VOLUME FIVE HUNDRED AND FIFTY SEVEN

METHODS IN ENZYMOLOGY

Membrane Proteins—Engineering, Purification and Crystallization

METHODS IN ENZYMOLOGY

Editors-in-Chief

JOHN N. ABELSON and MELVIN I. SIMON
Division of Biology
California Institute of Technology
Pasadena, California

ANNA MARIE PYLE
Departments of Molecular, Cellular and Developmental Biology and Department of Chemistry Investigator
Howard Hughes Medical Institute
Yale University

Founding Editors

SIDNEY P. COLOWICK and NATHAN O. KAPLAN

VOLUME FIVE HUNDRED AND FIFTY SEVEN

METHODS IN ENZYMOLOGY

Membrane Proteins—Engineering, Purification and Crystallization

Edited by

ARUN K. SHUKLA
Department of Biological Sciences and Bioengineering
Indian Institute of Technology
Kanpur, India

AMSTERDAM • BOSTON • HEIDELBERG • LONDON
NEW YORK • OXFORD • PARIS • SAN DIEGO
SAN FRANCISCO • SINGAPORE • SYDNEY • TOKYO
Academic Press is an imprint of Elsevier

Academic Press is an imprint of Elsevier
225 Wyman Street, Waltham, MA 02451, USA
525 B Street, Suite 1800, San Diego, CA 92101-4495, USA
125 London Wall, London, EC2Y 5AS, UK
The Boulevard, Langford Lane, Kidlington, Oxford OX5 1GB, UK

First edition 2015

Notices

Knowledge and best practice in this field are constantly changing. As new research and experience broaden our understanding, changes in research methods, professional practices, or medical treatment may become necessary.

Practitioners and researchers must always rely on their own experience and knowledge in evaluating and using any information, methods, compounds, or experiments described herein. In using such information or methods they should be mindful of their own safety and the safety of others, including parties for whom they have a professional responsibility.

To the fullest extent of the law, neither the Publisher nor the authors, contributors, or editors, assume any liability for any injury and/or damage to persons or property as a matter of products liability, negligence or otherwise, or from any use or operation of any methods, products, instructions, or ideas contained in the material herein.

ISBN: 978-0-12-802183-5
ISSN: 0076-6879

For information on all Academic Press publications
visit our website at store.elsevier.com

CONTENTS

CONTRIBUTORS

Enrique E. Abola
Department of Integrative Structural and Computational Biology, The Scripps Research Institute, La Jolla, California, USA

Bo Y. Baker
Department of Pharmacology, and Cleveland Center for Membrane and Structural Biology, School of Medicine, Case Western Reserve University, Cleveland, Ohio, USA

Lindsay A. Baker
NMR Spectroscopy, Bijvoet Center for Biomolecular Research, Department of Chemistry, Faculty of Science, Utrecht University, Utrecht, The Netherlands

Marc Baldus
NMR Spectroscopy, Bijvoet Center for Biomolecular Research, Department of Chemistry, Faculty of Science, Utrecht University, Utrecht, The Netherlands

Shibom Basu
Department of Chemistry and Biochemistry, Arizona State University, Tempe, Arizona, USA

Dominic Birth
Institute for Biochemistry and Molecular Biology, ZBMZ, BIOSS Centre for Biological Signalling Studies, and Faculty of Biology, University of Freiburg, Freiburg, Germany

Jani Reddy Bolla
Department of Chemistry, Iowa State University, Ames, Iowa, USA

Bernadette Byrne
Department of Life Sciences, Imperial College London, London, United Kingdom

Pil Seok Chae
Department of Bionanotechnology, Hanyang University, Ansan, South Korea

Sudha Chakrapani
Department of Physiology and Biophysics, School of Medicine, Case Western Reserve University, Cleveland, Ohio, USA

Amitabha Chattopadhyay
CSIR-Centre for Cellular and Molecular Biology, Hyderabad, and Academy of Scientific and Innovative Research, New Delhi, India

Vadim Cherezov
Department of Integrative Structural and Computational Biology, The Scripps Research Institute, La Jolla, California, USA

Kyung Ho Cho
Department of Bionanotechnology, Hanyang University, Ansan, South Korea

Ka Young Chung
School of Pharmacy, Sungkyunkwan University, Suwon, Republic of Korea

Jesse Coe
Department of Chemistry and Biochemistry, Arizona State University, Tempe, Arizona, USA

Linda Columbus
Department of Chemistry, University of Virginia, Charlottesville, Virginia, USA

Chelsie E. Conrad
Department of Chemistry and Biochemistry, Arizona State University, Tempe, Arizona, USA

Jared A. Delmar
Department of Physics and Astronomy, Iowa State University, Ames, Iowa, USA

Pawel K. Dominik
Department of Biochemistry and Molecular Biology, Institute for Biophysical Dynamics, The University of Chicago, Chicago, Illinois, USA

Mark E. Dumont
Department of Biochemistry and Biophysics, P.O. Box 712, University of Rochester School of Medicine and Dentistry, Rochester, New York, USA

Manish Dwivedi
Department of Biological Chemistry, Alexander Silberman Institute of Life Sciences, Hebrew University of Jerusalem, Jerusalem, Israel

Claudia Escher
Institute for Biochemistry and Molecular Biology, ZBMZ, BIOSS Centre for Biological Signalling Studies, University of Freiburg, Freiburg, Germany

Salem Faham
Center for Membrane Biology, Department of Molecular Physiology and Biological Physics, University of Virginia, Charlottesville, Virginia, USA

Gustavo Fenalti
Department of Integrative Structural and Computational Biology, The Scripps Research Institute, La Jolla, California, USA

Gert E. Folkers
NMR Spectroscopy, Bijvoet Center for Biomolecular Research, Department of Chemistry, Faculty of Science, Utrecht University, Utrecht, The Netherlands

Petra Fromme
Department of Chemistry and Biochemistry, Arizona State University, Tempe, Arizona, USA

Raimund Fromme
Department of Chemistry and Biochemistry, Arizona State University, Tempe, Arizona, USA

Sahil Gulati
Department of Pharmacology, and Cleveland Center for Membrane and Structural Biology, School of Medicine, Case Western Reserve University, Cleveland, Ohio, USA

Anjali Bansal Gupta
Department of Biological Sciences and Bioengineering, Indian Institute of Technology Kanpur, Kanpur, India

Charu Gupta
Department of Biological Sciences and Bioengineering, Indian Institute of Technology, Kanpur, India

Klaartje Houben
NMR Spectroscopy, Bijvoet Center for Biomolecular Research, Department of Chemistry, Faculty of Science, Utrecht University, Utrecht, The Netherlands

Carola Hunte
Institute for Biochemistry and Molecular Biology, ZBMZ, BIOSS Centre for Biological Signalling Studies, University of Freiburg, Freiburg, Germany

Md. Jafurulla
CSIR-Centre for Cellular and Molecular Biology, Hyderabad, India

Deepika Jaiman
Department of Biological Sciences and Bioengineering, Indian Institute of Technology, Kanpur, India

Zongchao Jia
Department of Biomedical and Molecular Sciences, Queen's University, Kingston, Ontario, Canada

Wei-Chun Kao
Institute for Biochemistry and Molecular Biology, ZBMZ, BIOSS Centre for Biological Signalling Studies, University of Freiburg, Freiburg, Germany

Mohammed Kaplan
NMR Spectroscopy, Bijvoet Center for Biomolecular Research, Department of Chemistry, Faculty of Science, Utrecht University, Utrecht, The Netherlands

Anthony A. Kossiakoff
Department of Biochemistry and Molecular Biology, Institute for Biophysical Dynamics, The University of Chicago, Chicago, Illinois, USA

Brett Kroncke
Department of Biochemistry and Center for Structural Biology, Vanderbilt University School of Medicine, Nashville, Tennessee, USA

Stefan Köster
Department of Structural Biology, Max Planck Institute of Biophysics, Frankfurt am Main, Germany

Christopher Kupitz
Department of Chemistry and Biochemistry, Arizona State University, Tempe, Arizona, USA

Morgan Lawrenz
Department of Chemistry, Stanford University, Stanford, California, USA

Su Youn Lee
School of Pharmacy, Sungkyunkwan University, Suwon, Republic of Korea

Sheng Li
Department of Medicine, University of California at San Diego, San Diego, California, USA

M. Gregor Madej
Department of Physiology, David Geffen School of Medicine, University of California Los Angeles, Los Angeles, California, USA

Elizabeth Mathew
Department of Biochemistry and Biophysics, P.O. Box 712, University of Rochester School of Medicine and Dentistry, Rochester, New York, USA

Syed H. Mir
Institute for Biochemistry and Molecular Biology, ZBMZ, BIOSS Centre for Biological Signalling Studies, University of Freiburg, Freiburg, Germany, and Department of Clinical Biochemistry, University of Kashmir, Srinagar, India

Jacob L.W. Morgan
Center for Membrane Biology, Department of Molecular Physiology and Biological Physics, University of Virginia, Charlottesville, Virginia, USA

Etana Padan
Department of Biological Chemistry, Alexander Silberman Institute of Life Sciences, Hebrew University of Jerusalem, Jerusalem, Israel

Aritra Pal
Department of Cellular and Molecular Physiology, Yale University School of Medicine, New Haven, Connecticut, USA

Krzysztof Palczewski
Department of Pharmacology, and Cleveland Center for Membrane and Structural Biology, School of Medicine, Case Western Reserve University, Cleveland, Ohio, USA

Vijay S. Pande
Department of Chemistry, and SIMBIOS NIH Center for Biomedical Computation, Stanford University, Stanford, California, USA

Sandra Poulos
Center for Membrane Biology, Department of Molecular Physiology and Biological Physics, University of Virginia, Charlottesville, Virginia, USA

Chelsy Prince
Department of Biomedical and Molecular Sciences, Queen's University, Kingston, Ontario, Canada

Bhagyashree D. Rao
CSIR-Indian Institute of Chemical Technology, Hyderabad, and Academy of Scientific and Innovative Research, New Delhi, India

Shatabdi Roy-Chowdhury
Department of Chemistry and Biochemistry, Arizona State University, Tempe, Arizona, USA

Ankita Roy
Department of Medicine, University of Pittsburgh, Pittsburgh, Pennsylvania, USA

Aiman Sadaf
Department of Bionanotechnology, Hanyang University, Ansan, South Korea

Ramasubbu Sankararamakrishnan
Department of Biological Sciences and Bioengineering, and Center of Excellence for Chemical Biology, Indian Institute of Technology Kanpur, Kanpur, India

Ingeborg Schmidt-Krey
School of Biology, and School of Chemistry and Biochemistry, Georgia Institute of Technology, Atlanta, Georgia, USA

Wuxian Shi
Center for Proteomics and Bioinformatics, Center for Synchrotron Biosciences, School of Medicine, Case Western Reserve University, Cleveland, Ohio, USA

Arun K. Shukla
Department of Biological Sciences and Bioengineering, Indian Institute of Technology, Kanpur, India

Diwakar Shukla
Department of Chemistry, and SIMBIOS NIH Center for Biomedical Computation, Stanford University, Stanford, California, USA

Satinder K. Singh
Department of Cellular and Molecular Physiology, Yale University School of Medicine, New Haven, Connecticut, USA

Tessa Sinnige
NMR Spectroscopy, Bijvoet Center for Biomolecular Research, Department of Chemistry, Faculty of Science, Utrecht University, Utrecht, The Netherlands

Ashish Srivastava
Department of Biological Sciences and Bioengineering, Indian Institute of Technology, Kanpur, India

Phoebe L. Stewart
Department of Pharmacology, and Cleveland Center for Membrane and Structural Biology, School of Medicine, Case Western Reserve University, Cleveland, Ohio, USA

Chih-Chia Su
Department of Physics and Astronomy, Iowa State University, Ames, Iowa, USA

Robert Tampé
Institute of Biochemistry, Biocenter, and Cluster of Excellence—Macromolecular Complexes, Goethe-University Frankfurt, Frankfurt/Main, Germany

Simon Trowitzsch
Institute of Biochemistry, Biocenter, Goethe-University Frankfurt, Frankfurt/Main, Germany

Yusuf M. Uddin
School of Biology, Georgia Institute of Technology, Atlanta, Georgia, USA

Elwin A.W. van der Cruijsen
NMR Spectroscopy, Bijvoet Center for Biomolecular Research, Department of Chemistry, Faculty of Science, Utrecht University, Utrecht, The Netherlands

Katharina van Pee
Department of Structural Biology, Max Planck Institute of Biophysics, Frankfurt am Main, Germany

Ravi Kumar Verma
Department of Biological Sciences and Bioengineering, Indian Institute of Technology Kanpur, Kanpur, India

Benlian Wang
Center for Proteomics and Bioinformatics, Center for Synchrotron Biosciences, School of Medicine, Case Western Reserve University, Cleveland, Ohio, USA

Chong Wang
Department of Integrative Structural and Computational Biology, The Scripps Research Institute, La Jolla, California, USA

Christophe Wirth
Institute for Biochemistry and Molecular Biology, ZBMZ, BIOSS Centre for Biological Signalling Studies, University of Freiburg, Freiburg, Germany

Beili Wu
Department of Integrative Structural and Computational Biology, The Scripps Research Institute, La Jolla, California, USA

Özkan Yildiz
Department of Structural Biology, Max Planck Institute of Biophysics, Frankfurt am Main, Germany

Edward W. Yu
Department of Physics and Astronomy, and Department of Chemistry, Iowa State University, Ames, Iowa, USA

Jochen Zimmer
Center for Membrane Biology, Department of Molecular Physiology and Biological Physics, University of Virginia, Charlottesville, Virginia, USA

PREFACE

Integral membrane proteins constitute a significant portion of the entire proteome in different organisms. They mediate a wide range of signal recognition and communication processes across the cell membranes. These proteins are one of the most important classes of drug targets, and more than half of the currently marketed drugs are aimed at membrane proteins. In spite of their crucial physiological roles, structural characterization, especially high-resolution structure determination, of membrane proteins lags significantly behind that of soluble proteins. There are numerous challenges encountered at every step in the process of membrane protein crystallography such as recombinant protein expression, homogenous purification, and crystallization. These two volumes (volume 556 & 557) of *Methods in Enzymology* aim to provide a comprehensive coverage of various steps involved in the process of membrane protein structural characterization through general protocols and case examples.

The very first step in the process of structural studies of membrane proteins is their recombinant expression in heterologous hosts for large-scale protein production. In Section I of Volume 556, we provide a collection of chapters that describe either a step-by-step protocol or a general overview for recombinant expression of various types of membrane proteins in different expression hosts. These chapters cover recent advances in conventional *E. coli*-based expression of membrane proteins, yeast-based expression systems for large-scale production of eukaryotic membrane proteins, and cell culture-based membrane protein overexpression in insect cells and mammalian cells. Furthermore, several chapters in this section also discuss relatively uncommon but promising strategies for expressing membrane proteins, e.g., *Drosophila melanogaster*, *Xenopus* oocytes, and *Wolinella succinogenes*.

Biochemical and functional characterization of recombinant membrane proteins is important to ensure their native-like behavior before structural studies can be undertaken. Section II of Volume 556 encompasses several chapters that cover various methods for characterizing membrane proteins such as reconstitution in lipid environment, cross-linking, fluorescence, and spectroscopy-based approaches to investigate conformational changes and surface plasmon resonance-based strategies to study ligand–protein interactions.

Once the recombinant membrane protein expression has been established and functional characterization reveals native-like properties, the next steps are to solubilize and purify them efficiently in functional forms. In Section I of Volume 557, we present a collection of chapters that provide generally applicable protocols and discussions on efficient solubilization and purification of recombinant membrane proteins. One of the major challenges in membrane protein crystallization is their conformational flexibility and limited polar surface area to make crystal contacts. In Section II of Volume 557, three chapters describe general protocols and successful examples of generating antibody fragments against membrane proteins using phage display technology to address these challenges.

Although crystallography of membrane proteins provides high-resolution structural information, it yields only a static snapshot of the protein architecture. Therefore, dynamic studies of membrane proteins are highly invaluable to obtain a complete understanding of their function. Chapters 13–16 in Section III of Volume 557 highlight various biophysical approaches such as electron paramagnetic resonance and nuclear magnetic resonance methodologies that yield dynamic insights into the structure and function of membrane proteins.

One of the major goals in membrane protein structural studies is their crystallization for high-resolution structure determination by X-ray crystallography. Chapters 17–22 in Section IV of Volume 557 present streamlined protocols and discussions on various methods for protein crystallization including a case example of protein crystallography by X-ray free-electron laser, the latest development in the area of protein crystallography.

Similar to dynamic methodologies, computational approaches can also provide extremely valuable insights into membrane protein functions. In Section V of Volume 557, authors present case examples of computational approaches that were applied to better understand three different classes of membrane proteins.

Overall, these two volumes provide an extensive and unique coverage of various aspects in membrane protein studies and they will be extremely useful to researchers engaged in the area of membrane proteins.

Arun K. Shukla

Membrane Protein Engineering, Solubilization and Purification

CHAPTER ONE

Multicolor Fluorescence-Based Screening Toward Structural Analysis of Multiprotein Membrane Complexes

Simon Trowitzsch*, Robert Tampé*,†,1

*Institute of Biochemistry, Biocenter, Goethe-University Frankfurt, Frankfurt/Main, Germany
†Cluster of Excellence—Macromolecular Complexes, Goethe-University Frankfurt, Frankfurt/Main, Germany
[1]Corresponding author: e-mail address: tampe@em.uni-frankfurt.de

Contents

Methods in Enzymology, Volume 557
ISSN 0076-6879
http://dx.doi.org/10.1016/bs.mie.2014.11.043

Abstract

Structures of membrane protein complexes provide a wealth of information on their biological function, the interplay among their subunits, and on ligand binding. Structural genomics of multiprotein membrane complexes seek to deliver structural information ideally of most of these complexes. Models of abundant native membrane protein complexes have been proposed from X-ray crystallography or single-particle cryo-EM. However, most of the remaining membrane protein complexes persist in very low copy numbers per cell and cannot be isolated from their native source without tremendous efforts. Therefore, heterologous expression systems are continually being developed to overproduce membrane protein complexes in various host cells of bacterial or eukaryotic origin. Still, only a small fraction of membrane proteins is suitable for structure determination due to poor expression levels, misfolding and aggregation, complex heterogeneity, imbalanced stoichiometry, and difficulties in solubilization as well as stabilization of the complexes. Powerful tools are therefore necessary to identify the correct expression host and to validate extraction and purification strategies for a given membrane protein complex at the earliest time point. Here, we discuss a fluorescence-based screening approach particularly tailored for the handy and sensitive analysis of the production and purification process for multiprotein membrane complexes. Multicolor fluorescence-detection size-exclusion chromatography provides a powerful readout system and allows quantitative monitoring of the production of critical single subunits of membrane protein complexes. The approach facilitates the tracking of improvements during sample optimization for monodispersity, balanced stoichiometry, and stability of multisubunit membrane protein complexes.

1. INTRODUCTION

Membrane proteins and membrane protein complexes perform a wide range of essential biological functions including metabolism, signal transduction, energy conversion and utilization. They represent with more than 50% the largest class of potential novel protein drug targets (Arinaminpathy, Khurana, Engelman, & Gerstein, 2009; Russell & Eggleston, 2000; Yildirim, Goh, Cusick, Barabasi, & Vidal, 2007). Although membrane proteins make up nearly 30% of all proteins in the cell, they only constitute ~2% of all structures deposited in the Protein Data Bank (PDB, www.rcsb.org). Only a few membrane proteins are abundant in cells, such as mammalian and bacterial rhodopsins, aquaporins, respiratory complexes, F/V/P-type ATPases, photosynthetic complexes, reaction centers, and light harvesting proteins (Bill et al., 2011). The first published structure of an integral membrane protein complex was the photosynthetic reaction center of *Rhodopseudomonas viridis* extracted from native sources in 1985

(Deisenhofer, Epp, Miki, Huber, & Michel, 1985). And still nowadays, significant numbers of structures solved from endogenous material are deposited in the PDB each year (Buschmann et al., 2010; Efremov, Baradaran, & Sazanov, 2010; Lee, Stewart, Donohoe, Bernal, & Stock, 2010; Mesa, Deniaud, Montoya, & Schaffitzel, 2013; von der Hocht et al., 2011).

However, low abundance of the vast majority of membrane proteins renders overexpression essential for large-scale production for structural studies, because milligram quantities of pure, monodisperse, and stoichiometric sample are typically needed. The production of high-quality samples is not trivial though. It took more than a decade after the structural characterization of the photosynthetic reaction center until the first crystal structure of a membrane protein heterologously expressed in *Escherichia coli* was solved (Doyle et al., 1998). Since then, rigorous efforts to prepare homogeneous and active recombinant samples for structural investigation have lead to the determination of around 500 unique structures of membrane proteins and membrane protein complexes at atomic or near atomic resolution (Fig. 1; http://blanco.biomol.uci.edu/mpstruc/).

A number of heterologous expression systems have been used and optimized for the production of membrane proteins, such as bacterial and yeast expression systems, baculovirus-based insect cell systems, and cultured mammalian cells (Andrell & Tate, 2013; Bonander & Bill, 2012;

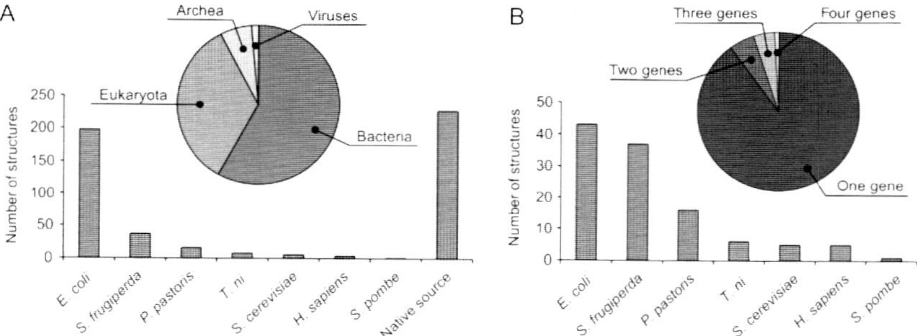

Figure 1 Classification of released unique membrane protein structures in the PDB based on the source organism for sample preparation (either recombinant or from a native source). (A) The histogram shows the distribution of unique membrane protein structures from all domains of life as a plot against the expression host, from which the sample was derived. Inset: a pie chart showing the distribution of the genetic origin of the membrane protein structures. (B) The histogram summarizes the distribution of unique eukaryotic, recombinant membrane protein structures as a plot against the expression host. Inset: a pie chart showing the distribution of the number of subunits for each unique eukaryotic, recombinant membrane protein structure.

Condreay & Kost, 2007; Frelet-Barrand, Boutigny, Kunji, & Rolland, 2010; Junge et al., 2008; Sahdev, Khattar, & Saini, 2008). Cell-free protein production is also being used and constantly developed for membrane protein synthesis for structural analysis (Harbers, 2014). Engineered baculoviruses are promising versatile vectors, not only for protein production in insect cells but also for transducing numerous types of mammalian cells, thus facilitating the search for the best expression host (Kost, Condreay, & Jarvis, 2005; Liu, Chen, & Chao, 2010; Sung et al., 2014). Not surprisingly, eukaryotic expression systems are becoming more and more important as expression hosts for the production of membrane proteins of eukaryotic origin (Fig. 1). Although approximately 38% of heterologously expressed eukaryotic membrane protein structures were determined using samples produced in *E. coli*, the insect cell expression system including cell lines from *Spodoptera frugiperda* and *Trichoplusia ni* is turning out to be a very attractive approach (He, Wang, & Yan, 2014). Yeast and mammalian systems rank third and fourth accounting for ~20% and 4% of the unique structures deposited, respectively. Identifying the best expression host for a given eukaryotic membrane protein or membrane protein complex is still an empirical process and usually takes a significant amount of time and resources (Bernaudat et al., 2011).

In order to overcome a critical bottleneck in structure determination of membrane proteins, high-throughput approaches have been developed and implemented but only show success rates of ~0.3% (Mancia & Love, 2010). The main problem is that conditions need to be optimized toward stability and homogeneity for each specimen in order to obtain a conformationally rigid sample, which can eventually intercalate into a regular crystal lattice for structure determination by X-ray crystallography. Recent technological advances in direct electron detection cameras and innovative image processing algorithms enabled structure determination by single-particle cryo-electron microscopy (cryo-EM) at near atomic resolution (Allegretti, Mills, McMullan, Kühlbrandt, & Vonck, 2014; Amunts et al., 2014; Bai, Fernandez, McMullan, & Scheres, 2013; Kim et al., 2014; Li et al., 2013). Of note is the first structure of a heterologously expressed homotetrameric ion channel, TRPV1, extracted from transiently transfected HEK293S GnTI$^-$ cells (Liao, Cao, Julius, & Cheng, 2013). Another milestone for structural analysis of heterologously produced multiprotein membrane complexes by single-particle cryo-EM is the 3D reconstruction of the heterotetrameric γ-secretase at 4.5 Å resolution and the heterodimeric ABC transporter TmrAB (Kim et al., 2014; Lu et al., 2014). One has yet to bear in mind that sample preparation and structure determination of these complexes took several years and that rigorous efforts were made to obtain homogeneous, active sample for structural investigation. Taking into account all the difficulties during production, purification, and stabilization already encountered for a single recombinant membrane protein, it is not surprising that only a few structures of hetero-oligomeric eukaryotic multiprotein membrane complexes exist (Fig. 1B). The development of methods providing a rapid way to identify membrane proteins that can be expressed and purified in sufficient amounts and of sufficient quality is therefore an area of intensive efforts (Drew, Lerch, Kunji, Slotboom, & de Gier, 2006; Drew, von Heijne, Nordlund, & de Gier, 2001; Kawate & Gouaux, 2006; Martinez Molina et al., 2008; Newstead, Kim, von Heijne, Iwata, & Drew, 2007).

1.1. Screening approaches for membrane proteins and membrane protein complexes

Besides choosing the appropriate expression host, construct engineering is key for the successful production of homogeneous and stable membrane protein complexes for structural analysis. Crystallization of G protein-coupled receptors, for example, was for the first time possible by generating T4 lysozyme fusions (Rosenbaum et al., 2007; Rosenbaum, Rasmussen, & Kobilka, 2009). However, assessing general stabilizing effects on membrane proteins is not trivial. Stability assays developed for soluble proteins rely on fluorescent probes, which change their fluorescent properties upon binding to hydrophobic patches exposed during the unfolding process of the target molecule (Niesen, Berglund, & Vedadi, 2007; Vedadi et al., 2006). Due to the intrinsic hydrophobicity of membrane proteins and the high concentration of detergents and lipids in membrane protein samples, this approach is not applicable for membrane proteins. Cysteines buried in the interior of membrane proteins were shown to be suitable as stability sensors (Alexandrov, Mileni, Chien, Hanson, & Stevens, 2008). These embedded residues become solvent exposed upon thermal denaturation of the target protein and are then attacked by a thiol-reactive fluorescent probe. The fluorescence of the probe in isolation is relatively low, but increases strongly, when the probe is covalently linked to cysteines. This tool was developed to improve stability of membrane proteins by optimizing expression constructs, identifying stabilizing ligands or lipids, and assessing detergents (Alexandrov et al., 2008; Liu, Hanson, Stevens, & Cherezov, 2010).

Fluorescent protein tags not only proved to be indispensable for studying protein–protein interactions and protein localization in living cells (Day & Davidson, 2009; Kremers, Gilbert, Cranfill, Davidson, & Piston, 2011), but also were successfully applied for protein production to monitor system performance and to track protein production (Drew et al., 2001; Fitzgerald et al., 2006; Geertsma, Groeneveld, Slotboom, & Poolman, 2008; Nie, Bellon-Echeverria, Trowitzsch, Bieniossek, & Berger, 2014; Trowitzsch, Palmberger, Fitzgerald, Takagi, & Berger, 2012). Conditions for the overproduction and purification of membrane proteins for functional and structural studies were significantly improved by the introduction of a C-terminal fusion of the green fluorescent protein (GFP), which was used to track protein production in *E. coli* and *Saccharomyces cerevisiae* (Drew et al., 2006, 2001; Hammon, Palanivelu, Chen, Patel, & Minor, 2009). When fused to the C terminus of a membrane protein, GFP can be used to quantitatively report on the folding state of the membrane protein (Newstead et al., 2007). GFP will only fold into a sodium dodecyl sulfate (SDS)-resistant fluorescent state if the preceding target protein is likewise correctly folded (Geertsma et al., 2008; Pedelacq et al., 2002). An extension of the fluorescence-based concept was introduced by the Gouaux lab for rapid screening of membrane protein stability and monodispersity by gel filtration and was named fluorescence-detection size-exclusion chromatography (FSEC; Kawate & Gouaux, 2006). In this approach, the C-terminal fluorescent protein tag is exploited to characterize a membrane protein's oligomeric state from only nanogram quantities of unpurified sample (Sonoda et al., 2010). Data about the yield, degree of monodispersity, and approximate molecular weight of a recombinant membrane protein can thus be obtained from small culture samples. The applicability of the FSEC method as a powerful precrystallization screening approach for integral membrane proteins was recently demonstrated by the crystallization and structure determination of two ion channels (Gonzales, Kawate, & Gouaux, 2009; Kawate, Michel, Birdsong, & Gouaux, 2009). Later on, this technique was developed into a thermostability assay (Hattori, Hibbs, & Gouaux, 2012). The FSEC method has only been used for proteins coded by a single gene, either as a monomer or assembled into homo-oligomers. We have recently established the FSEC approach to analyze multiprotein membrane complexes by introducing tags with different fluorescent properties (Parcej, Guntrum, Schmidt, Hinz, & Tampé, 2013). Tagged with different fluorescent proteins, individual subunits of multiprotein membrane complexes can be followed at nanogram quantities at each purification step *via* various methods as outlined below.

1.2. Multicolor fluorescence-based screening of multiprotein membrane complexes

Purification of macromolecular multiprotein membrane complexes requires as an initial step the isolation and enrichment of membranes by centrifugation and the subsequent solubilization of the membrane protein complex by detergents. Detergents are ubiquitous and necessary reagents in membrane protein biochemistry and the identification of the proper detergent for a given complex is critical for complex stability, stoichiometry, monodispersity, and structure determination (Privé, 2007). Once the ideal detergent for solubilization has been found, affinity and ion exchange chromatography are routinely used as further purification steps prior to polishing by size-exclusion chromatography. Each of these steps during production and purification of multiprotein membrane complexes can be monitored by the use of fused fluorescent protein reporters. Expression levels can be specifically and quantitatively assessed by in-gel fluorescence or fluorescence measurements on whole cell extracts (Drew et al., 2006). The purification process can be optimized by tracking fluorescence of the individual subunits using multicolor fluorescence-detection size-exclusion chromatography (MC-FSEC), including detergent screening and detergent exchange (Parcej et al., 2013). It is recommended to analyze purified membrane protein complexes by blue-native PAGE and clear-native PAGE (Wittig, Braun, & Schägger, 2006; Wittig & Schägger, 2008). The effect of small molecules and proteins, such as small peptidic inhibitors, nanobodies, Fab fragments, or DARPins, on the stability of the multiprotein membrane complex can be readily assessed by MC-FSEC. An overview of the multicolor fluorescence-based approach is shown in Fig. 2.

1.3. The major histocompatibility complex class I peptide-loading complex: A challenge for structural biology of membrane protein complexes

Cellular proteins are constantly converted into peptides and are presented on the cell surface *via* major histocompatibility complex class I (MHC I) molecules for policing by the adaptive immune system. Once virally or malignantly transformed, cells present antigenic peptides on their surface and are in turn recognized and eliminated by cytotoxic T cells. Display of these antigenic peptides on the cell surface is a highly regulated and compartmentalized process involving various soluble proteins and membrane protein complexes (Hulpke & Tampé, 2013). Central to peptide loading of MHC I molecules is the peptide-loading complex (PLC; Fig. 3). The

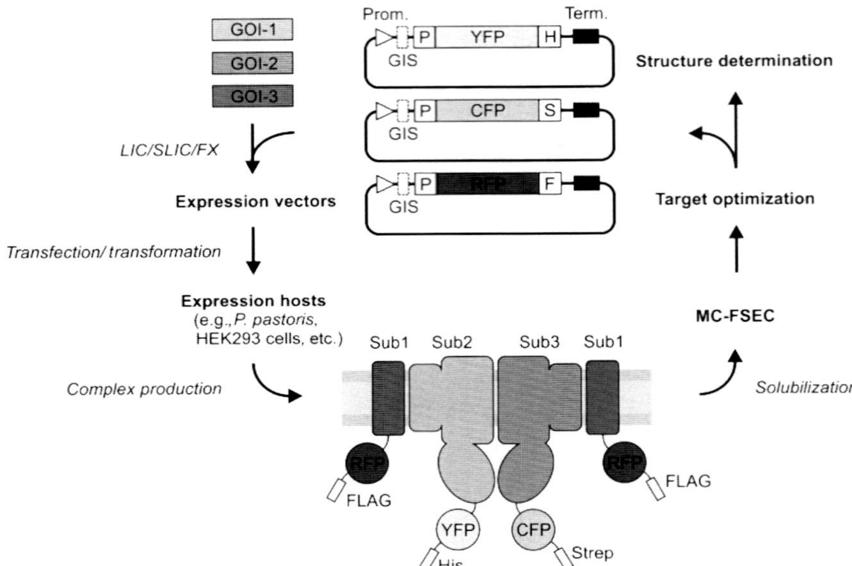

Figure 2 Flowchart for a multicolor fluorescence-based approach toward structure determination of integral multiprotein membrane complexes. Expression cassettes of three vectors for cloning of genes of interest (GOI) are schematically shown. GOIs are integrated into the vectors *via* gene integration sites (GIS) by ligation-independent cloning (LIC), sequence- and ligation-independent cloning (SLIC), or fragment exchange (FX) cloning (Aslanidis & de Jong, 1990; Geertsma & Dutzler, 2011; Li & Elledge, 2007). Host cells are transfected or transformed with the charged expression vectors and complexes are produced. A hetero-oligomeric membrane protein complex embedded in a lipid bilayer is depicted with fluorescence and purification tags. The recombinant protein complex is detergent solubilized and tested for stability using multicolor fluorescence-detection size-exclusion chromatography (MC-FSEC). After successful, iterative target optimization, the multiprotein membrane complex can be structurally characterized. P, protease cleavage site; YFP, yellow (light gray in the print version) fluorescent protein; CFP, cyan (gray in the print version) fluorescent protein; RFP, red (dark gray in the print version) fluorescent protein; H, His-tag; S, Strep-tag; F, FLAG-tag; Sub, subunit; Prom., promoter; Term., terminator.

PLC is organized around the transporter associated with antigen processing (TAP), which is the active player in shuttling proteasomal-derived peptides from the cytoplasm into the endoplasmic reticulum (ER; Abele & Tampé, 2011). With the help of auxiliary factors, these peptides are loaded onto MHC I molecules, which then travel from the ER to the cell surface for antigen presentation to cytotoxic T lymphocytes (Neefjes, Jongsma, Paul, & Bakke, 2011).

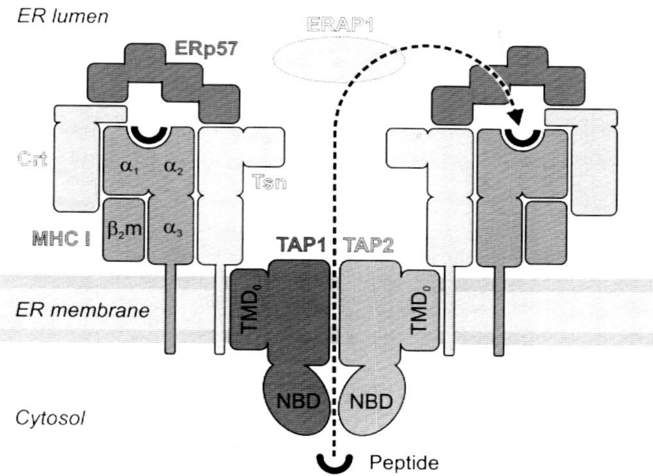

Figure 3 Structural organization of the PLC. Subunits are schematically depicted according to their domain organization. Transmembrane domains (TMD$_0$) and the nucleotide-binding domains (NBDs) of TAP1 and TAP2 are shown. The cartoon highlights multivalent intersubunit interactions, which stabilize the PLC (see text for details). Peptides are translocated from the cytosol to the ER lumen *via* TAP (dashed arrow), where they are further processed *via* the aminopeptidase ERAP1, prior to loading onto MHC I molecules. ER, endoplasmic reticulum; PLC, peptide-loading complex; Tsn, tapasin; Crt, calreticulin; MHC I, major histocompatibility complex class I; β_2m, β_2-microglobulin; TAP, transporter associated with antigen processing; ERAP1, endoplasmic reticulum aminopeptidase 1. (See the color plate.)

TAP is found in all nucleated cells of jawed vertebrates (Parcej & Tampé, 2010) and belongs to ATP-binding cassette (ABC) transporters, which constitute one of the largest transporter subfamily in all three kingdoms of life (Higgins, 1992). The basic blueprint of an ABC transporter consists of two transmembrane domains (TMDs) that provide a translocation pathway, and two cytoplasmic, solvent-exposed nucleotide-binding domains (NBDs) that hydrolyze ATP (Seyffer & Tampé, 2014). The heterodimeric TAP complex is composed of the two half-transporters, TAP1 and TAP2, each harboring a specialized N-terminal TMD$_0$.

TAP interacts *via* its TMD$_0$s with the adaptor chaperone tapasin (Tsn), which is a type I membrane glycoprotein (Bangia, Lehner, Hughes, Surman, & Cresswell, 1999; Hulpke et al., 2012). Tsn in turn connects TAP and MHC I molecules (Everett & Edidin, 2007; Ortmann et al., 1997; Rizvi & Raghavan, 2010; Sadasivan, Lehner, Ortmann, Spies, & Cresswell, 1996). Within human PLC, Tsn is covalently linked to the oxidoreductase ERp57 by an intermolecular disulfide bridge (Dick, Bangia, Peaper, & Cresswell, 2002; Peaper, Wearsch, & Cresswell, 2005). The PLC is stabilized by the soluble protein calreticulin, which tethers MHC I to ERp57 (Frickel et al., 2002; Oliver, Roderick, Llewellyn, & High, 1999). Interestingly, only monoglycosylated MHC I can be incorporated into the PLC pointing toward a key function of the lectin domain of calreticulin as a sensor for regulating the assembly/disassembly of the PLC (Wearsch, Peaper, & Cresswell, 2011; Zhang, Wearsch, Zhu, Leonhardt, & Cresswell, 2011). In brief, these numerous intermolecular interactions exemplify the significance of tight temporally and spatially controlled assembly of the PLC, which is furthermore regulated by posttranslational modifications. These various factors illustrate the difficulty of producing and assembling recombinant PLC and show the need for robust screening methods for protein production, complex assembly, stability, and stoichiometry. The following sections detail the expression, purification, and analysis of recombinant fluorescently labeled TAP complexes produced in *Pichia pastoris* and HEK293T cells.

2. PRODUCTION OF FLUORESCENTLY LABELED TAP COMPLEXES IN *P. PASTORIS*

2.1. Construct design

Two expression vectors for the production of fluorescently labeled TAP1 and TAP2 in *P. pastoris* were generated from the pPICZ C vector (Invitrogen). The pPICZ C vector was engineered to code for a C-terminal Tobacco etch virus (TEV) protease cleavage site, a fluorescent protein mVenus or mCerulean, and a purification tag under the control of the AOX1 promoter (Parcej et al., 2013). We have engineered two *Bsm*BI sites downstream of the AOX1 promoter in order to allow rapid insertion of genes of interest by ligation-independent cloning (LIC; Aslanidis & de Jong, 1990). To grant for orthogonal purification of the membrane protein complex, we have engineered a combination of mVenus with a His_{10}-tag and mCerulean with a StrepII-tag. For structural analysis of the membrane protein complex, the fluorescence, and purification tags can be eventually removed by TEV protease treatment. We typically use the protease-deficient *P. pastoris* cell strain SMD1163H (Invitrogen), which allows growth in minimal media. The following protocols describe transfection and screening procedures optimized for the production of mVenus-His_{10}-tagged TAP1 and mCerulean-StrepII-tagged TAP2. The next four sections (Sections 2.2–2.5) outline the steps from cloning, transformation, expression screening in small scale to large-scale production, and purification of fluorescently labeled TAP complex.

2.2. Transformation

1. Linearize 50 μg of each expression vector by *Mss*I cleavage. Purify the linearized fragments and elute the vector DNAs in water.
2. Mix 30 μg of each linearized vector and concentrate the solution to 10–20 μl in a Speedvac.
3. Grow *P. pastoris* cells over night to an optical density (OD_{600}) of 1.2–1.3 in 10 ml YPD media (1% yeast extract, 2% peptone, 2% dextrose).
4. Harvest the cells by centrifugation for 5 min at $1500 \times g$ and resuspend the cell pellet in 5 ml buffer containing 10 m*M* tris–HCl, pH 7.5, 100 m*M* lithium acetate, 600 m*M* sorbitol, and 10 m*M* dithiothreitol.
5. Incubate the cell suspension for 30 min at room temperature, pellet the cells, and transfer them into a 2-ml centrifuge tube.
6. Wash the cells three times with ice-cold 1 *M* sorbitol solution and bring the cells to a final OD_{600} of 10 (this corresponds to $\sim 5 \times 10^9$ cells/ml).
7. Mix 100 μl cell suspension with the concentrated linearized vector DNA and incubate for 5 min on ice.
8. Transfer the cell suspension to a prechilled electroporation cuvette (2 mm gap) and electroporate at 2.5 kV, 25 μF, and 400 Ω (BioRad Genepulser II).
9. Add 650 μl of a 1 *M* sorbitol solution immediately after the pulse, transfer the solution to a 15 ml conical tube, and let the cells recover for 2 h at 30 °C.
10. Add 750 μl YPD media to the cell suspension and shake the cells for another 2 h at 30 °C.
11. Plate the cells on YPDS plates (1% yeast extract, 2% peptone, 2% dextrose, 1 *M* sorbitol, 15 g/l agar) containing 2, 3, and 4 mg/ml Zeocin and incubate the plates for ~3 days at 30 °C.

2.3. Colony screening

1. Inoculate 2 ml MGY media (3.4 g/l nitrogen base, 10 g/l ammonium sulfate, 0.4 mg/l biotin, 1% [w/v] glycerol) with single colonies in 10 ml deep well plates.

2. Shake the plates for 1 day at 30 °C and inoculate 2 ml MGY media in a second deep well plate.
3. Grow the cultures to an OD_{600} of 5.0 and harvest the cells by centrifugation.
4. Resuspend each cell pellet in 10 ml MM media (3.4 g/l nitrogen base, 10 g/l ammonium sulfate, 0.4 mg/l biotin, 0.5% methanol), transfer the cell suspension to 50 ml conical tubes, and shake the cells for 24 h at 30 °C.
5. Harvest the cells by centrifugation at 1500 × *g* for 5 min and resuspend the pellet in 500 μl breaking buffer (50 m*M* KH_2PO_4, pH 7.4, 1 m*M* ethylenediaminetetraacetic acid, 5 m*M* 6-aminohexanoic acid, 5% [w/v] glycerol, 2.5 m*M* benzamidine, 0.1 mg/ml soybean trypsin inhibitor, and 1 m*M* phenylmethanesulfonylfluoride, PMSF).
6. Transfer the cell suspension to a 2 ml centrifuge tube and add an equal volume of ice-cold glass beads (0.5 mm diameter).
7. Break the cells using a FastPrep-24 (MP Biomedicals) for 5 × 20 s at 6 m/s, add 1 ml buffer, and centrifuge the mixture at 4500 × *g* for 5 min.
8. Centrifuge the supernatant at 100,000 × *g* for 30 min to collect the membrane fraction. One can freeze the pellets at this point or proceed immediately to the next step.
9. Resuspend the pellet in 150 μl solubilization buffer (20 m*M* HEPES–NaOH, pH 7.4, 200 m*M* NaCl, 50 m*M* KCl, 15% glycerol, 1 m*M* PMSF, 2.5 m*M* benzamidine) supplemented with 1% (w/w) *fos*-choline-12 (FC-12) and mix for 1 h at 4 °C.
10. Centrifuge the sample for 1 h at 100,000 × *g* and pass the supernatant through 0.2 μm Spin-X centrifuge filter (Corning Life Sciences). Use 50 μl filtrate per MC-FSEC run (see Section **4.1**).

2.4. Large-scale expression

1. Inoculate 50 ml of MYG media with a colony, which produces the protein complex of interest, and incubate the culture in a shaker at 30 °C over night.
2. The next day, inoculate 800 ml of MYG media with the overnight culture and proceed incubating at 30 °C until an OD_{600} of 5–6 has been reached.
3. Sediment the cells by centrifugation at 1500 × *g* for 5 min and resuspend the cell pellet in 2 l of MM media.

4. Incubate the culture in a shaker for another 24 h at 30 °C.
5. Harvest the cells by centrifugation, flash freeze the pellet in liquid nitrogen, and store the pellet at −80 °C until further use.

2.5. Orthogonal purification

1. Thaw 15 g of wet cell pellet and top up to 30 ml with breaking buffer.
2. Mix the cell suspension with 20 ml glass beads and break the cells using a FastPrep-24 (MP Biomedicals) for 4 × 45 s at 6 m/s.
3. Recover the supernatant by centrifugation for 15 min at 5000 × *g*, add 35 ml breaking buffer to the beads, and repeat the FastPrep-24 run.
4. Mix the two supernatants and centrifuge for 45 min at 150,000 × *g*.
5. Discard supernatant and resuspend membranes to a total protein concentration of 20 mg/ml with solubilization buffer.
6. (Optional) Flash freeze membranes in liquid nitrogen in aliquots of 25 ml and store aliquots at −80 °C.
7. Dilute 25 ml of membranes to 10 mg/ml with 25 ml solubilization buffer supplemented with 10 m*M* imidazole, add digitonin to a final concentration of 2% (w/v), and let the solution gently mix for 1 h.
8. Centrifuge the solution at 100,000 × *g* for 1 h and keep the supernatant (take crude extract sample).
9. Equilibrate 2 ml (bed volume, bv) of Ni^{2+}-NTA Sepharose 6 fast flow (GE Healthcare) in binding buffer (20 m*M* HEPES–NaOH, pH 7.4, 200 m*M* NaCl, 50 m*M* KCl, 0.1 m*M* PMSF, 0.25 m*M* benzamidine) containing 2–10 times the critical micelle concentration (CMC) of the detergent to be tested and supplement with 10 m*M* imidazole. Keep the same detergent concentration in all subsequent steps.
10. Add the supernatant to the equilibrated Ni^{2+}-NTA resin and mix for 2 h or overnight in batch.
11. Collect the Ni^{2+}-NTA resin in an empty gravity flow column (take immobilized metal affinity chromatography (IMAC) flow-through sample).
12. Wash the Ni^{2+}-NTA resin with 20 ml binding buffer supplemented with 40 m*M* imidazole.
13. Add 8 ml binding buffer supplemented with 200 m*M* imidazole onto the column and collect the eluate (take IMAC elution sample).
14. Load 2 ml (bv) *Strep*-Tactin Superflow high capacity (IBA, Göttingen) in an empty gravity flow column and equilibrate the resin with binding buffer.

15. Transfer the IMAC eluate to the equilibrated *Strep*-Tactin resin. For optimal binding and recovery, ensure that the sample passes very slowly trough the column (flow rate of ~0.1 ml/min).
16. Collect the flow through of the *Strep*-Tactin column (take Strep flow-through sample).
17. Wash the *Strep*-Tactin column with 10 ml of binding buffer and elute the sample with 10 ml binding buffer supplemented with 10 m*M* D-desthiobiotin (take Strep elution sample).
18. Concentrate the sample to 1 ml by ultrafiltration using a centrifugal filter unit with a molecular weight cut off of 100,000 (Millipore, Darmstadt) and take a final sample. In order to avoid protein precipitation, concentrate the sample in small steps and gently mix the solution in between. The procedure typically yields ~500 μg of purified membrane protein complex.
19. Analyze all samples by MC-FSEC (Section 4.1) and SDS-PAGE (Coomassie brilliant blue stain and in-gel fluorescence).

3. PRODUCTION OF FLUORESCENTLY LABELED TAP COMPLEXES IN MAMMALIAN CELLS

3.1. Construct design

In order to facilitate construct exchange, we have chosen the fragment exchange (FX) cloning strategy (Geertsma & Dutzler, 2011) to generate mammalian expression vectors for the production of fluorescently labeled TAP complexes in HEK293T HeLa, or TAP-deficient cell lines. Based on the pcDXC3GMS vector (Geertsma & Dutzler, 2011), which codes for a C-terminal human rhinovirus 3C protease, a GFP, a myc-tag, and a streptavidin-binding protein (SBP) tag, we generated two vectors with either mVenus- or mCerulean-tag. The vectors were constructed by replacing the GFP-myc-SBP coding sequence with a sequence coding for mVenus, a C8-tag, and a His_{10}-tag, or by replacing GFP with mCerulean. The constructs allow detection of chimeric fusion protein products by fluorescence or immunoblotting and orthogonal purification *via* the His_{10}-, C8-, or SBP-tag.

3.2. Transfection

1. Seed ~5 × 10^6 cells in Dulbecco's Modified Eagles Medium (DMEM) supplemented with 10% fetal calf serum (FCS) in 15 cm culture dishes

1 day prior to transfection. At the day of transfection, cells should be 70–80% confluent.

2. At the day of transfection, dilute 30 μg vector DNA and 90 μg polyethylenimine (PEI) separately in 700 μl serum-free DMEM.
3. Mix DNA and PEI solutions and incubate at room temperature for 30 min. Vortex the solution several times during incubation.
4. Replace media with 10 ml serum-free DMEM and add the DNA-PEI mixture dropwise into the dish.
5. Change media to 15 ml DMEM/10% FCS 6 h post transfection.
6. Two days after transfection, wash the cells in ice-cold PBS, scrape cells off using a rubber policeman and flash freeze the cell pellet in liquid nitrogen. Store the pellet at −80 °C until further use or proceed directly with orthogonal purification (Section **3.3**).

3.3. Orthogonal purification

1. Thaw the cell pellet on ice and solubilize membrane proteins in prechilled buffer A (50 m*M* tris–HCl, pH 8.0, 250 m*M* NaCl, 10% glycerol, 0.05% digitonin) supplemented with 2% digitonin for 1 h on an overhead shaker.
2. Centrifuge the suspension at 150,000 × *g* for 1 h.
3. Equilibrate 100 μl streptavidin agarose beads (Pierce, USA) in buffer A and incubate the supernatant with agarose beads for 1 h.
4. Wash the beads twice with 1 ml of buffer A and elute bound proteins in 1 ml buffer A supplemented with 2.5 m*M* biotin.
5. Bind 10 μg of C8 antibody to 50 μl sheep anti-mouse IgG Dynabeads (Life Technologies) for 1 h and wash the beads twice with buffer A.
6. Incubate the streptavidin eluate with the C8 antibody-coated beads for 1 h and wash the beads twice with 1 ml buffer A.
7. Elute bound proteins by adding 100 μl 2 × SDS-PAGE-loading dye and incubating the mixture for 30 min at 37 °C. The samples can now be analyzed by in-gel fluorescence and immunoblotting.

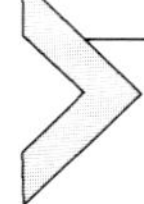

4. MULTICOLOR FLUORESCENCE-BASED SCREENING APPROACHES

This section describes various approaches, which benefit from using multicolor fluorescently labeled membrane protein complexes.

4.1. MC-FSEC analysis

For MC-FSEC experiments, we use an Agilent 1200 series high-performance liquid chromatography (HPLC) system equipped with an auto-sampler and a Shodex semi-micro KW404-4F (4.6 mm × 300 mm) size-exclusion chromatography column. We routinely perform our experiments in the cold room with a temperature-controlled column holder. A protocol for dual detection of mVenus and mCerulean fluorescence as well as for three-color experiments is outlined below. It is recommended to calibrate the fluorescence detectors of the system with defined amounts of recombinant fluorescent proteins.

1. Equilibrate the system in running buffer (20 m*M* HEPES–NaOH, pH 7.4, 200 m*M* NaCl, and 50 m*M* KCl) containing two to five times the CMC of the detergent to be tested.
2. Prefilter samples through a 0.2 μm Spin-X centrifuge filter prior to injection on the SEC column.
3. For each run, set a run time of 20 min at a flow rate of 0.3 ml/min. Detect mCerulean fluorescence at an excitation wavelength of 435 nm and an emission wavelength of 470 nm (detector gain 12) and mVenus fluorescence at an excitation wavelength of 515 nm and an emission wavelength of 535 nm (detector gain 10).
4. For three-color experiments, set the first fluorescence detector (FLD) to single emission detection at a wavelength of 520 nm (gain 12) and double excitation at wavelengths of 430 nm for mCerulean and 510 nm for mVenus. Use the second FLD for detection of a red fluorophore (e.g., Atto565) at excitation and emission wavelengths of 563 and 589 nm, respectively (gain 14).

4.2. In-gel fluorescence and immunoblotting

We routinely use in-gel fluorescence to analyze and quantify membrane protein complex production at the earliest time point. *Note*: to detect in-gel fluorescence, it is important not to boil the samples in SDS-PAGE-loading dye to prevent unfolding of the fluorescent tags.

1. Separate proteins by 8% tris/tricine-PAGE at 120 V.
2. Visualize fluorescently tagged proteins on a typhoon scanner (GE Healthcare). Detect fluorescence of mCerulean and mVenus tags at wavelengths of $\lambda_{ex/em}$ 457/526 nm (short-pass filter) and $\lambda_{ex/em}$ 488/555 nm ($\pm$20 nm band-pass filter), respectively.

3. For immunoblotting, heat the samples at 96 °C for 10 min to unfold the fluorescent tags and run out the samples on 8% tris/tricine gels.
4. Blot the proteins onto a PVDF membrane and stain with mouse monoclonal α-TAP1 (148.3) and α-TAP2 (435.3) primary antibodies. Alternatively, proteins can be detected with tag-specific primary antibodies.
5. Incubate the membrane with a horseradish peroxidase-conjugated α-mouse secondary antibody and record chemiluminescence with a LumiImager (Roche). Images can be further analyzed using the LumiAnalyst 3.1 software.

4.3. Monitoring peptide binding to TAP

As outlined in Section **1**, binding of small molecules or Fab fragments, for example, may help to stabilize membrane protein complexes for structural studies. In the case of TAP, various viral proteins have been identified, which lock the complex in certain states (Abele & Tampé, 2009). These viral factors may be instrumental to structurally characterize TAP in different fixed conformations, thereby expanding our understanding of the transport cycle of TAP. A particular example is the herpes simplex virus immediate early gene product ICP47, which inhibits TAP by blocking the peptide-binding site of TAP (Ahn et al., 1996; Neumann, Kraas, Uebel, Jung, & Tampé, 1997; Tomazin et al., 1996). The following protocol allows to monitor the binding of fluorescently labeled ICP47 peptide and an antigenic peptide substrate derived from Histone H3.3 to the TAP complex (Eggensperger, Fisette, Parcej, Schäfer, & Tampé, 2014; Parcej et al., 2013).

1. Express viral inhibitor His_{10}-ICP47 (S81C) in *E. coli* BL21(DE3)pLysS and purify as described (Ahn et al., 1996).
2. The antigenic epitope RRYCKSTEL is synthesized by Fmoc solid phase chemistry.
3. Label His_{10}-ICP47 polypeptide and the antigenic epitope peptide with maleimide-Atto565 dye (ATTO-TEC GmbH) as described (Neumann et al., 1997).
4. Purify labeled polypeptides *via* reverse-phase C18-HPLC and verify the fluorescently labeled product by mass spectrometry.
5. Mix purified fluorescently labeled TAP1/2 complex and labeled ICP47 peptide or antigenic peptide in a stoichiometric ratio to a final concentration of 1 μ*M* TAP1/2 in running buffer and let the mixture incubate for 20 min at 4 °C.
6. Analyze the mixture *via* MC-FSEC (see Section **4.1**).

7. To assess nonspecific binding of the fluorophore, add 100-fold molar excess of unlabeled peptides to the binding reaction prior to MC-FSEC analysis.

4.4. Measurement of MHC I surface presentation by flow cytometry

The TAP-deficient cell lines BRE-169 (TAP1$^-$) and STF-169 (TAP2$^-$) derived from fibroblasts of patients with bare lymphocyte syndrome (type I) are significantly impaired in MHC I peptide loading and surface expression due to the defect in one TAP subunit (de la Salle et al., 1994, 1999). These defects make the cells ideal systems to study the effects of mutations or truncations in TAP subunits by complementation assays. Transfection of these cells with the respective TAP subunit restores MHC I peptide loading and surface expression. MHC I antigen surface presentation and the presence of the transfected fluorescently labeled TAP subunit can be monitored and correlated by flow cytometry using fluorescently labeled anti-MHC I antibodies as recently described (Hinz et al., 2014).

1. Cultivate TAP-deficient cell lines in DMEM media supplemented with 10% FCS and 100 μg/ml Hygromycin B.
2. One day prior to transfection, seed 2.5×10^5 cells in 2 ml DMEM/10% FCS media.
3. Transfect TAP1-deficient BRE-169 or TAP2-deficient STF-169 cells with the respective TAP1-mVenus or TAP2-mCerulean constructs using X-tremeGENE (Roche).
4. Dilute 2 μg plasmid DNA and 6 μl X-tremeGENE separately in 150 μl serum-free DMEM media. Mix DNA and transfection reagent solutions and incubate for 30 min at room temperature.
5. Replace culture media with 500 μl serum-free DMEM media and add the transfection mix dropwise. Change culture media to 2 ml DMEM/ 10% FCS after 4 h.
6. Two days posttransfection, harvest 1×10^6 cells by trypsinizing and washing cells twice with ice-cold FACS buffer (phosphate-buffered saline, pH 7.3, supplemented with 2% (v/v) FCS).
7. Resuspend the cells in 250 μl FACS buffer supplemented with 5% (w/v) bovine serum albumin and incubate on ice for 20 min to block nonspecific binding.
8. Wash the cells twice with FACS buffer (centrifuge for 3 min at $300 \times g$) and add 0.2 μg polyclonal phycoerythrin-coupled anti-human HLA-ABC (W6/32) antibody (BioLegend) diluted in 250 μl FACS buffer.

In parallel, perform an isotype control using an unspecific phycoerythrin-coupled antibody. Incubate the cell suspension for 1 h on ice in the dark.

9. Wash the cells twice with FACS buffer to remove excess of antibody and resuspend the cell pellet in 500 μl FACS buffer. Directly proceed to Step 10.
10. Collect flow cytometry data at a FACSCanto II (BD Biosciences) and process the data with FlowJo software (www.flowjo.com). For a typical experiment at least 3×10^4 cells should be evaluated.

5. SUMMARY

Integrative approaches combining X-ray crystallography, single-particle cryo-EM, as well as crosslinking and mass spectrometry (CL-MS) will be instrumental to generate quasi atomic models of large multiprotein membrane complexes. Here, we outlined multicolor fluorescence-based screening approaches for the production of stoichiometric, homogeneous, and monodisperse specimen in high quality and quantity for structural studies. As an example, we detailed vector design together with expression and purification strategies for the production of a recombinant, heterodimeric membrane complex: TAP. The strategies described in this contribution for the ABC transporter TAP will be beneficial for structural biology projects of large hetero-oligomeric multiprotein membrane complexes in the future.

ACKNOWLEDGMENTS

We thank Valentina Herbring, Sabine Eggensperger, Christine Le Gal, as well as Drs. Rupert Abele, Inga Hänelt, Peter Mayerhofer, and Fabian Seyffer for helpful comments on the manuscript. The German Research Foundation (SFB 807–Transport and Communication across Biological Membranes to R. T.) and the European Drug Initiative on Channels and Transporters (EDICT to R. T.) funded by the European Commission Seventh Framework Program supported this work.

REFERENCES

Abele, R., & Tampé, R. (2009). Peptide trafficking and translocation across membranes in cellular signaling and self-defense strategies. *Current Opinion in Cell Biology*, *21*(4), 508–515.

Abele, R., & Tampé, R. (2011). The TAP translocation machinery in adaptive immunity and viral escape mechanisms. *Essays in Biochemistry*, *50*(1), 249–264.

Ahn, K., Meyer, T. H., Uebel, S., Sempé, P., Djaballah, H., Yang, Y., et al. (1996). Molecular mechanism and species specificity of TAP inhibition by herpes simplex virus ICP47. *EMBO Journal*, *15*(13), 3247–3255.

Alexandrov, A. I., Mileni, M., Chien, E. Y., Hanson, M. A., & Stevens, R. C. (2008). Microscale fluorescent thermal stability assay for membrane proteins. *Structure*, *16*(3), 351–359.

Allegretti, M., Mills, D. J., McMullan, G., Kuhlbrandt, W., & Vonck, J. (2014). Atomic model of the F420-reducing [NiFe] hydrogenase by electron cryo-microscopy using a direct electron detector. *eLife*, *3*, e01963.

Amunts, A., Brown, A., Bai, X. C., Llacer, J. L., Hussain, T., Emsley, P., et al. (2014). Structure of the yeast mitochondrial large ribosomal subunit. *Science*, *343*(6178), 1485–1489.

Andrell, J., & Tate, C. G. (2013). Overexpression of membrane proteins in mammalian cells for structural studies. *Molecular Membrane Biology*, *30*(1), 52–63.

Arinaminpathy, Y., Khurana, E., Engelman, D. M., & Gerstein, M. B. (2009). Computational analysis of membrane proteins: The largest class of drug targets. *Drug Discovery Today*, *14*(23–24), 1130–1135.

Aslanidis, C., & de Jong, P. J. (1990). Ligation-independent cloning of PCR products (LIC-PCR). *Nucleic Acids Research*, *18*(20), 6069–6074.

Bai, X. C., Fernandez, I. S., McMullan, G., & Scheres, S. H. (2013). Ribosome structures to near-atomic resolution from thirty thousand cryo-EM particles. *eLife*, *2*, e00461.

Bangia, N., Lehner, P. J., Hughes, E. A., Surman, M., & Cresswell, P. (1999). The N-terminal region of tapasin is required to stabilize the MHC class I loading complex. *European Journal of Immunology*, *29*(6), 1858–1870.

Bernaudat, F., Frelet-Barrand, A., Pochon, N., Dementin, S., Hivin, P., Boutigny, S., et al. (2011). Heterologous expression of membrane proteins: Choosing the appropriate host. *PLoS One*, *6*(12), e29191.

Bill, R. M., Henderson, P. J., Iwata, S., Kunji, E. R., Michel, H., Neutze, R., et al. (2011). Overcoming barriers to membrane protein structure determination. *Nature Biotechnology*, *29*(4), 335–340.

Bonander, N., & Bill, R. M. (2012). Optimising yeast as a host for recombinant protein production (review). *Methods in Molecular Biology*, *866*, 1–9.

Buschmann, S., Warkentin, E., Xie, H., Langer, J. D., Ermler, U., & Michel, H. (2010). The structure of cbb3 cytochrome oxidase provides insights into proton pumping. *Science*, *329*(5989), 327–330.

Condreay, J. P., & Kost, T. A. (2007). Baculovirus expression vectors for insect and mammalian cells. *Current Drug Targets*, *8*(10), 1126–1131.

Day, R. N., & Davidson, M. W. (2009). The fluorescent protein palette: Tools for cellular imaging. *Chemical Society Reviews*, *38*(10), 2887–2921.

de la Salle, H., Hanau, D., Fricker, D., Urlacher, A., Kelly, A., Salamero, J., et al. (1994). Homozygous human TAP peptide transporter mutation in HLA class I deficiency. *Science*, *265*(5169), 237–241.

de la Salle, H., Zimmer, J., Fricker, D., Angenieux, C., Cazenave, J. P., Okubo, M., et al. (1999). HLA class I deficiencies due to mutations in subunit 1 of the peptide transporter TAP1. *The Journal of Clinical Investigation*, *103*(5), R9–R13.

Deisenhofer, J., Epp, O., Miki, K., Huber, R., & Michel, H. (1985). Structure of the protein subunits in the photosynthetic reaction centre of Rhodopseudomonas viridis at 3A resolution. *Nature*, *318*(6047), 618–624.

Dick, T. P., Bangia, N., Peaper, D. R., & Cresswell, P. (2002). Disulfide bond isomerization and the assembly of MHC class I-peptide complexes. *Immunity*, *16*(1), 87–98.

Doyle, D. A., Morais Cabral, J., Pfuetzner, R. A., Kuo, A., Gulbis, J. M., Cohen, S. L., et al. (1998). The structure of the potassium channel: Molecular basis of K+ conduction and selectivity. *Science*, *280*(5360), 69–77.

Drew, D. E., Lerch, M., Kunji, E., Slotboom, D. J., & de Gier, J. W. (2006). Optimization of membrane protein overexpression and purification using GFP fusions. *Nature Methods*, *3*(4), 303–313.

Drew, D. E., von Heijne, G., Nordlund, P., & de Gier, J. W. (2001). Green fluorescent protein as an indicator to monitor membrane protein overexpression in Escherichia coli. *FEBS Letters*, *507*(2), 220–224.

Efremov, R. G., Baradaran, R., & Sazanov, L. A. (2010). The architecture of respiratory complex I. *Nature*, *465*(7297), 441–445.

Eggensperger, S., Fisette, O., Parcej, D., Schäfer, L. V., & Tampé, R. (2014). An annular lipid belt is essential for allosteric coupling and viral inhibition of the antigen translocation complex TAP (transporter associated with antigen processing). *The Journal of Biological Chemistry*, *289*(48), 33098–33108.

Everett, M. W., & Edidin, M. (2007). Tapasin increases efficiency of MHC I assembly in the endoplasmic reticulum but does not affect MHC I stability at the cell surface. *The Journal of Immunology*, *179*(11), 7646–7652.

Fitzgerald, D. J., Berger, P., Schaffitzel, C., Yamada, K., Richmond, T. J., & Berger, I. (2006). Protein complex expression by using multigene baculoviral vectors. *Nature Methods*, *3*(12), 1021–1032.

Frelet-Barrand, A., Boutigny, S., Kunji, E. R., & Rolland, N. (2010). Membrane protein expression in Lactococcus lactis. *Methods in Molecular Biology*, *601*, 67–85.

Frickel, E. M., Riek, R., Jelesarov, I., Helenius, A., Wuthrich, K., & Ellgaard, L. (2002). TROSY-NMR reveals interaction between ERp57 and the tip of the calreticulin P-domain. *Proceedings of the National Academy of Sciences of the United States of America*, *99*(4), 1954–1959.

Geertsma, E. R., & Dutzler, R. (2011). A versatile and efficient high-throughput cloning tool for structural biology. *Biochemistry*, *50*(15), 3272–3278.

Geertsma, E. R., Groeneveld, M., Slotboom, D. J., & Poolman, B. (2008). Quality control of overexpressed membrane proteins. *Proceedings of the National Academy of Sciences of the United States of America*, *105*(15), 5722–5727.

Gonzales, E. B., Kawate, T., & Gouaux, E. (2009). Pore architecture and ion sites in acid-sensing ion channels and P2X receptors. *Nature*, *460*(7255), 599–604.

Hammon, J., Palanivelu, D. V., Chen, J., Patel, C., & Minor, D. L., Jr. (2009). A green fluorescent protein screen for identification of well-expressed membrane proteins from a cohort of extremophilic organisms. *Protein Science*, *18*(1), 121–133.

Harbers, M. (2014). Wheat germ systems for cell-free protein expression. *FEBS Letters*, *588*(17), 2762–2773.

Hattori, M., Hibbs, R. E., & Gouaux, E. (2012). A fluorescence-detection size-exclusion chromatography-based thermostability assay for membrane protein precrystallization screening. *Structure*, *20*(8), 1293–1299.

He, Y., Wang, K., & Yan, N. (2014). The recombinant expression systems for structure determination of eukaryotic membrane proteins. *Protein & Cell*, *5*(9), 658–672.

Higgins, C. F. (1992). ABC transporters: From microorganisms to man. *Annual Review of Cell and Developmental Biology*, *8*, 67–113.

Hinz, A., Jedamzick, J., Herbring, V., Fischbach, H., Hartmann, J., Parcej, D., et al. (2014). Assembly and function of the major histocompatibility complex (MHC) I peptide-loading complex are conserved across higher vertebrates. *The Journal of Biological Chemistry*, *289*(48), 33109–33117.

Hulpke, S., & Tampé, R. (2013). The MHC I loading complex: A multitasking machinery in adaptive immunity. *Trends in Biochemical Sciences*, *38*(8), 412–420.

Hulpke, S., Tomioka, M., Kremmer, E., Ueda, K., Abele, R., & Tampé, R. (2012). Direct evidence that the N-terminal extensions of the TAP complex act as autonomous interaction scaffolds for the assembly of the MHC I peptide-loading complex. *Cellular and Molecular Life Sciences*, *69*(19), 3317–3327.

Junge, F., Schneider, B., Reckel, S., Schwarz, D., Dötsch, V., & Bernhard, F. (2008). Large-scale production of functional membrane proteins. *Cellular and Molecular Life Sciences*, *65*(11), 1729–1755.

Kawate, T., & Gouaux, E. (2006). Fluorescence-detection size-exclusion chromatography for precrystallization screening of integral membrane proteins. *Structure*, *14*(4), 673–681.

Kawate, T., Michel, J. C., Birdsong, W. T., & Gouaux, E. (2009). Crystal structure of the ATP-gated P2X(4) ion channel in the closed state. *Nature*, *460*(7255), 592–598.

Kim, J., Wu, S., Tomasiak, T. M., Mergel, C., Winter, M. B., Stiller, S. B., et al. (2014). Subnanometre-resolution electron cryomicroscopy structure of a heterodimeric ABC exporter. *Nature*. http://dx.doi.org/10.1038/nature13872.

Kost, T. A., Condreay, J. P., & Jarvis, D. L. (2005). Baculovirus as versatile vectors for protein expression in insect and mammalian cells. *Nature Biotechnology*, *23*(5), 567–575.

Kremers, G. J., Gilbert, S. G., Cranfill, P. J., Davidson, M. W., & Piston, D. W. (2011). Fluorescent proteins at a glance. *Journal of Cell Science*, *124*(Pt 2), 157–160.

Lee, L. K., Stewart, A. G., Donohoe, M., Bernal, R. A., & Stock, D. (2010). The structure of the peripheral stalk of Thermus thermophilus H+-ATPase/synthase. *Nature Structural & Molecular Biology*, *17*(3), 373–378.

Li, M. Z., & Elledge, S. J. (2007). Harnessing homologous recombination in vitro to generate recombinant DNA via SLIC. *Nature Methods*, *4*(3), 251–256.

Li, X., Mooney, P., Zheng, S., Booth, C. R., Braunfeld, M. B., Gubbens, S., et al. (2013). Electron counting and beam-induced motion correction enable near-atomic-resolution single-particle cryo-EM. *Nature Methods*, *10*(6), 584–590.

Liao, M., Cao, E., Julius, D., & Cheng, Y. (2013). Structure of the TRPV1 ion channel determined by electron cryo-microscopy. *Nature*, *504*(7478), 107–112.

Liu, C. Y., Chen, H. Z., & Chao, Y. C. (2010). Maximizing baculovirus-mediated foreign proteins expression in mammalian cells. *Current Gene Therapy*, *10*(3), 232–241.

Liu, W., Hanson, M. A., Stevens, R. C., & Cherezov, V. (2010). LCP-Tm: An assay to measure and understand stability of membrane proteins in a membrane environment. *Biophysical Journal*, *98*(8), 1539–1548.

Lu, P., Bai, X. C., Ma, D., Xie, T., Yan, C., Sun, L., et al. (2014). Three-dimensional structure of human gamma-secretase. *Nature*, *512*(7513), 166–170.

Mancia, F., & Love, J. (2010). High-throughput expression and purification of membrane proteins. *Journal of Structural Biology*, *172*(1), 85–93.

Martinez Molina, D., Cornvik, T., Eshaghi, S., Haeggstrom, J. Z., Nordlund, P., & Sabet, M. I. (2008). Engineering membrane protein overproduction in Escherichia coli. *Protein Science*, *17*(4), 673–680.

Mesa, P., Deniaud, A., Montoya, G., & Schaffitzel, C. (2013). Directly from the source: Endogenous preparations of molecular machines. *Current Opinion in Structural Biology*, *23*(3), 319–325.

Neefjes, J., Jongsma, M. L., Paul, P., & Bakke, O. (2011). Towards a systems understanding of MHC class I and MHC class II antigen presentation. *Nature Reviews. Immunology*, *11*(12), 823–836.

Neumann, L., Kraas, W., Uebel, S., Jung, G., & Tampé, R. (1997). The active domain of the herpes simplex virus protein ICP47: A potent inhibitor of the transporter associated with antigen processing. *Journal of Molecular Biology*, *272*(4), 484–492.

Newstead, S., Kim, H., von Heijne, G., Iwata, S., & Drew, D. (2007). High-throughput fluorescent-based optimization of eukaryotic membrane protein overexpression and purification in Saccharomyces cerevisiae. *Proceedings of the National Academy of Sciences of the United States of America*, *104*(35), 13936–13941.

Nie, Y., Bellon-Echeverria, I., Trowitzsch, S., Bieniossek, C., & Berger, I. (2014). Multiprotein complex production in insect cells by using polyproteins. *Methods in Molecular Biology*, *1091*, 131–141.

Niesen, F. H., Berglund, H., & Vedadi, M. (2007). The use of differential scanning fluorimetry to detect ligand interactions that promote protein stability. *Nature Protocols*, *2*(9), 2212–2221.

Oliver, J. D., Roderick, H. L., Llewellyn, D. H., & High, S. (1999). ERp57 functions as a subunit of specific complexes formed with the ER lectins calreticulin and calnexin. *Molecular Biology of the Cell, 10*(8), 2573–2582.

Ortmann, B., Copeman, J., Lehner, P. J., Sadasivan, B., Herberg, J. A., Grandea, A. G., et al. (1997). A critical role for tapasin in the assembly and function of multimeric MHC class I-TAP complexes. *Science, 277*(5330), 1306–1309.

Parcej, D., Guntrum, R., Schmidt, S., Hinz, A., & Tampé, R. (2013). Multicolour fluorescence-detection size-exclusion chromatography for structural genomics of membrane multiprotein complexes. *PLoS One, 8*(6), e67112.

Parcej, D., & Tampé, R. (2010). ABC proteins in antigen translocation and viral inhibition. *Nature Chemical Biology, 6*(8), 572–580.

Peaper, D. R., Wearsch, P. A., & Cresswell, P. (2005). Tapasin and ERp57 form a stable disulfide-linked dimer within the MHC class I peptide-loading complex. *EMBO Journal, 24*(20), 3613–3623.

Pedelacq, J. D., Piltch, E., Liong, E. C., Berendzen, J., Kim, C. Y., Rho, B. S., et al. (2002). Engineering soluble proteins for structural genomics. *Nature Biotechnology, 20*(9), 927–932.

Privé, G. G. (2007). Detergents for the stabilization and crystallization of membrane proteins. *Methods, 41*(4), 388–397.

Rizvi, S. M., & Raghavan, M. (2010). Mechanisms of function of tapasin, a critical major histocompatibility complex class I assembly factor. *Traffic, 11*(3), 332–347.

Rosenbaum, D. M., Cherezov, V., Hanson, M. A., Rasmussen, S. G., Thian, F. S., Kobilka, T. S., et al. (2007). GPCR engineering yields high-resolution structural insights into beta2-adrenergic receptor function. *Science, 318*(5854), 1266–1273.

Rosenbaum, D. M., Rasmussen, S. G., & Kobilka, B. K. (2009). The structure and function of G-protein-coupled receptors. *Nature, 459*(7245), 356–363.

Russell, R. B., & Eggleston, D. S. (2000). New roles for structure in biology and drug discovery. *Nature Structural Biology*, 7(Suppl.), 928–930.

Sadasivan, B., Lehner, P. J., Ortmann, B., Spies, T., & Cresswell, P. (1996). Roles for calreticulin and a novel glycoprotein, tapasin, in the interaction of MHC class I molecules with TAP. *Immunity, 5*(2), 103–114.

Sahdev, S., Khattar, S. K., & Saini, K. S. (2008). Production of active eukaryotic proteins through bacterial expression systems: A review of the existing biotechnology strategies. *Molecular and Cellular Biochemistry, 307*(1–2), 249–264.

Seyffer, F., & Tampé, R. (2014). ABC transporters in adaptive immunity. *Biochimica et Biophysica Acta*. http://dx.doi.org/10.1016/j.bbagen.2014.05.022.

Sonoda, Y., Cameron, A., Newstead, S., Omote, H., Moriyama, Y., Kasahara, M., et al. (2010). Tricks of the trade used to accelerate high-resolution structure determination of membrane proteins. *FEBS Letters, 584*(12), 2539–2547.

Sung, L. Y., Chen, C. L., Lin, S. Y., Li, K. C., Yeh, C. L., Chen, G. Y., et al. (2014). Efficient gene delivery into cell lines and stem cells using baculovirus. *Nature Protocols, 9*(8), 1882–1899.

Tomazin, R., Hill, A. B., Jugovic, P., York, I., van Endert, P., Ploegh, H. L., et al. (1996). Stable binding of the herpes simplex virus ICP47 protein to the peptide binding site of TAP. *EMBO Journal, 15*(13), 3256–3266.

Trowitzsch, S., Palmberger, D., Fitzgerald, D., Takagi, Y., & Berger, I. (2012). MultiBac complexomics. *Expert Review of Proteomics, 9*(4), 363–373.

Vedadi, M., Niesen, F. H., Allali-Hassani, A., Fedorov, O. Y., Finerty, P. J., Jr., Wasney, G. A., et al. (2006). Chemical screening methods to identify ligands that promote protein stability, protein crystallization, and structure determination. *Proceedings of the National Academy of Sciences of the United States of America, 103*(43), 15835–15840.

von der Hocht, I., van Wonderen, J. H., Hilbers, F., Angerer, H., MacMillan, F., & Michel, H. (2011). Interconversions of P and F intermediates of cytochrome c oxidase from Paracoccus denitrificans. *Proceedings of the National Academy of Sciences of the United States of America*, *108*(10), 3964–3969.

Wearsch, P. A., Peaper, D. R., & Cresswell, P. (2011). Essential glycan-dependent interactions optimize MHC class I peptide loading. *Proceedings of the National Academy of Sciences of the United States of America*, *108*(12), 4950–4955.

Wittig, I., Braun, H. P., & Schägger, H. (2006). Blue native PAGE. *Nature Protocols*, *1*(1), 418–428.

Wittig, I., & Schägger, H. (2008). Features and applications of blue-native and clear-native electrophoresis. *Proteomics*, *8*(19), 3974–3990.

Yildirim, M. A., Goh, K. I., Cusick, M. E., Barabasi, A. L., & Vidal, M. (2007). Drug-target network. *Nature Biotechnology*, *25*(10), 1119–1126.

Zhang, W., Wearsch, P. A., Zhu, Y., Leonhardt, R. M., & Cresswell, P. (2011). A role for UDP-glucose glycoprotein glucosyltransferase in expression and quality control of MHC class I molecules. *Proceedings of the National Academy of Sciences of the United States of America*, *108*(12), 4956–4961.

CHAPTER TWO

A Novel Screening Approach for Optimal and Functional Fusion of T4 Lysozyme in GPCRs

Elizabeth Mathew, Mark E. Dumont[1]
Department of Biochemistry and Biophysics, P.O. Box 712, University of Rochester School of Medicine and Dentistry, Rochester, New York, USA
[1]Corresponding author: e-mail address: mark_dumont@urmc.rochester.edu

Contents

Abstract

Determination of high-resolution, three-dimensional structures of transmembrane proteins (TMPs) has, in many cases, only been accomplished through the use of stabilized variant forms of the proteins being studied. For the important G protein-coupled receptor superfamily, this has most often been achieved by inserting a stable soluble protein, such as T4 lysozyme (T4L), in an internal loop of a receptor. However, creation of such fusion proteins generally results in loss of the ability of receptors to activate their cognate cytoplasmic G proteins. Furthermore, the criteria for designing fusions that minimally perturb receptor structure are not well established. We describe here a method for creating a library of receptor variants containing T4L inserted into an internal loop at varying positions and as replacements for varying amounts of the original receptor sequence. We also describe methods for screening for variants displaying maximal expression levels, ligand binding capacity, and signaling function. When applied to

Methods in Enzymology, Volume 557
ISSN 0076-6879
http://dx.doi.org/10.1016/bs.mie.2014.12.031

the yeast α-factor receptor, Ste2p, this approach allowed recovery of well-expressed receptor variants containing internally fused T4L that retained nearly normal signaling function. The approach we describe can be readily adapted to creation of stabilized fusions of other TMPs expressed in yeast or other expression systems.

1. INTRODUCTION

Genes for transmembrane proteins (TMPs) make up 20–30% of most genomes (Wallin & Heijne, 1998), and based on their critical roles in cell physiology, TMPs are the targets of a large fraction of clinically useful drugs. Thus, determination of their structures and modes of action is of considerable importance. However, the number of available high-resolution TMP structures remains relatively low (see Web page of Stephen White; http://blanco.biomol.uci.edu/mpstruc/). Two explanations that are frequently invoked for the difficulty of obtaining well-diffracting crystals of TMPs for use in X-ray diffraction studies are (1) TMPs are structurally dynamic with flexible regions that make it difficult for the protein to adopt the uniform conformation required for packing into a crystal; and (2) surfaces of transmembrane regions of TMPs do not participate as readily as the surfaces of globular soluble proteins in protein–protein interactions that are required for crystal packing. Introducing a stable soluble protein into a surface-exposed region of a TMP provides a way of potentially circumventing both of these problems, since such an insertion at an internal location in a TMP can reduce overall flexibility and since the presence of a readily crystallizable soluble domain can provide intermolecular contacts required for crystallization (Chun et al., 2012; Engel, Chen, & Privé, 2002).

Fusion of T4 lysozyme (T4L) at internal locations in TMPs has, to date, provided a successful approach for determination of the structures of diverse members of the important GPCR superfamily (Cherezov et al., 2007; Chien et al., 2010; Dore et al., 2014; Fenalti et al., 2014; Granier et al., 2012; Haga et al., 2012; Hanson et al., 2012; Hollenstein et al., 2013; Jaakola et al., 2008; Kruse et al., 2012; Manglik et al., 2012; Rosenbaum et al., 2007, 2011; Shimamura et al., 2011; Tan et al., 2013; Wacker et al., 2013; Wang et al., 2013; White et al., 2012; Wu et al., 2010, 2012; Xu et al., 2011; Zhang et al., 2012). Similar internal fusions of a thermally stabilized apocytochrome *b* (Wacker et al., 2013; Wang et al., 2013; Zhang, Zhang, Gao, Paoletta et al., 2014; Zhang, Zhang, Gao, Zhang, et al., 2014) and rubredoxin (Tan et al., 2013) have also yielded GPCR structures.

In addition, several GPCR structures have been obtained by crystallization of constructs in which the stabilizing protein partner was fused at the N-terminus of the receptor sequence, rather than at an internal position (Fenalti et al., 2014; Rasmussen, Choi, et al., 2011; Rasmussen, DeVree, et al., 2011; Siu et al., 2013; Thompson et al., 2012; Wu et al., 2014), suggesting that the role of fusion proteins in facilitation of formation of inter-molecular contacts may be more important for promoting crystallization than the reduction of TMP flexibility. Alternative approaches for obtaining GPCR structures that do not involve the use of fusion proteins have included cocrystallization with anti-receptor antibodies (Kruse et al., 2013; Rasmussen, Choi, et al., 2011; Rasmussen et al., 2007; Rasmussen, DeVree, et al., 2011; Ring et al., 2013) and the crystallization of GPCR variants that have been thermostabilized by the introduction of point mutations and deletions (Egloff et al., 2014; Scott, Kummer, Tremmel, & Pluckthun, 2013; Tate, 2012). In fact, most fusion protein-containing GPCR variants used for successful structure determination have also incorporated point mutations and truncations designed to further enhance protein stability and reduce flexibility.

Insertion of stable soluble proteins into TMPs for the purpose of promoting crystallization has generally been performed as an insertion or replacement for residues in the third intracellular (IC3) loop of GPCRs. This is based on the expectation that this region in receptors is flexible and may control the relative motion of surrounding helices (Rosenbaum et al., 2007), as well as reducing the likelihood that insertion of the fusion partner will disrupt the overall GPCR structure (Chun et al., 2012). While such insertions often do not disrupt the overall structures of receptors (based on retention of at least some ligand binding affinity by the fusion constructs), they do generally result in loss of the ability to activate the cognate trimeric G protein (Rosenbaum et al., 2007), which is not surprising since the IC3 loop is often the site of functional interactions between GPCRs and the trimeric G proteins that are their immediate downstream effectors. Furthermore, the criteria for establishing particular junction sites for insertion of fusion proteins and for determining the amount of receptor sequence that they replace are not well established. A successful fusion must provide an optimal match between the spacing and orientation of the N- and C-termini of the inserted protein and the junction sites in the GPCR (Chun et al., 2012; Engel et al., 2002). Thus, creation of a useful fusion may require synthesis or construction of multiple versions of fused genes (Rosenbaum et al., 2007).

We describe here a strategy for generating and screening a library of variant forms of a receptor containing insertions of T4L at different points in the third intracellular (IC3) loop as a replacement for different numbers of amino acid residues in the loop sequence (Mathew, Ding, Naider, & Dumont, 2013). By expressing receptors in yeast, it is possible to create randomized libraries of variants with lysozyme insertions and to directly screen for constructs with maximal overall expression levels, numbers of ligand binding sites at the cell surface, and even receptor function, assayed based on activation of the cognate G protein. The approaches were successful in identifying receptors with insertions in the IC3 loop that retain nearly full signaling function. This suggests that, when attached via appropriate linking sequences, the T4L molecule can be inserted into this loop without preventing access of the G protein to relevant surfaces of the activated receptor.

The described protocols were developed using Ste2p, an endogenous GPCR of the yeast *Saccharomyces cerevisiae* used as a tractable initial system for which no three-dimensional structure is yet available. However, similar approaches could be used to identify insertion-containing variants of the numerous mammalian GPCRs that have been reported to be capable of activating the yeast pheromone response pathway (Brown et al., 2000; Dowell & Brown, 2009; King, Dohlman, Thorner, Caron, & Lefkowitz, 1990; Pausch, 1997), or to other TMPs for which it is possible to assay for function in intact cells. Furthermore, yeast species have been used as expression systems for structure determination of GPCRs and other TMPs (Clark et al., 2010), including for the determination of the crystal structure of the human H1 histamine receptor (Shimamura et al., 2011; Shiroishi et al., 2011). Ste2p is the receptor for the yeast mating pheromone α-factor, a 13-residue peptide. Binding of this ligand results in activation of a cytoplasmic heterotrimeric G protein, which, in turn, activates a MAP kinase pathway, ultimately leading to transcriptional responses, changes in cell shape, cell cycle arrest, and fusion with yeast cells of the opposite mating type.

2. OVERALL STRATEGY

A library of mutant receptors was created in which T4L was inserted at different points in the predicted IC3 loop, accompanied by different extents of deletion of residues in this loop. (We refer to the region targeted for insertion and/or replacement by T4L as the "replacement region.") As shown in Fig. 1, this was accomplished by initially creating an allele of the

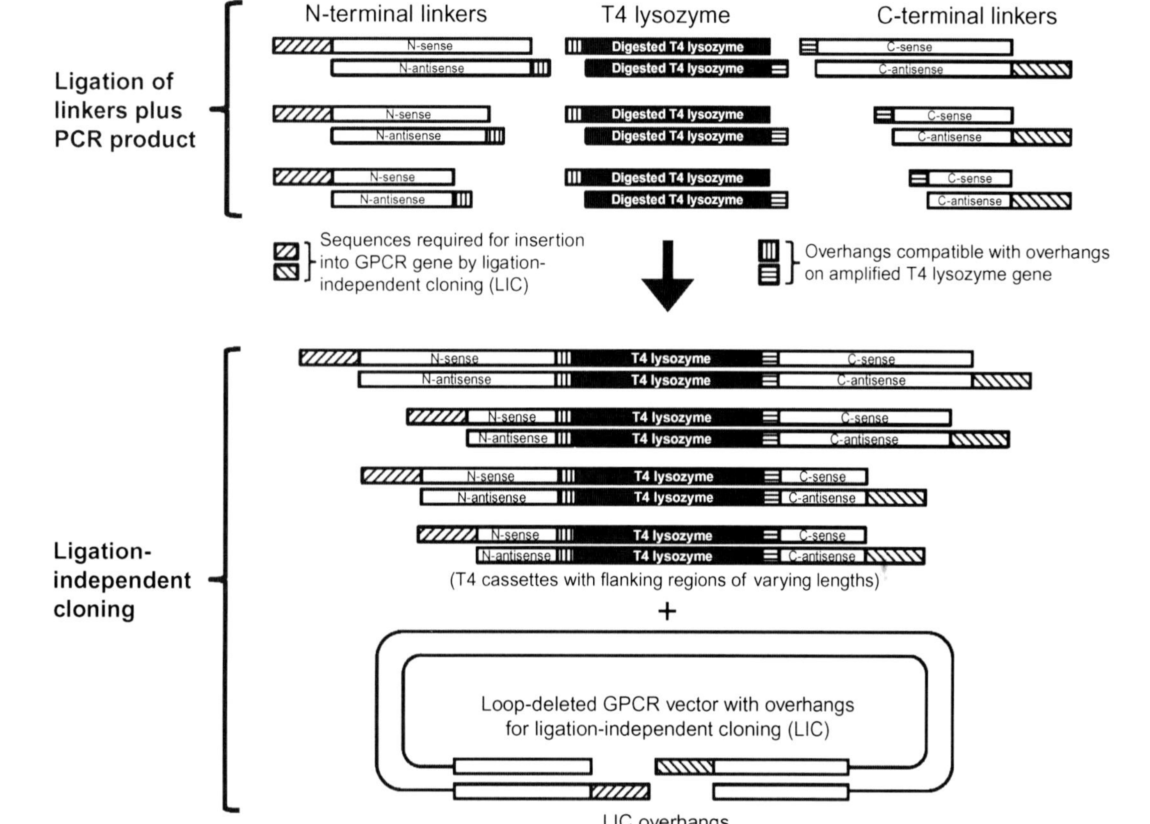

Figure 1 Overall strategy for library construction. Sense and antisense oligonucleotides are annealed to form linkers that have progressive amino acid deletions of the GPCR replacement region. The N- and C-terminal linkers have deletions from the ends that are ligated with the N- and C-terminal ends of T4L, respectively. Ligation of the linkers and the restriction-digested T4L PCR product results in a library of T4L cassettes that have varied lengths of the GPCR replacement region sequence flanking the T4L gene. These cassettes are then introduced into the GPCR gene that is lacking the replacement region.

receptor-encoding gene *STE2* in which most of the IC3 loop was deleted and replaced by a unique restriction site. The receptor-encoding gene was also fused at its C-terminus to green fluorescent protein (GFP). A cysteine-free version of T4L was PCR amplified, introducing restriction sites at the 5′ and 3′ ends of the coding region. A family of double-stranded DNA linkers corresponding to different lengths of sequence from the receptor third loop was prepared from synthetic oligonucleotides. The linkers contained overhangs that allow them to be annealed and ligated to the 5′ and 3′ ends of the amplified T4L gene that had been restriction digested to create complementary overhangs. The ligated products containing T4L flanked by various lengths of sequence corresponding to the IC3 loop can be conveniently inserted into the IC3 loop-deleted receptor allele via ligation-independent cloning (LIC) (Alexandrov et al., 2004; Aslanidis & de Jong, 1990; Dieckman, Gu, Stols, Donnelly, & Collart, 2002). Following transformation into bacteria, a library of plasmids is isolated and used to transform a yeast strain lacking the normal *STE2* gene, so that the T4L-containing variants are the only α-factor receptors present in the cells. Screening for overall expression is accomplished by flow cytometry of the library of transformed cells, detecting expression of the Ste2p–GFP fusion. Screening for ligand binding capability is also conducted using flow cytometry to detect binding of a fluorescently labeled ligand. Screening for receptors' signaling function is conducted based on expression of a *FUS1-HIS3* reporter gene that undergoes transcriptional induction in response to pheromone pathway activation, allowing growth of the host *his3-Δ* strain on medium lacking histidine. Further characterization of receptor expression levels and signaling function can be conducted using quantitative assays of the binding of fluorescent ligands and of reporter gene induction as well as by immunoblotting of individual isolates.

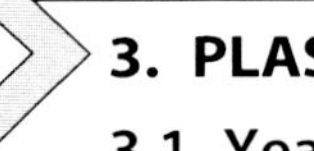

3. PLASMIDS, YEAST STRAINS, AND MEDIA

3.1 Yeast strains

A575—*Mat**a** ste2-Δ bar1-Δ far1 cry1*^{R} *ade2-1 his4-580 lys2*$_{oc}$ *trp1*$_{am}$ *tyr1-*$_{oc}$*SUP4-3*^{ts} *leu2 ura3 FUS1::*p*[FUS1-lacZ TRP1]* (Sommers et al., 2000)

A2953—*Mat**a** ste2-Δ bar1-Δ far1-Δ leu 2*$^{-}$ *his3-Δ200 ura3*$^{-}$ *FUS1-HIS3 FUS1-lacZ* (Taslimi, Mathew, Čelić, Wessel, & Dumont, 2012)

BJ5460—(ATCC® 208285) *Mat**a** ura3*-52 *trp1 lys2-801 leu2Δ1 his3Δ200 pep4::HIS3 prb1Δ1.6R can1 GAL* (Jones, 1991)

3.2 Plasmids

pMD1167—Full-length *STE2* with a C-terminal GFP fusion, under the control of the *STE2* promoter (confers ampicillin resistance in *Escherichia coli* and uracil auxotrophy in *S. cerevisiae*) (Gehret, Bajaj, Naider, & Dumont, 2006).

pMD1859—Created by site-directed mutagenesis of plasmid pMD1167, replacing the IC3 loop of *STE2* with nucleotide sequences that code for two *Bsa*BI restriction sites flanking a *Sma*I site (confers ampicillin resistance in *E. coli* and uracil auxotrophy in *S. cerevisiae*) (Mathew et al., 2013).

pSGP46—Expression vector that contains the alcohol dehydrogenase (*ADH2*) promoter (confers ampicillin resistance in *E. coli* and uracil auxotrophy in *S. cerevisiae*) (Clark et al., 2010).

3.3 Media

Yeast media were prepared as described by Sherman (2002).

4. LIBRARY CONSTRUCTION

4.1 Generation of T4L-containing cassettes with variable lengths of flanking regions

1. Design and purchase a set of synthetic oligonucleotides corresponding to the replacement region of the receptor according to the following criteria:
   - **i.** The oligonucleotides will consist of sense and antisense strands encoding progressively shortened (by one codon at a time) portions of the replacement sequence. The longest linkers encode the entire region while the smallest ones encode only one or two codons.
   - **ii.** Two distinct sets of oligonucleotides are designed for the N- and C-terminal sides of the insertion. In each case, the oligonucleotides are designed to provide progressive deletions at the site of junction to T4L.
   - **iii.** The two sets of oligonucleotides are designed to provide overlaps that will allow annealing and ligation, respectively, to two different unique restriction sites introduced at the N- and C-terminal sides of the T4L insert.
   - **iv.** The oligonucleotides are designed to introduce Gly-Ser sequences immediately adjacent to the restriction site overhang used for fusion

to T4L. This sequence is introduced to provide flexibility at the fusion junction to enhance the likelihood of correct folding of T4L and the insertion-containing receptor.

v. The oligonucleotides are designed to contain sequences derived from the receptor at the end where they are to be fused into the receptor. These receptor-derived sequences consist of 16–18 bases containing only three of the four types of DNA nucleotide bases, as required for the use of LIC.

2. Prepare double-stranded forms of the linkers by annealing pairs of complementary oligonucleotides (5 pmol of each oligo in 50 μl annealing buffer (100 m*M* NaCl, 50 m*M* Tris–HCl, 10 m*M* $MgCl_2$, 1 m*M* DTT, pH 7.9)). Annealing is performed by placing the reactions in a 95 °C water bath (400 ml) and then allowing them to cool on the benchtop to room temperature, followed by storage at 4 °C.

3. PCR amplify the T4L gene (minus the stop codon) with primers containing 5′ extensions that introduce restriction sites at both ends for in-frame, directional ligation with the linkers. The restriction sites, when digested, should create overhangs that will not ligate with each other (either by using a single enzyme that cuts outside of the recognition site or by using two different enzymes).

4. Digest the amplified T4L gene with the restriction enzymes corresponding to the sites introduced adjacent to the T4L gene. Gel purify using a commercially available kit.

5. Ligate the mixed linkers to the amplified T4L gene. Mix 0.1 pmol of each annealed linker such that the final concentration of total mixed linkers is 0.1 pmol/μl. Incubate 9 μl (0.9 pmol) of this annealed linker mix with 0.3 pmol of digested amplified T4L gene in the presence of 1 m*M* ATP and 700 units T4 DNA ligase (New England Biolabs) in a total volume of 35 μl annealing buffer. Incubate overnight at 16 °C.

6. Purify the ligation products by electrophoresis on an agarose gel, excising a gel slice encompassing the size range of the expected products. Extract DNA from the gel using a commercially available kit.

4.2 Insertion of cassettes containing the T4L gene into the replacement region of the receptor

1. Create an allele of the receptor gene that contains a deletion of the replacement region (the region of insertion or replacement by T4L) (Fig. 2). A unique restriction site should be inserted as a replacement

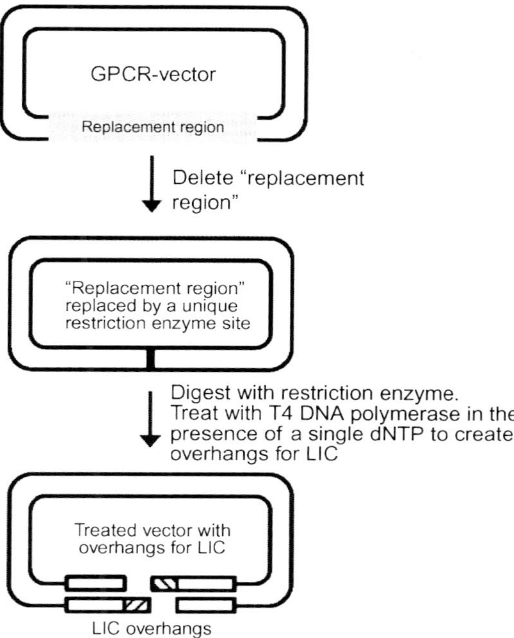

Figure 2 Construction of the GPCR vector lacking the replacement region.

for the deleted region. This can be performed by standard techniques of site-directed mutagenesis. If LIC is to be used to insert the T4L-containing cassettes, this restriction site should be flanked by sequences that contain three of the four types of nucleotide bases but lacking the fourth type of base. If screening for overall expression levels is planned, the starting allele should be constructed so as to encode a receptor fused at its C-terminal to GFP. It is anticipated that the GFP will subsequently be removed from forms of the receptor to be used for protein purification and crystallization.

2. If possible, the deletion should introduce a frameshift such that undesired self-ligation of vectors without insert will not result in read-through of the C-terminal region of the receptor. This will reduce background from plasmids in the library that lack an inserted T4L.
3. Perform LIC essentially as described by Clark et al. (2010). The individual steps are as follows:
   i. Digest the plasmid encoding the deletion-containing receptor gene with the restriction enzyme at the introduced site and gel purify the linearized plasmid.

ii. Use the exonuclease activity of T4 DNA polymerase to digest the 3′ ends of the cut vector in the presence of only a single dNTP corresponding to the nucleotide that is missing from the junction sequence. Under these conditions (listed below), the polymerase removes nucleotides until it reaches a base complementary to the one present in the reaction, creating 5′ overhangs.

Restriction enzyme-digested vector	400 ng
Single dNTP	2.5 m*M*
Dithiothreitol	5 m*M*
T4 DNA polymerase buffer (Novagen)	Final dilution 1×
T4 DNA polymerase (LIC-qualified; Novagen)	2 units
Total volume	40 μl
↓	
22 °C/40 min	
↓	
75 °C/15 min (to inactivate enzyme)	

iii. Incubate the T4-polymerase-treated vector (final concentration ~2.5 ng/μl) and T4L-containing cassettes (final concentration of assembled cassettes is ~0.5 ng/μl) in a total volume of 10 μl annealing buffer and anneal for 5–60 min at 22 °C. Add 2.5 μl of 25 m*M* EDTA and continue to incubate at 22 °C for an additional 5 min.

iv. Use 3 μl of the vector/insert mix to transform 25 μl competent NOVABlue cells following the manufacturer's instructions (Novagen). Enzymes present in the recipient cells carry out ligation of the vector and insert and can also fill in gaps formed in annealing overhangs that are not fully complementary. Plate the cells on an appropriate selection medium for the vector being used. (Typically, 6000–7500 transformants are obtained, compared with 4–6 in the absence of insert.)

4. Perform DNA sequencing on plasmids isolated from randomly selected *E. coli* transformants to evaluate the frequency of plasmids with inserts and the diversity of the library. Of the 40 plasmids that were sequenced from the *STE2* library, all contained inserted T4L with variable flanking regions.

5. Pool the bacterial transformants by adding 3–5 ml LB to the plate and gently scraping the surface of the agar with a sterile glass spreader. Pipette off as much of the LB plus cells as possible and transfer to a sterile microfuge tube. Culture the pooled colonies and libraries in LB supplemented with 100 μg/ml ampicillin and isolate plasmid DNA. This DNA library can now be introduced into the test cells of choice (bacteria, yeast, tissue culture, etc.) and screened as needed for function and expression. For the *STE2* library, the pooled plasmids were used to transform a yeast strain containing a chromosomal *STE2* deletion (A2953), creating a library of 2400 yeast transformants.

5. EXPRESSION SCREENING AND FUNCTIONAL ASSAYS

1. Flow cytometry is used to screen and quantitate the overall expression of the Ste2p–lysozyme fusions by detecting the emission from the C-terminally fused GFP.
2. Screening for fusions that are correctly folded and trafficked to the cell surface is conducted based on binding of a fluorescent derivative of the agonist α-factor essentially as described previously (Bajaj et al., 2004; Bajaj, Connelly, Gehret, Naider, & Dumont, 2007; Mathew et al., 2011). The α-factor analog, [K^7(bodipy-TR),Nle^{12}]α-factor (provided by Dr. Fred Naider of the College of Staten Island, City University of New York), with a red-shifted fluorophore, bodipy-TR (6-(((4-(4, 4-difluoro-5-(2-thienyl)-4-bora-3a,4a-diaza-*s*-indacene-3-yl)phenoxy) acetyl)amino)hexanoic acid), is used to minimize interference from the GFP fused to the receptor. Cells grown to an OD_{600} of ~1 in SD-ura (synthetic dextrose medium lacking uracil) (Sherman, 2002) are incubated at a density of ~3×10^6 cells/ml in 0.4 ml 20 m*M* acetate buffer (pH 4.6) with 100 n*M* fluorescent ligand, on ice (to prevent internalization), for 45 min. Fluorescence is then determined using a flow cytometer (i.e., a LSRII from BD Biosciences). The bodipy fluorophore was excited at 532 nm and detected with a 610/20 detector.
3. Screening for functional receptors is conducted based on induction of the pheromone-responsive *FUS1-HIS3* reporter gene, which is integrated into the chromosome of strain A2953. The library transformants are replica-plated onto solid SD-ura and SD-ura-his medium with and without ~300 n*M* α-factor. There should be no growth when histidine and α-factor are both absent.

4. Clones containing fusions with the desired expression levels and functional properties should be cultured for the reisolation of fusion-encoding plasmids. This allows for the determination of the sequences of the T4L-containing genes. It is also important to confirm apparent phenotypes of particular alleles by retransforming into a fresh host strain and reevaluating expression, folding, and signaling function to verify that the observed properties are plasmid dependent. Plasmids are isolated from *S. cerevisiae* using a modification of the manufacturer's protocol for the "Wizard® Plus SV Minipreps DNA Purification System" (Promega). To disrupt the yeast cell wall before plasmid isolation, approximately 0.5 ml of 0.5-mm zirconia/silica beads (BioSpec Products, Inc.) are added to cells pelleted from 1.5 ml of fresh culture grown overnight to an $OD_{600} > 1.5$. "Resuspension" and "Cell Lysis" solutions (250 μl each) are added, and the cells are vortexed for 5 min and centrifuged at 12,000 × *g* for 10 min to pellet the beads and unbroken cells. The supernatant from this spin is then pipetted into 350 μl "Neutralization" buffer. The manufacturer's instructions are followed from this point except for the final elution volume, which is reduced from 100 to 30 μl. Since plasmid yield is generally poor, the entire miniprep is used to transform competent *E. coli* and then plasmid is isolated from the bacteria for further manipulation.
5. To quantitate signaling function of T4L-containing receptors, plasmids of interest are used to transform the *ste2*-Δ yeast strain, A575 containing a chromosomally integrated *Fus1-lacZ* reporter (Sommers & Dumont, 1999). β-Galactosidase assays are performed as described previously (Sommers & Dumont, 1999).
6. As a first step toward high-level production of fusion proteins for use in crystallization, the *STE2* genes encoding fusion proteins of interest (but without the C-terminally fused GFP) are PCR amplified and inserted via LIC into the expression vector pSGP46 under the control of the inducible *ADH2* promoter, as described previously (Clark et al., 2010). Reading frames are expressed from this vector with C-terminal ZZ (protein A) and $RGSHis_{10}$ tags that are cleavable by rhinovirus 3C protease. Plasmids encoding the desired constructs are isolated from *E. coli* and transformed into the protease-deficient expression strain BJ5460 (Jones, 1991).
7. For immunoblotting, yeast expressing receptor–T4L fusions from plasmid pSGP46 are cultured in 10 ml SD-ura medium to an OD_{600} of ~3, then diluted fourfold in YPD (total volume is 40 ml), and recultured to

an OD_{600} of ~10, depleting glucose in order to induce the *ADH2* promoter. Cells are harvested by centrifuging at 1000 × *g* for 5 min at 4 °C, washed with 10 ml water, and resuspended in 300 μl of 20 m*M* sodium acetate (pH 4.6), 0.1 m*M* EDTA, and 1 m*M* PMSF. They are then lysed by vortexing at 4 °C together with 0.5-mm zirconia/silica beads (BioSpec Products, Inc.; use a volume of beads corresponding to ~250 μl in a microfuge tube) in 10 cycles of 20 s each, separated by 2 min of cooling on ice to prevent overheating. After centrifugation at 12,000 × *g* for 2 min to pellet the beads, the supernatant is transferred to a fresh tube. A total of 8 μl of lysate is then mixed with 22 μl of loading buffer (40 m*M* Tris–HCl (pH 6.8), 0.1 m*M* EDTA, 9 *M* urea, 5% sodium dodecyl sulfate, 0.02 mg/ml bromophenol blue) and incubated at 37 °C for 5 min. To prevent irreversible aggregation of TMPs, it is important to avoid overheating the samples at this stage. Following SDS–PAGE, proteins are transferred onto nitrocellulose and probed with an HRP-conjugated anti-RGS-His antibody (Qiagen) or an anti-enolase antibody (for visualizing enolase as a loading control; a gift of Michael Holland of the University of California, Davis) followed by an HRP-conjugated anti-rabbit secondary antibody (Sigma). Protein bands are visualized using SuperSignal reagent (Pierce) and exposed to X-ray film (CL-XPosure Film by Thermo Scientific).

6. RESULTS FOR THE YEAST α-FACTOR RECEPTOR STE2P

Application of the described procedures for creating a library of variant forms of the yeast α-factor receptor containing T4L inserted into the IC3 loop allowed efficient recovery of the full range of possible receptor variants, with some bias against variants containing insertions with the shortest loop sequences (Mathew et al., 2013). Based on GFP fluorescence, many of the variant receptors were expressed at levels that were similar to the levels of receptors with no insertions. Cells expressing T4L-containing receptors exhibited up to 80% of the ligand capacity of cells expressing normal receptors, with similar ligand affinities. Surprisingly, some receptor variants containing the inserted protein exhibited up to 66% of the maximal signaling response of normal receptors, based on induction of the *FUS1*-lacZ reporter, with similar EC_{50}s for activation by pheromone.

A screen was also conducted searching for insertions of T4L into the third loop of receptors with truncations of their cytoplasmic C-terminal tails. C-terminally truncated receptors are generally fully functional and present at the cell surface in greater numbers than similarly expressed full-length receptors. While random screening uncovered some C-terminally truncated Ste2p–T4L fusions that bound up to 18% of the high levels of ligand bound by similarly expressed truncated receptors with no insertion, none of the truncated receptors containing the T4L insertion exhibited more than ~5% of the signaling activity of the unfused receptors. The C-terminal tail may, therefore, provide stabilization to receptors containing perturbations such as inserted proteins (Mathew et al., 2013). This is an important point to be considered in view of the fact that GPCR constructs used for crystallography involve truncations of the C-terminal to remove regions presumed to have excess flexibility.

The most active T4L-containing variants of full-length Ste2p contained the inserted protein flanked by nearly complete IC3 loop sequences that were duplicated at both the N- and C-terminal sides of the insertion. We confirmed that duplication of the IC3 loop, even in the absence of any inserted T4L, is not detrimental to receptor signaling. Thus, it seems likely that the presence of extended sequences on both sides of the inserted protein provides the flexibility needed to move T4L out of the way so that the receptor can participate in the interactions with the cytoplasmic G protein that are required for signaling. The extended sequences present in the active constructs may not provide significant rigidity to the surrounding transmembrane helices, but recent successes in structure determination of GPCRs with fusions only at their extreme N-termini (Fenalti et al., 2014; Siu et al., 2013; Thompson et al., 2012; Wu et al., 2014; Zou, Weis, & Kobilka, 2012) suggest that such rigidification may not be an important aspect of the fusion protein approach to stabilization.

ACKNOWLEDGMENTS

We thank Dr. Fred Naider (CUNY, Staten Island) for providing fluorescent ligand and many helpful discussions. We also thank the University of Rochester Flow Cytometry Core facility for assistance with cell sorting and flow cytometry. This work was supported by National Institutes of Health grants R01GM084083 and U54GM09461 to M. E. D.

REFERENCES

Alexandrov, A., Vignali, M., LaCount, D. J., Quartley, E., de Vries, C., De Rosa, D., et al. (2004). A facile method for high-throughput co-expression of protein pairs. *Molecular & Cellular Proteomics*, *3*(9), 934–938.

Aslanidis, C., & de Jong, P. J. (1990). Ligation-independent cloning of PCR products (LIC-PCR). *Nucleic Acids Research*, *18*(20), 6069–6074.

Bajaj, A., Ćelić, A., Ding, F.-X., Naider, F., Becker, J. M., & Dumont, M. E. (2004). A fluorescent α-factor analogue exhibits multiple steps on binding to its G protein coupled receptor in yeast. *Biochemistry*, *43*(42), 13564–13578.

Bajaj, A., Connelly, S. M., Gehret, A. U., Naider, F., & Dumont, M. E. (2007). Role of extracellular charged amino acids in the yeast α-factor receptor. *Biochimica et Biophysica Acta (BBA)—Biomembranes*, *1773*(6), 707–717.

Brown, A. J., Dyos, S. L., Whiteway, M. S., White, J. H. M., Watson, M.-A. E. A., Marzioch, M., et al. (2000). Functional coupling of mammalian receptors to the yeast mating pathway using novel yeast/mammalian G protein α-subunit chimeras. *Yeast*, *16*(1), 11–22.

Cherezov, V., Rosenbaum, D. M., Hanson, M. A., Rasmussen, S. G. F., Thian, F. S., Kobilka, T. S., et al. (2007). High-resolution crystal structure of an engineered human β2-adrenergic G protein coupled receptor. *Science*, *318*(5854), 1258–1265.

Chien, E. Y. T., Liu, W., Zhao, Q., Katritch, V., Won Han, G., Hanson, M. A., et al. (2010). Structure of the human dopamine D3 receptor in complex with a D2/D3 selective antagonist. *Science*, *330*(6007), 1091–1095.

Chun, E., Thompson, A. A., Liu, W., Roth, C. B., Griffith, M. T., Katritch, V., et al. (2012). Fusion partner toolchest for the stabilization and crystallization of G protein-coupled receptors. *Structure (London, England: 1993)*, *20*(6), 967–976.

Clark, K. M., Fedoriw, N., Robinson, K., Connelly, S. M., Randles, J., Malkowski, M. G., et al. (2010). Purification of transmembrane proteins from *Saccharomyces cerevisiae* for X-ray crystallography. *Protein Expression and Purification*, *71*(2), 207–223.

Dieckman, L., Gu, M., Stols, L., Donnelly, M. I., & Collart, F. R. (2002). High throughput methods for gene cloning and expression. *Protein Expression and Purification*, *25*(1), 1–7.

Dore, A. S., Okrasa, K., Patel, J. C., Serrano-Vega, M., Bennett, K., Cooke, R. M., et al. (2014). Structure of class C GPCR metabotropic glutamate receptor 5 transmembrane domain. *Nature*, *511*(7511), 557–562.

Dowell, S. J., & Brown, A. J. (2009). Yeast assays for G protein-coupled receptors. *Methods in Molecular Biology*, *552*, 213–229.

Egloff, P., Hillenbrand, M., Klenk, C., Batyuk, A., Heine, P., Balada, S., et al. (2014). Structure of signaling-competent neurotensin receptor 1 obtained by directed evolution in *Escherichia coli*. *Proceedings of the National Academy of Sciences of the United States of America*, *111*(6), E655–E662.

Engel, C. K., Chen, L., & Privé, G. G. (2002). Insertion of carrier proteins into hydrophilic loops of the *Escherichia coli* lactose permease. *Biochimica et Biophysica Acta (BBA)—Biomembranes*, *1564*(1), 38–46.

Fenalti, G., Giguere, P. M., Katritch, V., Huang, X. P., Thompson, A. A., Cherezov, V., et al. (2014). Molecular control of delta-opioid receptor signalling. *Nature*, *506*(7487), 191–196.

Gehret, A. U., Bajaj, A., Naider, F., & Dumont, M. E. (2006). Oligomerization of the yeast α-factor receptor. *Journal of Biological Chemistry*, *281*(30), 20698–20714.

Granier, S., Manglik, A., Kruse, A. C., Kobilka, T. S., Thian, F. S., Weis, W. I., et al. (2012). Structure of the δ-opioid receptor bound to naltrindole. *Nature*, *485*(7398), 400–404.

Haga, K., Kruse, A. C., Asada, H., Yurugi-Kobayashi, T., Shiroishi, M., Zhang, C., et al. (2012). Structure of the human M2 muscarinic acetylcholine receptor bound to an antagonist. *Nature*, *482*(7386), 547–551.

Hanson, M. A., Roth, C. B., Jo, E., Griffith, M. T., Scott, F. L., Reinhart, G., et al. (2012). Crystal structure of a lipid G protein-coupled receptor. *Science*, *335*(6070), 851–855.

Hollenstein, K., Kean, J., Bortolato, A., Cheng, R. K., Dore, A. S., Jazayeri, A., et al. (2013). Structure of class B GPCR corticotropin-releasing factor receptor 1. *Nature*, *499*(7459), 438–443.

Jaakola, V.-P., Griffith, M. T., Hanson, M. A., Cherezov, V., Chien, E. Y. T., Lane, J. R., et al. (2008). The 2.6 angstrom crystal structure of a human A2A adenosine receptor bound to an antagonist. *Science*, *322*(5905), 1211–1217.

Jones, E. W. (1991). Tackling the protease problem in *Saccharomyces cerevisiae*. *Methods in Enzymology*, *194*, 428–453.

King, K., Dohlman, H. G., Thorner, J., Caron, M. G., & Lefkowitz, R. J. (1990). Control of yeast mating signal transduction by a mammalian beta 2-adrenergic receptor and Gs alpha subunit. *Science*, *250*(4977), 121–123.

Kruse, A. C., Hu, J., Pan, A. C., Arlow, D. H., Rosenbaum, D. M., Rosemond, E., et al. (2012). Structure and dynamics of the M3 muscarinic acetylcholine receptor. *Nature*, *482*(7386), 552–556.

Kruse, A. C., Ring, A. M., Manglik, A., Hu, J., Hu, K., Eitel, K., et al. (2013). Activation and allosteric modulation of a muscarinic acetylcholine receptor. *Nature*, *504*(7478), 101–106.

Manglik, A., Kruse, A. C., Kobilka, T. S., Thian, F. S., Mathiesen, J. M., Sunahara, R. K., et al. (2012). Crystal structure of the μ-opioid receptor bound to a morphinan antagonist. *Nature*, *485*(7398), 321–326.

Mathew, E., Bajaj, A., Connelly, S. M., Sargsyan, H., Ding, F.-X., Hajduczok, A. G., et al. (2011). Differential interactions of fluorescent agonists and antagonists with the yeast G protein coupled receptor Ste2p. *Journal of Molecular Biology*, *409*(4), 513–528.

Mathew, E., Ding, F.-X., Naider, F., & Dumont, M. E. (2013). Functional fusions of T4 lysozyme in the third intracellular loop of a G protein-coupled receptor identified by a random screening approach in yeast. *Protein Engineering, Design and Selection*, *26*(1), 59–71.

Pausch, M. H. (1997). G-protein-coupled receptors in *Saccharomyces cerevisiae*: High-throughput screening assays for drug discovery. *Trends in Biotechnology*, *15*(12), 487–494.

Rasmussen, S. G. F., Choi, H.-J., Fung, J. J., Pardon, E., Casarosa, P., Chae, P. S., et al. (2011). Structure of a nanobody-stabilized active state of the [bgr]2 adrenoceptor. *Nature*, *469*(7329), 175–180.

Rasmussen, S. G. F., Choi, H.-J., Rosenbaum, D. M., Kobilka, T. S., Thian, F. S., Edwards, P. C., et al. (2007). Crystal structure of the human β2 adrenergic G-protein-coupled receptor. *Nature*, *450*(7168), 383–387.

Rasmussen, S. G. F., DeVree, B. T., Zou, Y., Kruse, A. C., Chung, K. Y., Kobilka, T. S., et al. (2011). Crystal structure of the [bgr]2 adrenergic receptor-Gs protein complex. *Nature*, *477*(7366), 549–555.

Ring, A. M., Manglik, A., Kruse, A. C., Enos, M. D., Weis, W. I., Garcia, K. C., et al. (2013). Adrenaline-activated structure of beta2-adrenoceptor stabilized by an engineered nanobody. *Nature*, *502*(7472), 575–579.

Rosenbaum, D. M., Cherezov, V., Hanson, M. A., Rasmussen, S. G. F., Thian, F. S., Kobilka, T. S., et al. (2007). GPCR engineering yields high-resolution structural insights into β2-adrenergic receptor function. *Science*, *318*(5854), 1266–1273.

Rosenbaum, D. M., Zhang, C., Lyons, J. A., Holl, R., Aragao, D., Arlow, D. H., et al. (2011). Structure and function of an irreversible agonist-[bgr]2 adrenoceptor complex. *Nature*, *469*(7329), 236–240.

Scott, D. J., Kummer, L., Tremmel, D., & Pluckthun, A. (2013). Stabilizing membrane proteins through protein engineering. *Current Opinion in Chemical Biology*, *17*(3), 427–435.

Sherman, F. (2002). Getting started with yeast. *Methods in Enzymology*, *350*, 3–41.

Shimamura, T., Shiroishi, M., Weyand, S., Tsujimoto, H., Winter, G., Katritch, V., et al. (2011). Structure of the human histamine H1 receptor complex with doxepin. *Nature*, *475*(7354), 65–70.

Shiroishi, M., Kobayashi, T., Ogasawara, S., Tsujimoto, H., Ikeda-Suno, C., Iwata, S., et al. (2011). Production of the stable human histamine H1 receptor in *Pichia pastoris* for structural determination. *Methods*, *55*(4), 281–286.

Siu, F. Y., He, M., de Graaf, C., Han, G. W., Yang, D., Zhang, Z., et al. (2013). Structure of the human glucagon class B G-protein-coupled receptor. *Nature*, *499*(7459), 444–449.

Sommers, C. M., & Dumont, M. E. (1999). Genetic approaches for studying the structure and function of G protein-coupled receptors in yeast. In J. Wess (Ed.), *Structure-function analysis of G protein coupled receptors* (pp. 141–166). New York: Wiley-Liss.

Sommers, C. M., Martin, N. P., Akal-Strader, A., Becker, J. M., Naider, F., & Dumont, M. E. (2000). A limited spectrum of mutations causes constitutive activation of the yeast α-factor receptor. *Biochemistry*, *39*(23), 6898–6909.

Tan, Q., Zhu, Y., Li, J., Chen, Z., Han, G. W., Kufareva, I., et al. (2013). Structure of the CCR5 chemokine receptor-HIV entry inhibitor maraviroc complex. *Science*, *341*(6152), 1387–1390.

Taslimi, A., Mathew, E., Ćelić, A., Wessel, S., & Dumont, M. E. (2012). Identifying functionally important conformational changes in proteins: Activation of the yeast α-factor receptor Ste2p. *Journal of Molecular Biology*, *418*(5), 367–378.

Tate, C. G. (2012). A crystal clear solution for determining G-protein-coupled receptor structures. *Trends in Biochemical Sciences*, *37*(9), 343–352.

Thompson, A. A., Liu, W., Chun, E., Katritch, V., Wu, H., Vardy, E., et al. (2012). Structure of the nociceptin/orphanin FQ receptor in complex with a peptide mimetic. *Nature*, *485*(7398), 395–399.

Wacker, D., Wang, C., Katritch, V., Han, G. W., Huang, X.-P., Vardy, E., et al. (2013). Structural features for functional selectivity at serotonin receptors. *Science*, *340*(6132), 615–619.

Wallin, E., & Heijne, G. V. (1998). Genome-wide analysis of integral membrane proteins from eubacterial, archaean, and eukaryotic organisms. *Protein Science*, 7(4), 1029–1038.

Wang, C., Jiang, Y., Ma, J., Wu, H., Wacker, D., Katritch, V., et al. (2013). Structural basis for molecular recognition at serotonin receptors. *Science*, *340*(6132), 610–614.

White, J. F., Noinaj, N., Shibata, Y., Love, J., Kloss, B., Xu, F., et al. (2012). Structure of the agonist-bound neurotensin receptor. *Nature*, *490*(7421), 508–513.

Wu, B., Chien, E. Y. T., Mol, C. D., Fenalti, G., Liu, W., Katritch, V., et al. (2010). Structures of the CXCR4 chemokine GPCR with small-molecule and cyclic peptide antagonists. *Science*, *330*(6007), 1066–1071.

Wu, H., Wacker, D., Mileni, M., Katritch, V., Han, G. W., Vardy, E., et al. (2012). Structure of the human κ-opioid receptor in complex with JDTic. *Nature*, *485*(7398), 327–332.

Wu, H., Wang, C., Gregory, K. J., Han, G. W., Cho, H. P., Xia, Y., et al. (2014). Structure of a class C GPCR metabotropic glutamate receptor 1 bound to an allosteric modulator. *Science*, *344*(6179), 58–64.

Xu, F., Wu, H., Katritch, V., Han, G. W., Jacobson, K. A., Gao, Z.-G., et al. (2011). Structure of an agonist-bound human A2A adenosine receptor. *Science*, *332*(6027), 322–327.

Zhang, C., Srinivasan, Y., Arlow, D. H., Fung, J. J., Palmer, D., Zheng, Y., et al. (2012). High-resolution crystal structure of human protease-activated receptor 1. *Nature*, *492*(7429), 387–392.

Zhang, J., Zhang, K., Gao, Z. G., Paoletta, S., Zhang, D., Han, G. W., et al. (2014). Agonist-bound structure of the human P2Y12 receptor. *Nature*, *509*(7498), 119–122.

Zhang, K., Zhang, J., Gao, Z. G., Zhang, D., Zhu, L., Han, G. W., et al. (2014). Structure of the human P2Y12 receptor in complex with an antithrombotic drug. *Nature*, *509*(7498), 115–118.

Zou, Y., Weis, W. I., & Kobilka, B. K. (2012). N-terminal T4 lysozyme fusion facilitates crystallization of a G protein coupled receptor. *PLoS One*, 7(10), e46039.

CHAPTER THREE

Membrane Preparation and Solubilization

Ankita Roy[1]
Department of Medicine, University of Pittsburgh, Pittsburgh, Pennsylvania, USA
[1]Corresponding author: e-mail address: amr161@pitt.edu

Contents

Abstract

Membrane proteins play an essential role in several biological processes like ion transport, signal transduction, and electron transfer to name a few. For structural and functional studies of integral membrane proteins, it is critically important to isolate proteins from the membrane using biological detergents. Detergents disrupt the native lipid components of the native membrane and encase the membrane protein in an unnatural environment in aqueous solution. However, a particular membrane protein is best solubilized in a specific detergent; therefore, screening for the optimal detergent is essential. Apart from keeping the membrane protein monodispered in solution, the detergent has to be compatible with downstream processes to isolate and characterize a membrane protein. Over the past several years, a number of membrane proteins have been successfully isolated for structural and functional studies that allowed an outline of general strategies for isolating a novel membrane protein of interest.

Methods in Enzymology, Volume 557
ISSN 0076-6879
http://dx.doi.org/10.1016/bs.mie.2014.11.044

1. INTRODUCTION

Cell membranes act as a physiological barrier separating the external environment from the internal milieu. Membranes are rich in integral membrane proteins that play diverse role in cellular processes ranging from transport of nutrients, ions, electron transfer, and signal transduction to several other processes. Additionally, cells are packed with internal membranes that are rich in membrane proteins. As approximately 70% of current day medications are targeted against membrane proteins, it is essential to elucidate the structure of these proteins. Though 30% of the human genome codes for integral membrane proteins, only 30,000 structures deposited in the protein data bank, less than 1%, represent membrane protein. The paucity of the deposited structures can be attributed to the amphipathic nature of these proteins.

There are several limitations in purifying membrane proteins for structural and functional studies. Firstly, membrane proteins are present in minute amounts in native tissues. However for structural studies, milligram amounts of protein must be isolated. To overcome this problem, recombinant proteins are overexpressed in heterologous host organisms. Although *Escherichia coli* has been used extensively to overexpress and purify membrane proteins, other bacterial hosts such as *Lactococcus lactis*, yeast expression systems such as *Pichia pastoris* and *Saccharomyces cerevisiae*, and insect and mammalian expression systems have been routinely used for the purpose (Lundstrom et al., 2006; Morello et al., 2008; Sahdev, Khattar, & Saini, 2008; Shukla, Reinhart, & Michel, 2006). To heterologously express and purify a membrane protein, systematic screening should be performed which involves expression analysis using a variety of host organisms, expression vectors, and expression conditions (Surade et al., 2006).

Often overexpression of membrane proteins can be problematic; as large quantities of the protein are expressed using a strong promoter, due to lack of membranes, the expressed proteins fail to fold properly. To overcome this problem, membrane-rich expression hosts like *Rhodobacter sphaeroides* have been developed. These photosynthetic organisms are genetically engineered to provide excess membrane surface area in comparison to typical expression hosts for harboring overexpressed membrane proteins (Laible, Scott, Henry, & Hanson, 2004; Roy, Shukla, Haase, & Michel, 2008).

Secondly, membrane proteins must be solubilized and purified from the membranes using small amphipathic molecules called detergents. Several types of detergents have been used to isolate membrane proteins; however, it is important to select a detergent that not only maximally extracts the protein from the membrane but also stabilizes it. The detergent used to solubilize a membrane protein must also be compatible with the purification strategy and crystallization. Often detergent mixtures have been employed in the purification and crystallization stages to improve crystal quality for structural studies.

In recent years, high-throughput platforms like JCSG, MePNet, EMeP, etc., have generated structure determination pipelines which involve working with larger set of proteins. Large numbers of membrane proteins are processed in parallel; this is advantageous from the standpoint of optimization of strategies involved in the structure determination (Lundstrom, 2006).

This chapter describes a general protocol for membrane preparation and then shows how to screen for the most suitable detergent to solubilize a protein.

2. MEMBRANE PREPARATION

Membrane preparation is the first step in the isolation and purification of a membrane protein. Membrane preparation involves disruption of the cell membrane using mechanical methods and differential centrifugation, which helps to isolate the membranes from soluble proteins, nuclear material, and other internal membranes. Membrane preparation is an important step in purifying a membrane protein, as isolation of membranes, in itself, is a purification step. Soluble proteins and other membrane-associated proteins (like outer membrane proteins, mitochondria associated proteins, etc.) are removed from the membrane fraction. Moreover, a good membrane preparation ensures reproducibility amongst purification preps, which is a prerequisite for successful crystallization trials.

2.1. Choice of buffer

Membranes are prepared by breaking open cells in a suitable buffer such as Tris, HEPES, and phosphate buffer. The pH of the buffer is critical as it plays a role in protein solubility and stability. Additionally, buffers also affect binding and elution of a protein from a column matrix during purification. Care should also be taken to check if the properties of the buffer affect the enzymatic activity of a membrane protein. As detergents are required in subsequent steps of membrane protein isolation, the ionic strength of the buffer should be compatible with the detergent used for solubilization and purification. Postcell disruption, several proteases are released from the cell that degrades the target protein thereby reducing the yield during the isolation steps. Protease inhibitors such as Complete Protease Inhibitor Cocktails should be added to the buffer in each step of the protein isolation process. Reducing agents like dithiothreitol should be present in the buffer to prevent oxidation of disulfide bonds in a protein.

2.2. Cell lysis

Based on the source material, cells are lysed using a variety of methods. Cells with tough walls are broken using mechanical force such as ultrasonication, homogenization involving French Press, Microfluidizer, or mechanical lysis using glass beads-based agitation and nitrogen cavitation method. For smaller volumes of cells, osmotic shock, repeated cycles of freeze thaw or enzymatic lysis using lysozyme or zymolase can be employed (Table 1).

2.3. Membrane fractionation

Postcell lysis, membranes are fractionated using differential centrifugation. Membranes are composed of lipid–protein complexes, which have a specific

Table 1 Cell lysis methods for large-scale membrane preparation

Lysis method	Apparatus	Advantages	Disadvantages	Cells lysed
Ultrasonication	Sonicator (20–50 kHz in short pulses of 5–15 s)	Easy to use, applied to various cell types	Ear protection required, prevent generation of heat	Wide variety of cell types
Homogenization	French Press: 18,000–30,000 psi	Effective in small volumes	Prone to clogging	Wide variety of cell types
	Microfluidizer: 20,000–30,000 psi	Effective for larger volume, reproducible	Cooling required	Wide variety of cell types
Mechanical	Beads-based agitation	Effective in small volumes	Not reproducible, heat generated	Mammalian cells
	Nitrogen cell cavitation	Not effective for cells with tough wall	Care in handling	Mammalian cells

density; this allows them to be separated from soluble proteins and other organelles like mitochondria and nuclei using centrifugation techniques. Sucrose density centrifugation can be employed to separate different types of membranes. The inner membranes like the intracytoplasmic membranes of *R. sphaeroides* that houses the photosynthetic proteins, such as the reaction center and the light harvesting complexes can be separated from the cell membrane using sucrose density gradient (Rafferty & Clayton, 1997).

2.4. Methodology

Described below is a general approach to prepare membrane fragments for subsequent solubilization and purification of a membrane protein from *E. coli*.

Materials

1. Source material
2. Low-speed centrifuge
3. Ultracentrifuge
4. Instrument for lysis
5. Bicinchoninic acid (BCA) protein estimation kit
6. Spectrophotometer (UV/Vis)

Methods

1. *E. coli* cells were harvested when OD_{600} of about 1.5 was reached.
2. Washed bacterial cell pellet (5 g) was resuspended in 10 ml of ice-cold breaking buffer (20 m*M* HEPES, 100 m*M* NaCl, 1 m*M* phenylmethyl sulfonyl fluoride, EDTA-free protease inhibitors).
3. Lysozyme was added at a final concentration of 1 mg/ml.
4. Cells were disrupted by passage through French Press (18,000 psi pressure) at a flow rate of 10 ml/min at 4 °C. The process was repeated thrice.
5. Unbroken cells, nuclear material and cell debris were removed by low-speed centrifugation at 5000 × *g*, at 4 °C for 10 min.
6. The supernatant was subjected to ultracentrifugation at 100,000 × *g* at 4 °C for 60 min.
7. The pellet was resuspended in 10 ml of 20 m*M* HEPES, 100 m*M* NaCl, 1 m*M* phenylmethyl sulfonyl fluoride, EDTA-free protease inhibitors, and ultracentrifuged again at 100,000 × *g* at 4 °C for 30 min.
8. The washed membrane pellet was resuspended in 1 ml of 20 m*M* HEPES, 100 m*M* NaCl, 1 m*M* phenylmethyl sulfonyl fluoride, EDTA-free protease inhibitors. Smaller aliquots were flash-frozen and

stored at −80 °C. A small fraction of the membrane suspension was diluted and used to determine the protein concentration using BCA protein quantification method.

Notes

1. Yeast cells have very rigid cell walls that are difficult to break. Vigorous vortexing of yeast cells mixed with 0.5 mm glass beads for 30 min result in efficient cell lysis (André et al., 2006).
2. Outer membranes from *E. coli* can be isolated by two methods. The first one involves the use of density gradient centrifugation using a 45–70% sucrose gradient. The second method involves the use of an extraction strategy using 0.5% Sarkosyl NL97 for 30 min at 28 °C followed by ultracentrifugation at 10,000 × *g* for 60 min. The outer membrane is present in the pellet (Bachhawat & Ghosh, 1987).
3. To isolate internal membranes, like intracytoplasmic membranes of *R. sphaeroides* rate-zonal centrifugation can be employed to separate the cell membrane from the ICMs by layering the cell extract on a 5–35% (w/w) sucrose gradient in 1 m*M* Tris pH 7.5, 1 m*M* EDTA.
4. BCA method is an efficient method for estimation of membrane protein concentration as the assay is less sensitive to the detergents and buffer components. However, other assays like a modified method based on Lowry protein estimation assay can also be used. This method involves precipitating proteins from a detergent solubilized solution with trichloroacetic acid and deoxycholate prior to protein estimation.

3. SOLUBILIZATION

Membrane proteins must be purified to near homogeneity for structural or functional studies. Integral membrane proteins must be extracted from the native lipid environment of the membrane in a functional form using detergents. Detergents are amphipathic molecules; they contain a polar head group at one end and long hydrocarbon chain at the other end. In solution, at low concentration, detergents are present as monomers; however, with increase in the concentration of the detergent above a Critical Micellar Concentration (CMC), the detergent from micelles. Micelles are ordered thermodynamically stable spherical structures with polar head groups facing the exterior, while hydrophobic tails toward the interior. During solubilization, membrane proteins partition from the lipid-rich native membrane to the detergent micelles. The detergent molecules surround the hydrophobic regions of the membrane protein, mimicking the lipid bilayer. Usually, a suitable detergent to protein ratio is employed to solubilize an integral membrane protein, which is determined empirically.

3.1. Types of detergents

Detergents are amphipathic in nature, but structurally diverse. They are primarily characterized by the CMC, which is the optimal concentration of the detergent when micelles start to form. In order words, CMC is the minimal concentration of detergent that should be present in extraction buffers to keep the protein monodispersed in solution. CMC of a particular detergent also depends on several other factors like temperature of solubilization, buffer pH, and ionic strength. Detergents are classified based on either the hydrophilic head group or the hydrophobic tail group. Based on the head group, there are four categories of detergents—nonionic, anionic, cationic, and zwitterionic. Nonionic detergents are most commonly used for solubilization. Nonionic polyethylene detergents are characterized by the presence of a polymeric chain $(O–CH_2–CH_2)_N–OH$. This class of detergent is milder and is commonly used to extract peripheral membrane protein. These mild detergents help retain lipids bound to the protein, thereby retaining catalytic activity of the protein. Common detergents in this category include, $C_{12}E_9$ (Thesit), Brij series, Triton X-100, Tween 20, etc. Most common detergents used in solubilizing integral membrane proteins are sugar detergents, which are considered mild. Sugar detergents have a hydrophobic part linked to a carbohydrate moiety like glucoside, maltoside via a glycosidic bond. Dodecyl maltoside (DDM), also referred to as laurylmaltoside (LM), is considered one of the best detergents for purification and crystallization of membrane proteins. Digitonin is a detergent milder than DDM and it preserves mild interactions between membrane integral proteins, thus used extensively in protein interaction studies. However, digitonin cannot effectively solubilize membrane proteins, it is used in combination with other detergents for extraction. Ionic detergents are harsh than nonionic detergents; however, anionic detergents like cholate and deoxycholate have been used for extraction of mitochondrial respiratory proteins (Hatefi, 1978). However, such detergents can solubilize proteins at a higher pH. Anionic detergents like bile salts can be used at a lower pH. Zwitterionic detergents have both a positive and negative group. Lauryldimethylammoniumoxide (LDAO), a zwitterionic detergent, has two charged groups at higher pH. LDAO has been successfully used to purify and crystallize reaction center proteins of photosynthetic bacteria, some of the earliest structures of membrane proteins solved (Deisenhofer, Epp, Miki, Huber, & Michel, 1984).

Detergents have also been classified based on the size and flexibility of the tail. Longer the tail length, greater is the chance of denaturation. Triton series detergents are the most flexible while bile salts such as CHAPS and CHAPSO fall in the rigid category.

3.2. Choice of detergent

Choice of a suitable detergent involves selection of the detergent that most efficiently solubilizes a protein of interest. Firstly, the detergent should preserve the functionality of the target protein. Secondly, the most suitable detergent should have the highest efficiency of extraction in comparison to other detergents. Thirdly, the detergent should be compatible with the downstream processes like purification and crystallization. Many times, membrane proteins are solubilized in a detergent; however, in the later purification stages the detergent is replaced by a second detergent that can promote crystal lattice formation for structural studies. Detergent exchange can be performed when the protein is bound to the column matrix; alternatively, dialysis can be employed which is a time-consuming method. Several studies have shown that the intrinsic lipids are bound tightly to membrane proteins to retain their activity. Therefore, care should be taken to choose suitable detergents that will not delipidate a membrane protein (Palsdottir & Hunte, 2004).

3.3. Stability in detergent

Long-term stability of the purified membrane protein can be assessed by techniques such as size-exclusion chromatography (SEC), SDS-polyacrylamide gel electrophoresis (SDS-PAGE), circular dichroism (CD), and differential scanning calorimetry (DSC). However, prior to the purification, some techniques can be employed to determine the stability of a membrane protein in a particular detergent. To determine the stability of a protein in a given detergent, a method was designed whereby the target membrane protein was tagged with Green Fluorescent Protein (GFP) at the N- or the C-terminus. Detergent solubilized extracts of cells transfected or transformed with the GFP fusion constructs were directly loaded on an SEC column and the stability of the protein was analyzed based on GFP fluorescence. By this method, the stability of a GFP fusion protein can be determined in an unpurified heterogeneous sample (Kawate & Gouaux, 2006). In case of proteins like G protein-coupled receptors (GPCRs), radioligand-binding assays can be performed using the different detergents under varying conditions (Magnani, Shibata, Serrano-Vega, & Tate, 2008; Serrano-Vega, Magnani, Shibata, & Tate, 2008).

3.4. Alternative solubilization strategy

To maximally extract a membrane protein, detergents are used in excess. Often, excess detergent in solution obscures biophysical measurements. In the past decade, several attempts have been made to improve solubilization of membrane protein using alternative compounds. One such approach involves the use of amphipols such as PMAL-B-100, which are detergent-like amphipathic polymers. The advantage of using amphipols over detergents is that the amphipols are used in stoichiometric quantities and not in excess quantities like detergents; additionally, they are capable of preserving the function of most membrane proteins tested so far (Popot et al., 2003). However, till date there are no data on the use of amphipols in crystallization of a membrane protein.

3.5. Presolubilization screen

The best approach to select a suitable detergent to solubilize and purify a new target protein is to employ a presolubilization screen. A presolubilization screen involves a systematic approach to screen for some detergents (Table 2), which have been successfully used for functional solubilization of several membrane proteins. Apart from choosing the right detergent it is also important to identify the suitable solubilization conditions such as buffer pH, temperature, and time of solubilization.

Table 2 Commonly used detergents in membrane protein research

Detergent	**Type**	**Molecular weight**	**CMC (m*M*)**
DDM (LM)	Nonionic	511	0.2
DM	Nonionic	483	0.2
OG	Nonionic	308	15
NG	Nonionic	306.4	6.5
C12E8	Nonionic	540	<0.1
LDAO	Zwitterionic	229	1
CHAPS	Zwitterionic	615	3–10
Fos12	Ionic	351.5	1.5
Fos14	Ionic	395	0.12

The detergent that maximally extracts the membrane protein leaving little or no protein in the pellet fraction can be considered suitable for purification and crystallization. At this point, it is important to assess the stability of the protein. However, in most cases, testing stability before purification is not possible. In such cases, stability of the protein can be tested only after purification. Once the conditions of solubilization are standardized using the presolubilization screen, large-scale solubilization can be performed by scaling up the amount of the membranes and detergent using the most suitable solubilization conditions.

3.6. Methodology

Materials

1. Isolated membranes
2. Ultracentrifuge
3. Detergent stocks (stored at −20 °C)
4. BCA protein estimation kit
5. Spectrophotometer (UV/Vis)
6. SDS-polyacrylamide gel
7. Western blot apparatus
8. PVDF/nitrocellulose membrane
9. Protein molecular weight marker

Methods

1. Membrane pellet was resuspended in suitable buffer (20 m*M* HEPES, 100 m*M* NaCl, pH 7.4) with protease inhibitors at a total concentration of 10 mg/ml in 500 μl total volume. Detergents are used at a final concentration of 1% except for octylglucoside (OG), which is 2% (generally two times CMC).
2. The samples are incubated in rotating shaker at 4, 37 °C, or room temperature for 2 h.
3. The solubilized membranes are centrifuged at 100,000 × *g* at 4 °C for 1 h.
4. The supernatant is loaded on an SDS-PAGE and analyzed as described previously; the amount of protein solubilized is quantified based on the unsolubilized control protein used (Roy et al., 2008).

Notes

1. Following presolubilization screening, detergent titration is performed to ensure reproducible solubilization. This can be accomplished in two steps: in the first step, different amount of membrane sample 2.5–10 mg/ml is taken and solubilized with the same amount of

detergent under previously determined optimal conditions. The amount solubilized is detected using SDS-PAGE analysis and quantification. Similarly, different detergent concentration can also be tested using the same amount of starting material. These initial experiments provide sufficient information for maintaining reproducibility of solubilization from batch to batch. Based on all these parameters, large-scale solubilization can be performed prior to purification.

2. At this stage, the activity of the target protein is also tested to identify the detergent in which the protein is functionally stable. In case of GPCRs, radioligand-binding assays can be performed in different detergents. The advantage of this assay is that it can be performed using picomolar amounts of unpurified receptor proteins. The method is reproducible and easy to perform. It has been previously shown that ligand binding can also increase the stability of the GPCR during solubilization. However, in case of other membrane proteins such as transporters, such assays cannot be used, the stability of the protein can be determined by reconstitution in lipid vesicles only after purification.
3. GPCRs contain characteristic cholesterol binding clefts, which were identified based on the crystal structure of β1 and β2 adrenergic receptors (Hanson et al., 2008). Cholesterol analogs such as 0.2% (w/v) cholesterol hemisuccinate (CHS) are added to stabilize the protein of interest (Lopez et al., 2008). Dynamic light scattering and small-angle X-ray scattering have shown that CHS forms bicelle-like architecture when mixed with DDM (Thompson et al., 2011).
4. Integral lipids are required for stability of some membrane proteins; longer solubilization time should be avoided as the proteins might get delipidated.

REFERENCES

André, N., Cherouati, N., Prual, C., Steffan, T., Zeder-Lutz, G., Magnin, T., et al. (2006). Enhancing functional production of G protein-coupled receptors in *Pichia pastoris* to levels required for structural studies via a single expression screen. *Protein Science*, *15*(5), 1115–1126.

Bachhawat, A. K., & Ghosh, S. (1987). Isolation and characterization of outer membrane proteins of *Azospirillum brasilense*. *Journal of General Microbiology*, *133*, 1751–1758.

Deisenhofer, J., Epp, O., Miki, K., Huber, R., & Michel, H. (1984). X-ray structure analysis of a membrane protein complex. Electron density map at 3 A resolution and a model of the chromophores of the photosynthetic reaction center from *Rhodopseudomonas viridis*. *Journal of Molecular Biology*, *180*(2), 385–398.

Hanson, M. A., Cherezov, V., Griffith, M. T., Roth, C. B., Jaakola, V. P., Chien, E. Y., et al. (2008). A specific cholesterol binding site is established by the 2.8 A structure of the human beta2-adrenergic receptor. *Structure*, *16*(6), 897–905.

Hatefi, Y. (1978). Preparation and properties of dihydroubiquinone: Cytochrome c oxido-reductase (complex III). *Methods in Enzymology, 53*, 35–40.

Kawate, T., & Gouaux, E. (2006). Fluorescence-detection size-exclusion chromatography for precrystallization screening of integral membrane proteins. *Structure, 14*(4), 673–681.

Laible, P. D., Scott, H. N., Henry, L., & Hanson, D. K. (2004). Towards higher-throughput membrane protein production for structural genomics initiatives. *Journal of Structural and Functional Genomics, 5*(1–2), 167–172.

Lopez, J. J., Shukla, A. K., Reinhart, C., Schwalbe, H., Michel, H., & Glaubitz, C. (2008). The structure of the neuropeptide bradykinin bound to the human G-protein coupled receptor bradykinin B2 as determined by solid-state NMR spectroscopy. *Angewandte Chemie (International Ed. in English), 47*(9), 1668–1671.

Lundstrom, K. (2006). Structural genomics for membrane proteins. *Cellular and Molecular Life Science, 63*(22), 2597–2607.

Lundstrom, K., Wagner, R., Reinhart, C., Desmyter, A., Cherouati, N., Magnin, T., et al. (2006). Structural genomics on membrane proteins: Comparison of more than 100 GPCRs in 3 expression systems. *Journal of Structural and Functional Genomics, 2*, 77–91.

Magnani, F., Shibata, Y., Serrano-Vega, M. J., & Tate, C. G. (2008). Co-evolving stability and conformational homogeneity of the human adenosine A2a receptor. *Proceedings of the National Academy of Sciences of the United States of America, 105*(31), 10744–10749.

Morello, E., Bermúdez-Humarán, L. G., Llull, D., Solé, V., Miraglio, N., Langella, P., et al. (2008). *Lactococcus lactis*, an efficient cell factory for recombinant protein production and secretion. *Journal of Molecular Microbiology and Biotechnology, 14*(1–3), 48–58.

Palsdottir, H., & Hunte, C. (2004). Lipids in membrane protein structures. *Biochimica et Biophysica Acta, 1666*(1–2), 2–18.

Popot, J. L., Berry, E. A., Charvolin, D., Creuzenet, C., Ebel, C., Engelman, D. M., et al. (2003). Amphipols: Polymeric surfactants for membrane biology research. *Cellular and Molecular Life Sciences, 60*(8), 1559–1574.

Rafferty, C. N., & Clayton, R. K. (1997). The orientations of reaction center transition moments in the chromatophore membrane of *Rhodopseudomonas sphareroides*, bases on new linear dichroism and photoselection measurements. *Biochimica et Biophysica Acta, 546*(2), 189–206.

Roy, A., Shukla, A. K., Haase, W., & Michel, H. (2008). Employing *Rhodobacter sphaeroides* to functionally express and purify human G protein-coupled receptors. *Biological Chemistry, 389*(1), 69–78.

Sahdev, S., Khattar, S. K., & Saini, K. S. (2008). Production of active eukaryotic proteins through bacterial expression systems: A review of the existing biotechnology strategies. *Molecular and Cellular Biochemistry, 307*(1–2), 249–264.

Serrano-Vega, M. J., Magnani, F., Shibata, Y., & Tate, C. G. (2008). Conformational thermostabilization of the beta1-adrenergic receptor in a detergent-resistant form. *Proceedings of the National Academy of Sciences of the United States of America, 105*(3), 877–882.

Shukla, A. K., Reinhart, C., & Michel, H. (2006). Comparative analysis of the human angiotensin II type 1a receptor heterologously produced in insect cells and mammalian cells. *Biochemical and Biophysical Research Communications, 349*(1), 6–14.

Surade, S., Klein, M., Stolt-Bergner, P. C., Muenke, C., Roy, A., & Michel, H. (2006). Comparative analysis and "expression space" coverage of the production of prokaryotic membrane proteins for structural genomics. *Protein Science, 15*(9), 2178–2189.

Thompson, A. A., Liu, J. J., Chun, E., Wacker, D., Wu, H., Cherezov, V., et al. (2011). GPCR stabilization using the bicelle-like architecture of mixed sterol-detergent micelles. *Methods, 55*, 310–317.

CHAPTER FOUR

Amphipathic Agents for Membrane Protein Study

Aiman Sadaf*, Kyung Ho Cho*, Bernadette Byrne†, Pil Seok Chae*,[1]
*Department of Bionanotechnology, Hanyang University, Ansan, South Korea
†Department of Life Sciences, Imperial College London, London, United Kingdom
[1]Corresponding author: e-mail address: pchae@hanyang.ac.kr

Contents

Abstract

Membrane proteins (MPs) are insoluble in aqueous media as a result of incompatibility between the hydrophilic property of the solvent molecules and the hydrophobic nature of MP surfaces, normally associated with lipid membranes. Amphipathic compounds are necessary for extraction of these macromolecules from the native membranes and their maintenance in solution. The amphipathic agents surround the hydrophobic segments of MPs, thus serving as a membrane mimetic system. Of the available amphipathic agents, detergents are most widely used for MP manipulation. However, MPs encapsulated by conventional detergent micelles have a tendency to undergo structural degradation, hampering MP advance, and necessitating the development of novel detergents with enhanced efficacy for MP study. In this chapter, we will introduce both conventional and novel classes of detergents and discuss about the chemical structures, design principles, and efficacies of these compounds for MP solubilization and stabilization. The behaviors of those agents toward MP crystallization will be a primary topic in our discussion. This discussion highlights the common features of popular

Methods in Enzymology, Volume 557
ISSN 0076-6879
http://dx.doi.org/10.1016/bs.mie.2014.12.021

conventional/novel detergents essential for successful MP structural study. The conclusions reached by this discussion would not only enable MP scientists to rationally select a set of detergent candidates among a large number of detergents but also provide detergent inventors with useful guidelines in designing novel amphipathic systems.

1. INTRODUCTION

Membrane proteins (MPs) are amphipathic biomacromolecules with a central hydrophobic segment and two flanking hydrophilic moieties. Because of the presence of large hydrophobic surfaces, these membrane macromolecules are normally located in membrane bilayers comprised of amphipathic lipids. Conventional detergents share the amphipathic properties of MPs and lipids, and are popularly used to work with MPs as they form a membrane mimetic system, called micelles, above their individual critical micelle concentrations (CMCs). These self-assemblies have the ability to associate with and protect the hydrophobic segment of MPs from a polar aqueous medium (Moller & le Maire, 1993; White & Wimley, 1999). Detergent micelles formed by the so-called hydrophobic effect usually have a spherical, cylindrical, or elliptical shape, with a hydrophobic interior and a hydrophilic exterior (Fig. 1; Israelachvili, Mitchell, & Ninham, 1977; Tanford, 1980). Although a large number of detergents are available, one of the most challenging tasks for scientists is to select an optimal detergent for MP solubilization and stabilization. Unfortunately, there are few guidelines on effective detergent selection, mainly because little information is known on detergent structure–property–efficacy relationships, i.e., how detergent structural features correlate to their physical properties and efficacy for MP studies (Bowie, 2001; Garavito & Ferguson-Miller, 2001; Rosenbusch, 2001). The limited understanding of the properties that confer desirable detergent characteristics also makes it difficult for organic chemists to design and prepare novel amphiphiles with enhanced efficacy, thus significantly hampering advances in this area. In this chapter, we will focus on a number of conventional detergents and representatives of novel amphipathic systems and critically discuss their respective advantages and disadvantages for MP research in order to provide guidelines for the selection of detergent candidates and the design of novel agents for MP study.

2. MP STABILITY IN MEMBRANE ARCHITECTURE

As mentioned above, integral MPs exist within the highly hydrophobic environment of lipid membranes. However, for successful MP structural

Figure 1 Schematic representation of detergent micelles. Micelles are comprised of amphipathic molecules with a hydrophobic and a hydrophilic group within the molecules. These molecules are often called detergents but other names such as amphiphiles and surfactants are also used.

study, it is critical to maintain the MPs in a soluble, native, and stable state in aqueous media. In order to understand what maintains an MP in a stable state, a good place to start is the lipid membrane. The key feature of biological membranes resides in the bilayer architecture which provides lateral pressure on MPs (Anglin & Conboy, 2008; Lee, 2011). This MP-holding force is critical to prevent protein denaturation and aggregation, two mechanisms responsible for protein degradation (Bowie, 2001; Garavito & Ferguson-Miller, 2001). Many MPs undergo significant conformational changes essential for them to carry out their cellular functions. Due to their highly dynamic nature, MPs are prone to structural degradation. The lipid superassembly encompassing MPs is effective at preventing protein denaturation by providing a recovering force (i.e., lateral pressure) on partially unfolded MPs. This architecture also has the capability to prevent uncontrolled protein–protein interaction (i.e., protein aggregation) by favorable association of lipid molecules with the hydrophobic segments of MPs. Another feature of the native membranes for MP stabilization comes from the presence of lipid molecules specifically bound to the MP surface (Marsh, 2008). Similar to a ligand bound to the active site of an MP, these lipids have protein-stabilizing effect. Thus, the membrane provides stabilizing forces both from the association of individual lipids and from the lateral forces provided by the membrane bilayer.

3. CONVENTIONAL DETERGENTS

3.1 Detergent classification

Approximately 120 conventional detergents are available for MP study. This large number of detergents can be classified into three categories depending on the ionic nature of their hydrophilic groups: ionic, zwitterionic, and nonionic detergents. An ionic detergent contains a negative (anionic) or positive (cationic) charge within the headgroup. A sulfate group is the most popular as an anionic headgroup, as exemplified by sodium dodecyl sulfate (SDS), followed by sulfonate and carboxylate (e.g., sodium deoxycholate) (Fig. 2). In the case of cationic headgroups, a quaternary ammonium headgroup (e.g., trimethylammonium group) is most common (e.g., cetyltrimethylammonium bromide (CTAB)) (Fig. 2). In general, however, these ionic detergents are not ideal for MP study because of their high tendency to degrade MPs illustrated by the popular use of SDS for protein denaturation (Privé, 2007). With a few exceptions, zwitterionic detergents are more suitable for MP research than ionic detergents. The headgroups of zwitterionic detergents carry both positive and negative charges, although the overall charge is zero. A few zwitterionic headgroups such as *N*-oxides, phosphorylcholines, sulfobetaines, and carboxybetaines appear commonly in detergent structures. Representatives of these zwitterionic detergents include laurydimethylamine-*N*-oxide (LDAO), Anzergent 3–12, Fos-choline-12 (FC-12), 3-[(3-cholamidopropyl)dimethylammonio]-1-propanesulfonate (CHAPS), and 3-[(3-cholamidopropyl)dimethylammonio]-2-hydroxy-1-propanesulfonate (CHAPSO) (Fig. 2). These agents are popularly used for MP research, particularly for MP extraction (Privé, 2007). Among these detergents, LDAO is the most successful in terms of high-resolution MP structure determination. However, the charged character of these detergents is not ideal for the stabilization of most MPs, with only 20% MPs predicted to be stable in LDAO (Michel, 2001). Uncharged, nonionic, detergents are most widely used for both extraction and stabilization of MPs in solution. Representatives of nonionic hydrophilic groups include *N*-methylglucamides (MEGAs), *N*-hydroxymethylglucamides (HEGAs),

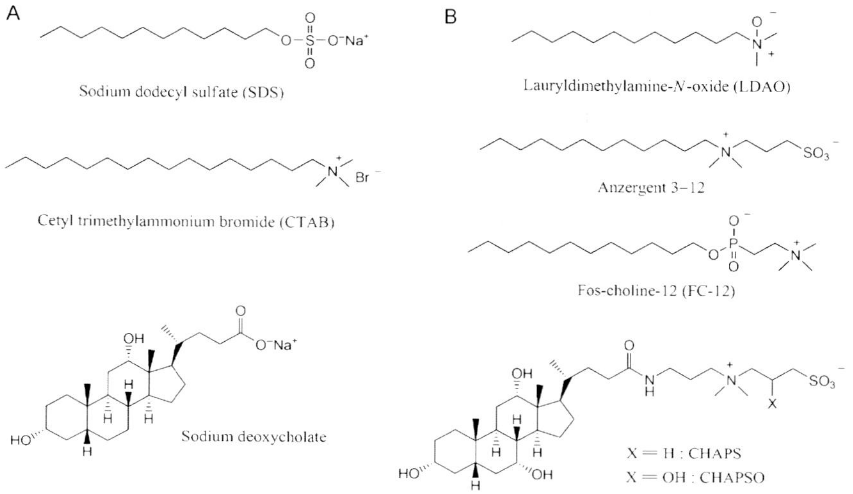

A

Sodium dodecyl sulfate (SDS)

Cetyl trimethylammonium bromide (CTAB)

Sodium deoxycholate

B

Lauryldimethylamine-*N*-oxide (LDAO)

Anzergent 3–12

Fos-choline-12 (FC-12)

X = H : CHAPS
X = OH : CHAPSO

Figure 2 Chemical structures of representative conventional detergents (ionic detergents (A) and zwitterionic detergents (B)). (A) Sodium dodecyl sulfate (SDS) and sodium deoxycholate have a negative charge in their hydrophilic groups, thus being classified as anionic detergents whereas cetyltrimethylammonium bromide (CTAB) with a positive charge is categorized as a cationic detergent. (B) These agents contain both positive and negative charge on their hydrophilic groups, making them overall neutral. Three headgroups are popular in these zwitterionic detergents: *N*-oxide (e.g., LDAO), sulfobetaines (e.g., Anzergent 3–12, CHAPS, and CHAPSO), and phosphorylcholine (e.g., FC-12).

polyoxyethylene glycols (C_mE_n), glucosides, and maltosides (Fig. 3). These agents are known to have mild properties in terms of MP denaturation and, thus are generally effective at stabilizing a variety of MPs (Privé, 2007).

3.2 Use of conventional detergents for MP study

Use of detergents starts with protein extraction from the native membranes. Next, a target MP solubilized by detergent treatment is isolated in a pure and functional form via several protein purification steps where detergent molecules play an essential role in maintaining protein solubility and stability. Such functional proteins purified in detergent micelles are directly used for MP crystallization using the vapor-diffusion method (i.e., *in surfo* method) or transferred to lipidic cubic phase (LCP) in LCP-based crystallization method (i.e., *in meso* method). Thus, detergents are indispensable elements for MP manipulation including protein extraction/solubilization, purification, and crystallization. During this whole process, maintaining the native conformation of a target MP is a prerequisite for the success of

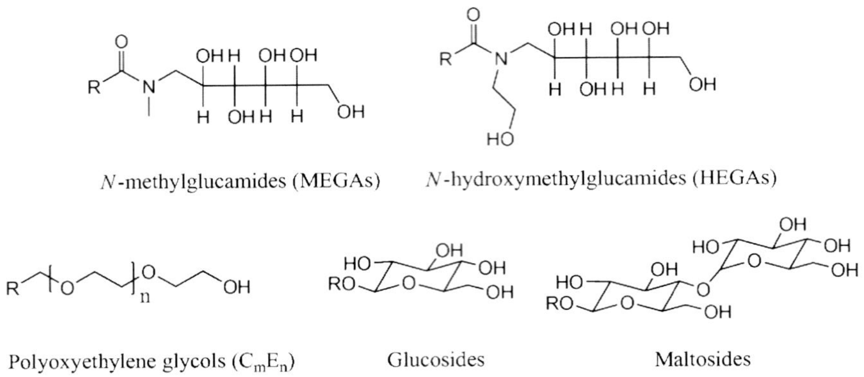

N-methylglucamides (MEGAs) *N*-hydroxymethylglucamides (HEGAs)

Polyoxyethylene glycols (C_mE_n) Glucosides Maltosides

Figure 3 General chemical structures of conventional nonionic detergents. Most of these agents utilize carbohydrates with multiple hydroxyl groups such as glucamide, glucose, and maltose although some members have polyoxyethylene glycol as their hydrophilic groups. Glucamide and polyoxyethylene glycol are open chain headgroups, while glucoside and maltoside are closed ring headgroups which may be related to the efficacy of these agents for membrane protein manipulation (R is an alkyl group).

protein structure determination, which is a reason why detergents with excellent MP stabilization efficacy are most widely used for MP study (Bowie, 2001; Garavito & Ferguson-Miller, 2001). It is noteworthy that the optimal detergent to use varies depending on the step of the process (Gohon & Popot, 2003). In other words, a promising detergent for MP solubilization is not necessarily suitable for MP stabilization or crystallization.

A generally accepted detergent order for MP solubilization efficiency would be ionic > zwitterionic > nonionic detergents. However, this order is typically reversed when considering MP stabilization efficacy; nonionic detergents are most stabilizing, while ionic detergents are least stabilizing. Thus, there is a compromise to make in terms of solubilization efficiency and stabilization efficacy. For example, use of an ionic detergent (e.g., SDS) usually extracts MPs very efficiently from the native membranes, but the extracted proteins typically undergo a complete loss of activity as a result of protein unfolding. In contrast, MPs extracted by nonionic detergents (e.g., DDM (*n*-dodecyl-β-D-maltopyranoside); Fig. 4) generally retain their activity, but the amount of solubilized proteins can be low. Because most MPs are produced at comparatively low levels even in recombinant expression systems, it is difficult to obtain a sufficient amount of MPs for structural and functional study (Junge et al., 2008; Midgett & Madeen, 2007). Thus, MP solubilization efficiency is critical, but the protein must still be maintained in a structurally and functionally intact form.

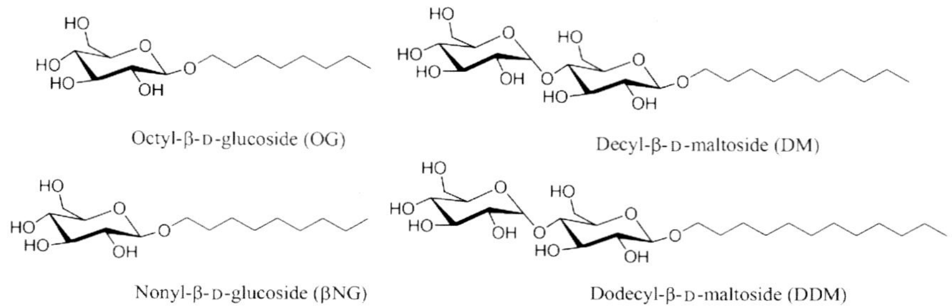

Figure 4 Chemical structures of nonionic detergents that are popularly used for membrane protein structural study. Glucoside detergents (OG and βNG) tend to form smaller PDCs than maltoside detergents (DM and DDM) presumably due to the presence of the small headgroup. On the other hand, maltoside detergents are superior to glucoside agents in maintaining the native structures of membrane proteins. Both glucoside and maltoside detergents are popularly used for MP structural study as detergent efficacy for both small PDC formation and MP stabilization are critical to facilitate membrane protein crystal formation.

Accordingly, zwitterionic detergents such as LDAO and FC-12 which are more efficient at MP solubilization are sometimes used in preference to the milder nonionic agents such as DDM and DM (*n*-decyl-β-D-maltopyranoside) (Privé, 2007). In the next step of protein purification, however, preserving the native structures of MPs (i.e., MP stabilization efficacy) becomes most important. Thus, nonionic detergents with promising stabilization efficacy are the best candidates for this process. As MP degradation could take place via two mechanisms, i.e., protein denaturation and aggregation, a detergent should have the ability to prevent both protein denaturation and aggregation to be effective for MP stabilization. In this context, mild detergents can be classified into two categories with different characteristics. For instance, cholate-based conventional detergents (e.g., CHAPS and CHAPSO) are excellent at minimizing the denaturation of MPs but tend to be poor at preventing protein aggregation. On the other hand, maltoside-bearing conventional agents (e.g., DDM and DM) are good at inhibiting both protein denaturation and aggregation, thereby showing enhanced behaviors relative to the former toward preserving the native structures of a variety of MPs. The most commonly used detergent is DDM which combines reasonable solubilization efficiency with good MP stabilization characteristics (Caffrey, Li, & Dukkipati, 2012). However, even if the same detergent is used for protein extraction and solubilization, it is often essential to explore other detergents for crystallization by the process of detergent exchange. Detergent properties ideal for MP crystallization are somewhat different from those for MP solubilization or purification. The protein sample is at a higher concentration than during isolation (~10 mg/ml) and is thus more prone to protein aggregation through non-specific hydrophobic interactions. In order to prevent protein aggregation, a detergent with a strong binding affinity to MP surfaces would be more relevant than a weakly binding agent. Consequently, a detergent that minimizes protein aggregation rather than protein denaturation is preferably used for MP crystallization. An additional factor important for MP crystallization is the size of protein–detergent complexes (PDCs). A small PDC is known to be beneficial for protein crystal formation as this exposes a larger hydrophilic protein surface, available to form all-important protein–protein contacts essential for crystal lattice formation, particularly using the *surfo* method (Kantardjieff & Rupp, 2003). Small headgroup-bearing detergents (e.g., LDAO) tend to form small PDCs (Privé, 2007) and thus are popularly used for MP crystallization.

The importance of detergent binding strength to MP surfaces and PDC size for successful MP crystallization can be gained through an analysis of results obtained for conventional detergents. Of more than 120 conventional detergents, a small number of agents are popularly used for MP structural study. For instance, a study from 2008 revealed that five conventional detergents, OG (*n*-octyl-β-D-glucopyranoside), βNG (*n*-nonyl-β-D-glucopyranoside), DM, DDM, and LDAO, facilitated crystal structure determinations of about 70% of α-helical MPs (Fig. 4; Newstead, Ferrandon, & Iwata, 2008). Of these five agents, DDM shows the greatest stabilizing efficacy and it is not surprising that it has a strong track record in MP structural determination. However, glucoside-bearing agents (OG and βNG) with much lower stabilizing efficacy are also commonly used for MP structural study (Chae, Rana, et al., 2013). Furthermore, LDAO, a rather harsh detergent, is also popularly used for MP crystallization, indicating that MP stability is neither the sole nor main determinant for successful MP crystallization. The formation of small PDCs and a strong binding affinity to MPs, common features of OG, βNG, and LDAO, are believed to have significant roles in successful MP structure study.

3.3 Limitation of conventional detergents

Even popular detergents, such as LDAO, OG, and DDM, have a serious limitation in MP study (Breyton, Tribet, Olive, Dubacq, & Popot, 1997; Brotherus, Jost, Griffith, & Hokin, 1979). As we described above, LDAO and OG tend to form small PDCs, but many MPs encapsulated by these detergent micelles undergo marked structural degradation. Due to their rather harsh nature, these agents are not good candidates for denaturation-sensitive MPs as the proteins are not viable in the micelles of these agents. Rather, glucoside or *N*-oxide detergents have compatibility with relatively robust MPs (e.g., cytochrome bo_3 ubiquinol oxidase in OG (Abramson et al., 2000) and bovine rhodopsin in LDAO (Li, Edwards, Burghammer, Villa, & Schertler, 2004). In contrast, DDM is clearly superior to OG and LDAO in terms of MP stabilization efficacy but tends to form large PDCs, unfavorable for MP crystallization (Kantardjieff & Rupp, 2003). Consequently, maltoside-bearing detergents such as DM and DDM are favorable for relatively fragile MPs and MP complexes with comparatively large hydrophilic surface areas available for crystal contact formation (e.g., cytochrome b_6f complex; Stroebel, Choquet, Popot, & Picol, 2003). These findings lead to the conclusion that each agent has strength and weakness, i.e., no one detergent is suitable for all MPs. Most MPs of known structure are conspicuous by their stability (Alexandrov, Mileni, Chien, Hanson, & Stevens, 2008). In the future, we will need to address the issues of more challenging MPs with far lower inherent stability such as eukaryotic MPs or MP complexes. Even DDM, the most-stabilizing conventional detergent, has displayed limited ability to stabilize MP complexes (e.g., human β_2AR:G-protein complex) for structural studies (Rasmussen, DeVree, et al., 2011). However, integral MP complexes remain key targets for structural studies. The low variability in detergent chemical structures is likely to be responsible for their limited application. Most conventional detergents have a single flexible alkyl chain and a single head group. Amphipathic agents with novel architectures have significant potential to facilitate future MP structural studies (Chae, Laible, & Gellman, 2010).

4. NOVEL AMPHIPATHIC SYSTEMS

4.1 Detergent properties for MP structural study

Before introducing a number of novel systems, it is valuable to summarize which detergent properties are beneficial for successful MP structural study. First, the detergent needs to keep the protein stable through extraction and isolation and beyond. Following isolation, MPs should retain their native structures for at least a few days, the minimum required to obtain protein crystals. Second, detergents with the ability to form small PDCs will facilitate crystal formation by increasing the regions of the protein available to form crystal contacts. A number of different approaches have been developed to increase the hydrophilic surface areas of MPs such as expression of a fusion protein (e.g., GPCR + T4 lysozyme) and complex formation with an antibody (e.g., Fab) or a nanobody (Rasmussen, Choi, et al., 2011; Rasmussen et al., 2007). However, the need for generic, widely applicable approaches means that there is still a demand for detergents that can not only stabilize the native structure of MPs but also form small PDCs to increase the chances of obtaining well-diffracting crystals. As can be seen for in DDM and LDAO cases, however, a detergent with high MP stabilization efficacy tends to form large PDCs, while a small PDC-forming detergent typically destabilizes MPs, indicating that conventional detergents fail to confer both favorable characteristics. In addition to these two properties, detergents that exhibit high solubilization efficiency will also be favorable for MP structural study since the use of these agents results in a greater amount of target protein for downstream isolation and crystallization. When novel detergents are evaluated for MP study, therefore, the three detergent properties should be considered together: MP stabilization efficacy, PDC size, and MP solubilization efficiency.

4.2 Small amphipathic compounds

4.2.1 Conventional detergent variants

Most conventional detergents are small molecules with a molecular weight in the range of 200–1000 Da. The majority of novel agents also have relatively low molecular weight, but their molecular weights are usually a little higher than those of conventional ones. Some of these novel agents were derived from conventional detergents, but showed improved detergent efficacy for MP stabilization. For example, the carbohydrate versions of Triton X-100 (TX-100), designated Chae's glyco-triton (CGT), were recently reported (Fig. 5; Chae, Wander, Cho, Laible, & Gellman, 2013). TX-100 is a nonionic detergent, with a polyethylene oxide as a hydrophilic group and a branched alkyl substituent-bearing benzene ring as a hydrophobic group. This agent serves as a popular solubilizing agent for a number of molecules with hydrophobic surface including MPs, presumably due to the presence of the branched alkyl chain in the detergent lipophilic region. Bearing in mind that popular detergents for MP crystallization (e.g., OG and DDM) contain glucose or maltose groups rather than a polyethylene oxide, a maltoside, or a branched diglucoside headgroup was connected to the hydrophobic group of TX-100 to generate the CGT amphiphiles. Thus, these new agents are the hydrophilic variants of TX-100. Maltoside bearing agents (DDM and

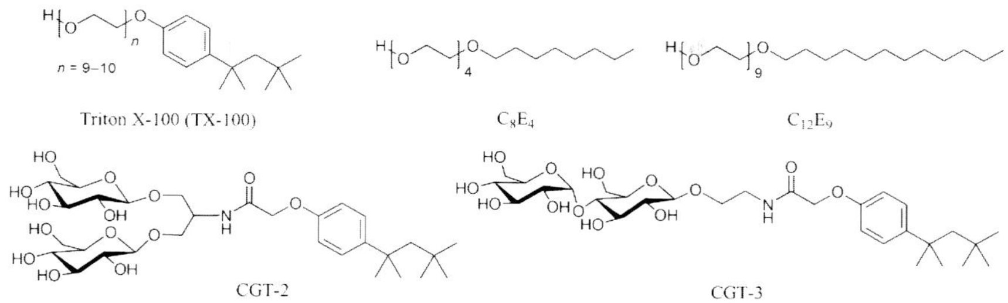

Figure 5 Chemical structures of conventional detergents (Triton X-100, C_8E_4, and $C_{12}E_9$) and novel amphiphiles (CGT-2 and CGT-3). As detergent lipophilic group, Triton X-100 (TX-100) has a benzene ring with a branched alkyl pendant while C_8E_4 and $C_{12}E_9$ have a straight alkyl chain. Chae's glyco-tritons (CGTs) are the carbohydrate versions of TX-100 with a branched diglucoside (CGT-2) or a maltoside (CGT-3) headgroup.

DM) are typically more stabilizing than their polyethylene oxide counterparts (e.g., C_4E_8 and C_9E_{12}) (Fig. 5). Consistent with this observation, CGTs (CGT-2 and CGT-3) were demonstrated to be superior to TX-100 in terms of stabilization of an MP complex, *Rhodobacter capsulatus* photosynthetic superassembly, comprised of the light harvesting complex I (LHI) and the reaction center (RC). Additionally, these novel agents are resistant to oxidation.

Other classes of novel amphiphiles that were derived from conventional detergents (CHAPS and sodium deoxycholate, respectively) include cholate- and deoxycholate-based *N*-oxide classes (CAOs or DCAOs) (Fig. 6; Chae, Sadaf, & Gellman, 2014). The *N*-oxide headgroup (such as in LDAO, the small PDC-forming detergent) is more popular than a sulfobetaine or carboxylate group for MP study. When the CAOs and DCAOs were evaluated for the solubilization and stabilization of LHI–RC complex, those agents were shown to be superior to LDAO or other *N*-oxide-containing synthetic agents, particularly for MP stabilization efficacy. In spite of the zwitterionic nature of the *N*-oxide headgroup, quite interestingly, CAOs and DCAOs turned out to be more effective at stabilizing LHI–RC complexes compared to DDM. The presence of nonhydrocarbon groups (e.g., hydroxy) in their lipophilic regions is thought to impart enhanced stabilization efficacy to these classes of amphiphiles.

4.2.2 Hemifluorinated surfactants

MPs with multiple subunits maintain their tertiary and quaternary structures via relatively weak intramolecular/intermolecular forces. Strong

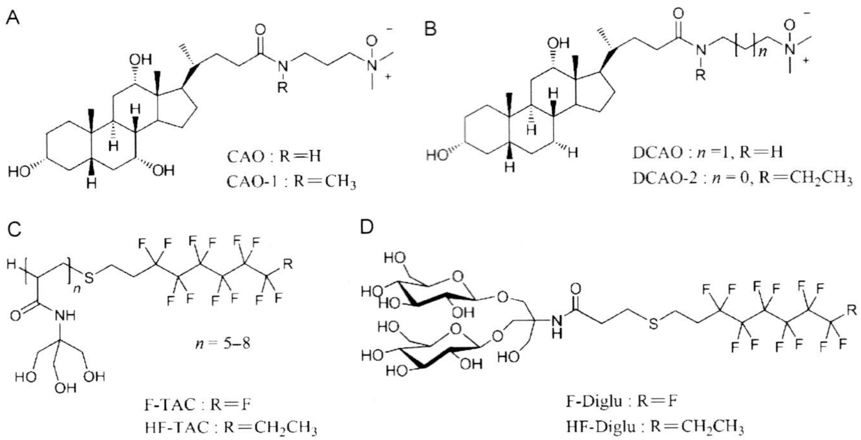

A

CAO : R=H
CAO-1 : R=CH_3

B

DCAO : n =1, R=H
DCAO-2 : n = 0, R=CH_2CH_3

C

n = 5–8

F-TAC : R=F
HF-TAC : R=CH_2CH_3

D

F-Diglu : R=F
HF-Diglu : R=CH_2CH_3

Figure 6 Chemical structures of cholate and deoxycholate-based amphiphiles (CAOs and DCAOs, respectively) (A and B) and (hemi)fluorinated surfactants ((H)FSs) (C and D). CAOs and DCAOs bear the multiple hydroxyl groups in the lipophilic region and have an *N*-oxide headgroup. Fully fluorinated alkyl chains were used to give fluorinated surfactants (FSs; F-TAC and F-Diglu), while alkyl tips such as ethyl and propyl groups were attached to the end of fluorinated alkyl groups to generate hemifluorinated surfactants (HFSs; HF-TAC and HF-Diglu).

interactions between detergent micelles and MPs are likely to interfere with these weak interactions, resulting in MP denaturation. In addition, a number of MPs contain hydrophobic cofactors in their interiors and lipids specifically bound to their surfaces. These cofactors and lipids confer enhanced stability on MPs. Detergent micelles could readily dissolve these hydrophobic or amphipathic molecules due to the presence of hydrophobic interior and amphipathic character (Chabaud, Barthélémy, Mora, Popot, & Pucci, 1998; Popot et al., 2011). Fluorinated alkyl chains are distinct from hydrocarbon alkyl chains in that they are hydrophobic *yet* lipophobic, rendering interactions between these two types of chains weak. Thus, amphipathic agents with a fluorinated alkyl chain would interfere much less with weak intramolecular or intermolecular forces essential for protein folding compared to their hydrocarbon counterparts, and are therefore much less likely to denature the protein. Furthermore, micelles comprised of fluorinated surfactants (FSs) have little affinity to hydrophobic cofactors and amphiphilic lipid molecules (Fig. 6). Thus, these FSs could be ideal solubilization and stabilization agents for MPs with weakly associated but essential hydrophobic cofactors and lipids. However, the fully FSs proved too lipophobic to effectively interact with MPs, thereby favoring irregular MP–MP association or aggregation, rather than MP–detergent micelle interaction. One strategy to solve this issue is to introduce a hydrocarbon alkyl tip at the end of the hydrophobic chain. This idea has produced hemifluorinated surfactants (HFSs) (e.g., HF-TAC; Fig. 6) with ethyl or propyl tips at the end of the chains (Abla, Durand, & Pucci, 2011; Breyton, Chabaud, Chaudier, Pucci, & Popot, 2004). These surfactants were shown to be not only mild enough to minimize protein denaturation but also lipophilic enough to prevent protein aggregation. Generally speaking, protein stability is dependent on detergent concentration with a high detergent concentration more likely to cause protein denaturation. However, the use of HFSs at a high concentration (e.g., 2 m*M*) was less detrimental to the stability of cytochrome b_6f complex than that of a conventional detergent, DDM, at the same concentration indicating that this class could provide a large window with respect to detergent concentration in MP crystallization (Breyton et al., 2004).

4.2.3 Tripod amphiphiles

Conventional detergents generally contain a single flexible alkyl tail group. The inherent flexibility of the tail groups may contribute to the disorder in the PDC which in turn may negatively influence the ability of the protein molecules to crystallize. As the name suggests, the tripod amphiphiles (TPAs) contain three alkyl chains projecting from a quaternary carbon (Fig. 7A). Thus, the lipophilic groups of TPAs are much less flexible than

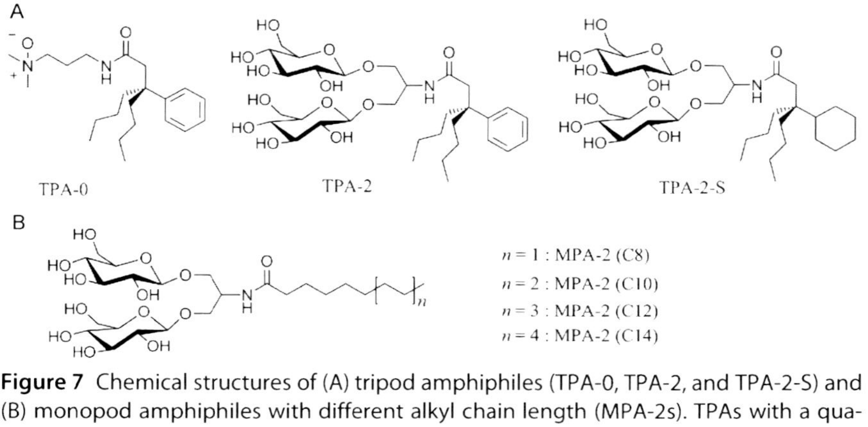

A

TPA-0 TPA-2 TPA-2-S

B

$n = 1$: MPA-2 (C8)
$n = 2$: MPA-2 (C10)
$n = 3$: MPA-2 (C12)
$n = 4$: MPA-2 (C14)

Figure 7 Chemical structures of (A) tripod amphiphiles (TPA-0, TPA-2, and TPA-2-S) and (B) monopod amphiphiles with different alkyl chain length (MPA-2s). TPAs with a quaternary carbon in the lipophilic region have either an *N*-oxide (TPA-0) or a branched diglucoside (TPA-2 and TPA-2-S) as a headgroup. MPA-2s share a branched diglucoside headgroup with TPA-2, but have linear and flexible alkyl chains.

those in conventional detergents. Such structural restriction was hypothesized to result in a less dynamic micelle and increase the likelihood of forming well-diffracting protein crystals (McQuade et al., 2000). Although the main expectation for these class members was to facilitate MP crystallization, TPAs are highly efficient at MP solubilization. For example, one TPA with an *N*-oxide headgroup, TPA-0 (commercial name, TRIPAO), extracted bacteriorhodopsin (BR) almost completely following an 30-min incubation. In contrast, TX-100 required 20 h for complete BR solubilization (Yu et al., 2000). The superior solubilization efficiency of TPAs was strongly supported by a comparative study of TPA-2 and its conventional analogs, monopod amphiphiles (MPA-2s), both of which bear a branched diglucoside headgroup (Fig. 7B). When glyco-TPAs (TPA-2 and TPA-2-S) were compared with MPA-2 variants with different carbon chain length for the solubilization of *R. capsulatus* superassembly, TPA-2 and TPA-2-S were substantially more efficient than MPA-2s at solubilizing the superassembly (50–70% vs. 30%) (Chae, Wander, Bowling, Laible, & Gellman, 2008). It is difficult to provide a precise reason why the TPA architecture showed favorable solubilization efficiency relative to MPAs. However, it is possible that the three hydrophobic tail groups enable TPAs to make multiple interactions with the MPs, which make the detergents more effective at disrupting the interactions between the proteins and the lipid bilayer. A number of glyco-TPAs with structural variation in the nature/number of head and tail groups have been generated by a modular synthetic approach. Ten hydrophobic variants of TPA-2 and TPA-2-S were all shown to solubilize LHI–RC without any detectable structural degradation (Chae, Bae, Muhammad, Hussain, & Kim, 2014; Chae, Cho, et al., 2014). Since all these agents contain a branched diglucoside headgroup, this result indicates that this headgroup is effective at stabilizing the native conformations of MPs. A TPA with a branched diglucoside headgroup was even more effective than a TPA with a maltoside group (e.g., TPA-4) at maintaining the integrity of LHI–RC complex. Interestingly, in contrast to the common superiority in MP stabilization, these hydrophobic variants showed a large variation in the solubilization efficiency for the complexes, ranging from 10% to 90%, suggesting that detergent hydrophobic groups mainly dictate MP solubilization efficiency, while detergent hydrophilic groups primarily determine MP stabilization efficacy (Chae, Cho, et al., 2014). This TPA class has been shown to have potential for crystallization having yielded well-diffracting BR crystals (Theisen, Potocky, McQuade, Gellman, & Chiu, 2005). It will be interesting to see if the TPAs can contribute to crystallization of more challenging MPs.

4.2.4 Facial amphiphiles

Lipid molecules in biological membranes facially interact with MPs. Facial amphiphiles (FAs) have garnered significant attention because these agents closely mimic membrane systems. The best-characterized examples are cholate-based FAs in which multiple headgroups project from one side of the steroidal lipophilic structure (Fig. 8A). Various headgroups such as glucose, maltose, and phosphorylcholine have been introduced to generate a number of FAs (Lee et al., 2013). These compounds are structurally similar to CHAPS and CHAPSO since they are derived from cholic acid. However, the FAs have two or three large hydrophilic groups and so have greater facial

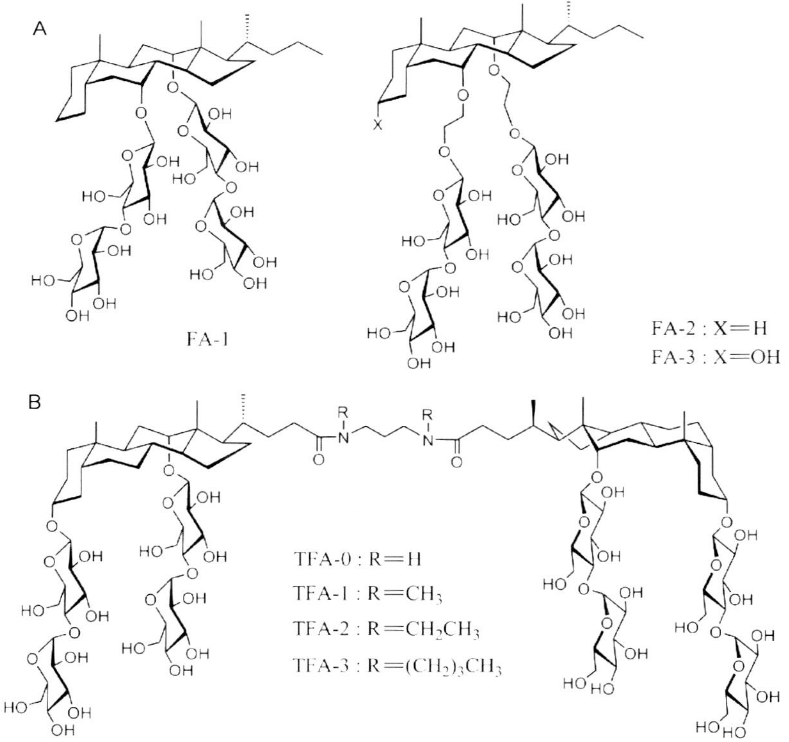

A

FA-1

FA-2 : X=H
FA-3 : X=OH

B

TFA-0 : R=H
TFA-1 : R=CH_3
TFA-2 : R=CH_2CH_3
TFA-3 : R=$(CH_2)_3CH_3$

Figure 8 Chemical structures of (A) facial amphiphiles with different headgroups (FA-1, FA-2, and FA-3) and (B) tandem facial amphiphiles with different alkyl chain length (TFA-0, TFA-1, TFA-2, and TFA-3). All of these amphiphiles were derived from cholic acid or deoxycholic acid. The hydroxyl groups in cholic acid were modified to position two or three maltosides on one side of the FAs, while two steroidal units are connected together via a diamine linker in the preparation of TFAs. TFAs were designed to match the length of the hydrophobic groups with the width of the hydrophobic segment of membrane proteins. TFA-1 (commercial name: TFA) is available from Anatrace.

amphiphilicity than CHAPS and CHAPSO with three hydroxy groups as the equivalent groups. In addition, the architecture of these agents is significantly different from most conventional amphiphiles because of the presence of the large hydrophobic surface. As this large surface is utilized to interact with the hydrophobic segment of MPs, these agents would bind to a target MP tightly and effectively, resulting in increased protein stability. The strong association of FAs with MPs was supported by ESI-mass spectra of an ATP-binding cassette (ABC) transporter (MsbA) solubilized in a conventional detergent (UDM) versus a facial amphiphile (FA-3; Fig. 8B) (Lee et al., 2013). In addition, several MPs including a BR, MsbA, and a human gap junction channel protein (connexin 26) have been shown to be better stabilized by FAs than conventional detergents (e.g., OG and DDM) (Lee et al., 2013). Furthermore, the tendency of FAs to form small PDCs may facilitate crystal formation of some MPs suitable for X-ray crystallographic analysis. Other FAs are the tandem facial amphiphiles (TFAs) (Chae, Gotfryd, et al., 2010). In this architecture, two steroidal units were connected together by a linker with different chain length via an amide coupling reaction so that the length of the rigid hydrophobic groups matches the width of the hydrophobic segment of MPs, ~30 nm (Fig. 8B). Interestingly, these agents form small micelles, comprised of only six to eight TFA molecules, in contrast to 50–100 molecules present in the micelles of conventional detergents. These TFAs were particularly superior to conventional detergents at stabilizing a number of MPs including LHI–RC complex and leucine transporter (LeuT) but not good at the stabilization of β_2-adrenergic receptor (β_2AR). The amide functional groups of these agents used to connect two steroidal units decreases their affinity for the hydrophobic surface of MPs, which could be a reason for their limited scope of utility. Thus, the ether or thio-ether version of the FAs may exhibit improved MP stabilization properties.

4.2.5 Rigid hydrophobic group-bearing amphiphiles

TPAs and FAs are rigid hydrophobic group-bearing amphiphiles due to the presence of a quaternary carbon and a fused ring system, respectively. Another well-known example is a class of amphiphiles with two consecutive hydrophobic rings (benzene and/or cyclohexane ring) in the lipophilic group (Fig. 9A). Some of these rigid hydrophobic group-bearing agents were shown to increase the stability of GPCRs and the human Patched protein receptor compared to DDM (Hovers et al., 2011). In addition, cytochrome b_6f complex from *Chlamydomonas reinhardtii* yielded well-diffracting

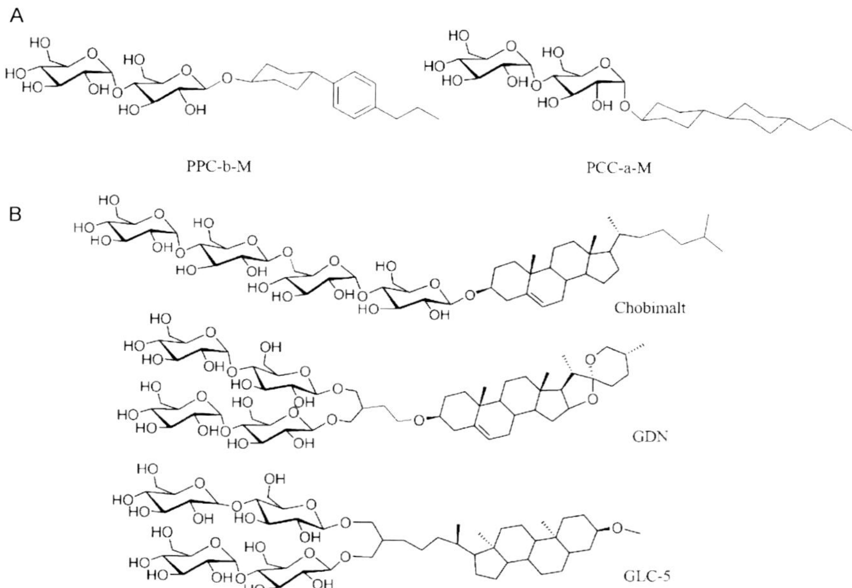

Figure 9 Chemical structures of rigid hydrophobic group-bearing amphiphiles. (A) PPC-b-M and PCC-a-M that shares a maltoside headgroup contain two ring systems (two cyclohexane rings or a cyclohexane and a benzene ring) positioned consecutively in the lipophilic region. (B) The rigid hydrophobic groups such as cholesterol (Chobimalt), diosgenin (GDN), and lithocholic acid (GLC-5) were utilized to generate these multifused ring-bearing agents. The hydrophilic groups vary from a linear carbohydrate (Chobimalt) to a branched dimaltoside (GDN and GLC-5). GDN is commercially available from Anatrace.

crystals in one of this class of detergents (PCC-a-M). These findings indicate that detergent rigidity maybe favor both MP stabilization and crystallization (Hovers et al., 2011). Cholate-related structures such as cholesterol and diosgenin have been utilized to produce amphiphiles with rigid architectures. For instance, cholesterol was used as a hydrophobic building block for the preparation of Chobimalt (Fig. 9B; Howell et al., 2010). However, Chobimalt alone, without addition of a classical detergent, exhibited only limited stabilization efficacy for the human kappa opioid receptor type I (hKOR1) (Howell et al., 2010). Most recently, glyco-lithocholate amphiphile (GLC-5) and glyco-diosgenin amphiphile (GDN) were developed. These amphiphiles both contain a hydrophobic group containing a fused ring system (Fig. 9B). Of the two, GDN conferred increased stability on several MPs including LHI–RC, LeuT, β_2AR, and melibiose permease

(MelB) compared to DDM (Chae et al., 2012). The hydrophobic group of this agent has a panel-like architecture that is suggested to promote favorable interactions with the MP surfaces, thus forming stable PDCs. Note that GDN, along with TFA-1, is commercially available.

4.2.6 Neopentyl glycol (NG) class amphiphiles

Most novel amphiphiles are difficult to prepare in large scale, which significantly reduces their practical utility. In contrast, the popular detergents such as OG, LDAO, DM, and DDM are easily prepared in large amounts. Furthermore, these conventional detergents have proved effective for a variety of MPs. In order to compete with these conventional agents, a novel amphiphile should possess both an excellent efficacy and practicability. In this respect, the neopentyl glycol (NG) class is the most outstanding because these class members could be prepared in four synthetic steps with an overall yield more than 80% (Chae, Rana, et al., 2013; Chae, Rasmussen, et al., 2010). This class is divided into two subclasses, glucose neopentyl glycol (GNG) and maltose neopentyl glycol (MNG), depending on their headgroup. Several NG class members are commercially available under the name of OGNG, DMNG, and LMNG in a price comparable to OG and DDM from Anatrace (Fig. 10). As a headgroup, OGNG has a branched diglucoside while DMNG and LMNG have a branched dimaltoside. These amphiphiles have a quaternary carbon between the hydrophobic and the hydrophilic groups, different from TPA architecture with a quaternary carbon within the lipophilic portion.

In terms of MP solubilization efficiency, GNGs and MNGs are almost comparable to their conventional counterparts, OG and DDM, respectively. For example, some GNG members could extract LHI–RC from the membrane with an efficiency almost comparable to OG while MNG-3 was comparable to DDM in the solubilization of LHI–RC, MelB, CMP-Sia (a human transporter), and LeuT (Chae, Rasmussen, et al., 2010; Cho, Bae, Das, Gellman, & Chae, 2014). The main merit of these novel agents comes from the ability to stabilize the native structures of MPs. All eight GNGs evaluated so far were superior to OG at stabilizing LHI–RC complex, with enhanced MP stabilization efficacy in some cases (e.g., GNG-1 and GNG-2) almost comparable to DDM (Fig. 10A) (Cho, Bae, et al., 2014). In addition, they produced PDCs as small as OG, suggesting that GNGs hold significant promise for MP structural study (Chae, Rana, et al., 2013). Indeed, the commercially available GNG, OGNG, has generated the high-resolution crystal structures of two MPs; Na^+-pumping

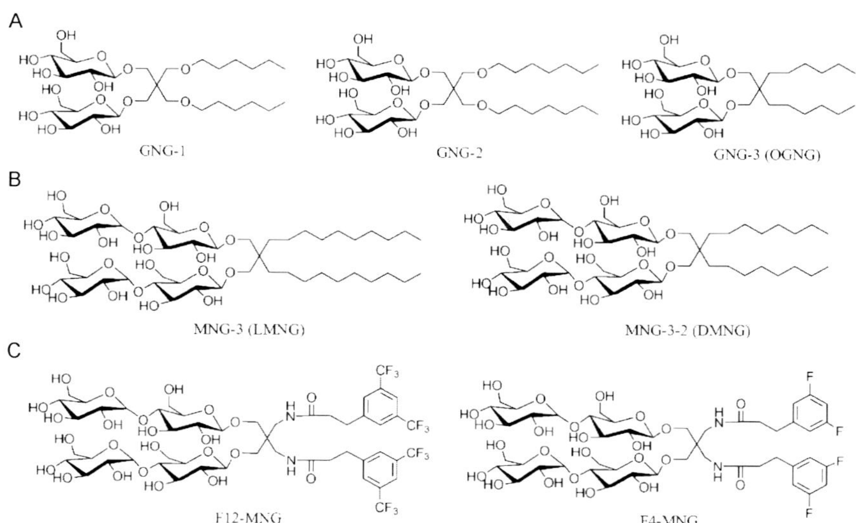

Figure 10 Chemical structures of neopentyl glycol (NG) class amphiphiles (GNG-1, GNG-2, GNG-3, MNG-3, MNG-3-2) and its fluorinated versions (F12-MNG and F4-MNG). The NG class amphiphiles can be categorized into two subclasses, (A) GNGs and (B) MNGs, depending on the hydrophilic groups. (C) Fluorinated MNG agents utilize the MNG scaffold and fluorinated aryl groups to give a class of hemifluorinated amphiphiles. GNG-3, MNG-3, and MNG-3-2 are commercially available from Anatrace under the name of OGNG, LMNG, and DMNG, respectively.

pyrophosphatase (Kellosalo, Kajander, Kogan, Pokharel, & Goldman, 2012) and human aquaporin 2 (AQP2) (Frick et al., 2014).

MNG agents bear characteristics quite different from those of GNGs although their overall architectures are similar, indicating that detergent efficacy is sensitive to small structural differences (Fig. 10B). The biggest advantage of MNG use for MP study is their excellent protein stabilization efficacy. Almost every MP that has been evaluated with these agents showed enhanced stability compared to DDM. Additionally, this architecture allowed us to conveniently prepare hemifluorinated version of MNGs, designated HF-MNGs, which showed outstanding stabilizing effect on LHI–RC complexes (Fig. 10C; Cho, Byrne, & Chae, 2013). Furthermore, the concentration range of MNG agents for effective MP stabilization is much larger than DDM, thus providing a wide window in terms of MP crystallization conditions. In terms of PDC size, MNGs were comparable to that formed by DDM, which is believed to be a sole limitation of this class. However, due to their remarkable stabilizing properties, MNGs have been very successfully applied to the functional and structural characterization of MPs.

During the last 4 years, LMNG and DMNG have facilitated the crystal structure determination of more than 15 MPs including a number of GPCRs. β_2AR structures with an irreversible agonist, a endogenous ligand (adrenaline), and G-protein-mimicking nanobody (NB-80) were successfully determined by the use of LMNG to high resolution (Rasmussen, DeVree, et al., 2011; Ring et al., 2013; Rosenbaum et al., 2011). Two muscarinic acetylcholine receptor subtypes (M2 and M3 AchR), two opioid receptors (μ-OR and δ-OR), and the neurotensin receptor (NTSR) were stabilized in LMNG micelles and successfully crystallized to produce high-resolution crystal structures (Granier et al., 2012; Haga et al., 2012; Kruse et al., 2012; Manglik et al., 2012; White et al., 2012). These GPCR structures were determined after transferring the receptor from LMNG micelles to LCP systems. The greater stability conferred by LMNG is a distinct advantage for crystallization using LCP as it typically takes a longer time to form crystals in this medium. The relative large PDCs formed by LMNG may hinder MP crystallization via the vapor-diffusion method. However, it is notable that LMNG has provided the crystal structure of the twin-arginine protein transport system via the vapor-diffusion method (Rollauer et al., 2012). In this crystal structure, one LMNG molecule was clearly resolved due to specific binding onto the protein surface. Very recently, the crystal structure of a human $GABA_A$ receptor was obtained by the same method utilizing DMNG micelles (Miller & Aricescu, 2014). These collective results clearly demonstrate the promising features of MNGs for two popular crystallization methods (both *in surfo* and *in meso*). LMNG has been shown to stabilize MP complexes such as β_2AR:G-protein and β_2AR:arrestin-1 (Rasmussen, DeVree, et al., 2011; Shukla et al., 2014). Stabilization of a GPCR is challenging but maintaining the stability of a receptor complex is extremely difficult. For example, the β_2AR:G-protein complex exhibited substantial dissociation in DDM micelles while the complex was stable enough in LMNG micelles to yield well-diffracting crystals, ultimately used to solve the structure. In addition to GPCRs, other MPs purified in LMNG yielded well-diffracting crystals via LCP method, as exemplified by hetero-tetrameric *N*-methyl-D-aspartate (NMDA) receptor ion channel (Karakas & Furukaya, 2014) and a tight junction (claudin) (Suzuki et al., 2014).

4.3 Peptide-based amphiphiles

Lipopeptide detergents (LPDs) are the best-characterized peptide-based amphiphiles and bear two alkyl chains at the end of peptide backbone

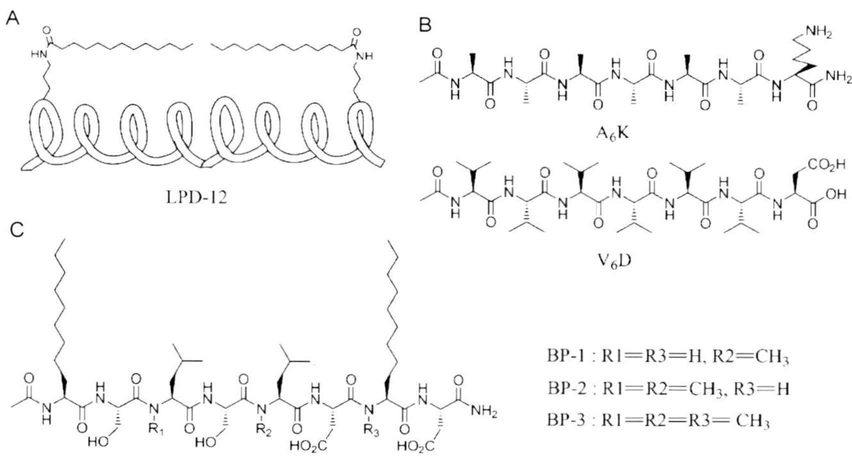

Figure 11 Schematic representation of (A) lipopeptide detergent-12 (LPD-12) and chemical structures of (B) short peptides (A_6K and V_6D) and (C) β-peptides (BP-1, BP-2, and BP-3). LPD-12 contains an α-helical backbone with two C12 alkyl chains conjugated to the side chains of ornithine residues via amide linkages. Short peptides contain hydrophobic amino acids such as alanine and valine, and hydrophilic amino acids such as lysine and aspartic acid, making them amphipathic. β-peptides (BPs) form β-pleated sheet in aqueous media and are thus well segregated between hydrophobic and hydrophilic faces.

(Fig. 11A). These agents were designed to mimic the architecture of the native membranes in which two alkyl chains in a lipid molecule facially interact with the hydrophobic segment of MPs (McGregor et al., 2003). In order to achieve such facial interaction between detergent molecules and MPs, an α-helix forming peptide comprised of 25 amino acids was utilized as a rigid backbone and the alkyl chains were attached to carboxylic acid side chains located at both ends of the helix via amide linkages. The length of the α-helical peptide was designed to match the hydrophobic width of lipid bilayers. This design principle was well supported by the crystal structure of LPD-12 cocrystallized with DDM (Privé, 2009). LPDs have been shown to maintain three MPs (BR, PagP, and Lac permease) in solution without aggregation and denaturation, and the structure of PagP in LPD-12 has been analyzed by nuclear magnetic resonance (NMR) spectroscopy. Predecessors of these agents are peptitergents with no long alkyl chains, the first peptide-based amphiphiles successfully used for MP study (Schafmeister, Miercke, & Stroud, 1993). A peptitergent was able to maintain two MPs (BR and rhodopsin) in a solubilized state, but was not effective at solubilizing PhoE porin protein. Although LPDs and peptitergents showed favorable behaviors in the solubilization and stabilization of some MPs, these agents have difficulty in the preparation of multigram scale, decreasing their practical utility. Furthermore, the multiple amide functional groups present in peptide backbone may be suboptimal for MP study because the relatively polar nature of this functional group could weaken the detergent interaction with MPs. Note that all popular conventional detergents such as DDM, OG, and LDAO do not contain amide functional groups in the lipophilic regions. Although more than 10 and 20 years passed since their invention, respectively, LPDs and peptitergents have yet to contribute to the crystal structures of any MPs.

To overcome the difficulty in large-scale preparation of LPDs, short peptides comprised of several amino acids were developed. These short peptides are comprised of hydrophilic amino acids such as aspartic acid and lysine and hydrophobic amino acids such as alanine and valine (Fig. 11B; Zhao et al., 2006). Despite the short length and no obvious presence of a secondary structure, a mixture of A_6K and V_6D was superior to conventional detergents (DDM and OG) both for solubilization of glycerol-3-phosphate dehydrogenase (GlpD) and for stability of photosystem I on a dry surface (Das et al., 2004; Yeh, Du, Tortajada, Paulo, & Zhang, 2005). However, to date, no successful MP crystallization has been reported using any of these agents.

Very recently, nanostructured β-sheet peptides (BPs) were invented. These differ from peptitergents and LPDs in terms of their secondary structures (Fig. 11C; Tao et al., 2013). These β-strand peptides self-assemble into filaments in solution but undergo molecular rearrangement to form PDCs in the presence of MPs. Because of the effective segregation of the hydrophilic and the hydrophobic surfaces, located on opposite sides of the strand, as well as a high affinity for hydrophobic surfaces, the β-strand prefers to interact with MPs rather than to form a nanosized self-assembly upon the addition of MPs. This rearrangement of the self-assemblies to PDCs is unprecedented for amphipathic compounds developed for MP research, but it is questionable that this dynamic nature is directly correlated to their favorable behaviors in MP study. These β-strands markedly increase the stability of several MP systems including the MsbA, BR, the tetrameric voltage-gated potassium channel (KcsA), the glucagon receptor and a class B GPCR (Tao et al., 2013). The enhanced MP stability achieved by the use of BPs is believed to originate from their ability to form hydrogen bonds between the BP peptide chains surrounding the MP surface. This is supported by the finding that BP-1, which is capable of forming more number of hydrogen bonds than BP-2 and BP-3, conferred greater stability on the MPs.

Electron microscopy studies of MsbA in BP-1 revealed details of the protein dynamics. However, like other peptide-based amphiphiles, these BPs are difficult to synthesize in large amounts and are yet to contribute to a high-resolution MP structure.

4.4 Large amphipathic systems

4.4.1 Amphipathic polymers (amphipols)

In the case of conventional detergents with low-molecular weights, dozens of detergent molecules generally assemble to give nanostructured micelles at concentrations above the CMCs of individual detergents in an aqueous medium. An innovative approach was made by the invention of amphipathic polymers (amphipols; APols) which differ significantly from conventional detergent architecture (Fig. 12; Tribet, Audebert, & Popot, 1996). These polymeric materials are probably one of the most extensively studied nonstandard detergents. A8-35 was the first successful APol and is now commercially available from Anatrace (Fig. 12A; Popot et al., 2011). This agent has a highly flexible polyacrylate backbone, comprised of ~70 acrylate residues. Octyl amine and isopropylamine were grafted to some portions of carboxylates present in the backbone to provide a hydrophobic surface.

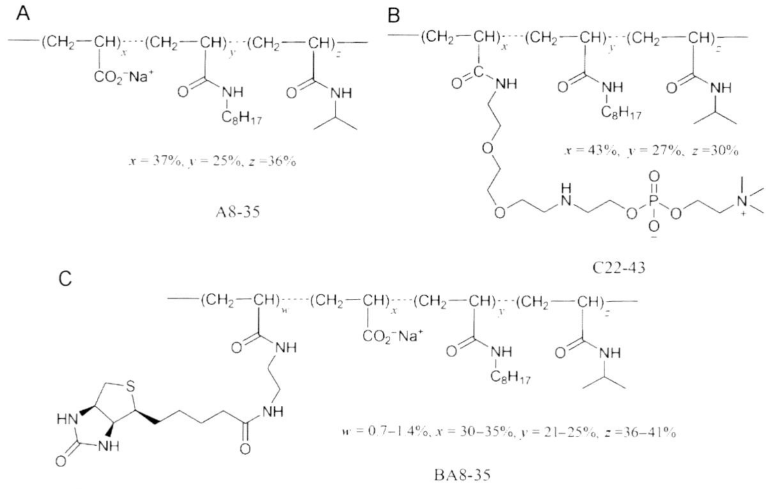

Figure 12 Chemical structures of amphiphilic polymers (amphipols; APols). An anionic Apol, (A) A8-35, is the first example of these class members proved effective at stabilizing a number of MPs. A number of new Apols were developed such as (B) phosphorylcholine-based amphiphiles (C22-43) and (C) biotinylated A8-35 (BA8-35). C22-43 is a zwitterionic version of A8-35 and BA8-35 was successfully used for the immobilization of a few MPs on the solid surface.

A typical A8-35 contains ~17 octylamine chains, ~28 isopropylamine groups, and ~25 remaining carboxylates, and is thus classified as an ionic surfactant. This polymeric material has an average molecular weight of about 9–10 kDa and forms well-defined small globular particles in aqueous solutions. The molecular weight of A8-35 particles are about 40 kDa, indicating that roughly four polymer chains combine together to form the aggregates (Gohon et al., 2006). Despite the ionic nature of the hydrophilic group, A8-35 has showed excellent stabilizing effects on dozens of proteins ranging from an MP containing a single small transmembrane helix (<5 kDa) to a large mitochondrial complex I (>1.1 MDa) (Popot et al., 2003). This agent exhibits different behaviors from standard detergents. First, A8-35 does not solubilize membranes (Popot et al., 2003). In order to use this agent for MP study, a target protein needs first to be extracted from the membrane by a conventional detergent (e.g., OG, DDM, or FC-12) and the resulting detergent-solubilized MP is subjected to detergent exchange into A8-35. Hydrophobic biobeads have been used to facilitate detergent removal and exchange with an APol in detergent-solubilized MP solutions. Second, this agent tends to form quasi-irreversible interactions with MPs (Tribet, Audebert, & Popot, 1997; Zoonens, Giusti, Zito, & Popot, 2007). After complex formation with MPs, it is almost impossible to remove A8-35 without extensive washing with a detergent solution. This property may prevent MP aggregation, but could hinder regular formation of the protein–protein contacts essential for crystal lattice formation. This may explain why A8-35 has not provided crystal structures of MPs yet, in spite of its remarkable MP stabilization efficacy.

Subsequent studies have generated a number of variants of A8-35 with zwitterionic or nonionic headgroups (Diab, Tribet, Gohon, Popot, & Winnik, 2007; Prata et al., 2001; Fig. 12B and C). Some of these variants showed improved MP-stabilizing properties relative to the original APol, A8-35. The polymeric nature of A8-35 has also been successfully utilized to immobilize MPs on solid surface (Charvolin et al., 2009). To this end, A8-35 was conjugated with a biotin to generate biotinylated A8-35 (BA8-35) used to make complex formations with MPs. The resulting MPs–APol complexes were fixed on a streptavidin-coated solid surface. As a proof of concept, four MPs (tOmpA, BR, cytochrome b_6f complex, and cytochrome bc_1 complex) were immobilized and binding events with their individual cognate ligands (antibodies or a snake toxin) studied via surface plasmon resonance and fluorescence microscopy. Thus, this protocol could provide a general method for MP immobilization without destruction of protein functions and also provide a platform for the identification of potential drugs. More detailed descriptions of APols can be found in a couple of recent review papers (Popot, 2010; Popot et al., 2011).

4.4.2 Nanodiscs and nanolipodisqs

Nanodiscs (NDs) are comprised of a patch of lipid assemblies stabilized by the association of two amphipathic polypeptide chains at the rim. The polypeptide chains, the so-called membrane scaffolding proteins (MSPs), were derived from human high density lipoprotein apoA-1 (Fig. 13; Nath, Atkins, & Sligar, 2007; Ritchie et al., 2009). This polypeptide–lipid superassembly generates a small island of native-like bilayer in the central region. The size of an ND is dependent on the chain length of the peptides as well as the surface area of the lipids, typically in the range of ~10 to ~20 nm, with molecular weights of more than 150 kDa (Borch & Hamann, 2009). Several MPs including P450 cytochromes, NADPH-cytochrome P450 reductase (Borch & Hamann, 2009), BR (Bayburt & Sligar, 2003), the bacterial Tar chemoreceptor (Boldog, Grimme, Li, Sligar, & Hazelbauer, 2006), the SecYEG translocon complex (Alami, Dalal, Lelj-Garolla, Sligar, & Duong, 2007), and GPCRs including the β_2AR (Leitz, Bayburt, Barnakov, Springer, & Sligar, 2006), rhodopsin (Bayburt, Leitz, Xie, Oprian, & Sligar, 2007), a opioid receptor (OR) (Kuszak et al., 2009) have been successfully reconstituted into NDs. The size of MPs that can be incorporated into NDs is variable, depending on the length of the associated

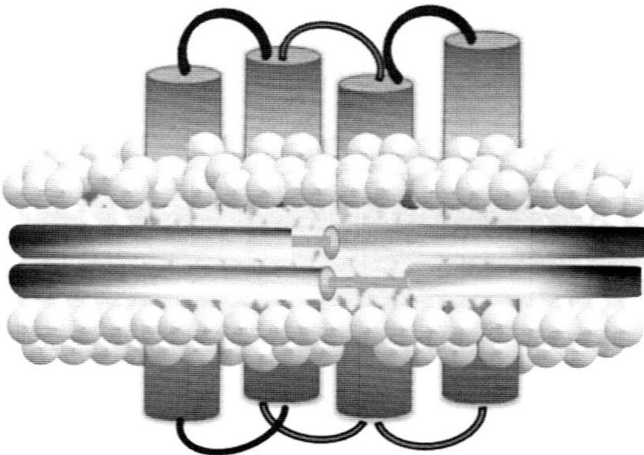

Figure 13 Schematic representation of a nanodisc (ND) incorporating a membrane protein. The central lipid assembly is stabilized by two amphipathic polypeptide chains derived from membrane scaffolding proteins (MSPs). This protein–lipid superassembly contains a patch of lipid bilayer, capable of incorporating a target membrane protein.

peptide chains. The largest MP system so far encapsulated by NDs is the BR trimer containing 21 transmembrane helices (Bayburt, Grinkova, & Sligar, 2006). A number of laboratories have taken up ND methodology, which appears to be excellent for functional studies of MPs. However, fusion events between NDs with an MP, that are a prerequisite for MP crystallization, are unlikely to occur because of the tight interaction between a patch of lipid bilayer and the peptide chains. In addition, the large size of the ND/MP complex is not favorable for MP crystallization.

Because of the presence of the polypeptide chains (MSPs), NDs may interfere with the absorbance properties of MPs incorporated. In addition, detergents are necessary for MP incorporation into NDs. In an attempt to overcome these drawbacks of NDs, the lipodisq system was developed (Orwick-Rydmark et al., 2012). This system is comprised of a patch of lipid molecules and a polymer obtained by using styrene and maleic acid (SMA). Thus, this SMA-lipid system and APols are similar in terms of the use of polymer chain(s) for MP stabilization, but different in terms of their dependency on lipid molecules for the formation of the systems; a patch of lipid molecules is only present in the former. The lipodisq system is similar to NDs because both systems have a central lipid patch stabilized by a polymeric material, but differ in the type of polymer used (SMA polymer vs. MSPs). As expected from the similar architectures, lipodisq particles confer high MP stabilization similar to that achieved by NDs. When BR was incorporated into the lipodisq system, for example, the protein was in the monomeric state and maintained the same dynamic and structural integrity as observed in the plasma membranes. The three membrane mimetic systems (APols, NDs, and lipodisq) have favorable aspects for the functional studies of MPs because these systems enable us to add ligands/cofactors to both sides of MPs, which is not possible using typical proteoliposomes, protein-inserted liposomes, where ligand accessibility is blocked on the liposome interior.

4.4.3 Bicelles and LCP

Bicelles with a discoid shape contain a central lipid patch, comprised of dimyristoyl phosphatidylcholine (DMPC) or ditridecanoyl phosphatidylcholine (DTPC), which is surrounded by short chain lipids (e.g., dihexanoyl-phosphatidylcholine (DHPC)) or a particular type of detergents (e.g., CHAPSO) (Fig. 14). The size of the bicelles is variable, depending on the ratio of lipid:detergent (q), and medium temperature

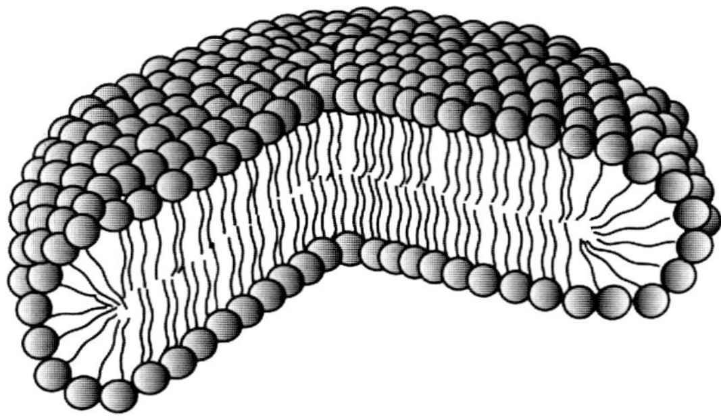

Figure 14 Schematic representation of a bicelle. This discoid structure is generated by the combination of lipid molecules and short chain lipids/detergents. The central bilayer, comprised of lipid molecules such as dimyristoyl phosphatidylcholine (DMPC) and ditridecanoyl phosphatidylcholine (DTPC), is surrounded by short chain lipids (e.g., dihexanoyl-phosphatidylcholine (DHPC)) or detergents (e.g., CHAPSO) at the rim.

(Tiburu, Moton, & Lorigan, 2001). The *in surfo* method involving crystallization of protein in detergent is the most popular for MP structural study, however, as mentioned earlier many MPs encapsulated by detergent micelles exhibit limited stability due in part to the lack of sufficient lateral pressure. In contrast, membrane mimetic systems utilizing membrane patches (e.g., proteoliposomes, NDs, and lipodisq particles) are superior to detergent micelles in terms of MP stabilization efficacy. They are excellent for MP functional study, but have yet to show utility for crystallization. The rigid architectures of these systems stabilize the MPs, but are likely to prevent MP–MP association essential for crystal lattice formation. In contrast, bicelles were shown to be highly suitable for MP structural study. Since the first success of BR crystallization in 2002 (Faham & Bowie, 2002), a number of MP structures including β_2AR (Rasmussen et al., 2007), voltage-dependent anion channel 1 (Ujwal et al., 2008), xanthorhodopsin (Luecke et al., 2008), and rhomboid protease (Vinothkumar, 2011) have been determined by this bicellar method. In every case, except for xanthorhodopsin, CHAPSO was used as the detergent component while, with the exception of β_2AR, DMPC was used as the central lipid component in bicelle formulation (Ujwal & Bowie, 2011). CHAPSO has a unique structure with three hydroxyl groups facially oriented in the lipophilic region and a sulfobetaine headgroup (Fig. 2). Because of the structural similarity to CHAPSO, FAs and TFAs possess potential as detergent components for bicelle formulation.

But why is the bicellar system superior to NDs and lipodisq particles in terms of MP crystallization? The bicellar system contains the favorable characteristics of both detergent micelles and lipid bilayers. Bicelles have structural similarities to NDs and lipodisq particles because of the presence of a lipid patch wrapped by amphipathic molecules at the rim of the discs. However, instead of using polymeric chain(s) in the cases of NDs and lipodisq particles, bicelles utilize the detergent molecules to generate the rim structure. Consequently, the rim structure of bicelles is expected to be much more flexible. This flexibility is likely to allow bicelles with individual MPs to fuse together, thereby providing increased chance of forming the protein–protein contacts essential for crystal lattice formation.

LCP is another membrane mimetic system that has proved very useful in MP crystallization (Landau & Rosenbusch, 1996). This system consists of a single lipid bilayer organized into a three-dimensional structure formed by an interconnected aqueous channel (Fig. 15). Monoolein and related compounds are utilized as components to generate lipidic cubic phase, capable of incorporating integral MPs. Many MPs have been successfully crystallized in LCP. BR crystal structure came first as a proof of concept, followed by the crystal structures of sensory rhodopsin II–transducer complex (Gordeliy et al., 2002) and β_2AR:Gs protein complex (Rasmussen, DeVree, et al., 2011). The LCP is only suitable as a crystallization matrix and cannot be used for MP extraction and purification.

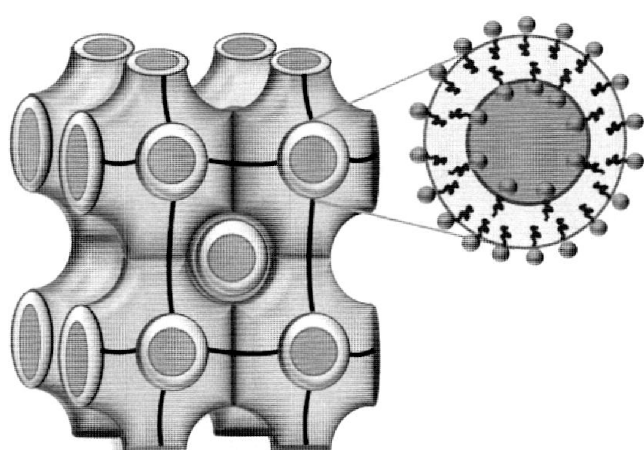

Figure 15 Schematic representation of lipidic cubic phase (LCP). In this architecture, the three-dimensional structure of a single lipid bilayer is associated with an interconnected aqueous channel. Monoolein with a single double bond within the alkyl chain is the most popular lipid component in LCP system. (See the color plate.)

5. SUMMARY OF DETERGENT PROPERTIES

In addition to widely used conventional detergents, a number of novel amphiphiles are now available for MP structural study. The availability of this growing number of tools means that it is often difficult for an MP scientist to select a workable number of approaches for a particular target MP. This difficulty is mainly due to the lack of information on detergent structure–property–efficacy relationships. Here, the general properties of a few classes of detergents are summarized in a comparative way, which may provide useful guidelines for MP scientists. First, the ionic nature of detergent hydrophilic groups is known to be critical for MP solubilization and stabilization as described in Section 1. Generally speaking, zwitterionic detergents can be favorably used for MP solubilization while nonionic detergents have a beneficial effect on MP stability. Second, the popular detergents such as LDAO, OG, βNG, DM and DDM are the first choice for MP structural study. These popular agents have no serious defect in all the processes of MP manipulation including extraction, purification, and crystallization. On the other hand, it is unlikely that CHAPS and CHAPSO have general utility for MP structural studies because the interactions of these agents with MPs are too weak to prevent protein aggregation. Third, small headgroup-bearing detergents such as OG and LDAO form small PDCs and this may be preferential to large headgroup-containing agents (e.g., DDM). Fourth, lipid bilayer systems such as liposomes, NDs, lipodisq particles, bicelles, and LCPs are better at stabilizing MPs than detergent micellar systems, being better mimics of the membrane. However, detergent micellar systems are superior to the lipid bilayer systems for MP extraction from the membranes and, in most cases, for MP crystallization.

Well-behaving agents for MP crystallization include conventional detergents, OG, LDAO, and DDM, as well as novel classes of amphiphiles, the FAs, TPAs, GNGs, and MNGs (Zhang, Tao, & Hong, 2011). The common features of these compounds could provide guidelines important for the direction of future detergent research, as currently such guidelines are severely limited. The first common feature of these agents is that they are highly efficient at solubilizing MPs from the native membranes. In contrast, their MP stabilization efficacies are variable. An efficacy order of these detergents for MP stabilization is dependent on the characteristics of the target MPs but the general order is as follows: MNGs ∼ FAs > DDM ∼ TPAs ∼ GNGs > OG > LDAO. This analysis indicates that MP solubilization efficiency could be more important than MP stabilization efficacy for the successful MP crystallization. Finally, the PDC size has an important but not essential role in MP crystal lattice formation. For instance, DDM and MNGs have both been used to successfully produce well-diffracting MP crystals by the *in surfo* method despite the fact that PDCs formed by these agents are relatively large. Furthermore, other alternatives such as bicelle- and LCP-based crystallization methods have been developed to avoid the direct usage of PDCs for MP crystallization. However, the vapor-diffusion method is still the most successful in terms of high-resolution structure determination of MPs. Thus, it would be valuable to know a general detergent trend in terms of PDC size. PDC size tends to be strongly dependent on the size of detergent headgroups. Thus, amphipathic agents with small headgroups such as *N*-oxide and glucoside (e.g., LDAO, OG, GNGs, and TPAs) tend to form small PDCs, while compounds with a large hydrophilic group, maltoside (e.g., DDM and MNGs) usually generate large PDCs. As small headgroup-bearing detergents (e.g., LDAO) are more destabilizing to MPs than large headgroup-bearing agents (e.g., DDM), detergents with greater stabilization efficacy usually produce rather large PDCs, necessitating a compromise between MP stability and PDC size.

6. DETERGENT SELECTION

Most researchers usually screen a set of popular or a large number of detergents rather than rationally selecting detergent candidates for the structural study of their target MPs. This is due to the general perception that detergent efficacy is variable from one MP to others. Thus, detergents favorable for a target MP are believed to be determined by experimental approach rather than rational analysis of detergent and MP properties. This seems obvious because MPs have a large variation in their structures and functions, thereby making it impossible to *a priori* determine a detergent set for a target MP without detailed structural information. In spite of the large number of highly variable MPs, it is possible to categorize them on their propensity for MP denaturation and aggregation; denaturation-sensitive, aggregation-sensitive, or both denaturation and aggregation-sensitive protein. Thus, depending on the tendency of a specific MP to denature and aggregate, we can rationally approach selecting detergent candidates for MP study. For example, denaturation-sensitive MPs necessitate the use of weakly binding detergents rather than strongly binding detergents for stabilization because the latter tend to disrupt the subtle folding interactions present within 3D structures of MPs. In contrast, due to their ability to prevent protein aggregation, strongly binding detergents would be more suitable for aggregation-sensitive MPs. Thus, it would be of importance to have information on the MP propensity to denature and aggregate before carrying out detergent screening using a large number of agents.

Another way to rationally select a set of detergents is to obtain some preliminary results for three conventional detergents, LDAO, OG, and DDM with a target MP. Depending on the results of MP stability in each detergent, we can adopt a different approach as follows. If a target MP is robust enough to be stable in LDAO, small PDC-forming detergents such as LDAO, OG, and βNG could be used for MP crystallization. In the case of target MPs unstable in LDAO, detergents with high MP stabilization efficacy and small PDC-formation property such as the glucoside-bearing agents (OG and βNG) may be more suitable. If the target MPs are not stable in the glucoside detergents, MP stability is the most important factor. Thus, DDM and other agents with high MP stabilization efficacy such as GNGs and TPAs would be the first choice. A number of MPs are stable in DDM micelles but there are still a number of MPs or MP complexes that are not sufficiently stabilized by this agent for downstream studies. In this case, novel classes of amphiphiles such as FAs, MNGs, and GDN which have been shown to result in higher MP stability than DDM could be tried. If a target MP is stable in DDM but does not form well-diffracting MP crystals, TPAs and more preferably GNGs which confer reasonable stability but form small PDCs may be suitable. Thus, a preliminary study with the three representative conventional detergents can provide a useful guide to decide which type of detergents are most likely to result in well-diffracting crystals. The laborious trial-error approach and high-throughput detergent screening scheme could be markedly reduced by following this protocol.

7. FUTURE DIRECTION

Advances in MP research have been significantly hampered by the absence of detailed information on detergent structure–property–efficacy relationships. This information will not only provide guidelines for the future design of novel amphiphiles but also offers criteria for choosing a set of detergents for a target MP. Obtaining such information, however, is challenging partly because of the requirement for a large number of amphipathic molecules with systematic structural variations. Many classes of amphiphiles developed so far have limited flexibility to introduce the systematic variations into their structures. In the case of polymeric amphiphiles such as APols, the introduction of structural modification into specific locations is relatively straightforward. However, the interpretation of results on detergent structure–property–efficacy relationships is often difficult because of the heterogeneous and complex structures of polymeric materials. Another hurdle to accessing information on detergent structure–property–efficacy relationships is the requirement for a rather large amount of detergent samples for evaluation. Measuring basic physical properties such as CMC and micelle size via conventional methods consumes a substantial amount of material. In addition, novel detergents need to be evaluated for their efficacy using a range of different MPs to show wide utility. Hundreds of milligrams of detergents are necessary for basic characterization of novel agents with MPs.

Detergent efficacy tends to strongly depend on the target MP and the evaluation method (e.g., assay conditions and detergent concentrations). As a result, despite a large number of detergent studies being available, it is very hard to extract detergent structure–property–efficacy relationships. However, as can be seen in a few TPA studies, some meaningful relationships could be found because it was possible to systematically change the structures of TPA molecules and evaluate these with MPs to find the detergent efficacy for MP solubilization, stabilization, and crystallization. Rather than development of a single detergent suitable for MP study, finding a detergent design principle is probably much more valuable. Thus, the future direction of detergent research should focus on discovery of detergent design principles, i.e., detergent structure–property–efficacy relationships.

REFERENCES

Abla, M., Durand, G., & Pucci, B. (2011). Propyl ended hemifluorinated surfactants: Synthesis and self-assembling properties. *The Journal of Organic Chemistry*, *76*, 2084–2093.

Abramson, J., Riistama, S., Larsson, G., Jasaitis, A., Svensson-Ek, M., Laakkonen, L., et al. (2000). The structure of the ubiquinol oxidase from *Escherichia coli* and its ubiquinone binding site. *Nature Structural Biology*, 7, 910–917.

Alami, M., Dalal, K., Lelj-Garolla, B., Sligar, S. G., & Duong, F. (2007). Nanodiscs unravel the interaction between the SecYEG channel and its cytosolic partner SecA. *The EMBO Journal*, *26*, 1995–2004.

Alexandrov, A. I., Mileni, M., Chien, E. Y., Hanson, M. A., & Stevens, R. C. (2008). Microscale fluorescent thermal stability assay for membrane proteins. *Structure*, *16*, 351–359.

Anglin, T. C., & Conboy, J. C. (2008). Lateral pressure dependence of the phospholipid transmembrane diffusion rate in planar-supported lipid bilayers. *Biophysical Journal*, *95*, 186–193.

Bayburt, T. H., Grinkova, Y. V., & Sligar, S. G. (2006). Assembly of single bacteriorhodopsin trimers in bilayer nanodiscs. *Archives of Biochemistry and Biophysics, 450*, 215–222.

Bayburt, T. H., Leitz, A. J., Xie, G., Oprian, D. D., & Sligar, S. G. (2007). Transducin activation by nanoscale lipid bilayers containing one and two rhodopsins. *The Journal of Biological Chemistry, 282*, 14875–14881.

Bayburt, T. H., & Sligar, S. G. (2003). Self-assembly of single integral membrane proteins into soluble nanoscale phospholipid bilayers. *Protein Science, 12*, 2476–2481.

Boldog, T., Grimme, S., Li, M., Sligar, S. G., & Hazelbauer, G. L. (2006). Nanodiscs separate chemoreceptor oligomeric states and reveal their signaling properties. *Proceedings of the National Academy of Sciences of the United States of America, 103*, 11509–11514.

Borch, J., & Hamann, T. (2009). The nanodisc: A novel tool for membrane protein studies. *Biological Chemistry, 390*, 805–814.

Bowie, J. U. (2001). Stabilizing membrane proteins. *Current Opinion in Structural Biology, 11*, 397–402.

Breyton, C., Chabaud, E., Chaudier, Y., Pucci, B., & Popot, J.-L. (2004). Hemifluorinated surfactants: A nondissociating environment for handling membrane proteins in aqueous solutions? *FEBS Letters, 564*, 312–318.

Breyton, C., Tribet, C., Olive, J., Dubacq, J.-P., & Popot, J.-L. (1997). Dimer to monomer transition of the cytochrome b6 f complex: Causes and consequences. *The Journal of Biological Chemistry, 272*, 21892–21900.

Brotherus, J. R., Jost, P. C., Griffith, O. H., & Hokin, L. E. (1979). Detergent inactivation of sodium- and potassium-activated adenosinetriphosphatase of the electric eel. *Biochemistry, 18*, 5043–5050.

Caffrey, M., Li, D., & Dukkipati, A. (2012). Membrane protein structure determination using crystallography and lipidic mesophases: Recent advances and successes. *Biochemistry, 51*, 6266–6288.

Chabaud, E., Barthélémy, P., Mora, N., Popot, J.-L., & Pucci, B. (1998). Stabilization of integral membrane proteins in aqueous solution using fluorinated surfactants. *Biochimie, 80*, 515–530.

Chae, P. S., Bae, H. E., Muhammad, E., Hussain, H., & Kim, J. W. (2014). New ganglio-tripod amphiphiles (TPAs) for membrane protein solubilization and stabilization: Implications for detergent structure–property relationships. *Organic & Biomolecular Chemistry, 12*, 8480–8487.

Chae, P. S., Cho, K. H., Wander, M. J., Bae, H. E., Gellman, S. H., & Laible, P. D. (2014). Hydrophobic variants of ganglio-tripod amphiphiles for membrane protein manipulation. *Biochimica et Biophysica Acta, 1838*, 278–286.

Chae, P. S., Gotfryd, K., Pacyna, J., Miercke, L. J. W., Rasmussen, S. G. F., Robbins, R. A., et al. (2010). Tandem facial amphiphiles for membrane protein stabilization. *Journal of the American Chemical Society, 132*, 16750–16752.

Chae, P. S., Laible, P. D., & Gellman, S. H. (2010). Tripod amphiphiles for membrane protein manipulation. *Molecular BioSystems, 6*, 89–94.

Chae, P. S., Rana, R. R., Gotfryd, K., Rasmussen, S. G. F., Kruse, A. C., Cho, K. H., et al. (2013). Glucose-neopentyl glycol (GNG) amphiphiles for membrane protein study. *Chemical Communications, 49*, 2287–2289.

Chae, P. S., Rasmussen, S. G. F., Rana, R. R., Gotfryd, K., Chandra, R., Goren, M. A., et al. (2010). Maltose-neopentyl glycol (MNG) amphiphiles for solubilization, stabilization and crystallization of membrane proteins. *Nature Methods, 7*, 1003–1008.

Chae, P. S., Rasmussen, S. G. F., Rana, R. R., Gotfryd, K., Kruse, A. C., Manglik, A., et al. (2012). A new class of amphiphiles bearing rigid hydrophobic groups for solubilization and stabilization of membrane proteins. *Chemistry: A European Journal, 18*, 9485–9490.

Chae, P. S., Sadaf, A., & Gellman, S. H. (2014). Hydrophobic variations of *N*-oxide amphiphiles for membrane protein manipulation: Importance of non-hydrocarbon groups in the hydrophobic portion. *Chemistry: An Asian Journal*, *9*, 110–116.

Chae, P. S., Wander, M. J., Bowling, A. P., Laible, P. D., & Gellman, S. H. (2008). Glycotripod amphiphiles for solubilisation and stabilization of a membrane-protein superassembly: Importance of branching in the hydrophilic portion. *ChemBioChem*, *9*, 1706–1709.

Chae, P. S., Wander, M. J., Cho, K. H., Laible, P. D., & Gellman, S. H. (2013). Carbohydrate-containing Triton X-100 analogues for membrane protein solubilization and stabilization. *Molecular BioSystems*, *9*, 626–629.

Charvolin, D., Perez, J.-B., Rouviere, F., Giusti, F., Bazzacco, P., Abdine, A., et al. (2009). The use of amphipols as universal molecular adapters to immobilize membrane proteins onto solid supports. *Proceedings of the National Academy of Sciences of the United States of America*, *106*, 405–410.

Cho, K. H., Bae, H. E., Das, M., Gellman, S. H., & Chae, P. S. (2014). Improved glucose-neopentyl glycol (GNG) amphiphiles for membrane protein solubilisation and stabilization. *Chemistry—An Asian Journal*, *9*, 632–638.

Cho, K. H., Byrne, B., & Chae, P. S. (2013). Hemifluorinated maltose-neopentyl glycol (HF-MNG) amphiphiles for membrane protein stabilisation. *ChemBioChem*, *14*, 452–455.

Das, R., Kiley, P. J., Segal, M., Norville, J., Yu, A., Wang, L., et al. (2004). Integration of photosynthetic protein molecular complexes in solid-state electronic devices. *Nano Letters*, *4*, 1079–1083.

Diab, C., Tribet, C., Gohon, Y., Popot, J.-L., & Winnik, F. M. (2007). Complexation of integral membrane proteins by phosphorylcholine-based amphipols. *Biochimica et Biophysica Acta*, *1768*, 2737–2747.

Faham, S., & Bowie, J. U. (2002). Bicelle crystallization: A new method for crystallizing membrane proteins yields a monomeric bacteriorhodopsin structure. *Journal of Molecular Biology*, *316*, 1–6.

Frick, A., Eriksson, U. K., de Mattia, F., Oberg, F., Hedfalk, K., Neutze, R., et al. (2014). X-ray structure of human aquaporin 2 and its implications for nephrogenic diabetes insipidus and trafficking. *Proceedings of the National Academy of Sciences of the United States of America*, *111*, 6305–6310.

Garavito, R. M., & Ferguson-Miller, S. (2001). Detergents as tools in membrane biochemistry. *The Journal of Biological Chemistry*, *276*, 32403–32406.

Gohon, Y., Giusti, F., Prata, C., Charvolin, D., Timmins, P., Ebel, C., et al. (2006). Well-defined nanoparticles formed by hydrophobic assembly of a short and polydisperse random terpolymer, amphipol A8-35. *Langmuir*, *22*, 1281–1290.

Gohon, Y., & Popot, J.-L. (2003). Membrane protein-surfactant complexes. *Current Opinion in Colloid and Interface Science*, *8*, 15–22.

Gordeliy, V. I., Labahn, J., Moukhametzianov, R., Efremov, R., Granzin, J., Schlesinger, R., et al. (2002). Molecular basis of transmembrane signalling by sensory rhodopsin II-transducer complex. *Nature*, *419*, 484–487.

Granier, S., Manglik, A., Kruse, A. C., Kobilka, T. S., Thian, F. S., Weis, W. I., et al. (2012). Structure of the δ-opioid receptor bound to naltrindole. *Nature*, *485*, 400–404.

Haga, K., Kruse, A. C., Asada, H., Kobayashi, T. Y., Shiroishi, M., Zhang, C., et al. (2012). Structure of the human M_2 muscarinic acetylcholine receptor bound to an antagonist. *Nature*, *482*, 547–551.

Hovers, J., Potschies, M., Polidori, A., Pucci, B., Raynal, S., Bonneté, F., et al. (2011). A class of mild surfactants that keep integral membrane proteins water-soluble for functional studies and crystallization. *Molecular Membrane Biology*, *28*, 171–181.

Howell, S. C., Mittal, R., Huang, L., Travis, B., Breyer, R. M., & Sanders, C. R. (2010). CHOBIMALT: A cholesterol-based detergent. *Biochemistry, 49*, 9572–9583.

Israelachvili, J. N., Mitchell, D. J., & Ninham, B. W. (1977). Theory of self-assembly of lipid bilayers and vesicles. *Biochimica et Biophysica Acta, 470*, 185–201.

Junge, F., Schneider, B., Reckel, S., Schwarz, D., Dotsch, V., & Bernhard, F. (2008). Large-scale production of functional membrane proteins. *Cellular and Molecular Life Sciences, 65*, 1729–1755.

Kantardjieff, K. A., & Rupp, B. (2003). Matthews coefficient probabilities: Improved estimates for unit cell contents of proteins, DNA, and protein–nucleic acid complex crystals. *Protein Science, 12*, 1865–1871.

Karakas, E., & Furukaya, H. (2014). Crystal structure of a heterotetrameric NMDA receptor ion channel. *Science, 344*, 992–997.

Kellosalo, J., Kajander, T., Kogan, K., Pokharel, K., & Goldman, A. (2012). The structure and catalytic cycle of a sodium-pumping pyrophosphatase. *Science, 337*, 473–476.

Kruse, A. C., Hu, J., Pan, A. C., Arlow, D. H., Rosenbaum, D. M., Rosemond, E., et al. (2012). Structure and dynamics of the M3 muscarinic acetylcholine receptor. *Nature, 482*, 552–556.

Kuszak, A. J., Pitchiaya, S., Anand, J. P., Mosberg, H. I., Walter, N. G., & Sunahara, R. K. (2009). Purification and functional reconstitution of monomeric μ-opioid receptors: Allosteric modulation of agonist binding by Gi2. *The Journal of Biological Chemistry, 284*, 26732–26741.

Landau, E. M., & Rosenbusch, J. P. (1996). Lipidic cubic phases: A novel concept for the crystallization of membrane proteins. *Proceedings of the National Academy of Sciences of the United States of America, 93*, 14532–14535.

Lee, A. G. (2011). How to understand lipid-protein interactions in biological membranes in biological membranes. In P. L. Yeagle (Ed.), *The structure of biological membranes* (3rd ed., pp. 273–313). Boca Raton, FL, USA: CRC Press, Taylor & Francis Group.

Lee, S. C., Bennett, B. C., Hong, W.-X., Fu, Y., Baker, K. A., Marcoux, J., et al. (2013). Steroid-based facial amphiphiles for stabilization and crystallization of membrane proteins. *Proceedings of the National Academy of Sciences of the United States of America, 110*, E1203–E1211.

Leitz, A. J., Bayburt, T. H., Barnakov, A. N., Springer, B. A., & Sligar, S. G. (2006). Functional reconstitution of β2-adrenergic receptors utilizing self-assembling nanodisc technology. *BioTechniques, 40*, 601–612.

Li, J., Edwards, P. C., Burghammer, M., Villa, C., & Schertler, G. F. X. (2004). Structure of bovine rhodopsin in a trigonal crystal form. *Journal of Molecular Biology, 343*, 1409–1438.

Luecke, H., Schobert, B., Stagno, J., Imasheva, E. S., Wang, J. M., Balashov, S. P., et al. (2008). Crystallographic structure of xanthorhodopsin, the light-driven proton pump with a dual chromophore. *Proceedings of the National Academy of Sciences of the United States of America, 105*, 16561–16565.

Manglik, A., Kruse, A. C., Kobilka, T. S., Thian, F. S., Mathiesen, J. M., Sunahara, R. K., et al. (2012). Crystal structure of the μ-opioid receptor bound to a morphinan antagonist. *Nature, 485*, 321–326.

Marsh, D. (2008). Protein modulation of lipids, and vice-versa, in membranes. *Biochimica et Biophysica Acta, 1778*, 1545–1575.

McGregor, C.-L., Chen, L., Pomroy, N. C., Hwang, P., Go, S., Chakrabartty, A., et al. (2003). Lipopeptide detergents designed for the structural study of membrane proteins. *Nature Biotechnology, 21*, 171–176.

McQuade, D. T., Quinn, M. A., Yu, S. M., Polans, A. S., Krebs, M. P., & Gellman, S. H. (2000). Rigid amphiphiles for membrane protein manipulation. *Angewandte Chemie International Edition, 39*, 758–761.

Michel, H. (2001). Crystallization of membrane proteins. In M. Rossmann & E. Arnold (Eds.), *International tables for crystallography* (pp. 94–99). Vol. F. Dordrecht: Kluwer Academic Publishers.

Midgett, C. R., & Madeen, D. R. (2007). Breaking the bottleneck: Eukaryotic membrane protein expression for high-resolution structural studies. *Journal of Structural Biology*, *160*, 265–274.

Miller, P. S., & Aricescu, A. R. (2014). Crystal structure of a human $GABA_A$ receptor. *Nature*, *512*, 270–275.

Moller, J. V., & le Maire, M. (1993). Detergent binding as a measure of hydrophobic surface area of integral membrane proteins. *The Journal of Biological Chemistry*, *268*, 18659–18672.

Nath, A., Atkins, W. M., & Sligar, S. G. (2007). Applications of phospholipid bilayer nanodiscs in the study of membranes and membrane proteins. *Biochemistry*, *46*, 2059–2069.

Newstead, S., Ferrandon, S., & Iwata, S. (2008). Rationalizing alpha-helical membrane protein crystallization. *Protein Science*, *17*, 466–472.

Orwick-Rydmark, M., Lovett, J. E., Graziadei, A., Lindholm, L., Hicks, M. R., & Watts, A. (2012). Detergent-free incorporation of a seven-transmembrane receptor protein into nanosized bilayer lipodisq particles for functional and biophysical studies. *Nano Letters*, *12*, 4687–4692.

Popot, J.-L. (2010). Amphipols, nanodiscs, and fluorinated surfactants: Three non-conventional approaches to studying membrane proteins in aqueous solutions. *Annual Review of Biochemistry*, *79*, 737–775.

Popot, J.-L., Althoff, T., Bagnard, D., Baneres, J.-L., Bazzacco, P., Billon-Denis, E., et al. (2011). Amphipols from A to Z. *Annual Review of Biophysics*, *40*, 379–408.

Popot, J.-L., Berry, E. A., Charvolin, D., Creuzenet, C., Ebel, C., et al. (2003). Amphipols: Polymeric surfactants for membrane biology research. *Cellular and Molecular Life Sciences*, *60*, 1559–1574.

Prata, C., Giusti, F., Gohon, Y., Pucci, B., Popot, J.-L., & Tribet, C. (2001). Non-ionic amphiphilic polymers derived from tris(hydroxymethyl)-acrylamidomethane keep membrane proteins soluble and native in the absence of detergent. *Biopolymers*, *56*, 77–84.

Privé, G. G. (2007). Detergents for the stabilization and crystallization of membrane proteins. *Methods*, *41*, 388–397.

Privé, G. G. (2009). Lipopeptide detergents for membrane protein studies. *Current Opinion in Structural Biology*, *19*, 1–7.

Rasmussen, S. G. F., Choi, H.-J., Fung, J. J., Pardon, E., Casarosa, P., Chae, P. S., et al. (2011). Structure of a nanobody-stabilized active state of the β_2 adrenoceptor. *Nature*, *469*, 175–180.

Rasmussen, S. G. F., Choi, H.-J., Rosenbaum, D. M., Kobilka, T. S., Thian, F. S., Edwards, P. C., et al. (2007). Crystal structure of the human β_2 adrenergic G-protein-coupled receptor. *Nature*, *450*, 383–387.

Rasmussen, S. G. F., DeVree, B. T., Zou, Y., Kruse, A. C., Chung, K. Y., Kobilka, T. S., et al. (2011). Crystal structure of the β_2 adrenergic receptor-Gs protein complex. *Nature*, *477*, 549–555.

Ring, A. M., Manglik, A., Kruse, A. C., Enos, M. D., Weis, W. I., Garcia, K. C., et al. (2013). Adrenaline-activated structure of β_2-adrenoceptor stabilized by an engineered nanobody. *Nature*, *502*, 575–579.

Ritchie, T. K., Grinkova, Y. V., Bayburt, T. H., Denisov, I. G., Zolnerciks, J. K., Atkins, W. M., et al. (2009). Reconstitution of membrane proteins in phospholipid bilayer nanodiscs. *Methods in Enzymology*, *464*, 211–231.

Rollauer, S. E., Tarry, M. J., Graham, J. E., Jääskeläinen, M., Jäger, F., Johnson, S., et al. (2012). Structure of the TatC core of the twin-arginine protein transport system. *Nature*, *492*, 210–214.

Rosenbaum, D. M., Zhang, C., Lyons, J. A., Holl, R., Aragao, D., Arlow, D. H., et al. (2011). Structure and function of an irreversible agonist-β_2 adrenoceptor complex. *Nature*, *469*, 236–240.

Rosenbusch, J. P. (2001). Stability of membrane proteins: Relevance for the selection of appropriate methods for high-resolution structure determinations. *Journal of Structural Biology*, *136*, 144–157.

Schafmeister, C. E., Miercke, L. J. W., & Stroud, R. A. (1993). Structure at 2.5 Å of a designed peptide that maintains solubility of membrane proteins. *Science*, *262*, 734–738.

Shukla, A. K., Westfield, G. H., Xiao, K., Reis, R. I., Huang, L.-Y., Tripathi-Shukla, P., et al. (2014). Visualization of arrestin recruitment by a G-protein-coupled receptor. *Nature*, *512*, 218–222.

Stroebel, D., Choquet, Y., Popot, J.-L., & Picol, D. (2003). An atypical haem in the cytochrome b_6f complex. *Nature*, *426*, 413–418.

Suzuki, H., Nishizawa, T., Tani, K., Yamazaki, Y., Tamura, A., Ishitani, R., et al. (2014). Crystal structure of a claudin provides insight into the architecture of tight junctions. *Science*, *344*, 304–307.

Tanford, C. (1980). *The hydrophobic effect: Formation of micelles and biological membranes*. New York: John Wiley & Sons, 233 pp.

Tao, H., Lee, S. C., Moeller, A., Roy, R. S., Siu, F. Y., Zimmermann, J., et al. (2013). Engineered nanostructured β-sheet peptides protect membrane proteins. *Nature Methods*, *10*, 759–761.

Theisen, M. J., Potocky, T. B., McQuade, D. T., Gellman, S. H., & Chiu, M. L. (2005). Crystallization of bacteriorhodopsin solubilized by a tripod amphiphile. *Biochimica et Biophysica Acta*, *1751*, 213–216.

Tiburu, E. K., Moton, D. M., & Lorigan, G. A. (2001). Development of magnetically aligned phospholipid bilayers in mixtures of palmitoylstearoylphosphatidylcholine and dihexanoylphosphatidylcholine by solid-state NMR spectroscopy. *Biochimica et Biophysica Acta*, *1512*, 206–214.

Tribet, C., Audebert, R., & Popot, J.-L. (1996). Amphipols: Polymers that keep membrane proteins soluble in aqueous solutions. *Proceedings of the National Academy of Sciences of the United States of America*, *93*, 15047–15050.

Tribet, C., Audebert, R., & Popot, J.-L. (1997). Stabilisation of hydrophobic colloidal dispersions in water with amphiphilic polymers: Application to integral membrane proteins. *Langmuir*, *13*, 5570–5576.

Ujwal, R., & Bowie, J. U. (2011). Crystallizing membrane proteins using lipidic bicelles. *Methods*, *55*, 337–341.

Ujwal, R., Cascio, D., Colletier, J.-P., Faham, S., Zhang, J., Toro, L., et al. (2008). The crystal structure of mouse VDAC1 at 2.3 A resolution reveals mechanistic insights into metabolite gating. *Proceedings of the National Academy of Sciences of the United States of America*, *105*, 17742–17747.

Vinothkumar, K. R. (2011). Structure of rhomboid protease in a lipid environment. *Journal of Molecular Biology*, *407*, 232–247.

White, J. F., Noinaj, N., Shibata, Y., Love, J., Kloss, B., Xu, F., et al. (2012). Structure of the agonist-bound neurotensin receptor. *Nature*, *490*, 508–513.

White, S. H., & Wimley, W. C. (1999). Membrane protein folding and stability: Physical principles. *Annual Review of Biophysics and Biomolecular Structure*, *28*, 319–365.

Yeh, J. I., Du, S., Tortajada, A., Paulo, J., & Zhang, S. (2005). Peptergents: Peptide detergents that improve stability and functionality of a membrane protein, glycerol-3-phosphate dehydrogenase. *Biochemistry, 44*, 16912–16919.

Yu, S. M., McQuade, D. T., Quinn, M. A., Hackenberger, C. P., Krebs, M. P., & Polans, A. S. (2000). An improved tripod amphiphile for membrane protein solubilization. *Protein Science, 9*, 2518–2527.

Zhang, Q., Tao, H., & Hong, W.-X. (2011). New amphiphiles for membrane protein structural biology. *Methods, 55*, 318–323.

Zhao, X., Nagai, Y., Reeves, P. J., Kiley, P., Khorana, H. G., & Zhang, S. (2006). Designer short peptide surfactants stabilize G protein-coupled receptor bovine rhodopsin. *Proceedings of the National Academy of Sciences of the United States of America, 103*, 17707–17712.

Zoonens, M., Giusti, F., Zito, F., & Popot, J.-L. (2007). Dynamics of membrane protein/amphipol association studied by Förster resonance energy transfer: Implications for in vitro studies of amphipol-stabilized membrane proteins. *Biochemistry, 46*, 10392–10404.

CHAPTER FIVE

Quantification of Detergent Using Colorimetric Methods in Membrane Protein Crystallography

Chelsy Prince, Zongchao Jia[1]
Department of Biomedical and Molecular Sciences, Queen's University, Kingston, Ontario, Canada
[1]Corresponding author: e-mail address: jia@queensu.ca

Contents

Abstract

Membrane protein crystallography has the potential to greatly aid our understanding of membrane protein biology. Yet, membrane protein crystals remain challenging to produce. Although robust methods for the expression and purification of membrane proteins continue to be developed, the detergent component of membrane protein samples is equally important to crystallization efforts. This chapter describes the development of three colorimetric assays for the quantitation of detergent in membrane

Methods in Enzymology, Volume 557
ISSN 0076-6879
http://dx.doi.org/10.1016/bs.mie.2014.12.002

protein samples and provides detailed protocols. All of these techniques use small sample volumes and have potential applications in crystallography. The application of these techniques in crystallization prescreening, detergent concentration modification, and detergent exchange experiments is demonstrated. It has been observed that the concentration of detergent in a membrane protein sample can be just as important as the protein concentration when attempting to reproduce crystallization lead conditions.

1. DETERGENT IN MEMBRANE PROTEIN CRYSTALLOGRAPHY

High-resolution structures of membrane proteins remain severely underrepresented in structural databases (Arinaminpathy, Khurana, Engelman, & Gerstein, 2009). Purification of membrane proteins requires the addition of detergent for both extraction of the protein from the lipid bilayer and stabilization of the protein *in vitro*. Unfortunately, the addition of detergent can be detrimental to crystallization efforts (Hitscherich, Kaplan, Allaman, Wiencek, & Loll, 2000). Protein solubilization results in the formation of protein–detergent complexes (PDC), in which detergent coats a large portion of the protein surface. The flexible and dynamic detergent belt surrounding the protein can dictate the properties of the PDC and impede the formation of well-ordered crystal lattices (Hitscherich et al., 2000; Prive, 2007; Sonoda et al., 2010).

Productive crystal contacts can only be formed on protein surfaces that are exposed to the solvent (Sonoda et al., 2010). Therefore, the protein to detergent ratio of the final PDC and the buried surface area of the protein are important parameters to consider (DaCosta & Baenziger, 2002; Gutmann et al., 2007). Practically, smaller detergents are desirable because they reduce the overall PDC size (Kunji, Harding, Butler, & Akamine, 2008; Marone, Thiyagarajan, Wagner, & Tiede, 1999) and allow tighter crystal packing (Garavito, Picot, Loll, & Daniel, 1996), which can produce PDC crystals with better than average diffraction (Prive, 2007; Sonoda et al., 2010). In contrast, smaller detergents are more likely to denature and destabilize proteins or dissociate physiologically relevant oligomers (Gan, Vararattanavech, Nordin, Eshaghi, & Torres, 2011; Prive, 2007). Typically, many detergents must be screened to find an appropriate stabilizing detergent that allows crystallization of a particular membrane protein.

The level of detergent and/or lipid present in solution is also a crucial factor in the success of crystallization trials (DaCosta & Baenziger, 2002;

Eriks, Mayor, & Kaplan, 2003; Prive, 2007; Strop & Brunger, 2005). In the past decade, many techniques have been proposed for the measurement of detergent concentrations. They include thin-layer chromatography combined with densitometry (Eriks et al., 2003), gas chromatography coupled with flame ionization detection or mass spectrometry (Shi, Han, Xiong, & Tian, 2009; Shi, Shao, Xiong, & Tian, 2008), nuclear magnetic resonance (Maslennikov et al., 2007), contact angle measurement (Kaufmann, Engel, & Rémigy, 2006), refractive index measurement (Strop & Brunger, 2005), or Fourier-transform infrared spectrometry (DaCosta & Baenziger, 2002). Many of the most popular detergents used in membrane protein crystallography can also be quantified using specific chemical reactions that produce a colored product. These colorimetric assays include the phenol assay for sugars (Prince & Jia, 2012; Urbani & Warne, 2005), the molybdate assay for total phosphate (Ames, 1966), the sulfuric acid assay for bile salts (Urbani & Warne, 2005), and the methylene blue/chloroform assay for the detection of sodium dodecyl sulfate (SDS) (Prive, 2007). These colorimetric methods require limited equipment and provide a quick assessment that is particularly advantageous to crystallographers.

2. 2,6-DIMETHYLPHENOL ASSAY FOR SUGAR-BASED DETERGENTS

The measurement of sugar concentrations using a phenol/sulfuric acid colorimetric reaction was originally proposed in the 1950s and has been used in a wide range of applications (Dubois, Gilles, Hamilton, Rebers, & Smith, 1956). This technique has been shown to effectively quantify sugar-based detergents in membrane protein samples (Urbani & Warne, 2005). More recently, a modification of this technique using 2,6-dimethylphenol as a colorimetric reagent with higher sensitivity has been applied specifically to the quantification of sugar-based detergents in protein samples for crystallography (Mallya & Pattabiraman, 1997; Prince & Jia, 2012). The three most popular sugar-based detergents in crystallography are *n*-dodecyl-β-D-maltoside (DDM), *n*-decyl-β-D-maltoside (DM), and *n*-octyl-β-D-glucoside (OG). Despite the wide variety of detergents available, these detergents are collectively used in approximately 45% of successful crystallizations (Membrane Protein Databank or MPDB—http://www.mpdb.tcd.ie). This assay can also be extended to an additional 12% of cases in which less common sugar-based detergents have been used (MPDB). Therefore, this assay addresses the most popular detergents for crystallography and could have broad applications.

2.1. Overall protocol

These reactions typically consist of combining three basic components: the detergent sample, the colorimetric reagent (phenol or its derivatives), and sulfuric acid under a fume hood. Using a sample tube specifically designed for high-temperature reactions, such as a SafeSeal 1.5-mL tube (Starstedt, Germany) is crucial, because regular 1.5 mL plastic tubes cannot withstand the heat generated during exothermic mixing and will leak. Concentrations of sulfuric acid always refer to the final proportion after mixing. After the reaction cools to room temperature, the absorbance of the product is measured using a sealed disposable plastic cuvette. This reaction functionalizes phenol at the *para*-position resulting in a new peak in the visible spectrum at 490 nm for phenol derivatives and at 510 nm for 2,6-dimethylphenol derivatives (Fig. 1). In the following sections, the optimization of this assay for the detection of sugar-based detergents in membrane protein samples is described. The objective was to minimize the quantity of sample consumed while maintaining measurement accuracy, which is critical given the small volume, concentrated, protein samples used in crystallography.

2.2. Assay optimization

Previous work has shown 2,6-dimethylphenol to be a superior prochromogen to phenol and many other substituted phenols (Mallya &

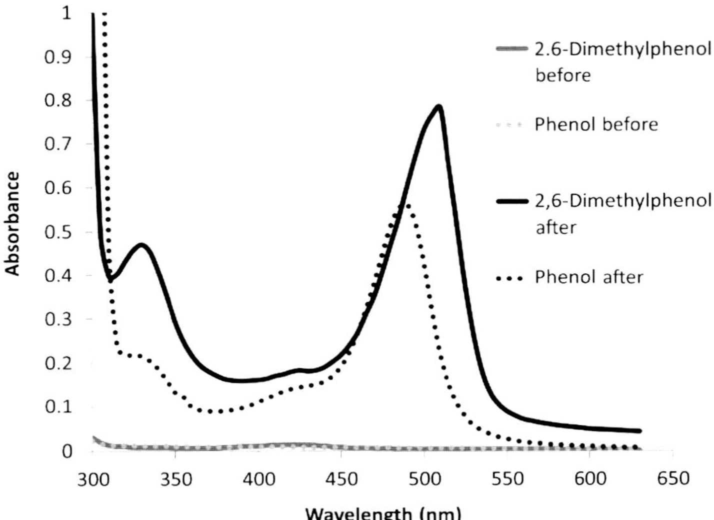

Figure 1 Absorbance peaks resulting from the reaction of phenol/2,6-dimethylphenol, sulfuric acid, and 0.2% DDM.

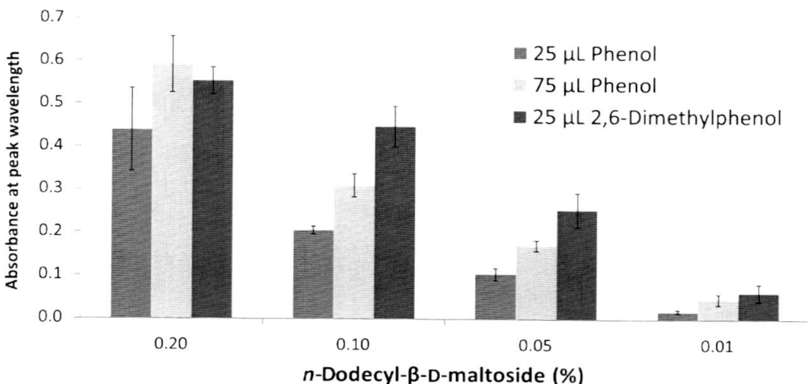

Figure 2 Comparison of reaction efficiency between phenol (490 nm) and 2,6-dimethylphenol (510 nm) as colorimetric reagents. Reactions were performed with 5 μL sample volumes and 75% sulfuric acid with a total reaction volume of 1 mL. Reported values are the average of three independent measurements and error bars represent 1 standard deviation.

Pattabiraman, 1997). Specifically, 2,6-dimethylphenol has been shown to be three times more efficient than phenol as a colorimetric reagent. In this first stage of optimization, the performance of phenol and 2,6-dimethylphenol is directly compared using the same reaction protocol with a total reaction volume of 1 mL. Both reagents were prepared as a 20% solution in absolute ethanol. Phenol produces a colorless solution, compared to the 2,6-dimethylphenol stock solution which is slightly yellow. All reactions were performed in triplicate using an one-step protocol with 750 μL sulfuric acid added last and then mixed by inversion. Three scenarios were compared in which 25 μL 20% phenol, 75 μL 20% phenol, or 25 μL 20% 2,6-dimethylphenol were used as the colorimetric reagent to detect the presence of 0.1% DDM (Fig. 2). The absorbance values reported are after subtraction of a blank reaction with water. It was immediately apparent that 2,6-dimethylphenol had a greater reaction efficiency and performed better, especially at low concentrations. For this reason, it was decided to perform further optimizations with 2,6-dimethylphenol, as this would allow measurement with the smallest possible sample volumes.

There are two common methods that have been proposed for phenol–sulfuric acid reactions, the one-step and the two-step procedures (Fig. 3A and B). In the one-step protocol, all reagents are mixed at the beginning, with sulfuric acid added last. A potential source of error in this method is the highly exothermic mixing of the reaction after sulfuric acid addition. In the two-step protocol, the sugar is reacted with sulfuric acid before the

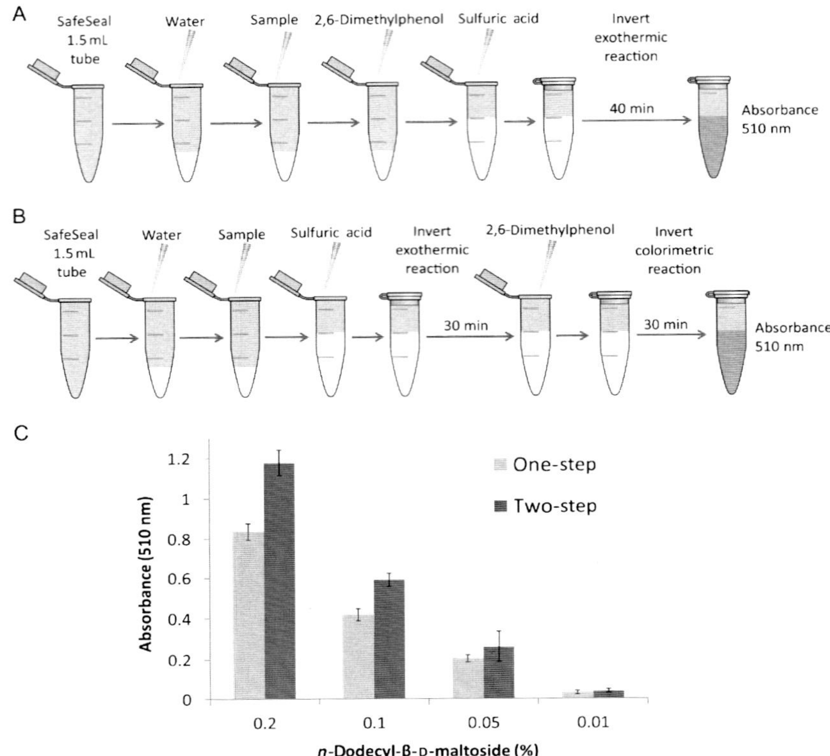

Figure 3 Reaction schematics for the one-step (A) and two-step (B) colorimetric 2,6-dimethylphenol/sulfuric acid reaction for the measurement of sugars in aqueous solution. (C) Comparison of the resulting absorbance from the one-step and two-step reaction procedures using 75% sulfuric acid and 25 μL of 2,6-dimethylphenol. Reactions were performed in triplicate, and error bars represent 1 standard deviation.

addition of 2,6-dimethylphenol. This protects the colorimetric reagent from the exothermic mixing of sulfuric acid with water, but increases the time required. Although the two-step procedure does increase the signal at most detergent concentrations, the increased signal at very low concentrations is minimal (due to increased absorbance from the blank), resulting in a minimal increase in sensitivity (Fig. 3C).

Previous work has also shown that the ratio of sulfuric acid and colorimetric reagent can affect the absorbance of the final product (Dubois et al., 1956; Mallya & Pattabiraman, 1997). The optimum sulfuric acid concentration is around 75% for the two-step procedure (Mallya & Pattabiraman, 1997). When sulfuric acid concentration in the one-step procedure is optimized to 65%, the "lost signal" is recovered (compare Figs. 3C and 4A). A comparison of the one-step and two-step procedures reveals that both effectively quantify DDM with no significant difference in accuracy. Therefore, the more efficient and convenient one-step procedure is selected. The volume of 20% 2,6-dimethylphenol required was also optimized with 25 μL proving sufficient to obtain good signal (Fig. 4B).

The length of the assay was also investigated to ensure the colorimetric reaction was complete before measurement. A reaction time course was performed to monitor the development and decay of absorbance at 510 nm. The standard deviation of the mean absorbance readings taken between 40 and 90 min, 0.013, is less than the average standard deviation in the triplicate measurements, 0.027 (Fig. 4C), so readings taken within this time period are stable and should provide an accurate detergent concentration.

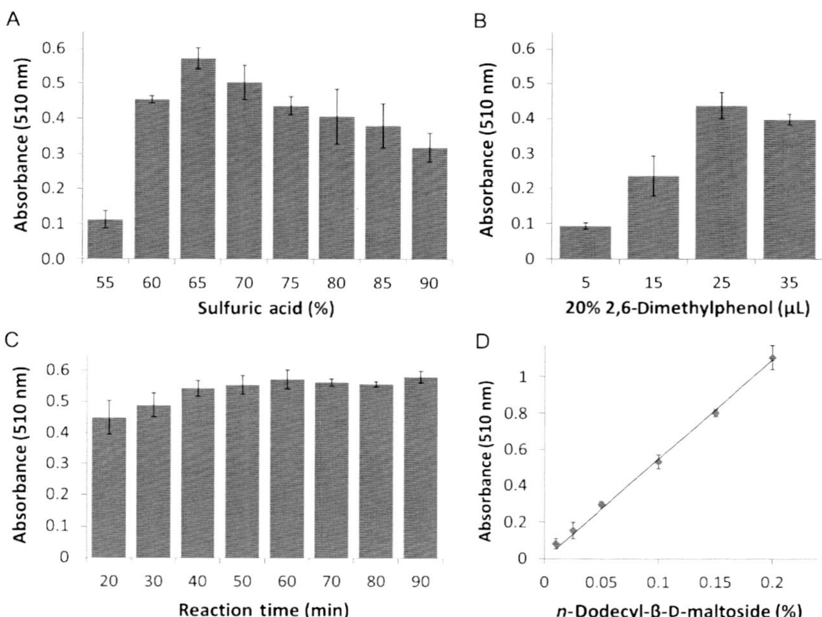

Figure 4 Optimization of the 2,6-dimethylphenol assay. (A) Optimization of absorbance signal from an one-step reaction with 5 μL 0.1% DDM, 25 μL 2,6-dimethylphenol, and varying concentrations of sulfuric acid. (B) Optimization of absorbance signal from an one-step reaction with 5 μL 0.1% DDM, 75% sulfuric acid, and varying amounts of 2,6-dimethylphenol. (C) Time course with 5 μL 0.1% DDM, 65% sulfuric acid, and 25 μL 2,6-dimethylphenol. (D) Standard curve generated by the reaction of 5 μL samples with 65% sulfuric acid and 25 μL 2,6-dimethylphenol using an one-step assay. Reactions were performed in triplicate and error bars represent 1 standard deviation.

The final optimized protocol for the 2,6-dimethylphenol assay employs the one-step procedure and requires the combination of 25 μL 20% 2,6-dimethylphenol, 650 μL sulfuric acid, 5 μL sample, and 320 μL water. Measurements are performed after 40 min. The optimized protocol has been used to generate a standard curve with linearity over a 20-fold range and a correlation coefficient of 0.99 (Fig. 4D). This is now an extremely sensitive assay and samples should be diluted so that the resultant absorbance is within the standard curve and a minimum of 5 μL is always added to the reaction, allowing consistent pipetting. When very dilute samples are measured, the sample volume can be increased (by reducing the water added) to further increase the sensitivity of the reaction. Reactions are always performed in triplicate and a blank assay is performed in parallel with all experimental assays. The standard curve is regenerated each time the 2,6-dimethylphenol stock solution is remade.

2.3. Interfering compounds

When tested with water alone, a resultant absorbance is observed. Therefore, subtraction of the absorbance from an appropriate blank is always required. To ensure that the 2,6-dimethylphenol assay can be used on complex protein samples, cross-reaction with common buffer components and several protein standards was also tested (Fig. 5). These tests used 5 μL samples of highly concentrated solutions. The buffers, salt, reducing agents, and protein tested were found to contribute negligible signal. The compound

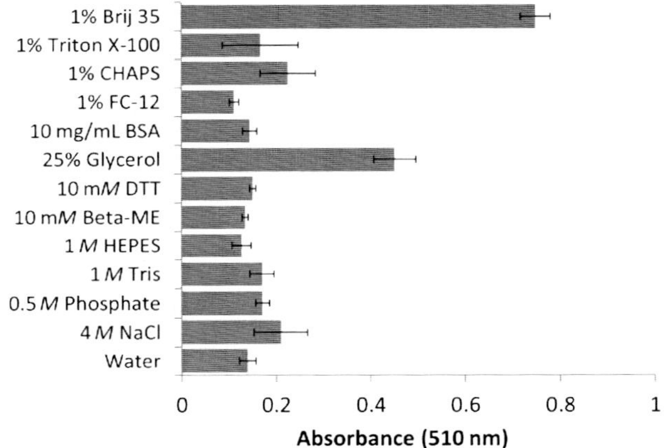

Figure 5 Reaction of 5 μL control samples with 65% sulfuric acid and 25 μL 2,6-dimethylphenol. Reactions were performed in triplicate. Error bars represent 1 standard deviation.

with the most interference, NaCl, gave only a small increase at 4 *M*; therefore, the more typical concentrations in protein buffers (100–300 m*M*) would be expected to give negligible signal. More caution must be applied with additives like glycerol, which is sometimes added to protein buffers to improve solubility. In these experiments, both glycerol and Brij 35 produced significant signal. When these compounds must be used, it is possible to subtract out their contribution using appropriate blank reactions.

3. MOLYBDATE ASSAY FOR TOTAL PHOSPHATE

This procedure was originally described by Bruce Ames in 1966 for the detection of organic phosphates (Ames, 1966). It has been suggested in multiple publications that this assay can be further applied to the fos-choline series of phospholipid-derived detergents. These detergents combine a very small head group with long aliphatic tail groups and are of interest to crystallographers because of their small micelle size. The common detergents in this class are fos-choline 12 and fos-choline 14, but thus far they represent less than 1% of the deposited structures in the MPDB. In the following sections, the effectiveness of a total-phosphate assay for the quantification of fos-choline 12 is discussed. This provides an additional advantage to the use of this detergent in membrane protein crystallization.

3.1. Protocol

This protocol is adapted from the original publication (Ames, 1966) with minor modifications to take advantage of modern equipment. This assay requires multiple stock solutions that should be prepared in advance (Table 1). The protocol consists of combining a 5-μL sample with magnesium nitrate and ethanol (30 μL) in a glass test tube followed by ashing over a strong flame. The resulting powder is then dissolved in 350 μL of 0.5 N hydrochloric acid by vortexing and 300 μL is transferred to a SafeSeal 1.5-mL tube (Starstedt). The addition of excess acid allows exactly 300 μL of sample to be accurately transferred. The sample is then boiled on a hot plate for 15 min. After the tube cools, 700 μL of the ascorbic acid/molybdate solution is added. Incubation at 45 °C on a second hot plate results in a phosphomolybdate complex that is reduced by ascorbic acid to produce the colorimetric product (Fig. 6A). Total phosphate in the sample can be quantified by measurement of the absorbance at its peak wavelength of 825 nm (Fig. 6B).

Table 1 Stock solutions required for the total-phosphate assay

Solution	Composition	Storage
10% Magnesium nitrate	1 g in 10 mL 95% ethanol	Room temperature
0.5 N Hydrochloric acid	1.8 mL concentrated HCl in 100 mL water	Room temperature
10% Ascorbic acid	1 g in 10 mL water	4 °C up to a month
Ammonium molybdate	0.42 g ammonium molybdate tetrahydrate 2.86 mL sulfuric acid in 100 mL water	Room temperature
Colorimetric mix	Mix one part 10% ascorbic acid to six parts ammonium molybdate	12 h on ice

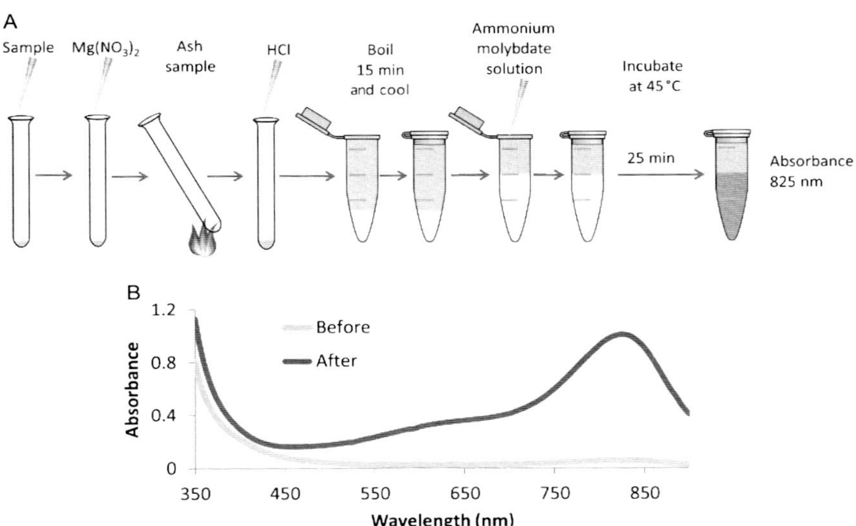

Figure 6 (A) Procedure for the total organic phosphate assay. (B) Absorbance spectrum before and after reaction of fos-choline 12 in the total-phosphate assay.

3.2. Assay optimization

The main parameter optimized for this assay was the incubation time for the colorimetric reaction with ammonium molybdate and ascorbic acid. A 45 °C incubation was performed using a heat block and it was found that the maximum absorbance signal was measured after 25 min (Fig. 7A). Given the variation in absorbance with incubation time, this becomes an important variable to ensure the accuracy of the measurement.

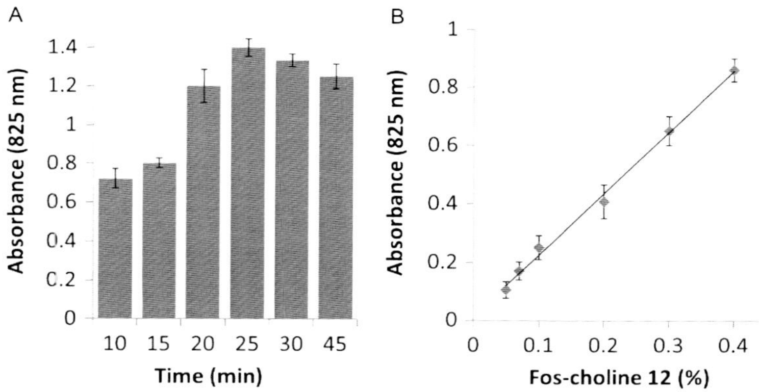

Figure 7 (A) Reaction time course of the ammonium molybdate/ascorbic acid reaction using 5 μL of 0.5% FC-12 as the substrate. (B) Standard curve with FC-12 and the total-phosphate assay. Reactions were performed in triplicate. Error bars represent 1 standard deviation.

Another potential source of error in this assay is the cleanliness of the glass test tubes used. Interference has been noted from silicates evolved from the glass during ashing and residue from trisodium phosphate detergents used during cleaning (Ames, 1966). Given the inexpensive nature of test tubes, new glassware from the same manufacturer is recommended. The ammonium molybdate/ascorbic acid reagent is present in excess and will readily produce a dark blue solution with absorbance >3 AU when high concentrations of phosphate are tested. Therefore, it is important to dilute the sample prior to measurement to fit within the dynamic range of the assay as demonstrated by the standard curve (Fig. 7B). The 5-μL sample volume of this assay is suitable for the small volume samples found in membrane protein crystallography.

3.3. Interfering compounds

Using a similar panel as the 2,6-dimethylphenol assay, the potential for interference from common protein buffer components as well as other detergents were assessed (Fig. 8). HEPES buffer and glycerol were incompatible with the ashing process, resulting in black flakes that did not dissolve in the hydrochloric acid and interfered with the final absorbance measurement. As expected, phosphate buffers also resulted in a reaction with absorbance greater than 3 and obscured the measurement of phosphate from the detergent. There was minimal absorbance generated from highly concentrated salt, reducing agent, or protein and therefore, this assay is applicable to membrane

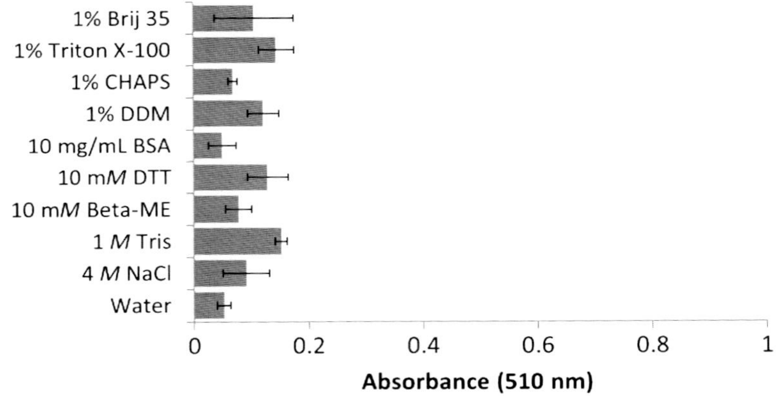

Figure 8 Total-phosphate assay with 5 μL of compatible control samples. Reactions were performed in triplicate. Error bars represent 1 standard deviation.

protein samples as long as an appropriate blank and a compatible buffer, such as Tris, are used. Caution must still be used when applying this technique to phosphorylated proteins as the additional phosphate will be detected.

4. ASSAY OF BILE SALT WITH SULFURIC ACID

The direct reaction of sulfuric acid with bile acid detergents such as CHAPS and cholate (Urbani & Warne, 2005) allows for the quantification of these detergents. The reaction of the sulfuric acid with the hydroxyl group in the cholate ring produces a colored product with an absorption peak at 390 nm that can be used for quantitation (Fig. 9A). CHAPS and CHAPSO are present in approximately 2% of deposited membrane protein structures (MPDB).

4.1. Protocol

In this assay, a 5-μL sample is diluted with water and then sulfuric acid is added to a final volume of 1 mL in a SafeSeal 1.5-mL tube (Starstedt). The reaction is allowed to cool and then the absorbance signal is measured (Fig. 9B).

4.2. Assay optimization and interfering compounds

Given the observed variation in absorbance with sulfuric acid concentration in the 2,6-dimethylphenol/sulfuric acid assay, a similar optimization strategy was applied here. In this case, the sulfuric acid was varied from 50% to 90% of the reaction volume and maximum absorbance occurred with 85% sulfuric acid (Fig. 10A). The reaction time for color development was also

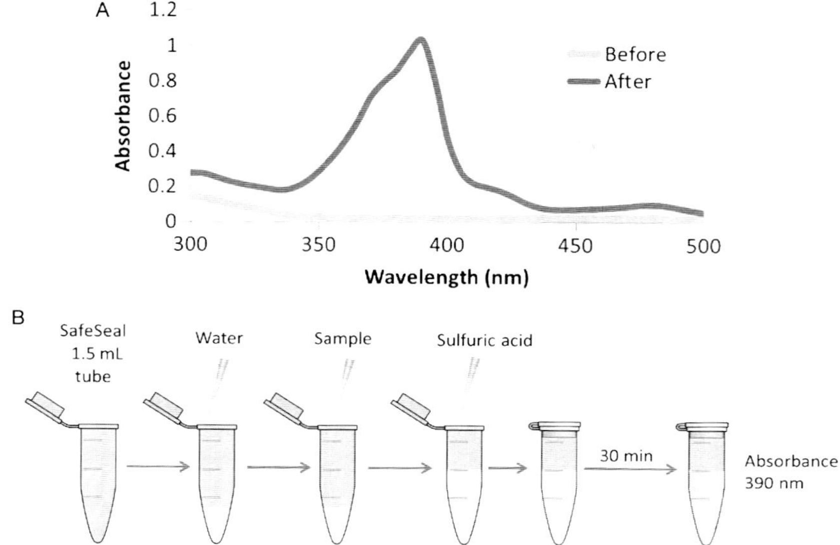

Figure 9 (A) Absorbance spectrum before and after reaction of CHAPS with sulfuric acid. (B) Schematic of CHAPS detergent assay with sulfuric acid.

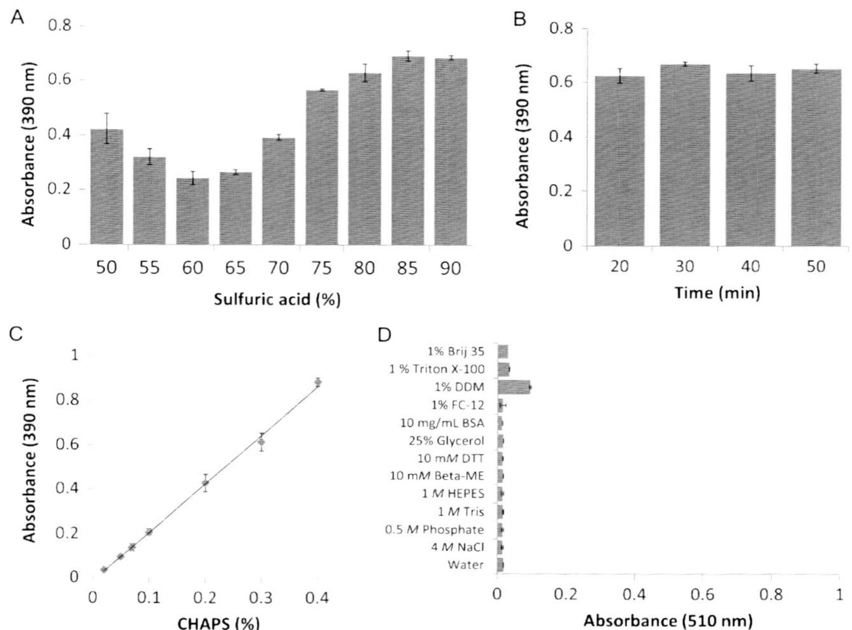

Figure 10 (A) Absorbance generated by the reaction of sulfuric acid with 5 μL 0.3% CHAPS detergent and a total volume of 1 mL. (B) Reaction time course of 0.3% CHAPS with 85% sulfuric acid. (C) Standard curve for the CHAPS assay with 85% sulfuric acid. (D) Reaction of 5 μL of control samples with 85% sulfuric acid. All reactions were performed in triplicate. Error bars represent 1 standard deviation.

considered and maximum absorbance was observed after 30 min (Fig. 10B). Overall, this assay also performs well with small sample volumes of 5 μL, producing a standard curve with a linear range from 0.02% to 0.4% detergent (Fig. 10C).

The panel of 12 control compounds was also tested with 85% sulfuric acid to assess potential side reactions (Fig. 10D). The sulfuric acid reaction with bile salts appears valid for application to membrane protein samples, with the only cross-reaction being with DDM.

5. APPLICATIONS OF DETERGENT MEASUREMENT

5.1. Detergent content and crystallizability

5.1.1 Detergent can affect the phase behavior of crystallization drops

During initial crystal screening of Etk NM2 (an inner membrane protein from *Escherichia coli*) solubilized in DM, a condition was identified which produced pseudo-crystals after phase separation. In an attempt to produce crystals from this condition, the effect of detergent concentration on the phase behavior of the crystallization drops was explored. Etk NM2 samples with a protein concentration of 5.2–5.8 mg/mL were adjusted to contain three divergent concentrations ofDM: 1.2%, 2.0%, and 4.9%. These samples were tested in parallel above a communal well solution of 0.1 *M* MES pH 5.5 and 30% PEG2000 MME. After 24 h, divergent phase behavior could be observed (Fig. 11), although the original pseudo-crystals were not reproduced. This experiment gives a direct example of the significant effect detergent concentration can have on a crystallization experiment, even when all other conditions are kept constant.

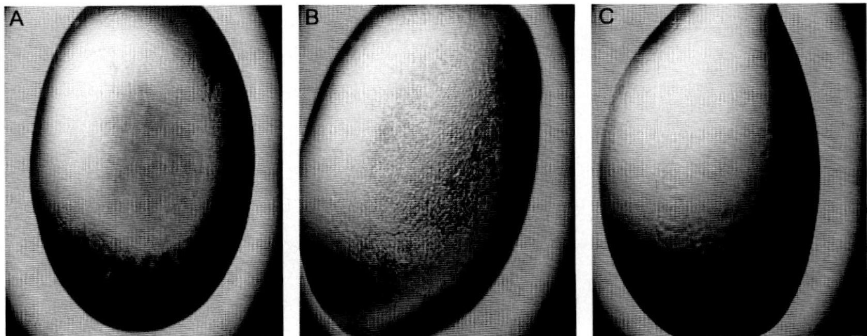

Figure 11 Effect of increasing concentrations of detergent on the phase behavior of Etk NM2 crystallization. Trials were set up with Etk NM2 containing (A) 1.2%, (B) 2.0%, and (C) 4.9% DM. Drops contained 2 μL protein mixed with 2 μL well solution.

5.1.2. Crystallizability of protein samples depends on both the protein and detergent content

In order for crystal nucleation to occur, it is essential that supersaturation is achieved. Even in cases where seeding is employed, the crystallization drop must be maintained in the meta-stable zone of the phase diagram for crystal growth to continue. These conditions are achieved using precipitants (i.e., PEG, alcohols, salts) mixed with the protein. It is commonly accepted that a minimum threshold for protein concentration exists in any specific crystallization experiment to ensure supersaturation and/or a meta-stable state are reached. The concentration of protein required is determined by the solubility of the protein, concentration of stabilizing additives, and the concentration of precipitant to be used. In membrane protein samples, the concentration of detergent is also starting to be appreciated as an equally important variable. The Etk NM2 construct solubilized in DDM produces small–medium hexagonal prism crystals (Fig. 12A). Crystallization of Etk

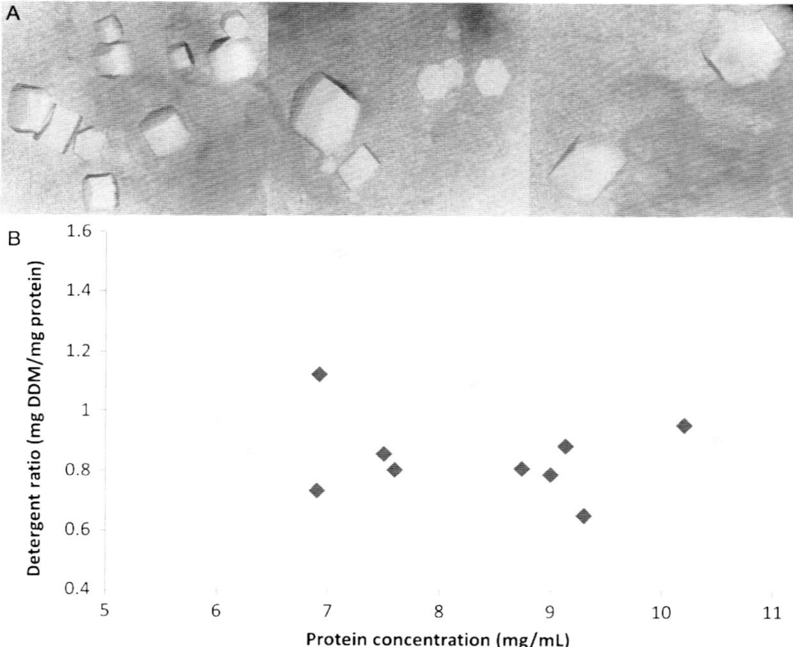

Figure 12 (A) Etk NM2 crystals grow as small to medium hexagonal prisms. (B) Plot of protein samples that successfully crystallized (red) and those that did not (blue) as a function of both the protein and detergent concentration in the sample. (See the color plate.)

NM2 was performed using the hanging drop method and a well solution containing a three-component mixture of 0.1 *M* MES (pH 6.0), 1–5% PEG3000, and 20–30% PEG200. A minimum of 24 wells were set up for each protein/detergent ratio tested. Without prior measurement of detergent concentration, the success rate for crystallization was less than 10%. Frozen samples from previous work, in addition to several new samples, were assayed for both detergent and protein concentration. By comparing detergent concentration to crystallization success, an optimal range of protein and detergent concentrations for crystallization was determined (Fig. 12B). Based on this analysis, it was determined that crystallization requires a minimum protein concentration of 6.9 mg/mL and a maximum of 10.5 mg/mL. In addition, the amount of detergent present in the sample has an important role, with crystallization most likely when the detergent concentration is between a detergent/protein ratio of 0.65 and 1.10 mg DDM/mg protein. Incorporation of this information has significantly increased the frequency of crystallization success.

5.2. Reducing detergent concentration in centrifugal concentrators

The usual school of thought is that variations in protein samples for crystallization can be minimized by careful application of a well-established and reproducible protocol. The difficulty with membrane protein samples is the protein concentration process, during which protein-free detergent micelles are also enriched (Prive, 2007; Shi et al., 2008; Urbani & Warne, 2005). The detergent concentration factor of centrifugal concentrators has been shown to vary between manufacturers and within units from the same manufacturer (Maslennikov et al., 2007). In order to minimize detergent variation, one could use the same concentrator repeatedly, but variation can still occur as the concentrator ages. Therefore, the concentration of detergent is often unknown after protein concentration and obtaining reproducible concentrations can be difficult. Despite this issue, ultrafiltration in centrifugal concentrators remains the most common technique for obtaining reproducible protein concentrations (Maslennikov et al., 2007).

The ability to quickly assess the detergent content and potential crystallizability of a sample is a valuable predictor of success. Even more compelling is the ability to correct the detergent content of samples that fail the criteria for crystallization. The addition of more detergent is relatively straightforward. More complicated is the removal of excess detergent from a premade sample. In this case, there are limited options, though the use of detergent absorbent beads is being offered as a possible solution. These beads have been successfully applied in 2D crystallization experiments and complete detergent removal for protein exchange into lipids or other lipophilic alternatives (Arunmanee, Harris, & Lakey, 2014; Rigaud et al., 1997). In this case study, it is demonstrated that the concentrators themselves and serial dilution can also be used to make minor adjustments to the amount of detergent present. Given that the protein content can be measured nondestructively using the optical density of the solution at 280 nm and the limited loss of sample in the 2,6-dimethylphenol assay, one can confidently adjust the sample to be within the range of expected crystallization success (Fig. 13). In this case, the starting sample of Etk NM2 solubilized in DDM contains 6.7 mg/mL protein and 1.0 mg DDM/mg protein, which is on the edge of the ideal zone for crystallization. Through four cycles of dilution and concentration the protein concentration is adjusted to 8.4 mg/mL and the detergent ratio is reduced to 0.81 mg DDM/mg protein, which makes this sample now a good candidate for crystallization success. Each cycle consists of diluting 2 mL of protein/detergent solution with 10 mL buffer without detergent and reconcentrating the protein in a 50-kDa concentrator (Millipore) back to 2 mL (about 25 min). When performing this analysis, the protein to detergent ratio is used as a more convenient guideline than absolute concentration, because slight variations in the volume after reconcentration will not negatively affect the outcome.

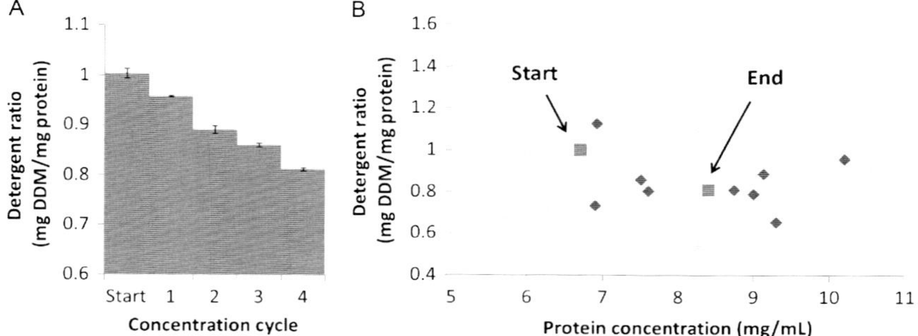

Figure 13 Detergent dilution using a centrifugal concentrator. (A) Detergent concentration is monitored as the ratio of detergent to protein to account for small variations in sample volume after each round of concentration. (B) Position of the protein batch on the crystallization prediction graph at the start and the end of detergent dilution. Protein samples with confirmed crystals are in red (dark gray in the print version) and samples that did not crystallize are in blue (light gray in the print version).

5.3. Detergent exchange

Detergents with long aliphatic side chains can be gentler and more appropriate for extracting proteins from the membrane. During purification, some researchers find it valuable to extract membrane proteins using an inexpensive, low CMC, detergent and later change to a potentially more expensive, specialized detergent that is more suitable for crystallization. A recent example of this application is crystallization of membrane-bound pyrophosphatases. In this case, crystals could be obtained from DDM, but crystals with better diffraction were obtained after exchange into alternative detergents, the best being octyl glucose neopentyl glycol (Kellosalo, Kajander, Honkanen, & Goldman, 2013). During these experiments, detergent measurement techniques that both quantitate and identify the detergent are especially valuable because they can be used to continually monitor the exchange for consistency. These techniques include thin-layer chromatography, gas chromatography, and nuclear magnetic resonance (Eriks et al., 2003; Maslennikov et al., 2007; Shi et al., 2009). NMR has been specifically used to monitor the exchange of fos-choline 14 for DDM and DM in samples of Etk (Maslennikov et al., 2007). In addition, TLC has been applied to monitor the exchange of sarkosyl with DDM in samples of the mitochondrial citrate transport protein (Eriks et al., 2003).

It is not uncommon for detergent exchange to be performed between detergents of different classes, in which case colorimetric assays could be applied to monitor the exchange and determine an appropriate endpoint when exchange is complete. For example, a protein is purified in DDM for its low CMC and high efficiency during extraction from the membrane, but then the protein is exchanged into a smaller, harsher detergent that is more favorable in crystallization. There are examples of small, harsh detergents that do not contain sugar groups. In any of these cases the 2,6-dimethylphenol/sulfuric acid assay can be applied to ensure that DDM is completely removed from the sample.

As an example, the exchange of DDM to FC-12 in a membrane protein sample bound to nickel resin is explored. By monitoring the detergent content of the wash fractions using colorimetric methods, we can determine when DDM is no longer being eluted from the column and use this to determine the number of washes required. In the case of the exchange to FC-12, we see that after seven washes (10 mL) with 0.05% FC-12 there is no detectable DDM being eluted from the column (Fig. 14). In this experiment, 50 μL samples were tested with the 2,6-dimethylphenol assay to increase

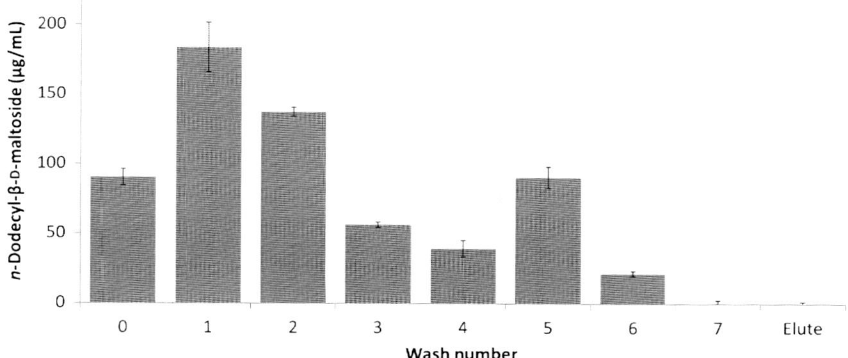

Figure 14 Exchange of DDM to FC-12 by consecutive washes of Ni-agarose bound protein. Wash 0 represents the initial loading and elution of excess DDM. Washes 2–7 each represent 10 mL washes with 0.05% FC-12. Average equilibration time between washes was 5 min, with the exception of wash 5, which was equilibrated for 1 h.

sensitivity. When the protein is subsequently eluted with imidazole, the eluted protein also does not contain a measurable quantity of DDM. In this example, the effect of equilibration time between washes is also evident. The average equilibration time between washes was 5 min, except between washes 4 and 5 when the column was equilibrated for 1 h instead. In wash 5, we see an increased elution of DDM, presumably from the increased time for FC-12 and DDM to exchange. This would suggest that equilibration time is extremely important and that longer equilibration times could allow exchange to be performed with fewer washes.

6. CONCLUDING REMARKS

The detergent concentration used during initial crystallization screens for membrane proteins remains an important consideration. It has been proposed that minimization of excess detergent and "empty" detergent micelles would be favorable to crystallization efforts (DaCosta & Baenziger, 2002; Wiener, 2004). It has also been suggested that concentrations near a detergent's cloud point (i.e., the phase boundary in which intermicellar attractive forces drive micelles into a separate phase) can show correlation with crystallization of the PDC (Hitscherich, Aseyev, Wiencek, & Loll, 2001; Wiener, 2004). In the Etk NM2 case study examined here, it was found that detergent concentrations between 0.65 and 1.10 mg DDM/mg protein were most likely to produce crystals. Overall, the detergent required for the crystallization of a specific membrane protein is likely case dependent, and the tracking of detergent concentration during crystallization efforts may be critical to success.

Many factors are worth considering when measuring detergent concentration, including the availability of equipment, measurement precision, processing time, and sample volume required. This ensures smooth incorporation of the technique into the membrane protein preparation and crystallization pipeline. It has been found that working quickly with membrane proteins *in vitro* can be beneficial for crystallization (Prive, 2007; Screpanti, Padan, Rimon, Michel, & Hunte, 2006). Therefore, the time frame required for colorimetric assays (40–90 min) is practical for incorporation into the timeline required by membrane protein crystallization. As a group, the three colorimetric assays presented in this document cover a range of detergent options for crystallization experiments. Yet, there are still popular detergents like lauryldimethylamine-*N*-oxide that have no specific chemical group which can be targeted by these techniques. In these cases, the more general techniques listed in the introduction will be needed. Moving forward, the ability to measure detergent will be one more factor to consider when choosing a detergent for crystallography experiments.

ACKNOWLEDGMENTS

This work was supported by a research grant to Z. J. (Canadian Institutes of Health Research). Z. J. holds a Canada Research Chair in Structural Biology. C. P. holds a Frederick Banting and Charles Best Canada Graduate Scholarship.

REFERENCES

Ames, B. (1966). Assay of inorganic phosphate, total phosphate and phosphatase. *Methods in Enzymology*, *8*, 115–118.

Arinaminpathy, Y., Khurana, E., Engelman, D. M., & Gerstein, M. B. (2009). Computational analysis of membrane proteins: The largest class of drug targets. *Drug Discovery Today*, *14*(23–24), 1130–1135. http://dx.doi.org/10.1016/j.drudis.2009.08.006.

Arunmanee, W., Harris, J. R., & Lakey, J. H. (2014). Outer membrane protein F stabilised with minimal amphipol forms linear arrays and LPS-dependent 2D crystals. *The Journal of Membrane Biology*, *247*(9–10), 949–956. http://dx.doi.org/10.1007/s00232-014-9640-5.

DaCosta, C. J. B., & Baenziger, J. E. (2002). A rapid method for assessing lipid:protein and detergent:protein ratios in membrane-protein crystallization. *Acta Crystallographica. Section D, Biological Crystallography*, *59*(1), 77–83. http://dx.doi.org/10.1107/S0907444902019236.

Dubois, M., Gilles, K. A., Hamilton, J. K., Rebers, P. A., & Smith, F. (1956). Colorimetric method for determination of sugars and related substances. *Analytical Chemistry*, *28*(3), 350–356. http://dx.doi.org/10.1021/ac60111a017.

Eriks, L. R., Mayor, J. A., & Kaplan, R. S. (2003). A strategy for identification and quantification of detergents frequently used in the purification of membrane proteins. *Analytical Biochemistry*, *323*(2), 234–241. http://dx.doi.org/10.1016/j.ab.2003.09.002.

Gan, S. W., Vararattanavech, A., Nordin, N., Eshaghi, S., & Torres, J. (2011). A cost-effective method for simultaneous homo-oligomeric size determination and monodispersity conditions for membrane proteins. *Analytical Biochemistry, 416*(1), 100–106. http://dx.doi.org/10.1016/j.ab.2011.05.007.

Garavito, R. M., Picot, D., Loll, P. J., & Daniel, P. (1996). Strategies for crystallizing membrane proteins. *Journal of Bioenergetics and Biomembranes, 28*(1), 13–27. Retrieved from, http://www.ncbi.nlm.nih.gov/pubmed/8786233.

Gutmann, D. A. P., Mizohata, E., Newstead, S., Ferrandon, S., Henderson, P. J. F., Veen, H. W. V., et al. (2007). A high-throughput method for membrane protein solubility screening: The ultracentrifugation dispersity sedimentation assay. *Protein Science, 16*(7), 1422–1428. http://dx.doi.org/10.1110/ps.072759907.1422.

Hitscherich, C., Aseyev, V., Wiencek, J., & Loll, P. J. (2001). Effects of PEG on detergent micelles: Implications for the crystallization of integral membrane proteins. *Acta Crystallographica. Section D, Biological Crystallography, 57*(7), 1020–1029. http://dx.doi.org/10.1107/S0907444901006242.

Hitscherich, C., Kaplan, J., Allaman, M., Wiencek, J., & Loll, P. J. (2000). Static light scattering studies of OmpF porin: Implications for integral membrane protein crystallization. *Protein Science, 9*(8), 1559–1566. http://dx.doi.org/10.1110/ps.9.8.1559.

Kaufmann, T. C., Engel, A., & Rémigy, H.-W. (2006). A novel method for detergent concentration determination. *Biophysical Journal, 90*(1), 310–317. http://dx.doi.org/10.1529/biophysj.105.070193.

Kellosalo, J., Kajander, T., Honkanen, R., & Goldman, A. (2013). Crystallization and preliminary X-ray analysis of membrane-bound pyrophosphatases. *Molecular Membrane Biology, 30*(1), 64–74. http://dx.doi.org/10.3109/09687688.2012.712162.

Kunji, E. R. S., Harding, M., Butler, P. J. G., & Akamine, P. (2008). Determination of the molecular mass and dimensions of membrane proteins by size exclusion chromatography. *Methods, 46*(2), 62–72. http://dx.doi.org/10.1016/j.ymeth.2008.10.020.

Mallya, H. M. M., & Pattabiraman, T. N. N. (1997). Effect of acid concentration on chromogen formation from hexoses in sulfuric acid-based reactions. *Analytical Biochemistry, 301*(1988), 299–301. Retrieved from http://www.ncbi.nlm.nih.gov/pubmed/9299033.

Marone, P. A., Thiyagarajan, P., Wagner, A. M., & Tiede, D. M. (1999). Effect of detergent alkyl chain length on crystallization of a detergent-solubilized membrane protein: Correlation of protein-detergent particle size and particle-particle interaction with crystallization of the photosynthetic reaction center from R. *Journal of Crystal Growth, 207*, 214–225.

Maslennikov, I., Kefala, G., Johnson, C., Riek, R., Choe, S., & Kwiatkowski, W. (2007). NMR spectroscopic and analytical ultracentrifuge analysis of membrane protein detergent complexes. *BMC Structural Biology*, 7, 74. http://dx.doi.org/10.1186/1472-6807-7-74.

Prince, C., & Jia, Z. (2012). Measurement of detergent concentration using 2,6-dimethylphenol in membrane-protein crystallization. *Acta Crystallographica. Section D, Biological Crystallography, 68*, 1694–1696. http://dx.doi.org/10.1107/S0907444912040176.

Prive, G. G. (2007). Detergents for the stabilization and crystallization of membrane proteins. *Methods, 41*(4), 388–397. http://dx.doi.org/10.1016/j.ymeth.2007.01.007.

Rigaud, J. L., Mosser, G., Lacapere, J. J., Olofsson, A., Levy, D., & Ranck, J. L. (1997). Bio-Beads: An efficient strategy for two-dimensional crystallization of membrane proteins. *Journal of Structural Biology, 118*(3), 226–235. http://dx.doi.org/10.1006/jsbi.1997.3848.

Screpanti, E., Padan, E., Rimon, A., Michel, H., & Hunte, C. (2006). Crucial steps in the structure determination of the Na+/H+ antiporter NhaA in its native conformation. *Journal of Molecular Biology, 362*(2), 192–202. http://dx.doi.org/10.1016/j.jmb.2006.07.019.

Shi, C., Han, F., Xiong, Y., & Tian, C. (2009). A gas chromatography–mass spectrometry method to monitor detergents removal from a membrane protein sample. *Protein Expression and Purification*, *68*(2), 221–225. http://dx.doi.org/10.1016/j.pep.2009.07.012.

Shi, C., Shao, W., Xiong, Y., & Tian, C. (2008). A gas chromatographic method for quantification of detergents frequently used in membrane protein structural studies. *Analytical Biochemistry*, *383*(2), 326–328. http://dx.doi.org/10.1016/j.ab.2008.08.028.

Sonoda, Y., Cameron, A., Newstead, S., Omote, H., Moriyama, Y., Kasahara, M., et al. (2010). Tricks of the trade used to accelerate high-resolution structure determination of membrane proteins. *FEBS Letters*, *584*(12), 2539–2547. http://dx.doi.org/10.1016/j.febslet.2010.04.015.

Strop, P., & Brunger, A. T. (2005). Refractive index-based determination of detergent concentration and its application to the study of membrane proteins. *Protein Science*, *14*, 2207–2211. http://dx.doi.org/10.1110/ps.051543805.

Urbani, A., & Warne, T. (2005). A colorimetric determination for glycosidic and bile salt-based detergents: Applications in membrane protein research. *Analytical Biochemistry*, *336*(1), 117–124. http://dx.doi.org/10.1016/j.ab.2004.09.040.

Wiener, M. C. (2004). A pedestrian guide to membrane protein crystallization. *Methods*, *34*(3), 364–372. http://dx.doi.org/10.1016/j.ymeth.2004.03.025.

CHAPTER SIX

Solubilization of G Protein-Coupled Receptors: A Convenient Strategy to Explore Lipid–Receptor Interaction

Amitabha Chattopadhyay[*,‡,1], **Bhagyashree D. Rao**[†,‡], **Md. Jafurulla**[*]
[*]CSIR-Centre for Cellular and Molecular Biology, Hyderabad, India
[†]CSIR-Indian Institute of Chemical Technology, Hyderabad, India
[‡]Academy of Scientific and Innovative Research, New Delhi, India
[1]Corresponding author: e-mail address: amit@ccmb.res.in

Contents

Abstract

G protein-coupled receptors (GPCRs) are the largest class of molecules involved in signal transduction across cell membranes and are major drug targets. Since GPCRs are integral membrane proteins, their structure and function are modulated by membrane lipids. In particular, membrane cholesterol is an important lipid in the context of GPCR function. Solubilization of integral membrane proteins is a process in which the proteins and lipids in native membranes are dissociated in the presence of a suitable amphiphilic detergent. Interestingly, solubilization offers a convenient approach to monitor lipid–receptor interaction as it results in differential extents of lipid solubilization, thereby allowing to assess the role of specific lipids on receptor function. In this review, we highlight how this solubilization strategy is utilized to decipher novel information about the structural stringency of cholesterol necessary for supporting the function of the serotonin$_{1A}$ receptor. We envision that insight in GPCR–lipid interaction would result in better understanding of GPCR function in health and disease.

Methods in Enzymology, Volume 557
ISSN 0076-6879
http://dx.doi.org/10.1016/bs.mie.2015.01.001

1. G PROTEIN-COUPLED RECEPTORS

The G protein-coupled receptors (GPCRs) represent the largest and most diverse group of proteins in mammals, involved in information transfer (signal transduction) from outside the cell to the cellular interior (Chattopadhyay, 2014; Perez, 2003; Pierce, Premont, & Lefkowitz, 2002; Rosenbaum, Rasmussen, & Kobilka, 2009). GPCRs are seven transmembrane domain proteins and include more than 800 members which are encoded by approximately 5% of human genes (Zhang, DeVries, & Skolnick, 2006). They transmit extracellular signals to the cellular interior by concerted changes in the transmembrane domain structure (Deupi & Kobilka, 2010; Nygaard et al., 2013). GPCRs respond to a variety of physiological stimuli that include endogenous ligands (such as biogenic amines) and exogenous ligands (e.g., odorants, pheromones, and photons) for sensory perception. As a consequence, GPCRs regulate a large number of physiological processes such as neurotransmission, secretion, cellular differentiation, growth, entry of pathogens into host cells, and inflammatory and immune responses. For this reason, GPCRs represent major drug targets in all clinical areas (Ellis & The Nature Reviews Drug Discovery GPCR Questionnaire Participants, 2004; Heilker, Wolff, Tautermann, & Bieler, 2009; Insel, Tang, Hahntow, & Michel, 2007; Jacoby, Bouhelal, Gerspacher, & Seuwen, 2006). It is estimated that approximately 50% of clinically prescribed drugs and 25 of the 100 top-selling drugs target GPCRs (Schlyer & Horuk, 2006; Thomsen, Frazer, & Unett, 2005).

2. MEMBRANE LIPIDS IN GPCR ORGANIZATION AND FUNCTION

GPCRs are integral membrane proteins with seven passes across the membrane, and as a result, a considerable portion of GPCRs remains in contact with the membrane lipid environment. Membrane lipids therefore act as important modulators of GPCR structure and function. Cells possess the ability to vary their membrane lipid composition in response to a variety of stress and stimuli, thereby changing the environment and the activity of the membrane receptors. Such interplay between the function of a given GPCR and its immediate lipid environment in the membrane is physiologically relevant. Results from our laboratory and others have shown that the interaction of GPCRs with membrane lipids is crucial for their structure and function (Burger, Gimpl, & Fahrenholz, 2000; Jafurulla & Chattopadhyay, 2013a; Oates & Watts, 2011; Paila & Chattopadhyay, 2010; Pucadyil & Chattopadhyay, 2006; Soubias & Gawrisch, 2012). It has recently been reported that even the interaction between GPCRs and G-proteins could be modulated by membrane lipids (Inagaki et al., 2012). Interestingly, the membrane lipid environment of GPCRs has been implicated in disease progression during aging (Alemany et al., 2007). The most studied lipid in the context of GPCR–lipid interaction is cholesterol.

3. CHOLESTEROL: AN IMPORTANT MODULATOR OF GPCR FUNCTION

Cholesterol is a crucial membrane lipid in higher eukaryotes. It plays an important role in membrane organization, dynamics, function, and sorting (Mouritsen & Zuckermann, 2004; Simons & Ikonen, 2000). Typically, membrane cholesterol is distributed in a nonrandom fashion in domains in biological and model membranes (Chaudhuri & Chattopadhyay, 2011; Lingwood & Simons, 2010; Mukherjee & Maxfield, 2004; Xu & London, 2000). These membrane domains are believed to play a key role in membrane sorting and trafficking (Simons & van Meer, 1988), signal transduction (Simons & Toomre, 2000), and the entry of pathogens into host cells (Chattopadhyay & Jafurulla, 2012; Pucadyil & Chattopadhyay, 2007; Roy, Kumar, Jafurulla, Mandal, & Chattopadhyay, 2014).

The role of membrane cholesterol in the organization and function of membrane proteins in general, and GPCRs in particular is an exciting and contemporary area of research (Burger et al., 2000; Jafurulla & Chattopadhyay, 2013a; Oates & Watts, 2011; Paila & Chattopadhyay, 2010; Pucadyil & Chattopadhyay, 2006; Soubias & Gawrisch, 2012). The mechanism underlying the effect of membrane cholesterol on the structure and function of membrane receptors is not straightforward and still emerging (Lee, 2011; Paila & Chattopadhyay, 2009, 2010). Membrane cholesterol could modulate the function of membrane proteins by direct (specific) interaction, which could induce local conformational change(s) in the receptor. Another mechanism proposes an indirect effect by altering the physical properties of the membrane in which the protein is embedded. A third possibility could be a combination of both types of effects.

As stated above, membrane cholesterol has been reported to influence the function of a number of GPCRs. A representative GPCR in the context of cholesterol sensitivity of receptor organization, dynamics, and function is the serotonin$_{1A}$ receptor (Jafurulla & Chattopadhyay, 2013a; Paila & Chattopadhyay, 2010; Pucadyil & Chattopadhyay, 2006). The serotonin$_{1A}$ receptor is an important neurotransmitter receptor, which acts as a drug target for neuropsychiatric disorders (Celada, Bortolozzi, & Artigas, 2013; Kalipatnapu & Chattopadhyay, 2007; Müller, Carey, Huston, & De Souza Silva, 2007; Pucadyil, Kalipatnapu, & Chattopadhyay, 2005; Savitz, Lucki, & Drevets, 2009). The receptor is implicated in the generation and modulation of various cognitive, behavioral, and developmental functions. Previous work from our laboratory has shown that the organization, dynamics, and function of the serotonin$_{1A}$ receptor are critically dependent on membrane cholesterol (reviewed in Jafurulla & Chattopadhyay, 2013a, 2013b; Paila & Chattopadhyay, 2010; Pucadyil & Chattopadhyay, 2006). Utilizing a number of approaches, we showed that membrane cholesterol plays an important role in the ligand-binding activity and G-protein coupling of the receptor. These approaches include: (i) physical depletion of membrane cholesterol using MβCD; (ii) treatment with agents such as nystatin and digitonin, which complex cholesterol and modulate the availability of membrane cholesterol without physically depleting it; (iii) oxidation of cholesterol to cholestenone (chemical modification) using cholesterol oxidase; and (iv) metabolic inhibition of cholesterol biosynthesis using inhibitors such as statin and AY 9944. Another important approach used by us to monitor the effect of specific lipids on receptor function was solubilization of the receptor using suitable detergents. Solubilization offers a convenient way to explore lipid–receptor interaction as it results in differential extents of lipid solubilization, thereby allowing to assess the role of specific lipids on receptor function (see below).

4. MEMBRANE PROTEIN SOLUBILIZATION: AN ESSENTIAL STEP TOWARD PURIFICATION

Biological membranes represent a complex milieu of a large variety of lipids and proteins, the organization of which allows the membrane to carry out its function. A commonly used approach to study membranes is to dissociate the membrane into its components. An important step in this direction is purification of membrane proteins, an area of considerable experimental challenge (Anson, 2009). Experiments performed using purified and reconstituted membrane receptors have helped significantly in our current understanding of the function of membrane receptors. An essential criterion for purification of a transmembrane protein is that the protein must be carefully removed from the native membrane environment and dispersed in solution. This is carried out using suitable amphiphilic detergents and the process is known as solubilization (Duquesne & Sturgis, 2010; Helenius & Simons, 1975; Hjelmeland & Chrambach, 1984; Jones, Earnest, & McNamee, 1987; Kalipatnapu & Chattopadhyay, 2005; Kubicek, Block, Maertens, Spriestersbach, & Labahn, 2014; Madden, 1986; Privé, 2007; Seddon, Curnow, & Booth, 2004).

Solubilization of membrane proteins could be defined as a process in which proteins and lipids, held together in native membranes, are suitably dissociated in a buffered solution containing an appropriate detergent. The dissociation of the native membrane leads to the formation of small clusters of protein, lipid, and detergent that remain dissolved in the aqueous solution (see Fig. 1). An important criterion for effective solubilization and purification of membrane proteins is that the function of the protein should be retained to the maximum possible extent. This poses a considerable challenge since many detergents irreversibly denature membrane proteins (Garavito & Ferguson-Miller, 2001), which is responsible for the modest list of membrane proteins solubilized with retention of function. In case of GPCRs, solubilization and purification from natural sources is still rare due to low amounts of the receptor present in the native tissue. Since solubilization constitutes the crucial first step toward purification of any transmembrane receptor, it is important to identify factors responsible for achieving successful solubilization. We outline below some crucial aspects of membrane receptor solubilization.

4.1 Choice of an appropriate detergent

Efficient solubilization of functional GPCRs utilizing a suitable detergent constitutes the first step in their molecular characterization. Detergents are soluble amphiphiles with critical micelle concentrations (CMCs) typically in the range of millimolar. The ability of a detergent to solubilize membranes is related to its hydrophile–lipophile balance (HLB), especially for solubilization by nonionic detergents (Helenius & Simons, 1975; Neugebauer, 1990). This principle has been utilized earlier in order to achieve optimum solubilization of membrane proteins (Slinde & Flatmark, 1976). HLB is an empirical parameter and is a measure of the hydrophilic character of a detergent. It is calculated as the weight percentage of hydrophilic versus lipophilic groups present in a detergent. Detergents

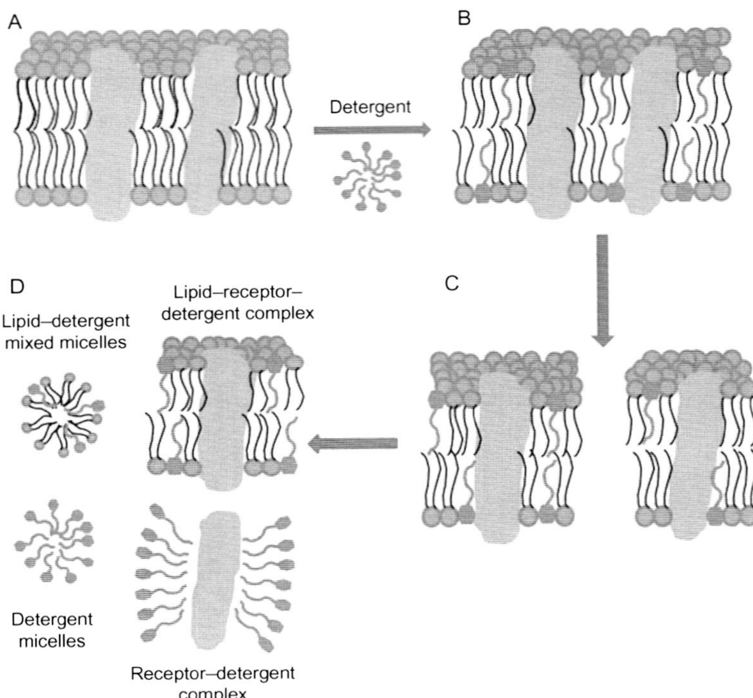

Figure 1 A schematic representation of different stages of solubilization of biological membranes by detergents. When detergents are added to biological membranes (shown in (A)), the detergent monomers (shown in maroon (dark gray in the print version) with single tails) bind to the membrane and cause minimum perturbation at low concentrations (B). With increasing detergent concentration, the membrane bilayer gets further perturbed (C). At even higher detergent concentrations, complexes of detergent, lipid, and receptor of varying compositions are formed. These complexes include lipid–detergent mixed micelles, lipid–receptor–detergent complex, receptor–detergent complex, and detergent micelles (D).

with a relatively high HLB value of 12–20 are recommended for efficient solubilization of membrane proteins without denaturation (Bhairi & Mohan, 2001).

Detergents that belong to the class of nonionic and zwitterionic detergents are particularly popular for their ability to solubilize membrane proteins with retention of function. An important member of this class of detergents is CHAPS (3-[(3-cholamidopropyl)dimethylammonio]-1-propanesulfonate; see Fig. 2A), which is a mild, nondenaturing, and zwitterionic detergent (Hjelmeland, 1980). CHAPS is a synthetic detergent that combines useful features of both the bile salt hydrophobic group and the

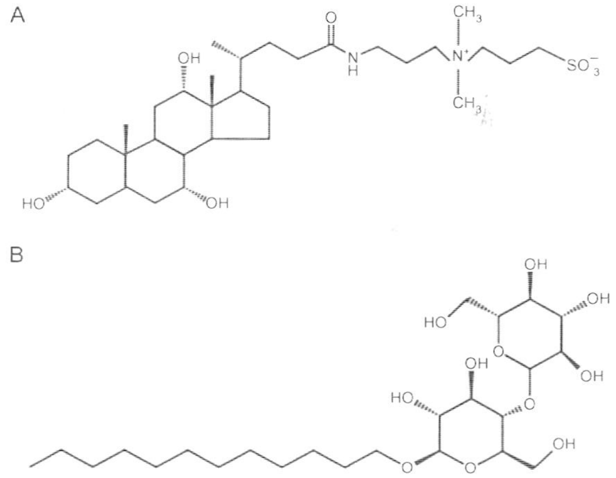

Figure 2 Chemical structures of detergents commonly used for solubilization of GPCRs: (A) CHAPS and (B) DDM. See text for more details.

N-alkyl sulfobetaine-type polar group. It is more efficient in solubilizing membrane proteins than its parent bile acid (such as cholate, which is anionic) due to the absence of net charge. The advantages of using CHAPS for solubilization include its low absorbance at 280 nm and lack of circular dichroic signature in the far-UV region, thereby making it ideal for studies of membrane proteins using optical spectroscopy. For these reasons, CHAPS is widely used in solubilization of membrane proteins and receptors (Banerjee, Joo, Buse, & Dawson, 1995; Chattopadhyay & Harikumar, 1996; Chattopadhyay, Harikumar, & Kalipatnapu, 2002; Cladera, Rigaud, Villaverde, & Duñach, 1997; Kline, Park, & Meyerson, 1989; Locatelli-Hoops, Gorshkova, Gawrisch, & Yeliseev, 2013; Talmont, Moulédous, Mollereau, & Zajac, 2014; Vukoti, Kimura, Macke, Gawrisch, & Yeliseev, 2012; White et al., 2012).

Another detergent that has been used extensively in the last few years to solubilize and crystallize GPCRs is DDM (*n*-dodecyl-β-D-maltoside) (see Fig. 2B). DDM is a mild, nonionic detergent and has been found to be effective in solubilizing several membrane proteins, due to its gentle nature and favorable properties for maintaining the function of aggregation-prone membrane proteins in solution (Privé, 2007). It forms large micelles, which offer the advantage of preventing membrane protein aggregation. However, this could be a limitation in structural studies since DDM masks the protein to a large extent in protein–detergent complexes (Privé, 2007). DDM is known to occlude hydrophilic regions of the protein that are essential to form crystal contacts, which is not conducive for crystallization of GPCRs (Tate, 2012). This was avoided in later studies by increasing the hydrophilic regions of the GPCRs using antibodies or fusion proteins. DDM has been used to effectively solubilize GPCRs such as β_2-adrenergic receptor (Cherezov et al., 2007; Rasmussen et al., 2011), A_{2A} adenosine receptor (Liu et al., 2012), μ-opioid receptor (Manglik et al., 2012), κ-opioid receptor (Wu et al., 2012), β_1-adrenergic receptor (Huang, Chen, Zhang, & Huang, 2013), serotonin$_{1B}$ receptor (Wang et al., 2013), serotonin$_{2B}$ receptor (Wacker et al., 2013), and metabotropic glutamate type 1 receptor (Wu et al., 2014). Some GPCRs such as neurotensin receptor (White et al., 2012) and CB_2 cannabinoid receptor (Locatelli-Hoops et al., 2013; Vukoti et al., 2012) have been solubilized utilizing a combination of DDM and CHAPS.

It should be noted that the choice of a suitable detergent for optimal solubilization of a given membrane protein has to be worked out on an individual basis (Privé, 2007). For example, efficient solubilization of the IgE receptor has been shown to occur with the anionic detergent cholate but not with the nonionic detergent octyl glucoside (Rivnay & Metzger, 1982). Compatibility of the detergent in biochemical assays is another important factor to be considered.

4.2 CMC of detergents

Detergents are soluble amphiphiles and above a critical concentration (strictly speaking, a narrow concentration range), referred to as the CMC, they self-associate to form thermodynamically stable, noncovalent aggregates called micelles (Tanford, 1978). The concept of micelle formation is important in the context of solubilization and reconstitution of membrane receptors. There is a certain correlation between micelle formation and detergent concentration necessary for solubilization (Rivnay & Metzger, 1982). In case of receptors such as the insulin receptor, opioid receptor, and angiotensin II receptor, efficient solubilization is achieved only with high (>1 m*M*) CMC detergents such as CHAPS and octyl glucoside at concentrations below the CMC (Hjelmeland & Chrambach, 1984). Detergents used at concentrations above their CMC invariably resulted in loss of receptor function. The mechanism by which detergents solubilize membranes at concentrations below the CMC, and the related loss of function above the CMC is not clear. This has given rise to the useful concept of "effective CMC" (Chattopadhyay & Harikumar, 1996; Chattopadhyay et al., 2002; Jones et al., 1987; Rivnay & Metzger, 1982; Schürholz, 1996), which is the concentration of detergent existing as monomers at a given condition (such as lipids, proteins, ionic strength, pH, and temperature). Solubilization could therefore be carried out below the CMC if the effective CMC is lower than literature CMC. Another key parameter is the critical solubilization concentration (CSC), which is the minimal detergent concentration required to disrupt a given membrane into micellar dispersion (Privé, 2007). Selective solubilization of membrane proteins at detergent concentrations below CSC could be an effective purification strategy.

4.3 Detergent–lipid–protein ratio

Membrane solubilization by detergents is a multistep process (Helenius & Simons, 1975; Hjelmeland & Chrambach, 1984; Jones et al., 1987; le Maire, Champeil, & Møller, 2000; see Fig. 1). The relative detergent–lipid–protein ratio is an important factor for optimal solubilization of membrane proteins. At a given protein or lipid concentration, with increasing detergent concentration, an increase in solubilized lipid (Pucadyil & Chattopadhyay, 2004) or protein (Demoliou-Mason & Barnard, 1984) is observed until saturation is reached. However, it is not advisable to use high detergent concentrations since membrane protein function is often compromised under such conditions. To overcome this, a mild concentration of detergent could be used which may balance these two aspects, that is, maximize solubilization yet preserve protein function. Arriving at an optimal detergent, lipid, and protein ratio involves trial and error by carrying out solubilization over a wide range of detergent–lipid ratios.

An empirical relationship between these experimental parameters was developed in which the parameter (ρ) was defined as the molar ratio of detergent to lipid optimal for functional solubilization (Rivnay & Metzger, 1982).

$$\rho = \frac{[\text{Detergent}] - \text{CMC}_{\text{eff}}}{[\text{Phospholipid}]}$$

where CMC_{eff} represents the effective CMC determined under specific experimental conditions (as mentioned above). An increase in solubilization is expected with increase in the value of the ρ parameter (generally up to ~2).

5. SOLUBILIZATION AS A STRATEGY TO MONITOR LIPID–PROTEIN INTERACTIONS

As mentioned earlier, solubilization provides a convenient approach to explore lipid–receptor interaction since it results in differential extents of lipid solubilization, thereby allowing to assess the role of specific lipids on receptor function. A common feature often associated with membrane solubilization is delipidation (loss of lipids). This results in loss of protein function since lipid–protein interactions play a crucial role in maintaining the structure and function of integral membrane proteins and receptors (Lee, 2003). For example, displacement of annular lipids from the receptor was shown to be an integral feature of detergent-induced inactivation in case of the nicotinic acetylcholine receptor (Jones, Eubanks, Earnest, & McNamee, 1988). Interestingly, the phenomenon of delipidation caused by solubilization and the subsequent loss of membrane protein function has been effectively utilized to gain molecular insight into the specific lipid requirements of membrane proteins (Jones et al., 1988; Kirilovsky & Schramm, 1983).

This strategy has been successfully utilized for exploring lipid–GPCR interaction. It was previously reported that solubilization of the native hippocampal serotonin$_{1A}$ receptors using CHAPS results in loss of receptor activity and membrane cholesterol (Banerjee, Buse, & Dawson, 1990; Banerjee et al., 1995; Chattopadhyay, Jafurulla, Kalipatnapu, Pucadyil, & Harikumar, 2005). We previously demonstrated that specific ligand binding of the serotonin$_{1A}$ receptor could be restored upon replenishment of cholesterol into solubilized membranes (Chattopadhyay et al., 2005). Utilizing this experimental strategy, we were able to examine the degree of stringency required by closely related analogs of cholesterol, necessary for restoring receptor activity. In order to explore the structural stringency of cholesterol necessary for supporting receptor function, we replaced cholesterol with its close structural analogs with minor differences (see Fig. 3). In one set of experiments, solubilized membranes were replenished with 7-dehydrocholesterol (7-DHC) and desmosterol, which are immediate biosynthetic precursors of cholesterol in the Kandutsch–Russell and Bloch pathways, respectively, both of which differ with cholesterol merely in an additional double bond. While 7-DHC differs with cholesterol only in a double bond at the seventh position in the sterol ring, desmosterol differs with cholesterol only in a double bond at the 24th position in its flexible alkyl side chain (see Fig. 3). Accumulation of either 7-DHC or desmosterol due to defective sterol biosynthesis has been shown to result in fatal neurological disorders (Porter & Herman, 2011). Figure 4A shows that while desmosterol could support receptor function, 7-DHC could not restore receptor activity (Chattopadhyay et al., 2007; Singh et al., 2011). This brings out the fine stringency of cholesterol requirement for receptor function

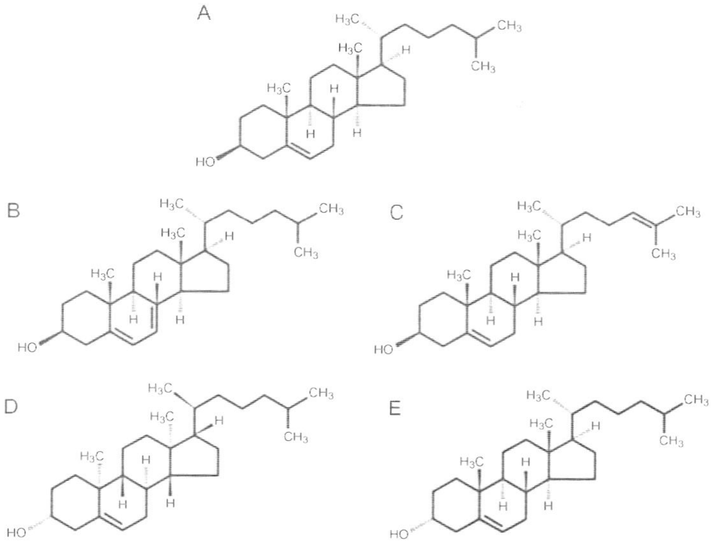

Figure 3 Chemical structures of (A) cholesterol, (B) 7-dehydrocholesterol, (C) desmosterol, (D) *ent*-cholesterol, and (E) *epi*-cholesterol. Both 7-dehydrocholesterol (7-DHC) and desmosterol are immediate biosynthetic precursors of cholesterol in Kandutsch–Russell and Bloch pathways, respectively, differing with cholesterol only in a double bond. While 7-dehydrocholesterol differs with cholesterol only in a double bond at the 7th position in the sterol ring, desmosterol differs with cholesterol only in a double bond at the 24th position of the flexible alkyl side chain. Patients with mutations in enzymes that catalyze the final step in these pathways exhibit low levels of serum cholesterol and accumulation (high levels) of the respective immediate precursor (7-DHC or desmosterol) leading to diseases such as the Smith–Lemli–Opitz syndrome (SLOS) and desmosterolosis. Both *ent*-cholesterol and *epi*-cholesterol are stereoisomers of cholesterol. *ent*-Cholesterol is the enantiomer of cholesterol and is a nonsuperimposable mirror image of cholesterol, whereas *epi*-cholesterol is a diastereomer of cholesterol which differs with cholesterol in the orientation of hydroxyl group at carbon-3 position. *ent*-Cholesterol shares similar physicochemical properties with cholesterol but *epi*-cholesterol does not.

since the presence of an additional double bond in the sterol ring (7-DHC) appears more detrimental to receptor function than the presence of an extra double bond in the alkyl side chain (desmosterol).

The degree of structural stringency was explored further by examining whether stereoisomers of cholesterol (*ent*-cholesterol and *epi*-cholesterol) could support receptor function. While *ent*-cholesterol is the enantiomer of cholesterol and is a nonsuperimposable mirror image of cholesterol, *epi*-cholesterol is a diastereomer (not a mirror image of cholesterol) (see

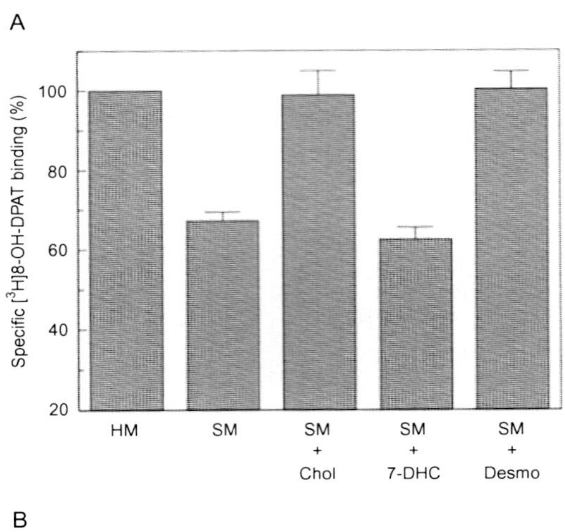

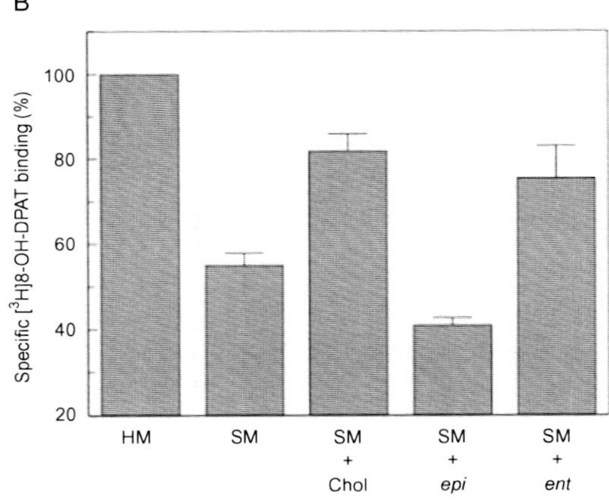

Figure 4 Effect of replenishment of (A) 7-DHC and desmosterol, and (B) *ent*-cholesterol and *epi*-cholesterol into solubilized membranes (SM) on specific binding of the agonist [^{3}H]8-OH-DPAT to the serotonin$_{1A}$ receptor. Data for solubilized membranes replenished with cholesterol are shown in all cases for comparison. Native hippocampal membranes (HM) without any treatment served as control. See text for more details. *Data shown in panel (A) are from Chattopadhyay et al. (2007) and Singh, Jafurulla, Paila, and Chattopadhyay (2011). Panel (B) is adapted and modified from Jafurulla et al. (2014).*

Fig. 3). Figure 4B shows that *ent*-cholesterol could replace cholesterol in supporting the function of the serotonin$_{1A}$ receptor, but *epi*-cholesterol could not (Jafurulla et al., 2014). These results clearly show that the requirement of membrane cholesterol for the serotonin$_{1A}$ receptor function is *diastereospecific*, yet not *enantiospecific*. Taken together, we were able to

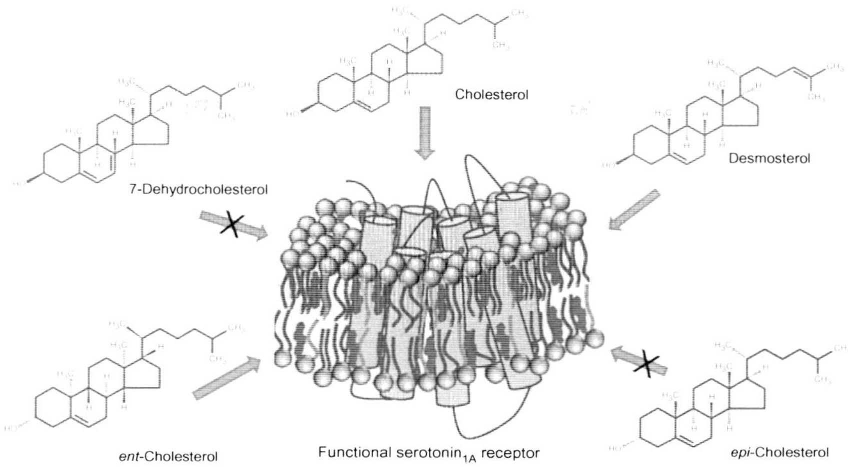

Figure 5 A schematic representation showing the role of various sterols in supporting the function of the reconstituted serotonin$_{1A}$ receptor. The serotonin$_{1A}$ receptor is shown in purple, and the replenished sterol molecules are shown in maroon. The ligand-binding activity of the reconstituted serotonin$_{1A}$ receptor was supported upon replenishment with cholesterol, desmosterol, and *ent*-cholesterol, whereas replenishment with 7-DHC and *epi*-cholesterol could not support the function of the receptor. This brings out the important message that even subtle changes in sterol structure could be detrimental for receptor function, thereby implying stringent receptor–lipid interaction. (See the color plate.)

decipher the subtle details of structural stringency of cholesterol necessary for serotonin$_{1A}$ receptor function, utilizing solubilization strategy. A comprehensive representation of the role of various cholesterol analogs on serotonin$_{1A}$ receptor activity is shown in Fig. 5.

6. CONCLUSIONS AND FUTURE PERSPECTIVES

GPCRs represent one of the evolutionarily conserved families of membrane receptors dating back more than a billion years (Schöneberg, Hofreiter, Schulz, & Römpler, 2007). GPCRs occupy a unique position in contemporary biology due to their ability to transduce a variety of information across the cell membrane and as drug targets (Chattopadhyay, 2014). Interaction of membrane lipids with GPCRs leading to functional modulation of the receptor is an increasingly emerging area of interest. The recent reports of high-resolution crystal structures of several GPCRs with cholesterol bound to various positions of the receptor (reviewed in

Chattopadhyay, 2014; Jafurulla & Chattopadhyay, 2013a) have provided more impact to this field. Unfortunately, a majority of GPCRs are not available in purified form from native sources. This is a severe limitation in attempts to reconstitute the purified receptor into a defined lipid environment, thereby preventing lipid–receptor studies using established approaches. In this overall scenario, solubilization using suitable detergents which allow selective depletion of membrane lipids offers a window of opportunity to assess the lipid specificity of GPCRs. The knowledge gained from these studies will provide a better understanding of specific lipid dependence of receptor function. Such advances in deciphering molecular details of receptor–lipid interaction would lead to better understanding of GPCR function in health and disease.

ACKNOWLEDGMENTS

Work in A.C.'s laboratory was supported by the Council of Scientific and Industrial Research, Govt. of India. B.D.R. thanks the University Grants Commission for the award of a Junior Research Fellowship. A.C. gratefully acknowledges support from J.C. Bose Fellowship (Department of Science and Technology, Govt. of India). A.C. is an Adjunct Professor of Jawaharlal Nehru University (New Delhi), Indian Institute of Science Education and Research (Mohali), Indian Institute of Technology (Kanpur), and Honorary Professor of the Jawaharlal Nehru Centre for Advanced Scientific Research (Bangalore). We thank G. Aditya Kumar for help in making figures, and members of the Chattopadhyay laboratory for their comments and discussions.

REFERENCES

Alemany, R., Perona, J. S., Sánchez-Dominguez, J. M., Montero, E., Cañizares, J., Bressani, R., et al. (2007). G protein-coupled receptor systems and their lipid environment in health disorders during aging. *Biochimica et Biophysica Acta*, *1768*, 964–975.

Anson, L. (2009). Membrane protein biophysics. *Nature*, *459*, 343.

Banerjee, P., Buse, J. T., & Dawson, G. (1990). Asymmetric extraction of membrane lipids by CHAPS. *Biochimica et Biophysica Acta*, *1044*, 305–314.

Banerjee, P., Joo, J. B., Buse, J. T., & Dawson, G. (1995). Differential solubilization of lipids along with membrane proteins by different classes of detergents. *Chemistry and Physics of Lipids*, *77*, 65–78.

Bhairi, S. M., & Mohan, C. (2001). *Detergents—A guide to the properties and uses of detergents in biological systems*. San Diego, CA: Calbiochem-Novabiochem.

Burger, K., Gimpl, G., & Fahrenholz, F. (2000). Regulation of receptor function by cholesterol. *Cellular and Molecular Life Sciences*, *57*, 1577–1592.

Celada, P., Bortolozzi, A., & Artigas, F. (2013). Serotonin 5-HT_{1A} receptors as targets for agents to treat psychiatric disorders: Rationale and current status of research. *CNS Drugs*, *27*, 703–716.

Chattopadhyay, A. (2014). GPCRs: Lipid-dependent membrane receptors that act as drug targets. *Advances in Biology*, *2014*, 143023.

Chattopadhyay, A., & Harikumar, K. G. (1996). Dependence of critical micelle concentration of a zwitterionic detergent on ionic strength: Implications in receptor solubilization. *FEBS Letters*, *391*, 199–202.

Chattopadhyay, A., Harikumar, K. G., & Kalipatnapu, S. (2002). Solubilization of high affinity G-protein-coupled serotonin$_{1A}$ receptors from bovine hippocampus using pre-micellar CHAPS at low concentration. *Molecular Membrane Biology, 19*, 211–220.

Chattopadhyay, A., & Jafurulla, M. (2012). Role of membrane cholesterol in leishmanial infection. *Advances in Experimental Medicine and Biology, 749*, 201–213.

Chattopadhyay, A., Jafurulla, M., Kalipatnapu, S., Pucadyil, T. J., & Harikumar, K. G. (2005). Role of cholesterol in ligand binding and G-protein coupling of serotonin$_{1A}$ receptors solubilized from bovine hippocampus. *Biochemical and Biophysical Research Communications, 327*, 1036–1041.

Chattopadhyay, A., Paila, Y. D., Jafurulla, M., Chaudhuri, A., Singh, P., Murty, M. R. V. S., et al. (2007). Differential effects of cholesterol and 7-dehydrocholesterol on ligand binding of solubilized hippocampal serotonin$_{1A}$ receptors: Implications in SLOS. *Biochemical and Biophysical Research Communications, 363*, 800–805.

Chaudhuri, A., & Chattopadhyay, A. (2011). Transbilayer organization of membrane cholesterol at low concentrations: Implications in health and disease. *Biochimica et Biophysica Acta, 1808*, 19–25.

Cherezov, V., Rosenbaum, D. M., Hanson, M. A., Rasmussen, S. G. F., Thian, F. S., Kobilka, T. S., et al. (2007). High-resolution crystal structure of an engineered human β_2-adrenergic G protein-coupled receptor. *Science, 318*, 1258–1265.

Cladera, J., Rigaud, J.-L., Villaverde, J., & Duñach, M. (1997). Liposome solubilization and membrane protein reconstitution using Chaps and Chapso. *European Journal of Biochemistry, 243*, 798–804.

Demoliou-Mason, C. D., & Barnard, E. A. (1984). Solubilization in high yield of opioid receptors retaining high-affinity delta, mu and kappa binding sites. *FEBS Letters, 170*, 378–382.

Deupi, X., & Kobilka, B. K. (2010). Energy landscapes as a tool to integrate GPCR structure, dynamics, and function. *Physiology (Bethesda), 25*, 293–303.

Duquesne, K., & Sturgis, J. N. (2010). Membrane protein solubilization. *Methods in Molecular Biology, 601*, 205–217.

Ellis, C., & The Nature Reviews Drug Discovery GPCR Questionnaire Participants (2004). The state of GPCR research in 2004. *Nature Reviews Drug Discovery, 3*, 577–626.

Garavito, R. M., & Ferguson-Miller, S. (2001). Detergents as tools in membrane biochemistry. *The Journal of Biological Chemistry, 276*, 32403–32406.

Heilker, R., Wolff, M., Tautermann, C. S., & Bieler, M. (2009). G-protein-coupled receptor-focused drug discovery using a target class platform approach. *Drug Discovery Today, 14*, 231–240.

Helenius, A., & Simons, K. (1975). Solubilization of membranes by detergents. *Biochimica et Biophysica Acta, 415*, 29–79.

Hjelmeland, L. M. (1980). A nondenaturing zwitterionic detergent for membrane biochemistry: Design and synthesis. *Proceedings of the National Academy of Sciences of the United States of America, 77*, 6368–6370.

Hjelmeland, L. M., & Chrambach, A. (1984). Solubilization of functional membrane-bound receptors. In J. C. Venter & L. C. Harrison (Eds.), *Membranes, detergents, and receptor solubilization* (pp. 35–46). New York: Alan R. Liss.

Huang, J., Chen, S., Zhang, J. J., & Huang, X.-Y. (2013). Crystal structure of oligomeric β_1-adrenergic G protein-coupled receptors in ligand-free basal state. *Nature Structural & Molecular Biology, 20*, 419–425.

Inagaki, S., Ghirlando, R., White, J. F., Gvozdenovic-Jeremic, J., Northup, J. K., & Grisshammer, R. (2012). Modulation of the interaction between neurotensin receptor NTS1 and Gq protein by lipid. *Journal of Molecular Biology, 417*, 95–111.

Insel, P. A., Tang, C.-M., Hahntow, I., & Michel, M. C. (2007). Impact of GPCRs in clinical medicine: Monogenic diseases, genetic variants and drug targets. *Biochimica et Biophysica Acta, 1768*, 994–1005.

Jacoby, E., Bouhelal, R., Gerspacher, M., & Seuwen, K. (2006). The 7TM G-protein-coupled receptor target family. *ChemMedChem*, *1*, 760–782.

Jafurulla, M., & Chattopadhyay, A. (2013a). Membrane lipids in the function of serotonin and adrenergic receptors. *Current Medicinal Chemistry*, *20*, 47–55.

Jafurulla, M., & Chattopadhyay, A. (2013b). Application of quantitative fluorescence microscopic approaches to monitor organization and dynamics of the serotonin$_{1A}$ receptor. In Y. Mely & G. Duportail (Eds.), Fluorescent Methods to Study Biological Membranes. *Springer series on fluorescence: Vol. 13*. (pp. 417–437). Heidelberg: Springer.

Jafurulla, M., Rao, B. D., Sreedevi, S., Ruysschaert, J.-M., Covey, D. F., & Chattopadhyay, A. (2014). Stereospecific requirement of cholesterol in the function of the serotonin$_{1A}$ receptor. *Biochimica et Biophysica Acta*, *1838*, 158–163.

Jones, O. T., Earnest, J. P., & McNamee, M. G. (1987). Solubilization and reconstitution of membrane proteins. In J. B. C. Findlay & W. H. Evans (Eds.), *Biological membranes: A practical approach* (pp. 139–177). Oxford: IRL press.

Jones, O. T., Eubanks, J. H., Earnest, J. P., & McNamee, M. G. (1988). A minimum number of lipids are required to support the functional properties of the nicotinic acetylcholine receptor. *Biochemistry*, *27*, 3733–3742.

Kalipatnapu, S., & Chattopadhyay, A. (2005). Membrane protein solubilization: Recent advances and challenges in solubilization of serotonin$_{1A}$ receptors. *IUBMB Life*, *57*, 505–512.

Kalipatnapu, S., & Chattopadhyay, A. (2007). Membrane organization and function of the serotonin$_{1A}$ receptor. *Cellular and Molecular Neurobiology*, *27*, 1097–1116.

Kirilovsky, J., & Schramm, M. (1983). Delipidation of a β-adrenergic receptor preparation and reconstitution by specific lipids. *The Journal of Biological Chemistry*, *258*, 6841–6849.

Kline, T., Park, H., & Meyerson, L. R. (1989). CHAPS solubilization of a G-protein sensitive 5-HT$_{1A}$ receptor from bovine hippocampus. *Life Sciences*, *45*, 1997–2005.

Kubicek, J., Block, H., Maertens, B., Spriestersbach, A., & Labahn, J. (2014). Expression and purification of membrane proteins. *Methods in Enzymology*, *541*, 117–140.

le Maire, M., Champeil, P., & Møller, J. V. (2000). Interaction of membrane proteins and lipids with solubilizing detergents. *Biochimica et Biophysica Acta*, *1508*, 86–111.

Lee, A. G. (2003). Lipid-protein interactions in biological membranes: A structural perspective. *Biochimica et Biophysica Acta*, *1612*, 1–40.

Lee, A. G. (2011). Biological membranes: The importance of molecular detail. *Trends in Biochemical Sciences*, *36*, 493–500.

Lingwood, D., & Simons, K. (2010). Lipid rafts as a membrane-organizing principle. *Science*, *327*, 46–50.

Liu, W., Chun, E., Thompson, A. A., Chubukov, P., Xu, F., Katritch, V., et al. (2012). Structural basis for allosteric regulation of GPCRs by sodium ions. *Science*, *337*, 232–236.

Locatelli-Hoops, S. C., Gorshkova, I., Gawrisch, K., & Yeliseev, A. A. (2013). Expression, surface immobilization, and characterization of functional recombinant cannabinoid receptor CB$_2$. *Biochimica et Biophysica Acta*, *1834*, 2045–2056.

Madden, T. D. (1986). Current concepts in membrane protein reconstitution. *Chemistry and Physics of Lipids*, *40*, 207–222.

Manglik, A., Kruse, A. C., Kobilka, T. S., Thian, F. S., Mathiesen, J. M., Sunahara, R. K., et al. (2012). Crystal structure of the μ-opioid receptor bound to a morphinan antagonist. *Nature*, *485*, 321–326.

Mouritsen, O. G., & Zuckermann, M. J. (2004). What's so special about cholesterol? *Lipids*, *39*, 1101–1113.

Mukherjee, S., & Maxfield, F. R. (2004). Membrane domains. *Annual Review of Cell and Developmental Biology*, *20*, 839–866.

Müller, C. P., Carey, R. J., Huston, J. P., & De Souza Silva, M. A. (2007). Serotonin and psychostimulant addiction: Focus on 5-HT$_{1A}$-receptors. *Progress in Neurobiology*, *81*, 133–178.

Neugebauer, J. M. (1990). Detergents: An overview. *Methods in Enzymology, 182*, 239–253.

Nygaard, R., Zou, Y., Dror, R. O., Mildorf, T. J., Arlow, D. H., Manglik, A., et al. (2013). The dynamic process of β_2-adrenergic receptor activation. *Cell, 152*, 532–542.

Oates, J., & Watts, A. (2011). Uncovering the intimate relationship between lipids, cholesterol and GPCR activation. *Current Opinion in Structural Biology, 21*, 802–807.

Paila, Y. D., & Chattopadhyay, A. (2009). The function of G-protein coupled receptors and membrane cholesterol: Specific or general interaction? *Glycoconjugate Journal, 26*, 711–720.

Paila, Y. D., & Chattopadhyay, A. (2010). Membrane cholesterol in the function and organization of G-protein coupled receptors. *Subcellular Biochemistry, 51*, 439–466.

Perez, D. M. (2003). The evolutionarily triumphant G-protein-coupled receptor. *Molecular Pharmacology, 63*, 1202–1205.

Pierce, K. L., Premont, R. T., & Lefkowitz, R. J. (2002). Seven-transmembrane receptors. *Nature Reviews Molecular Cell Biology, 3*, 639–650.

Porter, F. D., & Herman, G. E. (2011). Malformation syndromes caused by disorders of cholesterol synthesis. *Journal of Lipid Research, 52*, 6–34.

Privé, G. G. (2007). Detergents for the stabilization and crystallization of membrane proteins. *Methods, 41*, 388–397.

Pucadyil, T. J., & Chattopadhyay, A. (2004). Exploring detergent insolubility in bovine hippocampal membranes: A critical assessment of the requirement for cholesterol. *Biochimica et Biophysica Acta, 1661*, 9–17.

Pucadyil, T. J., & Chattopadhyay, A. (2006). Role of cholesterol in the function and organization of G-protein coupled receptors. *Progress in Lipid Research, 45*, 295–333.

Pucadyil, T. J., & Chattopadhyay, A. (2007). Cholesterol: A potential therapeutic target in *Leishmania* infection? *Trends in Parasitology, 23*, 49–53.

Pucadyil, T. J., Kalipatnapu, S., & Chattopadhyay, A. (2005). The serotonin$_{1A}$ receptor: A representative member of the serotonin receptor family. *Cellular and Molecular Neurobiology, 25*, 553–580.

Rasmussen, S. G. F., DeVree, B. T., Zou, Y., Kruse, A. C., Chung, K. Y., Kobilka, T. S., et al. (2011). Crystal structure of the β_2 adrenergic receptor-Gs protein complex. *Nature, 477*, 549–555.

Rivnay, B., & Metzger, H. (1982). Reconstitution of the receptor for immunoglobulin E into liposomes. Conditions for incorporation of the receptor into vesicles. *The Journal of Biological Chemistry, 257*, 12800–12808.

Rosenbaum, D. M., Rasmussen, S. G. F., & Kobilka, B. K. (2009). The structure and function of G-protein-coupled receptors. *Nature, 459*, 356–363.

Roy, S., Kumar, G. A., Jafurulla, M., Mandal, C., & Chattopadhyay, A. (2014). Integrity of the actin cytoskeleton of host macrophages is essential for *Leishmania donovani* infection. *Biochimica et Biophysica Acta, 1838*, 2011–2018.

Savitz, J., Lucki, I., & Drevets, W. C. (2009). 5-HT$_{1A}$ receptor function in major depressive disorder. *Progress in Neurobiology, 88*, 17–31.

Schlyer, S., & Horuk, R. (2006). I want a new drug: G-protein-coupled receptors in drug development. *Drug Discovery Today, 11*, 481–493.

Schöneberg, T., Hofreiter, M., Schulz, A., & Römpler, H. (2007). Learning from the past: Evolution of GPCR functions. *Trends in Pharmacological Sciences, 28*, 117–121.

Schürholz, T. (1996). Critical dependence of the solubilization of lipid vesicles by the detergent CHAPS on the lipid composition. Functional reconstitution of the nicotinic acetylcholine receptor into preformed vesicles above the critical micellization concentration. *Biophysical Chemistry, 58*, 87–96.

Seddon, A. M., Curnow, P., & Booth, P. J. (2004). Membrane proteins, lipids and detergents: Not just a soap opera. *Biochimica et Biophysica Acta, 1666*, 105–117.

Simons, K., & Ikonen, E. (2000). How cells handle cholesterol. *Science, 290*, 1721–1726.

Simons, K., & Toomre, D. (2000). Lipid rafts and signal transduction. *Nature Reviews Molecular Cell Biology*, *1*, 31–39.

Simons, K., & van Meer, G. (1988). Lipid sorting in epithelial cells. *Biochemistry*, *27*, 6197–6202.

Singh, P., Jafurulla, M., Paila, Y. D., & Chattopadhyay, A. (2011). Desmosterol replaces cholesterol for ligand binding function of the serotonin$_{1A}$ receptor in solubilized hippocampal membranes: Support for nonannular binding sites for cholesterol? *Biochimica et Biophysica Acta*, *1808*, 2428–2434.

Slinde, E., & Flatmark, T. (1976). Effect of the hydrophile-lipophile balance of non-ionic detergents (Triton X-series) on the solubilization of biological membranes and their integral *b*-type cytochromes. *Biochimica et Biophysica Acta*, *455*, 796–805.

Soubias, O., & Gawrisch, K. (2012). The role of the lipid matrix for structure and function of the GPCR rhodopsin. *Biochimica et Biophysica Acta*, *1818*, 234–240.

Talmont, F., Moulédous, L., Mollereau, C., & Zajac, J.-M. (2014). Solubilization and reconstitution of the mu-opioid receptor expressed in human neuronal SH-SY5Y and CHO cells. *Peptides*, *55*, 79–84.

Tanford, C. (1978). The hydrophobic effect and the organization of living matter. *Science*, *200*, 1012–1018.

Tate, C. G. (2012). A crystal clear solution for determining G-protein-coupled receptor structures. *Trends in Biochemical Sciences*, *37*, 343–352.

Thomsen, W., Frazer, J., & Unett, D. (2005). Functional assays for screening GPCR targets. *Current Opinion in Biotechnology*, *16*, 655–665.

Vukoti, K., Kimura, T., Macke, L., Gawrisch, K., & Yeliseev, A. (2012). Stabilization of functional recombinant cannabinoid receptor CB_2 in detergent micelles and lipid bilayers. *PLoS One*, *7*, e46290.

Wacker, D., Wang, C., Katritch, V., Han, G. W., Huang, X.-P., Vardy, E., et al. (2013). Structural features for functional selectivity at serotonin receptors. *Science*, *340*, 615–619.

Wang, C., Jiang, Y., Ma, J., Wu, H., Wacker, D., Katritch, V., et al. (2013). Structural basis for molecular recognition at serotonin receptors. *Science*, *340*, 610–614.

White, J. F., Noinaj, N., Shibata, Y., Love, J., Kloss, B., Xu, F., et al. (2012). Structure of the agonist-bound neurotensin receptor. *Nature*, *490*, 508–513.

Wu, H., Wacker, D., Mileni, M., Katritch, V., Han, G. W., Vardy, E., et al. (2012). Structure of the human κ-opioid receptor in complex with JDTic. *Nature*, *485*, 327–332.

Wu, H., Wang, C., Gregory, K. J., Han, G. W., Cho, H. P., Xia, Y., et al. (2014). Structure of a class C GPCR metabotropic glutamate receptor 1 bound to an allosteric modulator. *Science*, *344*, 58–64.

Xu, X., & London, E. (2000). The effect of sterol structure on membrane lipid domains reveals how cholesterol can induce lipid domain formation. *Biochemistry*, *39*, 843–849.

Zhang, Y., DeVries, M. E., & Skolnick, J. (2006). Structure modeling of all identified G protein-coupled receptors in the human genome. *PLoS Computational Biology*, *2*, e13.

CHAPTER SEVEN

Overexpression, Isolation, Purification, and Crystallization of NhaA

Etana Padan[1], Manish Dwivedi

Department of Biological Chemistry, Alexander Silberman Institute of Life Sciences, Hebrew University of Jerusalem, Jerusalem, Israel

[1]Corresponding author: e-mail address: etana@vms.huji.ac.il

Contents

Abstract

Living cells are critically dependent on processes that regulate intracellular pH, Na^+ content, and volume. Na^+/H^+ antiporters play a primary role in these homeostatic mechanisms. They are found in the cytoplasmic and intracellular membranes of most organisms from bacteria to humans and have long been human drug targets.

Methods in Enzymology, Volume 557
ISSN 0076-6879
http://dx.doi.org/10.1016/bs.mie.2014.12.003

NhaA, the principal Na^+/H^+ antiporter in *Escherichia coli*, plays an essential role in homeostasis of Na^+ and H^+. It constitutes a paradigm for the study of its numerous prokaryotic homologs and of several human Na^+/H^+ antiporters. The crystal structure of NhaA, determined at pH 4, has provided the first structural and functional insights into the antiport mechanism and pH regulation of an Na^+/H^+ antiporter. Remarkably, the NhaA structure revealed a new and unique fold (the "NhaA fold") that has since been observed in four additional bacterial secondary transporters. The NhaA structure has facilitated the rational interpretation of mutational data obtained in NhaA, revealing the antiporter's functional organization. Nevertheless, the crystal structure is a single snapshot, determined at acidic pH, when NhaA is downregulated; NhaA is activated at pH 6.5 and reaches maximal activity at pH 8.5. Therefore, it is crucial to crystallize the active conformations of NhaA. Herein, we present a procedure for determining the structure of NhaA.

1. INTRODUCTION

Na^+/H^+ antiporters are instrumental to cell homeostasis, participating in several critical mechanisms, including regulation of intracellular pH, Na^+ content, and volume (see Padan, 2014; Padan, Kozachkov, Herz, & Rimon, 2009 for recent reviews). They are found in the cytoplasmic and intracellular membranes of most organisms from bacteria to humans (Orlowski & Grinstein, 2007; Padan et al., 2004) and are well-known human drug targets (Fliegel, 2008). Determination of the crystal structure of NhaA, the principal Na^+/H^+ antiporter in *Escherichia coli* (crystallized at acidic pH; Hunte et al., 2005), has constituted an important milestone in the study of these antiporters, elucidating key aspects of the structure and function of their antiport and pH regulation mechanisms (Padan, 2008).

NhaA appears in numerous types of enterobacteria (Padan, Venturi, Gerchman, & Dover, 2001) and has orthologs throughout the biological kingdoms, including humans (Brett, Donowitz, & Rao, 2005). It is largely responsible for the ability of bacteria to adapt to high salinity, challenge Li^+ toxicity, and grow at alkaline pH (in the presence of Na^+; Padan, Bibi, Masahiro, & Krulwich, 2005). These functions are facilitated by the antiporter's unique biochemical properties, which include very high turnover (10^5 min^{-1}; Taglicht, Padan, & Schuldiner, 1991), electrogenicity with a stoichiometry of $2H^+/Na^+$ (Taglicht, Padan, & Schuldiner, 1993), and strong pH dependence (Taglicht et al., 1991), a property it shares with other prokaryotic (Padan et al., 2005) as well as eukaryotic Na^+/H^+ antiporters (reviewed in Orlowski & Grinstein, 2004, 2007; Putney, Denker, & Barber, 2002; Wakabayashi, Hisamitsu, Pang, & Shigekawa, 2003).

The crystal structure of NhaA crystallized at acidic pH (Hunte et al., 2005) has provided the first structural insights into the antiport mechanism and pH regulation of a Na^+/H^+ antiporter (Padan, 2008). NhaA consists of 12 transmembrane helices (TMs), whose N and C termini are oriented to the cytoplasm. The TMs are organized in a unique fold: TMs III, IV, and V are topologically inverted with respect to TMs X, XI, and XII. In each repeat, one TM (IV and XI, respectively) is interrupted by an extended chain, and these chains cross each other in the middle of the membrane. This creates a structure in which two short helices are oriented toward the cytoplasm (c) (IVc and XIc) and two are oriented toward the periplasm (p) (IVp and XIp; Hunte et al., 2005; Fig. 1A). The N- or C-terminal partial dipoles of the short helices face each other. This noncanonical TM assembly creates a delicately balanced electrostatic environment in the middle of the membrane at the ion-binding site(s), which probably plays a critical role in the cation exchange activity of the antiporter.

Remarkably, since the determination of the NhaA structure, additional bacterial secondary transporters have been shown to share the so-called "NhaA fold": NapA of *Thermus thermophilus* (Lee, Kang, et al., 2013); ASBT, which is found in *Neisseria meningitidis* (Hu, Iwata, Cameron, &

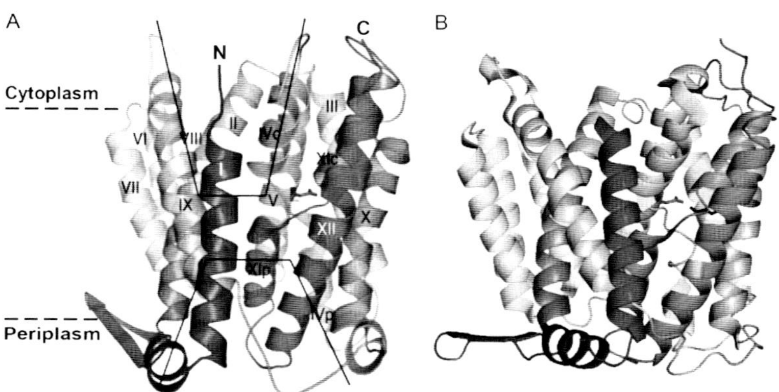

Figure 1 The crystal structure of NhaA (A). NhaA was originally crystallized in 2005 (Hunte et al., 2005). (B) In 2007, NhaA was crystallized again, with very similar results, but with different assignment of TM X (Lee, Yashiro, et al. in PDB:4AU5, 2014). It is not yet known whether TM X was misassigned in the original structure or whether the second structure represents a different conformation. The figures were illustrated using Pymol (rainbow, ribbon presentation). In (A), black lines indicate funnels, and the TMs are marked with roman numerals. (See the color plate.)

Drew, 2011) and *Yersinia frederiksenii* (Zhou et al., 2014), is a bacterial homologue of the human cholic acid transporter and NhaP1 of *Methanocaldococcus jannaschii* (Paulino & Kuhlbrandt, 2014).Overall, there are three different types of folds that are known to characterize secondary transporters: LacY, LeuT, and NhaA (Shi, 2013).

The NhaA structure has enabled mutational data to be interpreted in a rational way (Padan, 2008), revealing the functional organization of NhaA: (a) a cluster of amino acyl side chains, about 9 Å away from the catalytic site, is proposed to modulate the response to pH, and (b) a catalytic region contains the ion-binding sites. Electrophysiological analysis supports these propositions (Mager, 2013).

Although the crystal structure of NhaA has provided valuable insights, it is only a single snapshot. Moreover, it was determined at pH 4, at which NhaA is downregulated (Hunte et al., 2005); NhaA is activated at pH 6.5 and reaches maximal activity at pH 8.5 (Taglicht et al., 1991). Therefore, great efforts are currently being invested in crystallizing an active conformation of NhaA and in studying its functional dynamics both *in vitro* and *in vivo* at physiological pH (Kozachkov & Padan, 2011; Mager, Rimon, Padan, & Fendler, 2011; Rimon, Kozachkov-Magrisso, & Padan, 2012). Herein, we present a procedure for determining the structure of NhaA. The composition of the solutions used is provided in detail in Section 8.

2. OVEREXPRESSION

2.1 Plasmid encoding His-tagged NhaA

In general, transporters, like many other membrane proteins, are present in small quantities in the membrane. NhaA proteins make up about 0.1–0.2% (Taglicht et al., 1991) of the proteins in the *E. coli* membrane. Therefore, it is necessary to overexpress NhaA in order to obtain a sufficiently high concentration for crystallography. However, overexpression of NhaA—like the overexpression of other membrane proteins—harms the membrane and kills the cells (Taglicht et al., 1991). To circumvent this problem, it is preferable to use a plasmid that overexpresses the protein under strict transcriptional control. With overexpression under transcriptional control, the proportion of NhaA proteins (among the proteins in the membrane) can reach about 10% (Taglicht et al., 1991).

The plasmid constructed for overexpression of NhaA is called pAXH (previously called pYG10; Olami, Rimon, Gerchman, Rothman, & Padan, 1997). It is a pET20b (Novagen) derivative (ampicillin-resistant), which uses the tac promoter to encode NhaA, fused in frame in its C-terminus to two factor Xa cleavage sites followed by six histidines (His-tagged NhaA, for further details see Olami et al., 1997). However, because the tac promoter is leaky and pAXH is a multicopy plasmid, the NhaA produced in the TA15 *E. coli* strain (Table 1) also kills the cells. Therefore, for successful overexpression, the LacYsuper-repressor I^Q is introduced into the host cells either on another compatible plasmid pI^Q or on the bacterial chromosome; this process yields the TA16 *E. coli* strain (Taglicht et al., 1991). The plasmid pI^Q is a pACYC184 derivative carrying $lacI^Q$ encoding the lac repressor I^Q, needed for controlled induction by IPTG (isopropyl-β-D-thiogalactopyranoside), and is tetracycline resistant.

2.2 Growth of *E. coli* for production of His-tagged NhaA

The *E. coli* strains used for His-tagged NhaA overexpression are given in Table 1. RK20/pAXH/pI^Q (Padan et al., 1989; Screpanti et al., 2006) is the best strain because it is completely deleted of chromosomal NhaA and carries two compatible plasmids, one expressing His-tagged NhaA and the other the super-repressor of the tac promoter (I^Q).

For strain storage, grow the cells in modified L broth (LBK) in which NaCl has been replaced with KCl (67 m*M*, pH 7.5; Padan et al., 1989b), prepare glycerol stocks by adding 20% glycerol, and store at −80 °C. For large-scale production in a 10-l fermenter, prepare minimal medium A without sodium citrate (Davies & Mingioli, 1950; Taglicht et al., 1991; see Section 8), supplemented with 0.5% glycerol, 0.01% $MgSO_4$, and

Table 1 Bacterial strains

Bacterial strain	Genotype	Reference
TA15	*mel*BLid, Δ*lac*ZY	Goldberg et al. (1987)
TA16	*mel*BLid, Δ*lac*ZY, *lac*I^Q	Taglicht et al. (1991)
EP432	*mel*BLid, Δ*nha*A1::kan, Δ*nha*B1::cat, Δ*lac*ZY, *thr*1	Pinner, Kotler, Padan, and Schuldiner (1993)
RK20	*mel*BLid, Δ*nha*A2::kan, Δ*lac*ZY, *thr*1	Padan, Maisler, Taglicht, Karpel, and Schuldiner (1989a) and Screpanti, Padan, Rimon, Michel, and Hunte (2006)

All strains are K12 derivatives. The respective antibiotic concentrations are as follows: 100 μg/ml ampicillin and/or 50 μg/ml kanamycin and/or 6.5 μg/ml tetracycline, 20 μg/ml chloramphenicol. RK20 bears a complete deletion of NhaA and therefore is used in His-tagged NhaA overexpression for crystallization (Screpanti et al., 2006).

thiamine (2.5 μg/ml). For growth of strains EP432 and RK20, also add threonine (100 μg/ml).

The growth procedure is a slight modification of that presented in Venturi and Padan (2003). Prepare a small preculture by directly inoculating 5 ml LBK with 20 μl of the glycerol stock and incubate at 37 °C (250 r.p.m.) overnight. Centrifuge (15,000 × *g*) the cells and wash once in minimal medium A. Then, prepare a large preculture by inoculating 500 ml of minimal medium with the washed cells and incubate (200 r.p.m.) overnight at 30 °C. Inoculate the large preculture into 10 l prewarmed minimal medium A and incubate at 37 °C (stirring at 470–480 r.p.m. with continuous aeration under pressurized air stream (10 l/h)) and grow to OD_{600nm} of 0.7–0.8. Induce by addition of 1 m*M* IPTG and further incubate for 1.5–2 h for overexpression. Stop incubation if the OD decreases, which indicate cells lysis. Chill the cells rapidly on ice and harvest by centrifugation (14,000 × *g*) in weighted test tubes. Wash the pellet once with 500 ml TSC (see Section 8) and weigh again to evaluate the pellet wet weight. The pellet can be stored overnight at 4 °C or flash frozen in liquid nitrogen and stored longer at −80 °C. Cells for overexpression of NhaA can also be grown in 2 l flasks, 0.5 l medium per flask, and incubate at 37 °C (250 r.p.m.) until the OD is reached for induction. Then, the procedure is continued as elaborated in the preceding text.

3. ISOLATION OF MEMBRANES FROM STRAIN EXPRESSING HIS-TAGGED NhaA

The remainder of the procedure outlined herein must be conducted at 4 °C (cold room or on ice). Resuspend the washed cells in TSC, 5 volumes per 1 g wet weight. Before breakage, supplement the cell suspension with the additives for cell disruption (see Section 8). Disrupt the cells three times in a French-press cell at 20,000 psi or in a Microfluidizer® Processor (M-110EHI, Microfluidics) at 14,000 psi. In a successful disruption, the color of the suspension turns from milky white to brownish. The DNase rapidly digests the DNA, which otherwise would cause the extract to be highly viscous. Remove the debris and intact cells by centrifugation (14,000 × *g*). Collect the membrane fraction by ultracentrifugation (260,000 × *g*). Resuspend the pellet containing the membranes in 10–20 ml TSC, using a 10-ml syringe with a blunt needle (No. 18). Centrifuge and wash again as above. Determine the membrane protein concentration using the Bradford assay (Bradford, 1976). With TSC, adjust the membrane protein concentration to 10 mg protein/ml. Make 9 ml aliquots in 15-ml Falcon tubes, flash freeze in liquid nitrogen, and store at −80 °C. From 10 l cells, about 50 ml of membrane solution (10 mg membrane protein/ml) is obtained.

4. AFFINITY PURIFICATION OF HIS-TAGGED NhaA

4.1 Membrane solubilization

As noted earlier, this step of the procedure must be conducted at 4 °C. The procedure given here is for 9 ml membranes (10 mg membrane protein/ml) derived from TA16/pAXH. This preparation yields about 5 mg pure His-tagged NhaA. The protocol can easily be scaled up or down.

Defrost 9 ml aliquot of membrane suspension. Solubilize the membranes by mixing with 18 ml solubilizing buffer (see Section 8). Stir gently at 4 °C for 1 h. Ultracentrifuge (260,000 × *g*, 1 h, 4 °C) and collect the supernatant, which contains NhaA.

4.2 Immobilized metal affinity chromatography purification of His-tagged NhaA

Immobilized metal affinity chromatography purification (IMAC) is used to purify His-tagged NhaA. Prepare 5-ml Ni^{2+}-NTA affinity resin (Qiagen) in batch by washing twice in 80 ml ddH_2O and twice in 150 ml binding buffer. Get rid of the extra fluid, mix the solution containing NhaA with the prepared resin, and mix gently (4 °C, 1 h). Transfer the mixture into a column (Bio-Rad) and allow to pack by gravity. Connect to a fraction collector with controlled flow rate. Wash with binding buffer (120 ml, 1.4 ml/min) and then with washing buffer (120 ml, 2.2 ml/min). Exchange the detergent using 150 ml exchange buffer (1.4 ml/min). In this step, the detergent, 0.015% DDM (*n*-dodecyl-β-D-maltopyranoside), is exchanged for 0.015% α-DDM (*n*-dodecyl-α-D-maltopyranoside; or another detergent for detergent exchange screening). Elute with about 50 ml elution buffer (2.2 ml/min) but collect only the peak (about 7–10 ml). Add sucrose immediately (80%) to obtain a final concentration of 20% sucrose. Dialyze the eluate for 1 h in 1 l dialysis buffer and then overnight in a fresh 1 l of buffer. Ultracentrifuge (260,000 × *g*) and determine NhaA concentration using an empirical extinction coefficient: 1.56 OD_{280nm} =1 mg NhaA protein. This procedure yields about 4–5 mg NhaA per liter of cell culture. The protein is virtually pure, stable, and homogenous (Screpanti et al., 2006).

Concentrate the protein (Centriprep YM-50 or Centricon MWCO 50,000;EMD Millipore) to about 12 mg/ml, supplemented with 20% sucrose. Freeze in liquid nitrogen and store at −80 °C. It is important to know the final detergent concentration and to ensure that it is reproducible. Therefore, the elution volume is kept constant and as small as possible, enabling the detergent concentration to be estimated after concentration of the protein.

Before crystallization, use mini-gel filtration through a bio-spin column (Bio-gel P-6, Bio-Rad) to further decrease the extra detergent concentration and ultracentrifuge (260,000 × *g*, 4 °C, 1 h).

5. ANALYSIS OF PROTEIN QUALITY

5.1 SDS-PAGE

The matrix gel is composed of a polyacrylamide gradient of 4–12%, in a Bis-Tris System buffered at pH 6.4. The running buffer composition is 50 m*M* 2-(*N*-morpholino) ethane sulfonic acid (MES), 50 m*M* Tris base, 3.5 m*M* sodium dodecyl sulfate (SDS), 1 m*M* ethylenediaminetetraacetic acid (EDTA), pH 7.3.

5.2 Blue native gel electrophoresis

Blue native gel electrophoresis is carried out as described in Rimon, Tzubery, and Padan (2007). The main gel and the overlayer are made of 10%and 4% polyacrylamide, respectively, in 0.015% DDM, 2.5 m*M* imidazole/HCl, pH 7, 0.5 *M* 6-aminocaproic acid, and main gel 10% glycerol. The sample buffer contains 50 m*M* NaCl, 1 m*M* Na/EDTA, 50 m*M* imidazole/HCl (pH 7), 0.015% DDM, 5%glycerol, and 5 m*M* aminocaproic acid. The cathode buffer contains 50 m*M* Tricine/NaOH, pH 7, 7.5 m*M* imidazole, pH 7, Coomassie blue 0.02%(G-250, Merck), and 0.015% DDM. The anode buffer contains 25 m*M* imidazole, pH 7, and 0.015% DDM. The electrophoresis is conducted at 15 mA for 1 h. The gel is stained by Coomassie Blue or silver staining and is dried, and the band densities are determined.

5.3 Gel filtration

For each analysis, 50 μg of total protein is loaded onto a Superose 6 PC 3.2/3.0 column, on a SMART™ system (Pharmacia). The column is equilibrated in the same buffer as that used for the protein. The flow rate is 40 μl/min. Analytical gel filtration is also carried out on an Äktasystem (GE Healthcare Life Sciences) with a Superose 12 column equilibrated in the same buffer as for the protein but without sucrose. The flow rate is 40 μl/min.

6. CRYSTALLIZATION OF HIS-TAGGED NhaA

Before crystallization, ultracentrifuge (260,000 × *g*) the concentrated protein solution. Crystals are grown with the hanging drop vapor diffusion technique at 6 °C. One-ml reservoir solution is used in each well of the 24-well VDXm Plate Greased™ (Hampton Research). The drop is prepared by mixing the reservoir solution and protein solution (4–6 mg/ml) in equal volumes (2 μl + 2 μl). For details see Screpanti et al. (2006).

7. FUNCTIONAL ASSAY

It is crucial to determine the functionality (ligand binding and antiporter activity) of the protein used for crystallization.

7.1 Ligand binding

Isothermal titration calorimetry (ITC) is used to determine binding of Li^+ to purified NhaA in detergent micelles (Maes, Rimon, Kozachkov-Magrisso, Friedler, & Padan, 2012). The IMAC-purified protein is eluted in a buffer contained 300 m*M* imidazole, 25 m*M* citric acid, 100 m*M* KCl, 5 m*M* $MgCl_2$, and 0.015% β-DDM (final pH 7.9). After addition of sucrose (10%), the eluted protein solution is dialyzed overnight at 4 °C in acidic elution buffer containing 10% sucrose, 100 m*M* KCl, 25 m*M* citric acid, 5 m*M* $MgCl_2$, and 0.015% β-DDM (pH 4.0) and is frozen at −80 °C.

To prepare the frozen NhaA solution for ITC, melt it (about 0.7 ml containing ~650 μg protein) at 4 °C, concentrating it 2.5-fold by filtration (Amicon Ultra, 30 K), and wash three times with the original volume of the reaction buffer containing 50 m*M* BTP, 150 m*M* choline chloride, 5 m*M* $MgCl_2$, 10% sucrose (pH 8.5). These washing cycles yield a reaction mixture for ITC, containing 40–50 μ*M* of the protein and 0.04% β-DDM. ITC experiments are performed using the micro-calorimeter ITC_{200} (MicroCal, GE Healthcare Life Sciences). The reaction mixture (300 μl) is loaded into the sample cell. For titration, 40 m*M* LiCl (dissolved in the ITC reaction buffer) is loaded into the injection syringe. Before data collection, the system is equilibrated to 10 °C with the stirring speed set to 500 r.p.m. Titration curves for binding Li^+ are initiated by injection of 0.8 μl followed by successive 2 μl injections of the ligand every 200 s. Injections of ligands into reaction buffer without protein, or injection of reaction buffer into reaction buffer, are performed to determine background corrections. The integrated heats from each injection, normalized to the moles of ligand per injection, are fitted to a single-site binding isotherm using ORIGIN 7 (OriginLab). The integrated peak of the first injection is excluded from the fit due to the large errors in the first step.

7.2 Na^+/H^+ antiport activity in proteoliposomes

Na^+/H^+ antiport activity is measured in reconstituted NhaA proteoliposomes according to Taglicht et al. (1991). For NhaA reconstitution in proteoliposomes, add His-tagged NhaA (30–60 μg) in 200 μl dialysis buffer to 200 μl of resuspension buffer in glass vials (12 × 75 mm) and sonicate to clarity (2 × 10 s, see Section 8). This mixture can be frozen in liquid nitrogen and stored at −80 °C.

Otherwise, proceed to reconstitution: prepare 300-μl reconstitution buffer and sonicate to clarity as above in glass tubes. Then add to each 150 μl of the reconstitution buffer 200 μl of the prepared protein (in resuspension buffer) and sonicate again to clarity. To reconstitute the proteoliposomes and load them with NH_4Cl, dilute the mixture into 36 volumes of dilution buffer, stir well for 5 min, and incubate for 20 min at room temperature without stirring. Collect the proteoliposomes by centrifugation (Ti70 Beckman, 100,000 × *g*, 1 h, 4 °C). Resuspend the pellet in 60 μl of the dilution buffer. The proteoliposome suspension can be flash-frozen in liquid nitrogen and stored at −80 °C, where it is stable for at least 6 months. The NH_4Cl-loaded proteoliposomes generate a proton gradient upon dilution into the reaction mixture of the activity assay (see Section 8). For ^{22}Na uptake determination, it is advisable to freeze aliquots of 40–60 μl. Before the assay, resonicate the samples as above. Reconstitution by biobeads is also possible (see Venturi & Padan, 2003).

For transport assay, ammonium-chloride loaded proteoliposomes (20 μl) are diluted into 0.5 ml reaction buffer and incubated at room temperature for 30 s up to 5 min. The transport reaction is stopped by dilution into 3 ml (at 4 °C) washing buffer, is filtered through 0.22 μm filters (EMD Millipore) using a vacuum device, and is washed again. The radioactivity trapped on the filters is measured in a gamma counter. The total radioactivity is measured for data evaluation (typically, 5 μl of transport reaction mixture contains 5000–6000 c.p.m.).

For a negative control (no ΔpH), NH_4Cl is added to the reaction mixture at time 0 to a final concentration of 150 m*M* instead of choline chloride.

8. PREPARATION OF SOLUTIONS

8.1 Medium A

60 m*M* K_2HPO_4, 33 m*M* KH_2PO_4, 7.5 m*M* $(NH_4)_2SO_4$, 0.01% $MgSO_4.7H_2O$, 0.5% glycerol, 2.5 μg/ml thiamine, 100 μg/ml threonine if needed.

8.2 Membrane isolation and His-tagged NhaA purification

TSC 10 m*M* Tris/HCl, pH 7.5, 250 m*M* sucrose, 15 m*M* choline chloride.

Additives for membranes disruption 1 m*M* DTT (freshly prepared), 2.5 m*M* $MgCl_2$, 1.0 micro g/ml DNAse, and a protease inhibitor cocktail, freshly prepared (1 m*M* PMSF, 1 m*M* aminocaproic acid, 1 m*M* benzamidine hydrochloride).

Affinity purification Membrane solubilization buffer for 9 ml membranes (10 mg membrane protein/ml): 100 m*M* MOPS/KOH, pH 7, 28% glycerol, 1% β-DDM, total volume 28 ml).

Binding buffer 20 m*M* Tris, pH 7.9, 500 m*M* NaCl, 5 m*M* imidazole, pH 7.9, 0.015% β-DDM.

Washing buffer 20 m*M* Tris is pH 7.9, 500 m*M* NaCl, 30 m*M* imidazole, pH 7.9, 0.015% β-DDM.

Exchange buffer 20 m*M* Bis-Tris, pH 6.5, 100 m*M* KCl, 5 m*M* $MgCl_2$, 0.015% α-DDM.

Elution buffer *For crystallization*. 25 m*M* MES, pH 6, 500 m*M* KCl, 300 m*M* imidazole, pH 4, 5 m*M* $MgCl_2$, 0.015% α-DDM. *For ligand binding or reconstitution*. 25 m*M* citric acid, 100 m*M* choline chloride, 300 m*M* imidazole, 5 m*M* $MgCl_2$, 10% glycerol, final pH 7,9.

Dialysis buffer (to get rid of the imidazole) *For crystallization*. 25 m*M* MES, pH 6, 100 m*M* KCl, 5 m*M* $MgCl_2$, 20% sucrose, 0.015% α-DDM. *For ligand binding or reconstitution*. 25 m*M* citric acid, 100 m*M* choline chloride, 5 m*M* $MgCl_2$, 10% sucrose, 0.015% α-DDM, final pH 4.

8.3 Reconstitution of NhaA proteoliposomes by detergent dilution

Resuspension buffer 1.2% (W/V) octyl-β-D-glucoside (OG, Sigma), 0.5 mg/ml *E. coli* phospholipids (50% phosphatydilethanolamine, PE, Sigma), 50 m*M* MOPS/KOH (pH 7.0), 32% (v/v) glycerol, 1 m*M* DTT.

Reconstitution buffer 1.2% (W/V) OG, 27 mg/ml *E. coli* phospholipids, 100 m*M* MOPS/KOH (pH 7.0). This solution is prepared and frozen in 150 μl aliquots. Before use, it must be sonicated to clarity (two times, 5 s at maximum power, Branson Sonifier250 (G. Heinemann GmbH)).

Dilution buffer 150 m*M* NH_4Cl, 15 m*M* Tris/Cl (pH 7.5), 1 m*M* DTT.

8.4 Na^+/H^+ exchange activity by NhaA

Reaction mixture. 150 m*M* choline chloride, 10 m*M* Tris/Hepes/HCl (pH 8.6), 2 m*M* $MgS0_4$, 50 μ*M* NaCl, ^{22}Na (0.5–1 μCi/ml). Washing buffer: Same as the reaction buffer but contains 100 instead of 150 μ*M* choline chloride

ACKNOWLEDGMENTS

E.P. thanks the Israel Science Foundation (Grant No. 284/12). M.D. thanks the PBC, Council for Higher Education and the Hebrew University Program for Fellowships for Outstanding Postdoctoral Fellows from China and India (2015). Thanks are also due to Sofia Hollschwandner and Dr. Rimon Abraham for discussions.

REFERENCES

Bradford, M. M. (1976). A rapid and sensitive method for the quantitation of microgram quantities of protein utilizing the principle of protein-dye binding. *Analytical Biochemistry*, *72*, 248–254.

Brett, C. L., Donowitz, M., & Rao, R. (2005). Evolutionary origins of eukaryotic sodium/proton exchangers. *American Journal of Physiology Cell Physiology*, *288*, C223–C239.

Davies, B., & Mingioli, E. (1950). Mutants of *Escherichia Coli* requiring methionine or vitamin B12. *Journal of Bacteriology*, *60*, 17–28.

Fliegel, L. (2008). Molecular biology of the myocardial Na+/H+ exchanger. *Journal of Molecular and Cellular Cardiology*, *44*, 228–237.

Goldberg, E. B., Arbel, T., Chen, J., Karpel, R., Mackie, G. A., Schuldiner, S., et al. (1987). Characterization of a Na^+/H^+ antiporter gene of *Escherichia coli*. *Proceedings of the National Academy of Sciences of the United States of America*, *84*, 2615–2619.

Hu, N. J., Iwata, S., Cameron, A. D., & Drew, D. (2011). Crystal structure of a bacterial homologue of the bile acid sodium symporter ASBT. *Nature*, *478*, 408–411.

Hunte, C., Screpanti, M., Venturi, M., Rimon, A., Padan, E., & Michel, H. (2005). Structure of a Na^+/H^+ antiporter and insights into mechanism of action and regulation by pH. *Nature*, *534*, 1197–1202.

Kozachkov, L., & Padan, E. (2011). Site-directed tryptophan fluorescence reveals two essential conformational changes in the Na+/H+ antiporter NhaA. *Proceedings of the National Academy of Sciences of the United States of America*, *108*, 15769–15774.

Lee, C., Kang, H. J., von Ballmoos, C., Newstead, S., Uzdavinys, P., Dotson, D. L., et al. (2013). A two-domain elevator mechanism for sodium/proton antiport. *Nature*, *501*, 573–577.

Lee, C., Yashiro, S., Dotson, D. L., Uzdavinys, P., Iwata, S., Sansom, M. S., et al. (2014). Crystal structure of the sodium-proton antiporter NhaA dimer and new mechanistic insights. *Journal of General Physiology*, *144*, 529–544.

Maes, M., Rimon, A., Kozachkov-Magrisso, L., Friedler, A., & Padan, E. (2012). Revealing the ligand binding site of NhaA Na+/H+ antiporter and its pH dependence. *The Journal of Biological Chemistry, 287*, 38150–38157.

Mager, T. (2013). Differential effect of mutations on the transport properties of NhaA from *Escherichia coli*. *The Journal of Biological Chemistry, 288*, 24666–24675.

Mager, T., Rimon, A., Padan, E., & Fendler, K. (2011). Transport mechanism and pH regulation of the Na+/H+ antiporter NhaA from Escherichia coli: An electrophysiological study. *The Journal of Biological Chemistry, 286*, 23570–23581.

Olami, Y., Rimon, A., Gerchman, Y., Rothman, A., & Padan, E. (1997). Histidine 225, a residue of the NhaA-Na^+/H^+ antiporter of *Escherichia coli* is exposed and faces the cell exterior. *The Journal of Biological Chemistry, 272*, 1761–1768.

Orlowski, J., & Grinstein, S. (2004). Diversity of the mammalian sodium/proton exchanger SLC9 gene family. *Pflügers Archiv, 447*, 549–565.

Orlowski, J., & Grinstein, S. (2007). Emerging roles of alkali cation/proton exchangers in organellar homeostasis. *Current Opinion in Cell Biology, 19*, 483–492.

Padan, E. (2008). The enlightening encounter between structure and function in the NhaA Na(+)-H(+) antiporter. *Trends in Biochemical Sciences, 33*, 435–443.

Padan, E. (2014). Functional and structural dynamics of NhaA, a prototype for Na^+ and H^+ antiporters, which are responsible for Na^+ and H^+ homeostasis in cells. *Biochimica et Biophysica Acta, 1837*, 1047–1062.

Padan, E., Bibi, E., Masahiro, I., & Krulwich, T. A. (2005). Alkaline pH homeostasis in bacteria: New insights. *Biochimica et Biophysica Acta, 1717*, 67–88.

Padan, E., Kozachkov, L., Herz, K., & Rimon, A. (2009). NhaA crystal structure: Functional-structural insights. *The Journal of Experimental Biology, 212*, 1593–1603.

Padan, E., Maisler, N., Taglicht, D., Karpel, R., & Schuldiner, S. (1989). Deletion of ant in *Escherichia coli* reveals its function in adaptation to high salinity and an alternative Na+/H+ antiporter system(s). *The Journal of Biological Chemistry, 264*, 20297–20302.

Padan, E., Tzubery, T., Herz, K., Kozachkov, L., Rimon, A., & Galili, L. (2004). NhaA of *Escherichia coli*, as a model of a pH-regulated Na^+/H^+ antiporter. *Biochimica et Biophysica Acta, 1658*, 2–13.

Padan, E., Venturi, M., Gerchman, Y., & Dover, N. (2001). Na^+/H^+ antiporters. *Biochimica et Biophysica Acta, 1505*, 144–157.

Paulino, C., & Kuhlbrandt, W. (2014). pH- and sodium-induced changes in a sodium/proton antiporter. *eLife, 3*, e01412.

Pinner, E., Kotler, Y., Padan, E., & Schuldiner, S. (1993). Physiological role of *nhaB*, a specific Na^+/H^+ antiporter in *Escherichia coli*. *The Journal of Biological Chemistry, 268*, 1729–1734.

Putney, L. K., Denker, S. P., & Barber, D. L. (2002). The changing face of the Na+/H+ exchanger, NHE1: Structure, regulation, and cellular actions. *Annual Review of Pharmacology and Toxicology, 42*, 527–552.

Rimon, A., Kozachkov-Magrisso, L., & Padan, E. (2012). The unwound portion dividing helix IV of NhaA undergoes a conformational change at physiological pH and lines the cation passage. *Biochemistry, 51*, 9560–9569.

Rimon, A., Tzubery, T., & Padan, E. (2007). Monomers of nhaa NA+/H+ antiporter of *Escherichia coli* are fully functional yet dimers are beneficial under extreme stress conditions at alkaline ph in the presence of NA+ or LI+. *The Journal of Biological Chemistry, 282*, 26810–26821.

Screpanti, E., Padan, E., Rimon, A., Michel, H., & Hunte, C. (2006). Crucial steps in the structure determination of the Na+/H+ antiporter NhaA in its native conformation. *Journal of Molecular Biology, 362*, 192–202.

Shi, Y. (2013). Common folds and transport mechanisms of secondary active transporters. *Annual Review of Biophysics, 42*, 51–72.

Taglicht, D., Padan, E., & Schuldiner, S. (1991). Overproduction and purification of a functional Na^+/H^+ antiporter coded by *nhaA* (*ant*) from *Escherichia coli*. *The Journal of Biological Chemistry*, *266*, 11289–11294.

Taglicht, D., Padan, E., & Schuldiner, S. (1993). Proton-sodium stoichiometry of NhaA, an electrogenic antiporter from *Escherichia coli*. *The Journal of Biological Chemistry*, *268*, 5382–5387.

Venturi, M. P., & Padan, E. (2003). In C. Hunte, G. von Jagow, & H. Schagger (Eds.), *Purification of NhaA Na+/H+ antiporter of Escherichia coli for 3D and 2D crystallization*. San Diego, CA, USA: Academic Press/Elsevier Science.

Wakabayashi, S., Hisamitsu, T., Pang, T., & Shigekawa, M. (2003). Kinetic dissection of two distinct proton binding sites in Na^+/H^+ exchangers by measurement of reverse mode reaction. *The Journal of Biological Chemistry*, *278*, 43580–43585.

Zhou, X., Levin, E. J., Pan, Y., McCoy, J. G., Sharma, R., & Kloss, B. (2014). Structural basis of the alternating-access mechanism in a bile acid transporter. *Nature*, *505*, 569–573.

CHAPTER EIGHT

Purification, Refolding, and Crystallization of the Outer Membrane Protein OmpG from *Escherichia coli*

Stefan Köster[1], Katharina van Pee, Özkan Yildiz[2]

Department of Structural Biology, Max Planck Institute of Biophysics, Frankfurt am Main, Germany

[2]Corresponding author: e-mail address: oezkan.yildiz@biophys.mpg.de

Contents

Abstract

OmpG is a pore-forming protein from *E. coli* outer membranes. Unlike the classical outer membrane porins, which are trimers, the OmpG channel is a monomeric β-barrel made of 14 antiparallel β-strands with short periplasmic turns and longer extracellular loops. The channel activity of OmpG is pH dependent and the channel is gated by the extracellular loop L6. At neutral/high pH, the channel is open and permeable for substrate molecules with a size up to 900 Da. At acidic pH, loop L6 folds across the channel and blocks the pore. The channel blockage at acidic pH appears to be triggered by the

[1] Present address: Division of Infectious Diseases, Department of Medicine, New York University School of Medicine, 522 First Avenue, Smilow 901, New York, NY 10016, USA.

Methods in Enzymology, Volume 557
ISSN 0076-6879
http://dx.doi.org/10.1016/bs.mie.2015.01.018

protonation of a histidine pair on neighboring β-strands, which repel one another, resulting in the rearrangement of loop L6 and channel closure. OmpG was purified by refolding from inclusion bodies and crystallized in two and three dimensions. Crystallization and analysis by electron microscopy and X-ray crystallography revealed the fundamental mechanisms essential for the channel activity.

1. INTRODUCTION

The outer membrane of Gram-negative bacteria is a selective permeability barrier and prevents uncontrolled exchange of substrates (Benz, Janko, Boos, & Lauger, 1978; Nikaido, 2003). Various proteins are involved in transport, uptake, or efflux of a large variety of compounds, nutrients, or toxic molecules (sugars, small peptides, fatty acids, nucleotides, drugs, chemicals). Influx is largely controlled by porins, which are water-filled open channels that span the outer membrane and allow the passive transport of hydrophilic molecules. Porins are also involved in pathogenicity since they determine the bacterial surface structure. This makes them important targets for development of therapies (Achouak, Heulin, & Pages, 2001; Galdiero, Vitiello, & Galdiero, 2003; Naumann, Rudel, & Meyer, 1999). The majority of outer membrane porins involved in solute uptake are membrane-spanning β-barrels of 16 or 18 strands surrounding an aqueous pore. They conform to a common pattern whereby three monomers associate into trimers as shown in structures solved by two-dimensional (2D) (Jap, Downing, & Walian, 1990; Sass et al., 1989) and three-dimensional (3D) crystallography (Benz, Schmid, & Vos-Scheperkeuter, 1987).

The outer membrane protein G (OmpG) is a 14-stranded β-barrel porin from *Escherichia coli*. Both the condition under which it is expressed in *E. coli* and its biological function are unknown. In contrast to most porins found in the outer membrane of bacteria, no oligomerization is required for channel formation in OmpG as shown in electrophysiological measurements (Conlan, Zhang, Cheley, & Bayley, 2000) and in 2D crystals (Behlau, Mills, Quader, Kühlbrandt, & Vonck, 2001). OmpG is a simple and very robust transmembrane β-barrel protein that is permeable to molecules with a molecular mass of up to 900 Da (Fajardo et al., 1998). Due to its excellent stability, its monomeric nature, high availability by expression into inclusion bodies, and the high-resolution structures of OmpG (Fig. 1) determined

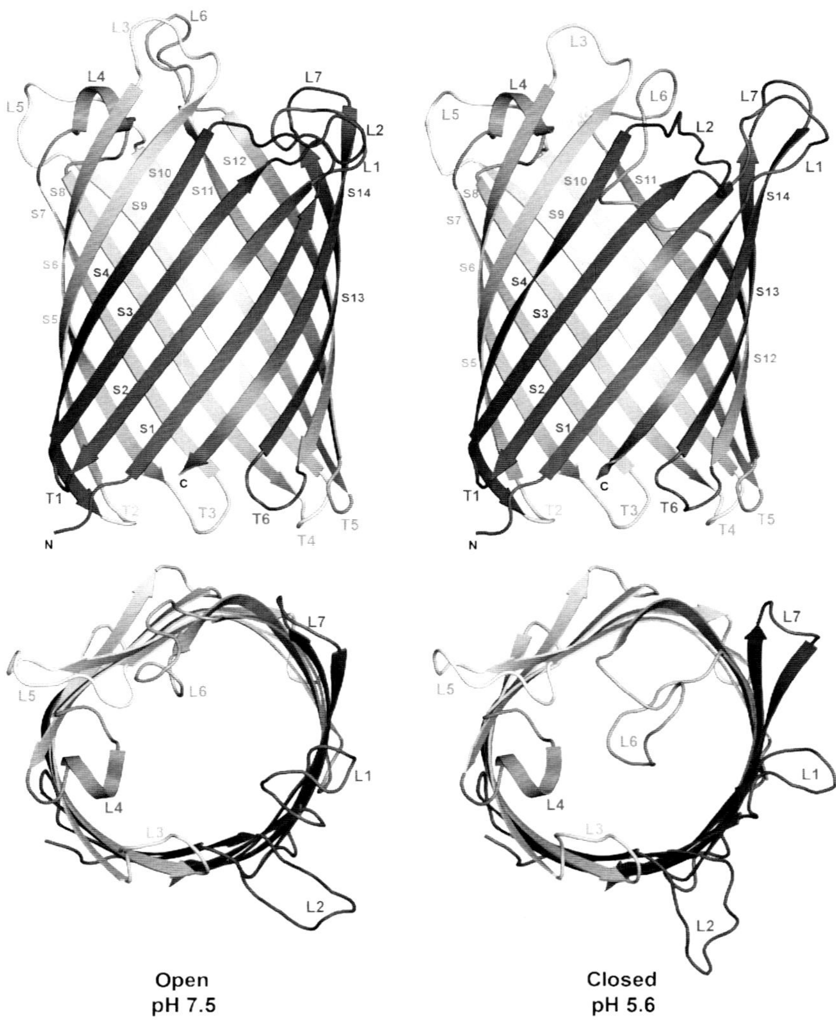

Figure 1 Overall structure of OmpG. The X-ray structure shows the 14-stranded, monomeric β-barrel in the open and closed conformation (Yildiz, Vinothkumar, Goswami, & Kühlbrandt, 2006) viewed along the membrane plane (top) and from the extracellular side (bottom). *Adapted from Yildiz et al. (2006) with permission from John Wiley & Sons.* (See the color plate.)

both in the open state at high pH and in the closed state at low pH (Yildiz et al., 2006), OmpG is a promising target as template in biotechnological applications including biosensors, substrate-specific channels, and for engineering of switchable nanopores with tailor-made properties.

2. OmpG PRODUCTION AND PURIFICATION

Plasmid-based overproduction in *E. coli* is possible and results in membrane-integrated OmpG (Subbarao & van den Berg, 2006). However, the yield of 4–5 mg OmpG out of 12 l of culture is very low. Here we describe a protocol in which the deletion of the amino-terminal region including the signal sequence (amino acids 1–20) leads to overexpression of OmpG in inclusion bodies (Conlan & Bayley, 2003) from where it is purified in high quantities by unfolding, ion-exchange (IE) chromatography, and refolding in presence of detergents (Yildiz et al., 2006) with a final yield of 20–30 mg refolded OmpG per liter of culture.

2.1 Preparation of Inclusion Bodies and Unfolding in Urea

1. OmpG was expressed without the N-terminal signal sequence and fusion tags using the expression plasmid pET26b in *E. coli* expression strain C43 (DE3) (Miroux & Walker, 1996).
2. 10 ml of an overnight LB culture containing 50 μg/ml kanamycin was transferred to 1000 ml TB medium containing kanamycin and incubated at 37 °C and 160 rpm.
3. Upon reaching OD_{600} of ≈ 2, the protein overproduction was induced with 1 m*M* isopropyl β-D-1-thiogalactopyranoside.
4. After 4 h, cells were harvested by a 15 min centrifugation at 4 °C and $5000 \times g$ using a Beckman J6-MI centrifuge and JS 4.2 rotor.
5. Harvested cells were resuspended in lysis buffer containing 25 m*M* Tris–HCl pH 8.0.
6. The cell suspension was filtered to remove cell clumps.
7. For cell lysis, the cell suspension was passed six times through a microfluidizer (model M-110L; Microfluidics Corp., Newton, Massachusetts, USA) with a 100-ml interaction chamber and cooling coil both immersed in ice.
8. Cell debris and inclusion bodies were pelleted for 20 min at $15{,}200 \times g$ using a Sorvall RC-5B centrifuge and Thermo Scientific SLA-1500 rotor.
9. Inclusion bodies were resuspended in 25 m*M* Tris pH 8.0, 1 *M* urea, 1% Triton X-100.
10. The inclusion bodies were pelleted for 20 min at $15{,}200 \times g$.
11. The pelleted inclusion bodies were resuspended in 25 m*M* Tris–HCl pH 8.0, 1 *M* urea to remove Triton X-100.

12. The inclusion bodies were pelleted again by a centrifugation at 15,200 × *g* for 20 min.
13. The inclusion bodies were dissolved in 25 m*M* Tris–HCl pH 8.0, 8 *M* urea at room temperature for at least 2 h or overnight on a stirrer.
14. To remove the nondissolved inclusion bodies, the suspension was centrifuged for 20 min at 15,200 × *g*.
15. The supernatant contains the unfolded OmpG.

2.2 Ion-Exchange Chromatography and Refolding

1. IE chromatography was done with the FPLC System Äkta Purifier or Äkta Prime, both from GE Healthcare.
2. A 30 ml Q-Sepharose (QS) column (GE Healthcare) was preequilibrated with unfolding buffer containing 25 m*M* Tris–HCl pH 8.0 and 8 *M* urea.
3. The solution containing unfolded OmpG was loaded onto the column while the flow-through monitored in the UV cell at 280 nm.
4. The column was washed with the unfolding buffer until the baseline has been reached.
5. Impurities were washed with unfolding buffer containing 165 m*M* NaCl until the absorbance reached baseline.
6. OmpG was eluted with unfolding buffer containing 250 m*M* NaCl and the elution was collected in 4-ml fractions.
7. Fractions were analyzed by UV absorption and sodium dodecyl sulfate polyacrylamide gel electrophoresis (SDS–PAGE; Fig. 2).
8. Fractions with the highest OmpG purity were pooled.
9. OmpG was refolded by diluting the pooled OmpG fractions into the refolding buffer containing 25 m*M* Tris–HCl pH 8.0, 3 *M* urea, and 1% (w/v) β-D-octylglucoside (OG) at room temperature on a stirrer.
10. The QS column, which has been thoroughly washed with water, was preequilibrated with refolding buffer.
11. The solution with the refolded OmpG was loaded onto the preequilibrated QS column.
12. The QS column was washed with the refolding buffer.
13. After reaching the baseline, the column was washed with urea-free buffer containing 25 m*M* Tris–HCl pH 8.0 and 0.5% (w/v) OG.
14. Refolded OmpG was eluted with 25 m*M* Tris–HCl pH 8.0, 0.5% (w/v) OG, and 500 m*M* NaCl, and the elution was collected in 4-ml fractions.

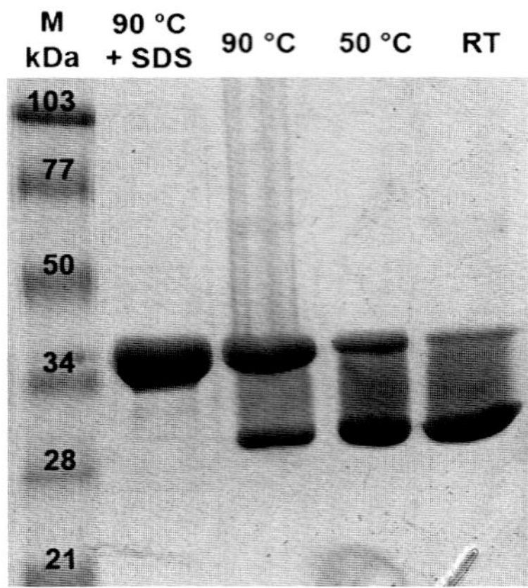

Figure 2 SDS–PAGE analysis of OmpG. 12% SDS–PAGE of OmpG incubated 10 min at 50 and 90 °C and at 90 °C in presence of SDS prior to SDS–PAGE. At room temperature, OmpG runs at an apparent mass of 28 kDa. Heating at 50 °C for 10 min results in partial unfolding indicated by an intensity increase of the protein band at 34 kDa. Higher temperature leads to further unfolding which is still not sufficient to completely unfold OmpG. Complete unfolding is observed upon incubating at 90 °C in presence of SDS in the sample buffer. The first lane (M) is loaded with protein molecular weight marker. The gel was stained with Coomassie Blue.

15. Fractions were analyzed by UV absorption at 280 nm and SDS–PAGE.

16. OmpG containing elution fractions were pooled and concentrated to ~50 mg/ml using ultrafiltration devices (Centricon) with a molecular weight cut-off (MWCO) of 30 kDa. The OG micelle has a molecular weight of 8–30 kDa resulting in 38–60 kDa for OmpG/micelle complex, which will not pass the 30 kDa ultrafiltration.

17. Alternatively, the protein was concentrated by dialysis against buffer containing 20% PEG-35000 in dialysis tubes with a 12 kDa cut-off membranes. This procedure can also be used in combination with detergent exchange (Section 2.3). In contrast to ultrafiltration, the detergent concentration does not increase during concentration procedure by dialysis.

18. The desired final protein concentration was adjusted by adding an appropriate amount of buffer to the concentrated protein stock.

19. The final yield was 20–30 mg of refolded OmpG per liter of culture. Protein purity was greater than 95%, as estimated by SDS–PAGE and Coomassie staining.

2.3 Detergent Exchange by Dialysis

1. A dialysis tube with a MWCO of 12 kDa was cut according to the volume of OmpG.
2. One end of the dialysis tube was closed with a clamp.
3. The dialysis tube was filled ~80% with the OmpG protein solution and the other end was closed with a clamp.
4. The dialysis tube was transferred into a glass beaker with 25 m*M* Tris–HCl pH 8.0 and the detergent of choice at twofold critical micelle concentration.
5. The dialysis buffer was exchanged three times with a dialysis time of at least 4 h each time. Alternatively, the dialysis was also performed overnight which was possible due to the high stability of OmpG.
6. After the dialysis, the OmpG solution was transferred from the dialysis tube in an Eppendorf tube, and the protein concentration was determined by UV absorption at 280 nm or by the Bradford assay (Bradford, 1976).

3. PROTEIN ANALYSIS

The efficiency of OmpG refolding was verified using various experiments including SDS–PAGE, circular dichroism (CD) spectroscopy, and analytical size-exclusion chromatography (SEC).

3.1 Electrophoresis

Outer membrane porins with a β-barrel structure are very stable. Trimeric porins do not monomerize (Martinez-Martinez, 2008) or denature in SDS–PAGE unless the samples are not heated upon addition of SDS sample buffer (Bolla, Loret, Zalewski, & Pages, 1995; Martinez-Martinez, 2008; Nakamura & Mizushima, 1976; Rosenbusch, 1974). Heating of the protein samples for 5–10 min at 97 °C results in monomerization and migration of the protein at its calculated molecular mass. Similarly, the monomeric OmpG with a calculated mass of 34 kDa migrates in SDS–PAGE at ~28 kDa (Fig. 2). Upon heating, it migrates at its calculated molecular mass. To use this heat modifiability for a half-quantitative evaluation of the refolding, it is advisable to run the SDS–PAGE in a cooled environment and at lower currents, to prevent partial unfolding during the electrophoresis, which results in a smeary band between 28 and 34 kDa (Fig. 2). Due to the absence of cysteines in OmpG, it is not necessary to add reducing agents like β-mercaptoethanol to the sample buffer.

1. 10 μl of elution fraction was mixed with 10 μl of 2 × concentrated SDS sample buffer.
2. The precast gel was mounted (Mini-PROTEAN system; Bio-Rad).
3. 5 μl of the molecular weight standard was loaded in the first well.
4. 5–10 μl of the protein samples were loaded in the other wells.
5. Applying a constant current of 40 mA started the electrophoresis run.
6. The electrophoresis was stopped when the blue dye front reached the bottom of the gel.
7. The gel cassette was opened and the gel was transferred into the Coomassie staining solution containing 0.1% (w/v) of Coomassie Brilliant Blue R250/G250 dissolved in 45% (v/v) ethanol, 5% (v/v) acetic acid.
8. The gel within the staining solution was heated using a microwave until the solution started to boil.
9. The staining solution with the gel was incubated for 5 min.
10. The staining solution was replaced by the destaining solution (10% (v/v) ethanol, 5% (v/v) acetic acid).
11. The gel was heated in the microwave until the solution starts to boil and incubated for 5 min in the hot destaining solution.
12. The destaining solution was replaced by water that was heated in the microwave until it was boiling.
13. The gel was incubated in hot water until the background was nearly clear.
14. Destaining was repeated with water if the gel still had Coomassie Blue background.

3.2 CD Spectroscopy

CD spectroscopy is used to estimate the β-sheet content in the OmpG protein solution after refolding. Buffer solutions, 8 *M* urea and unfolded OmpG (in 8 *M* urea), show no CD signals in the wavelength region 200–240 nm (Fig. 3). In contrast, proteins with high α-helix content show a strong double CD signal peak at 208 and 222 nm and β-sheet-rich proteins show a weaker single CD signal peak at 215 nm. Therefore, an increase of the CD signal at 215 nm is indicative of a successful refolding of OmpG. The CD spectra of OmpG were recorded using a JASCO J-810 spectrometer.

1. OmpG samples were prepared by diluting the protein in 25 m*M* Tris–HCl pH 8.0 to a final protein concentration of 0.5 mg/ml.

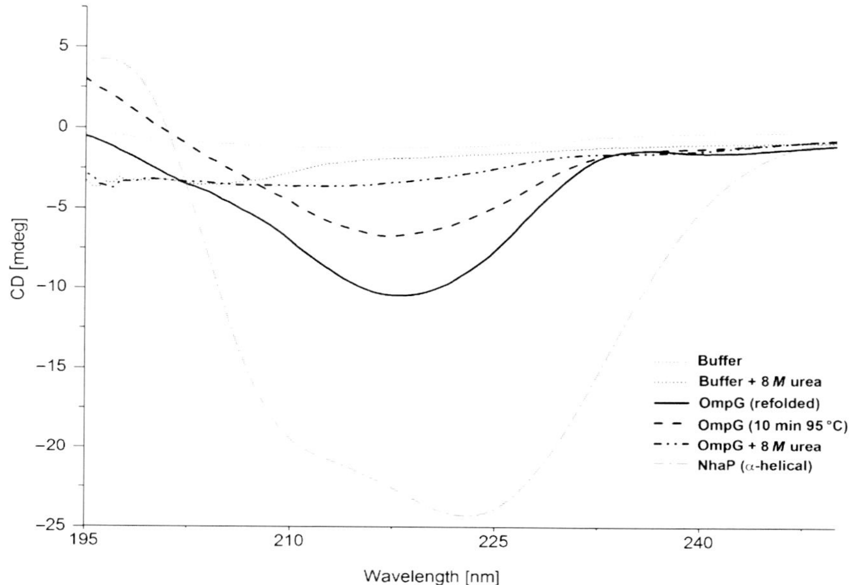

Figure 3 CD spectrum of OmpG in different buffers. The circular dichroism (*Y*-axis) is shown for the wavelength region between 195 and 250 nm (*X*-axis). The negative controls (buffer solution and buffer solution with 8 *M* urea, orange (gray in the print version) and cyan (light gray in the print version)) do not show any significant CD signals. Refolded OmpG (blue (dark gray in the print version)) with high β-sheet content shows a CD signal at 218 nm. Heating of OmpG for 10 min at 95 °C (green (light gray in the print version)) reduces the CD signal for β-sheet structure. Unfolded OmpG in urea (black) shows a similar CD signal like the negative controls. Proteins with high α-helix content exhibit CD signals at 208 and 222 nm as shown for the sodium proton antiporter NhaP (purple (light gray in the print version)) from *Methanococcus jannaschii* (Goswami et al., 2011).

2. The protein solution was pipetted into a 0.2-mm cuvette that was placed in the spectrometer.
3. The spectra were recorded at 298 K every 0.5 nm in the wavelength range of 190–250 nm and analyzed by the neural network program CDNN.

3.3 Analytical SEC

The monodispersity of a protein solution by analytical SEC is often a prerequisite for successful structural characterization. Ideally, the size-exclusion chromatogram for OmpG shows a single peak with increased Stokes radius that results from the β-barrel shape and detergent micelle around OmpG. The analytical SEC runs were performed with the Ettan LC

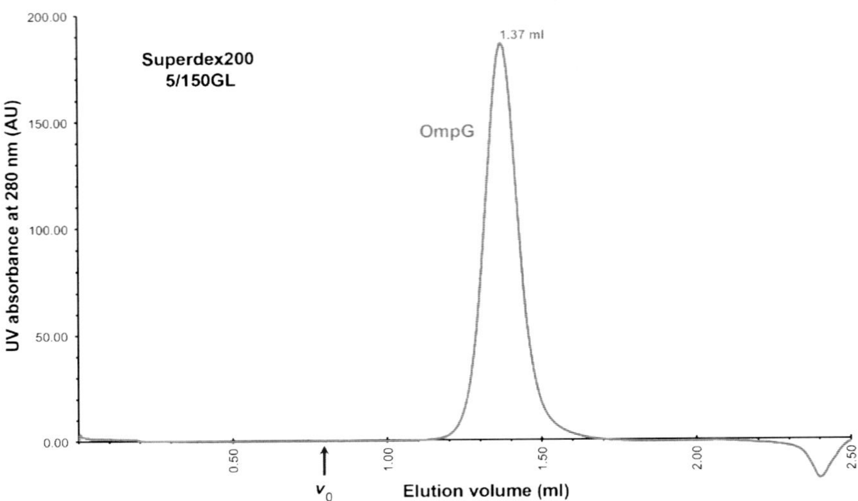

Figure 4 Analytical size-exclusion chromatography of OmpG. 200 μg of OmpG was loaded to a Superdex200 5/150GL column (GE Healthcare) equilibrated with 25 m*M* Tris pH 8.0 and 20 m*M* OG. The arrow indicates the void volume of the column at 0.8 ml. The elution from the column was analyzed photometrical at a wavelength of 280 nm. The elution volume of OmpG in the detergent OG is indicated by the absorption maximum at 1.37 ml.

chromatography system (GE Healthcare) and Superdex200 3.2/150 size-exclusion columns (Fig. 4).

1. The column was equilibrated with running buffer containing 25 m*M* Tris–HCl pH 8.0, 0.5% (w/v) OG.
2. OmpG at a protein concentration of 5 mg/ml was injected into a 25-μl loop.
3. The analysis was started by loading the loop content onto the column at a flow rate of 30 μl/min.
4. The chromatogram was recorded at 280 nm on the connected computer.

SEC as additional purification step was not critical for crystallization since the second IE chromatography step removed all small amounts of remaining, unfolded or partially refolded OmpG.

4. CRYSTALLIZATION OF OmpG

Two methods were used for the structural characterization of OmpG. One is the cryo-electron microscopy (cryo-EM) of 2D crystals (Behlau et al., 2001) and the other is the X-ray crystallography of 3D crystals (Subbarao & van den Berg, 2006; Yildiz et al., 2006).

4.1 2D Crystallization

Cryo-EM of 2D crystals is a powerful method in investigation of membrane proteins in their native-like, lipid-embedded environment to produce a medium-resolution map. Depending on the quality of the crystals, this method can provide important information about secondary structure and domain organization in the proteins. With this method, it is also possible to observe structural changes within the lipid-embedded protein upon incubation in buffer solutions with varying conditions like pH or substrate. Native OmpG solubilized from the *E. coli* membranes using OG and refolded OmpG could be successfully crystallized in two dimensions (Fig. 5) and yielded a projection map at 6 Å resolution (Yildiz et al., 2006). The following section describes the 2D crystallization of refolded OmpG purified from inclusion bodies.

1. Expression into inclusion bodies and purification by unfolding and refolding as described in Sections 2.1 and 2.2.

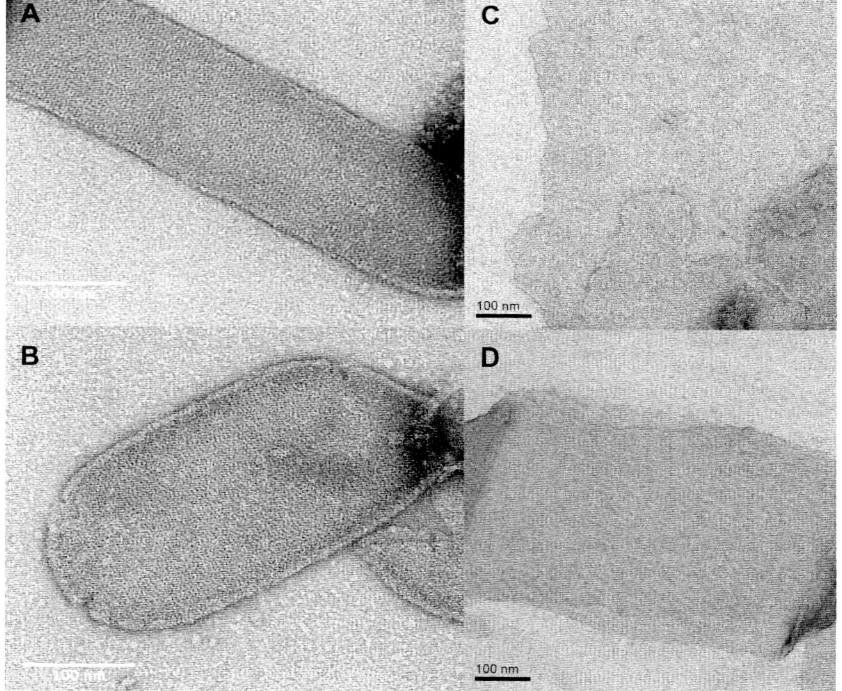

Figure 5 2D crystals of OmpG. Negatively stained tubular 2D crystals of OmpG (A, B) were 0.1–0.2 μm wide and had a length of up to 1 μm. Sheet-like 2D crystals (C, D) were bigger with an overall dimension up to 2 μm × 2 μm. Layers of the sheet-like crystals were often stacked.

2. 20 mg native *E. coli* polar lipids (Avanti) were dried under a nitrogen stream for 2 h at room temperature and resuspended in 1 ml 1% *n*-decyl-β-D-maltoside (DM) to a final lipid concentration of 20 mg/ml.
3. 10 μl OmpG solution (10 mg/ml) and 2 μl lipid solubilized in 1% DM were mixed to get a lipid/protein ratio of 0.4 (w/w).
4. The OmpG/DM/lipid mixture was incubated for 1 h at room temperature before it was transferred into a dialysis device (Pierce) with 100 μl volume. The mixture was supplemented with 88 μl buffer containing 25 m*M* Tris–HCl pH 8.0 and 0.5% OG.
5. The dialysis device was placed in a beaker with 400 ml dialysis buffer containing 25 m*M* Tris–HCl pH 7.0, 300 m*M* NaCl, 25 m*M* $MgCl_2$, and 2 m*M* NaN_3.
6. The beaker with the dialysis device was placed in an incubator at 37 °C.
7. After every week, the dialysis buffer was exchanged.
8. After 6–8 weeks, the 2D crystals were transferred from the dialysis device into an Eppendorf tube and stored at 4 °C.

4.1.1 Negative Stain EM

1. Copper grids carrying a carbon support film were produced with a carbon-coating unit (Edwards Auto 306 Turbo; Edwards Vacuum Ltd., UK).
2. Carbon-coated copper grids were glow-discharged for 30 s at a plasma current of about 10 mA (Balzers Union Ltd., Liechtenstein).
3. The morphology and quality of 2D crystals were assessed by negative stain EM.
4. 1–3 μl of the OmpG 2D crystal suspension was adsorbed to the grid that was blotted with the tip of a No. 4 Whatman filter paper (Whatman Ltd.).
5. 2 μl of 1% uranyl acetate was added, immediately blotted, replaced by fresh 2 μl of 1% uranyl acetate, and incubated on the grid for 1–2 min.
6. The grid was blotted and allowed to dry and inspected in an FEI G^2 Tecnai Spirit electron microscope (FEI, The Netherlands).

4.1.2 Sample Preparation for Electron Cryo-Microscopy

1. Specimens for electron cryo-microscopy were prepared by the back injection method (Wang & Kühlbrandt, 1991).
2. A fresh square piece of carbon was cut to the approximate size of a grid and detached from the mica by immersing it into a ~1-ml drop of embedding medium of choice containing 4% trehalose.

3. The floating carbon film was picked up with the darker side of a washed copper grid (400 mesh) and held with tweezers.
4. 1.5-μl sample was applied from the opposite side to which the carbon film was attached and mixed several times with the embedding medium.
5. After an incubation time of 1 min, excessive buffer was removed with a pipette.
6. The grid was blotted on a double No. 4 Whatman filter paper and air-dried for 5 s before plunge-freezing it in liquid nitrogen.
7. Grids were stored in liquid nitrogen.
8. 2D crystals were analyzed by cryo-EM using a JEOL 3000 SFF electron microscope equipped with a field emission gun and liquid-helium-cooled top-entry stage, with an accelerating voltage of 300 kV at 4 K specimen temperature.
9. Images were recorded at a magnification of 53K using spot-scan procedure with a 35-ms exposure time per spot.
10. Images were recorded on Kodak SO-163 electron emulsion film and were developed for 12 min in full-strength Kodak D19 developer.
11. The quality of the crystals was evaluated by optical diffraction, and regions showing strong reflections were scanned by 4000 × 4000 pixel areas with a pixel size of 7 μm on a Zeiss SCAI scanner.
12. The MRC image-processing programs were used for correcting the lattice distortions and contrast transfer function (Crowther, Henderson, & Smith, 1996; Henderson, Baldwin, Downing, Lepault, & Zemlin, 1986). The program ALLSPACE was used to determine the phase residuals (Valpuesta, Carrascosa, & Henderson, 1994). Phases and scaled amplitudes of structure factors obtained from projection maps at pH 8.0 and 4.0 were used to calculate the difference map.

4.2 3D Crystallization

For structural information at atomic level, refolded OmpG was crystallized in three dimensions using the method of vapor diffusion. For 3D crystallization, OmpG was dialyzed against buffers with different detergents including OG, DM, lauryldimethylamine-*N*-oxide (LDAO), polyoxyethylene (C_8E_4), and mixtures of these at final protein concentrations of 2.5, 5, 7.5, 10, and 15 mg/ml. Initial crystallization trials were performed with commercially available sparse matrix screen solutions from Hampton Research (Screen I/II, MemFac, Index, and PEG-Ion), from Jena

Bioscience (JBScreen Classic and JBScreen Membrane), and from NeXtal/QIAGEN (MBClass Suite, MBClass II Suite, $AmSO_4$ Suite, and MPD Suite).

1. 100 μl of screen solutions were pipetted into the reservoir well of a 96-well sitting-drop crystallization plate (Greiner).
2. For the initial screening, 0.2–0.6 μl of the screen solutions were mixed using a pipetting robot (Cartesian Technologies) with the same volume of the protein solution and pipetted into sitting-drop well of the crystallization plate.
3. The plates were sealed with a crystal clear sealing tape (Hampton Research).
4. Immediately after setting up the drops, the plates were checked under the stereomicroscope and stored in a vibration-free incubator (Rumed) at 18 °C.
5. The drops were analyzed once a day under the stereomicroscope.
6. Initial crystals were obtained (Fig. 6) in a wide range of conditions. Ammonium sulfate and PEG-3350 were frequent precipitation compounds in these crystallization conditions.

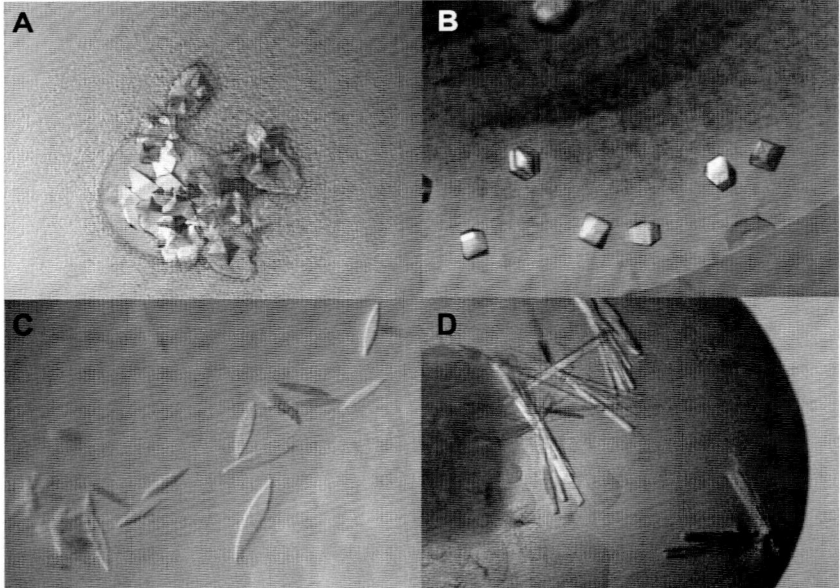

Figure 6 Initial 3D crystals of OmpG. Initial 3D crystals grown in 96-well plates were typically small (A–C) or clustered (D). Often, the crystals were surrounded by viscose reservoir (A), precipitated protein (B), or phase separation (C, D). (See the color plate.)

7. The initial crystallization conditions were reproduced and optimized (Fig. 7) in a 96-well sitting-drop well plates or in pregreased 24-well VDX hanging drop plates using the vapor diffusion method.
8. The reservoir volume for the 24-well plate was 500 µl and the hanging drop size was 2 µl (1 µl protein + 1 µl reservoir solution).
9. In one optimized condition, OmpG crystals grew over 3 days in hanging drops in 1 µl of 10 mg/ml OmpG in 25 m*M* Tris at pH 8.0 with 5 m*M* LDAO mixed with 1 µl of 100 m*M* HEPES pH 7.5, 30% PEG-4000, and 200 m*M* $CaCl_2$. These crystals were very fragile and resembled a flattened rice grain of dimensions 700 µm × 400 µm × 40 µm (Fig. 7).
10. Another crystal form grew in sitting drop of a 96-well plate for 4–5 days in 0.4 µl of 7.5 mg OmpG in 25 m*M* Tris at pH 8.0 and 20 m*M* OG mixed with 0.4 µl of 100 m*M* Na-citrate at pH 5.6, 150 m*M* NaCl, and 12% PEG-3350.

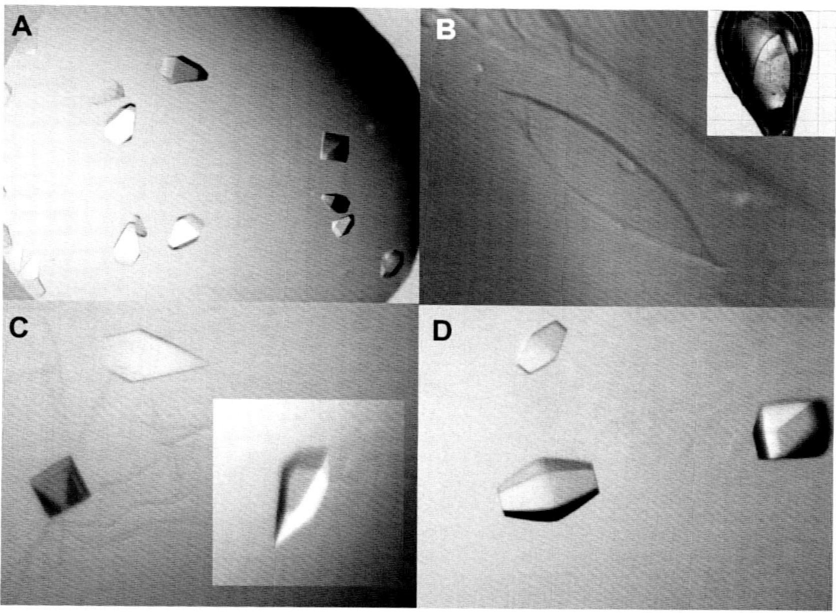

Figure 7 Optimized 3D crystals of OmpG. The initial conditions were optimized in 24-well plates and were bigger in size and were nicely separated. The crystals belong to the tetragonal (A, C), orthorhombic (D), and triclinic (B) space group.

11. For data collection, the crystals were transferred to cryoprotectant solutions and flash-frozen in liquid nitrogen.

12. Native data were collected at the beamline PXI and PXII at the Swiss Light Source and processed with the XDS package (Kabsch, 1993).

5. SUMMARY

The monomeric outer membrane protein OmpG from *E. coli* was produced in high amounts by overexpression into inclusion bodies. By unfolding in urea, IE chromatography, and refolding in presence of detergents, it could be efficiently purified to homogeneity. The protein quality was evaluated by SDS–PAGE, CD spectroscopy, and SEC. After detergent exchange, OmpG could be crystallized in both 2D and 3D that allowed the structural analysis by electron microscopy and X-ray crystallography, respectively. These crystals enabled a detailed description of the gating mechanism of OmpG as described in further studies. The projection maps from the electron microscopic analysis (Behlau et al., 2001; Yildiz et al., 2006) and the atomic structures from the X-ray analysis of two different crystal forms (Yildiz et al., 2006) enlightened the mechanisms of channel opening and closure. Further, the 2D crystals of OmpG were investigated by Fourier transform infrared spectroscopy (Korkmaz, Köster, Yildiz, & Mantele, 2008; Korkmaz-Özkan, Köster, Kühlbrandt, Mäntele, & Yildiz, 2010) and atomic force microscopic methods (Damaghi, Bippes, et al., 2010; Damaghi, Sapra, et al., 2010; Mari et al., 2010). The unique properties of OmpG, its easy production, and the structural accessibility predestinate OmpG in biotechnological applications.

REFERENCES

Achouak, W., Heulin, T., & Pages, J. M. (2001). Multiple facets of bacterial porins. *FEMS Microbiology Letters*, *199*(1), 1–7.

Behlau, M., Mills, D. J., Quader, H., Kühlbrandt, W., & Vonck, J. (2001). Projection structure of the monomeric porin OmpG at 6 A resolution. *Journal of Molecular Biology*, *305*(1), 71–77. http://dx.doi.org/10.1006/jmbi.2000.4284.

Benz, R., Janko, K., Boos, W., & Lauger, P. (1978). Formation of large, ion-permeable membrane channels by the matrix protein (porin) of Escherichia coli. *Biochimica et Biophysica Acta*, *511*(3), 305–319. http://dx.doi.org/10.1016/0005-2736(78)90269-9.

Benz, R., Schmid, A., & Vos-Scheperkeuter, G. H. (1987). Mechanism of sugar transport through the sugar-specific LamB channel of Escherichia coli outer membrane. *The Journal of Membrane Biology*, *100*(1), 21–29.

Bolla, J. M., Loret, E., Zalewski, M., & Pages, J. M. (1995). Conformational analysis of the Campylobacter jejuni porin. *Journal of Bacteriology*, *177*(15), 4266–4271.

Bradford, M. M. (1976). A rapid and sensitive method for the quantitation of microgram quantities of protein utilizing the principle of protein-dye binding. *Analytical Biochemistry, 72*, 248–254.

Conlan, S., & Bayley, H. (2003). Folding of a monomeric porin, OmpG, in detergent solution. *Biochemistry, 42*(31), 9453–9465. http://dx.doi.org/10.1021/bi0344228.

Conlan, S., Zhang, Y., Cheley, S., & Bayley, H. (2000). Biochemical and biophysical characterization of OmpG: A monomeric porin. *Biochemistry, 39*(39), 11845–11854. http://dx.doi.org/10.1021/bi001065h.

Crowther, R. A., Henderson, R., & Smith, J. M. (1996). MRC image processing programs. *Journal of Structural Biology, 116*(1), 9–16. http://dx.doi.org/10.1006/jsbi.1996.0003.

Damaghi, M., Bippes, C., Koster, S., Yildiz, O., Mari, S. A., Kuhlbrandt, W., et al. (2010). pH-dependent interactions guide the folding and gate the transmembrane pore of the beta-barrel membrane protein OmpG. *Journal of Molecular Biology, 397*(4), 878–882. http://dx.doi.org/10.1016/j.jmb.2010.02.023.

Damaghi, M., Sapra, K. T., Koster, S., Yildiz, O., Kuhlbrandt, W., & Muller, D. J. (2010). Dual energy landscape: The functional state of the beta-barrel outer membrane protein G molds its unfolding energy landscape. *Proteomics, 10*(23), 4151–4162. http://dx.doi.org/10.1002/pmic.201000241.

Fajardo, D. A., Cheung, J., Ito, C., Sugawara, E., Nikaido, H., & Misra, R. (1998). Biochemistry and regulation of a novel Escherichia coli K-12 porin protein, OmpG, which produces unusually large channels. *Journal of Bacteriology, 180*(17), 4452–4459.

Galdiero, M., Vitiello, M., & Galdiero, S. (2003). Eukaryotic cell signaling and transcriptional activation induced by bacterial porins. *FEMS Microbiology Letters, 226*(1), 57–64.

Goswami, P., Paulino, C., Hizlan, D., Vonck, J., Yildiz, O., & Kuhlbrandt, W. (2011). Structure of the archaeal Na+/H+ antiporter NhaP1 and functional role of transmembrane helix 1. *EMBO Journal, 30*(2), 439–449. http://dx.doi.org/10.1038/emboj.2010.321.

Henderson, R., Baldwin, J. M., Downing, K. H., Lepault, J., & Zemlin, F. (1986). Structure of purple membrane from Halobacterium halobium—Recording, measurement and evaluation of electron micrographs at 3.5 A resolution. *Ultramicroscopy, 19*(2), 147–178.

Jap, B. K., Downing, K. H., & Walian, P. J. (1990). Structure of PhoE porin in projection at 3.5 A resolution. *Journal of Structural Biology, 103*(1), 57–63.

Kabsch, W. (1993). Automatic processing of rotation diffraction data from crystals of initially unknown symmetry and cell constants. *Journal of Applied Crystallography, 26*, 795–800.

Korkmaz, F., Köster, S., Yildiz, O., & Mantele, W. (2008). The role of lipids for the functional integrity of porin: An FTIR study using lipid and protein reporter groups. *Biochemistry, 47*(46), 12126–12134.

Korkmaz-Özkan, F., Köster, S., Kühlbrandt, W., Mäntele, W., & Yildiz, Ö. (2010). Correlation between the OmpG secondary structure and its pH-dependent alterations monitored by FTIR. *Journal of Molecular Biology, 401*, 56–67. http://dx.doi.org/10.1016/j.jmb.2010.06.015.

Mari, S. A., Koster, S., Bippes, C. A., Yildiz, O., Kuhlbrandt, W., & Muller, D. J. (2010). pH-induced conformational change of the beta-barrel-forming protein OmpG reconstituted into native E. coli lipids. *Journal of Molecular Biology, 396*(3), 610–616. http://dx.doi.org/10.1016/j.jmb.2009.12.034.

Martinez-Martinez, L. (2008). Extended-spectrum beta-lactamases and the permeability barrier. *Clinical Microbiology and Infection, 14*(Suppl. 1), 82–89. http://dx.doi.org/10.1111/j.1469-0691.2007.01860.x.

Miroux, B., & Walker, J. E. (1996). Over-production of proteins in Escherichia coli: Mutant hosts that allow synthesis of some membrane proteins and globular proteins at high levels. *Journal of Molecular Biology, 260*(3), 289–298. http://dx.doi.org/10.1006/jmbi.1996.0399.

Nakamura, K., & Mizushima, S. (1976). Effects of heating in dodecyl sulfate solution on the conformation and electrophoretic mobility of isolated major outer membrane proteins from Escherichia coli K-12. *Journal of Biochemistry, 80*(6), 1411–1422.

Naumann, M., Rudel, T., & Meyer, T. F. (1999). Host cell interactions and signalling with Neisseria gonorrhoeae. *Current Opinion in Microbiology, 2*(1), 62–70.

Nikaido, H. (2003). Molecular basis of bacterial outer membrane permeability revisited. *Microbiology and Molecular Biology Reviews, 67*(4), 593–656.

Rosenbusch, J. P. (1974). Characterization of the major envelope protein from Escherichia coli. Regular arrangement on the peptidoglycan and unusual dodecyl sulfate binding. *Journal of Biological Chemistry, 249*(24), 8019–8029.

Sass, H. J., Buldt, G., Beckmann, E., Zemlin, F., van Heel, M., Zeitler, E., et al. (1989). Densely packed beta-structure at the protein-lipid interface of porin is revealed by high-resolution cryo-electron microscopy. *Journal of Molecular Biology, 209*(1), 171–175. http://dx.doi.org/10.1016/0022-2836(89)90180-0.

Subbarao, G. V., & van den Berg, B. (2006). Crystal structure of the monomeric porin OmpG. *Journal of Molecular Biology, 360*(4), 750–759.

Valpuesta, J. M., Carrascosa, J. L., & Henderson, R. (1994). Analysis of electron microscope images and electron diffraction patterns of thin crystals of phi 29 connectors in ice. *Journal of Molecular Biology, 240*(4), 281–287.

Wang, D. N., & Kühlbrandt, W. (1991). High-resolution electron crystallography of light-harvesting chlorophyll a/b-protein complex in three different media. *Journal of Molecular Biology, 217*, 691–699. http://dx.doi.org/10.1016/0022-2836(91)90526-C.

Yildiz, Ö., Vinothkumar, K. R., Goswami, P., & Kühlbrandt, W. (2006). Structure of the monomeric outer-membrane porin OmpG in the open and closed conformation. *EMBO Journal, 25*(15), 3702–3713.

CHAPTER NINE

Biophysical Approaches to the Study of LeuT, a Prokaryotic Homolog of Neurotransmitter Sodium Symporters

Satinder K. Singh[1], Aritra Pal
Department of Cellular and Molecular Physiology, Yale University School of Medicine, New Haven, Connecticut, USA
[1]Corresponding author: e-mail address: satinder.k.singh@yale.edu

Contents

Abstract

Ion-coupled secondary transport is utilized by multiple integral membrane proteins as a means of achieving the thermodynamically unfavorable translocation of solute molecules across the lipid bilayer. The chemical nature of these molecules is diverse and

Methods in Enzymology, Volume 557
ISSN 0076-6879
http://dx.doi.org/10.1016/bs.mie.2015.01.002

includes sugars, amino acids, neurotransmitters, and other ions. LeuT is a sodium-coupled, nonpolar amino acid symporter and eubacterial member of the solute carrier 6 (SLC6) family of Na^+/Cl^--dependent neurotransmitter transporters. Eukaryotic counterparts encompass the clinically and pharmacologically significant transporters for γ-aminobutyric acid (GABA), glycine, serotonin (5-hydroxytryptamine, 5-HT), dopamine (DA), and norepinephrine (NE). Since the crystal structure of LeuT was first solved in 2005, subsequent crystallographic, binding, flux, and spectroscopic studies, complemented with homology modeling and molecular dynamic simulations, have allowed this protein to emerge as a remarkable mechanistic paradigm for both the SLC6 class as well as several other sequence-unrelated SLCs whose members possess astonishingly similar architectures. Despite yielding groundbreaking conceptual advances, this vast treasure trove of data has also been the source of contentious hypotheses. This chapter will present a historical scientific overview of SLC6s; recount how the initial and subsequent LeuT structures were solved, describing the insights they each provided; detail the accompanying functional techniques, emphasizing how they either supported or refuted the static crystallographic data; and assemble these individual findings into a mechanism of transport and inhibition.

1. INTRODUCTION

Communication across chemical synapses is the principal mode by which electrical signals are transmitted among neurons in the brain (Vanhatalo & Sohila, 1998). Subsequent to Ca^{2+}-dependent release of neurotransmitters from the presynaptic neuron and activation of fast-acting ionotropic or slower-acting metabotropic postsynaptic receptors, these small molecules are cleared from the synaptic cleft primarily by presynaptic sodium-dependent neurotransmitter transporters. These integral membrane proteins rely on preexisting ion gradients to catalyze the thermodynamically unfavorable movement of their substrates across the phospholipid bilayer (Masson, Sagné, Hamon, & El Mestikawy, 1999).

Members of the SLC6 group, also known as neurotransmitter sodium symporters, depend on Na^+/Cl^- to transport an expansive spectrum of small molecules such as amino acids (glycine and GABA); osmolytes; and the monoamines (serotonin (5-hydroxytryptamine, 5-HT), dopamine (DA), and norepinephrine (NE); Kristensen et al., 2011). They are of particular clinical significance because their dysfunction has been implicated in multiple debilitating neurological and neuropsychiatric illnesses (Hahn & Blakely, 2007), and they are the target of numerous psychoactive agents (Kristensen et al., 2011).

Prior to cloning of any SLC6 members, seminal experiments detecting NE transport in the heart (Iversen, 1963) as well as nerve endings (Hetting & Axelrod, 1961); and unearthing the uptake's sodium-dependence (Iversen &

Kravitz, 1966), stereospecificity (Iversen, Jarrott, & Simmonds, 1971), saturability (Iversen, 1971), and inhibition by cocaine (Iversen, 1965) as well as tricyclic antidepressants (TCAs; Iversen, 1965) reinforced the concept that an integral membrane protein was responsible. These early studies were further enhanced by exploiting membrane vesicles enriched in transporter, prepared from brain synaptosomes (Kanner, 1978) and blood platelets (Rudnick, 1977), as well as partially purified transporter reconstituted into lipid vesicles (Radian & Kanner, 1985). Properties such as substrate/inhibitor specificity, ion selectivity, and substrate–ion stoichiometry were soon characterized.

Cloning of the rat GABA type 1 transporter (GAT1; Guastella et al., 1990) followed by the human norepinephrine transporter (NET; Pacholczyk, Blakely, & Amara, 1991) and others each represented breakthroughs in SLC6 research. At the most basic level, they showed that seemingly functionally diverse transporters all belonged to the same family. They enabled hydropathy analyses with the prediction of 12 transmembrane (TM) segments and intracellularly located amino- and carboxy-termini, a topology that was later verified by site-directed chemical labeling (Chen, Liu-Chen, & Rudnick, 1998). Notably, recombinant expression in heterologous systems permitted scientists to examine the consequences of mutating distinct amino acids on activity, trafficking, and regulation.

As more data amassed, the need for a three-dimensional template became evident. Although attempts to solubilize and purify representative eukaryotic SLC6 members from heterologously expressing cells and some native tissues were sufficient for functional assays, they fell far short of the milligram quantities demanded for crystallography. Comprehensive efforts began by targeting mammalian proteins, such as the rat serotonin transporter (SERT). However, this protein has proven problematic (Tate et al., 2003), being stable and active exclusively in the heterogeneous detergent digitonin (Fig. 1A), from which no membrane protein has ever been *directly* crystallized (Tate, 2010).

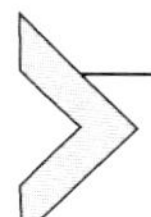

2. LeuT EXPRESSION, PURIFICATION, CRYSTALLIZATION, AND STRUCTURE DETERMINATION

2.1 Background

Structure determination of integral membrane proteins remains a daunting task. Before attempts at crystallization are even made, obtaining enough pure monodisperse protein with which to work remains a major bottleneck.

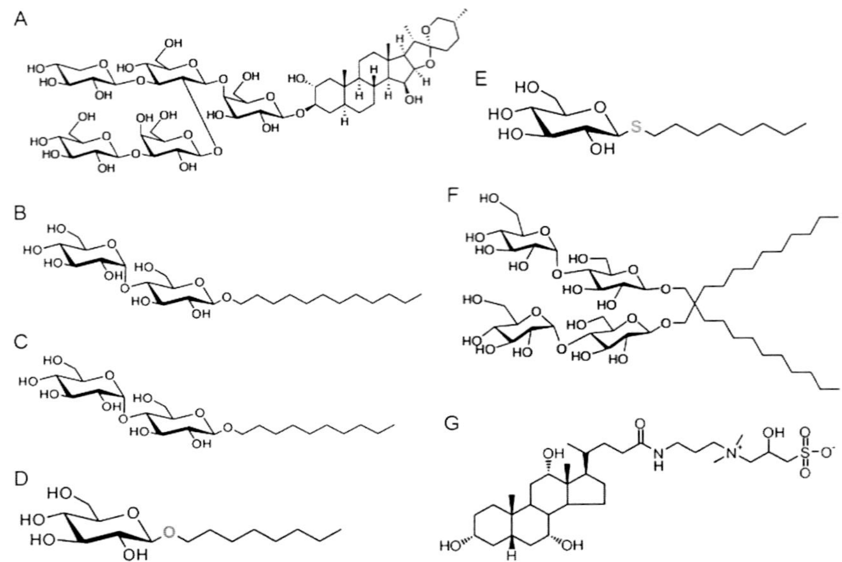

Figure 1 Detergents used in structural and functional studies of some eukaryotic SLC6 members and the prokaryotic counterpart LeuT. (A) Digitonin. (B) *n*-Dodecyl-β-D-maltoside (DDM). (C) *n*-Decyl-β-D-maltoside. (D) *n*-Octyl-β-D-glucoside (β-OG). (E) *n*-Octyl-β-D-thioglucoside (C_8SG). (F) Lauryl maltose neopentyl glycol (LMNPG). (G) 3-[(3-Cholamidopropyl)dimethylammonio]-2-hydroxy-1-propanesulfonate (CHAPSO). Note that C_8SG is somewhat larger than β-OG due to sulfur's slightly longer atomic radius. *Permission provided by Anatrace.*

Prerequisites include adequate expression levels, monodispersity, long-term stability, large hydrophilic surface area, and activity. Probably the most crucial variable is detergent. There is a diverse array now available whose properties vary by headgroup, charge, and alkyl tail length (Moraes, Evans, Sanchez-Weatherby, Newstead, & Stewart, 2014). Generally, nonionic shorter chain-length ones promote formation of crystals capable of diffracting to higher resolution, but they can also disrupt quaternary structure and/or induce aggregation. Thus, the goal is to identify the shortest chain-length detergent that preserves stability, function, and oligomeric state.

Transporters are particularly challenging because they are almost completely buried in the hydrophobic bilayer, are conformationally heterogeneous, and frequently have long, flexible termini and loops. One way of improving the odds of success is to evaluate monodispersity and stability of homologs from multiple organisms via fluorescence-detection size-exclusion chromatography (FSEC), a high-throughput method in which the target protein is fused to a fluorescent tag, expressed, and behavior monitored on a gel filtration column (Kawate & Gouaux, 2006). The fact that FSEC requires only nanograms of unpurified protein versus micrograms of purified protein greatly expedites the process. Even among thermophilic prokaryotes, whose proteins are inherently more stable than those of their eukaryotic cousins and generally express well in inexpensive hosts like *Escherichia coli* and/or *Lactococcus lactis* NZ9000 (Kunji, Slotboom, & Poolman, 2003), FSEC screening can make an enormous difference.

For the SLC6 family, a pivotal discovery was that bacterial counterparts exist (Nelson, 1998). The tryptophan transporter (TnaT) from *Symbiobacterium thermophilum* was the first to be functionally characterized (Androutsellis-Theotokis et al., 2003), suggesting that it might be the first homolog to unveil its atomic secrets. However, despite the fact that the protein expressed to high levels and could be purified to homogeneity via metal affinity chromatography, most of it was aggregated when expressed in standard *E. coli* strains and solubilized in the mild, nonionic detergent *n*-dodecyl-β-D-maltoside (DDM) (Fig. 1B). For many prokaryotic membrane proteins, expression in *L. lactis* or the *E. coli* "Walker" variants, C41(λDE3) and C43(λDE3) (Miroux & Walker, 1996) has proven more beneficial. After exhaustive trials, *L. lactis* and C41(λDE3) were found to be the best strains for overexpressing two other prokaryotic SLC6s, Tyt1 (Quick & Javitch, 2007) and LeuT (Yamashita, Singh, Kawate, Jin, & Gouaux, 2005), respectively, in fully functional monodisperse form.

2.2 Toward the first SLC6 crystal structure: The LeuT Na^+/Leu-bound outward-occluded state

LeuT is from the hyperthermophilic, chemolithotrophic eubacterium *Aquifex aeolicus* (Deckert et al., 1998) and was one of seven bacterial homologs selected from a PSI-BLAST search against the rat glycine transporter 1 (GlyT1). It, along with one from *Methanococcus jannachi*, expressed extremely well in C41(λDE3) cells, but LeuT crystallized much more readily. Hanging-drop vapor diffusion trays were initially set up with protein that had been solubilized and NiNTA-purified in DDM, subjected to gel filtration chromatography to remove any aggregates and exchange to the shorter-chain detergent *n*-decyl-β-D-maltoside (DM; Fig. 1C), and then concentrated. Crystals grew, but they diffracted to only ~15 Å, suggesting that molecular packing was likely mediated by detergent rather than protein. Because DDM is extremely difficult to remove, having a low critical micelle concentration (CMC) of only 0.2 m*M*, and detergent exchange to DM was achieved on a size-exclusion column rather than an affinity matrix, it is conceivable that the LeuT-detergent micelle was actually mixed, consisting of both DDM and DM, but this possibility was never explored.

The adjustment that dramatically improved diffraction was exchange from DDM to the much smaller *n*-octyl-β-D-glucoside (β-OG; Fig. 1D). As with DM, LeuT eluted as a sharp Gaussian peak with little or no aggregate. Although the protein was somewhat less stable in β-OG, with visible precipitate appearing in 1–2 days at 4 °C, trays set up immediately yielded crystals within a few days at ~20 °C and in HEPES buffer with PEG 550 monomethylether as the precipitant. These crystals were best cryoprotected by gradually increasing the PEG concentration to 35% instead of adding glycerol, the latter of which actually deteriorated resolution. The final structure was solved with native and selenomethionine multiwavelength anomalous diffraction (Se-MAD) X-ray data (Hendrickson, Horton, & LeMaster, 1990) extending to 1.65 and 1.90 Å, respectively. Because of the high resolution and excellent starting phases, model building was trivial, greatly assisted by the automatic tracing algorithm ARP/wARP (Langer, Cohen, Lazmin, & Perrakis, 2008). Despite the fortuitous appearance of Leu and two putative ions sitting at the core, there was no functional evidence beforehand that this protein was even a transporter, let alone a sodium-dependent Leu symporter. Nevertheless, it was designated "LeuT," and the name has remained ever since.

Compared with the electron density for Leu, which was unequivocal, that for the two Na^+ ions was less so because, even at such high resolution, these latter peaks could have easily been modeled as water molecules or other cations. Therefore, parameters, including valence calculations (Nayal & Di Cera, 1996); coordinating distances/geometry; and refined atomic positions, temperature factors, and residual peaks in F_o–F_c maps were employed to validate their identity (Yamashita et al., 2005).

2.3 Initial revelations (Fig. 13B)

The first LeuT structure afforded unprecedented glimpses into SLC6 architecture, substrate/Na^+-binding sites, putative extracellular/intracellular gates, and a speculative transport mechanism. Although it confirmed the 12-TM topology (Fig. 2) as well as juxtaposition of EL2 and EL4 (Fig. 3), it unexpectedly revealed an internal structural repeat relating TMs 1–5 and 6–10 by an antiparallel pseudo twofold in the membrane plane (Fig. 2). LeuT has a central substrate/ion-binding site (S1 hereafter) and another cavity, the extracellular vestibule (EV), located approximately

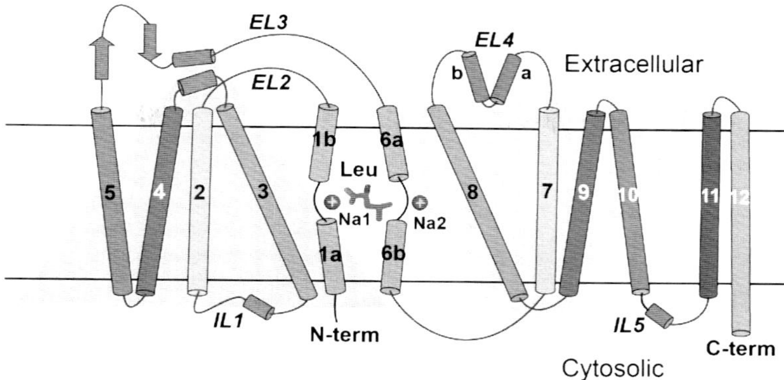

Figure 2 LeuT topology. Leu and two sodium ions (purple (gray in the print version) circles) are drawn halfway across the bilayer. Membrane boundaries are demarcated by two thick black horizontal lines. TM helices are depicted as cylinders, with TMs 1 and 6 unwound close to Leu and the two sodiums ions. Faint triangles illustrate the component helices of the 5+5 inverted repeat, each pair of which is shown in the same color.

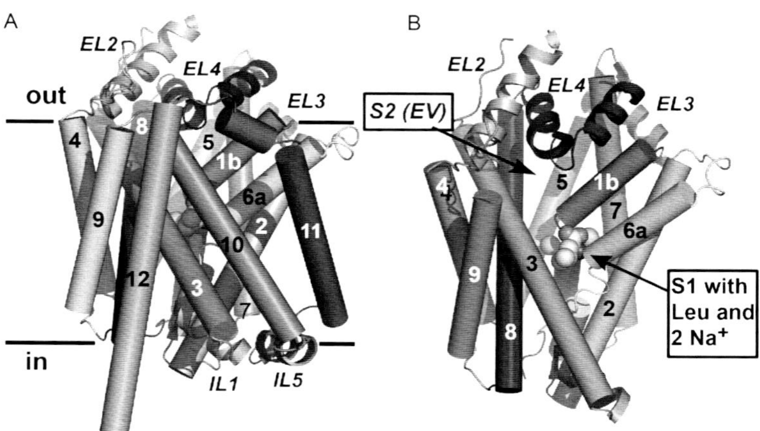

Figure 3 (A) LeuT in the membrane plane. Transmembrane helices (TM) and extracellular/intracellular loops are depicted as cylinders and coils, respectively. Heavy black lines demarcate the membrane boundaries. (B) Same view as (A) except tilted ~15° toward the reader and with TMs 10–12 removed to expose the substrates Leu (in yellow [carbon]/red [oxygen] spheres) and the two sodium ions (cyan spheres), all bound near the unwound sections of TMs 1 and 6. *Adapted from Singh (2008).* (See the color plate.)

11 Å above. Although the EV, specifically a site within known as S2, was subsequently shown to bind a diverse array of hydrophobic molecules, including β-OG (Quick et al., 2009), only water molecules were originally built into the weak, discontinuous electron density. Nonetheless, the presence of β-OG was later validated by selenium-containing *n*-heptyl seleno-β-D-glucoside (β-SeHG), a β-OG analog with an anomalous signal (Wang, Elferich, & Gouaux, 2012).

The two central TMs, 1 and 6, are unwound halfway across the lipid bilayer, unmasking backbone carbonyl oxygens and amide nitrogens for Na^+ and leucine coordination (Fig. 4A). The two sodiums, Na1 and Na2, bind in a completely dehydrated site, also formed by the unwound sections of TMs 1 and 6, in addition to residues in the middle of TMs 3 and 8. Coordination is provided by five or six precisely arranged oxygen ligands with defined geometry (octahedral for Na1 and trigonal bipyramidal for Na2) and within a defined distance (2.28 Å) of each of the dehydrated sodium ions. Only Na1 is in direct contact with Leu, specifically its carboxylate, suggesting it plays a vital role in forming the substrate-binding pocket. The universally conserved Y108 interacts with Leu's carboxylate and stabilizes TM1 near the unwound region. In the monoamine transporters, whose respective substrates possess a primary amine but lack a carboxylate, G24 in

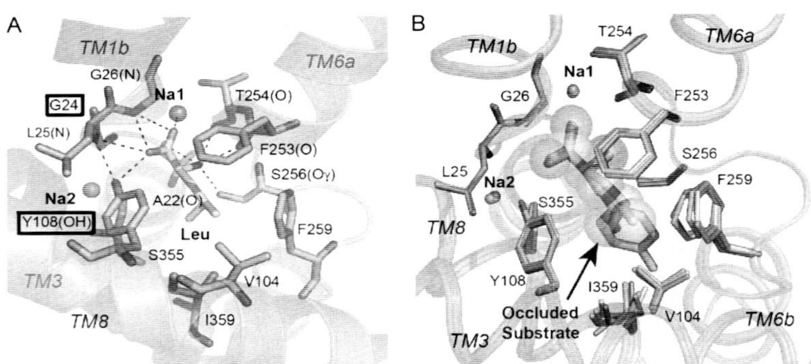

Figure 4 Close-up of S1. (A) Leu and sodium ions are primarily coordinated by backbone amide nitrogens (blue (gray in the print version)), backbone carbonyl oxygens (red (dark gray in the print version)), and side chain hydroxyls (red (dark gray in the print version)). Labels for the invariant tyrosine in TM3 and glycine in TM1, an aspartate in the monoamine transporters, are boxed. Hydrogen bonds are shown as dashed lines. (B) Overlay of LeuT substrates (sticks) Leu (gray), Ala (green (gray in the print version)), Gly (magenta (dark gray in the print version)), Met (blue (gray in the print version)), and L-4-F-Phe (orange (gray in the print version)). Atoms of bound Leu are shown as semi-transparent van der Waals spheres. *Left and right images respectively reproduced from Singh (2008) and Singh, Piscitelli, Yamashita, and Gouaux (2008), with permission from Landes Bioscience and AAAS, respectively.*

LeuT is replaced with Asp whose side-chain carboxylate likely coordinates Na1 and hydrogen bonds with both Y's hydroxyl as well as the primary amine (Fig. 4A). Although these suppositions are supported by human SERT (Celik et al., 2008) and DAT (Huang & Zhan, 2007) homology models, they have yet to be substantiated by the structure of a Na^+/substrate-bound eukaryotic monoamine transporter. For example, the recent structure of a thermostabilized *Drosophila melanogaster* dopamine transporter (dDAT), incapable of transport, complexed with Na^+, Cl^-, and presumed inhibitor, demonstrates that Na1 indirectly interacts with Asp's side-chain carboxylate through a water molecule, but it bares no information about how the substrate DA is coordinated.

In LeuT's outward-occluded state, S1 is solvent inaccessible from both sides of the membrane, with extracellular access obstructed to a lesser degree (Figs. 5 [right panel] and 7A). Near the extracellular gate, at the bottom of the EV, is a water-mediated salt bridge between D404 (TM10) and R30 (TM1), the latter of which is indirectly coupled to S1 via a hydrogen-bond network (Fig. 5 [left panel]). At the opposite end of the bilayer, comprising part of the cytoplasmic gate, is another salt bridge, an ionic latch between D369 (TM8) and R5 (N-terminal domain/TM1), the latter of which is partly stabilized by a hydrogen bond to Y268 (TMs 6–7). W8 (TM8) fits snugly into a hydrophobic pocket formed by TMs 1 and 6, further anchoring the N-terminus and R5 in their positions (Fig. 6).

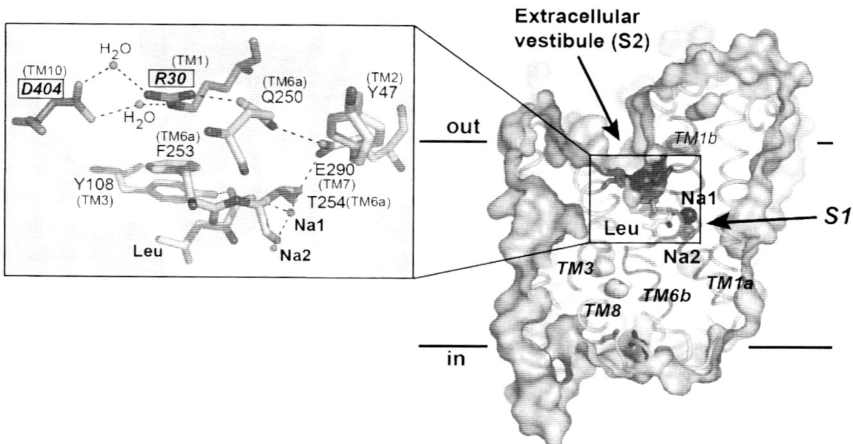

Figure 5 LeuT's extracellular gate. Both S1 and S2 are labeled. On the left panel, note the water-mediated salt bridge between R30 and D404, the hydrogen-bond network linking R30 with S1, and the layering of F253 and Y108 on top of Leu. Residues involved in the gating conformational changes in SLC6 members are italicized, boldfaced, and boxed. *Reproduced from Singh (2008) with permission from Landes Bioscience.*

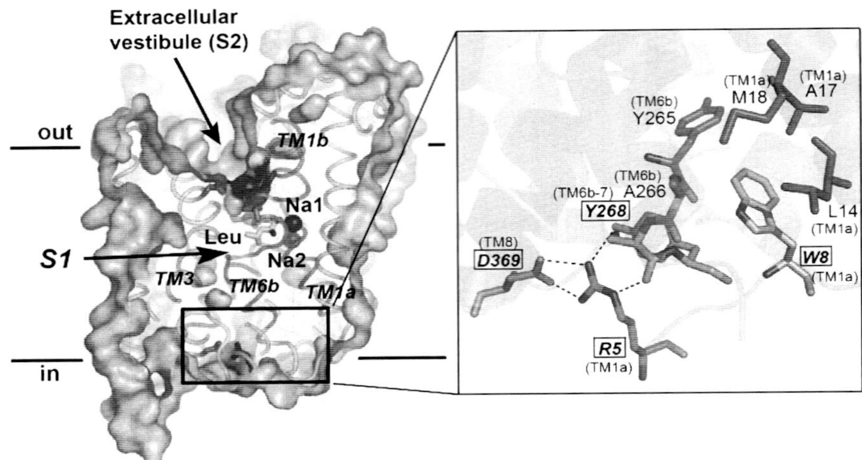

Figure 6 LeuT's intracellular gate. (Left panel) Note the direct salt bridge between D369 and R5. Labels for residues involved in the gating conformational changes in SLC6 members are italicized, boldfaced, and boxed. *Reproduced from Singh (2008) with permission from Landes Bioscience.*

The LeuT-fold was novel back in 2005, but, surprisingly, crystal structures of numerous transporters from diverse, sequence-unrelated families have since been found to exhibit the 5 + 5 inverted repeat. These comprise members of the amino acid–polyamine–organocation (APC) (AdiC, ApcT, and GadC); betaine/carnitine/choline (BCCT) (BetP and CaiT); nucleobase:cation symporter-1 (NCS1) (Mhp1); solute–sodium symporter (SSS or SLC5) (vSGLT) (Shi, 2013); and NRAMP (SLC11; ScaDMT) families (Ehrnstorfer, Geertsma, Pardon, Steyaert, & Dutzler, 2014).

2.4 Noncompetitive inhibitor-bound structures (Fig. 13G)

Akin to the outward-occluded structure, crystallizing subsequent cocomplexes with the noncompetitive tricyclic antidepressants (TCAs; Singh, Yamashita, & Gouaux, 2007; Zhou et al., 2007) and later, the selective serotonin reuptake inhibitors (SSRIs), sertraline (SRT) and fluoxetine (FLX; Zhou et al., 2009), was straightforward. It merely involved purifying LeuT as usual and then adding 10–40 m*M* clomipramine (CMI), imipramine (IMI), desipramine (DMI), SRT, or FLX to the protein for cocrystallization trials. Crystals grew within 1 week under comparable conditions and diffracted to 1.7–1.9 Å (Singh et al., 2007), 2.9 Å (Zhou et al., 2007), or 2.1–2.5 Å (Zhou et al., 2009). The cocomplexes were solved via

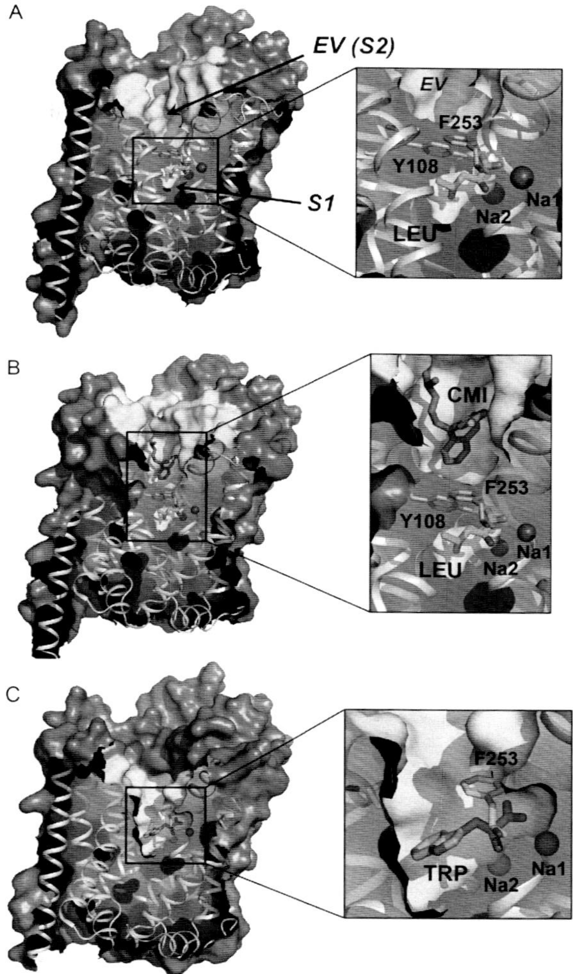

Figure 7 Surface renderings of LeuT–Na^+/Leu and inhibitor complexes with S1 and S2 labeled in (A). Na^+ ions are purple (dark gray in the print version) spheres in all panels. Magnified sections with F253 and Y108 are indicated, colored orange (gray in the print version) in (A) and (B) and green (gray in the print version) in (C). (A) LeuT–Na^+/Leu complex (outward-occluded state); Leu is colored by element, with carbon, oxygen, and nitrogen shown in green (light gray in the print version), red (gray in the print version), and blue (dark gray in the print version), respectively. (B) LeuT–Na^+/Leu/CMI complex (noncompetitively inhibited stabilized outward-occluded state); CMI is colored by element, with carbon, nitrogen, and chlorine depicted in red (gray in the print version), blue (dark gray in the print version), and green (light gray in the print version), respectively. (C) Leu–Na^+/Trp complex (competitively inhibited, trapped outward-open state); Trp atoms are colored by element, with carbon, oxygen, and nitrogen shown in orange (light gray in the print version), red (gray in the print version), and blue (dark gray in the print version), respectively.

difference Fourier analysis or molecular replacement and refined well. All LeuT–TCA/SSRI complexes assume the same outward-occluded conformation as the original LeuT structure with Leu and Na^+ bound in S1 and a single antidepressant molecule bound in S2 (Figs. 7B and 8A), directly above the R30–D404 ion pair (Fig. 8B), displacing β-OG. The drug's presence triggers an almost 180° flip in R30's guanidium ring to expel two water molecules and form a direct salt bridge with D404 (Fig. 8B), a presumably stronger interaction than one mediated by water.

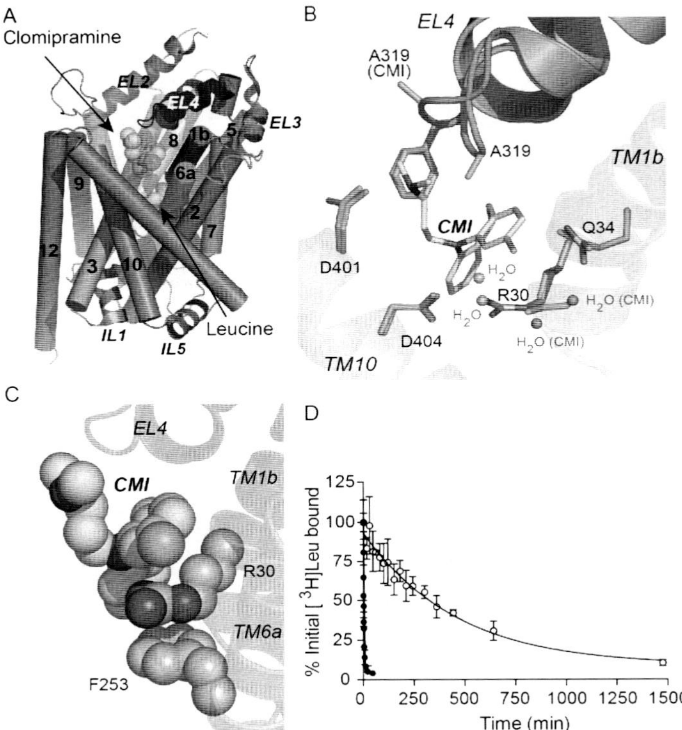

Figure 8 (A) The TCA clomipramine (CMI) binds to LeuT in S2 about 11 Å above S1 and just above the R30–D404 salt bridge. (B) Magnification of the CMI-binding site from LeuT–Na^+/Leu/CMI complex (pink (gray in the print version)) overlaid onto that of the –Na^+/Leu, depicting the flip of the guanidium group of R30 to form a direct salt bridge with D404 and the displacement of two water molecules. (C) Space-filling model of CMI (yellow (light gray in the print version)), R30 (blue (gray in the print version)) and F253 (gray), illustrating how CMI seems to stabilize R30 in its flipped position via a "layered" cation–π interaction with F253. (D) Dissociation of [^{3}H]Leu from LeuT in the absence (filled circles) and presence (open circles) of 3 m*M* CMI. *Reproduced from Singh (2008) with permission from Landes Bioscience.*

2.5 Substrate- and competitive inhibitor-bound (Fig. 13F) structures

Relative to LeuT–Na^+/Leu, crystallizing other substrate-bound complexes was more difficult because these other amino acids had to supplant the endogenously bound Leu. LeuT–Ala and –Trp were the easiest, simply requiring the inclusion of saturating Ala or Trp, respectively, to all buffers. The others were marginally more complicated, demanding the same high concentrations of Ala during solubilization and then gradual replacement with the desired amino acid during purification and dialysis. Because of Tyr's extremely low solubility, the more soluble isosteric Tyr analog, L-4-fluorophenylalanine (L-4-F-Phe), was used instead. All cocrystals diffracted to 1.8–2.3 Å and the structures were solved via molecular replacement using outward-occluded LeuT–Na^+/Leu as a phase probe. This strategy was successful for the substrate-bound complexes but failed for LeuT–Na^+/Trp, signifying that a structural rearrangement had occurred. Thus, the search model was modified to delete parts of TM1, TM6, EL4a, and EL4b. Molecular replacement with this new probe revealed displaced density for the missing helices, indicating they had indeed moved. To corroborate this new conformation, experimental phases from a selenomethionine-substituted LeuT–Na^+/Trp complex were obtained, and the resulting structure was indistinguishable from the refined molecular replacement solution (Singh et al., 2008).

What did the LeuT–Na^+/substrate and –Na^+/Trp structures uncover? As predicted, all of the substrate complexes adopt the same outward-occluded state, with Y108's hydroxyl maintaining its critical interaction with the substrate carboxylate and L25's amide nitrogen. The only variations are localized to F259 and I359 abutting the substrates' R group. Depending on the size of this moiety, F259 and I359 pivot into or out from S1, with the greatest torsion in and out observed with the Gly and Tyr complexes, respectively (Fig. 4B). The Trp complex, on the other hand, adopts an outward-open conformation, in which TMs 1b, 2a, and 6a rotate approximately 9° outward, accompanied by substantial displacement of TM11 and ELs 2, 3, and 4a (Figs. 7C and 9 [left panel]), all of which open a solvent-accessible channel from the extracellular side to S1 (Fig. 7C). These movements can be traced to Trp's broad rigid indole ring, which separates Y108 from the Trp carboxylate by ~5.1 Å to preclude formation of a critical hydrogen bond. It also increases the distance between Y108 and F253 by 3 Å (Fig. 9 [lower right panel]; Singh et al., 2008). Curiously, as model

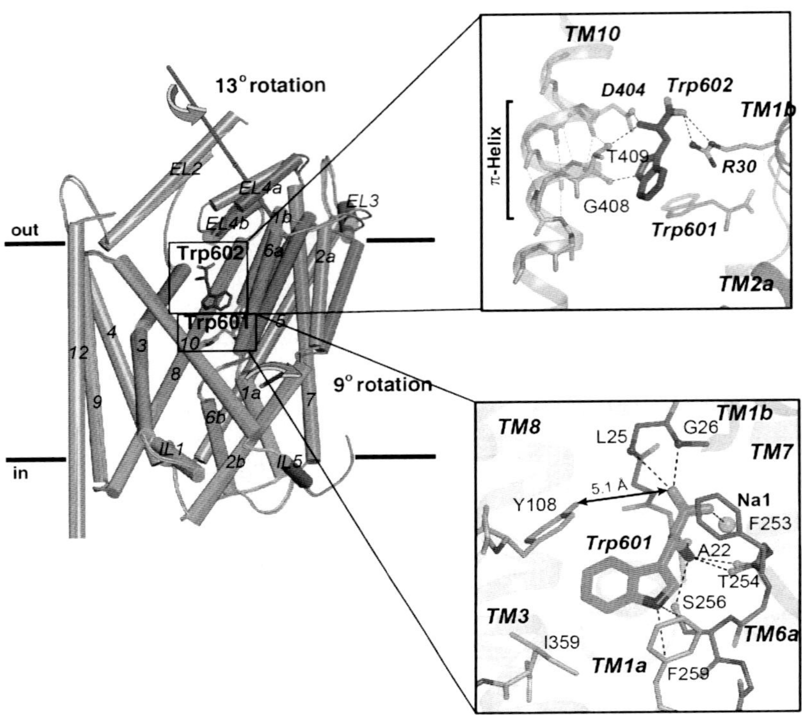

Figure 9 (Left panel) Cα superposition of the LeuT–Na^+/Leu (gray cylinders) and –Na^+/Trp (sand (light gray in the print version), red (dark gray in the print version), and magenta (dark gray in the print version) cylinders) complexes. Helices engaged in the domain shift (TMs 1b, 2a, and 6a) and EL4a are colored red (dark gray in the print version) and magenta (dark gray in the print version), respectively, as are their rotation axes. TM11 and ELs 2 and 3 are also involved but not highlighted. The bound Trps are shown as sticks, with the carbon atoms of 601 and 602 colored green (gray in the print version) and dark green (dark gray in the print version), respectively. (Lower right panel) Magnification of the Cα superposition depicting the hydrogen-bond network in S1 of LeuT–Na^+/Trp. A double-headed arrow indicates disruption of the critical hydrogen-bond between Y108 and the Trp601 carboxylate. For clarity, view has been rotated approximately 90° toward the reader relative to that in the larger left panel. (Upper right panel) A second Trp molecule (602) is bound between R30 and D404, flanked by a π-helix in TM10, just below S2. *Reproduced from Singh et al. (2008) with permission from AAAS.*

building and refinement progressed, density consistent with a second Trp molecule (Trp602) appeared in the more open permeation pathway, between R30 and D404, and this position, distinct from S2, might represent a low-affinity, transiently occupied site (Fig. 9 [upper right panel]).

2.6 Outward-open, Na^+-bound, substrate-free state (Fig. 13A)

Unlike the substrate- and inhibitor-bound complexes, crystallizing the substrate-free, sodium-bound outward-open state was more challenging because LeuT is not as stable in β-OG in the absence of ligands. Four components were critical for success: (1) use of a Y108F mutant to dramatically impair Leu binding by eliminating the stabilizing hydrogen bond with the Leu carboxylate (Piscitelli, Krishnamurthy, & Gouaux, 2010); (2) extensive washing of LeuT-Y108F membranes with sodium-free buffer supplemented with the sodium-specific chelator, 15-crown-5 (Christensen, Hill, & Izatt, 1971); (3) complexation with a conformation-specific Fab antibody fragment (2B12); and (4) detergent exchange to the slightly larger sulfur-containing *n*-octyl-β-D-thioglucoside (C_8SG; Fig. 1E) rather than the smaller oxygen-containing β-OG. Although these crystals diffracted to only 3.1 Å, there was clearly no extra density in S1 except that for sodium, and LeuT had adopted an outward-open conformation (Fig. 10A), akin to the LeuT–Na^+/Trp competitively inhibited complex, except for an additional 90° upward rotation of F253 (Krishnamurthy & Gouaux, 2012).

2.7 Inward-open "apo" state (Fig. 13D)

This structure was the most challenging and required an exhaustive search of the literature for clues. Two proved instrumental: First, molecular dynamic simulations with "LeuT-fold" transporters Mhp1 (Shimamura et al., 2010) and vSGLT1 (Watanabe et al., 2010), both of which share LeuT's Na2 site (Shi, 2013), had implied that disrupting selected Na2-coordinating residues might promote intracellular substrate release. Second, mutation of the DAT intracellular gating residue Y335, analogous to Y268 in LeuT, had previously been shown to favor the DAT cytoplasmic-facing conformation (Kniazeff, Shi, Loland, Weinstein, & Gether, 2008). Based on these data, a new LeuT construct incorporating three mutations, two at Na2 (T354V and S355A) and one at the intracellular gate (Y268A), was generated.

The final modifications employed to crystallize the inward-open conformation included (1) use of this new mutant, LeuT-TSY; (2) utilization of another conformation-specific Fab (6A10); and (3) supplementation with the stabilizing lipid, 1,2-dimyristoyl-*sn*-glycero-3-phosphoethanolamine (DMPE; Fig. 11A). These steps were in addition to washing of membranes with sodium-free, 15-crown-5-supplemented buffer and exchange of LeuT-TSY into C_8SG instead of β-OG. Like LeuT-Y108A-Fab crystals,

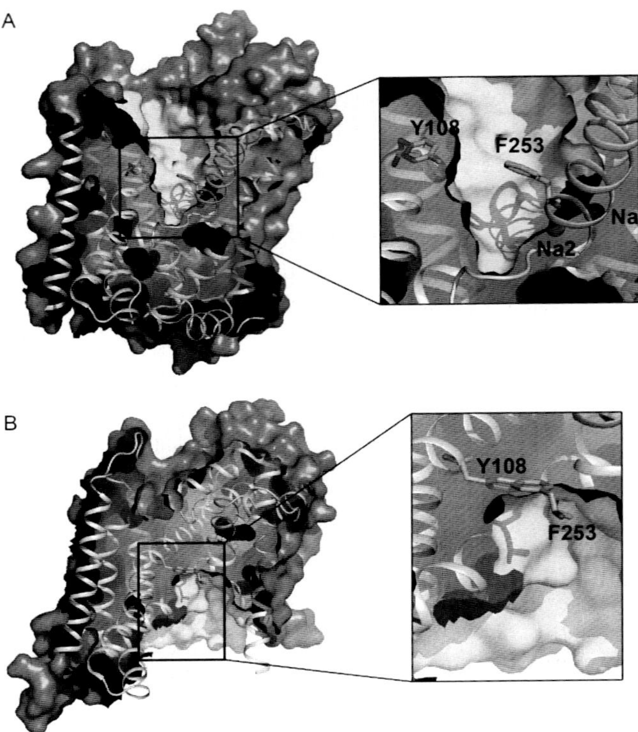

Figure 10 Surface representation of the (A) Na^+-bound/substrate-free, outward-open state and (B) apo inward-open conformation. Right panels magnify S1, with gating residues F253 and Y108 depicted as orange (gray in the print version) sticks. Panel A also includes the two sodium ions (purple (dark gray in the print version) spheres) partly behind the yellow (light gray in the print version) surface.

these only diffracted to 3.2 Å. The structure was solved by molecular replacement, but the phase probe had to be heavily altered, using a combination of a previously solved Fab structure and a partial LeuT model. Initial maps definitively indicated that sizable rearrangements had occurred. First, TMs 1b and 6a flex to similar extents to partly close the extracellular pathway, while TM6b concomitantly bends away from the intracellular pathway by 17°. Second, TMs bracing TMs 1 and 6, namely TMs 2, 5, and 7, flex rather than tilt about their midpoints, facilitated by centrally located Gly or Pro residues. The bending of TM7 permits EL4 to "plug" the EV, mostly collapsing S2. The most striking movement is an almost 45° bend in TM1a away from the cytoplasmic side (Fig. 12) to open a solvent-accessible channel to S1 (Krishnamurthy & Gouaux, 2012; Fig. 10B). Such an enormous bend is startling, and, as the authors note, it is debatable if such a shift actually occurs in the membrane.

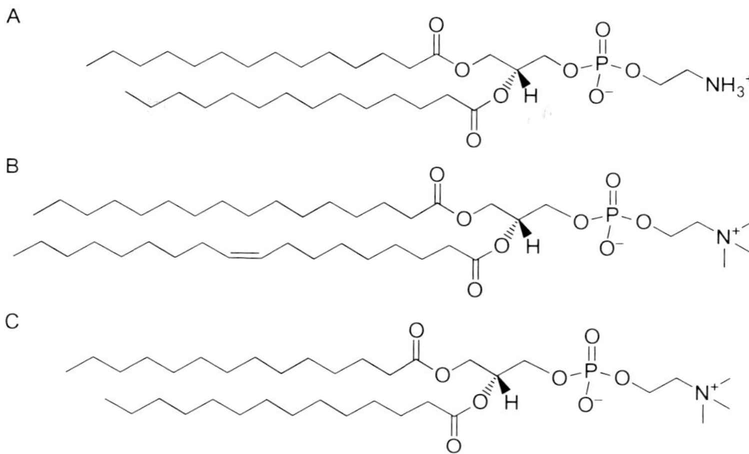

Figure 11 Lipids used in LeuT structure/function experiments. (A) 1,2-Dimyristoyl-*sn*-glycero-3-phosphoethanolamine (DMPE). (B) 1-Palmitoyl-2-oleoyl-*sn*-glycero-3-phosphocholine (POPC). (C) 1,2-Dimyristoyl-*sn*-glycero-3-phosphocholine (DMPC). *Permission provided by Avanti Polar Lipids, Inc.*

2.8 Inward-occluded Na^+/substrate-bound state

Although structures of LeuT have been solved in multiple conformations along the translocation cycle, there are still two missing states—Na^+/-substrate-bound inward-occluded (Fig. 13C) and Na^+/substrate-free outward-open (Fig. 13E). Recently, however, the crystal structure of another bacterial SLC6 homolog from *Bacillus halodurans* (MhsT), a sodium-dependent aromatic amino acid transporter, was solved in the Na^+/Trp-bound inward-occluded state in two separate lipidic environments (Fig. 13C; Malinauskaite et al., 2014). Unlike the LeuT apo inward-open structure, which was not crystallized in a membranous milieu. TM1a does not move nearly as far into the membrane and TM5 unwinds, enabling solvent access to Na2. Notably, a leucine at position 29 in LeuT is replaced with the larger tryptophan, as is the case with every other SLC6 member, and this Trp, in conjunction with EL4, completely collapses the EV, including S2.

3. FUNCTIONAL CHARACTERIZATION

Despite the undeniable power of crystallography, it alone is rarely a panacea for addressing every mechanistic question. Its full potential is realized only when combined with complementary biochemical, biophysical,

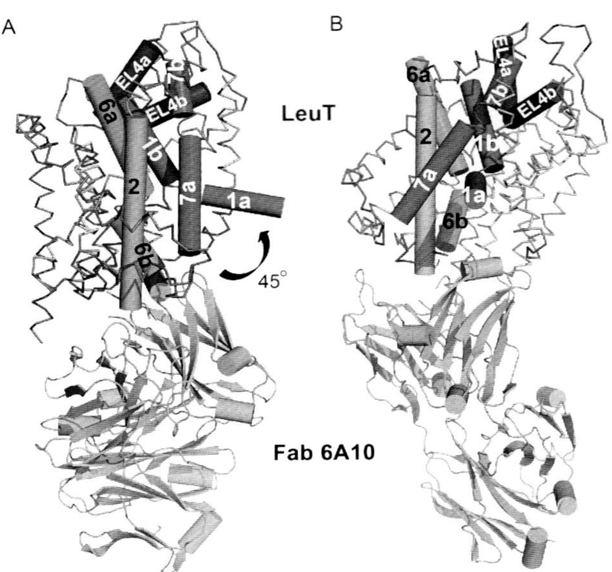

Figure 12 (A) Cα trace of the apo inward-open-Fab complex, with the "bundle" helices and EL4 depicted as cylinders. TMs 1, 2, 6, 7, and EL4 are colored red (dark gray in the print version), pink (gray in the print version), green (gray in the print version), magenta (dark gray in the print version), and blue (darkest gray in the print version), respectively. The 45° bend of TM1a into the predicted membrane is indicated. (B) Same image in (A) except rotated 90° toward the reader.

and pharmacological techniques. Indeed, these methods were indispensable for delineating LeuT substrate specificity, elucidating TCA/Trp inhibition mechanisms, and characterizing solution conformational dynamics.

3.1 Flux assays

Measuring ion-dependent radiolabeled substrate uptake in intact cells is often the first avenue traveled when investigating function. Surmising that LeuT could transport Leu based on this amino acid's serendipitous presence in the original crystal structure, assays were initially attempted with [^{3}H]Leu in LeuT-expressing cultures of the *E. coli* transposon knockout strain FB21219, which exhibits considerably reduced endogenous transport of branched amino acids (Adams et al., 1990). Unfortunately, the signal-to-noise was only twofold over untransformed knockouts. Therefore, DDM-purified LeuT was reconstituted into liposomes composed of *E. coli* total lipid extract and egg phosphatidylcholine, the predominant species of which is 1-palmitoyl-2-oleoyl-*sn*-glycero-3-phosphocholine

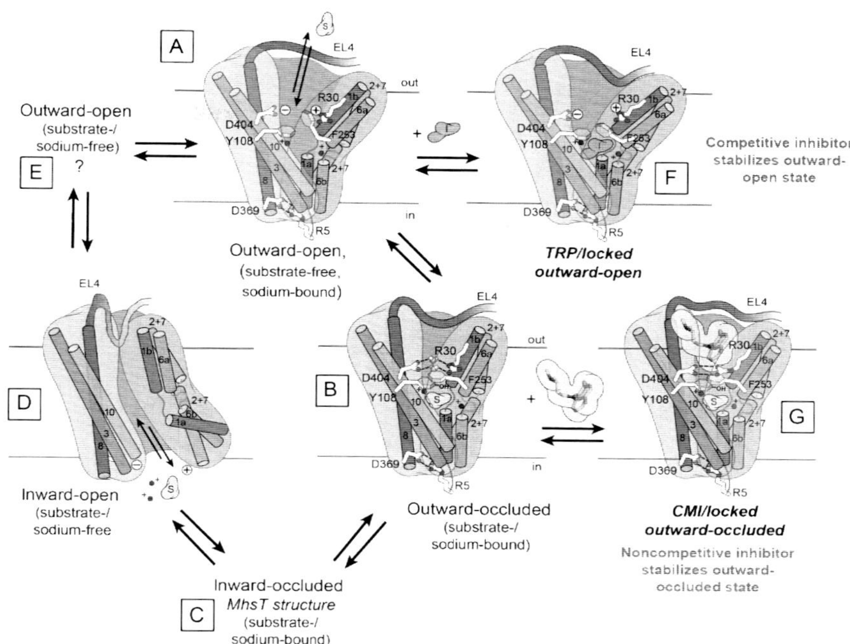

Figure 13 Schematic of transport and inhibition based exclusively on crystal structures and steady-state kinetic data of LeuT except for the substrate/sodium-bound inward-occluded state, which is based on MhsT, and the apo outward-open state, for which no representative structure is yet available. Conformational changes associated with isomerization from (A) sodium-bound, substrate-free outward-open to (B) substrate/sodium-bound outward-occluded state to (C) sodium/substrate-bound inward-occluded (MhsT) to (D) apo inward-open to (E) sodium/substrate-free outward-open. Panel (F) and (G), respectively, depict structures of the competitively inhibited, stabilized outward-open Trp/Na^+-bound and the noncompetitively inhibited, stabilized outward-occluded Leu/Na^+/CMI-bound states. TM5 is not shown for clarity, although, as mentioned, it does play an important role in the translocation cycle. (See the color plate.)

(POPC; Fig. 11B), at a protein:lipid ratio of 1:400 (Yamashita et al., 2005) and later 1:100 (Singh et al., 2007, 2008). These assays were successful and demonstrated that [^{3}H]Leu transport is electrogenic and Na^+- but not Cl^--dependent, like other bacterial SLC6 members but unlike its eukaryotic cousins.

Nevertheless, sequence comparisons between LeuT and eukaryotic SLC6 members permitted identification of a putative Cl^--binding site in SERT (Forrest, Tavoulari, Zhang, Rudnick, & Honig, 2007), GAT1, GAT4, and DAT (Zomot et al., 2007) and demonstration that Cl^-'s negative charge (E290 in LeuT) promotes Na^+ binding and substrate binding/translocation. The proposed site was later confirmed with crystal structures of 1) LeuT-E290S complexed with Na^+ and Cl^- or Br^-, the either anion coordinated by Y47, Q250, T254, and S290 (Kantcheva et al., 2013); 2) LeuBAT, a hybrid between LeuT and SERT/DAT/NET (Wang et al., 2013); and 3) the aforementioned thermostabilized dDAT mutant (Penmatsa, Wang, & Gouaux, 2013).

Successive flux assays illustrated that LeuT transports a wide array of aliphatic/aromatic amino acids, including Gly, Ala, Met, and Tyr, in addition to its namesake Leu, with catalytic efficiencies roughly corresponding to the inverse of substrate volume. Steady-state kinetics established that Ala is actually a much better substrate than Leu, with a catalytic rate constant (k_{cat}) at least sixfold greater, and hence far superior for examining inhibition (Singh et al., 2007, 2008). These experiments indicated that although Trp can bind LeuT, it cannot be transported, and is instead a competitive inhibitor that stabilizes the outward-open conformation. Competitive inhibitory behavior was manifested by an obvious increase in Michaelis constant (K_m) but no change in maximum velocity (V_{max}) when concentrations of [^{3}H]Ala and Trp were varied as substrate and inhibitor, respectively, graphically portrayed on an Eadie-Hofstee (EH) plot as nonparallel lines intersecting on the y-axis (Singh et al., 2008).

Since LeuT is a homolog of the clinically important SERT, DAT, NET, GAT1, and GlyTs, an immediate goal was to determine if any of the psychoactive agents that target these proteins can inhibit LeuT. A comprehensive screen with LeuT proteoliposomes revealed that the TCA CMI is the most potent, with IC_{50}s of 250 and 5 μM, for inhibition of [^{3}H]Leu and [^{3}H]Ala uptake, respectively. Unlike the aliphatic and aromatic amino acids, none of the TCAs (CMI, IMI, or DMI) could displace bound [^{3}H]Leu, provisionally excluding competitive inhibition as the mechanism. To conclusively define the mode of inhibition, steady-state kinetics with varying concentrations of [^{3}H]Ala as the substrate and CMI as the inhibitor were conducted. Contrary to Trp inhibition kinetics, there was no change in K_m but a noticeable reduction in V_{max}, graphically illustrated by parallel lines on an EH plot, unambiguous indicators of noncompetitive inhibition (Singh et al., 2007). As mentioned above, all LeuT–TCA structures suggested that these drugs stabilize the occluded state. To provide functional support for this postulate, dissociation assays with bound [^{3}H]Leu in the presence and absence of 3 mM CMI were performed. [^{3}H]Leu instead of [^{3}H]Ala was used for these experiments because the latter dissociates too quickly to accurately measure its off-rate. Data showed that CMI attenuates [^{3}H]Leu release by almost 700-fold (Fig. 8D). Thus, the structure in combination with steady-state kinetic and dissociation data demonstrated that a noncompetitive inhibitor binding at S2 works by strengthening the R30/D404 salt bridge at the extracellular gate.

While it is tantalizing to surmise that the TCA/SSRI site observed in LeuT is present in NET and SERT, the actual targets in humans, almost all functional data on mammalian SERT, NET, and DAT indicate that these compounds inhibit monoamine transport competitively and bind at or near S1 (Kristensen et al., 2011). Moreover, alanine mutants of analogous residues within 5 Å of S2 in human SERT decreased CMI potency by only two- to fivefold (Singh, 2008; Zhou et al., 2007), far less than the values reported for S1 mutants and incompatible with molecular docking (Kristensen et al., 2011). The only region that seemed to have a more pronounced effect, at least for the SSRI sertraline, was termed the "halogen binding pocket" (Zhou et al., 2009), especially residue I111 in LeuT (I179, I155, and F168 in human SERT, NET, and DAT, respectively).

A recent structure of the thermostabilized dDAT variant complexed with Na^+, Cl^-, and the TCA nortriptyline (Penmatsa et al., 2013), along with several of LeuBAT, each complexed with Na^+, Cl^-, and a clinically relevant antidepressant (Wang et al., 2013), may seem to have resolved the issue, but there are serious caveats. The transporters are presumably trapped in an outward-open conformation, with all of the drugs bound in S1 except for the serotonin–norepinephrine reuptake inhibitor desvenlafaxine, which is bound in both S1 and S2. Although the authors contend that they have conclusively resolved the controversy over where these compounds bind and how they competitively inhibit substrate transport, their constructs cannot transport substrate and perhaps are perpetually frozen in the outward-open state. Furthermore, there is no evidence that these proteins can even bind substrate despite the fact that a simple displacement assay was already established to gauge relative antagonist potencies. Thus, it is impossible for the authors to directly correlate kinetic inhibition patterns with substrate/antagonist-binding sites and associated conformational changes, as was accomplished with both the noncompetitively inhibited LeuT–CMI and competitively inhibited LeuT–Trp complexes. Knowledge of both structure *and* kinetics is crucial for designing more selective pharmacological agents with fewer side effects.

Note that the ambiguity does not mitigate the significance of definitive drug binding in S2 of LeuT. First, the data advance a tangible, testable hypothesis for the *general* phenomenon of noncompetitive inhibition that may be applicable to eukaryotic SLC6 members. Second, they pinpoint a potential antagonist-binding site that may exist in eukaryotic cousins and be exploited in rational drug design efforts. Two possible examples of such inhibitors include the NET-specific χ-conotoxin MrIA (Paczkowski, Sharpe, Dutertre, & Lewis, 2007) and the GlyT2-specific *N*-arachidonylglycine (Edington et al., 2009). In addition, S2 cannot be excluded as the low-affinity allosteric site reported in SERT (Wennogle & Meyerson, 1982). Ibogaine, a noncompetitive inhibitor of SERT (Jacobs, Zhang, Campbell, & Rudnick, 2007) and DAT (Bulling et al., 2012) does not fit into the "S2 category" because it stabilizes the inward-open conformation and binds to an unknown site, ostensibly accessible from the extracellular side.

3.2 Binding via scintillation proximity, equilibrium dialysis, and isothermal titration calorimetry

Another method of assessing transporter function is via binding of a cognate ligand. For LeuT, this can be any one of the amino acid substrates, the competitive inhibitor Trp, or the noncompetitive TCAs/SSRIs. Multiple means have been applied to evaluate LeuT function other than standard filter binding, including scintillation proximity assay (SPA), equilibrium dialysis (ED), and isothermal titration calorimetry (ITC). Probably the most convenient but also contentious is SPA. It utilizes fluoromicrospheres coated with a capture molecule (often copper chelate) to which a protein of interest (usually histidine-tagged) can adhere. The beads are filled with scintillant that emits light only when excited by a radiolabel bound either directly to the bead or to an attached target protein (Harder & Fotiadis, 2012; Quick & Javitch, 2007). Despite its advantages, one serious limitation is the unreliability of scintillation counting efficiency, potentially leading to errors when converting from cpm to moles substrate and thus, to substrate: protein molar binding stoichiometry. The "SPA" controversy has arisen from this very issue, which has led to divergence in published stoichiometry measurements, i.e., 2:1 versus 1:1 (Lim & Miller, 2012). That said, complementary binding methods such as ED have reinforced the disparity (see below).

What could be causing the discrepancy? One problem may stem from the fact that almost all crystallographic and binding experiments have been performed in two distinct detergents, β-OG versus DDM, the former of which binds in S2 (Wang et al., 2012) to conceivably displace a second substrate and inhibit transport (Quick et al., 2009). A second complication may focus on DDM concentration itself. SPA and ED conducted with LeuT at two different DDM concentrations (2 versus 6 m*M*) reported a substrate: protein binding stoichiometry of 1:1 in 6 m*M* (Piscitelli et al., 2010;

Quick, Shi, Zehnpfennig, Weinstein, & Javitch, 2012) but 2:1 in 2 m*M* DDM (Quick et al., 2012). Even single mutations in S1 (F253A) and S2 (L400C, L400S, L400A), supposedly designed to obstruct substrate at these respective sites, have yielded conflicting outcomes. Moreover, whether these mutants actually obstruct the respective sites is also a source of debate (Piscitelli et al., 2010; Wang & Gouaux, 2012).

Although crystallography tends to negate the two-substrate theory, it is not irrefutable. First, despite the perception that LeuT purified in DDM and crystallized in bicelles composed of the detergent CHAPSO (Fig. 1G) and the lipid DMPC (Fig. 11C), revealed no substrate-like density in S2, DDM associates/dissociates extremely slowly (Quick et al., 2012) so it is possible that some DDM remained in the bicelles, invisible because of disorder (Lim & Miller, 2012). Second, although the inward-open LeuT conformation showed a collapsed S2, with presumably no room for a second substrate, there is some unspecified electron density present which has yet to be unequivocally identified (Krishamurthy & Gouaux, 2012). This finding directly contradicts the MhsT structure, which, with its bulkier Trp instead of Leu at position 29, possesses no mysterious density in this area (Malinauskaite et al., 2014).

3.3 SPA and LeuT-nanodiscs

The above binding methods were all performed in either DDM or the newly synthesized lauryl maltose neopentyl glycol (LMNPG; Fig. 1F), in which many membrane proteins, including LeuT, are more stable (Chae et al., 2010). Still, ligand binding should ideally take place with protein reconstituted into lipid bilayers. Proteoliposomes are an option and were employed in one investigation (Quick et al., 2012). Nanodiscs are another and have recently been tested in combination with SPA (Nasr & Singh, 2014). Equally crucial is the use of BioBeads to remove detergent, which were utilized in both proteoliposome and nanodisc preparations but not with bicelles.

SPA with both [^{3}H]Leu and [^{3}H]Ala showed that LeuT is almost twice as active in nanodiscs versus DDM, as would be expected for a membrane protein in a lipid bilayer, but the increase in B_{max} values for both substrates is intriguing given the clash over the number of substrate-binding sites. It is plausible that the twofold rise is due to simultaneous binding of two substrates, but this interpretation will not be convincing until binding to the S1/S2 knockout mutant proteins is also gauged in the same hydrophobic nanodisc environment. Therefore, existence of dual substrate binding in LeuT remains uncertain at this juncture. Curiously, the protein-to-substrate molar binding stoichiometry in MhsT, which, as mentioned above, possesses the larger Trp instead of Leu in S2, like all other SLC6 members, is 1:1 (Malinauskaite et al., 2014) versus the 2:1 reported for LeuT. Thus, if the 2:1 stoichiometry is correct, it might hold true for LeuT exclusively and not the other members of the SLC6 family.

3.4 Spectroscopy

Two spectroscopic methods have been elegantly applied to LeuT, complementing static crystallographic experiments as well as bulk, steady-state binding/flux assays: pulsed electron paramagnetic resonance (EPR), along with double electron–electron resonance (DEER), and single-molecule Förster resonance energy transfer (smFRET). These techniques are fundamental for unraveling protein-mediated reaction mechanisms due to their intrinsic ability to track each step in real-time. Both require strategically placed cysteines on which to attach either nitroxide spin labels (EPR and DEER) or appropriate fluorophores (smFRET). These in turn report intervening distances and angles between two probes at various points along a reaction pathway, a reflection of associated conformational changes in response to the presence/absence of substrates, ions, and/or antagonists. LeuT is ideal for this work because it has no endogenous cysteines.

Four groundbreaking papers have capitalized on these spectroscopies, which have collectively validated, explained, and/or rebutted observations from bulk binding/transport and/or direct structural studies. One EPR study centered on the dynamics associated with Na^+, substrate, and inhibitor binding (Claxton et al., 2010), the results of which are largely concordant with solved crystal structures. In the apo state (no Na^+, substrate, or inhibitor), the nature of which remains impervious to crystallography, two populations emerged, one major (~26 Å) and one minor (~30–40 Å), interpreted as open and closed conformers, respectively. As anticipated, Na^+ and Na^+/Trp stabilized the outward-open, while Na^+/Leu and Na^+/Leu/CMI or Na^+/Leu/β-OG stabilized the outward-occluded conformations. These results are nearly congruous with previously solved structures. Interestingly, distance distribution shits, coupled with accessibility measurements, suggested that the Na^+-bound outward-open state is more conformationally heterogeneous than either the Na^+/Leu-bound outward-occluded or Na^+/Trp-bound outward-open states, which may clarify why LeuT–Na^+ required Fab complexation to crystallize.

One smFRET study (Zhao et al., 2011) concentrated on ligand-induced gating dynamics and the relative activation energies required to transition from outward- to inward-open states. LeuT variants with cysteine substitutions on nonconserved, intracellular residues (H7C [distal N-terminus] and R86C [IL1] or T515C [TM12 C-terminal end]), were labeled with Cy3 or Cy5 maleimide, respectively. In the absence of Na^+, two FRET states were detected, differing by ~13 Å between labeled pairs and thus inferred as outward- and inward-open conformations, consistent with EPR data. Increasing Na^+ concentrations diminished the two FRET distributions, selectively stabilizing the inward-closed state by approximately sevenfold.

Aiming to discern conformational changes associated with the transport process, Leu was then added at a Na^+ concentration sufficient for Leu binding, and the inward-closed state was stabilized by ~3.5-fold under these conditions. By contrast, when Ala was the substrate, rates between the two FRET states increased by approximately fourfold. Detailed transition state analysis indicates that whereas Ala substantially reduces, Leu raises the activation energy barrier between the outward- and inward-open states, providing a satisfying, time-resolved thermodynamic explication for Ala's higher catalytic rate constant (Singh et al., 2008). The nature of the coupling cation and presence of inhibitors also impacted intracellular gating dynamics. Substituting Na^+ with saturating concentrations of Li^+, a cation with a ~0.3 Å smaller ionic radius (Shannon, 1976), permitted Ala binding to LeuT, but not transport, steady-state data buttressed by the fact that Ala/Li^+ stabilizes the inward-closed state by twofold and thus is unable to enhance intracellular opening. As expected, the inhibitors CMI and β-OG, in the presence of Ala and Na^+, completely blocked intracellular opening, stabilizing LeuT in a single high-FRET peak representing the inward-closed conformation. These time-resolved results nicely recapitulate the IC_{50} values reported previously (Singh et al., 2007) and again illustrate how time-resolved single-molecule spectroscopic data can furnish gratifying molecular explanations for bulk measurements.

Another smFRET study (Zhao et al., 2010) also focused on intracellular motions but this time further evaluated the degree of conformational coupling between both sides. It used the same three LeuT mutants delineated above and added two extracellular cysteine substitutions (K239C [EL3] and H480C [EL6]) as well as two more cytoplasmic ones (R185C [IL2] and K271C [IL3]). The second pair did not exhibit the Na^+- or Na^+/Leu-dependent changes of H7C/R86C or H7C/T515C, consistent with the view that IL2/IL3 does not move as much during intracellular gating. These mutations were made in otherwise WT LeuT, the results of which, predictably paralleled those of the 2011 smFRET study. To investigate the nature of coupling, the same series of mutants was also generated in

LeuT-R5A, -Y268A, and R30A backgrounds, the former two of which disrupt the *intracellular gate* to favor the inward-open state (Kniazeff et al., 2008) and the latter of which putatively disturbs the *extracellular gate* by precluding formation of both the salt bridge with D404 and the cation–π interaction with F253. Interestingly, without Na^+, the presence of R5A or Y268A markedly affected FRET distributions for both intracellular H7C/R86C as well as extracellular K239C/H490C, hinting that intracellular outward and extracellular inward movements occur concomitantly. By contrast, the presence of R30A affected only H7C/R86C.

This pioneering smFRET work thus argues that conformational transitions between the two sides are communicated through a linked series of small local, flexible rearrangements instead of one concerted tilt of a rigid bundle and is also partly suggestive of large movements in TM1a. Nevertheless, there was no indication that TM1a moves as much as 45° into the membrane. Indeed, evidence from a subsequent series of EPR experiments supports this contention (Kazmier et al., 2014). Although the broad EPR line distributions were consonant with LeuT's innately dynamic character, and TM1a of the Y268A mutant did move far into the presumed lipid bilayer, it did not exhibit such a large amplitude change in the wild-type protein, as implied by the structure of the inward-open, highly mutagenized Fab-complexed LeuT. The fact that R5A also mimicked the anomalous behavior of Y268A reinforces the notion that tampering with intracellular gating residues skews conformation and is likely an artifact. An additional complication with the inward-open structure is the fact that the 6A10 Fab abuts the intracellular face and protrudes into the intracellular-facing cavity (Fig. 12). Its presence on the same side as the one that moves in the inward-open conformation may prevent TM1a from resting there and possibly force it to flip up as much as it does. It is also conceivable that the drastic movement is a synergistic combination of the Y268A mutant and Fab. Regardless, such sizable bending is incompatible with solution EPR data of WT LeuT.

4. TRANSPORT MECHANISM UNVEILED FROM STRUCTURE, FUNCTION, AND COMPUTATIONAL BIOLOGY

The most prevalent theory to describe mechanism in secondary transporters is one in which substrates and ions bind to unique sites on the protein and an ensuing isomerization permits alternating access to either the extracellular or cytoplasmic side. This thought was first drafted in outline form by Peter Mitchell over 50 years ago with his vision of a "translocase," a moving "barrier" that was alternately accessible from either side of the lipid bilayer (Mitchell, 1957). Some variations and details were added to this basic two-state concept over the next few years, reformulated and dubbed "gate-type non-carrier" by Patlak (1957), "two-shape allosteric" by Vidaver (1966), and "alternating access" by Jardetsky (1966).

The atomic framework for these nebulous ideas has emerged from dozens of crystal structures from many different SLC families. One model is based on the intriguing outward-occluded state of the original LeuT structure, which signifies that, in addition to outward- and inward-open configurations of the paradigmatic alternating access model, there is at least one more state where both gates are closed. The unwound helices of TMs 1 and 6 first portended that they might bend like "fingers" during the transport cycle. Movement of EL4 was implied by the lid-like structure it forms over EV, and movement of EL2 was suggested by its proximal position on top of EL4. The fact that access to the substrate-binding site from the extracellular side is blocked by only a few residues indicated that only modest rotations in key helices such as TMs 1b, 6a, 3, and 8, along with more dramatic shifts in EL2 and EL4, would be required for the transporter to open to the outside. By contrast, the 25 Å of ordered protein structure between the intracellular gate and S1 hinted that this entire region would have to rearrange considerably in order for the transporter to open to the inside. Indeed, LeuT's Na^+-bound/substrate-free outward-open, competitively inhibited outward-open, and apo inward-open as well as MhsT's Na^+/Trp inward-occluded states have largely confirmed such movements. Specifically, concomitant rotations of TMs 2, 5, and 7 along with considerable rearrangements of ELs 2, 3, and 4, complemented with spectroscopic data, all bolster these conclusions. However, the pronounced movement of TM1a into the bilayer is not supported by solution smFRET and EPR data and stands in stark contrast to the simple rotation of the structurally related transporters Mhp1 and BetP.

A second model is based on the 5 + 5 architecture and pseudosymmetry astutely noticed in the inverted repeats, which were "swapped" to create a model of the inward-occluded state (Forrest et al., 2008). This "rocking bundle" proposal has two elements, a "scaffold" comprised of TMs 3–5 and 8–10 and a "bundle" comprised of TMs 1–2 and 6–7, the latter of which rocks back and forth relative to the "scaffold" to alternately expose S1 to the external and internal milieu. Accessibility measurements in SERT as well as the inward-facing vSGLT, Mhp1, and several BetP structures mostly support the premise. Although the "rocking bundle" hypothesis emphasizes tilting of a rigid bundle and cannot inherently predict EL and IL movements, it does not exclude the possibility of internal rearrangements and/or bending of individual helices envisioned by the first model (Forrest & Rudnick, 2009).

A third model also posits that the occluded state is an essential intermediate but that binding of a second substrate molecule in S2 acts as an allosteric trigger for release of S1-bound substrate and sodium to the intracellular medium (Shi, Quick, Zhao, Weinstein, & Javitch, 2008). S2-bound substrate is hypothesized to act as a "symport effector," whereas hydrophobic molecules that bind in S2 such as β-OG, TCAs, and SSRIs inhibit transport by serving as "symport uncouplers." This unorthodox concept, derived from SPA data and "steered" molecular dynamics, deviates significantly from the traditional alternating access mechanism and has evolved into a contentious issue (Lim & Miller, 2012). Nevertheless, the idea may be applicable just to LeuT, being the only SLC6 member with published evidence for a 2:1 molar substrate:protein binding stoichiometry. Notably, LeuT is the only homolog with Leu at position 29 in S2, whereas all others possess the larger Trp, which in MhsT completely collapses the EV, including S2.

5. SUMMARY

Biophysical studies on LeuT have catapulted our knowledge of both SLC6 members and structurally related SLCs, unraveling molecular details about transport mechanism, inhibition, substrate/ion-binding sites, and conformational dynamics. More importantly, they have provided an experimental platform on which the clinically significant eukaryotic SLC6 members can eventually be pursued via complementary crystallographic, steady-state kinetic, and real-time spectroscopic methods.

ACKNOWLEDGMENTS

The authors thank Gary Rudnick for insightful comments and Suraj Adhikary for assistance with figures. Research in the corresponding author's laboratory was/is funded by Yale University (Goodman-Gilman Scholar Award), the Alfred P. Sloan Foundation, the US-Israel Binational Science Foundation (BSF 2013198), and the NIH (R00MH083050, R21MH098180, and R01MH100688).

REFERENCES

Adams, M. D., Wagner, L. M., Graddis, T. J., Landick, R., Antonucci, T. K., Gibson, A. L., et al. (1990). Nucleotide sequence and genetic characterization reveal six essential genes for the LIV-I and LS transport systems of *Escherichia coli*. *The Journal of Biological Chemistry*, *265*, 11436–11443.

Androutsellis-Theotokis, A., Goldberg, N. R., Ueda, K., Beppu, T., Beckman, M. L., Das, S., et al. (2003). Characterization of a functional bacterial homologue of sodium-dependent neurotransmitter transporters. *The Journal of Biological Chemistry*, *278*, 12703–12709.

Bulling, S., Schicker, K., Zhang, Y. W., Steinkeller, T., Stockner, T., Gruber, C. W., et al. (2012). The mechanistic basis for noncompetitive ibogaine inhibition of serotonin and dopamine transporters. *The Journal of Biological Chemistry*, *287*, 18524–18534.

Celik, L., Sinning, S., Severinsen, K., Hansen, C. G., Moller, M. S., Bols, M., et al. (2008). Binding of serotonin to the human serotonin transporter. Molecular modeling and experimental validation. *Journal of the American Chemical Society*, *130*, 3853–3865.

Chae, P. S., Rasmussen, S. G. F., Rana, R. R., Gotfryd, K., Chandra, R., Goren, M. A., et al. (2010). Maltose-neopentyl glycol (MNG) amphiphiles for solubilization, stabilization, and crystallization of membrane proteins. *Nature Methods*, *7*, 1003–1008.

Chen, J. G., Liu-Chen, S., & Rudnick, G. (1998). Determination of external loop topology in the serotonin transporter by site-directed chemical labeling. *The Journal of Biological Chemistry*, *273*, 12675–12681.

Christensen, J. J., Hill, J. O., & Izatt, R. M. (1971). Ion binding by synthetic macrocyclic compounds. *Science*, *174*, 459–467.

Claxton, D. P., Quick, M., Shi, L., de Carvalho, F. D., Weinstein, H., Javitch, J. A., et al. (2010). Ion/substrate-dependent conformational dynamics of a bacterial homolog of neurotransmitter:sodium symporters. *Nature Structural & Molecular Biology*, *17*, 822–829.

Deckert, G., Warren, P. V., Gaasterland, T., Young, W. G., Lenox, A. L., Graham, D. E., et al. (1998). The complete genome of the hyperthermophilic bacterium *Aquifex aeolicus*. *Nature*, *392*, 353–358.

Edington, A. R., McKinzie, A. A., Reynolds, A. J., Kassiou, M., Ryan, R. M., & Vanderberg, R. J. (2009). Extracellular loops 2 and 4 of GLYT2 are required for N-arachidonylglycine inhibition of glycine transport. *The Journal of Biological Chemistry*, *284*, 36424–36430.

Ehrnstorfer, I. A., Geertsma, E. R., Pardon, E., Steyaert, J., & Dutzler, R. (2014). Crystal structure of a SLC11 (NRAMP) transporter reveals the basis for transition-metal ion transport. *Nature Structural & Molecular Biology*, *21*, 990–996.

Forrest, L. R., & Rudnick, G. (2009). The rocking bundle: A mechanism for ion-coupled solute flux by symmetrical transporters. *Physiology*, *24*, 377–386.

Forrest, L. R., Tavoulari, S., Zhang, Y. W., Rudnick, G., & Honig, B. (2007). Identification of a chloride ion binding site in Na^+/Cl^--dependent neurotransmitter transporters. *Proceedings of the National Academy of Sciences of the United States of America*, *104*, 12761–12766.

Forrest, L. R., Zhang, Y. W., Jacobs, M. T., Gesmonde, J., Xie, L., Honig, B. H., et al. (2008). Mechanism for alternating access in neurotransmitter transporters. *Proceedings of the National Academy of Sciences of the United States of America*, *105*, 10338–10343.

Guastella, J., Nelson, N., Nelson, H., Czyzyk, L., Keynan, S., Miedel, M. C., et al. (1990). Cloning and expression of a rat brain GABA transporter. *Science*, *249*, 1303–1306.

Hahn, M. K., & Blakely, R. D. (2007). The functional impact of SLC6 transporter genetic variation. *Annual Review of Pharmacology and Toxicology*, *47*, 401–441.

Harder, D., & Fotiadis, D. (2012). Measuring substrate binding and affinity of membrane transport proteins using the scintillation proximity assay. *Nature Protocols*, *7*, 1569–1578.

Hendrickson, W. A., Horton, J. R., & LeMaster, D. M. (1990). Selenomethionyl proteins produced for analysis by multiwavelength anomalous diffraction (MAD): A vehicle for direct determination of three-dimensional structure. *The EMBO Journal*, *9*, 1665–1672.

Hetting, G., & Axelrod, J. (1961). Fate of tritiated noradrenalin at the sympathetic nerve-endings. *Nature*, *192*, 172–173.

Huang, X., & Zhan, C. G. (2007). How dopamine transporter interacts with dopamine: Insights from molecular modeling and simulation. *Biophysical Journal*, *93*, 3627–3639.

Iversen, L. L. (1963). The uptake of noradrenalin by the isolated perfused heart. *British Journal of Pharmacology and Chemotherapy*, *21*, 523–537.

Iversen, L. L. (1965). Inhibition of noradrenaline uptake by drugs. *The Journal of Pharmacy and Pharmacology*, *17*, 62–64.

Iversen, L. L. (1971). Role of transmitter uptake mechanisms in synaptic neurotransmission. *British Journal of Pharmacology*, *41*, 571–591.

Iversen, L. L., Jarrott, B., & Simmonds, M. A. (1971). Differences in the uptake, storage and metabolism of (+)- and (−)-noradrenaline. *British Journal of Pharmacology*, *43*, 845–855.

Iversen, L. L., & Kravitz, E. A. (1966). Sodium dependence of transmitter uptake at adrenergic nerve terminals. *Molecular Pharmacology*, *2*, 360–362.

Jacobs, M. T., Zhang, Y. W., Campbell, S. D., & Rudnick, G. (2007). Ibogaine, a noncompetitive inhibitor of serotonin transport, acts by stabilizing the cytoplasmic-facing of the transporter. *The Journal of Biological Chemistry*, *282*, 29441–29447.

Jardetsky, O. (1966). Simple allosteric model for membrane pumps. *Nature*, *211*, 969–970.

Kanner, B. I. (1978). Active transport of gamma-aminobutyric acid by membrane vesicles isolated from rat brain. *Biochemistry*, *17*, 1207–1211.

Kantcheva, A. K., Quick, M., Shi, L., Winther, A. M., Stolzenberg, S., Weinstein, H., et al. (2013). Chloride binding site of neurotransmitter transporters. *Proceedings of the National Academy of Sciences of the United States of America*, *110*, 8489–8494.

Kawate, T., & Gouaux, E. (2006). Fluorescence-detection size-exclusion chromatography for precrystallization screening of integral membrane proteins. *Structure*, *14*, 673–681.

Kazmier, K., Sharma, S., Quick, M., Islam, S. M., Roux, B., Weinstein, H., et al. (2014). Conformational dynamics of ligand-dependent alternating access in LeuT. *Nature Structural & Molecular Biology*, *21*, 472–479.

Kniazeff, J., Shi, L., Loland, C. J., Weinstein, H., & Gether, U. (2008). An intracellular interaction network regulates conformational transitions in the dopamine transporter. *The Journal of Biological Chemistry*, *283*, 17691–17701.

Krishnamurthy, H., & Gouaux, E. (2012). X-ray structures of LeuT in substrate-free outward-open and apo inward-open states. *Nature*, *481*, 469–474.

Kristensen, A. S., Andersen, J., Jorgensen, T. N., Sorensen, L., Eriksen, J., Loland, C. J., et al. (2011). SLC6 neurotransmitter transporters: Structure, function, and regulation. *Pharmacy Review*, *63*, 585–640.

Kunji, E. R., Slotboom, D. J., & Poolman, B. (2003). *Lactococcus lactis* as host for overproduction of functional membrane proteins. *Biochimica et Biophysica Acta*, *1610*, 97–108.

Langer, G., Cohen, S. X., Lazmin, V. S., & Perrakis, A. (2008). Automated macromolecular model building for X-ray crystallography using ARP/wARP version 7. *Nature Protocols*, *3*, 1171–1179.

Lim, H. H., & Miller, C. (2012). It takes two to transport, or is it one? *Nature Structural & Molecular Biology*, *19*, 129–130.

Malinauskaite, L., Quick, M., Reinhard, L., Lyons, J. A., Yano, H., Javitch, J. A., et al. (2014). A mechanism for intracellular release of Na^+ by neurotransmitter/sodium symporters. *Nature Structural & Molecular Biology*, *21*, 1006–1012.

Masson, J., Sagné, C., Hamon, M., & El Mestikawy, S. (1999). Neurotransmitter transporters in the central nervous system. *Pharmacy Review*, *51*, 439–464.

Miroux, B., & Walker, J. E. (1996). Over-production of proteins in *Escherichia coli*: Mutant hosts that allow synthesis of some membrane proteins and globular proteins at high levels. *Journal of Molecular Biology*, *260*, 289–298.

Mitchell, P. (1957). A general theory of membrane transport from studies of bacteria. *Nature*, *180*, 134–136.

Moraes, I., Evans, G., Sanchez-Weatherby, J., Newstead, S., & Stewart, P. D. S. (2014). Membrane protein structure determination—The next generation. *Biochimica et Biophysica Acta*, *1838*, 78–97.

Nasr, M. L., & Singh, S. K. (2014). Radioligand binding to nanodisc-reconstituted membrane transporters assessed by the scintillation proximity assay. *Biochemistry*, *53*, 4–6.

Nayal, M., & Di Cera, E. (1996). Valence screening of water in protein crystals reveals potential Na^+ binding sites. *Journal of Molecular Biology*, *256*, 228–243.

Nelson, N. (1998). The family of Na^+/Cl^- neurotransmitter transporters. *Journal of Neurochemistry*, *71*, 1785–1803.

Pacholczyk, T., Blakely, R. D., & Amara, S. G. (1991). Expression cloning of a cocaine- and antidepressant-sensitive human noradrenaline transporter. *Nature*, *350*, 350–354.

Paczkowski, F. A., Sharpe, I. A., Dutertre, S., & Lewis, R. J. (2007). chi-Conopeptide and tricyclic antidepressant interactions at the norepinephrine transporter define a new transporter model. *The Journal of Biological Chemistry*, *282*, 17837–17844.

Patlak, C. S. (1957). Contributions to the theory of active transport: II. The gate type non-carrier mechanism and generalizations concerning tracer flow, efficiency, and measurement of energy expenditure. *The Bulletin of Mathematical Biophysics*, *19*, 209–235.

Penmatsa, A., Wang, K. H., & Gouaux, E. (2013). X-ray structure of dopamine transporter elucidates antidepressant mechanism. *Nature*, *503*, 85–90.

Piscitelli, C. L., Krishnamurthy, H., & Gouaux, E. (2010). Neurotransmitter/sodium symporter orthologue LeuT has a single high-affinity substrate site. *Nature*, *468*, 1129–1132.

Quick, M., & Javitch, J. A. (2007). Monitoring the function of membrane transport proteins in detergent-solubilized form. *Proceedings of the National Academy of Sciences of the United States of America*, *104*, 3603–3608.

Quick, M., Shi, L., Zehnpfennig, B., Weinstein, H., & Javitch, J. A. (2012). Experimental conditions can obscure the second high-affinity site in LeuT. *Nature Structural & Molecular Biology*, *19*, 207–211.

Quick, M., Winther, A. M., Shi, L., Nissen, P., Weinstein, H., & Javitch, J. A. (2009). Binding of an octylglucoside detergent molecule in the second substrate (S2) site of LeuT establishes an inhibitor bound conformation. *Proceedings of the National Academy of Sciences of the United States of America*, *106*, 5563–5568.

Radian, R., & Kanner, B. I. (1985). Reconstitution and purification of the sodium- and chloride-coupled gamma-aminobutyric acid transporter from rat brain. *The Journal of Biological Chemistry*, *260*, 11859–11865.

Rudnick, G. (1977). Active transport of 5-hydroxytryptamine by plasma membrane vesicles isolated from human blood platelets. *The Journal of Biological Chemistry*, *252*, 2170–2174.

Shannon, R. D. (1976). Revised effective ionic radii and systematic studies of interatomic distances in halides and chalcogenics. *Acta Crystallographica*, *A32*, 751–767.

Shi, Y. (2013). Common folds and transport mechanisms of secondary active transporters. *Annual Review of Biophysics*, *42*, 51–72.

Shi, L., Quick, M., Zhao, Y., Weinstein, H., & Javitch, J. A. (2008). The mechanism of a neurotransmitter:sodium symporter—Inward release of Na^+ and substrate is triggered by substrate in a binding site. *Molecular Cell*, *30*, 667–677.

Shimamura, T., Weyand, S., Beckstein, O., Rutherford, N. G., Hadden, J. M., Sharples, D., et al. (2010). Molecular basis of alternating access membrane transport by the sodium-hydantoin transporter Mhp1. *Science*, *328*, 470–473.

Singh, S. K. (2008). LeuT: A prokaryotic stepping stone on the way to a eukaryotic neurotransmitter transporter structure. *Channels (Austin, Tex)*, *2*, 380–389.

Singh, S. K., Piscitelli, C. L., Yamashita, A., & Gouaux, E. (2008). A competitive inhibitor stabilizes LeuT in an open-to-out conformation. *Science*, *322*, 1655–1661.

Singh, S. K., Yamashita, A., & Gouaux, E. (2007). Antidepressant binding site in a bacterial homologue of neurotransmitter transporters. *Nature*, *448*, 952–956.

Tate, C. G. (2010). Practical considerations of membrane protein instability during purification and crystallization. *Methods in Molecular Biology*, *601*, 187–203.

Tate, C. G., Haase, J., Baker, C., Boorsma, M., Magnani, F., Vallis, Y., et al. (2003). Comparison of seven different heterologous protein expression systems for the production of the serotonin transporter. *Biochimica et Biophysica Acta*, *1610*, 141–153.

Vanhatalo, S., & Sohila, S. (1998). The concept of chemical neurotransmission—Variations on a theme. *Annals of Medicine*, *20*, 151–158.

Vidaver, G. A. (1966). Inhibition of parallel flux and augmentation of counter flux by transport models not involving a mobile carrier. *Journal of Theoretical Biology*, *10*, 301–306.

Wang, H., Elferich, L., & Gouaux, E. (2012). Structures of LeuT in bicelles define conformation and substrate binding in a membrane-like context. *Nature Structural & Molecular Biology*, *19*, 212–219.

Wang, H., Goehring, A., Wang, K. H., Penmasta, A., Ressler, R., & Gouaux, E. (2013). Structural basis for action by diverse antidepressants on biogenic amine transporters. *Nature*, *503*, 141–145.

Wang, H., & Gouaux, E. (2012). Substrate binds in the S1 site of the F253A mutant of LeuT, a neurotransmitter sodium symporter homologue. *EMBO Reports*, *13*, 861–866.

Watanabe, A., Choe, S., Chaptal, V., Rosenberg, J. M., Wright, E. M., Grabe, M., et al. (2010). The mechanism of sodium and substrate release from the binding pocket of vSGLT. *Nature*, *468*, 988–991.

Wennogle, L. P., & Meyerson, L. R. (1982). Serotonin modulates the dissociation of [3H] imipramine from human platelet recognition sites. *European Journal of Pharmacology*, *86*, 303–307.

Yamashita, A., Singh, S. K., Kawate, T., Jin, Y., & Gouaux, E. (2005). Crystal structure of a bacterial homologue of Na^+/Cl^--dependent neurotransmitter transporters. *Nature*, *437*, 215–223.

Zhao, Y., Terry, D. S., Shi, L., Quick, M., Weinstein, H., Blanchard, S. C., et al. (2011). Substrate-modulated gating dynamics in a Na^+-coupled neurotransmitter transporter homologue. *Nature*, *474*, 109–113.

Zhao, Y., Terry, D., Shi, L., Weinstein, H., Blanchard, S. C., & Javitch, J. A. (2010). Single-molecule dynamics of gating in a neurotransmitter transporter homologue. *Nature*, *465*, 188–193.

Zhou, Z., Zhen, J., Karpowich, N. K., Goetz, R. M., Law, C. J., Reith, M. E., et al. (2007). LeuT-desipramine structure reveals how antidepressants block neurotransmitter reuptake. *Science*, *317*, 1390–1393.

Zhou, Z., Zhen, J., Karpowich, N. K., Law, C. J., Reith, M. E., & Wang, D. N. (2009). Antidepressants specificity of serotonin transporter suggested by three LeuT-SSRI complex structures. *Nature Structural & Molecular Biology*, *16*, 652–657.

Zomot, E., Bendahan, A., Quick, M., Zhao, Y., Javitch, J. A., & Kanner, B. I. (2007). Mechanism of chloride interaction with neurotransmitter:sodium symporters. *Nature*, *449*, 726–730.

Generation and Use of Antibody Fragments Against Membrane Proteins

CHAPTER TEN

Generation of Recombinant Antibody Fragments for Membrane Protein Crystallization

Syed H. Mir*[,†], Claudia Escher*, Wei-Chun Kao*, Dominic Birth*[,‡], Christophe Wirth*, Carola Hunte*[,1]

*Institute for Biochemistry and Molecular Biology, ZBMZ, BIOSS Centre for Biological Signalling Studies, University of Freiburg, Freiburg, Germany
[†]Department of Clinical Biochemistry, University of Kashmir, Srinagar, India
[‡]Faculty of Biology, University of Freiburg, Freiburg, Germany
[1]Corresponding author: e-mail address: carola.hunte@biochemie.uni-freiburg.de

Contents

Abstract

Membrane proteins are challenging targets for crystallization and structure determination by X-ray crystallography. Hurdles can be overcome by antibody-mediated crystallization. More than 25 unique structures of membrane protein:antibody complexes have already been determined. In the majority of cases, hybridoma-derived antibody fragments either in Fab or Fv fragment format were employed for these complexes. We will briefly introduce the background and current status of the strategy and describe in detail the current protocols of well-established methods for the immunization, the selection, and the characterization of antibodies, as well as the cloning, the production, and the purification of recombinant antibodies useful for structural analysis of membrane proteins.

Methods in Enzymology, Volume 557
ISSN 0076-6879
http://dx.doi.org/10.1016/bs.mie.2014.12.029

1. INTRODUCTION

Antibody-mediated crystallization of membrane proteins has been instrumental to obtain X-ray structures of challenging protein targets (Hino, Iwata, & Murata, 2013; Hunte & Michel, 2002). It was introduced in 1995 with the rationale that enlarging the limited hydrophilic surface of detergent-solubilized membrane proteins will enhance the chances to obtain ordered crystals, as hydrophilic surfaces can mediate stable contacts between molecules in the crystal lattice and the extra domains provide more space for the detergent micelle (Iwata, Ostermeier, Ludwig, & Michel, 1995; Ostermeier, Iwata, Ludwig, & Michel, 1995). Binding of an antibody may also help to overcome inherent properties of membrane proteins that hamper crystallization as the complex can trap a defined conformation and stabilize flexible loops, domains, and thermolabile proteins. Antibodies may also permit selection for defined covalent modifications. A growing number of laboratories employ this strategy, and a number of methods have been developed. It is important to note that only antibody fragments are used for cocrystallization, as the bivalent binding mode and the flexible hinge region of native antibody molecules are detrimental for crystallization.

For structural studies, recombinant Fv and Fab fragments, proteolytically generated Fab fragments, and nanobodies are in use. So far, antibody fragments used for cocrystallization and structure determination of membrane proteins are mainly derived from hybridoma cells (Table 1). This includes the initial approach to clone the genes of the variable regions of heavy and light chains from the parental hybridoma cell line and to produce the Fv fragment by periplasmic expression in *Escherichia coli* (Kleymann, Ostermeier, Ludwig, Skerra, & Michel, 1995). Alternatively, Fab fragments obtained by proteolytic digestion of purified monoclonal antibodies (Zhou et al., 2001) are frequently used. More recently, alternative approaches with promising potential arose. Structures were also obtained with Fab fragments, which were selected from synthetic Fab libraries using phage display methods (Shukla et al., 2013; Uysal et al., 2009). In addition, recombinant single-domain antibodies or nanobodies, the VHH fragment of camelid heavy-chain antibodies, which were generated from immunized lama, were successfully used (Rasmussen, Choi, et al., 2011; Rasmussen, DeVree, et al., 2011; Kruse et al., 2013). Membrane protein structures in defined conformations were obtained for mutation-stabilized outward and inward facing conformation of the transporter LeuT in complex with specific antibodies

Table 1 Structures of membrane proteins obtained by antibody-mediated crystallization

Types of integral membrane protein (IMP)	Organism	PDB	Immunization	Year	References
Crystallized with hybridoma (mouse)-derived recombinant Fv fragments[a]					
Cytochrome *c* oxidase:7E2 Fv[b]	*P. denitrificans*	1AR1	Detergent-solubilized IMP	1997	Ostermeier, Harrenga, Ermler, and Michel (1997)
Cytochrome bc_1 complex:18E11 Fv[b]	*S. cerevisiae*	1EZV	Detergent-solubilized IMP	2000	Hunte, Koepke, Lange, Rossmanith, and Michel (2000)
KvAP voltage-gated potassium channel:33H1 Fv	*A. pernix*	2A0L	n.d.	2005	Lee, Lee, Chen, and MacKinnon (2005)
Crystallized with hybridoma (mouse) derived proteolytic Fab fragments[c]					
Potassium channel KcsA:Fab[b]	*S. lividans*	1K4D	n.d.	2001	Zhou, Morais-Cabral, Kaufman, and MacKinnon (2001)
Chloride channel EcClC:Fab[b]	*E. coli*	1OTS	n.d.	2003	Dutzler, Campbell, and MacKinnon (2003)
Voltage-dependent potassium channel KvAP:6E1 Fab[b]	*A. pernix*	1ORQ	n.d.	2003	Jiang et al. (2003)
KvAP potassium channel voltage sensor:33H1 Fab[b]	*A. pernix*	1ORS	n.d.	2003	Jiang et al. (2003)
β_2 Adrenergic G protein-coupled receptor:Fab5	*H. sapiens*	2R4R	Purified IMP reconstituted in liposomes	2007	Rasmussen et al. (2007)
SecYE translocon:Fab	*T. thermophilus*	2ZJS	n.d.	2008	Tsukazaki et al. (2008)

Continued

Table 1 Structures of membrane proteins obtained by antibody-mediated crystallization—cont'd

Types of integral membrane protein (IMP)	Organism	PDB	Immunization	Year	References
Virtual proton pump AdiC:Fab	*S. typhimurium*	3NCY	Detergent solubilised IMP	2009	Fang et al. (2009)
Na^+-independent amino acid transporter ApcT:7F11 Fab	*M. jannaschii*	3GI9	n.d.	2009	Shaffer, Goehring, Shankaranarayanan, and Gouaux (2009)
Disulfide bond formation protein DsbB: Fab	*E. coli*	2ZUQ	n.d.	2009	Inaba et al. (2009)
Nitric oxide reductase:Fab25	*P. aeruginosa*	3O0R	n.d.	2010	Hino et al. (2010)
Inhibitory anion-selective Cys-loop receptor GluCl:Fab[b]	*C. elegans*	3RHW	n.d.	2011	Hibbs and Gouaux (2011)
LeuT (outward-open state):2B12 Fab	*A. aelicus*	3TT1	Detergent-solubilized IMP	2012	Krishnamurthy and Gouaux (2012)
LeuT (inward-open state):6A10 Fab	*A. aelicus*	3TT3	Detergent-solubilized IMP	2012	Krishnamurthy and Gouaux (2012)
Adenosine receptor $A_{2A}AR$:Fab	*H. sapiens*	3VG9	Detergent-solubilized IMP	2012	Hino et al. (2012)
TRAAK K^+ channel:13E9 Fab	*H. sapiens*	4I9W	Detergent-solubilized IMP	2013	Brohawn, Campbell, and MacKinnon (2013)
Structure of dopamine transporter (DAT):9D5 Fab	*D. melanogaster*	4M48	n.d.	2013	Penmatsa, Wang, and Gouaux (2013)
Ca^{2+} activated chloride channel BEST1:10D19 Fab	*G. gallus*	4RDQ	Detergent-solubilized IMP	2014	Dickson, Pedi, and Long (2014)

Crystallized with recombinant Fab derived from minimal synthetic-Fab phage displayed library					
Potassium channel KcsA:Fab	*S. lividans*	3EFF	No immunization	2009	Uysal et al. (2009)
Crystallized with recombinant nanobody derived from camelid VHH-phage library (immunized lama)					
β_2-Adrenoceptor (nanobody stabilized active state):Nb80-6His nanobody	*H. sapiens*	3P0G	Purified IMP reconstituted in liposomes	2011	Rasmussen, Choi, et al. (2011)
β_2-Adrenergic receptor–Gs protein complex:Nb35-6His nanobody	*H. sapiens*	3SN6	Purified IMP reconstituted in liposomes	2011	Rasmussen, DeVree, et al. (2011)
Serotonin receptor:VHH15-6His nanobody	*M. musculus*	4PIR	Detergent-solubilized IMP	2014	Hassaine et al. (2014)
Crystallized with recombinant nanobody derived from camelid VHH-yeast library (immunized lama)					
Muscarinic acetylcholine receptor: 9-8 nanobody	*H. sapiens*	4MQS	Purified IMP reconstituted in liposomes	2013	Kruse et al. 2013

[a]Fv fragments produced by periplasmic expression in *E. coli*.
[b]A representative structure is included for protein:antibody pairs, for which several structures are available.
[c]Fab fragments were generated by papain proteolysis of IgG harvested from hybridoma supernatants except for the IgG for Fab25 which was purified from ascites fluid.
n.d., not described.

(Krishnamurthy & Gouaux, 2012). Furthermore, antibody complexes were critical for generating structures of signaling assemblies (Rasmussen, DeVree, et al., 2011; Shukla et al., 2013). The use of antibodies as crystallization chaperones of membrane proteins bears a great potential that can be harnessed with current methods and ongoing developments.

Further dissemination of the approach and concomitant development of methods will help to obtain high-resolution membrane protein structures important for mechanistic understanding as well as for drug development. It will aid exploration of the yet poorly described structure space of membrane proteins. Here, we will report current robust methods for the generation of antibodies suitable for cocrystallization of membrane proteins and for further structural and functional studies.

2. GENERATION OF ANTIBODIES FOR STRUCTURAL CHARACTERIZATION OF MEMBRANE PROTEINS

2.1 Immunization

A well-established strategy to generate an immune response against native membrane proteins is the immunization with detergent-solubilized purified protein. The following long-term immunization protocol was successfully used to generate native binders against diverse membrane proteins. The respective antibodies and antibody-derived fragments facilitated structural characterization of multisubunit respiratory complex III from *Saccharomyces cerevisiae* (Birth, Kao, & Hunte, 2014; Hunte et al., 2000; Lange & Hunte, 2002; Lange, Nett, Trumpower, & Hunte, 2001; Solmaz & Hunte, 2008) and complex I from *Yarrowia lipolytica* (Zickermann et al., 2003) and of the Na^+/H^+ antiporter NhaA from *E. coli* (Padan, Venturi, Michel, & Hunte, 1998; Venturi et al., 2000).

2.1.1 Method

One hundred micrograms of protein in detergent-containing buffer is mixed with complete adjuvant (trehalose dimycolate, monophosphoryl lipid A, 2% oil–Tween-80; e.g., ABM 2, Pan Biotech GmbH, Aidenbach, Germany) at a final concentration of 50% (v/v) and a final volume of 250 μl. 10-week-old female BALB/c mice are immunized by intraperitoneal injection with 250 μl of the mixture. Three additional intraperitoneal injections in 30-days intervals are given with adjuvant-supplemented protein suspension (50 μg of protein in 50% (v/v) adjuvant (trehalose dimycolate, 2% oil–Tween-80; e.g., ABM 1, same supplier)) with a final volume of 250 μl per mouse. After

30 additional days, the same injection is repeated on 3 consecutive days and the mouse is sacrificed the following day for removal of the spleen. Blood sera of mice are taken prior to immunization and 10 days after each injection. Typically, pronounced immune responses are detected after the third immunization with positive signals in a standard enzyme-linked immunosorbent assay (ELISA) at serum dilutions of 10^5. In addition to this long-term protocol, shorter immunization schemes can be considered.

Whereas the adjuvant supplementation is apparently sufficient to stabilize the detergent-solubilized proteins and generate immune response to native epitopes, additional strategies for immunization were employed to enhance structural and functional integrity for unstable membrane protein targets. Monoclonal antibodies against a G protein-coupled receptor were obtained by immunizing mice with purified receptor reconstituted into phospholipid vesicles at high protein-to-lipid ratio (Day et al., 2007). In that case, the reconstituted receptor was further stabilized with a bound high-affinity inverse agonist. In order to generate monoclonal antibodies against poorly immunogenic mammalian membrane proteins, the use of immunodeficient mice (MRL/lpr) with protein reconstituted in phosphatidyl choline liposomes, which are supplemented with the adjuvant Lipid A was recommended to overcome immune tolerance (Hino et al., 2013).

2.2 Selection of antibodies

Monoclonal antibodies are readily obtained using the classical hybridoma technique (Kohler & Milstein, 1975) starting with the spleen of an immunized mouse and applying standard protocols. The selection of the antibodies is the crucial step to obtain native binders suitable for crystallization of a membrane protein or other structural studies. The antibodies should bind the protein of interest in its native conformation forming a stable and stoichiometric complex and avoiding the introduction of flexible domains. Therefore, the selection procedure should present the membrane protein in its native conformation using appropriate detergents and buffer conditions. As a guideline, conditions established for size-exclusion chromatography (SEC) of the protein of interest are a good choice, as the final complex is typically SEC-purified prior to crystallization. A robust method is the HIS-TAG ELISA, in which His-tagged recombinant membrane proteins are presented in native conformation via immobilization on Ni^{2+}-NTA coated ELISA plates (Padan et al., 1998). An updated protocol is described below. Presentation on polysterene surfaces as used in standard ELISAs should be avoided, as adsorption to the solid phase can cause partial denaturation of the protein. Antibodies obtained by this selection procedure are cross-checked whether they recognize the target protein in Western immunoblots after SDS-PAGE separation (Western negative). These linear-epitope binders are avoided for cocrystallization trials, as amino- and carboxy-termini of polypeptides as well as exposed loops are often flexible and binding of an antibody may introduce a mobile domain hindering crystallization.

2.2.1 Method

For HIS-TAG ELISA, detergent-solubilized purified His-tagged membrane protein (~10 μg/ml in binding buffer) is applied (100 μl/well) to Ni^{2+}-NTA 96-well plates (e.g., Ni-NTA HisSorb Plates; QIAGEN) and incubated overnight at 4 °C or for 1 h at room temperature. Binding buffer has to ensure stability of the protein. TBS/DDM (20 m*M* Tris/HCl, pH 7.5, 150 m*M* NaCl, 0.03% (w/v) dodecyl-β-D-maltoside (DDM)) or phosphate-buffered saline (PBS)/DDM (0.03% w/v) work well but might need protein-specific adaptations especially in respect to choice and concentration of detergent. The solution is discarded and wells are blocked with 350 μl 3% (w/v) protease-free bovine serum albumin (BSA) in binding buffer (incubation 1 h at room temperature). Excess protein is washed out by replacing the liquid three times (350 μl, 30 s incubation) with binding buffer. The decorated wells are then incubated with hybridoma supernatants (supplemented with 5 m*M* imidazole, pH 7.9, 0.03% (w/v) DDM, 1.5% (w/v) BSA, or diluted 1:1 up to 1:10 in imidazole and BSA-supplemented binding buffer) and incubated for 1 h at room temperature. Recombinant antibodies produced by periplasmic expression in *E. coli* can be probed by diluting periplasmic extracts 1:1 in binding buffer supplemented with 1.5% BSA (incubation 1 h at room temperature). Excess protein is washed out by replacing the liquid three times (350 μl, 5 min incubation) with binding buffer. Antibody binding is detected with a secondary antibody (anti-mouse IgG) conjugated to alkaline phosphatase (1:2000 dilution in binding buffer supplemented with 1.5% (w/v) BSA, incubation 1 h at room temperature). The previous washing step is repeated. For detection, AP buffer (1 mg/ml *p*-nitrophenylphosphate in 10% (v/v) diethanolamine, pH 9.8, 0.5 m*M* $MgCl_2$) is applied (100 μl/well) and incubated at room temperature for about 15–20 min. The signal is quantified with a plate reader at 405 nm (450 nm reference).

Alternative approaches for the selection of native binders for membrane proteins are in use or in development. Crucial is the presentation of the protein in its native form, and strategies employ protein in detergent micelles, reconstituted in liposomes or in a cellular or membrane context. Native presentation with proteoliposomes prepared with biotinylated phosphatidylethanolamine and presented on streptavidin-coated surfaces were indicated (Hino et al., 2013). Distinct conformations may be stabilized with specific ligands or mutations. Clearly, screening is the key step to select suitable antibodies from hybridoma fusions as well as from recombinant antibody libraries.

2.3 Characterization of antibodies

Strategies for the characterization of antibodies depend on the intended application and the number of antibodies available. If crystallization is the exclusive goal, only binders of conformational epitopes are usually considered. Some native binders may recognize linear epitopes, so that they can be of interest for topological characterization or other structural studies. For instance, the pH-dependent conformational changes of the NH_2 terminus of the bacterial Na^+/H^+ antiporter NhaA were demonstrated with mAb1F6, which binds a linear peptide that was identified with synthetic peptides (Venturi et al., 2000). Also, peptide mapping for conformational epitopes can be pursued, if information is needed. Yet, the main criterion is stable complex formation. For the latter, SEC analysis is usually the method of choice as the complex formed between membrane protein and antibody should be stable and in the best case monodisperse during this chromatographic step. As a faster alternative for probing a larger number of samples, we use electrophoretic mobility shift assays in blue native PAGE (BN-PAGE). In our hands, results for stable complex formation compare well with data obtained by SEC analysis. For large protein complexes and small binders, the resolution can be even better as compared to SEC. For instance, binding of two Fv fragments (28 kDa each) to the homodimeric multisubunit cytochrome bc_1 complex from *S. cerevisiae* (460-kDa protein mass alone, in DDM) leads to an electrophoretic mobility shift (Fig. 1), whereas the retention time in SEC analysis probed on a high-resolution column (TSK4000; TosoHaas) remained unchanged (data not shown; complex in SEC was shown by subsequent SDS-PAGE analysis as well as by crystallization and structure determination). Apparently, displacement of DDM by the dye Coomassie diminishes the extra mass in BN-PAGE thus improving

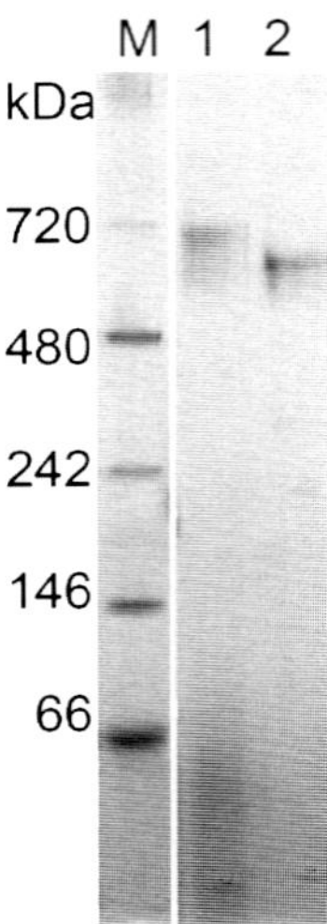

Figure 1 Binding of Fv fragment to yeast cytochrome *bc*$_1$ complex probed by BN-PAGE electrophoretic mobility shift assay. Shown is the Coomassie-stained BN-PAGE gel (4–16% Bis–Tris gel) after separation of 5 μg cytochrome *bc*$_1$ complex (homodimer of 460 kDa) incubated with 5 μg Fv fragment (lane 1). Five micrograms of cytochrome *bc*$_1$ complex were separated in lane 2. The band shift in lane 1 indicates complex formation. Standard proteins are shown in lane M (5 μl NativeMark™ Unstained Protein Ladder).

resolution, whereas the additional mass of bound Fv cannot be resolved in SEC for this large protein:detergent complex. Readymade gels are meanwhile available for BN-PAGE and accelerate the protocol. The results are clear-cut for antibody fragments. Native antibodies and oligomeric membrane proteins may form multiple complexes.

2.3.1 Methods

2.3.1.1 BN-PAGE electrophoretic mobility shift assay

For complex formation, 10 μg of purified membrane protein are mixed with antibody fragment in molar excess and incubated on ice for 1 h (10 μl final volume in detergent-containing buffer). Four microliters of 4 × sample buffer (NativePAGE™; Life Technologies) and 2 μl of 20 × cathode buffer additive (NativePAGE™) are added, and the solution is mixed well. The mixture is briefly centrifuged at 10,000 × *g*, 4 °C for 5 min to remove aggregates. Eight microliters of the sample is loaded per lane on a BN-PAGE gel (e.g., NativePAGE™ Novex®, 4–16% Bis–Tris Gel). The anode chamber is filled with 1 × running buffer (NativePAGE™), while the cathode chamber filled with the same buffer supplemented with 1 × cathode buffer additive

(NativePAGE™). After separation at 150 V for about 100 min at 4 °C, the gel is stained with Coomassie Brilliant Blue (overnight staining at 4 °C is recommended).

2.4 Generation of recombinant antibodies

Fv or Fab fragments are required for cocrystallization, as entire antibodies would introduce flexible domains and potentially form nonstoichiometric complexes. Antibody fragments can be produced as recombinant protein or, in case of the Fab fragment, also be generated by proteolysis of purified antibodies (Table 1). We will provide here a detailed description of cloning and production of antibody fragments from hybridoma cells, as recombinant fragments are versatile and render antibody production independent from hybridoma cell culture or animal experiments. Genes are amplified from hybridoma RNA by reverse transcriptase-polymerase chain reaction and transferred into the pASK68 vector for periplasmic expression of Fv fragments in *E. coli* (Kleymann et al., 1995; Ostermeier et al., 1995). Its bicistronic expression cassette introduces OmpA and PhoA signal sequences for periplasmic targeting of heavy and light chains, respectively as well as strep-tag and myc-tag fusion for purification and quality control.

2.4.1 Methods

2.4.1.1 Cloning of antibody fragments

2.4.1.1.1 RNA preparation All materials have to be RNA grade, and stock solutions are prepared with RNAse-free water. Total RNA is extracted from $\sim 5 \times 10^6$ hybridoma cells harvested from growing and antibody-producing cultures (five to six wells of a six-well plate that are covered to ~75% with cells). RNA extraction can be also performed from frozen cells, but DMSO needs to be washed out before using PBS. Kit-based RNA extraction works well (e.g., RNeasy Qiagen with RNAlater supplement). Concentrate RNA solution to 1–5 μg/μl.

2.4.1.1.2 First-strand cDNA synthesis In general, an oligo-thymidine primer (oligo-dT; stock solution 10 pmol/μl dT15 or oligo-dT12-18) is used for first-strand synthesis. Subtype specific first-strand synthesis is an alternative if amplification of pseudogenes needs to be suppressed (Hunte & Münke, 2004). One to five micrograms of total RNA (alternatively 50–500 ng of mRNA) and 40 pmol oligo-dT primer (H_2O added to a final volume of 11 μl) are heated for 10 min at 70 °C and immediately put on ice. One microliter of 10 m*M* dNTP, 4 μl

5 × first-strand buffer, 2 μl 0.1 M DTT, and 1 μl RNAse inhibitor (30 U/μl, e.g., Rnasin, MBI Fermentas or Promega) are added (final reaction volume 20 μl). After gentle mixing and preheating for 2 min at 42 °C, 1 μl (200 U) of reverse transcriptase (e.g., SuperScript II Rnase H- Reverse Transcriptase; GIBCO BRL/Invitrogen) is added, and the sample is gently mixed by pipetting. After incubation for 1 h at 42 °C, the reaction is stopped by heating for 15 min at 70 °C. The mixture is cooled on ice and then cleared by centrifugation at 4 °C, 14,000 rpm for 5 min.

2.4.1.1.3 Amplification of VH and VL genes and construction of Fv fragment expression vector Cloning of VH and VL genes takes advantage of the highly conserved sequences of the framework regions (Kabat et al., 1991). The protocol can be adapted for Fab fragment cloning. Here, we describe PCR amplification of VH and VL genes from first-strand cDNA with degenerated primers, which match the 5′-end of framework 1 and the 3′-end of framework 4. Primer pairs contain unique restriction sites to clone into the pASK68 vector (Table 2). They were designed for subclasses 1,2,3,5 of V_κ-chain, subclasses 4,6 of V_κ-chain, and for VH (Kleymann et al., 1995). Subtyping of antibodies can be performed beforehand with hybridoma supernatants or purified antibodies using commercial ELISA-based kits. In some cell lines, nonfunctional mRNA transcripts are abundant. They originate from the myeloma cell line used for hybridoma fusion. Amplification can be suppressed by antisense-directed RNase H digestion (Ostermeier & Michel, 1996). The VH and VL genes are cloned into the pASK68 vector following standard protocols using the respective restriction sites (VH gene: *Pst*I, *Bst*EII; VL gene: *Sst*I, *Xho*I). This ensures reconstruction of complete framework sequences. Genes are sequenced and expression products are tested for stability and binding. Hybrid Fab fragments can be constructed by insertion of VH and VL genes into a vector, which already contains constant domains of an antibody molecule (Venturi, Seifert, & Hunte, 2002). An alternative PCR strategy for cloning of antibody genes is the rapid amplification of cDNA ends. One should note that amplification and sequencing is generally also required for structure determination in cases in which proteolytic Fab fragments were used for crystallization.

2.5 Production and purification of recombinant antibodies

Several expression systems are in use for the production of recombinant antibody fragments including *E. coli*, *Pichia pastoris*, or mammalian cells. A well-established protocol is the periplasmic expression of Fv fragments in *E. coli*

Table 2 Degenerated primers for PCR amplification of murine VH and VL genes for insertion in pASK68 expression vector

Primer	Chain	Framework region	Primer sequence[a] respective residue number of chain												Restriction enzyme[b]	Name
Forward	VL	FR1 subtypes I, II, III, and V	GAC	ATT	GAG	CTC	AC**M**	CAG	**W**CT	CCA	**KY**C	TCC	CTG		*Sst*I	VKC1T
			1	2	3	4	5	6	7	8	9	10	11			
Forward	VL	FR1 subtypes IV and VI	GAC	ATT	GAG	CTC	ACC	CAG	TCT	CCA	GCA	ATC	ATG		*Sst*I	VKC2T
			1	2	3	4	5	6	7	8	9	10	11			
Backward	VL	FR4	CCG	TTT	CAG	CTC	GAG	CTT	GGT	**S**CC	**WS**C	**W**CC	GAA	CGT	*Xho*I	VKR1T
			108	107	106	105	104	103	102	101	100	99	98	97		
Forward	VH	FR1	GAG	GT**K**	**M**AG	CTG	CAG	**S**AG	TC**W**	GGG	**S**CT	GGC	**Y**TG		*Pst*I	VHC1T
			1	2	3	4	5	6	7	8	9	10	11			
Backward	VH	FR4	TGA	GGA	GAC	GGT	GAC	CGT	GGT	CCC	TTG	GCC	CCA		*Bst*EII	VHR1T
			113	112	111	110	109	108	107	106	105	104	103			

[a] **M** stands for A or C, **W** for A or T, **K** for G or T, **Y** for C or T, and **S** for C or G, and numbering is according to Kabat et al. (1991).

[b] Restriction site is underlined.

Adapted from Hunte and Münke (2004) and Kleymann et al. (1995).

using the pASK68 vector. Protocols for expression and purification were previously published (Hunte & Kannt, 2003; Kleymann et al., 1995), and several membrane protein structures were obtained with these antibodies (Birth et al., 2014; Hunte & Michel, 2002; Ostermeier et al., 1997; Palsdottir, Lojero, Trumpower, & Hunte, 2003; Solmaz & Hunte, 2008). We here describe the optimized protocol, which typically yields 0.5–1 mg pure Fv fragment per liter of bacterial cell culture.

2.5.1 Method

2.5.1.1 Periplasmic expression of Fv fragment

First preculture of *E. coli* strain JM83 transformed with pASK68-Fv (10 ml LB medium, 100 μg/ml carbenicillin, and 50 μg/ml streptomycin) is incubated during the day at 30 °C with shaking at 200 rpm. It is transferred to 400 ml medium (LB medium, 100 μg/ml carbenicillin) for overnight (o/n) growth at 30 °C with shaking at 200 rpm. Main cultures (6 × 2 l prewarmed LB^{Amp} medium) in 5-l flasks are started by addition of ~50 ml of o/n culture per flask (starting $OD_{600}=0.15–0.2$). Cells are grown at 22.5 °C with shaking at 200 rpm until OD_{600} of 0.5 (~4.5 h). Expression is induced with 1 m*M* IPTG and incubation continued for 3 h. Low growth temperature and short expression time work best. Cultures are transferred to precooled centrifugation buckets and cells are sedimented by centrifugation for 12 min at 4000 × *g* at 4 ° C. Sedimented cells are very gently resuspended in ice-cold extraction buffer (100 m*M* Tris/HCl, pH 8.0, 500 m*M* sucrose, 1 m*M* EDTA) at 1/100 of the culture volume and incubated for 30 min on ice. The periplasmic extract is cleared from cells and cell debris by centrifugation (5000 × *g* for 30 min at 4 °C). Cleared extract can be stored in appropriate aliquots at −20 °C for several months. Strep-tag and myc-tag at heavy and light chains, respectively, permit quality control by Western immunoblot using streptavidin–alkaline phosphatase and mAB 9E10.

2.5.1.2 Streptavidin affinity purification of Fv fragments

Periplasmic extract is cleared from biotin and biotin-carrying proteins by avidin treatment. Sixty milliliters of extract supplemented with 120 μg of avidin is incubated for 30 min on ice. Aggregated proteins are removed by centrifugation (40 min at 35,000 × *g* at 4 °C). The supernatant can be frozen and stored at −20 °C if required. Sixty milliliters of avidin-treated periplasmic extract is filtered (0.2 μm) and loaded onto a 5-ml streptavidin affinity column (e.g., *Strep*-Tactin® Superflow Cartridge H-PR; IBA, Göttingen) equilibrated with 10 column volumes of Tris buffer (100 m*M*

Tris/HCl, pH 8.0). A flow rate of 0.7 ml/min is used throughout purification. The same buffer is applied until the signal of the UV recorder reaches approximately 10% of the base level (5 × 5 ml). Fv fragments are eluted with 6 × 2.5 ml 2.5 m*M* Desthiobiotin in equilibration buffer. The eluate is concentrated by ultrafiltration (10 kDa cutoff) to a final protein concentration of ~5 mg/ml for storage at 4 °C. Desthiobiotin content is typically decreased by dilution and reconcentration. Many Fv fragments are stable in this buffer, some need addition of salt. Buffer exchange for freezing or downstream processes can be appended. Regular cleaning of the column with 2-(4-hydroxyphenylazo)benzoic acid (HABA) treatment (100 ml of 100 m*M* Tris, pH 8.0, 150 m*M* NaCl, 1 m*M* HABA) is recommended.

2.6 Crystallization

The common strategy for crystallization of membrane proteins with bound antibody fragments is to use stoichiometric complexes, which are prepared by mixing the membrane protein with little surplus of the binder followed by SEC purification of the complex. We consider these complexes as "new" proteins and subject them to comprehensive screening of crystallization space. So far, there is a lack of data as to which binding affinity is required for successful cocrystallization. As a rule of thumb, complexes stable in SEC can be readily put forward to crystallization trials.

ACKNOWLEDGMENTS

This work was supported by the Deutsche Forschungsgemeinschaft (CRC 746, CRC 992, RTG 1976) and by the Excellence Initiative of the German Federal and State Governments (EXC 294, BIOSS) and received funding from the European Community's Seventh Framework Programme FP7/2007-2013 under grant agreement no. HEALTH-F4-2007-201924, EDICT Consortium.

REFERENCES

Birth, D., Kao, W. C., & Hunte, C. (2014). Structural analysis of atovaquone-inhibited cytochrome bc1 complex reveals the molecular basis of antimalarial drug action. *Nature Communications*, *5*, 4029.

Brohawn, S. G., Campbell, E. B., & MacKinnon, R. (2013). Domain-swapped chain connectivity and gated membrane access in a Fab-mediated crystal of the human TRAAK K^+ channel. *Proceedings of the National Academy of Sciences of the United States of America*, *110*(6), 2129–2134.

Day, P. W., Rasmussen, S. G. F., Parnot, C., Fung, J. J., Masood, A., Kobilka, T. S., et al. (2007). A monoclonal antibody for G protein-coupled receptor crystallography. *Nature Methods*, *4*(11), 927–929.

Dickson, V. K., Pedi, L., & Long, S. B. (2014). Structure and insights into the function of a Ca^{2+}-activated Cl^- channel. *Nature*, *516*, 213–218. http://dx.doi.org/10.1038/nature13913.

Dutzler, R., Campbell, E. B., & MacKinnon, R. (2003). Gating the selectivity filter in ClC chloride channels. *Science*, *300*(5616), 108–112.

Fang, Y., Jayaram, H., Shane, T., Kolmakova-Partensky, L., Wu, F., Williams, C., et al. (2009). Structure of a prokaryotic virtual proton pump at 3.2 Å resolution. *Nature*, *460*(7258), 1040–1043.

Hassaine, G., Deluz, C., Grasso, L., Wyss, R., Tol, M. B., Hovius, R., et al. (2014). X-ray structure of the mouse serotonin 5-HT3 receptor. *Nature*, *512*(7514), 276–281.

Hibbs, R. E., & Gouaux, E. (2011). Principles of activation and permeation in an anion-selective Cys-loop receptor. *Nature*, *474*(7349), 54–60.

Hino, T., Arakawa, T., Iwanari, H., Yurugi-Kobayashi, T., Ikeda-Suno, C., Nakada-Nakura, Y., et al. (2012). G-protein-coupled receptor inactivation by an allosteric inverse-agonist antibody. *Nature*, *482*(7384), 237–240.

Hino, T., Iwata, S., & Murata, T. (2013). Generation of functional antibodies for mammalian membrane protein crystallography. *Current Opinion in Structural Biology*, *23*(4), 563–568.

Hino, T., Matsumoto, Y., Nagano, S., Sugimoto, H., Fukumori, Y., Murata, T., et al. (2010). Structural basis of biological N_2O generation by bacterial nitric oxide reductase. *Science*, *330*(6011), 1666–1670.

Hunte, C., & Kannt, A. (2003). Antibody fragment mediated crystallization of membrane proteins. In C. Hunte, G. von Jagow, & H. Schägger (Eds.), *Membrane protein purification and crystallization: A practical guide* (2nd ed., pp. 205–218). Academic Press.

Hunte, C., Koepke, J., Lange, C., Rossmanith, T., & Michel, H. (2000). Structure at 2.3 Å resolution of the cytochrome bc(1) complex from the yeast *Saccharomyces cerevisiae* co-crystallized with an antibody Fv fragment. *Structure*, *8*(6), 669–684.

Hunte, C., & Michel, H. (2002). Crystallisation of membrane proteins mediated by antibody fragments. *Current Opinion in Structural Biology*, *12*(4), 503–508.

Hunte, C., & Münke, C. (2004). Application of antibody fragments as crystallization enhancers. In T. Dingermann, D. Steinhilber, & G. Folkers (Eds.), *Molecular biology in medicinal chemistry: Vol. 21* (pp. 300–322). Weinheim: Wiley-VCH Verlag GmbH & Co. KGaA.

Inaba, K., Murakami, S., Nakagawa, A., Iida, H., Kinjo, M., Ito, K., et al. (2009). Dynamic nature of disulphide bond formation catalysts revealed by crystal structures of DsbB. *The EMBO Journal*, *28*(6), 779–791.

Iwata, S., Ostermeier, C., Ludwig, B., & Michel, H. (1995). Structure at 2.8 Å resolution of cytochrome c oxidase from *Paracoccus denitrificans*. *Nature*, *376*(6542), 660–669.

Jiang, Y., Lee, A., Chen, J., Ruta, V., Cadene, M., Chait, B. T., et al. (2003). X-ray structure of a voltage-dependent K^+ channel. *Nature*, *423*(6935), 33–41.

Kabat, E. A., Wu, T. T., Reid-Miller, M., Perry, H. M., Gottesmann, K. S., & Foeller, C. (1991). *Sequences of proteins of immunological interest* (5th ed.). Bethesda: US Department of Health and Human Services, Public Service, NIH.

Kleymann, G., Ostermeier, C., Ludwig, B., Skerra, A., & Michel, H. (1995). Engineered Fv fragments as a tool for the one-step purification of integral multisubunit membrane protein complexes. *Biotechnology (N. Y)*, *13*(2), 155–160.

Kohler, G., & Milstein, C. (1975). Continuous cultures of fused cells secreting antibody of predefined specificity. *Nature*, *256*(5517), 495–497.

Krishnamurthy, H., & Gouaux, E. (2012). X-ray structures of LeuT in substrate-free outward-open and apo inward-open states. *Nature*, *481*(7382), 469–474.

Kruse, A. C., Ring, A. M., Manglik, A., Hu, J., Hu, K., Eitel, K., et al. (2013). Activation and allosteric modulation of a muscarinic acetylcholine receptor. *Nature*, *504*(12735), 101–106.

Lange, C., & Hunte, C. (2002). Crystal structure of the yeast cytochrome bc(1) complex with its bound substrate cytochrome c. *Proceedings of the National Academy of Sciences of the United States of America, 99*(5), 2800–2805.

Lange, C., Nett, J. H., Trumpower, B. L., & Hunte, C. (2001). Specific roles of protein–phospholipid interactions in the yeast cytochrome bc(1) complex structure. *The EMBO Journal, 20*(23), 6591–6600.

Lee, S. Y., Lee, A., Chen, J. Y., & MacKinnon, R. (2005). Structure of the KvAP voltage-dependent K^+ channel and its dependence on the lipid membrane. *Proceedings of the National Academy of Sciences of the United States of America, 102*(43), 15441–15446.

Ostermeier, C., Harrenga, A., Ermler, U., & Michel, H. (1997). Structure at 2.7 Å resolution of the *Paracoccus denitrificans* two-subunit cytochrome c oxidase complexed with an antibody FV fragment. *Proceedings of the National Academy of Sciences of the United States of America, 94*(20), 10547–10553.

Ostermeier, C., Iwata, S., Ludwig, B., & Michel, H. (1995). F-V fragment mediated crystallization of the membrane-protein bacterial cytochrome-c-oxidase. *Nature Structural Biology, 2*(10), 842–846.

Ostermeier, C., & Michel, H. (1996). Improved cloning of antibody variable regions from hybridomas by an antisense-directed RNase H digestion of the P3-X63-Ag8.653 derived pseudogene mRNA. *Nucleic Acids Research, 24*(10), 1979–1980.

Padan, E., Venturi, M., Michel, H., & Hunte, C. (1998). Production and characterization of monoclonal antibodies directed against native epitopes of NhaA, the Na^+/H^+ antiporter of *Escherichia coli*. *FEBS Letters, 441*(1), 53–58.

Palsdottir, H., Lojero, C. G., Trumpower, B. L., & Hunte, C. (2003). Structure of the yeast cytochrome bc(1) complex with a hydroxyquinone anion Q(o) site inhibitor bound. *Journal of Biological Chemistry, 278*(33), 31303–31311.

Penmatsa, A., Wang, K. H., & Gouaux, E. (2013). X-ray structure of dopamine transporter elucidates antidepressant mechanism. *Nature, 503*(7474), 85–90.

Rasmussen, S. G., Choi, H. J., Fung, J. J., Pardon, E., Casarosa, P., Chae, P. S., et al. (2011). Structure of a nanobody-stabilized active state of the beta(2) adrenoceptor. *Nature, 469*(7329), 175–180.

Rasmussen, S. G., Choi, H. J., Rosenbaum, D. M., Kobilka, T. S., Thian, F. S., Edwards, P. C., et al. (2007). Crystal structure of the human beta2 adrenergic G-protein-coupled receptor. *Nature, 450*(7168), 383–387.

Rasmussen, S. G., DeVree, B. T., Zou, Y., Kruse, A. C., Chung, K. Y., Kobilka, T. S., et al. (2011). Crystal structure of the beta2 adrenergic receptor–Gs protein complex. *Nature, 477*(7366), 549–555.

Shaffer, P. L., Goehring, A., Shankaranarayanan, A., & Gouaux, E. (2009). Structure and mechanism of a Na^+-independent amino acid transporter. *Science, 325*(5943), 1010–1014.

Shukla, A. K., Manglik, A., Kruse, A. C., Xiao, K., Reis, R. I., Tseng, W. C., et al. (2013). Structure of active beta-arrestin-1 bound to a G-protein-coupled receptor phosphopeptide. *Nature, 497*(7447), 137–141.

Solmaz, S. R., & Hunte, C. (2008). Structure of complex III with bound cytochrome c in reduced state and definition of a minimal core interface for electron transfer. *Journal of Biological Chemistry, 283*(25), 17542–17549.

Tsukazaki, T., Mori, H., Fukai, S., Ishitani, R., Mori, T., Dohmae, N., et al. (2008). Conformational transition of Sec machinery inferred from bacterial SecYE structures. *Nature, 455*(7215), 988–991.

Uysal, S., Vasquez, V., Tereshko, V., Esaki, K., Fellouse, F. A., Sidhu, S. S., et al. (2009). Crystal structure of full-length KcsA in its closed conformation. *Proceedings of the National Academy of Sciences of the United States of America, 106*(16), 6644–6649.

Venturi, M., Rimon, A., Gerchman, Y., Hunte, C., Padan, E., & Michel, H. (2000). The monoclonal antibody 1F6 identifies a pH-dependent conformational change in the hydrophilic NH_2 terminus of NhaA Na^+/H^+ antiporter of *Escherichia coli*. *Journal of Biological Chemistry*, *275*(7), 4734–4742.

Venturi, M., Seifert, C., & Hunte, C. (2002). High level production of functional antibody Fab fragments in an oxidizing bacterial cytoplasm. *Journal of Molecular Biology*, *315*(1), 1–8.

Zhou, Y., Morais-Cabral, J. H., Kaufman, A., & MacKinnon, R. (2001). Chemistry of ion coordination and hydration revealed by a K^+ channel-Fab complex at 2.0 Å resolution. *Nature*, *414*(6859), 43–48.

Zickermann, V., Bostina, M., Hunte, C., Ruiz, T., Radermacher, M., & Brandt, U. (2003). Functional implications from an unexpected position of the 49-kDa subunit of NADH: ubiquinone oxidoreductase. *Journal of Biological Chemistry*, *278*(31), 29072–29078.

CHAPTER ELEVEN

Phage Display Selections for Affinity Reagents to Membrane Proteins in Nanodiscs

Pawel K. Dominik, Anthony A. Kossiakoff[1]

Department of Biochemistry and Molecular Biology, Institute for Biophysical Dynamics, The University of Chicago, Chicago, Illinois, USA

[1]Corresponding author: e-mail address: koss@bsd.uchicago.edu

Contents

Abstract

Phage display selections generate high-affinity synthetic reagents that can be used as tools in structural characterization of membrane proteins. Currently, most selection protocols are performed with membrane protein targets in detergents. However, there are numerous technical issues associated with this, primarily that detergents are poor mimics of the native lipid environment. Here, we describe a set of protocols for phage display selection that involves reconstituting membrane proteins in nanodiscs, which

Methods in Enzymology, Volume 557
ISSN 0076-6879
http://dx.doi.org/10.1016/bs.mie.2014.12.032

are small discoidal particles consisting of lipids enclosed by membrane scaffold proteins. The nanodisc format enabled us to expand the capabilities of competitive and subtractive phage display selection steps, and generation of high-quality synthetic reagents for membrane proteins in native-like lipid environment.

1. INTRODUCTION

The use of crystallization chaperones has been the enabling factor for determining a number of paradigm-shifting membrane protein structures. There are different types of crystallization chaperones, each with their own strengths and weaknesses. These chaperones vary in size and structure, and can be produced in several different formats (Bukowska & Grütter, 2013). However, they all function to promote crystallization by reducing conformational heterogeneity, masking hydrophobic surfaces to increase solubility, and providing primary contact points between molecules in the crystal lattice.

Our work focuses on generating chaperones based on the Herceptin Fab framework that has been engineered for high stability. We use phage display mutagenesis selections to identify Fabs that have desired characteristics, such as high affinity and specificity to the target protein antigen. These Fabs are derived from synthetic phage display libraries *in vitro* (Chen & Sidhu, 2014; Fellouse, Wiesmann, & Sidhu, 2004; Miller et al., 2012). Therefore, they are fully recombinant and renewable. Importantly, the phage display selection conditions can be precisely modified for different targets and specific uses.

The application of phage display to generate binders to myriad types of soluble proteins is well established (Fellouse et al., 2007; Hoogenboom, 2005; Koenig & Fuh, 2014; and others); however, there are a number of factors that make membrane proteins much more challenging. Principal among these is that membrane proteins are generally less stable than their soluble counterparts, and this is exacerbated by the fact that they need to be removed from their native membrane environments and solubilized in detergent before undergoing the phage display selection experiment. To circumvent this problem, we have developed an antibody phage display selection protocol that utilizes nanodiscs to keep the membrane proteins stabilized in a native-like environment during the selection process.

Nanodiscs are discoidal nanoparticles formed by synthetic lipids surrounded by a belt consisting of amphipathic domains of α-helical proteins, called membrane scaffold proteins (MSPs) (Nath, Atkins, & Sligar, 2007). Since the initial development of this technology by the Sligar laboratory, multiple proteins of different architecture were readily incorporated into nanodiscs (reviewed in Schuler, Denisov, & Sligar, 2013). Nanodiscs provide a unique advantage because their size and lipid composition can be modified. Besides providing a native-like environment for the membrane protein, they have numerous other advantages. Once embedded in the nanodisc, the membrane proteins are generally quite stable and easier to handle. For phage display selections in detergents, it is required to put a covalent affinity tag on the membrane protein, which, in many cases, significantly destabilizes the protein. However, with nanodiscs, the affinity tag can be placed on the MSP directly, leaving the membrane protein unchanged. Additionally, immobilization through the MSP provides for access to binding of the Fab to both sides of the embedded membrane protein. Therefore, the chaperones can be positioned to either the extra- or intracellular sides of the protein or both sides together. Finally, the use of nanodiscs simplifies the selection procedure because the membrane protein can be released from the nanodisc simply by adding detergent.

In this chapter, we describe how to exploit nanodiscs for selecting high-performance binders to membrane proteins. Besides generating Fabs that simply bind to the membrane proteins, the selections can be designed to capture and stabilize desired conformational states or target-specific surface epitopes of the protein. Although the protocol emphasizes the use of Fabs as the source scaffold for the antigen binders, most of the procedures described here can be modified to accommodate the use of other scaffolds.

1.1 Overview of the method

The set of protocols outlined in this chapter provide the basis for the generation of high-performance affinity reagents to membrane proteins. The complete phage display process is encompassed in the term "selection," which includes (1) the initial nanodisc preparation, antigen insertion, and quality control, (2) the phage display sorting procedures to generate binders with desired properties, and (3) validation and characterization of the binders to establish their effectiveness in a variety of applications (Fig. 1). This set of protocols has been optimized and validated using several different classes of membrane protein systems, and has been particularly successful in addressing challenging problems such as trapping conformational states and epitope targeting. Several aspects of the selection process have been published by

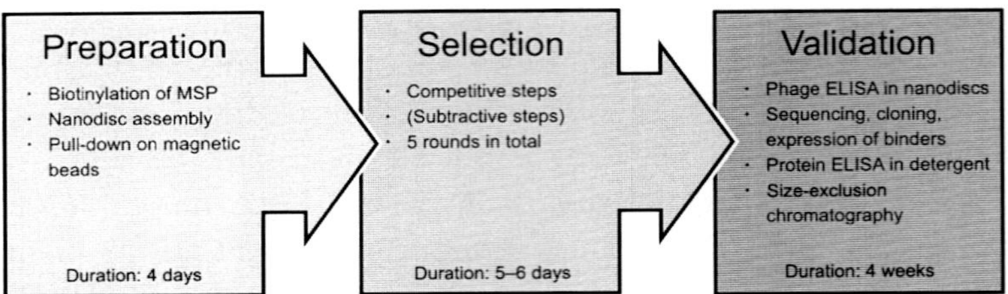

Figure 1 Overview of the method. Phage display selection with membrane proteins in nanodiscs is composed of three major segments: preparation (left), selection (middle), and validation (right). Typically, preparation step is the shortest and consists of MSP biotinylation, assembly of target membrane protein in nanodiscs, and quality control pull-down on streptavidin beads. The selection protocol itself consists of five consecutive rounds in which competitive (and optionally subtractive) steps are performed. Validation of potential synthetic binding reagents occurs in two steps: primary (phage ELISA with protein in nanodiscs, followed by sequencing, cloning, expression, and purification of binders), and secondary (which depending on reagent application may involve protein ELISA in detergent, size-exclusion chromatography, and others).

us and others (Gao, Sidhu, & Wells, 2009; Kim, Stroud, & Craik, 2011; Paduch et al., 2013). In those cases, we reference the initial work and focus on describing the modifications and nuances that differentiate the nanodisc selection pipeline from the original applications.

2. MATERIALS AND EQUIPMENT

2.1 Chemical biotinylation

1. Purified MSP, MSP1D1 or MSP1E3D1 without His-Tag; plasmids can be obtained via Addgene (#20061 and #20066)
2. EZ-Link NHS-PEG4-Biotin, No-Weight Format (Thermo Scientific)
3. PD-10 columns (GE Healthcare)
4. Amicon Centrifugal Units 10,000 MWCO (Millipore)

2.2 Nanodisc reconstitution

1. Target membrane protein in detergent of choice with affinity tag (His-tag protein is used for the purpose of this protocol)
2. Biotinylated and nonbiotinylated MSP (MSP1D1 or MSP1E3D1, Addgene) (biotinylation of MSP as described in Section 3.1)
3. Chicken egg phosphatidylcholine; eggPC (Avanti)
4. Glass vials with chloroform-resistant caps
5. Vacuum desiccator

6. Sonicator (with microtip or water bath)
7. *n*-Dodecyl-β-D-maltopyranoside; DDM (Affymetrix) or other detergent compatible with target membrane protein
8. Bio-Beads SM-2 Adsorbents (Bio-Rad)
9. Ni-NTA Superflow Resin (Qiagen)
10. Amicon Centrifugal Units 50,000 MWCO (Millipore)
11. Superdex 200 10/200 GL column (GE Healthcare) or equivalent

2.3 Pull-down on streptavidin-coated magnetic beads

1. Streptavidin magnetic beads: Streptavidin MagneSphere Paramagnetic Particles (Promega)
2. Magnetic stand: MagneSphere Magnetic Separation Stand (Promega) or small earth magnets
3. 10% fos-choline-12 (FC-12) (Affymetrix)
4. SDS-PAGE running system (Bio-Rad)

2.4 Competitive and subtractive phage display selection

1. Biotinylated nanodiscs with and without target membrane protein (see Section 3.2)
2. Phage display library
3. Selection buffer of choice supplemented with 1% BSA (Fisher)
4. PEG 8000 (Promega)
5. XL1-Blue *E. coli* strain (Stratagene)
6. Tetracycline (Acros Organics)
7. Kanamycin (Amresco)
8. Ampicillin (Gold Biotechnology)
9. Streptavidin magnetic beads: Streptavidin MagneSphere Paramagnetic Particles (Promega)
10. D-Biotin (Sigma-Aldrich)
11. 2YT media (Fisher)
12. M13 KO7 helper phage (NEB)
13. KingFisher Magnetic Particle Processor with disposable reservoir strips and combs (Thermo Scientific)

2.5 Single-point phage ELISA

1. Biotinylated nanodiscs (with and without target membrane protein) (see Section 3.2)
2. 2YT media (Fisher)

3. 96-Well Deep Well Plates, Square (Greiner)
4. Breathable Sealing Film (Corning)
5. M13 KO7 helper phage (NEB)
6. ELISA High-Binding 96-Well Flat Bottom MICROLON 600 Plate (Greiner)
7. Coating buffer: 50 m*M* sodium carbonate, pH 9.6
8. Neutravidin (Thermo Scientific)
9. HRP/anti-M13 antibody conjugate (GE Healthcare, #27-9421-01)
10. TMB substrate (Thermo Scientific)
11. Quenching agent: 1 *M* HCl or 1 *M* H_3PO_4
12. 96-Well Sterile Non-Binding U-shape plates (Greiner) (for dilutions)
13. 50 ml Reagent Reservoirs (Denville Scientific)

2.6 Size-exclusion chromatography

1. Amicon Centrifugal Units of various cut-off range (Millipore)
2. Superdex 75 or 200 10/200 GL column (GE Healthcare) or equivalent

3. METHODS

3.1 Chemical biotinylation of membrane scaffold protein

Generally, phage display library sorting requires introducing a high-affinity immobilization tag on the antigen to facilitate efficient pull-down during the phage capture step onto a substrate surface. However, in selections with membrane proteins embedded in nanodiscs, it is preferable to add a biotinylation tag directly to the MSP as this spares the membrane protein from any deleterious effects that might be induced by direct biotinylation. There are several different variants of MSP that can be used to produce nanodiscs of various diameters (Bayburt & Sligar, 2009; Hagn, Etzkorn, Raschle, & Wagner, 2013). Protocols describing expression and purification of MSP isoforms are reported elsewhere (Ritchie et al., 2009). The choice of an optimum nanodisc size depends on the antigen's oligomeric state, stability, and lipid composition, all of which have to be optimized independently. For the majority of membrane protein targets smaller than 7.5 nm in diameter, we suggest starting with either the MSP1D1 or MSP1E3D1 variants. Below, we briefly describe the protocol for the direct biotinylation of MSP for purposes of nanodisc immobilization.

1. Dialyze MSP (with His-tag cleaved off) into buffer compatible with the target membrane protein. The biotinylation reaction should be

performed in the buffer without primary amine groups (e.g., phosphate or HEPES buffers are compatible, while Tris is not) in 7.0–8.0 pH range. If the target membrane protein requires lower or higher pH, it can be adjusted prior to nanodiscs reconstitution, but after the biotinylation step.

2. Dissolve biotinylation reagent EZ-Link NHS-PEG4-Biotin in buffer to 10 mg/ml final concentration. Reagent has to be used immediately and cannot be stored. Alternatively, dry DMSO can be used to dissolve the reagent, and it can be then kept at −20 °C for around 10–14 days without significant loss of activity.
3. Dilute MSP to a final concentration of 0.5–3 mg/ml and mix it with EZ-Link biotinylation reagent at fivefold molar excess over final protein concentration. The reaction is efficient at this ratio for both MSP1D1 and MSP1E3D1 variants. For other MSPs, the amount of biotinylation reagent has to be optimized independently. Keep the reaction at room temperature for 60 min; no additional stirring or shaking is required.
4. Quench the reaction with 5 m*M* final Tris at pH 7.0–8.0 and remove the excess biotinylation reagent from MSP by desalting in PD-10 columns according to manufacturer's protocol. Alternatively, it is possible to dialyze the sample overnight in a desired buffer using a membrane with a 10 kDa cut-off. Note that neither short desalting nor dialysis completely removes the excess of biotinylation reagent; however, since the reconstitution of nanodiscs involves two additional purification steps, size-exclusion chromatography (SEC) is not an obligatory step.
5. Hard spin the biotinylated MSP at benchtop centrifuge for 10 min at 15,000 × *g* and 4 °C and transfer the supernatant to a fresh tube. Measure the MSP concentration by the method of your choice. For the purpose of nanodiscs reconstitution, concentrate the MSP to about ~100 μ*M* using Amicon Centrifugal Units with a 10 kDa cut-off (this step may take some time). Dispense biotinylated MSP in 50–100 μl aliquots and flash freeze in liquid nitrogen. Store at −80 °C.

3.2 Reconstitution of target membrane protein into biotinylated nanodiscs

Methodologies for incorporation of membrane protein into nanodiscs (Fig. 2) have been described (Ritchie et al., 2009). For the purpose of phage display selections, the main difference is the use of biotinylated MSP (see Section 3.1) and a detergent optimal for the target membrane protein (rather than sodium cholate) if the protein is marginally stable. To optimize the

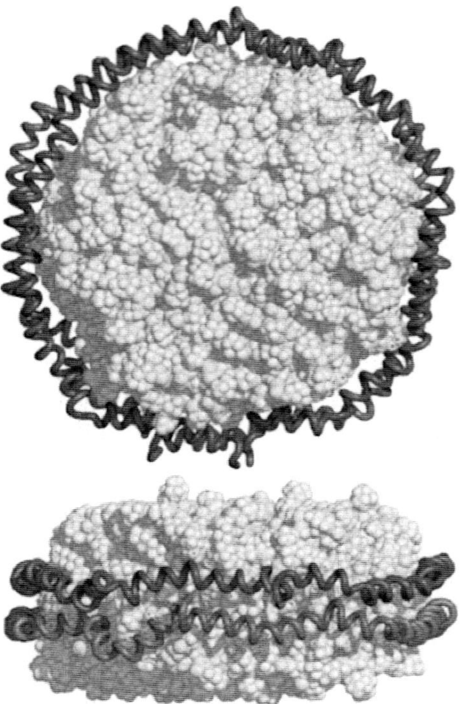

Figure 2 Structure of nanodisc. Nanodisc consists of lipid bilayer (white space filling model, here POPC are shown) and a double belt of membrane scaffold protein, MSP (gray ribbon). *Adapted from Ritchie et al. (2009), with permission from Elsevier.*

nanodisc reconstitution for a particular membrane protein, we suggest starting with MSP1D1 or MSP1E3D1 (10 and 13 nm nanodiscs diameter, respectively), and a mixture of phosphatidylcholine isolated from chicken egg (eggPC). This provides bilayers large enough to accommodate at least one copy of the protein per nanodisc in the majority of the membrane protein assemblies, as well as a neutral lipid environment. If the composition of lipids that support biological function of membrane proteins is established for liposome or nanodiscs reconstitution, we recommend using that for the assembly and phage display selection. It should also be noted that the incorporation efficiency of a membrane protein into a nanodisc varies. It is dependent on target protein stability, type of lipid, and MSP used and needs to be optimized on a case-by-case basis. Whereas low incorporation efficiency is problematic in other cases, amounts as low as 200 μl of 2 μM membrane protein in nanodiscs are sufficient for a single successful phage display selection experiment.

Note: To separate the empty nanodiscs from the nanodiscs reconstituted with His-tagged membrane proteins by affinity chromatography (Ni-NTA), it is necessary to cleave off the His-tag from MSP prior to reconstitution so that only the nanodiscs loaded with His-tagged targets bind to the column and the empty nanodiscs are washed away.

Steps 8 and 9 should be performed concurrently, Step 13 is optional.

1. Transfer total of 23.1 mg of eggPC (in chloroform) with a Hamilton syringe into a glass tube with a cap. Remove as much chloroform as you can with the stream of nitrogen gas under the hood. Make sure you rotate the vial as chloroform evaporates so that the lipids form a thin and evenly spread layer without any bubbles. Once no more liquid is observed, leave the vial for at least 4 h under a vacuum-closed container (e.g., desiccator) to remove traces amount of chloroform. Such films of dried lipids can be stored up to 1 week at −80 °C prior to nanodiscs reconstitution.
2. Resuspend dried eggPC lipids in 2 ml of reconstitution buffer (compatible with your protein), vortex until the film disappears and suspension turns opaque. Add 450 μl of 200 m*M* DDM (30 m*M* final) or other detergent in which target membrane protein is stable and vortex again. If the protein is stable in sodium cholate you can use it at up to 40 m*M* final to resuspend the lipids.
3. Sonicate using a tapered microtip that fits into the vial at 50% of amplitude with a 1-s pulse ON and 1-s pulse OFF until the solution becomes clear (around 3–5 min). Alternatively, run water bath sonicator for 1–2 h. At this point, the solution should be viscous and completely transparent indicating formation of mixed micelles between lipids and detergent. If this is not the case, most likely the chosen detergent is less effective in lipid solubilization and larger volumes have to be added to obtain clear mixed micelles suspension. Keep adding the detergent and sonicate until the solution is clear.
4. Add the remaining buffer to obtain a total volume of 3 ml. At this point, the final concentration of eggPC should be 10 m*M*, while DDM 30 m*M* (in case that different amount of detergent was necessary to obtain micelles, write down the final concentration for further reference). Mixed micelles composed of lipids:detergent can be stored at −80 °C for up to 1–2 months.
5. Calculate the amount of His-tagged membrane protein, reconstitution buffer, detergent, mixed micelles, and biotinylated MSP in the assembly reaction. Keep in mind that there are two MSP molecules per one

assembled nanodisc. For a 2 ml reaction with monomeric protein in MSP1D1 nanodiscs, prepare no more than 1.5 ml of assembly mix as follows—membrane protein:MSP:lipids in 1:10:650 molar ratio. Assuming 130 copies of eggPC per MSP1D1 nanodisc, you will roughly obtain one loaded nanodisc with target membrane protein and four empty ones. For MSP1E3D1, use a 1:10:1300 ratio instead. If target protein has a different oligomeric state, perform assembly at ratios where a single biological unit is incorporated in one out of five nanodiscs; for example, for tetrameric protein in MSP1D1 nanodiscs use a ratio of 4:10:650. Note that this is only a rough estimation and does not take into account the number of lipids displaced by the membrane protein. For more detailed protocols of reconstitution ratios, refer to Ritchie et al. (2009). Last, calculate the remaining volume of reconstitution buffer to be added up to 2 ml.

6. Add membrane protein in detergent into the calculated volume of reconstitution buffer supplemented with the detergent of your choice at concentrations above the CMC, and mix gently. This step is essential as it keeps the membrane protein in solution prior to mixing with lipids.
7. Add mixed micelles so that the amount of lipids corresponds to the molar ratio calculated in Step 6. Pipette gently a few times and leave on ice for 10 min. Make sure the final concentration of the detergent in the reconstitution mix (detergent from mixed micelles, your protein sample and buffer) is at least 10 m*M*. When working with marginally stable proteins, nanodisc assembly works at final concentrations of DDM as low as 4 m*M* (to avoid unfolding), albeit less efficiently.
8. Add the calculated amount of biotinylated MSP into the protein reconstitution mix. Pipette up and down several times and incubate on ice for 5 min.
9. Transfer 1 g of wet activated Bio-Beads into a new tube (0.5 g for every 1 ml of assembly mix). Remove any residual liquid from the Bio-Beads. To prepare 250 mg/ml activated Bio-Beads suspension, wash 5 g of Bio-Beads with methanol in 50-ml conical tube, followed by thorough washes with sterile water (~10 times 50 ml); fill up to 20 ml with sterile water. Such a suspension of activated Bio-Beads can be stored for prolonged periods at 4 °C.
10. Transfer nanodisc assembly mix onto Bio-Beads, mix up and down, and leave in the coldroom on a nutator overnight.
11. Collect as much of the assembly suspension as possible into new tube(s) and wash Bio-Beads with 0.5 ml of reconstitution buffer. Combine the

wash with assembly solution. Hard spin the sample on the benchtop centrifuge for 10 min at 15,000 × *g* and 4 °C, and transfer the supernatant to a fresh tube. From this point, do not use any detergents in your buffer as the majority of the target protein is in the lipid bilayer and an excess of detergents would readily destroy the nanodiscs.

12. Perform standard His-tag purification step using Ni-NTA Superflow Resin according to manufacturer's protocol. This step removes the excess of empty nanodiscs, micelles, and leftover protein that did not reconstitute properly. Note that other affinity purification methods can be used, depending on the tag available for membrane protein. Run a SDS-PAGE gel to check for the presence of MSP and target protein band in elution fraction(s).

13. (Optional): Remove the affinity tag by standard cleavage reaction as optimized for your protein in detergent. In our hands, the efficiency of thrombin, TEV, and 3C protease digestion in eggPC nanodiscs is similar or better when compared to reactions performed in DDM (Vergis & Wiener, 2011).

14. Concentrate biotinylated nanodisc with membrane protein using a 50 kDa cut-off Amicon Centrifugal Units and inject it on Superdex 200 column in reconstitution buffer without detergent. This step allows separation of nanodisc with reconstituted membrane protein from empty nanodiscs, higher order nanodisc assemblies, and leftover protein/lipid aggregates.

15. Run SEC fractions on a SDS-PAGE gel and concentrate the fractions corresponding to the monodisperse peak of reconstituted nanodiscs with 50 kDa cut-off Amicon Centrifugal Units.

16. Evaluate the protein concentration of nanodiscs with the method of choice, assuming that 1 biological unit of target membrane protein is incorporated per disc. Note that this is usually an oversimplification as multiple insertion events per disc are possible; however, for phage display selection purposes, it is not detrimental as long as the concentration is evaluated always the same way.

17. Dispense biotinylated nanodiscs with membrane protein into small aliquots and flash freeze in liquid nitrogen. Avoid adding glycerol to concentrations above 5% as it negatively impacts phage solubility. As an alternative 5% sucrose can be added as cryoprotectant, but for the majority of the proteins, no additives are required for short-term storage. Such aliquots can be kept at −80 °C for up to 6 months (depending on protein stability).

18. For selection purposes, empty nonbiotinylated nanodiscs composed of the same lipids will be required. While it is possible to recover the unbound fraction after affinity purification (Step 11), we recommend performing a separate and larger scale reconstitution of empty nanodiscs following Steps 7–17 of this section.

3.3 Pull-down and release of the membrane protein in biotinylated nanodiscs captured on streptavidin-coated magnetic beads

Prior to the phage display sorting, the efficiencies of the three following processes have to be evaluated: (1) the degree of the chemical biotinylation of the MSP, (2) the efficiency of the capture step by a pull-down assay of the membrane protein in the biotinylated nanodisc using streptavidin-coated magnetic beads, and (3) the efficiency of the release of the membrane protein from the nanodisc. Note that we employ magnetic beads for the capture step, but other solid substrates could be used with little modification of the protocol. Generally, capture is very efficient, even when the biotinylation of MSP is suboptimal because there are always two copies of the MSP per nanodisc, and thus the probability of having at least one copy biotinylated is high. Efficient release of the membrane protein from the beads is achieved by rapid disruption of nanodisc with high concentrations of zwitterionic detergent, FC-12.

1. Equilibrate 100–200 μl of Streptavidin MagneSphere Paramagnetic Particles with at least 3 bead volumes of buffer of your choice compatible with biotinylated nanodiscs (pull-down buffer). After the last wash, resuspend the beads in 50–100 μl final (further referred to as reaction volume). To wash and handle magnetic beads in an easy way, purchase a magnetic stand for small test tubes. Alternatively, you can use small earth magnets—by keeping them on the side of the tube it is easy to wash and collect the beads.
2. Remove the buffer and add up to 0.5–1.0 μg total of biotinylated nanodiscs resuspended in one reaction volume of pull-down buffer. Note that due to the size of the nanodiscs, the capacity of the magnetic beads is lower than what manufacturer reports for soluble proteins and membrane proteins in detergents. Prepare the same amount of the sample in a separate tube as a control for input.
3. Mix thoroughly and allow binding, mildly shaking on nutator at room temperature for at least 25 min.

4. Collect the supernatant and save in a separate tube for quality control. Wash the beads with 3 reaction volumes of pull-down buffer, keeping the first and last wash in separate tubes.
5. Resuspend the magnetic beads in elution buffer composed of 0.25–1.0% of FC-12 in the pull-down buffer and allow for mild shaking on nutator for at least 20 min. Keep the elution fraction in a separate tube. The exact concentration of FC-12 required for release of membrane protein must be optimized, and is dependent on the protein itself as well as the lipids used for nanodiscs reconstitution. In our hands, incubation with 1% FC-12 rapidly disassembles all variants of nanodiscs.
6. Wash the beads once with pull-down buffer and resuspend them in 1 reaction volume.
7. Take an equal volume of all fractions (input, flow-through, washes, elution, beads) and add SDS-PAGE loading dye. Boil all samples for 5 min at 95 °C. The boiling step is required to detach streptavidin from the magnetic beads; if membrane protein samples cannot be boiled, the amount of protein seen in the bead fraction might be lower.
8. Run SDS-PAGE, visualize the gel, and assess the efficiency of the pull-down. Usually ~80% or higher efficiency of capture and elution of target membrane protein from streptavidin beads is seen. A typical result is shown in Fig. 3. In case of lower efficiencies, troubleshooting of MSP biotinylation is advised as low efficiency may negatively impact the outcome of phage display library selection.

3.4 Competitive and subtractive phage display selection with nanodiscs

In this section, we describe the main steps of the phage display sorting procedure. We utilize a semi-automated phage display selection pipeline that was established under the auspices of the NIH (within Protein Structure Initiative). This protocol parallels what has been published for soluble proteins and therefore, we place particular emphasis on the differences in the protocols with regard to membrane proteins incorporated into nanodiscs. As previously stated, the significant advantage of nanodiscs is the presence of modular native-like lipid environments—eliminating the need for chemical modification of membrane protein for immobilization purposes. These are especially important aspects when working with marginally stable proteins or when there is no predetermined optimal detergent available to support function. On the other hand, phage display sorting with nanodiscs usually

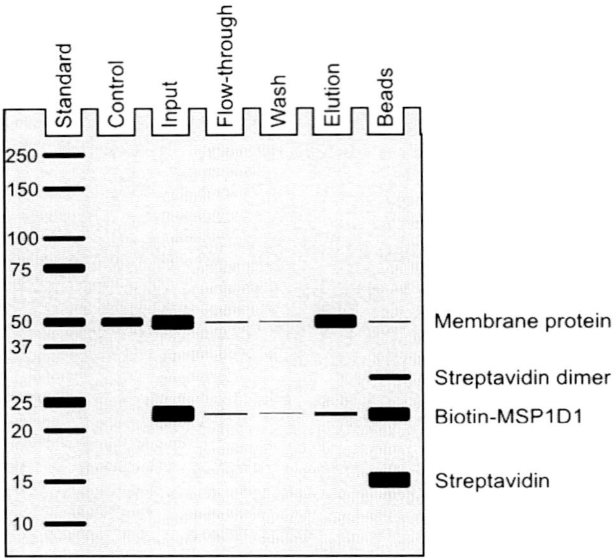

Figure 3 Pull-down with biotinylated nanodiscs. Graphical representation of desired outcome of procedure described in Section 3.3. A hypothetical 50-kDa membrane protein reconstituted into biotinylated MSP1D1 nanodiscs is efficiently captured on streptavidin-coated magnetic beads. The target membrane protein is then eluted from the nanodiscs by incubation with 1% fos-choline-12. The detergent readily releases the membrane protein into the solution, while the majority of biotinylated MSP1D1 remains attached to the beads. High efficiency of capture and release of target membrane protein on and off the beads is a prerequisite of a successful phage display selection experiment.

require in total 5 rounds versus 3 and 4 when performed with soluble proteins. This is due to the absence of detergent throughout the selection process, which removes nonspecific phage particles efficiently. We combat this by using excess empty nanodiscs as a soluble competitor throughout the selection to deplete the phage pool of nanodisc-specific binders.

General considerations for phage display in nanodiscs including the type and amount of membrane protein sample, buffer compositions, competitive and subtractive selection steps, as well as stringency of selection rounds are described in detail in Section 4 (see Note 3.4.a–d).

The sorting protocol described below (Fig. 4) is suitable for one to six independent membrane protein targets in biotinylated nanodiscs and is optimized to use with Library E (Miller et al., 2012) based on Herceptin Fab scaffold, similar to previously described protocol for competitive selection with soluble proteins by Paduch et al. (2013). With appropriate modifications, the protocol is compatible with any phage display library.

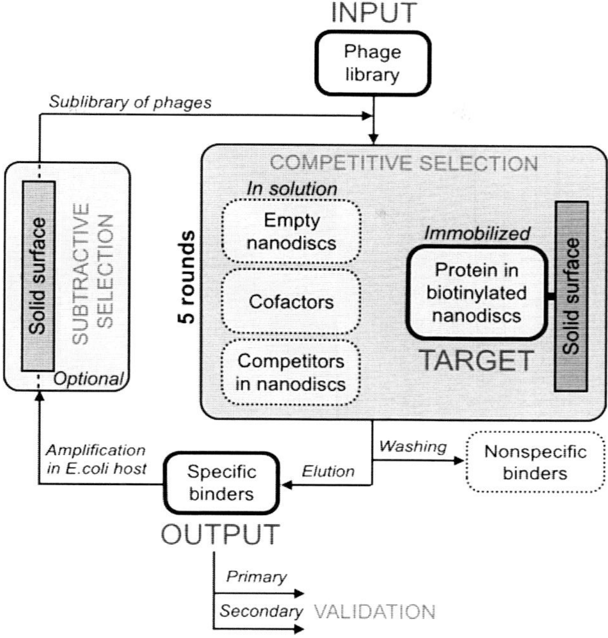

Figure 4 Competitive and subtractive phage display selection. The input phage library is incubated with the target membrane protein embedded in biotinylated nanodiscs, which are immobilized on a solid surface. In solution, the excess of empty nanodiscs, competitors (undesired conformations or regions of target protein), as well as necessary cofactors are used to create a desired competitive selection environment. Nonspecific or competitor-specific phage particles are washed away, while specific "binders" are recovered by elution in detergent. Subsequently, the specific phage particles are propagated in *E. coli* for another selection round. Optionally, prior to the next round, a subtractive selection step is performed by short incubation of sublibrary of phage particles with biotinylated competitors or solid surface alone to remove additional nonspecific binders. The leftover phage pool is used as a new input. In total, five consecutive rounds of binding selection are performed using decreasing concentrations of specific antigen as a selection pressure. As a result, specific phage clones are amplified and used for primary and secondary validation. (See the color plate.)

3.4.1 Selection—Round 1 (manual round)

Steps 2–3, 4, and 5–6 should be performed concurrently.

1. (Day before) Inoculate XL1-Blue cells into 1–3 ml of 2YT media containing 10 μg/ml tetracycline. Grow overnight at 37 °C shaking at 220 rpm.
2. (Day of experiment) Prepare phage library: transfer 1 ml of 10^{12} pfu/ml phage library (amount for six membrane protein targets in nanodiscs) in storage buffer into 16 ml of selection buffer (with BSA) and 3.5 ml of

20% PEG 8000/2.5 *M* NaCl. Mix several times by inverting gently and incubate for 1 h on ice. At this point phage particles are precipitating out of solution.

3. Spin precipitated phage particles for 20 min at 8000 × *g* at 4 °C, decant supernatant, remove excess precipitation buffer, and resuspended in 1 ml of selection buffer. Keep suspension of phage particles on ice.
4. Prepare cells: dispense 15 ml of 2YT media with 10 μg/ml tetracycline and inoculate with 100 μl XL1 cells from overnight starting culture. In parallel, prepare 2 × 2 ml of 2YT controls, one with 50 μg/ml ampicillin and the other with 50 μg/ml kanamycin. Inoculate controls with 100 μl of overnight XL1 culture to test for any possible contamination with phagemid or helper phage, respectively. Monitor cell density of 15 ml culture, keep shaking at 37 °C until OD_{600} ranges between 0.5 and 0.6 (log phase, usually it takes 2–3 h).
5. Prepare streptavidin magnetic beads: dispense 250 μl of magnetic beads per target and wash them three times with selection buffer (use magnetic stand or earth magnet to separate beads from the buffer). Incubate the beads with 500–1000 n*M* of biotinylated nanodiscs with each target in a separate tube for 25 min on orbital shaker/nutator (see Note 3.4.d). Wash three times in selection buffer.
6. Block magnetic beads with 5 m*M* D-biotin, incubate for 5 min on shaker/nutator, and wash three times in selection buffer. Keep on ice until phage suspension is ready.
7. Perform round 1 of binding selection by mixing 150 μl of precipitated phage library with each bead sample from Step 6 for 1 h on orbital shaker/nutator at room temperature. Keep excess of empty nanodiscs and additional competitors/cofactors in solution to counter-select undesired phage particles (see Note 3.4.c). In comparison to selection performed with soluble or membrane proteins in detergents, it is not recommended to mix the beads with different targets together and perform a single incubation with phages to prevent bias in the selection process for clones specific to nanodiscs and lipids. This is particularly important if no competitive steps are performed during round 1.
8. Wash beads three times with selection buffer. At this point, beads from separate targets can be combined into a single amplification pool. Alternatively, all protein targets can be kept separate.
9. To each amplification, add beads with bound target into 5 ml of log phase XL1 cells and incubate mildly shaking for 20 min at 37 °C. Meanwhile, prepare 30 ml of 2YT media with 50 μg/ml ampicillin

and 10^9 pfu/ml M13 KO7 helper phage; keep the media at 37 °C. Add media to infected cells and shake at 250–280 rpm in 37 °C for 18–20 h. Propagated phage particles are used as input for selection round 2.

10. Go to Step 1 to prepare cells for the next round of selection.

3.4.2 Selection—Round 2 (semi-automatic round using KingFisher robot)

Steps 1, 2, and 3 should be performed concurrently. Step 4 is optional.

1. Prepare phage particles: Transfer amplification culture from previous day to a 50-ml conical tube, spin down the cells at 6000 × *g* in 4 °C for 10 min. Pour supernatant into a fresh tube and add 6 ml of ice-cold PEG/NaCl solution. Incubate and spin like in Steps 2 and 3 of Section 3.4.1. Resuspend in 500 μl of buffer, spin at 6000 × *g* to remove residual bacteria, transfer supernatant to a fresh small tube and keep on ice. For each target, 20 μl of phage particles is used for selection; the rest of the phage suspension can be kept at −80 °C in 20% glycerol solution.
2. Prepare cells: see Step 4 of Section 3.4.1.
3. Prepare streptavidin magnetic beads: dispense 40 μl of magnetic beads per target and wash them three times with selection buffer like in Step 5 of Section 3.4.1.
4. (Optional): Perform subtractive step on precipitated phage particles from Step 1 (see Note 3.4.c). Incubate 100 μl of phage particles with 100 μl of streptavidin magnetic beads washed three times for 25 min on shaker/nutator. Use supernatant for further steps.
5. Prepare KingFisher binding plate—binding selection round 2: For each target, pipette the following solutions into separate KingFisher reservoir strips:
   a. Well A: biotinylated target nanodiscs at twofold lower molar concentration than in round 1 (see Note 3.4.d), nonbiotinylated empty nanodiscs (plus other competitors and cofactors, see Note 3.4.c), 20 μl of precipitated phage particles from Step 1 (or Step 4), selection buffer up to 100 μl.
   b. Well B: 40 μl of streptavidin magnetic beads from Step 3, 60 μl of selection buffer.
   c. Well C: 5 μl of 100 μM D-biotin, nonbiotinylated empty nanodiscs (plus other competitors and cofactors, see Note 3.4.c), selection buffer up to 100 μl.
   d. Wells D–G: nonbiotinylated empty nanodiscs (plus other competitors and cofactors, see Note 3.4.c), selection buffer up to 100 μl.
   e. Well H: 100 μl of 1% FC-12 in selection buffer.

6. Load binding plate into KingFisher drawer, use fresh plastic comb and run KingFisher selection protocol:
   a. Move the comb and collect the beads from well B by mixing vigorously for 30 s.
   b. Transfer beads into well A and mix vigorously for 15 min.
   c. Transfer beads into well C and mix vigorously for 20 s.
   d. Transfer beads into wells D to G, each time mixing vigorously for 20 s.
   e. Transfer beads into well H and mix vigorously for 10 min.
   f. Move comb with beads into well G, release the beads.
7. Discard the comb and transfer each elution into a fresh tube. For each target, infect 500 μl of log phase XL1 cells with 50 μl of elution, and incubate mildly shaking for 20 min at 37 °C. Add 20% glycerol to the leftover elution and store at −80 °C. Meanwhile, prepare 2.5 ml of 2YT media with 50 μg/ml ampicillin and 10^9 pfu/ml M13 KO7 helper phage; keep the media at 37 °C. Add media to infected cells and shake at 250–280 rpm in 37 °C for 18–20 h. Propagated phage particles are used as input for the next selection round.
8. Inoculate XL1-Blue cells like in Step 1 of Section 3.4.1.

3.4.3 Selection—Rounds 3–5

1. Repeat Steps 1–8 from Section 3.4.2 for three consecutive rounds with following differences:
   a. Precipitate supernatant from amplification and resuspend in 150 μl of selection buffer.
   b. For KingFisher binding plate, use lower concentrations of biotinylated nanodiscs with target membrane protein (see Note 3.4.d). Starting from round 3, include a separate control of empty beads (no target nanodisc added). For round 5, add second control of biotinylated empty nanodiscs. Starting from round 4, a visible enrichment in number of colonies should be observed for the membrane protein target-specific elutions when compared to colonies obtained from control elutions.
   c. During round 3, use 20 μl of washed streptavidin magnetic beads, for rounds 4 and 5, drop the amount to 10 μl.
   d. For amplifications prior to the next selection round, infect 100 μl of log phase XL1 with 10 μl of phage elution and add 2 ml of media with ampicillin and helper phage. No amplification is required after round 5.

2. Titer phage outputs after selection rounds 3–5: Do three to four 10-fold serial dilutions of infected cells by transferring 10 μl of infected XL1 cells from Step 1.d into 90 μl of 2YT media. For each target, plate 5 μl of each of the four consecutive dilutions (including controls) onto LB/agar/ampicillin50 plates. Grow in the incubator for 14–16 h at 37 °C.
3. After rounds 4 and 5, spread the remaining infected cells from the second dilution onto LB/agar/ampicillin50 plate (separate each target and do not plate cells infected with control phage particles). Use reduced streaking method to obtain single phage colonies that will be further used for initial validation with phage ELISA. Plates with single colonies can be kept for up to 1 week prior to amplification of phage particles (see Step 1 of Section 3.5).

3.5 Single-point phage ELISA

As an early validation step to evaluate the quality of the binders, we recommend performing a single-point phage ELISA assay. It is important to assess the binding of each single phage clones, not only to biotinylated nanodiscs with incorporated protein but also to empty nanodiscs and to uncoated wells as controls. Phage ELISA with nanodiscs is performed in the absence of detergents (e.g., Tween), which in many instances result in higher backgrounds. To circumvent this issue, we recommend increasing the concentration of BSA in all buffers (except for signal development and quenching) to 2% (w/v) final. It is also important to alter the type of capture agent through which the biotinylated target was attached during selection to eliminate the binders to the capture agent itself. For instance, neutravidin can be used as a surrogate of streptavidin to eliminate streptavidin-specific phage binders. Lastly, whereas for initial validation of phage clones where the selections are done in detergent competitive sandwich ELISA experiments are the method of choice, the method is impractical when the target protein is in nanodiscs. Because the amount of nonbiotinylated nanodisc that is required to perform the assay is relatively high, we recommend using direct phage ELISA experiments, without competitors, as described below.

Steps 6 and 7 should be performed concurrently.

1. Pick single colonies of XL1 cells infected with phage elution after the 4th and 5th rounds of selection (48 colonies each) and inoculate into 96-deep-well block plate containing 400 μl 2YT media with 50 μg/ml ampicillin and 10^9 pfu/ml of KO7 helper phage. Seal with breathable cap and grow for 16–20 h at 37 °C shaking at 250–280 rpm.

2. Spin phage plates for 20 min at 4000 × *g* and transfer 300 μl of phage particles suspension into a fresh plate. Save the leftover XL1 bacterial pellets. They can be frozen with the addition of 10% glycerol and used in further propagation of phage particles and/or phage DNA.
3. Coat ELISA binding plate with 50 μl of 2 μg/ml per well of Neutravidin resuspended in coating buffer (50 m*M* sodium carbonate pH 9.6). Incubate for 2 h at room temperature on orbital shaker, alternatively keep without shaking overnight at 4 °C.
4. Remove coating solution and add 100 μl of ELISA buffer (compatible with your protein, containing 2% BSA). Note that no detergent can be used at any point in this ELISA assay. Incubate for 1 h at room temperature shaking. Blocked plates can be stored sealed for approximately 2 weeks at 4 °C without significant increase of the average background levels.
5. In a separate tube, prepare the dilution of biotinylated targets (separately nanodiscs with and without protein) in ELISA buffer at concentrations equal to the final concentration of target used in the last round of selection—typically 20–50 n*M*.
6. For each phage screen prepare three ELISA wells: (1) buffer alone, (2) biotinylated empty nanodiscs, and (3) biotinylated nanodiscs with target protein embedded. Add 50 μl of corresponding solution into each well and incubate at room temperature shaking for 15–30 min.
7. For every well assayed prepare 60 μl of 5–10 × diluted overnight phage suspension in ELISA buffer on a separate plate.
8. Remove immobilization solution and wash each well three times with 100 μl of ELISA buffer to remove excess of biotinylated nanodiscs.
9. Transfer 50 μl of diluted phage particles into corresponding wells and shake for 15–30 min.
10. Remove solution and wash each well three times with 100 μl of ELISA buffer to remove excess of unbound phage particles.
11. For each plate prepare 5 ml + 2 ml of HRP-conjugated anti-M13 antibody at 1:5000 dilution in ELISA buffer. Add 50 μl into each well and incubate with shaking for 30 min.
12. Remove the solution and wash each well three times with 100 μl of ELISA buffer to remove excess of the secondary antibody.
13. Prepare TMB developing solution by mixing equal volumes of H_2O_2 and TMB substrate (always prepare it freshly) and add 50 μl to each well (you will need 5 ml/plate + 2 ml). Develop for 3–10 min until the signal in wells with biotinylated nanodiscs with target protein becomes blue (usually 1–3 min).

14. Quench the reaction with 50 μl of 1 M HCl (or H_3PO_4), mildly shake for even mixing. Be sure to remove any residual bubbles that might appear from washing steps.
15. Read the absorbance at 450 nm in plate reader.
16. Save and analyze the data. For sequencing analysis, pick those phage clones that exhibit low signal for empty nanodiscs, and >4 times greater signal for the biotinylated nanodisc with target protein than average background signal. In a typical phage display selection, the number of phage clones that are specific to empty nanodiscs should be no greater than 10 out of 96 screened clones.
17. Inoculate leftover spun bacteria from positive phage clones into 3 ml of 2YT media with 50 μg/ml ampicillin for phagemid propagation and mini-prep isolation.

3.6 Secondary validation: Characterizing positive "binders"

The secondary validation involves determining the affinity and specificity of the binders to their protein targets. For this, the evaluation of the phage ELISA positive clones is divided into several steps. First, the pool of positive clones is sequenced to determine the number of unique "binders." Depending upon the number of clones sequenced, there would be an expectation that a few of them will be duplicates. However, if one or two clones dominate the pool, it could be an indication that they possess some type of growth advantage and not necessarily be superior to others in the pool. Further analyses are performed in protein format. Because this requires expression and purification of the binders that may require additional subcloning steps, it is important to judiciously select the number of binders to study further. In cases where a relatively limited number of binders are identified, all should be further characterized. But, in cases where there are a large number of unique binders in the binding positive pool, we generally pick the top 24 to carry forward.

In our standard setup where the binders are in the Fab format, we use multipoint saturation ELISA assays and/or surface plasmon resonance to establish binding affinities to the target protein in nanodiscs. Alternatively, binding of Fabs can be evaluated to the target protein in detergent. This is particularly important if one considers using selected binders for pull-down based assays from detergent lysates, as well as for complex formation by SEC for crystallization purposes. Details of protocols on sequencing, small and large-scale purification of the Fabs, as well as protein ELISA analysis for Fabs were reported previously (Paduch et al., 2013). For the purposes of highlighting the difference between Fabs generated from phage display selections against membrane protein in detergent versus nanodiscs, we focus on the important steps required for efficient complex formation between Fabs and target membrane protein in detergent by SEC.

3.7 Fab-membrane protein complex formation in detergent by SEC

Fabs generated from phage display selections with soluble targets and membrane proteins in detergent were shown to be effective crystallization chaperones (Li et al., 2014; Rizk et al., 2011; Shukla et al., 2013; Uysal et al., 2009; and others). One of the prerequisites for high-throughput crystallization trials is the formation of stable and monodisperse complexes between Fab and membrane protein target in detergent using SEC. From our experience, this step of validation is as important as the selection process itself; and in this section, we comment on common procedures that lead to stable complex formation. While the common protocol for SEC with Fabs and membrane proteins in detergent is straightforward, it is the small details that allow for fast and efficient identification of the best chaperone candidates (see Notes 3.7.a–c).

1. Mix target membrane protein in detergent with 15–20% molar excess of Fab in a buffer of choice (see Note 3.7.a). For membrane proteins of higher oligomeric states, the excess should be calculated for each monomer separately to allow for monodisperse complexes, assuming that binding occurs at 1:1 (per monomer) stoichiometry. Incubate for 10–15 min on ice.
2. Meanwhile start equilibration of the column in buffer and detergent of choice.
3. Concentrate the complex using Amicon Centrifugal Units with 10 kDa molecular weight cut-off until desired injection volume is reached. Hard spin the sample at benchtop centrifuge for 10 min at 15,000 × *g* and 4 °C and transfer the supernatant to a fresh tube. At this point, free detergent micelles will be concentrated as well; however, the SEC step should efficiently remove their excess.
4. Inject the sample and run SEC (see Note 3.7.b), collecting 0.5–1 ml fractions.
5. Run fractions after SEC on an SDS-PAGE gel and concentrate the fractions corresponding to the complex with Amicon Centrifugal Units. Avoid excessive concentration of free detergent micelles, especially when sample is to be used for crystallization. If no complex formation is present, then refer to Note 3.7.c.

4. NOTES

Note 3.4.a—*Sample requirements*: As mentioned in Section 3.2, the type of nanodisc (MSP variant, hence diameter of nanodiscs after the reconstitution step) and the lipids used in the assembly should be optimized for each target. It is recommended to avoid a high percentage of positively charged lipids to minimize the nonspecific interaction between surface of nanodiscs and phage coat. For a single 5-round phage display sorting protocol, about 500 μl of 5 μ*M* of the membrane protein in biotinylated nanodiscs is ideal, as any leftover sample can be used for primary and secondary validation. However, it may be possible to use as little as 200 μl of 2 μ*M*, but this usually results in curtailing some of the secondary validation. Additionally, binding, blocking with biotin and the washing steps require at least 5–10 molar excess of nonbiotinylated empty nanodiscs assembled with the same lipids as incorporated in the nanodiscs with target membrane protein (see Note 3.4.c). For membrane protein targets that are marginally stable or contain little surface exposed surface area, it is recommended to use two to three times larger initial concentrations of the biotinylated nanodiscs to provide sufficient solvent-exposed epitopes for effective interaction. The immobilization and elution steps on and off the streptavidin-coated magnetic beads should be evaluated every time a new batch of biotinylated nanodiscs is prepared as described in Section 3.3.

Note 3.4.b—*Buffer requirements*: Protocols of phage display selections with nanodiscs described in this chapter were designed and tested with buffers around neutral pH (7.0–8.0 depending on buffering capacity); however, we recommend to start with the buffer and pH that is the most compatible with target membrane protein, including 100–200 m*M* NaCl (or other compatible salt) and supplemented with 1% BSA. Higher concentrations of BSA (0.1% used in standard protocols) reduce nonspecific background binding from "sticky" phage particles in the absence of detergent. Additionally, phage display selection buffer should not include any components that interfere with (1) phage solubility (glycerol above 5%, sucrose, above 5%, high-molecular PEGs), (2) phage infectivity (e.g., concentrations of EDTA above 2 m*M*), (3) library scaffold stability (e.g., reducing agents above 50 μ*M* for Fabs), and (4) nanodisc integrity (Tween and other detergents can be only present in the elution step), unless they are absolutely required to maintain the function of membrane protein during selection process.

Note 3.4.c—*Competitive and subtractive selection steps*: To eliminate the majority of nanodisc- and lipid-specific Fab binders, a 5- to 10-fold molar excess of empty nanodiscs over immobilized target protein in nanodiscs is recommended during the binding and washing steps starting from round 1 of selection (competitive strategy). Additionally, if conformation- or regio-specific binders are sought, any undesired mutants or isoforms of target membrane proteins in nanodiscs or as soluble domains can be used as competitors in similar fashion to empty nanodiscs. As an optional step, to remove bead-specific binders, prior to each round, the input library can be incubated with large amounts of streptavidin-coated magnetic beads and the resulting supernatant is then used as input (subtractive strategy). Note that subtractive selection steps are not 100% efficient; therefore, potential binders have to be further screened for binding to empty beads and/or streptavidin analogues are recommended for validation purposes at phage and protein level (see Section 3.5 for Step 3).

Note 3.4.d—*Stringency of selection rounds*: The optimal concentration of the target protein in nanodiscs between subsequent selection rounds depends on type of membrane protein and the desired affinity of binders. Usually, the concentration of the biotinylated nanodiscs with target should be dropped by 50–70% between four consecutive rounds, and with one additional round at concentrations similar to round 4. For instance, typical concentration values between subsequent phage display selection rounds would be 500 n*M*, then 250, 100, 40 n*M* and again 40 n*M* for rounds 1–5, respectively. For monomeric membrane proteins that are smaller than 30 kDa, we recommend less stringent selection protocols, i.e., 800 n*M*, and then 400, 200, 100, and 100 n*M*. It should be noted that this is very much dependent on the antigenicity, architecture and solvent-exposed surface area of the membrane protein itself.

Note 3.7.a—*SEC buffer conditions*: Usually, determining stable complex formation between the target protein in detergent and the Fabs obtained from phage display selection in nanodiscs starts with performing SEC experiments in the buffer, salt, pH, and detergent conditions in which the target membrane protein is the most stable or is known to retain its function. Initial conditions mimicking those in which phage display selection was performed (without BSA) and/or in which the ELISA binding signal was evaluated are both good starting points. When working with antibodies and their derivatives (Fabs, scFvs), it is generally advised to avoid reducing agents, unless they are absolutely required for membrane protein stability.

Note 3.7.b—*Separation of complexes*: Complexes containing the membrane protein with a Fab are usually well separated on Superdex 75/200 columns at their resolution ranges (depending on protein size). However, we have found that Fabs in the absence of their target membrane protein tend to interact with the resin, which appears to be a property of the hydrophobic content of the randomized CDR loops. As a result, when run alone, the Fabs tend to "stick" to Superdex or elute at much later elution volumes. If no elution peak is observed after 2 column volumes, extensive cleaning of the column according to the manufacturer's protocol is recommended. If there is a pressure increase, even after standard cleaning procedures, a solution of 1 mg/ml pepsin in 0.1 *M* acetic acid and 0.5 *M* NaCl should be passed through the column overnight to completely digest the residual Fab. Alternatively, for crude separations at preparative grade, Sephacryl columns can be used; however, the resolution is significantly lower, but is counterbalanced by the fact that the nonspecific binding of excess of unbound Fab is also significantly decreased.

Note 3.7.c—*Common problems with complex formation*: If there is no apparent stable complex formation between the membrane protein in detergent and the Fab generated from phage display selections in nanodiscs, this can be due to several factors. Below are the most common solutions to successful complex formation by SEC:

1. Compare the elution profile of "complex" to the membrane protein only control, do not run Fab by itself unless it is absolutely required (see Note 3.7.b). The vast majority of Fabs do shift the elution peak significantly upon binding, but this has to be confirmed by SDS-PAGE (co-elution of both protein components). If possible, during complex formation the concentrations of the membrane protein and Fab should be ~10 times above predicted K_D to shift the equilibrium toward complex formation even upon dilution of the sample on the column.
2. Change the binding conditions (see Note 3.7.a). Careful inspection of randomized regions of the library in a particular Fab might provide some insights about the character of interaction with the target (hydrophobicity, charge). For example, when large number of charged residues is present in randomized CDR loops, pH, and ionic strength of the buffer might play a role in robust complex formation between Fab and target protein.
3. There will certainly be cases where the Fab generated from phage display selection in a nanodisc does not bind to target membrane protein in

a particular detergent. This might be caused by several factors: slight differences in protein conformation, stability and flexibility, or epitope accessibility in a detergent micelle of a larger size. The latter can be tested using detergents that have similar head groups, but shorter hydrocarbon chains, e.g., *n*-octyl-β-D-glupyranoside instead of *n*-nonyl-β-D-glucopyranoside, or *n*-decyl-β-D-maltopyranoside in place of DDM.

4. In the instances where none of the Fabs bind in a detergent of a different micelle size, it is advised to change the purification strategy. For instance, when washing on an affinity column with detergent, fewer wash steps should be performed to allow for co-purification of essential structural lipids. For high-resolution structure determination, binding in bicelles, amphipols, or in lipidic cubic phases can be tested. Switching back to nanodiscs is a viable option only in the case of cryo-electron microscopy (Wu et al., 2012), as nanodiscs are too heterogeneous to crystallize on their own. Some structural epitopes available during selection in nanodiscs are highly dependent on lipids and are not present in detergent micelles. This is particularly true for smaller membrane proteins that lack extensive solvent-exposed surfaces.
5. If the Fab is incompatible with all viable detergents, sometimes the best solution is to simply try another Fab. We have found that a small fraction of the Fabs generated from selections in nanodiscs (and detergents), while showing significant binding in phage and protein format by ELISA or SPR, nevertheless, do not produce complex formation by SEC. This might be due to some paratopes at the interface of interaction that consist of hydrophobic portions of lipids/detergent molecules.

ACKNOWLEDGMENTS

We thank M. L. Paduch for providing advice on the competitive selection strategy, and B. E. Zalisko, and A. Koide for their input into various steps of the protocols. The DNA for Library E was provided by S. Koide. Initial optimization of elution of nanodisc with detergents was performed by members of the B. G. Fox laboratory (personal communication). We thank L. J. Bailey and S. S. Kim for helpful comments on the chapter.

This work is supported by National Institutes of Health Grants GM087519 and GM094588 to A. A. K., and the Chicago Biomedical Consortium.

REFERENCES

Bayburt, T. H., & Sligar, S. G. (2009). Membrane protein assembly into nanodiscs. *FEBS Letters*, *584*, 1721–1727.

Bukowska, M. A., & Grütter, M. G. (2013). New concepts and aids to facilitate crystallization. *Current Opinion in Structural Biology*, *23*, 409–416.

Chen, G., & Sidhu, S. S. (2014). Design and generation of synthetic antibody libraries for phage display. *Methods in Molecular Biology, 1131*, 113–131.

Fellouse, F. A., Esaki, K., Birtalan, S., Raptis, D., Cancasci, V. J., Koide, A., et al. (2007). High-throughput generation of synthetic antibodies from highly functional minimalist phage-displayed libraries. *Journal of Molecular Biology, 373*, 924–940.

Fellouse, F. A., Wiesmann, C., & Sidhu, S. S. (2004). Synthetic antibodies from a four-amino-acid code: A dominant role for tyrosine in antigen recognition. *Proceedings of the National Academy of Sciences of the United States of America, 101*, 12467–12472.

Gao, J., Sidhu, S. S., & Wells, J. A. (2009). Two-state selection of conformation-specific antibodies. *Proceedings of the National Academy of Sciences of the United States of America, 106*, 3071–3076.

Hagn, F., Etzkorn, M., Raschle, T., & Wagner, G. (2013). Optimized phospholipid bilayer nanodiscs facilitate high-resolution structure determination of membrane proteins. *Journal of the American Chemical Society, 135*, 1919–1925.

Hoogenboom, H. R. (2005). Selecting and screening recombinant antibody libraries. *Nature Biotechnology, 23*, 1105–1116.

Kim, J., Stroud, R. M., & Craik, C. S. (2011). Rapid identification of recombinant Fabs that bind to membrane proteins. *Methods, 55*, 303–309.

Koenig, P., & Fuh, G. (2014). Selection and screening using antibody phage display libraries. *Methods in Molecular Biology, 1131*, 133–149.

Li, Q., Wanderling, S., Paduch, M., Medovoy, D., Singharoy, A., McGreevy, R., et al. (2014). Structural mechanism of voltage-dependent gating in an isolated voltage-sensing domain. *Nature Structural and Molecular Biology, 21*, 244–252.

Miller, K. R., Koide, A., Leung, B., Fitzsimmons, J., Yoder, B., Yuan, H., et al. (2012). T cell receptor-like recognition of tumor in vivo by synthetic antibody fragment. *PLoS One, 7*, e43746.

Nath, A., Atkins, W. M., & Sligar, S. G. (2007). Applications of phospholipid bilayer nanodiscs in the study of membranes and membrane proteins. *Biochemistry, 46*, 2059–2069.

Paduch, M., Koide, A., Uysal, S., Rizk, S. S., Koidem, S., & Kossiakoff, A. A. (2013). Generating conformation-specific synthetic antibodies to trap proteins in selected functional states. *Methods, 60*, 3–14.

Ritchie, T. K., Grinkova, Y. V., Bayburt, T. H., Denisov, I. G., Zolnerciks, J. K., Atkins, W. M., et al. (2009). Reconstitution of membrane proteins in phospholipid bilayer nanodiscs. *Methods in Enzymology, 464*, 211–231.

Rizk, S. S., Paduch, M., Heithaus, J. H., Duguid, E. M., Sandstrom, A., & Kossiakoff, A. A. (2011). Allosteric control of ligand-binding affinity using engineered conformation-specific effector proteins. *Nature Structural and Molecular Biology, 18*, 437–442.

Schuler, M. A., Denisov, I. G., & Sligar, S. G. (2013). Nanodiscs as a new tool to examine lipid-protein interactions. *Methods in Molecular Biology, 974*, 415–433.

Shukla, A. K., Manglik, A., Kruse, A. C., Xiao, K., Reis, R. I., Tseng, W. C., et al. (2013). Structure of active β-arrestin-1 bound to a G-protein-coupled receptor phosphopeptide. *Nature, 497*, 137–141.

Uysal, S., Vásquez, V., Tereshko, V., Esaki, K., Fellouse, F. A., Sidhu, S. S., et al. (2009). Crystal structure of full-length KcsA in its closed conformation. *Proceedings of the National Academy of Sciences of the United States of America, 106*, 6644–6649.

Vergis, J. M., & Wiener, M. C. (2011). The variable detergent sensitivity of proteases that are utilized for recombinant protein affinity tag removal. *Protein Expression and Purification, 78*, 139–142.

Wu, S., Avila-Sakar, A., Kim, J., Booth, D. S., Greenberg, C. H., Rossi, A., et al. (2012). Fabs enable single particle cryoEM studies of small proteins. *Structure, 20*, 582–592.

CHAPTER TWELVE

Antibody Fragments for Stabilization and Crystallization of G Protein-Coupled Receptors and Their Signaling Complexes

Arun K. Shukla[1], Charu Gupta, Ashish Srivastava, Deepika Jaiman
Department of Biological Sciences and Bioengineering, Indian Institute of Technology, Kanpur, India
[1]Corresponding author: e-mail address: arshukla@iitk.ac.in

Contents

Abstract

G protein-coupled receptors (GPCRs) are one of the key players in extracellular signal recognition and their subsequent communications with cellular signaling machinery. Crystallization and high-resolution structure determination of GPCRs has been one of the major advances in the area of GPCR biology over the last 7–8 years. There have primarily been three approaches to GPCR crystallization till date. These are fusion protein strategy, thermostabilization, and antibody fragment-mediated crystallization. Of these, antibody fragment-mediated crystallization has not only provided the first breakthrough in structure determination of a non-rhodopsin GPCR but it has also assisted in obtaining structures of fully active conformations of GPCRs. Antibody fragment approach has also been crucial in obtaining structural information on GPCR signaling

Methods in Enzymology, Volume 557
ISSN 0076-6879
http://dx.doi.org/10.1016/bs.mie.2015.01.010

complexes. Here, we highlight the specific examples of GPCR crystal structures that have utilized antibody fragments for promoting crystallogenesis and structure solution. We also discuss emerging powerful technologies such as the nanobody technology and the synthetic phage display libraries in the context of GPCR crystallization and underline how these tools are likely to propel key GPCR structural studies in future.

1. INTRODUCTION

G protein-coupled receptors (GPCRs), also referred to as seven transmembrane receptors, bind to an incredibly diverse array of stimuli and mediate a wide range of cellular and physiological processes (Bockaert & Pin, 1999). They also represent the largest class of cell surface receptors with about 800 members in the superfamily and constitute the largest group of drug targets (Ma & Zemmel, 2002). Although their functional, biochemical, and physiological characterization has flourished a great deal over a several decades-long history of GPCR research, their significantly lagging direct structural characterization has started to surface only in the last two decades or so (Cherezov, Abola, & Stevens, 2010). Specifically, crystallization and X-ray structure determination of GPCRs had eluded the scientists until very recently. The first crystal structure of a GPCR, rhodopsin, was published in the year 2000 (Palczewski et al., 2000). The next crystal structure of a GPCR, β_2AR, was determined in 2007 (Rasmussen et al., 2007). However, in the past 7 years, since the first breakthrough of a non-rhodopsin GPCR structure, there have been crystal structures of 27 different GPCRs including several receptor now in complex with chemically and functionally distinct ligands (Ghosh, Kumari, Jaiman, & Shukla, 2015).

Antibody fragment-mediated crystallization and conformational stabilization approach has played an important role in the recent structural coverage of GPCRs in distinct conformations. The limitations that conventional hybridoma method poses in obtaining highly specific antibodies against challenging targets such as membrane proteins are being overcome now to some extent through the recent development of synthetic phage display antibody libraries and robust *in vitro* screening protocols (Shukla et al., 2013, 2014). Furthermore, remarkable success of nanobodies from llamas in trapping specific conformations of GPCRs and their ability to stabilize signaling complexes have further broadened the options available for this approach (Steyaert & Kobilka, 2011). In this chapter, we discuss examples of GPCRs where antibody fragments have been utilized successfully to stabilize them or their signaling complex, to obtain either high-resolution X-ray crystal structure or direct structural information using complementary approaches.

2. THE GENESIS OF ANTIBODY FRAGMENT-MEDIATED MEMBRANE PROTEIN CRYSTALLIZATION

The feasibility of antibody fragment-mediated crystallization of membrane proteins was first demonstrated by co-crystallization of cytochrome *c* oxidase in complex with a single chain variable fragment (ScFv) of an antibody (Ostermeier, Iwata, Ludwig, & Michel, 1995). Since then, several other membrane proteins including the cytochrome bc1 complex (Hunte, Koepke, Lange, Rossmanith, & Michel, 2000) and the potassium channel (Doyle et al., 1998) have been successfully crystallized in complex with either ScFv or antigen-binding fragment (Fab) of antibodies (Hino, Iwata, & Murata, 2013).

The potential advantage of the antibody fragment-mediated crystallization approach is twofold (Hunte & Michel, 2002). First, binding of a high-affinity antibody fragment is expected to reduce the conformational flexibility of membrane proteins which is a major limitation in their crystallization. Second, antibody fragments bound to purified membrane proteins extend the polar (hydrophilic) surface area which is then available to make crystal contacts. In case of GPCRs, both of these parameters are of critical importance to promote crystallogenesis as purified GPCRs are not only highly conformationally dynamic but they also contain relatively smaller hydrophilic loops.

2.1 First example of antibody fragment-assisted GPCR crystallization—β_2AR

β_2AR, the first non-rhodopsin GPCR to be cloned, has been one of the most intensively studied GPCRs from structural and functional angles. Not surprisingly, it has also been one of the favorite targets for crystallization attempts. A series of biochemical and biophysical studies revealed that in addition to the N-and the C-termini of the receptor, the third intracellular loop of the β_2AR is highly dynamic (Ghanouni, Gryczynski, et al., 2001; Ghanouni, Steenhuis, Farrens, & Kobilka, 2001) and therefore, it was conceived that it must be conformationally stabilized to obtain diffraction quality crystals. To achieve this, purified β_2AR reconstituted in lipid vesicles was injected in to a mouse to generate monoclonal antibodies. The resulting antibodies were characterized for their ability to specifically recognize the

β_2AR. The antigen-binding fragment (Fab) of one of these antibodies, referred to as Fab5, was found to form a stable complex with purified β_2AR (Day et al., 2007; Rasmussen et al., 2007). Furthermore, Fab5 recognized a three-dimensional epitope on β_2AR and did not bind to the denatured receptor protein. More importantly, this Fab was found to bind to the third intracellular loop of the β_2AR as revealed by a limited proteolysis assay and was therefore, considered to be a suitable stabilizing chaperone for β_2AR crystallogenesis. Indeed, the Fab5–β_2AR complex yielded diffracting crystals and resulted in 3.4/3.7 Å crystal structures of the carazolol-bound β_2AR—the first non-rhodopsin GPCR crystal structure (Fig. 1A). This crystal structure of the Fab5–β_2AR complex also revealed that Fab5 binds primarily to

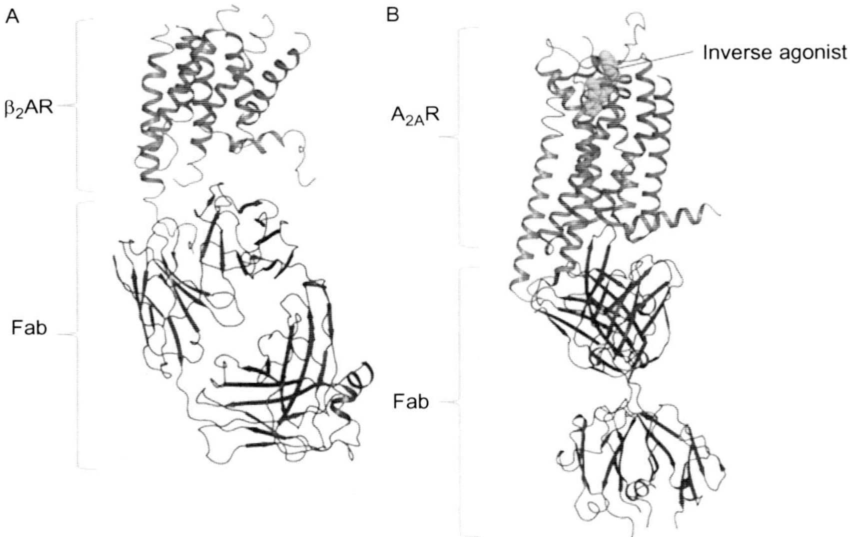

Figure 1 Antigen-binding fragment (Fab) assisted crystallization of inactive GPCR conformations. (A) Crystal structure of carazolol (not shown in figure) bound human β2 adrenergic receptor (β_2AR) in complex with a Fab corresponding to a mouse monoclonal antibody. The diagram represents the cartoon depiction of the structure (PDB ID 2R4R) generated using PyMol software. The β_2AR is shown in red (light gray in the print version) and the Fab is in blue (dark gray in the print version). (B) Crystal structure of an inverse agonist (ZM241385) bound human adenosine A_{2A} receptor (A_{2A}R) in complex with a Fab corresponding to a mouse monoclonal antibody. The diagram represents the cartoon depiction of the structure (PDB ID 3VG9) generated using PyMol software. The A_{2A}R is shown in red (light gray in the print version) while the Fab is in blue (dark gray in the print version) and the inverse agonist ZM241385 is shown in cyan colors (light gray in the print version).

the proximal region of the third intracellular loop of the receptor and therefore, reduces its flexibility to promote crystallogenesis.

2.2 Antibody fragment-assisted crystallization of the adenosine A_{2A} receptor ($A_{2A}R$)

The adenosine A_{2A} receptor represents the second GPCR to be crystallized using an antibody which was generated by conventional hybridoma method and its corresponding antigen-binding fragment that was used as a chaperone for crystallization and structure determination of the receptor (Hino et al., 2012). Similar to β_2AR, purified and lipid-reconstituted antagonist-bound receptor was used as an immunogen to raise conformationally sensitive monoclonal antibodies in mice. Several antibodies were obtained which specifically recognize conformational epitopes on the folded A_{2A}R and not the denatured protein. Subsequently, the Fab fragment of one of these antibodies, referred to as Fab2838, was bound to the purified A_{2A}R with low nanomolar affinity as determined by surface plasmon resonance and it was used for crystallization of antagonist-bound A_{2A}R. Interestingly, this Fab blocked the binding of agonist to the receptor, however, it did not alter antagonist binding. This striking observation suggests that the Fab used here stabilizes an inactive conformation of the receptor that prevents agonist occupancy of the receptor and in fact this is further corroborated by the crystal structure of the receptor which displays typical features of its inactive conformation (Fig. 1B). Unlike Fab5, Fab2838 exhibits a broader binding interface on the intracellular surface of the A_{2A}R and its binding to A_{2A}R appears to involve interactions with all three intracellular loops and the cytoplasmic end of several TM helices. Similar to the example of Fab5-bound β_2AR crystal structure, Fab2838-bound crystal structure of the A_{2A}R allowed the visualization of the native third intracellular loop of the receptor which was deleted in previously determined crystals structures of this receptor.

Interestingly, these two examples of β_2AR and A_{2A}R that highlight the successful application of mouse monoclonal antibody fragments in GPCR crystallization remain the only two cases among a total of more than 100 GPCR crystal structures where this strategy has been used. This trend not only highlights the tremendous success of the alternative approaches of GPCR crystallization, i.e., fusion protein strategy and thermostabilization but perhaps it also indicates the challenges and logistics involved in generating mouse monoclonal antibodies through hybridoma methods especially for structural studies.

3. NANOBODY TECHNOLOGY—TRAPPING ACTIVE GPCR CONFORMATIONS

Nanobodies are the variable fragments of the heavy chain-only antibodies produced by the members of the camelid family such as llamas and camels. Nanobodies are significantly smaller (approximately 12 kDa) compared to the conventional Fab (approximately 50 kDa) and ScFv (approximately 28 kDa), and appear to exhibit robust and high affinity binding to their targets. Nanobodies, in the context of GPCR crystallization, are primarily used for conformational stabilization that has resulted in crystallographic trapping of the active GPCR conformations (Steyaert & Kobilka, 2011).

3.1 Structure determination of active β_2AR conformation stabilized by a nanobody

Agonist binding to GPCRs leads to an active conformation of the receptor which is suitable for the coupling of heterotrimeric G proteins. Interestingly, crystallization of the β_2AR even bound to a covalent agonist resulted in only an inactive-like conformation of the receptor. This observation suggested that additional stabilization of the active conformation is required at least to visualize it crystallographically. Accordingly, purified β_2AR bound to a high-affinity agonist was injected in to llamas with the hope that they will generate antibodies that can selectively recognize and stabilize the active conformation of the receptor for crystallography (Rasmussen, Choi, et al., 2011). Subsequently, a phage display library of nanobody-coding genes was prepared and multiple rounds of *in vitro* selections were done to identify nanobodies that specifically recognize agonist-bound conformation of the β_2AR. Several nanobodies exhibited robust interaction only with the agonist-bound receptor and not antagonist-bound receptor. Further, binding of such nanobodies also enhanced the affinity of agonists for β_2AR suggesting their ability to stabilize an active conformation.

One of these nanobodies, referred to as Nb80, was used to crystallize high-affinity agonist BI-167107-bound β_2AR and it resulted in visualization of a fully active receptor conformation via X-ray crystallography (Fig. 2A). The Nb80-bound β_2AR crystal structure revealed a major outward movement of TM5 and 6 which were proposed earlier based on biophysical studies to be a signature of active GPCR conformation. Nb80

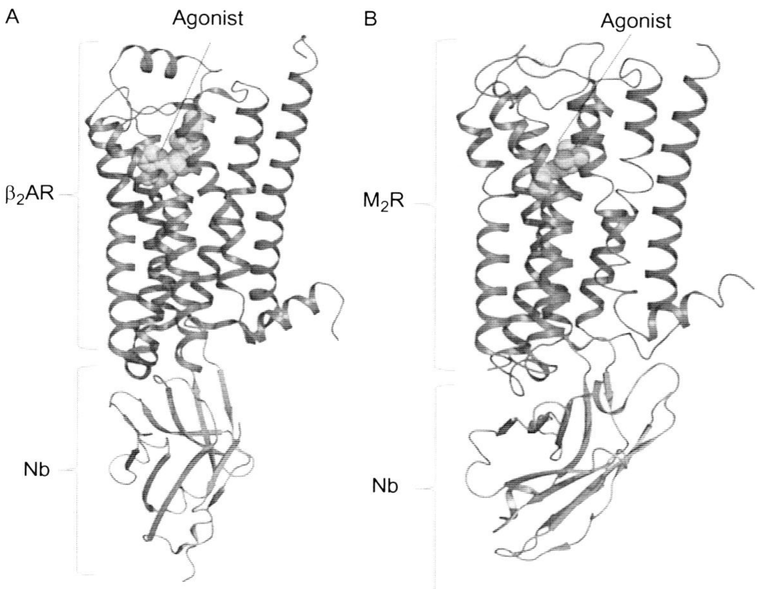

Figure 2 Nanobody-mediated trapping of active GPCR conformations. (A) Crystal structure of a high-affinity agonist-bound β_2AR in a fully active conformation in complex with a nanobody. The diagram represents the cartoon depiction of the structure (PDB ID 3POG) generated using PyMol software. The β_2AR is shown in red color, nanobody (Nb) in green and the agonist (BI-167107) is shown in cyan. (B) Crystal structure of the agonist-bound human muscarinic M_2 receptor (M_2R) in complex with an active state selective nanobody. The diagram represents the cartoon depiction of the structure (PDB ID 4MQS) generated using PyMol software. The M_2R is shown in red color, nanobody (Nb) in green, and the agonist (iperoxo) is shown in cyan. (See the color plate.)

bound to the intracellular face of the receptor and its complementarity determining region 3 protruded in to the cavity formed by the cytoplasmic surface of the receptor.

More recently, a modified variant of Nb80, referred to as Nb6B9 that binds to agonist occupied receptor with at least 10-fold better affinity compared to Nb80 was generated through a yeast display method (Ring et al., 2013). This Nb6B9 enhances the affinity of adrenaline for β_2AR by about 100-fold and indicates its robust ability to stabilize the adrenaline-bound active receptor conformation. Nb6B9 was used to determine the crystal structure of the β_2AR bound to adrenaline, the endogenous ligand of β_2AR and revealed similar features of active conformation as Nb80-bound receptor.

3.2 Structure of activated M_2R in complex with a nanobody

Another successful example of nanobody-mediated trapping of active GPCR conformation is the human muscarinic acetylcholine receptor subtype 2 (M_2R) (Kruse et al., 2013). Repeated crystallization trials of agonist-bound M_2R failed to yield diffracting crystals and therefore, agonist-bound receptor was immunized in to llamas to generate stabilizing nanobodies. Subsequently, a yeast display nanobody library was generated and then used to perform *in vitro* selections on agonist-bound M_2R. Several nanobodies were identified that selectively bound agonist-occupied M_2R and one of these nanobodies, referred to as Nb8-9 was used to crystallize an agonist-bound active conformation of the M_2R. Similar to Nb80 against β_2AR as described earlier, nanobody Nb8-9 also enhanced the agonist affinity for M_2R and the crystal structure of Nb8-9–M_2R complex (Fig. 2B) revealed typical features of active conformation similar to the nanobody-bound β_2AR crystal.

These examples underline the power of nanobodies to selectively recognize and stabilize specific GPCR conformations and going forward, provide a reliable option for capturing and structurally visualizing ligand-specific conformations of GPCRs, in general. It is also important to mention that nanobodies used for both, the β_2AR and M_2R are referred to as G protein mimetic because they enhance the agonist affinity as G protein does in ternary agonist-receptor–G protein complex and their binding interface also overlaps to G protein-binding site to a significant extent.

4. ANTIBODY FRAGMENTS IN VISUALIZING GPCR SIGNALING COMPLEXES

In addition to crystallizing inactive and active conformations, antibody fragments have also been proven to be invaluable tools in capturing and high-resolution structure determination of GPCR–effector signaling complexes as described below.

4.1 Nanobody stabilization of the β_2AR–Gs complex

Crystallization and structure determination of the β_2AR–G protein complex was a major milestone in the area of GPCR research (Rasmussen, DeVree, et al., 2011). Although β_2AR–G protein complex could be crystallized without any antibody fragment, the diffraction on the crystals did not reach beyond 6–7 Å. At this point, single-particle electron microscopy revealed highly dynamic nature of the alpha helical domain of the Gαs (GαsAH) and highlighted a potential bottleneck in obtaining better diffracting crystals (Westfield et al., 2011). To overcome this, llamas were immunized with a cross-linked β_2AR–G protein complex and nanobodies were raised against the complex. One of these nanobodies, referred to as Nb35, which binds to the interface of the Gα and Gβ subunits of G protein, improved the diffraction of the crystals to about 3 Å. This resulted in a crystal structure of β_2AR–Gs complex that revealed unprecedented insights in to GPCR–G protein coupling and downstream signaling.

A fascinating question is whether the same nanobody is capable of stabilizing complexes of other GPCRs with Gs protein for crystallization and high-resolution structure determination. Also, it would be interesting to see whether crystallization and structure determination of GPCRs with other subtypes of G proteins might also require nanobodies or similar stabilization tools for crystallization.

4.2 A synthetic Fab to stabilize GPCR–β-arrestin complex

Binding of β-arrestins to the activated and phosphorylated GPCRs not only terminates the G protein signaling but it also initiates a parallel signaling cascade downstream of the receptors (Shukla, 2014). Therefore, structural insights in to GPCR–β-arrestin interaction are likely to reveal fundamental insights in to GPCR regulation and signaling. GPCR–β-arrestin interaction is thought to be a two-step process wherein β-arrestin first binds to the phosphorylated carboxyl-terminus of GPCRs and then subsequently communicates with the 7TM bundle of the receptor. Accordingly, a phosphopeptide corresponding to the carboxyl-terminus of the arginine vasopressin type 2 receptor (V_2Rpp) has been utilized as a surrogate of phosphorylated GPCR to study the first step of GPCR–β-arrestin interaction. However, despite a range of optimization and engineering, crystallization of V_2Rpp–β-arrestin complex was not successful. At this point, a Fab (referred to as Fab30) against the V_2Rpp–β-arrestin 1 complex was selected from a synthetic phage display library to stabilize this complex further. This V_2Rpp–β-arrestin–Fab complex yielded well-diffracting crystals that resulted in structure determination of this complex at 2.6 Å (Shukla et al., 2013). This structure revealed not only a potentially conserved docking interface for the phosphorylated GPCRs on the N-domain of β-arrestin 1 but also allowed the visualization of activation dependent conformational changes in the β-arrestin 1 at high resolution. Although this Fab has not been tested for complex of β-arrestin to the phosphopeptide corresponding to the carboxyl-terminus of other GPCRs, it is likely that depending on the length, specific pattern, and extent of phosphorylation, other GPCRs might sample somewhat nonoverlapping conformational space on β-arrestins.

Interestingly, this Fab30 also turned to be absolutely essential to stabilize a β_2AR–β-arrestin complex for subsequent structural characterization (Shukla et al., 2014). Here, a chimeric β_2 receptor that contains the carboxyl-terminus of the arginine vasopressin subtype 2 receptor (referred to as β_2V_2R) was chosen as a model system because it displays higher affinity for β-arrestin. Moreover, as Fab30 was selected against the V_2Rpp-activated conformation of β-arrestin 1, it was expected to robustly stabilize the β_2V_2R–β-arrestin 1 complex. Indeed, Fab30 helped assembly of a highly stable, functional, and pharmacologically relevant β_2V_2R–β-arrestin 1 complex in detergent micelles. This complex was subjected to several biophysical studies including hydrogen/deuterium exchange (HDX), cross-linking, and single-particle electron microscopy and these complementary approaches revealed many novel structural insights in to the receptor-β–arrestin interaction. This study also provided the first direct visualization of a two-step-interaction mechanism of GPCR with β-arrestin, an observation that is likely to be a general mechanism in the GPCR superfamily. Similar to the nanobody approach in β_2AR-G protein crystallization, it remains to be investigated whether this particular Fab or a similar strategy might be required to stabilize and capture other GPCR–arrestin complexes.

5. CONCLUSION AND FUTURE PERSPECTIVE

As reflected in the examples described in this chapter, use of antibody fragment approach has not only provided the first breakthrough in crystallography of non-rhodopsin GPCRs but it has also turned out to be critical to overcome the bottlenecks in capturing fully active GPCR conformations and visualizing GPCR signaling complexes. Going forward, it would not be surprising if antibody fragments, especially nanobodies, became a general tool and play a major role in trapping different signaling conformations of GPCRs and their signaling assemblies (Ghosh, Nidhi, & Shukla, 2014; Shukla, Singh, & Ghosh, 2014). Recently described streamlined strategies for generation and selection of robust nanobody repertoires are likely to make the technology more accessible in future and benefit many researchers in this domain (Fridy et al., 2014; Pardon et al., 2014). Furthermore, major advances in the area of combinatorial biology such as design and optimization of synthetic phage display libraries and robust screening protocols are also emerging as powerful approaches to overcome the exclusive dependence on conventional and tedious hybridoma-based generation of antibody fragments for GPCR structural studies (Fellouse et al., 2007; Persson et al., 2013).

ACKNOWLEDGMENTS

We thank the members of Shukla laboratory for critical reading of the chapter and stimulating discussions. Research in Shukla laboratory is supported by the India Institute of Technology-Kanpur, Department of Science and Technology (DST, Government of India), Council for Scientific and Industrial Research (CSIR), and the Wellcome Trust DBT Indian Alliance. Dr. Shukla is an Intermediate Fellow of the Wellcome Trust DBT India Alliance.

REFERENCES

Bockaert, J., & Pin, J. P. (1999). Molecular tinkering of G protein-coupled receptors: An evolutionary success. *The EMBO Journal*, *18*, 1723–1729.

Cherezov, V., Abola, E., & Stevens, R. C. (2010). Recent progress in the structure determination of GPCRs, a membrane protein family with high potential as pharmaceutical targets. *Methods in Molecular Biology*, *654*, 141–168.

Day, P. W., Rasmussen, S. G., Parnot, C., Fung, J. J., Masood, A., Kobilka, T. S., et al. (2007). A monoclonal antibody for G protein-coupled receptor crystallography. *Nature Methods*, *4*, 927–929.

Doyle, D. A., Cabral, J. M., Pfuetzner, R. A., Kuo, A. L., Gulbis, J. M., Cohen, S. L., et al. (1998). The structure of the potassium channel: Molecular basis of K+ conduction and selectivity. *Science*, *280*, 69–77.

Fellouse, F. A., Esaki, K., Birtalan, S., Raptis, D., Cancasci, V. J., Koide, A., et al. (2007). High-throughput generation of synthetic antibodies from highly functional minimalist phage-displayed libraries. *Journal of Molecular Biology*, *373*, 924–940.

Fridy, P. C., Li, Y., Keegan, S., Thompson, M. K., Nudelman, I., Scheid, J. F., et al. (2014). A robust pipeline for rapid production of versatile nanobody repertoires. *Nature Methods*, *11*, 1253–1260.

Ghanouni, P., Gryczynski, Z., Steenhuis, J. J., Lee, T. W., Farrens, D. L., Lakowicz, J. R., et al. (2001). Functionally different agonists induce distinct conformations in the G protein coupling domain of the beta 2 adrenergic receptor. *The Journal of Biological Chemistry*, *276*, 24433–24436.

Ghanouni, P., Steenhuis, J. J., Farrens, D. L., & Kobilka, B. K. (2001). Agonist-induced conformational changes in the G-protein-coupling domain of the beta 2 adrenergic receptor. *Proceedings of the National Academy of Sciences of the United States of America*, *98*, 5997–6002.

Ghosh, E., Kumari, P., Jaiman, D., & Shukla, A. K. (2015). Methodological advances: The unsung heroes of the GPCR structural revolution. *Nature Reviews Molecular Cell Biology*, *16*(2), 69–81.

Ghosh, E., Nidhi, K., & Shukla, A. K. (2014). SnapShot: GPCR-ligand interactions. *Cell*, *159*, 1712–1712.e1.

Hino, T., Arakawa, T., Iwanari, H., Yurugi-Kobayashi, T., Ikeda-Suno, C., Nakada-Nakura, Y., et al. (2012). G-protein-coupled receptor inactivation by an allosteric inverse-agonist antibody. *Nature*, *482*, 237–240.

Hino, T., Iwata, S., & Murata, T. (2013). Generation of functional antibodies for mammalian membrane protein crystallography. *Current Opinion in Structural Biology*, *23*, 563–568.

Hunte, C., Koepke, J., Lange, C., Rossmanith, T., & Michel, H. (2000). Structure at 2.3 A resolution of the cytochrome bc(1) complex from the yeast Saccharomyces cerevisiae co-crystallized with an antibody Fv fragment. *Structure*, *8*, 669–684.

Hunte, C., & Michel, H. (2002). Crystallisation of membrane proteins mediated by antibody fragments. *Current Opinion in Structural Biology*, *12*, 503–508.

Kruse, A. C., Ring, A. M., Manglik, A., Hu, J., Hu, K., Eitel, K., et al. (2013). Activation and allosteric modulation of a muscarinic acetylcholine receptor. *Nature*, *504*, 101–106.

Ma, P., & Zemmel, R. (2002). Value of novelty? *Nature Reviews Drug Discovery*, *1*, 571–572.

Ostermeier, C., Iwata, S., Ludwig, B., & Michel, H. (1995). Fv fragment-mediated crystallization of the membrane protein bacterial cytochrome c oxidase. *Nature Structural Biology*, *2*, 842–846.

Palczewski, K., Kumasaka, T., Hori, T., Behnke, C. A., Motoshima, H., Fox, B. A., et al. (2000). Crystal structure of rhodopsin: A G protein-coupled receptor. *Science*, *289*, 739–745.

Pardon, E., Laeremans, T., Triest, S., Rasmussen, S. G., Wohlkonig, A., Ruf, A., et al. (2014). A general protocol for the generation of nanobodies for structural biology. *Nature Protocols*, *9*, 674–693.

Persson, H., Ye, W., Wernimont, A., Adams, J. J., Koide, A., Koide, S., et al. (2013). CDR-H3 diversity is not required for antigen recognition by synthetic antibodies. *Journal of Molecular Biology*, *425*, 803–811.

Rasmussen, S. G., Choi, H. J., Fung, J. J., Pardon, E., Casarosa, P., Chae, P. S., et al. (2011). Structure of a nanobody-stabilized active state of the beta(2) adrenoceptor. *Nature*, *469*, 175–180.

Rasmussen, S. G., Choi, H. J., Rosenbaum, D. M., Kobilka, T. S., Thian, F. S., Edwards, P. C., et al. (2007). Crystal structure of the human beta2 adrenergic G-protein-coupled receptor. *Nature*, *450*, 383–387.

Rasmussen, S. G., DeVree, B. T., Zou, Y., Kruse, A. C., Chung, K. Y., Kobilka, T. S., et al. (2011). Crystal structure of the beta2 adrenergic receptor-Gs protein complex. *Nature*, *477*, 549–555.

Ring, A. M., Manglik, A., Kruse, A. C., Enos, M. D., Weis, W. I., Garcia, K. C., et al. (2013). Adrenaline-activated structure ofbeta2-adrenoceptor stabilized by an engineered nanobody. *Nature*, *502*, 575–579.

Shukla, A. K. (2014). Biasing GPCR signaling from inside. *Science Signaling*, *7*, pe3.

Shukla, A. K., Manglik, A., Kruse, A. C., Xiao, K., Reis, R. I., Tseng, W. C., et al. (2013). Structure of active beta-arrestin-1 bound to a G-protein-coupled receptor phosphopeptide. *Nature*, *497*, 137–141.

Shukla, A. K., Singh, G., & Ghosh, E. (2014). Emerging structural insights into biased GPCR signaling. *Trends in Biochemical Sciences*, *39*, 594–602.

Shukla, A. K., Westfield, G. H., Xiao, K., Reis, R. I., Huang, L. Y., Tripathi-Shukla, P., et al. (2014). Visualization of arrestin recruitment by a G-protein-coupled receptor. *Nature*, *512*, 218–222.

Steyaert, J., & Kobilka, B. K. (2011). Nanobody stabilization of G protein-coupled receptor conformational states. *Current Opinion in Structural Biology*, *21*, 567–572.

Westfield, G. H., Rasmussen, S. G., Su, M., Dutta, S., DeVree, B. T., Chung, K. Y., et al. (2011). Structural flexibility of the G alpha s alpha-helical domain in the beta2-adrenoceptor Gs complex. *Proceedings of the National Academy of Sciences of the United States of America*, *108*, 16086–16091.

SECTION III

Biophysical Studies of Membrane Proteins

CHAPTER THIRTEEN

Conformational Analysis of G Protein-Coupled Receptor Signaling by Hydrogen/Deuterium Exchange Mass Spectrometry

Sheng Li*, Su Youn Lee†, Ka Young Chung†,1

*Department of Medicine, University of California at San Diego, San Diego, California, USA
†School of Pharmacy, Sungkyunkwan University, Suwon, Republic of Korea
[1]Corresponding author: e-mail address: kychung2@skku.edu

Contents

Abstract

Conformational change and protein–protein interactions are two major mechanisms of membrane protein signal transduction, including G protein-coupled receptors (GPCRs). Upon agonist binding, GPCRs change conformation, resulting in interaction with downstream signaling molecules such as G proteins. To understand the precise signaling mechanism, studies have investigated the structural mechanism of GPCR signaling using X-ray crystallography, nuclear magnetic resonance (NMR), or electron paramagnetic resonance. In addition to these techniques, hydrogen/deuterium exchange mass spectrometry (HDX-MS) has recently been used in GPCR studies. HDX-MS measures the rate at which peptide amide hydrogens exchange with deuterium in the solvent. Exposed or flexible regions have higher exchange rates and excluded or ordered regions have lower exchange rates. Therefore, HDX-MS is a useful tool for studying

Methods in Enzymology, Volume 557
ISSN 0076-6879
http://dx.doi.org/10.1016/bs.mie.2014.12.004

protein–protein interfaces and conformational changes after protein activation or protein–protein interactions. Although HDX-MS does not give high-resolution structures, it analyzes protein conformations that are difficult to study with X-ray crystallography or NMR. Furthermore, conformational information from HDX-MS can help in the crystallization of X-ray crystallography by suggesting highly flexible regions. Interactions between GPCRs and downstream signaling molecules are not easily analyzed by X-ray crystallography or NMR because of the large size of the GPCR-signaling molecule complexes, hydrophobicity, and flexibility of GPCRs. HDX-MS could be useful for analyzing the conformational mechanism of GPCR signaling. In this chapter, we discuss details of HDX-MS for analyzing GPCRs using the β_2AR-G protein complex as a model system.

1. INTRODUCTION

1.1 Structural study of GPCRs with related signaling proteins

In 2012, the Nobel Prize in Chemistry was awarded to two American scientists, Brian K. Kobilka and Robert J. Lefkowtiz, for "studies of G-protein-coupled receptors (GPCRs)." GPCRs are the largest family of membrane proteins with approximately 800 identified in the human genome. GPCRs perform vital signaling functions in vision and olfactory perception and are involved in signal transduction in the metabolic, endocrine, neuromuscular, and CNS systems. Many GPCRs are involved in various diseases, and 40% of drugs on the market targets GPCRs to treat diseases including heart failure, peptic ulcers, prostate carcinoma, hypertension, pain, and bronchial asthma (Garland, 2013; Shoichet & Kobilka, 2012).

Classical drug discovery approaches targeting GPCRs select and optimize compounds that interact with specific receptors using high-throughput screening (HTS) assays. However, HTS drug discovery efforts often fail because of the high degree of homology in the orthosteric ligand-binding sites among GPCRs, which makes finding a selective drug for a specific receptor subtype difficult (Shoichet & Kobilka, 2012). An alternative approach to regulating specific GPCRs is to design drugs based on the structural information of GPCRs (Shoichet & Kobilka, 2012).

Characterization of the conformation and structural dynamics of proteins is critical for understanding the molecular mechanism of proteins in normal physiological and abnormal pathological processes. All GPCRs share a common seven-transmembrane α-helical structure with an extracellular N-terminus and an intracellular C-terminus connected by three extracellular loops (ECLs) and three intracellular loops (ICLs; Katritch, Cherezov, & Stevens, 2013; Venkatakrishnan et al., 2013).

Upon activation by agonists, GPCRs undergo conformational changes that induce interactions between the intracellular parts of GPCRs and downstream signaling molecules such as G proteins and β-arrestins (Katritch et al., 2013; Lefkowitz, Rajagopal, & Whalen, 2006; Venkatakrishnan et al., 2013). The physiological or pathological outcome of GPCR-mediated signaling depends on the signaling molecules with which the GPCRs interact. Therefore, to design a drug targeting a GPCR, it is important to understand which structural characteristics provide the selective binding of ligands to a specific GPCR (ligand specificity) and the selective interaction of GPCRs with specific signaling molecules (signaling specificity; Garland, 2013; Shoichet & Kobilka, 2012).

Therefore, an effort has been made to define the mechanism of the ligand specificity in the structural studies of various GPCRs. Other studies have tried to define the signaling specificity by analyzing the conformational mechanisms of GPCRs interacting with downstream signaling molecules such as G proteins or β-arrestins. Tremendous efforts have focused on defining high-resolution GPCR structures (Zhao & Wu, 2012). Tools and techniques have been developed for better crystallization as discussed in this book: protein engineering (Chapter 2. A novel screening approach for optimal and functional fusion of T4 lysozyme in GPCRs), improved detergents (Chapter 4. Amphiphatic agents for membrane protein study), addition of antibodies (Section II. Generation and use of antibody fragments against membrane proteins), and improved crystallization conditions (Section IV. Crystallization of membrane proteins). As a result, great progress has been made in defining X-ray crystal structures of GPCRs during the past 7 years (Flight, 2013). A few GPCR crystal structures have been determined including agonist/antagonist-bound forms, a G protein-bound form, and different subtypes (Katritch et al., 2013; Rasmussen et al., 2011).

However, the precise mechanism of GPCR-signaling specificity is not fully understood because of several limitations of X-ray crystallography. Intracellular loop 3 (ICL3) of GPCRs is not well defined in the X-ray crystal structures. ICL3 is reported to be involved in the activation, internalization, desensitization, and oligomerization of GPCRs and is an important binding site for signaling molecules including G proteins, β-arrestins, G protein-coupled receptor kinase (GRK), and regulator of G protein signaling (Bernstein et al., 2004; Borroto-Escuela, Correia, et al., 2010; Borroto-Escuela, Garcia-Negredo, Garriga, Fuxe, & Ciruela, 2010; Hashimoto et al., 2008; Hu et al., 2012; Kang et al., 2005; Lee, Ptasienski, Pals-Rylaarsdam, Gurevich, & Hosey, 2000; Pao & Benovic, 2005; Peverelli et al., 2008; Singer-Lahat, Liu, Wess, & Felder, 1996). ICL3 is highly variable in length and sequence, and this might contribute to the diverse signaling mechanism of GPCRs (Katritch, Cherezov, & Stevens, 2012), but ICL3 in X-ray crystal structures of GPCRs is often missing or replaced by other proteins such as T4 lysozyme (Katritch et al., 2012; Venkatakrishnan et al., 2013). Another limitation is that only one X-ray crystal structure of a GPCR-signaling molecule complex (β_2Adrenoceptor-Gs complex) is available (Rasmussen et al., 2011). This makes it hard to identify the structural mechanism of signaling specificity. Moreover, dynamic conformational changes during GPCR-signaling protein complex formation and dissociation, which are not easily analyzed by X-ray crystallography, have not yet been systemically studied. Studying the sequential structural events of GPCR-signaling protein complex formation would contribute to an in-depth understanding of the structural mechanism of GPCR-signaling activation. Therefore, studying the structure of GPCRs with signaling molecules by not only X-ray crystallography but also other methods is necessary.

1.2 HDX-MS for conformational studies of GPCRs with related signaling proteins

Although X-ray crystallography and NMR spectroscopy are standard techniques for obtaining high-resolution protein structures, both have limitations. A major limitation of X-ray crystallography is the crystallization process. Some proteins such as membrane proteins that have large hydrophobic regions are not suitable for crystallization. Moreover, GPCRs are dynamic, which hampers the crystallization process. Another limitation is that flexible regions are often missing in defined structures such as ECL or ICL regions of GPCRs. X-ray crystallography provides structural information of only static protein states, and the crystallization environment is often not physiological. This is particularly true when studying GPCR-signaling protein complexes, which are unstable under the extreme precipitating conditions required for protein crystallography. On the other hand, NMR spectroscopy allows analysis of protein structures in more physiological, aqueous conditions. Current NMR spectroscopy technology is, however, not suitable for determining high-resolution structures of proteins larger than 40 kDa, although NMR is useful for tracking local conformational changes in specific regions of GPCRs by labeling the regions (Liu, Horst, Katritch, Stevens, & Wuthrich, 2012; Ratnala et al., 2007; Tapaneeyakorn, Goddard, Oates, Willis, & Watts, 2011). Both X-ray crystallography and NMR spectroscopy require large amounts of proteins, and obtaining structures of heterogeneous complexes, particularly in dynamic conformational states, is challenging.

Hydrogen/deuterium exchange mass spectrometry (HDX-MS) measures the rates at which peptide amide hydrogens exchange with deuterium in solvent. In folded proteins, the exchange rate depends on the position of the amide hydrogen (Marcsisin & Engen, 2010; Percy, Rey, Burns, & Schriemer, 2012). Exposed or highly dynamic regions show rapid exchange rates while excluded, and rigid regions show slow exchange rates (Marcsisin & Engen, 2010; Percy et al., 2012). Step 2 in Fig. 1 illustrates this phenomenon showing that the deuterium uptake occurs earlier in the flexible and exposed regions than in the ordered and excluded regions. Thus, the exchange rate provides information on the structural properties of a folded protein and contact sites between two proteins or a protein and small molecules (Marcsisin & Engen, 2010; Percy et al., 2012).

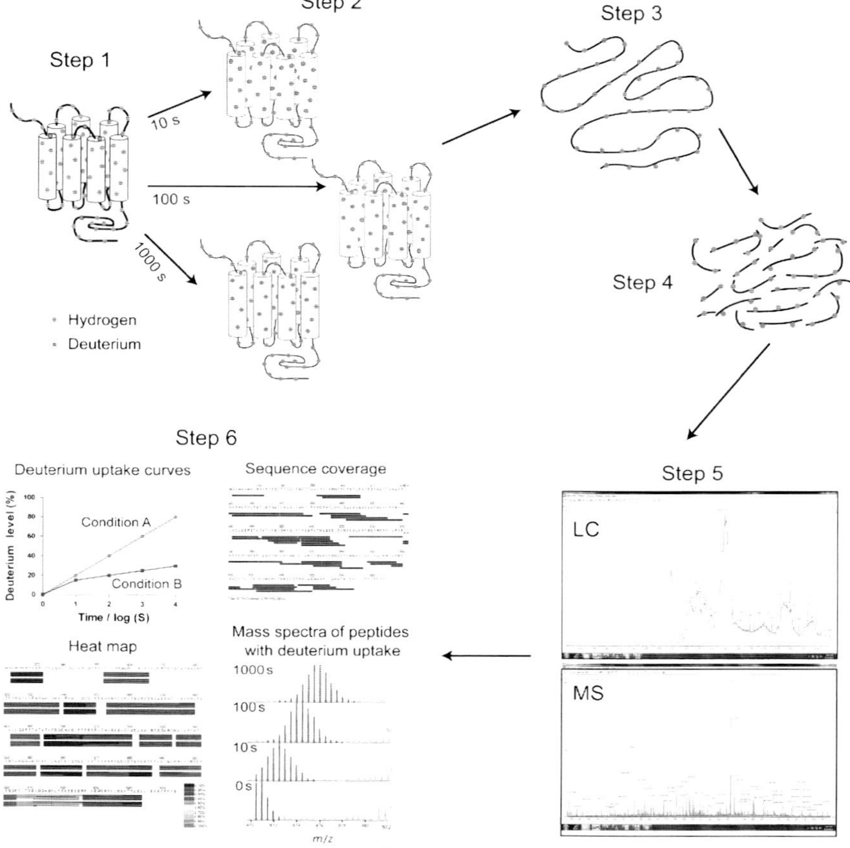

Figure 1 Theory and overall step-by-step procedure for HDX-MS regarding GPCR analysis (See the color plate.)

Although HDX-MS does not provide high-resolution structures, there are several advantages of HDX-MS over other techniques. HDX-MS requires relatively small amounts of proteins and can analyze large protein assemblies that are not amenable to other techniques. Sample preparation can be done under more physiological conditions, making the investigation of dynamic structural changes possible. For this reason, HDX-MS was adopted to analyze GPCR-G protein or GPCR-β-arrestin structures (Chung et al., 2011; Orban et al., 2012; Shukla et al., 2014).

For HDX-MS, the hydrogen/deuterium exchange is initiated by adding deuterated buffer to the target protein samples, and the reaction is stopped by adding quench buffer to the reaction mixture. Because deuterium is 1 Da heavier than hydrogen, the incorporation of deuterium is measured by mass spectrometry (MS). Hydrogen/deuterium exchange is performed under physiological conditions in deuterated buffer. The effects of protein conformational change are probed during this step.

GPCRs are difficult to analyze by MS, similar to other membrane proteins (Rabilloud, 2009). However, a few studies have analyzed GPCR conformations using HDX-MS, specifically the conformation of β_2-adrenergic receptor (β_2AR) and rhodopsin (Chung et al., 2011; Orban et al., 2012; Shukla et al., 2014; West et al., 2011; Zhang et al., 2010). The Griffin group and the Stevens group conducted a well-planned study to analyze β_2AR by optimizing the experimental conditions of the HDX-MS procedure (West et al., 2011; Zhang et al., 2010). The Kobilka group and the Palczewski group analyzed the conformation of the β_2AR-Gs complex and rhodopsin-transducin complex, respectively (Chung et al., 2011; Orban et al., 2012). The Kobilka group and the Lefkowitz group analyzed the conformation of the β_2AR-β-arrestin1 complex (Shukla et al., 2014). Here, we discuss detailed HDX-MS procedures and factors to consider for better HDX-MS results for GPCRs. We use the β_2AR-G protein complex as a model system.

2. EXPERIMENTAL PROCEDURE

2.1 Overall HDX-MS procedure (Fig. 1)

A general procedure for HDX-MS is divided into six steps (Fig. 1): (1) protein sample preparation; (2) hydrogen/deuterium exchange; (3) quenching and denaturation; (4) fragmentation by pepsin; (5) liquid chromatography and tandem mass spectrometry (LC/MS); and (6) data analysis. For

HDX-MS, proteins are prepared in a buffer condition that preserves the protein conformation. For example, to analyze conformational changes upon β_2AR–Gs complex formation, the β_2AR–Gs complex, β_2AR alone, and Gs alone are prepared in buffer conditions in which the β_2AR–Gs complex is stabilized. Prepared proteins are subjected to hydrogen/deuterium exchange by adding a large volume of deuterated buffer. The exchange rate is monitored after stopping hydrogen/deuterium exchange at various time points by adding quench buffer. The quench buffer not only stops the exchange reaction but also denatures the proteins, and quenched denatured proteins are fragmented by enzymatic digestion by pepsin. Fragmented peptides are separated by LC, and their molecular weight is measured by MS. Peptides from nondeuterated proteins are identified by matching experimental masses with theoretical masses of peptic fragments from the protein. Once peptides are identified, deuterium uptake levels are calculated for each peptide from the deuterated proteins using commercially available or home-made software. The first four steps of HDX-MS greatly affect data quality because of the challenge of maintaining the functional conformation of the β_2AR–Gs complex and getting quality peptides from membrane proteins such as GPCRs. We will not discuss how to make the stable β_2AR–Gs complex as it is outside the scope of this chapter. Instead, we will discuss how to get good-quality HDX-MS data from well-prepared protein samples, focusing on GPCRs. Detailed procedures and special cautions for GPCR analysis by HDX-MS are discussed below.

2.2 Protein sample preparation (Fig. 1, step 1)

1. *Protein stocks.* β_2AR, Gs, or β_2AR–Gs complex is solubilized in protein buffer (20 m*M* HEPES, pH 7.5, 100 m*M* NaCl, 100 μ*M* tris(2-carboxyethyl)phosphine (TCEP)) supplemented with either 0.1% *n*-dodecyl-*b*-D-maltopyranoside (DDM) or 0.02% maltose neopentyl glycol-3 (MNG-3). The protein concentration is optimized by testing proteins at various concentrations and adjusted to 70 μ*M*.

   Caution 2-1: Protein stocks and all buffers for HDX-MS should be filtered because particles can block LC/MS columns and tubing. Frozen/thawed protein stocks should be filtered before hydrogen/deuterium exchange because freezing can induce protein aggregation.

   Caution 2-2: Many membrane proteins including GPCRs have glycosylation in the extracellular parts of the protein. Glycosylation increases the peptide molecular weight to a various extent and adds complexity in the data analysis. Thus, deglycosylation is recommended for GPCRs before making protein stocks. This simplifies data analysis of MS-based peptide identification and increases the likelihood of higher sequence coverage of the extracellular part of GPCRs. GPCR deglycosylation can be done with peptide-*N*-glycosidase F, an endoglycosidase that releases N-linked oligosaccharides. Upon deglycosylation, asparagine becomes aspartate, so asparagine should be changed to aspartate when uploading protein sequences to search engines for the identification of peptides from nondeuterated proteins using theoretical masses. Released oligosaccharides can be removed by size-exclusion chromatography or affinity-tag purification.

Caution 2-3: Stock concentrations or protein amounts for HDX-MS should be optimized, depending on the sensitivity of the mass spectrometer.

Caution 2-4: GPCRs are hydrophobic membrane proteins, and detergents are used to extract and solubilize membrane proteins into an aqueous buffer. For GPCR solubilization, DDM has been widely used in a number of structural analyses including X-ray crystallography and NMR studies (Chung et al., 2012; Rasmussen et al., 2007). Recently, new detergents such as MNG-3 have been developed for more efficient stabilization of GPCRs (Chapter 5; Chae, Gotfryd, et al., 2010, Chae, Rasmussen, et al., 2010; Chung et al., 2012). These new detergents have been successfully used for X-ray crystallography, NMR, and other biochemical studies of GPCRs (Chung et al., 2012, 2011; Kruse et al., 2012; Rasmussen et al., 2011; Westfield et al., 2011).

Detergents, however, often cause low sequence coverage of membrane proteins because of various reasons, including reduction of the sensitivity of MS (Rabilloud, 2009). Therefore, the compatibility of detergents for HDX-MS should be tested. For the study of the β_2AR-G protein complex, we compared the sequence coverage of β_2AR prepared in DDM with the sequence coverage of β_2AR in MNG-3. MNG-3 stabilizes the β_2AR-G protein complex better than DDM, but DDM gave better sequence coverage than MNG-3 (Fig. 2A and B).

Caution 2-5: When protein solutions are concentrated by a spin concentrator, detergents will also be concentrated. Increased detergent

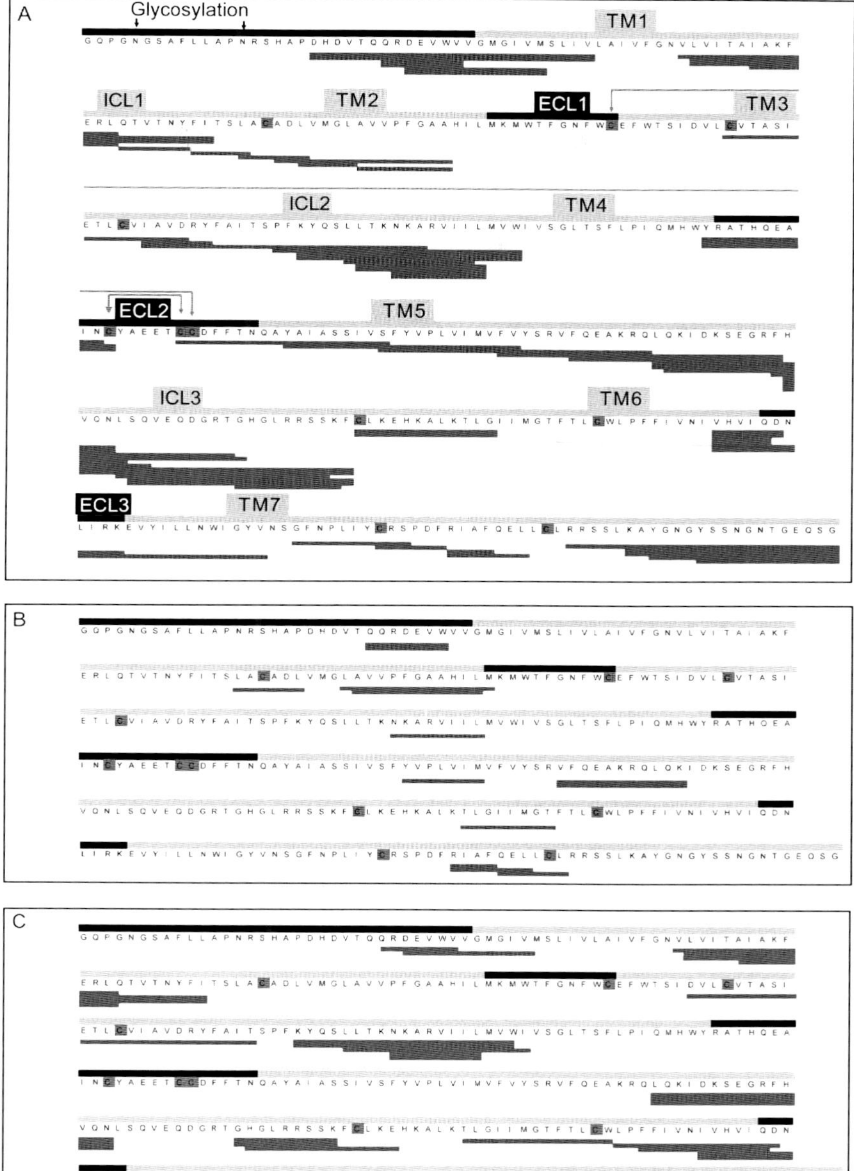

Figure 2 Optimization of sequence coverage of β_2AR. (A) Sequence coverage when β_2AR is solubilized in DDM. (B) Sequence coverage when β_2AR is solubilized in MNG-3. (C) Sequence coverage when β_2AR is solubilized in DDM, and guanidine is added in the quench buffer. Dark blue (black color in the print version) bars, peptic fragments; TM, transmembrane region; ICL, intracellular loop region, ECL; extracellular loop region. Red (gray color in the print version) arrows, connection between cysteines.

concentrations might affect the quality of mass spectra. Therefore, attention is required when proteins are concentrated. The details about this issue are discussed in Caution 3-2.

2.3 Hydrogen/deuterium exchange procedure (Fig. 1, step 2)

1. Hydrogen/deuterium exchange buffer (D_2O buffer) and nonexchange buffer (H_2O buffer).

An advantage of HDX-MS is that the hydrogen/deuterium exchange process can be done in the buffer in which the protein stock is prepared unless this buffer severely interferes with LC/MS analysis. Thus, D_2O and H_2O buffers are similar to the protein buffer described earlier except with lower detergent concentrations. D_2O buffer contains 20 m*M* HEPES, pH 7.5, 100 m*M* NaCl, 100 μ*M* TCEP with either 0.05% DDM or 0.0015% MNG in D_2O. H_2O buffer contains 20 m*M* HEPES, pH 7.5, 100 m*M* NaCl, 100 μ*M* TCEP with either 0.05% DDM or 0.0015% MNG in H_2O.

Caution 3-1: Because pH meters give inaccurate p*D*, the pH of the D_2O buffer is adjusted with highly concentrated HEPES stock solution (1 *M* HEPES, pH 7.5). The same concentrated HEPES stock solution is used to adjust the pH of the H_2O buffer. Another way to measure p*D* using a pH meter is by adding 0.4, the p*D* correction factor, to the pH meter reading.

$$pD = \text{pH meter reading} + 0.4$$

Caution 3-2: Factors that affect protein conformation should be considered when making the D_2O and H_2O buffers. The detergent concentration should be above the critical micelle concentration (CMC) to keep the GPCRs solubilized. The detergent concentration in D_2O or H_2O buffer should be lower than in the protein buffer but above the CMC because detergents tend to have deleterious effects on LC/MS analysis such as reducing MS sensitivity. If a protein stock is concentrated in a spin concentrator, detergents can be omitted from D_2O and H_2O buffers to minimize the deleterious effects of detergents because spin concentrators often increase detergent concentration.

If ligands are used to form a specific GPCR conformation, the ligands might need to be added to D_2O and H_2O buffers to maintain the specific active or inactive conformation. The amount of ligand can be adjusted by considering the K_d value of the ligand and the protein dilution factor after addition of D_2O or H_2O buffers.

2. Hydrogen/deuterium exchange.

Several automated systems such as H/D-X PAL (LEAP technologies) mix and incubate protein stocks with D_2O or H_2O buffers and quench buffer. Automated systems are convenient but relatively expensive for small academic laboratories. Hydrogen/deuterium exchange can also be done manually by the following procedure. Hydrogen/deuterium exchange is initiated by mixing 1.5-μl protein stock (70 μ*M*) with 4.5-μl ice-cold D_2O buffer on ice. To obtain the deuterium uptake rate, mixed samples are incubated for 10, 100, 1000, or 10,000 s on ice, and reactions are stopped with 15-μl ice-cold quench buffer. The quench step is described in detail in Section 2.4. For nondeuterated samples, 1.5-μl protein stock is mixed with 4.5-μl H_2O buffer and 15-μl quench buffer.

Caution 3-3: Hydrogen/deuterium exchange should be on ice in a cold room (4 °C) to minimize back exchange. Exchange rate decreases as temperature is lowered, and thus back exchange during and after quenching is reduced at lower temperatures. Pipette tips and pipettes should be cooled in the cold room before use.

Caution 3-4: Exchange duration can be optimized after initial experiments. If the deuterium uptake level in highly exchangeable regions is not saturated at 10,000 s, extend the exchange duration. If most regions show quick saturation, shorten the exchange duration.

Caution 3-5: The volume ratio between protein stock and D_2O buffer that is described here is 1:3. Therefore, the maximum deuterium uptake should not be more than 75%. When analyzing deuterium uptake mass spectra, the dilution factor should be considered.

2.4 Quench and denaturation (Fig. 1, step 3)

1. Quench and denaturation.

Hydrogen/deuterium exchange rate is dependent on temperature and pH (Marcsisin & Engen, 2010); the lower the temperature, the slower the exchange rate. Hydrogen/deuterium exchange is catalyzed by both acid and base. The exchange rate is lowest at around pH 2.5 and increases as pH increases or decreases. To slow the exchange, the temperature can be lowered to 0 °C, and the pH of the quench buffer can be adjusted to pH 2.5. The exchange reaction is stopped by addition of ice-cold quench buffer at pH 2.5.

Quenched proteins might need to be denatured for effective peptic digestion. The denaturing step should be quick to minimize possible back-exchange. The denaturing step, therefore, can be done during the quench step by adding denaturants (e.g., urea or guanidine) to the quench buffer. Most GPCRs have disulfide bonds, which should be broken by reducing agents such as dithiothreitol or TCEP for effective digestion and simplification of data analysis for peptide identification. The composition of quench buffer is described later.

For β_2AR–Gs analysis, 15-μl ice-cold quench buffer was added to 6-μl exchange reaction mixture (1.5-μl protein stock, 4.5 μl D_2O or H_2O buffer) at 10, 100, 1000, or 10,000 s.

Caution 4-1: If quenched samples will not immediately be analyzed, samples can be snap frozen with dry ice or liquid nitrogen and stored at −80 °C until use.

Caution 4-2: GPCRs tend to aggregate with urea or guanidine, so the effects of denaturants on sequence coverage should be tested. The addition of guanidine reduced the sequence coverage of β_2AR (Fig. 2A and C) but not Gs. Therefore, guanidine was not added in the quench buffer for β_2AR–Gs analysis.

2. Quench buffer.

Quench buffer is composed of 0.1 *M* NaH_2PO_4, pH 2.4, 20 m*M* TCEP, and 16.6% glycerol.

Caution 4-3: The quench buffer pH should be adjusted to a final pH of about 2.5 after addition of quench buffer to the protein reaction sample.

2.5 Fragmentation by pepsin (Fig. 1, step 4)

1. *Pepsin digestion*. Pepsin is used as the digestion enzyme because quenched samples should be kept at pH 2.5. Quenched samples or frozen/thawed quenched samples are immediately passed through an immobilized pepsin column (16 μl bed volume, with porcine pepsin [Sigma]) at a flow rate of 20 μl/min of 0.05% trifluoroacetic acid (TFA).

2.6 LC–MS/MS procedure (Fig. 1, step 5)

1. *HDX-MS instrument setup* (Fig. 3).

The instrumental configuration consists of microelectrically actuated high-pressure switching valves (EMHA, from VICI Valco Instruments,

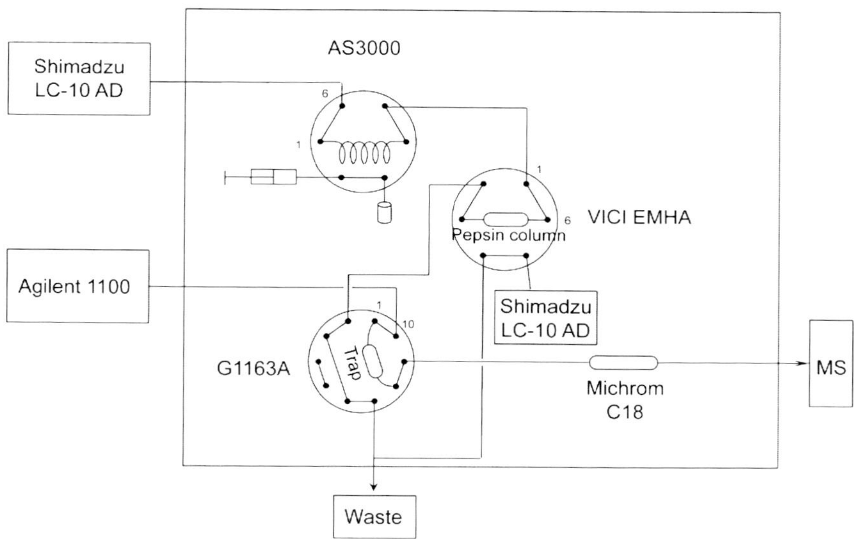

Figure 3 Instrumental configuration of pepsin-column-conjugated LC/MS.

Houston, TX; G1163A 2port/10 position microvalve) connected to pumps, columns, and other parts with PEEK tubing (Upchurch Scientific), in-house-made pepsin columns (pepsin or fungal protease XIII), C18 Trap and analytical columns, and an extensively modified AS3000 autosampler (Spectra Physics) for low-temperature operation. All but 10 sample vial positions are filled with powdered dry ice to keep the autosampler basin at −42 °C for 10 h. Two Shimadzu LC-10AD HPLC pumps and one Agilent 1100 series binary nanopump are operated by Chemstation software (Agilent Tech Inc., Santa Clara, CA). One LC-10AD pump delivers 0.05% TFA aqueous buffer to pass protein samples over the pepsin column and onto a C18 Trap column. The other pump delivers the same buffer to back flush the pepsin column after sample digestion. An Agilent binary pump delivers solvents for linear gradient elution from the reverse-phase HPLC column (Pump A: 0.05% TFA in water; pump B: 20% water, 80% acetonitrile, 0.01% TFA). The timing and sequence of pump and valve operations are controlled by programs run in Chemstation, with an interface provided by an external contact interface board (Agilent).

To minimize—back exchange, precise temperature control of the instrument is achieved by putting the autosampler, valves, connecting plumbing, and columns in a high thermal-capacity refrigerator maintained at 3.8 °C. Components with no contact with pure D_2O are immersed in ice water. MS analysis is carried out by a Finnigan LCQ Classic mass spectrometer with ESI voltage at 4.5 kV and capillary temperature of 200 °C.

2. General operational procedure.

Sets of autosampler vials containing quenched, functionally deuterated samples (stored at −80 °C) are placed in the dry ice-containing sample basin of the autosampler. Samples are kept at dry ice temperature until they are melted over 1 min by the autosampler at 4–5 °C during sample preparation. Samples are injected into the immobilized pepsin column coupled to 20AL support material from PerSeptive Biosystems. Samples are passed over the pepsin column (0.05% TFA at 20 μl/min) for 30–40 s. Digested peptides are collected on a C18 Trap column (Magic C18AQ, 3u, 200 Å, 0.2 × 2 mm; Michrom BioResources Inc.) for desalting (1 min at 60 μl/min) and separated on an analytical C18 column (Magic C18AQ, 3u, 200 Å, 0.2 × 50 mm) by a linear gradient of 8–48% B over 30 min, with solvent A, 0.05% TFA, and solvent B: 80% acetonitrile, 20% water, and 0.01% TFA. Eluant is directed to the mass spectrometer in positive ion mode over the m/z window 20–2000. Source parameters are optimized to minimize back exchange while maintaining ionization efficiency as spray voltage 4.5 kV and heated capillary temperature 200 °C (Walters, Ricciuti, Mayne, & Englander, 2012). Data are acquired in either MS profile mode or data-dependent MS/MS mode. A dynamic exclusion list is applied to the three most intense peaks in each MS/MS scan to allow additional data to be obtained on more peptide ions.

Caution 6-1: Some GPCR ligands are not soluble in aqueous solution and aggregate, which can block LC tubing.

2.7 Data analysis (Fig. 1, step 6)

HDX-MS data processing includes sequence-specific peptide identification and determination of deuterium incorporation for each peptide.

1. Sequence-specific peptide identification.

To identify pepsin-generated peptides, CID-based MS/MS runs are acquired using nondeuterated control samples, and lists of candidate peptides are generated by SEQUEST software. All tentative peptides are confirmed using specialized data reduction software (DXMS Explorer, Sierra Analytics Inc., Modesto, CA).

Caution 7-1: Even with well-controlled sample preparation and instrumental setup, identified peptides vary from sample to sample. Combining peptide lists from different sets of nondeuterated samples increases the chance of higher sequence coverage for deuterated samples. For β_2AR-Gs analysis, peptide lists are generated by combining peptide lists from three nondeuterated samples.

2. Determination of deuterium incorporation

   Currently, the rate-limiting step in HDX-MS studies is deuterium incorporation analysis. Software has been developed to accelerate data analysis (Pascal et al., 2012; Slysz et al., 2009; Weis, Engen, & Kass, 2006; Zhang, Zhang, & Xiao, 2012); however, calculation of the centroid mass for each identified peptide and extraction of deuterium incorporation information from the large quantity of data still relies on manual intervention, especially for low-yield peptides. In β_2AR-Gs analysis, the centroids of isotopic envelopes of all peptides were measured using DXMS Explorer, which fully automates the centroiding procedure. Corrections for back exchange use the formula of Zhang and Smith (1993).

$$\text{Deuterium incorporation} = (m(P) - m(N))/(m(F) - m(N)) \times \text{MaxD}$$

   where m(P), m(N), and m(F) are the centroid value of partially deuterated, nondeuterated, and fully deuterated peptides, respectively. MaxD is the calculated maximum deuterium incorporation for the corresponding peptides. Deuteration levels of peptides are further sublocalized using overlapping peptides.

3. CONCLUSION AND PERSPECTIVES

HDX-MS has been proved to be a nice tool to study the conformation of GPCR-downstream signaling molecule complexes such as β2AR-Gs and β2AR-β-arrestin1 (Chung et al., 2011; Shukla et al., 2014). We provided detailed experimental procedures for HDX-MS for GPCRs. However, HDX-MS for studying GPCRs has room to improve. In previous studies (Chung et al., 2011; Shukla et al., 2014), information about a GPCR (e.g., β2AR) is limited because of low sequence coverage. However, conformational information about downstream signaling molecules (e.g., Gs and β-arrestin1) is rich. One of the main reasons for poor sequence coverage of GPCRs is choice of detergents as discussed in Fig. 2. β2AR-Gs study used MNG-3, and β2AR-β-arrestin1 study used lauryl maltose neopentyl glycol for receptor solubilization because these detergents are better than DDM at stabilizing β2AR-downstream signaling molecule complexes. However, DDM is better for good sequence coverage of β2AR (Fig. 2). Therefore, new detergents that are compatible with HDX-MS and stabilize GPCR-downstream signaling molecule complexes are needed. With optimal detergents and conditions, as discussed in this chapter, HDX-MS could be a powerful tool for studying the structures of GPCR-downstream signaling molecule complexes.

ACKNOWLEDGMENTS

This work was supported by the basic science research program (NRF-2012R1A1A1039220) and medical research center program (NRF-2012R1A5A2A28671860) of the National Research Foundation of Korea.

REFERENCES

Bernstein, L. S., Ramineni, S., Hague, C., Cladman, W., Chidiac, P., Levey, A. I., et al. (2004). RGS2 binds directly and selectively to the M1 muscarinic acetylcholine receptor third intracellular loop to modulate Gq/11alpha signaling. *Journal of Biological Chemistry*, *279*, 21248–21256.

Borroto-Escuela, D. O., Correia, P. A., Perez Alea, M., Narvaez, M., Garriga, P., Fuxe, K., et al. (2010). Impaired M(3) muscarinic acetylcholine receptor signal transduction through blockade of binding of multiple proteins to its third intracellular loop. *Cellular Physiology and Biochemistry*, *25*, 397–408.

Borroto-Escuela, D. O., Garcia-Negredo, G., Garriga, P., Fuxe, K., & Ciruela, F. (2010). The M(5) muscarinic acetylcholine receptor third intracellular loop regulates receptor function and oligomerization. *Biochimica et Biophysica Acta*, *1803*, 813–825.

Chae, P. S., Gotfryd, K., Pacyna, J., Miercke, L. J., Rasmussen, S. G., Robbins, R. A., et al. (2010). Tandem facial amphiphiles for membrane protein stabilization. *Journal of the American Chemical Society*, *132*, 16750–16752.

Chae, P. S., Rasmussen, S. G., Rana, R. R., Gotfryd, K., Chandra, R., Goren, M. A., et al. (2010). Maltose-neopentyl glycol (MNG) amphiphiles for solubilization, stabilization and crystallization of membrane proteins. *Nature Methods*, *7*, 1003–1008.

Chung, K. Y., Kim, T. H., Manglik, A., Alvares, R., Kobilka, B. K., & Prosser, R. S. (2012). Role of detergents in conformational exchange of a G protein-coupled receptor. *Journal of Biological Chemistry*, *287*, 36305–36311.

Chung, K. Y., Rasmussen, S. G., Liu, T., Li, S., DeVree, B. T., Chae, P. S., et al. (2011). Conformational changes in the G protein Gs induced by the beta2 adrenergic receptor. *Nature*, *477*, 611–615.

Flight, M. H. (2013). Drug discovery: Structure-led design. *Nature*, *502*, S50–S52.

Garland, S. L. (2013). Are GPCRs still a source of new targets? *Journal of Biomolecular Screening*, *18*, 947–966.

Hashimoto, Y., Morisawa, K., Saito, H., Jojima, E., Yoshida, N., & Haga, T. (2008). Muscarinic M4 receptor recycling requires a motif in the third intracellular loop. *Journal of Pharmacology and Experimental Therapeutics*, *325*, 947–953.

Hu, J., Thor, D., Zhou, Y., Liu, T., Wang, Y., McMillin, S. M., et al. (2012). Structural aspects of M(3) muscarinic acetylcholine receptor dimer formation and activation. *FASEB Journal*, *26*, 604–616.

Kang, H., Lee, W. K., Choi, Y. H., Vukoti, K. M., Bang, W. G., & Yu, Y. G. (2005). Molecular analysis of the interaction between the intracellular loops of the human serotonin receptor type 6 (5-HT6) and the alpha subunit of GS protein. *Biochemical and Biophysical Research Communications*, *329*, 684–692.

Katritch, V., Cherezov, V., & Stevens, R. C. (2012). Diversity and modularity of G protein-coupled receptor structures. *Trends in Pharmacological Sciences*, *33*, 17–27.

Katritch, V., Cherezov, V., & Stevens, R. C. (2013). Structure-function of the G protein-coupled receptor superfamily. *Annual Review of Pharmacology and Toxicology*, *53*, 531–556.

Kruse, A. C., Hu, J., Pan, A. C., Arlow, D. H., Rosenbaum, D. M., Rosemond, E., et al. (2012). Structure and dynamics of the M3 muscarinic acetylcholine receptor. *Nature*, *482*, 552–556.

Lee, K. B., Ptasienski, J. A., Pals-Rylaarsdam, R., Gurevich, V. V., & Hosey, M. M. (2000). Arrestin binding to the M(2) muscarinic acetylcholine receptor is precluded by an inhibitory element in the third intracellular loop of the receptor. *Journal of Biological Chemistry*, *275*, 9284–9289.

Lefkowitz, R. J., Rajagopal, K., & Whalen, E. J. (2006). New roles for beta-arrestins in cell signaling: Not just for seven-transmembrane receptors. *Molecular Cell*, *24*, 643–652.

Liu, J. J., Horst, R., Katritch, V., Stevens, R. C., & Wuthrich, K. (2012). Biased signaling pathways in beta2-adrenergic receptor characterized by 19F-NMR. *Science*, *335*, 1106–1110.

Marcsisin, S. R., & Engen, J. R. (2010). Hydrogen exchange mass spectrometry: What is it and what can it tell us? *Analytical and Bioanalytical Chemistry*, *397*, 967–972.

Orban, T., Jastrzebska, B., Gupta, S., Wang, B., Miyagi, M., Chance, M. R., et al. (2012). Conformational dynamics of activation for the pentameric complex of dimeric G protein-coupled receptor and heterotrimeric G protein. *Structure*, *20*, 826–840.

Pao, C. S., & Benovic, J. L. (2005). Structure/function analysis of alpha2A-adrenergic receptor interaction with G protein-coupled receptor kinase 2. *Journal of Biological Chemistry*, *280*, 11052–11058.

Pascal, B. D., Willis, S., Lauer, J. L., Landgraf, R. R., West, G. M., Marciano, D., et al. (2012). HDX workbench: Software for the analysis of H/D exchange MS data. *Journal of the American Society for Mass Spectrometry*, *23*, 1512–1521.

Percy, A. J., Rey, M., Burns, K. M., & Schriemer, D. C. (2012). Probing protein interactions with hydrogen/deuterium exchange and mass spectrometry—A review. *Analytica Chimica Acta*, *721*, 7–21.

Peverelli, E., Mantovani, G., Calebiro, D., Doni, A., Bondioni, S., Lania, A., et al. (2008). The third intracellular loop of the human somatostatin receptor 5 is crucial for arrestin binding and receptor internalization after somatostatin stimulation. *Molecular Endocrinology*, *22*, 676–688.

Rabilloud, T. (2009). Membrane proteins and proteomics: Love is possible, but so difficult. *Electrophoresis*, *30*(Suppl. 1), S174–S180.

Rasmussen, S. G., Choi, H. J., Rosenbaum, D. M., Kobilka, T. S., Thian, F. S., Edwards, P. C., et al. (2007). Crystal structure of the human beta2 adrenergic G-protein-coupled receptor. *Nature*, *450*, 383–387.

Rasmussen, S. G., DeVree, B. T., Zou, Y., Kruse, A. C., Chung, K. Y., Kobilka, T. S., et al. (2011). Crystal structure of the beta2 adrenergic receptor-Gs protein complex. *Nature*, *477*, 549–555.

Ratnala, V. R., Kiihne, S. R., Buda, F., Leurs, R., de Groot, H. J., & DeGrip, W. J. (2007). Solid-state NMR evidence for a protonation switch in the binding pocket of the H1 receptor upon binding of the agonist histamine. *Journal of the American Chemical Society*, *129*, 867–872.

Shoichet, B. K., & Kobilka, B. K. (2012). Structure-based drug screening for G-protein-coupled receptors. *Trends in Pharmacological Sciences*, *33*, 268–272.

Shukla, A. K., Westfield, G. H., Xiao, K., Reis, R. I., Huang, L. Y., Tripathi-Shukla, P., et al. (2014). Visualization of arrestin recruitment by a G-protein-coupled receptor. *Nature*, *512*(7513), 218–222.

Singer-Lahat, D., Liu, J., Wess, J., & Felder, C. C. (1996). The third intracellular domain of the m3 muscarinic receptor determines coupling to calcium influx in transfected Chinese hamster ovary cells. *FEBS Letters*, *386*, 51–54.

Slysz, G. W., Baker, C. A., Bozsa, B. M., Dang, A., Percy, A. J., Bennett, M., et al. (2009). Hydra: Software for tailored processing of H/D exchange data from MS or tandem MS analyses. *BMC Bioinformatics*, *10*, 162.

Tapaneeyakorn, S., Goddard, A. D., Oates, J., Willis, C. L., & Watts, A. (2011). Solution- and solid-state NMR studies of GPCRs and their ligands. *Biochimica et Biophysica Acta*, *1808*, 1462–1475.

Venkatakrishnan, A. J., Deupi, X., Lebon, G., Tate, C. G., Schertler, G. F., & Babu, M. M. (2013). Molecular signatures of G-protein-coupled receptors. *Nature*, *494*, 185–194.

Walters, B. T., Ricciuti, A., Mayne, L., & Englander, S. W. (2012). Minimizing back exchange in the hydrogen exchange-mass spectrometry experiment. *Journal of the American Society for Mass Spectrometry*, *23*, 2132–2139.

Weis, D. D., Engen, J. R., & Kass, I. J. (2006). Semi-automated data processing of hydrogen exchange mass spectra using HX-express. *Journal of the American Society for Mass Spectrometry*, *17*, 1700–1703.

West, G. M., Chien, E. Y., Katritch, V., Gatchalian, J., Chalmers, M. J., Stevens, R. C., et al. (2011). Ligand-dependent perturbation of the conformational ensemble for the GPCR beta2 adrenergic receptor revealed by HDX. *Structure*, *19*, 1424–1432.

Westfield, G. H., Rasmussen, S. G., Su, M., Dutta, S., DeVree, B. T., Chung, K. Y., et al. (2011). Structural flexibility of the G alpha s alpha-helical domain in the beta2-adrenoceptor Gs complex. *Proceedings of the National Academy of Sciences of the United States of America*, *108*, 16086–16091.

Zhang, X., Chien, E. Y., Chalmers, M. J., Pascal, B. D., Gatchalian, J., Stevens, R. C., et al. (2010). Dynamics of the beta2-adrenergic G-protein coupled receptor revealed by hydrogen-deuterium exchange. *Analytical Chemistry*, *82*, 1100–1108.

Zhang, Z., & Smith, D. L. (1993). Determination of amide hydrogen exchange by mass spectrometry: A new tool for protein structure elucidation. *Protein Science*, *2*, 522–531.

Zhang, Z., Zhang, A., & Xiao, G. (2012). Improved protein hydrogen/deuterium exchange mass spectrometry platform with fully automated data processing. *Analytical Chemistry*, *84*, 4942–4949.

Zhao, Q., & Wu, B. L. (2012). Ice breaking in GPCR structural biology. *Acta Pharmacologica Sinica*, *33*, 324–334.

CHAPTER FOURTEEN

EPR Studies of Gating Mechanisms in Ion Channels

Sudha Chakrapani[1]
Department of Physiology and Biophysics, School of Medicine, Case Western Reserve University, Cleveland, Ohio, USA
[1]Corresponding author: e-mail address: sudha.chakrapani@case.edu

Contents

Abstract

Ion channels open and close in response to diverse stimuli, and the molecular events underlying these processes are extensively modulated by ligands of both endogenous and exogenous origin. In the past decade, high-resolution structures of several channel types have been solved, providing unprecedented details of the molecular architecture of these membrane proteins. Intrinsic conformational flexibility of ion channels critically governs their functions. However, the dynamics underlying gating mechanisms and modulations are obscured in the information from crystal structures. While nuclear magnetic resonance spectroscopic methods allow direct measurements of protein dynamics, they are limited by the large size of these membrane protein assemblies in detergent micelles or lipid membranes. Electron paramagnetic resonance (EPR) spectroscopy has emerged as a key biophysical tool to characterize structural dynamics of ion channels and to determine stimulus-driven conformational transition between functional states in a physiological environment. This review will provide an overview of the recent advances in the field of voltage- and ligand-gated channels and highlight some of the challenges and controversies surrounding the structural information available. It will discuss general methods used in site-directed spin labeling and EPR spectroscopy and illustrate how findings from these studies have narrowed the gap between high-resolution structures and gating mechanisms in membranes, and have thereby helped reconcile seemingly disparate models of ion channel function.

Methods in Enzymology, Volume 557
ISSN 0076-6879
http://dx.doi.org/10.1016/bs.mie.2014.12.030

A complete mechanistic description of ion channel function requires both the structure of the channel in multiple conformational states and the information of protein dynamics during transitions between these states. The functioning of ion channels is governed by their ability to undergo stimulus-driven movements that enable opening of the gate, and in most cases, these processes are accompanied by global protein motions. From a crystallographic point of view, "moving parts" and intrinsic flexibility of ion channels pose serious hurdles in crystallization and structure determination. Further, stabilizing channels in multiple conformational states under crystallographic conditions is yet another barrier. As a result, the structural basis for gating mechanisms in many channel types has been inferred from comparisons across homologues trapped in crystallographically distinct states. Although this strategy has proved informative, it is clearly plagued by two major issues: first, uncertainties in assigning functional states to crystallographically captured conformations, and second, issues with distinguishing intrinsic structural variations among homologues from differences between the conformational states. In some cases, truncation, limited proteolysis, and antibodies have been used to stabilize proteins with considerable success (Hassaine et al., 2014; Hibbs & Gouaux, 2011; Jiang et al., 2003; Miller & Aricescu, 2014; Zhou, Morais-Cabral, Kaufman, & MacKinnon, 2001), although these strategies suffer from limiting conformational transitions. Quite often, detergent solubilization, lattice forces, and crystal contacts have led to stabilization of states that are not part of the functional scheme or constitute only a minor population of the channels in the membrane (Cross, Sharma, Yi, & Zhou, 2011; Jiang, Lee, et al., 2003; Kim, Xu, Murray, & Cafiso, 2008). These limitations have spurred much of the discrepancies between various gating mechanisms and models. This necessitates strategies that allow structural and functional measurements under similar conditions. In this aspect, site-directed spin labeling (SDSL) and electron paramagnetic resonance (EPR) spectroscopy offer a unique perspective to multistep gating schemes (Columbus & Hubbell, 2002; Fanucci & Cafiso, 2006; Hubbell, Cafiso, & Altenbach, 2000; McHaourab, Lietzow, Hideg, & Hubbell, 1996; Sahu, McCarrick, & Lorigan, 2013). In comparison, comparison to X-ray and nuclear magnetic resonance, information from EPR is of relatively low resolution, however there are several advantages of this technique that surpass its limitations and make it highly complementary to other structural methods. Most notably, EPR is not constrained by the molecular size or the optical property of the system, and thereby allows measurements of the full-length constructs in a native lipid environment. In addition, this technique is extremely sensitive and the minimum sample requirements are in the pico- to micromolar range.

Structural understanding of ion channels was greatly facilitated by the discovery and cloning of prokaryotic homologues (Bocquet et al., 2007; Ruta, Jiang, Lee, Chen, & MacKinnon, 2003; Schrempf et al., 1995; Tasneem, Iyer, Jakobsson, & Aravind, 2005). These channels reveal remarkable conservation of the overall scaffold and basic channel functions, representing a minimal functional core of the channel. These model systems have therefore offered new avenues for determination of high-resolution structures and uncovering protein dynamics of the previously elusive channel class (Bocquet et al., 2009; Doyle et al., 1998; Hilf & Dutzler, 2008, 2009; Jiang, Lee, et al., 2003; Zhou et al., 2001). Advantages of using these channels include functionality in isolation from their native membranes, a smaller number of native cysteines, and strong resilience to mutagenesis (Gross, Columbus, Hideg, Altenbach, & Hubbell, 1999; Jiang, Lee, et al., 2003; Perozo, Cortes, & Cuello, 1998; Ruta et al., 2003; Velisetty & Chakrapani, 2012; Velisetty, Chalamalasetti, & Chakrapani, 2012, 2014). Thus, a complete understanding of structure, function, and dynamics has been made feasible.

1. SDSL AND EPR SPECTROSCOPY TO STUDY GATING MECHANISMS IN ION CHANNELS

The use of EPR spectroscopy in combination with SDSL was pioneered by Hubbell and colleagues and has been adopted for studying the molecular architecture and protein motions in a number of proteins (Altenbach, Marti, Khorana, & Hubbell, 1990; Bordignon, 2012; Columbus & Hubbell, 2002; Drescher, 2012; Fanucci & Cafiso, 2006; Hubbell et al., 2000; Hubbell, Lopez, Altenbach, & Yang, 2013; Hubbell, McHaourab, Altenbach, & Lietzow, 1996; Klare & Steinhoff, 2009; McHaourab et al., 1996; McHaourab, Steed, & Kazmier, 2011; Perozo, Cuello, Cortes, Liu, & Sompornpisut, 2002; Sahu et al., 2013). These techniques are best suited for studying the conformational dynamics of membrane proteins, especially in a native membrane environment. This method involves engineering cysteine substitutions at target positions by site-direct mutagenesis. Since SDSL requires a unique cysteine for site-specific introduction of the spin label, native cysteines within the protein need to be substituted with another amino acid to create a cysteine-less template. Work in numerous systems, including membrane proteins, has established that cysteine mutagenesis is generally well-tolerated in most parts of the protein without loss of function. In addition, the relatively small size of the spin probe causes minimal perturbation, and in many cases, the functionality of labeled proteins has been confirmed. Individual cysteine mutants are expressed, purified, and labeled with a nitroxide spin probe. To maximize reactivity, the labeling reaction of the protein is carried out in the detergent-solubilized form. After separation of the excess free spin label by size-exclusion chromatography, the protein is reconstituted into liposomes or bilayer-mimicking systems of defined lipid composition. The most commonly used spin label is a thiol-specific methanethiosulfonate derivative (1-oxyl-2,2,5,5-tetramethyl-Δ3-pyrroline-3-methyl) that attaches to the protein through a disulfide bond bridge. Due to its high specificity, relatively small size (slightly larger than tryptophan), and flexibility of the linker region, this spin label has been shown to have excellent reactivity even with buried cysteine. Continuous wave (CW) is the most commonly used EPR technique where the sample is subjected to varying magnetic fields in the presence of a fixed microwave frequency. The unpaired electron from the nitroxide group absorbs microwave radiation as it undergoes reorientation to the higher energy level. Spin-flip transitions occur when the energy separation between the two electron spin states matches the energy of the applied microwave radiation. EPR spectrum is usually represented as a first derivative of the absorption as a function of the magnetic field strength. EPR analyses reveal three different pieces of structural information (Altenbach, Froncisz, Hemker, McHaourab, & Hubbell, 2005; Altenbach, Greenhalgh, Khorana, & Hubbell, 1994; Altenbach et al., 1990; Farahbakhsh, Altenbach, & Hubbell, 1992; Hubbell et al., 2000, 1996): (1) *Line-shape analysis* of motional freedom or the dynamics of the spin probe. At room temperature, the spin probe can adopt multiple rotameric conformations, and the spectral line shape is sensitive to the motion of spin label side chain relative to the peptide backbone, backbone fluctuations, and global rotational motion of the protein. The mobility of the probe can be reasonably quantified as an empirical parameter $\left(\Delta H_0^{-1}\right)$, which is measured as an inverse of the central linewidth of the first-derivative absorption spectra (McHaourab et al., 1996). For a nitroxide spin probe in a motionally restricted environment, the linewidth is larger (and hence smaller ΔH_0^{-1}) compared to that of a spin probe in a location with high freedom of movement. This is a useful parameter to detect protein conformational changes or protein–protein interactions. (2) *Solvent accessibility* of the spin probe evaluated by collisional relaxation methods. Paramagnetic agents such as lipid-soluble O_2 (lipid-soluble) or water-soluble metal complex NiEDDA have different diffusional profiles across the membrane. Collision with these paramagnetic relaxing agents enhances the spin-lattice relaxation rate (T1) of the nitroxide by Heisenberg spin exchange (Altenbach et al., 2005). The accessibility parameter (Π) is estimated from power saturation experiments in which the vertical peak-to-peak amplitude of the central line of the first-derivative EPR spectrum is measured as a function of increasing incident microwave power (Farahbakhsh et al., 1992). A spin label at a specific position on the protein can be classified as one of the following: buried within the protein core inaccessible to either water or lipid, on the surface exposed to aqueous solution, or on the surface exposed to lipids. The accessibility parameters obtained for a series of consecutive sites can then be used to determine the secondary structure of a region, probe areas of protein–protein contact, measure the tilt of a protein within the membrane, estimate the immersion depth of interfacial loops, and identify water vestibules and lipid-bound pockets within the transmembrane helices (Altenbach et al., 1994). (3) *Interspin distances* between pairs of nitroxide. In systems with two or more spin labels, electron–electron dipolar interactions depend on the interspin distance and the angle between the paramagnetic centers relative to the external magnetic field. Measurements of these distances involve either CW or pulsed (double-electron electron resonance or DEER) EPR techniques. In CW, through-space dipolar interactions lead to spectral broadening and are a function of proximity between the labels. It is also influenced by relative orientation of the spin labels, and the frequency and amplitude of the motion averaging the orientation of the spin vector relative to the magnetic field. Line broadening is accompanied by a decrease in signal amplitude, and for distances up to ~15–20 Å, a decrease in peak height for spectra normalized to the same spin concentration can often be used as a qualitative indication of spin–spin interaction. Extent of broadening in comparison to the sum of singly labeled mutant spectra can also be used to semi-quantitatively estimate distances in 8–20 Å range (Rabenstein & Shin, 1995). While the information from the methods described above is intrinsically local in nature, mobility and accessibility data from a series of spin-labeled positions in combination with distance constraints allow modeling of the protein structure with a backbone level spatial resolution. In addition, more advanced pulsed EPR methods (DEER) have been developed to determine global changes in distances (Borbat, McHaourab, & Freed, 2002; Chiang, Borbat, & Freed, 2005; Jeschke, 2012; Jeschke & Polyhach, 2007; Jeschke, Wegener, Nietschke, Jung, &

Steinhoff, 2004). In DEER, the oscillations in the spin-echo amplitude are used to determine distances and distance distribution in the 20–70 Å range. Taken together, sequence-specific secondary structure, tertiary contacts, and distance constraints can be used to measure conformational changes under various conditions. The following examples of voltage- and ligand-gated channels will highlight how EPR studies have helped evaluate crystal structures, determine the structure of missing regions or distorted elements in the crystal structures, and finally, validate structural models for gating.

2. VOLTAGE-GATED ION CHANNELS

Voltage-gated channels play a critical role in cellular excitability and thereby form the basis for initiation and propagation of nerve impulses (Hille, 2001). In addition, they are involved in muscle contraction, neuronal activity, hormone secretion, osmotic balance, and immune response. Mutations in voltage-gated channels have been associated with cardiac arrhythmias, episodic ataxia, and epilepsy. These channels are primarily composed of two interacting yet modular units: a voltage sensor domain (VSD) formed from the first four transmembrane segments (S1–S4), and a pore domain (PD) formed from the remaining two transmembrane segments (S5 and S6). During voltage-dependent activation, gating charges on the S4 segment translocate from the "down" to the "up" conformation. Movement of S4 in turn is coupled with opening of the pore through the S4–S5 linker. The structure of the voltage sensor and the mechanisms underlying S4 movement leading to the opening channel gate have been the subject of intense debate for the past decade.

Much of this controversy originated with the first crystal structure of a voltage-gated channel from the archaea *Aeropirum pernix*, KvAP (Jiang, Lee, et al., 2003). The crystal structure, obtained in detergent and in the presence of Fab fragments, was found to be largely distorted and in a non-native conformation at the level of the VSD. With respect to the PD, S1 and S2 were oriented parallel to the membrane, and S3 and S4 were positioned close to the membrane inner-leaflet and perpendicular to the pore axis. However, transmembrane orientation of S1 relative to the rest of the VSD helices was partly reconciled based on the structure of the isolated VSD of KvAP (also solved in the presence of antibody) (Jiang, Lee, et al., 2003). Based on this structure, it was proposed that the S3b–S4 forms a "paddle" motif that is exposed to the membrane and moves as a hydrophobic cation upon membrane depolarization (Jiang, Ruta, Chen, Lee, & MacKinnon, 2003). This structure essentially challenged two of the long-held ideas in the field. The first was that the S1, S2, and S3 segments shield S4 and its charges from the lipid bilayer. The second was that the S4 segment undergoes translation or rotation within a canaliculi or gating pore (Goldstein, 1996; Yang, George, & Horn, 1996).

To address this controversy, Perozo's group sought to determine the architecture of the KvAP voltage sensor in a native environment by carrying out an extensive EPR analysis of spin probes at ~140 positions covering the entire VSD (Cuello, Cortes, & Perozo, 2004). EPR measurements were carried out in the absence of a membrane potential (0 mV), which in principle would favor the activated or "up" conformation of the VSD. Clear boundaries of probe dynamics and NiEDDA accessibility of the S1–S2/S3–S4 loops interspaced between the TM helices allowed the authors to conclude that the VSD domain indeed adopts a transmembrane orientation. While the overall architecture of the VSD predicted by EPR was incompatible with the KvAP full-length structure, it was in agreement with the KvAP-isolated VSD structure. A major finding of this work was that the spin labels on one face of the S4 helix showed high O_2 accessibility revealing exposure to the membrane and thereby provided the first direct evidence for a peripheral location of S4 in the lipid bilayer. Interestingly, the spin probes at positions corresponding to the S4 gating charges showed low mobility and limited accessibility to O_2 or NiEDDA suggesting that they lie on the protected face, away from the membrane lipid, and presumably interact with the rest of the TM helices. In addition, the pattern of mobility and high NiEDDA accessibility of several positions within the S3 segments were indicative of water crevices penetrating the VSD. These crevices have been implicated in reducing the energetic cost of placing the gating charges in a hydrophobic environment while at the same time focusing the membrane field which will allow an efficient transduction of the electric field into S4 movements. Another surprising result from this study was that the spin probes at positions along S1 were motionally restricted and solvent inaccessible. This was in contrast to findings from perturbation analysis in eukaryotic channels, which position S1 in the channel periphery with minimal interaction with the rest of the protein.

To unequivocally define the interacting footprint of PD on VSD, EPR analysis was then carried out at the same positions in the isolated VSD (Chakrapani, Cuello, Cortes, & Perozo, 2008). The expectation was that the region of VSD in contact with PD in the full-length channel would show an increase in probe mobility and O_2 accessibility in the isolated VSD, while the membrane-exposed region would be minimally perturbed. Indeed, spectra from S3 and S4 were essentially identical and there were no changes in the overall O_2 accessibility trends (Fig. 1). On the other hand, profound increases in spin label dynamics and lipid exposure were noted in the S1 and S2 segments, and more so for residues facing away from the VSD

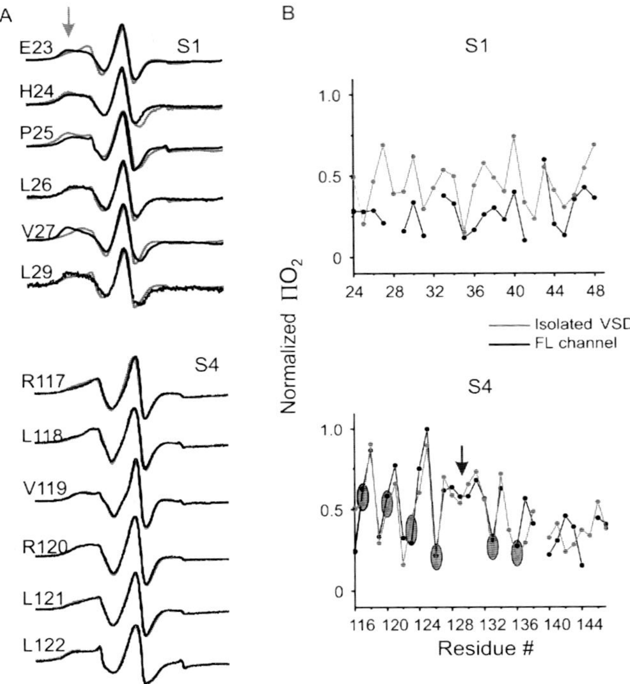

Figure 1 A comparison of spectral line-shape and residue-specific environmental parameters in the KvAP full-length (FL) channel, and the isolated VSD. (A) An overlay of representative X-band CW EPR spectra of spin-labeled mutants from S1 and S4 in the full-length channel (black) and the isolated VSD (red (light gray in the print version)). The green (gray in the print version) arrow head shows the location of immobile component of the spectra. Distinct differences in line shapes can be noted for positions in S1 that reflect increased mobility for side chains in the isolated VSD. The spectra for S4 are identical in the two constructs. (B) Profile of changes in the normalized O_2 accessibility. The arrow head marks the break in the α-helical periodicity of the accessibility. Positions of the gating charges in S4 are highlighted by ovals. While there are minimal changes in the membrane exposure in S4, dramatic differences in S1 clearly proves that interaction of the VSD with the pore occurs via S1 rather than S4. *This figure is adapted from the originally published data in Chakrapani et al. (2008).*

helical bundle. These results clearly showed that in the activated conformation of the KvAP sensor, S3 and S4 segments are at the periphery of the channel and make minimal contact with the PD. In contrast, the S1 and S2 segments are in close proximity to the PD. This finding was in agreement with DEER measurements in full-length KvAP, which showed shorter intersubunit distances for probes placed on S1 in comparison to those on S4 (Vamvouka, Cieslak, Van Eps, Hubbell, & Gross, 2008). These data were compared with the VSD orientation in the open-state structure of Kv1.2–Kv2.1 chimera (Long, Tao, Campbell, & MacKinnon, 2007). While there was a remarkable agreement with the arrangement of individual side chains, a discrepancy in the positioning of the VSD with respect to the PD was noted. The structure shows that VSD in the chimera interacts with the PD through S1 and S4, while the EPR data in KvAP indicated that the interaction with the PD was via S1 and S2, with S3 and S4 clearly placed in the periphery (Fig. 2). This arrangement of the two domains requires a rotation of ~70–100° (from the intracellular side) about an axis perpendicular to the membrane plane relative to the PD such that S1 and S2 are more closely placed to the PD.

Interestingly, water crevices were also identified in the isolated VSD and they were not considerably different from those in the full-length channel. It was thereby proposed that these cavities could be a common feature underlying the conductance of protons and other cations through the VSD, and that the VSD scaffold fully supports this pathway in the absence of auxiliary proteins and even in the monomeric state. Consistent with idea, EPR studies in a voltage-gated sodium channel (NaChBac) revealed conservation of fundamental features of the VSD in K_v and Na_v channel families (Chakrapani, Sompornpisut, Intharathep, Roux, & Perozo, 2010), although there were a few notable differences particularly at the packing of the helices and location of the S4 charges. While the periodicity of S1 and S2 accessibilities suggested that they are straight helices, a break in periodicity for S3 and S4 was reflective of a bending or kinking of the α-helix. In addition, patterns of ΠO_2 in NaChBac revealed that the S4 segment is not a simple α-helix; positions at the upper half of S4 show low values every third or fourth position, while at the bottom end it is every third position. The derived power spectrum of the N-terminal half of S4 shows a distinct peak centered at 99.8° reflecting its α-helical nature while the C-terminal end reveals a peak at 118.1°, which is characteristic of a 3_{10} helix. The occurrence of a 3_{10} helical hydrogen-bonding pattern in S4 was observed in the K_v1.2–2.1 chimera (Long et al., 2007), MlotiK (Clayton, Altieri, Heginbotham,

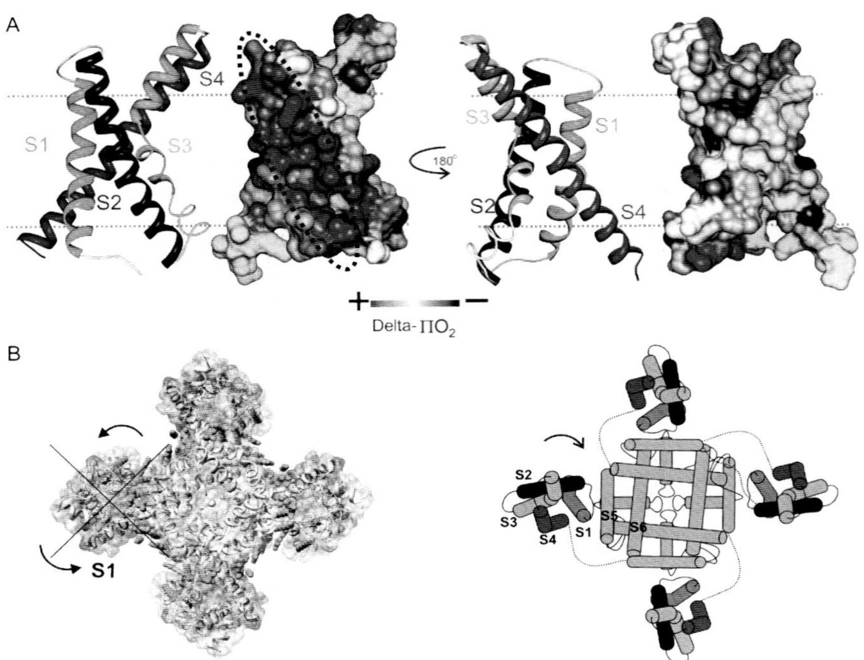

Figure 2 Mapping the molecular interacting surface of the VSD with the PD. Fractional differences in ΠO_2 mapped on the sensor structure and color coded with a gradient from red–white–blue. Red denotes increase and blue represents decrease in the environmental parameter. The surface of highest impact (S1 and S2) is marked by dotted black line. (B) Evaluation of the open-state structures for the K_v channels. The delta values mapped on to the coordinates of Kv1.2–Kv2.1 chimera structure (2R9R.pdb). A view from the intracellular part of the channel is shown. The EPR data from KvAP in the open-inactivated state predict a rotation of the VSD (~70–100°) to place S1 and S2 in the proximity of the pore as shown in the cartoon. *This figure is adapted from the originally published data in Chakrapani et al. (2008).* (See the color plate.)

Unger, & Morais-Cabral, 2008), and later in the Na_v structures (Payandeh, Scheuer, Zheng, & Catterall, 2011; Zhang et al., 2012). A varying length of 3_{10} stretched has been observed in these structures. In Na_VAb, the entire S4 is in a 3_{10}-helix conformation (Payandeh et al., 2011), while in Na_VRh structure the S4 segment is further outward and the 3_{10}-helix is partially relaxed to an α helix (Zhang et al., 2012). There is no evidence of 3_{10} in the KvAP sensor crystal structure (Jiang, Lee, et al., 2003) or in the EPR accessibility parameters of the KvAP sensor in the membrane (Chakrapani et al., 2008; Cuello et al., 2004). A 3_{10} helical conformation of S4 positions the series of charged arginine in a linear fashion pointing toward a water-filled crevice at the center of the VSD core, which would be energetically favorable (Long et al., 2007). An α-helix-to-3_{10} transition might allow S4 to stretch and spring back, changing the position of the gating charges within the electric field, as previously suggested (Long et al., 2007). Based on the extent of immobilization and NiEDDA accessibility, it was proposed that in the "up" conformation R4 and R5 occupy the narrowest constriction, separating the inner and outer vestibules in KvAP, while this position is occupied by R3 in NaChbac. These findings were in remarkable agreement with the NavAb structure (Payandeh et al., 2011) and the gating pore current measurements in NaChBac (El-Din, Scheuer, & Catterall, 2014). Mutation of R3 to a smaller hydrophilic residue in skeletal muscle Na_v creates gating pore currents in the "up" state of the voltage sensor and underlies potassium-sensitive normokalemic periodic paralysis (Sokolov, Scheuer, & Catterall, 2007). These findings collectively reinforce the modular nature of the VSD and PD in the voltage-gated channels.

Understanding the architecture of the voltage sensor completes just one piece of the giant puzzle, and another major part is the determination of S4 movement during voltage-dependent activation. Based on multitudes of experimental findings and predictions from molecular dynamic simulations (reviewed in (Bezanilla, 2008; Catterall, 2010; Swartz, 2008; Tombola, Pathak, & Isacoff, 2006), at least three different mechanistic models have been proposed: first, the voltage sensor paddle model, where the S3b–S4 paddle moves 15–20 Å across the membrane as a hydrophobic cation (Jiang, Ruta, Chen, Lee, & MacKinnon, 2003); second, the transporter-like model (Cha, Snyder, Selvin, & Bezanilla, 1999; Chanda, Asamoah, Blunck, Roux, & Bezanilla, 2005; Starace & Bezanilla, 2004), where a highly focused field across the membrane necessitates only small-scale (2–4 Å) movement of charges; and third, the helical-translocation/helical-screw model, where the S4 segment undergoes a vertical displacement/rotation to varying degrees (Catterall, 1986; Glauner, Mannuzzu, Gandhi, & Isacoff, 1999; Guy & Seetharamulu, 1986; Yang et al., 1996).

Understanding S4 movements and the associated gating charge translocation requires the VSD in its two end states: the "up" and "down" conformations. While crystal structures and EPR studies at 0 mV have provided a detailed understanding of the "up" conformation of the VSD from diverse sources (Clayton et al., 2008; Jiang, Lee, et al., 2003; Li, Wanderling, Paduch, et al., 2014; Long, Campbell, & Mackinnon, 2005; Long et al., 2007; Payandeh et al., 2011; Zhang et al., 2012), the challenge thus far has been to identify conditions that would stabilize the "down" state in the absence of a membrane potential. There is growing evidence that lipids influence the gating transitions in voltage-gated channels. Electrophysiological studies have shown that protein–lipid interaction plays a key role in shifting the conformational equilibrium between multiple functional states (Ramu, Xu, & Lu, 2006; Schmidt, Jiang, & MacKinnon, 2006; Xu, Ramu, & Lu, 2008; Zheng, Liu, Anderson, & Jiang, 2011). In Kv2.1, hydrolysis of sphingomyelin to ceramide by sphingomyelinase (SMase) D treatment led to a left shift in voltage-dependent activation and thereby an increase in open probability at resting membrane potentials (Ramu et al., 2006). Similarly, reconstitution of KvAP in nonphospholipid bilayers (DOTAP) led to a significant right shift in the activation curves, suggesting that the gating charge interaction with surrounding phospholipids is essential for the transition to the "up" conformation (Schmidt et al., 2006). Antibodies specific to the "down" state bind to KvAP in DOTAP membranes, further confirming that the conformation of the VSD in DOTAP is similar to the physiological hyperpolarization-driven resting state (Zheng et al., 2011). Perozo's group used this strategy to investigate the dynamics of the "down" state of isolated VSD of KvAP using EPR (Li, Wanderling, Sompornpisut, & Perozo, 2014). Probe mobility, solvent accessibility, and interhelical distance determinations from spin probes attached to the VSD in both phospho (POPC/POPG) and nonphospholipids (DOTAP) were used to elucidate the conformational changes associated with voltage-dependent activation of the VSD. In comparison to PC membranes, a dramatic increase in NiEDDA accessibility was observed for residues lining the intracellular crevasse. As a consequence, the water-excluded region was limited to 5 Å roughly at the center of the bilayer, in contrast to 15 Å for the VSD in PC membranes. Furthermore, the extracellular end of S4 showed a decrease in dynamics and NiEDDA accessibility, while the intracellular end of S4 showed an opposite effect consistent with an inward motion of S4 during transition to the "down" state. The extent of S4 movements with respect to relatively rigid ends of S1 and S2 was quantified by DEER measurement using bifunctional spin labels (reduce intrinsic flexibility) at positions in S4 and S1/S2. Overall, a 3 Å downward tilt away from S1/S2 and a 2 Å downward movement for S4 was proposed during the "down" state transition. A new "tilt-shift" model was proposed, suggesting a modest $\sim$3 Å S4 displacement. An extensive analysis of NiEDDa accessibility in the "up" and "down" states suggested that the S4 reorientation to the "down" state led to a large increase in intracellular water penetration within the VSD, thus putatively refocusing the electric field. These studies show how lipid–channel interactions can be used as a key tool in stabilizing specific channel conformation for structure determination by EPR.

3. LIGAND-GATED ION CHANNELS

Signal transduction at the postsynaptic membranes is mediated by neurotransmitter-gated channels. Pentameric ligand-gated channels (LGIC), also called Cys-loop receptors, mediate fast synaptic transmissions in the central and peripheral nervous systems and include both excitatory and inhibitory neurotransmitter receptors. The vertebrate members of this family consist of nicotinic acetylcholine receptor (nAChR), the γ-aminobutyric acid ($GABA_{A/C}R$) receptors, glycine receptors (GlyR), and serotonin receptors ($5\text{-}HT_3R$). Dysfunctions in LGIC underlie a number of neurological disorders, including Alzheimer's disease, Parkinson's disease, schizophrenia, epilepsies, myasthenia gravis, and congenital myasthenic syndromes, and are implicated in nicotine addiction and lung cancer. LGIC are targets for several classes of clinically used drugs, including general anesthetics, alcohols, benzodiazepines, neurosteroids, and barbiturates. In addition to these channels in humans, several orthologs within the LGIC are widely expressed in invertebrates and were recently discovered to be present in bacteria and archaea (Tasneem et al., 2005). While the invertebrate channels are activated by GABA, 5-HT3, glutamate, histamine, or protons, the bacterial channels are shown to respond to primary amines (including GABA) or protons (Bocquet et al., 2007; Lynagh & Pless, 2014; Miller & Smart, 2010; Zimmermann & Dutzler, 2011). Members within the family are known to have evolved from a single ancestral gene and share a common overall architecture with significant homology. These channels have served as important structural surrogates for understanding mechanisms in humans.

LGIC are comprised of five subunits (identical or homologous subunits) arranged around a central channel axis in a pseudo-symmetric manner. Each subunit in turn is composed of an extracellular domain (ECD) that hosts the neurotransmitter binding site, a transmembrane domain (TMD) containing four transmembrane helices (M1–M4) that provides the channel gating machinery, and an intracellular domain consisting of the M3–M4 linker and governs assembly, localization, and membrane trafficking through interactions with scaffolding proteins. Ligand binding within the ECD initiates conformational changes that traverse 50–60 Å across the channel and result in channel opening at the TMD. The allosteric transition is believed to involve large-scale global change in quaternary structure (Bocquet et al., 2009; Hilf & Dutzler, 2009; Taly et al., 2005; Unwin & Fujiyoshi, 2012). The molecular details of protein motions are not completely clear. Most channels within the family undergo desensitization subsequent to opening, which renders them refractory to the ligand. These gating mechanisms are modulated by a range of allosteric ligands of both endogenous and exogenous origin, and some of which have important therapeutic potential. Overall, there is tremendous interest in terms of understanding the molecular details of channel activation and desensitization, and, in turn, the basis of allosteric modulation in these channels.

Structural insights for human Cys-loop receptors have come from members of both prokaryotic and eukaryotic origin (Corringer et al., 2010; Labriola et al., 2013; Sauguet et al., 2014). Our first comprehensive view of the full-length channel in a membrane comes from a 4-Å resolution structural model of a heteromeric *Torpedo* nAChR generated using cryo-electron microscopy data (Unwin, 2005; Unwin & Fujiyoshi, 2012). First atomic-resolution snapshots of the agonist/antagonist-bound ECD came from structures of the acetylcholine-binding protein (AChBP), a naturally truncated homologue of the ECD of nAChR (Brejc et al., 2001; Celie et al., 2004; Hansen et al., 2005). The past decade has seen several landmarks in the LGIC field with structure determination of prokaryotic homologues GLIC and ELIC (Bocquet et al., 2009; Hilf & Dutzler, 2008, 2009), an invertebrate glutamate-gated chloride channel GluCl (Hibbs & Gouaux, 2011), and more recently, the mammalian $GABA_A$ β3 receptor (Miller & Aricescu, 2014) and the mouse serotonin $5\text{-}HT_3$ receptor (Hassaine et al., 2014). In spite of limited overall sequence identity, these solved structures superimpose well, revealing a remarkable conservation of the overall fold and architecture. In all these structures, the ECD has an archetypal arrangement of antiparallel β-sheet with the subunits arranged around a central ion pathway perpendicular to the membrane and lined by the M2 segments from individual subunits. However, interesting differences can be noted in the extent of rotation of the β-sandwich core of the ECD, the positioning of the helical segments within the TMD, and the relative orientation of the ECD with respect to the TMD (~20°). LGIC homologues have been crystallized under different conditions that, in principle, stabilize the channel in the closed, open, or desensitized conformation, and structural differences among them have been used to infer gating mechanisms. In general, it is believed that ELIC, $GLIC_c$ (at pH 7.0), and GluCl (Apo) structures may represent closed/resting states, while $GLIC_o$ (at pH 4.0) (Fig. 3) and GluCl

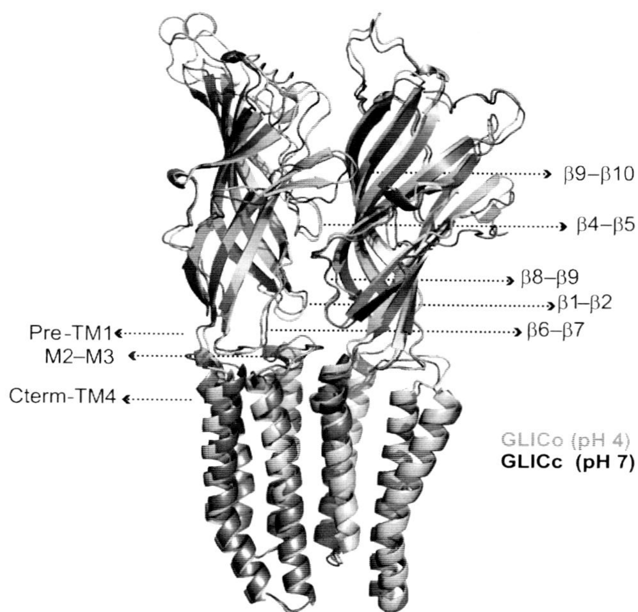

Figure 3 GLIC structures at neutral and acidic pH. An alignment of two adjacent subunits from GLIC structures representing putative closed and open states. The regions contributing toward the ECD–ECD and ECD–TMD interfaces are labeled.

(ivermectin bound) are in the putative open conformation (Althoff, Hibbs, Banerjee, & Gouaux, 2014; Bocquet et al., 2009; Hibbs & Gouaux, 2011; Hilf & Dutzler, 2008, 2009; Sauguet et al., 2014). The 5-HT_3 structure in the presence of nanobodies (Hassaine et al., 2014), which are potent inhibitors of channel activation, is proposed to be in an intermediate conformation, while the $GABA_A$ β3 bound to benzamidine, a novel activator of the channel, might be in a desensitized conformation (Miller & Aricescu, 2014). An overlay of these structures suggests that gating may involve rigid body motions of ECD and TMD relative to each other.

Overall, these structures have provided us atomic details of structural organization in LGIC, and there is reasonable consensus about the general architecture of the family. However, drawing mechanistic information from these structures is a goal far from achieved. Besides, LGIC are well known to be extensively modulated by membrane lipid composition; particularly in nAChR, transitions between closed-to-open-to-desensitized conformations are abrogated in the absence of negative lipids and cholesterol (Fong & McNamee, 1986). Molecular dynamic simulations predict that cholesterol binding at the lipid–protein interfacial sites of nAChR stabilizes ECD–TMD contacts and thereby regulates channel activation (Brannigan, Henin, Law, Eckenhoff, & Klein, 2008). Interestingly, GLIC structure reveals bound lipid molecules within the TMD cavity that are displaced in the presence of a general anesthetic (propofol), suggesting the modulatory role of tightly associated lipids (Nury et al., 2011). Similarly, POPC molecules were found to insert into the intersubunit crevice of GluCl, and quite remarkably this structure is stabilized in an expanded open-like conformation (Althoff et al., 2014). These observations clearly underscore the importance of membranes in channel dynamics.

The ability to use defined lipid composition in membrane reconstitution allows one to study the role of specific membrane constituent such as cholesterol, PIP2, or negative phospholipids at the functional level by patch-clamp measurement in proteoliposomes and monitor conformational changes by EPR spectroscopy. GLIC is chosen as an example to illustrate how this approach can be used to study gating mechanisms in this channel class, and particularly to probe the organization and packing of the TM helices in different conformational states, determine the extent of M2 movements during gating, and define the nature of conformational changes at the ECD–TMD interfaces.

GLIC is a pH-activated channel from *Gloeobacter violaceus* which was initially crystallized under acidic conditions (Bocquet et al., 2009; Hilf & Dutzler, 2009). The structure was proposed to be in an open conformation based on functional measurements in heterologous expression systems where channels were shown to display nondecaying macroscopic currents in response to pH changes (Bocquet et al., 2009, 2007; Hilf & Dutzler, 2009), although later studies showed evidence for GLIC desensitization (Gonzalez-Gutierrez & Grosman, 2010; Parikh, Bali, & Akabas, 2011). From the structure point of view, the pore-lining M2 region of GLIC was clearly in a more open conformation than the unliganded ELIC structure. However, this region was essentially identical to unliganded AChR model from *Torpedo* leading to questions about the mechanism of channel opening and the extent of M2 movement underlying activation. To directly study structural and functional correlates of GLIC gating in a membrane environment, we measured the functional properties of purified and reconstituted channels in proteoliposomes through patch-clamp recordings and investigated underlying conformational transitions by EPR spectroscopy (Velisetty & Chakrapani, 2012; Velisetty et al., 2012, 2014). Interestingly, GLIC currents showed robust desensitization within 1–3 min of

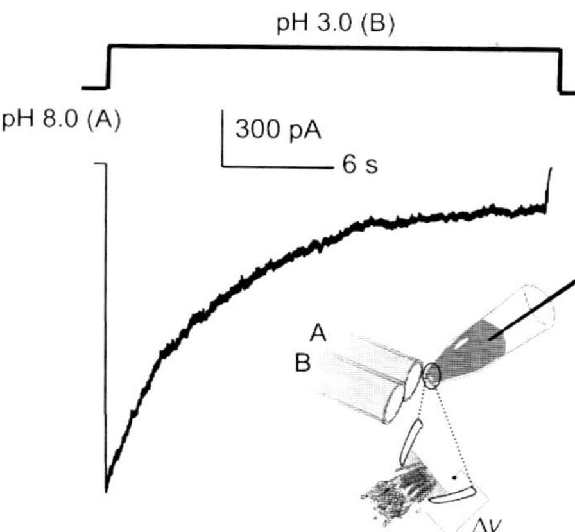

Figure 4 Functional reconstitution of GLIC in asolectin membranes. Macroscopic current elicited in response to pH jumps from excised inside-out patch under symmetrical 150 m*M* Na^+ at −50 mV holding membrane potential. GLIC shows rapid activation (10–30 ms) followed by desensitization (1–3 s) that results in 10–15% steady-state currents. *This figure is adapted from the originally published data in Velisetty and Chakrapanii (2012).*

exposure to protons (Fig. 4). In addition, GLIC desensitization had several hallmark features of the mechanism in eukaryotic channels, allowing us to investigate this far less understood mechanism in LGIC (Velisetty & Chakrapani, 2012).

We carried out extensive cysteine mutagenesis and SDSL at over 100 positions covering key regions of the protein, namely the M2 helix and the intra- and intersubunit interfaces (Velisetty & Chakrapani, 2012; Velisetty et al., 2012, 2014). Functionality of cysteine mutants at several key positions in the presence of spin label was confirmed by patch-clamp measurements of macroscopic currents. Since EPR measurements were carried out under steady-state conditions, GLIC mutants are expected to reside predominantly in the closed (at neutral pH) and desensitized conformations (at acidic pH). In the closed conformation, pore-facing positions in M2 revealed line-shape characteristic of motionally restricted spin labels, suggesting an overall sterically packed environment of M2. Furthermore, a number of pore-facing residues revealed broadening by short-range dipolar coupling arising from intersubunit proximity. Under acidic conditions, residues beyond the 9′ position showed a dramatic decrease in broadening and concomitant increase in amplitude, demonstrating that residues in this region move farther away from each other (Fig. 5). To quantify through-space

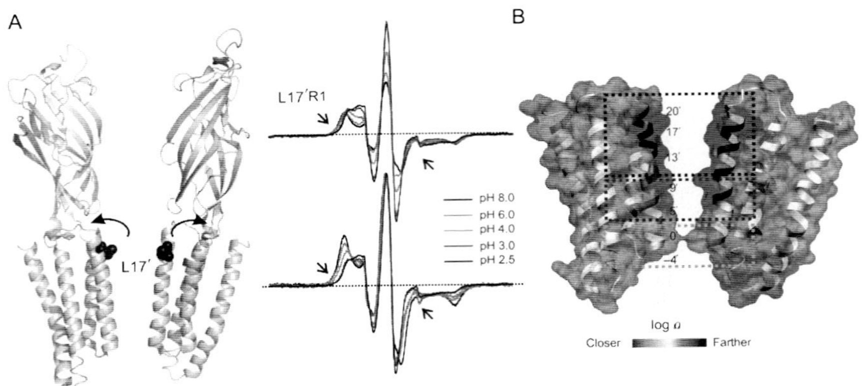

Figure 5 Changes in the spin label proximity during gating. (A) pH-dependent conformational changes at the putative activation gate L241 (highlighted in the GLIC crystal structure. Only diagonal subunits are shown). Spin-normalized EPR spectra show changes in amplitude and extent of broadening (indicated by arrows) revealing pH-dependent changes in proximity at this position in M2. (B) Profile of changes in the amplitude of the EPR signal (Ω) reveals differences in the spin–spin dipolar coupling in the closed and desensitized conformation. The log Ω parameter mapped on to the GLIC crystal structure. Residues displayed in red (gray in the print version) and blue (dark gray in the print version) show positive and negative log Ω values, respectively. Residues colored in white show values closer to 0. The boxed areas highlight regions moving away from the pore axis (log $\Omega > 0$), getting closer toward the pore axis (log $\Omega < 0$), and regions of no significant movement (log $\Omega \sim 0$). *This figure is adapted from the originally published data in Velisetty et al. (2012).*

intersubunit spin–spin dipolar coupling, we calculated the **Ω** parameter (Perozo, Cortes, & Cuello, 1999) as the ratio of spin-normalized amplitudes of the central resonance line in the desensitized and closed conformations. The **Ω** values lower than 1 indicate that the spin-labeled residues move closer to the symmetry axis, while values larger than 1 reflect motions away from the axis. The largest change in **Ω** was observed at the extracellular end of M2 in the region lined by hydrophobic residues (10′–20′) and suggests that this region tilts away radially during channel opening. In clear contrast, residues at the intracellular end (-4′-1′), comprising the selectivity filter region showed no major change in **Ω** values, suggesting that this region does not undergo significant movement. Interestingly, the region immediately above (residues 4′–9′, consisting mostly of polar residues) shows a decrease in **Ω**, presumably resulting from these residues moving closer to the pore axis. The polar region may therefore undergo a tighter packing during desensitization and thereby contribute to the desensitization gate of the channel. Consistent with this idea, bulky pore blockers such as lidocaine slow desensitization in GLIC presumably by a "foot-in-the-door" effect (Velisetty & Chakrapani, 2012), and small hydrophilic substitution at the 9′ position also decreases the rate and extent of desensitization (Gonzalez-Gutierrez & Grosman, 2010). Interestingly, in the GABA structure, M2 tapers downward toward -2′, where the narrowest point is 3.15 Å (Miller & Aricescu, 2014). The compression of the intracellular portion has been proposed to be a region of pore occlusion in the desensitized state, although this mechanism might be unique to $GABA_A$ β3 receptors. In contrast to ELIC, the helical packing in GLIC in the closed conformation creates a deep water-accessible vestibule behind M2 that extends roughly to the middle of the membrane (closer to the 2′ position) (Fig. 6). These vestibules have been implicated in anesthetic and alcohol modulation in this class of channels (Howard et al., 2011; Mascia, Trudell, & Harris, 2000; Nury et al., 2011; Weng, Yang, Corringer, & Sonner, 2010;

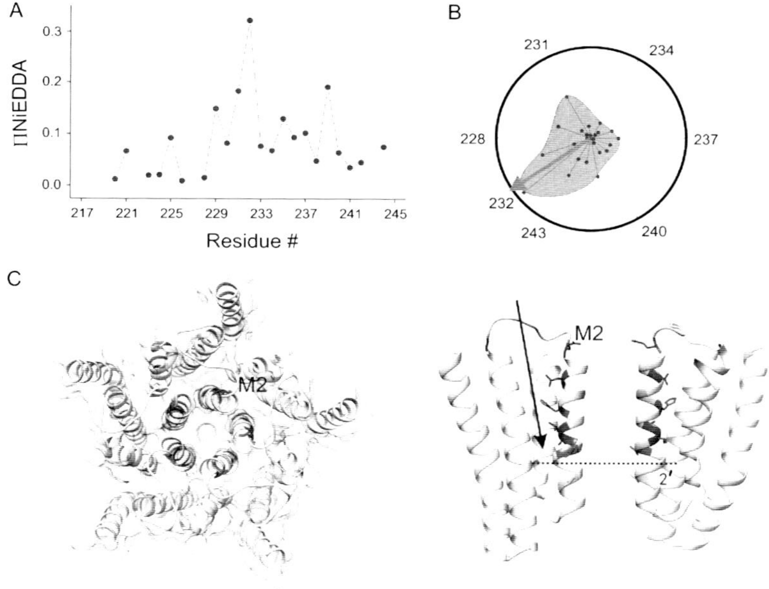

Figure 6 Aqueous vestibules in the GLIC closed state. (A) Periodicity of ΠNiEDDA values for the M2 residues in the closed state. (B) Helical wheel representation of ΠNiEDDA values superimposed in a polar coordinate. The resultant vector points toward the more solvent accessible face of the helix; in this case, the surface of M2 away from the pore lumen. The shaded area within the dashed lines highlights the projection for complete set of accessibility data relative to the maximal accessibility vector. (C) ΠNiEDDA values mapped on the GLIC closed structure. *This figure is adapted from the originally published data in Velisetty et al. (2012).*

Williams & Akabas, 1999). The EPR data of GLIC at acidic pH show general similarities with the overall arrangement of M2, as seen in the crystal structures, especially in the wide-open extracellular and tightly packed intracellular end (Bocquet et al., 2009; Hilf & Dutzler, 2009). A similar pore profile is also seen in $GluCl_o$, $GABA_A$ β3, and 5-HT_3. Overall, these results indicate that the activation gate is at the extracellular end of M2 and channel activation involves an outward motion of this region. This model for activation was in remarkable agreement with the subsequently solved GLIC structure at neutral pH that showed a narrow extracellular vestibule (Sauguet et al., 2014).

Ligand-binding events at the ECD are transduced to the M2 gate through the ECD–ECD and ECD–TMD interfacial regions (Fig. 3). The ECD–TMD interface is framed by tightly coupled interactions between the β6–β7 (Cys-loop), the M2–M3 linker, and the β1–β2 loop. In the closed conformation, residues within these loops were found to be sterically constrained and solvent occluded (Velisetty et al., 2014). In the desensitized conformation, these interactions were replaced by extensive accessibility to O_2 and NiEDDA. A decrease in dipolar broadening for labels at the N-terminal end of M2–M3 was indicative of large outward movement of the M2–M3 linker away from the pore axis. An outward motion of the M2–M3 segment (5 Å away from the channel pore) is evident in GLIC structures (Sauguet et al., 2014). Consistent with this idea, the tip of the M2–M3 segment was found to be disordered in 5HT_3 structures, indicating a high degree of flexibility of this region (Hassaine et al., 2014). In addition to these movements, there was an overall increase in probe dynamics and O_2 accessibility for residues lining the intrasubunit cavity (four-helix TM bundle), which would indicate an expansion of the intrasubunit cavity upon activation. The strategic position and increased membrane interaction of the β6–β7 linker suggested a downward movement into this cavity interacting with tightly associated membrane lipids. These protein motions may allow general anesthetics to access the intrasubunit upon channel activation and thereby play a role in the state-dependent binding of allosteric modulators.

The widely probed regions of the ECD–ECD interface are the β9–β10 (loop C) and β8–β9 (loop F). Studies in AChBP and LGIC have shown that agonists affect the mobility of loop C at the binding site and elicit closure of the binding cavity via capping this loop. This movement has been proposed to be the initial conformational change during channel activation (Celie et al., 2004; Hansen et al., 2005; Mukhtasimova, Lee, Wang, & Sine, 2009). Even though GLIC does not have a canonical ligand-binding site,

EPR line-shape changes and accessibility profiles are suggestive of a pH-dependent immobilization of loop C (Velisetty & Chakrapani, 2012). Such enhanced interaction across the ECD subunit interface was also noted in the β8–β9 loop on the complimentary face. Extensive movement of β8–β9 has been detected in AChBP and members of the LGIC (Hansen et al., 2002; Hansen, Sulzenbacher, Huxford, Marchot, Taylor, & Bourne, 2005; Hanson & Czajkowski, 2008; Khatri, Sedelnikova, & Weiss, 2009; Lyford, Sproul, Eddins, McLaughlin, & Rosenberg, 2003; Pless & Lynch, 2009; Thompson, Padgett, & Lummis, 2006). Intersubunit distance measurements by DEER for a key position in β8–β9 revealed a large inward motion (~9 Å) during activation (Dellisanti et al., 2013). Interestingly, the study noted the absence of β8–β9 movements when the channel was in detergents. Overall, these studies provide a detailed view of the dynamic changes with the TM helices and the subunit interfaces during gating, and how they relate to the conformations in the currently available structural models.

4. FUTURE DIRECTIONS

In combination with other structural techniques, SDSL–EPR spectroscopy has enhanced our understanding of conformational dynamics in large and complex systems. Several recent technological innovations have been geared toward improving sensitivity and extending the upper limit of distance determination (~80–100 Å) (Borbat, Georgieva, & Freed, 2013; Ghimire, McCarrick, Budil, & Lorigan, 2009; Hubbell et al., 2013; Lovett, Lovett, & Harmer, 2012; Ward et al., 2010). DEER measurements with a Q-band (35 GHz) spectrometer have substantially improved sensitivity with smaller sample requirements and shorter data collection times (Ghimire et al., 2009). Improved pulse protocols have helped achieve longer dipolar evolution times and hence an increase in distance resolution and measurable distances (Borbat et al., 2013; Ward et al., 2010). Multifrequency EPR approaches have the capabilities to resolve fast and slow dynamics corresponding to internal and overall motions, respectively (Nesmelov & Thomas, 2010; Zhang et al., 2010). Recent use of Nanodiscs (nanoscale bilayers surrounded by a belt of amphipathic protein derived by apolipoprotein A1; Bayburt & Sligar, 2010) has shown remarkable improvement in DEER sensitivity by decreasing background contribution that arises from intermolecular dipolar interactions. The next rapidly emerging areas are to measure dynamics over the entire range of protein motions (picosecond to millisecond timescales) and to further extend the determination to physiological temperatures (Hubbell et al., 2013). Furthermore, for systems where mutating native cysteines is not feasible (as is the case for LGIC), alternative nitroxide side chains based on site-specific incorporation of unnatural amino acids have been introduced (Fleissner et al., 2009). The recent structures of mammalian channels have ushered in a new era of structural biology, and the stage is set for an in-depth understanding of structural dynamics of these complex channel systems.

ACKNOWLEDGMENTS

I thank the Chakrapani lab members for helpful discussions and comments on the manuscript. This work was supported by the American Heart Association (NCRP Scientist Development Grant 12SDG12070069) and NIGMS (1R01GM108921) to S. C.

REFERENCES

Altenbach, C., Froncisz, W., Hemker, R., McHaourab, H., & Hubbell, W. L. (2005). Accessibility of nitroxide side chains: Absolute Heisenberg exchange rates from power saturation EPR. *Biophysical Journal*, *89*(3), 2103–2112.

Altenbach, C., Greenhalgh, D. A., Khorana, H. G., & Hubbell, W. L. (1994). A collision gradient method to determine the immersion depth of nitroxides in lipid bilayers: Application to spin-labeled mutants of bacteriorhodopsin. *Proceedings of the National Academy of Sciences of the United States of America*, *91*(5), 1667–1671.

Altenbach, C., Marti, T., Khorana, H. G., & Hubbell, W. L. (1990). Transmembrane protein structure: Spin labeling of bacteriorhodopsin mutants. *Science*, *248*(4959), 1088–1092.

Althoff, T., Hibbs, R. E., Banerjee, S., & Gouaux, E. (2014). X-ray structures of GluCl in apo states reveal a gating mechanism of Cys-loop receptors. *Nature*, *512*(7514), 333–337. http://dx.doi.org/10.1038/nature13669.

Bayburt, T. H., & Sligar, S. G. (2010). Membrane protein assembly into Nanodiscs. *FEBS Letters*, *584*(9), 1721–1727. http://dx.doi.org/10.1016/j.febslet.2009.10.024.

Bezanilla, F. (2008). How membrane proteins sense voltage. *Nature Reviews. Molecular Cell Biology*, *9*(4), 323–332. http://dx.doi.org/10.1038/nrm2376.

Bocquet, N., Nury, H., Baaden, M., Le Poupon, C., Changeux, J. P., Delarue, M., et al. (2009). X-ray structure of a pentameric ligand-gated ion channel in an apparently open conformation. *Nature*, *457*(7225), 111–114.

Bocquet, N., Prado de Carvalho, L., Cartaud, J., Neyton, J., Le Poupon, C., Taly, A., et al. (2007). A prokaryotic proton-gated ion channel from the nicotinic acetylcholine receptor family. *Nature*, *445*(7123), 116–119.

Borbat, P. P., Georgieva, E. R., & Freed, J. H. (2013). Improved sensitivity for long-distance measurements in biomolecules: Five-pulse double electron-electron resonance. *Journal of Physical Chemistry Letters*, *4*(1), 170–175. http://dx.doi.org/10.1021/jz301788n.

Borbat, P. P., McHaourab, H. S., & Freed, J. H. (2002). Protein structure determination using long-distance constraints from double-quantum coherence ESR: Study of T4 lysozyme. *Journal of the American Chemical Society*, *124*(19), 5304–5314.

Bordignon, E. (2012). Site-directed spin labeling of membrane proteins. *Topics in Current Chemistry*, *321*, 121–157. http://dx.doi.org/10.1007/128_2011_243.

Brannigan, G., Henin, J., Law, R., Eckenhoff, R., & Klein, M. L. (2008). Embedded cholesterol in the nicotinic acetylcholine receptor. *Proceedings of the National Academy of Sciences of the United States of America*, *105*(38), 14418–14423.

Brejc, K., van Dijk, W. J., Klaassen, R. V., Schuurmans, M., van Der Oost, J., Smit, A. B., et al. (2001). Crystal structure of an ACh-binding protein reveals the ligand-binding domain of nicotinic receptors. *Nature*, *411*(6835), 269–276.

Catterall, W. A. (1986). Molecular properties of voltage-sensitive sodium channels. *Annual Review of Biochemistry*, *55*, 953–985. http://dx.doi.org/10.1146/annurev.bi.55.070186.004513.

Catterall, W. A. (2010). Ion channel voltage sensors: Structure, function, and pathophysiology. *Neuron*, *67*(6), 915–928. http://dx.doi.org/10.1016/j.neuron.2010.08.021.

Celie, P. H., van Rossum-Fikkert, S. E., van Dijk, W. J., Brejc, K., Smit, A. B., & Sixma, T. K. (2004). Nicotine and carbamylcholine binding to nicotinic acetylcholine receptors as studied in AChBP crystal structures. *Neuron*, *41*(6), 907–914.

Cha, A., Snyder, G. E., Selvin, P. R., & Bezanilla, F. (1999). Atomic scale movement of the voltage-sensing region in a potassium channel measured via spectroscopy. *Nature*, *402*(6763), 809–813.

Chakrapani, S., Cuello, L. G., Cortes, D. M., & Perozo, E. (2008). Structural dynamics of an isolated voltage-sensor domain in a lipid bilayer. *Structure*, *16*(3), 398–409.

Chakrapani, S., Sompornpisut, P., Intharathep, P., Roux, B., & Perozo, E. (2010). The activated state of a sodium channel voltage sensor in a membrane environment. *Proceedings of the National Academy of Sciences of the United States of America*, *107*(12), 5435–5440.

Chanda, B., Asamoah, O. K., Blunck, R., Roux, B., & Bezanilla, F. (2005). Gating charge displacement in voltage-gated ion channels involves limited transmembrane movement. *Nature*, *436*(7052), 852–856.

Chiang, Y. W., Borbat, P. P., & Freed, J. H. (2005). The determination of pair distance distributions by pulsed ESR using Tikhonov regularization. *Journal of Magnetic Resonance*, *172*(2), 279–295. http://dx.doi.org/10.1016/j.jmr.2004.10.012.

Clayton, G. M., Altieri, S., Heginbotham, L., Unger, V. M., & Morais-Cabral, J. H. (2008). Structure of the transmembrane regions of a bacterial cyclic nucleotide-regulated channel. *Proceedings of the National Academy of Sciences of the United States of America*, *105*(5), 1511–1515. http://dx.doi.org/10.1073/pnas.0711533105.

Columbus, L., & Hubbell, W. L. (2002). A new spin on protein dynamics. *Trends in Biochemical Sciences*, *27*(6), 288–295.

Corringer, P. J., Baaden, M., Bocquet, N., Delarue, M., Dufresne, V., Nury, H., et al. (2010). Atomic structure and dynamics of pentameric ligand-gated ion channels: New insight from bacterial homologues. *The Journal of Physiology*, *588*(Pt. 4), 565–572.

Cross, T. A., Sharma, M., Yi, M., & Zhou, H. X. (2011). Influence of solubilizing environments on membrane protein structures. *Trends in Biochemical Sciences*, *36*(2), 117–125. http://dx.doi.org/10.1016/j.tibs.2010.07.005.

Cuello, L. G., Cortes, D. M., & Perozo, E. (2004). Molecular architecture of the KvAP voltage-dependent K+ channel in a lipid bilayer. *Science*, *306*(5695), 491–495.

Dellisanti, C. D., Ghosh, B., Hanson, S. M., Raspanti, J. M., Grant, V. A., Diarra, G. M., et al. (2013). Site-directed spin labeling reveals pentameric ligand-gated ion channel gating motions. *PLoS Biology*, *11*(11), e1001714. http://dx.doi.org/10.1371/journal.pbio.1001714.

Doyle, D. A., Morais Cabral, J., Pfuetzner, R. A., Kuo, A., Gulbis, J. M., Cohen, S. L., et al. (1998). The structure of the potassium channel: Molecular basis of K+ conduction and selectivity. *Science*, *280*(5360), 69–77.

Drescher, M. (2012). EPR in protein science: Intrinsically disordered proteins. *Topics in Current Chemistry*, *321*, 91–119. http://dx.doi.org/10.1007/128_2011_235.

El-Din, T. M., Scheuer, T., & Catterall, W. A. (2014). Tracking S4 movement by gating pore currents in the bacterial sodium channel NaChBac. *The Journal of General Physiology*, *144*(2), 147–157. http://dx.doi.org/10.1085/jgp.201411210.

Fanucci, G. E., & Cafiso, D. S. (2006). Recent advances and applications of site-directed spin labeling. *Current Opinion in Structural Biology*, *16*(5), 644–653.

Farahbakhsh, Z. T., Altenbach, C., & Hubbell, W. L. (1992). Spin labeled cysteines as sensors for protein-lipid interaction and conformation in rhodopsin. *Photochemistry and Photobiology*, *56*(6), 1019–1033.

Fleissner, M. R., Brustad, E. M., Kalai, T., Altenbach, C., Cascio, D., Peters, F. B., et al. (2009). Site-directed spin labeling of a genetically encoded unnatural amino acid. *Proceedings of the National Academy of Sciences of the United States of America*, *106*(51), 21637–21642. http://dx.doi.org/10.1073/pnas.0912009106.

Fong, T. M., & McNamee, M. G. (1986). Correlation between acetylcholine receptor function and structural properties of membranes. *Biochemistry*, *25*(4), 830–840.

Ghimire, H., McCarrick, R. M., Budil, D. E., & Lorigan, G. A. (2009). Significantly improved sensitivity of Q-band PELDOR/DEER experiments relative to X-band is observed in measuring the intercoil distance of a leucine zipper motif peptide (GCN4-LZ). *Biochemistry*, *48*(25), 5782–5784. http://dx.doi.org/10.1021/bi900781u.

Glauner, K. S., Mannuzzu, L. M., Gandhi, C. S., & Isacoff, E. Y. (1999). Spectroscopic mapping of voltage sensor movement in the Shaker potassium channel. *Nature*, *402*(6763), 813–817.

Goldstein, S. A. (1996). A structural vignette common to voltage sensors and conduction pores: Canaliculi. *Neuron*, *16*(4), 717–722.

Gonzalez-Gutierrez, G., & Grosman, C. (2010). Bridging the gap between structural models of nicotinic receptor superfamily ion channels and their corresponding functional states. *Journal of Molecular Biology*, *403*(5), 693–705.

Gross, A., Columbus, L., Hideg, K., Altenbach, C., & Hubbell, W. L. (1999). Structure of the KcsA potassium channel from *Streptomyces lividans*: A site-directed spin labeling study of the second transmembrane segment. *Biochemistry*, *38*(32), 10324–10335.

Guy, H. R., & Seetharamulu, P. (1986). Molecular model of the action potential sodium channel. *Proceedings of the National Academy of Sciences of the United States of America*, *83*(2), 508–512.

Hansen, S. B., Radic, Z., Talley, T. T., Molles, B. E., Deerinck, T., Tsigelny, I., et al. (2002). Tryptophan fluorescence reveals conformational changes in the acetylcholine binding protein. *The Journal of Biological Chemistry*, *277*(44), 41299–41302.

Hansen, S. B., Sulzenbacher, G., Huxford, T., Marchot, P., Taylor, P., & Bourne, Y. (2005). Structures of Aplysia AChBP complexes with nicotinic agonists and antagonists reveal distinctive binding interfaces and conformations. *The EMBO Journal*, *24*, 3635–3646.

Hanson, S. M., & Czajkowski, C. (2008). Structural mechanisms underlying benzodiazepine modulation of the GABA(A) receptor. *The Journal of Neuroscience*, *28*(13), 3490–3499. http://dx.doi.org/10.1523/JNEUROSCI.5727-07.2008.

Hassaine, G., Deluz, C., Grasso, L., Wyss, R., Tol, M. B., Hovius, R., et al. (2014). X-ray structure of the mouse serotonin 5-HT3 receptor. *Nature*, *512*(7514), 276–281. http://dx.doi.org/10.1038/nature13552.

Hibbs, R. E., & Gouaux, E. (2011). Principles of activation and permeation in an anion-selective Cys-loop receptor. *Nature*, *474*(7349), 54–60.

Hilf, R. J., & Dutzler, R. (2008). X-ray structure of a prokaryotic pentameric ligand-gated ion channel. *Nature*, *452*(7185), 375–379.

Hilf, R. J., & Dutzler, R. (2009). Structure of a potentially open state of a proton-activated pentameric ligand-gated ion channel. *Nature*, *457*(7225), 115–118.

Hille, B. (2001). *Ion channels of excitable membranes* (3rd ed.). Sunderland, MA, USA: Sinauer Associates Inc.

Howard, R. J., Murail, S., Ondricek, K. E., Corringer, P. J., Lindahl, E., Trudell, J. R., et al. (2011). Structural basis for alcohol modulation of a pentameric ligand-gated ion channel. *Proceedings of the National Academy of Sciences of the United States of America, 108*(29), 12149–12154.

Hubbell, W. L., Cafiso, D. S., & Altenbach, C. (2000). Identifying conformational changes with site-directed spin labeling. *Nature Structural Biology*, 7(9), 735–739.

Hubbell, W. L., Lopez, C. J., Altenbach, C., & Yang, Z. (2013). Technological advances in site-directed spin labeling of proteins. *Current Opinion in Structural Biology, 23*(5), 725–733. http://dx.doi.org/10.1016/j.sbi.2013.06.008.

Hubbell, W. L., McHaourab, H. S., Altenbach, C., & Lietzow, M. A. (1996). Watching proteins move using site-directed spin labeling. *Structure, 4*(7), 779–783.

Jeschke, G. (2012). DEER distance measurements on proteins. *Annual Review of Physical Chemistry, 63*, 419–446. http://dx.doi.org/10.1146/annurev-physchem-032511-143716.

Jeschke, G., & Polyhach, Y. (2007). Distance measurements on spin-labelled biomacromolecules by pulsed electron paramagnetic resonance. *Physical Chemistry Chemical Physics, 9*(16), 1895–1910. http://dx.doi.org/10.1039/b614920k.

Jeschke, G., Wegener, C., Nietschke, M., Jung, H., & Steinhoff, H. J. (2004). Interresidual distance determination by four-pulse double electron-electron resonance in an integral membrane protein: The Na+/proline transporter PutP of *Escherichia coli*. *Biophysical Journal, 86*(4), 2551–2557. http://dx.doi.org/10.1016/S0006-3495(04)74310-6.

Jiang, Y., Lee, A., Chen, J., Ruta, V., Cadene, M., Chait, B. T., et al. (2003). X-ray structure of a voltage-dependent K+ channel. *Nature, 423*(6935), 33–41.

Jiang, Y., Ruta, V., Chen, J., Lee, A., & MacKinnon, R. (2003). The principle of gating charge movement in a voltage-dependent K+ channel. *Nature, 423*(6935), 42–48.

Khatri, A., Sedelnikova, A., & Weiss, D. S. (2009). Structural rearrangements in loop F of the GABA receptor signal ligand binding, not channel activation. *Biophysical Journal, 96*(1), 45–55. http://dx.doi.org/10.1016/j.bpj.2008.09.011.

Kim, M., Xu, Q., Murray, D., & Cafiso, D. S. (2008). Solutes alter the conformation of the ligand binding loops in outer membrane transporters. *Biochemistry, 47*(2), 670–679. http://dx.doi.org/10.1021/bi7016415.

Klare, J. P., & Steinhoff, H. J. (2009). Spin labeling EPR. *Photosynthesis Research, 102*(2–3), 377–390. http://dx.doi.org/10.1007/s11120-009-9490-7.

Labriola, J. M., Pandhare, A., Jansen, M., Blanton, M. P., Corringer, P. J., & Baenziger, J. E. (2013). Structural sensitivity of a prokaryotic pentameric ligand-gated ion channel to its membrane environment. *The Journal of Biological Chemistry, 288*(16), 11294–11303.

Li, Q., Wanderling, S., Paduch, M., Medovoy, D., Singharoy, A., McGreevy, R., et al. (2014). Structural mechanism of voltage-dependent gating in an isolated voltage-sensing domain. *Nature Structural & Molecular Biology, 21*(3), 244–252. http://dx.doi.org/10.1038/nsmb.2768.

Li, Q., Wanderling, S., Sompornpisut, P., & Perozo, E. (2014). Structural basis of lipid-driven conformational transitions in the KvAP voltage-sensing domain. *Nature Structural & Molecular Biology, 21*(2), 160–166. http://dx.doi.org/10.1038/nsmb.2747.

Long, S. B., Campbell, E. B., & Mackinnon, R. (2005). Crystal structure of a mammalian voltage-dependent Shaker family K+ channel. *Science, 309*(5736), 897–903.

Long, S. B., Tao, X., Campbell, E. B., & MacKinnon, R. (2007). Atomic structure of a voltage-dependent K+ channel in a lipid membrane-like environment. *Nature, 450*(7168), 376–382.

Lovett, J. E., Lovett, B. W., & Harmer, J. (2012). DEER-Stitch: Combining three- and four-pulse DEER measurements for high sensitivity, deadtime free data. *Journal of Magnetic Resonance, 223*, 98–106. http://dx.doi.org/10.1016/j.jmr.2012.08.011.

Lyford, L. K., Sproul, A. D., Eddins, D., McLaughlin, J. T., & Rosenberg, R. L. (2003). Agonist-induced conformational changes in the extracellular domain of alpha 7 nicotinic acetylcholine receptors. *Molecular Pharmacology*, *64*(3), 650–658.

Lynagh, T., & Pless, S. A. (2014). Principles of agonist recognition in Cys-loop receptors. *Frontiers in Physiology*, *5*, 160. http://dx.doi.org/10.3389/fphys.2014.00160.

Mascia, M. P., Trudell, J. R., & Harris, R. A. (2000). Specific binding sites for alcohols and anesthetics on ligand-gated ion channels. *Proceedings of the National Academy of Sciences of the United States of America*, *97*(16), 9305–9310.

McHaourab, H. S., Lietzow, M. A., Hideg, K., & Hubbell, W. L. (1996). Motion of spin-labeled side chains in T4 lysozyme. Correlation with protein structure and dynamics. *Biochemistry*, *35*(24), 7692–7704.

McHaourab, H. S., Steed, P. R., & Kazmier, K. (2011). Toward the fourth dimension of membrane protein structure: Insight into dynamics from spin-labeling EPR spectroscopy. *Structure*, *19*(11), 1549–1561. http://dx.doi.org/10.1016/j.str.2011.10.009.

Miller, P. S., & Aricescu, A. R. (2014). Crystal structure of a human GABAA receptor. *Nature*, *512*(7514), 270–275. http://dx.doi.org/10.1038/nature13293.

Miller, P. S., & Smart, T. G. (2010). Binding, activation and modulation of Cys-loop receptors. *Trends in Pharmacological Sciences*, *31*(4), 161–174.

Mukhtasimova, N., Lee, W. Y., Wang, H. L., & Sine, S. M. (2009). Detection and trapping of intermediate states priming nicotinic receptor channel opening. *Nature*, *459*(7245), 451–454.

Nesmelov, Y. E., & Thomas, D. D. (2010). Protein structural dynamics revealed by site-directed spin labeling and multifrequency EPR. *Biophysical Reviews*, *2*(2), 91–99. http://dx.doi.org/10.1007/s12551-010-0032-5.

Nury, H., Van Renterghem, C., Weng, Y., Tran, A., Baaden, M., Dufresne, V., et al. (2011). X-ray structures of general anaesthetics bound to a pentameric ligand-gated ion channel. *Nature*, *469*, 428–431.

Parikh, R. B., Bali, M., & Akabas, M. H. (2011). Structure of the M2 transmembrane segment of GLIC, a prokaryotic Cys loop receptor homologue from *Gloeobacter violaceus*, probed by substituted cysteine accessibility. *The Journal of Biological Chemistry*, *286*(16), 14098–14109.

Payandeh, J., Scheuer, T., Zheng, N., & Catterall, W. A. (2011). The crystal structure of a voltage-gated sodium channel. *Nature*, *475*(7356), 353–358. http://dx.doi.org/10.1038/nature10238.

Perozo, E., Cortes, D. M., & Cuello, L. G. (1998). Three-dimensional architecture and gating mechanism of a K+ channel studied by EPR spectroscopy. *Nature Structural Biology*, *5*(6), 459–469.

Perozo, E., Cortes, D. M., & Cuello, L. G. (1999). Structural rearrangements underlying K+-channel activation gating. *Science*, *285*(5424), 73–78.

Perozo, E., Cuello, L. G., Cortes, D. M., Liu, Y. S., & Sompornpisut, P. (2002). EPR approaches to ion channel structure and function. *Novartis Foundation Symposium*, *245*, 146–158 (discussion 158–164, 165–148).

Pless, S. A., & Lynch, J. W. (2009). Ligand-specific conformational changes in the alpha1 glycine receptor ligand-binding domain. *The Journal of Biological Chemistry*, *284*(23), 15847–15856. http://dx.doi.org/10.1074/jbc.M809343200.

Rabenstein, M. D., & Shin, Y. K. (1995). Determination of the distance between two spin labels attached to a macromolecule. *Proceedings of the National Academy of Sciences of the United States of America*, *92*(18), 8239–8243.

Ramu, Y., Xu, Y., & Lu, Z. (2006). Enzymatic activation of voltage-gated potassium channels. *Nature*, *442*(7103), 696–699.

Ruta, V., Jiang, Y., Lee, A., Chen, J., & MacKinnon, R. (2003). Functional analysis of an archaebacterial voltage-dependent K+ channel. *Nature*, *422*(6928), 180–185.

Sahu, I. D., McCarrick, R. M., & Lorigan, G. A. (2013). Use of electron paramagnetic resonance to solve biochemical problems. *Biochemistry, 52*(35), 5967–5984. http://dx.doi.org/10.1021/bi400834a.

Sauguet, L., Shahsavar, A., Poitevin, F., Huon, C., Menny, A., Nemecz, A., et al. (2014). Crystal structures of a pentameric ligand-gated ion channel provide a mechanism for activation. *Proceedings of the National Academy of Sciences of the United States of America, 111*(3), 966–971. http://dx.doi.org/10.1073/pnas.1314997111.

Schmidt, D., Jiang, Q. X., & MacKinnon, R. (2006). Phospholipids and the origin of cationic gating charges in voltage sensors. *Nature, 444*(7120), 775–779.

Schrempf, H., Schmidt, O., Kummerlen, R., Hinnah, S., Muller, D., Betzler, M., et al. (1995). A prokaryotic potassium ion channel with two predicted transmembrane segments from *Streptomyces lividans*. *The EMBO Journal, 14*(21), 5170–5178.

Sokolov, S., Scheuer, T., & Catterall, W. A. (2007). Gating pore current in an inherited ion channelopathy. *Nature, 446*(7131), 76–78.

Starace, D. M., & Bezanilla, F. (2004). A proton pore in a potassium channel voltage sensor reveals a focused electric field. *Nature, 427*(6974), 548–553.

Swartz, K. J. (2008). Sensing voltage across lipid membranes. *Nature, 456*(7224), 891–897. http://dx.doi.org/10.1038/nature07620.

Taly, A., Delarue, M., Grutter, T., Nilges, M., Le Novere, N., Corringer, P. J., et al. (2005). Normal mode analysis suggests a quaternary twist model for the nicotinic receptor gating mechanism. *Biophysical Journal, 88*(6), 3954–3965.

Tasneem, A., Iyer, L. M., Jakobsson, E., & Aravind, L. (2005). Identification of the prokaryotic ligand-gated ion channels and their implications for the mechanisms and origins of animal Cys-loop ion channels. *Genome Biology, 6*(1), R4.

Thompson, A. J., Padgett, C. L., & Lummis, S. C. (2006). Mutagenesis and molecular modeling reveal the importance of the 5-HT3 receptor F-loop. *The Journal of Biological Chemistry, 281*(24), 16576–16582.

Tombola, F., Pathak, M. M., & Isacoff, E. Y. (2006). How does voltage open an ion channel? *Annual Review of Cell and Developmental Biology, 22*, 23–52.

Unwin, N. (2005). Refined structure of the nicotinic acetylcholine receptor at 4A resolution. *Journal of Molecular Biology, 346*(4), 967–989.

Unwin, N., & Fujiyoshi, Y. (2012). Gating movement of acetylcholine receptor caught by plunge-freezing. *Journal of Molecular Biology, 422*(5), 617–634.

Vamvouka, M., Cieslak, J., Van Eps, N., Hubbell, W., & Gross, A. (2008). The structure of the lipid-embedded potassium channel voltage sensor determined by double-electron-electron resonance spectroscopy. *Protein Science, 17*(3), 506–517. http://dx.doi.org/10.1110/ps.073310008.

Velisetty, P., & Chakrapani, S. (2012). Desensitization mechanism in prokaryotic ligand-gated ion channel. *The Journal of Biological Chemistry, 287*(22), 18467–18477.

Velisetty, P., Chalamalasetti, S. V., & Chakrapani, S. (2012). Conformational transitions underlying pore opening and desensitization in membrane-embedded *Gloeobacter violaceus* ligand-gated ion channel (GLIC). *The Journal of Biological Chemistry, 287*(44), 36864–36872.

Velisetty, P., Chalamalasetti, S. V., & Chakrapani, S. (2014). Structural basis for allosteric coupling at the membrane-protein interface in *Gloeobacter violaceus* ligand-gated ion channel (GLIC). *The Journal of Biological Chemistry, 289*(5), 3013–3025. http://dx.doi.org/10.1074/jbc.M113.523050.

Ward, R., Bowman, A., Sozudogru, E., El-Mkami, H., Owen-Hughes, T., & Norman, D. G. (2010). EPR distance measurements in deuterated proteins. *Journal of Magnetic Resonance, 207*(1), 164–167. http://dx.doi.org/10.1016/j.jmr.2010.08.002.

Weng, Y., Yang, L., Corringer, P. J., & Sonner, J. M. (2010). Anesthetic sensitivity of the *Gloeobacter violaceus* proton-gated ion channel. *Anesthesia and Analgesia, 110*(1), 59–63.

Williams, D. B., & Akabas, M. H. (1999). Gamma-aminobutyric acid increases the water accessibility of M3 membrane-spanning segment residues in gamma-aminobutyric acid type A receptors. *Biophysical Journal*, 77(5), 2563–2574.

Xu, Y., Ramu, Y., & Lu, Z. (2008). Removal of phospho-head groups of membrane lipids immobilizes voltage sensors of K+ channels. *Nature*, *451*(7180), 826–829.

Yang, N., George, A. L., Jr., & Horn, R. (1996). Molecular basis of charge movement in voltage-gated sodium channels. *Neuron*, *16*(1), 113–122.

Zhang, Z., Fleissner, M. R., Tipikin, D. S., Liang, Z., Moscicki, J. K., Earle, K. A., et al. (2010). Multifrequency electron spin resonance study of the dynamics of spin labeled T4 lysozyme. *The Journal of Physical Chemistry. B*, *114*(16), 5503–5521. http://dx.doi.org/10.1021/jp910606h.

Zhang, X., Ren, W., DeCaen, P., Yan, C., Tao, X., Tang, L., et al. (2012). Crystal structure of an orthologue of the NaChBac voltage-gated sodium channel. *Nature*, *486*(7401), 130–134. http://dx.doi.org/10.1038/nature11054.

Zheng, H., Liu, W., Anderson, L. Y., & Jiang, Q. X. (2011). Lipid-dependent gating of a voltage-gated potassium channel. *Nature Communications*, *2*, 250.

Zhou, Y., Morais-Cabral, J. H., Kaufman, A., & MacKinnon, R. (2001). Chemistry of ion coordination and hydration revealed by a K+ channel-Fab complex at 2.0 A resolution. *Nature*, *414*(6859), 43–48.

Zimmermann, I., & Dutzler, R. (2011). Ligand activation of the prokaryotic pentameric ligand-gated ion channel ELIC. *PLoS Biology*, *9*(6), e1001101.

CHAPTER FIFTEEN

Magic-Angle-Spinning Solid-State NMR of Membrane Proteins

Lindsay A. Baker, Gert E. Folkers, Tessa Sinnige, Klaartje Houben, Mohammed Kaplan, Elwin A.W. van der Cruijsen, Marc Baldus[1]

NMR Spectroscopy, Bijvoet Center for Biomolecular Research, Department of Chemistry, Faculty of Science, Utrecht University, Utrecht, The Netherlands

[1]Corresponding author: e-mail address: m.baldus@uu.nl

Contents

Abstract

Solid-state NMR spectroscopy (ssNMR) provides increasing possibilities to examine membrane proteins in different molecular settings, ranging from synthetic bilayers to whole cells. This flexibility often enables ssNMR experiments to be directly correlated with membrane protein function. In this contribution, we discuss experimental aspects of such studies starting with protein expression and labeling, leading to membrane protein isolation or to membrane proteins in a cellular environment. We show that optimized procedures can depend on aspects such as the achieved levels of expression, the stability of the protein during purification or proper refolding. Dealing with native membrane samples, such as isolated cellular membranes, can alleviate or entirely remove such biochemical challenges. Subsequently, we outline ssNMR experiments that involve the use of magic-angle-spinning and can be used to study membrane protein structure and their functional aspects. We pay specific attention to spectroscopic

Methods in Enzymology, Volume 557
ISSN 0076-6879
http://dx.doi.org/10.1016/bs.mie.2014.12.023

issues such as sensitivity and spectral resolution. The latter aspect can be controlled using a combination of tailored preparation procedures with solid-state NMR experiments that simplify the spectral analysis using specific filtering and correlation methods. Such approaches have already provided access to obtain structural views of membrane proteins and study their function in lipid bilayers. Ongoing developments in sample preparation and NMR methodology, in particular in using hyperpolarization or proton-detection schemes, offer additional opportunities to study membrane proteins close to their cellular function. These considerations suggest a further increase in the potential of using solid-state NMR in the context of prokaryotic or eukaryotic membrane protein systems in the near future.

1. INTRODUCTION

Membrane proteins (MPs) are involved in a diverse range of biological functions but pose unique challenges for structural biologists. Their amphipathic, heterogeneous native environment is challenging to mimic *in vitro*, complicating not only isolation of these proteins but also the interpretation of data obtained in nonnative settings. The choice of environmental mimetic, such as detergents or synthetic bilayers, can have a significant impact on the structure, function, and stability of MPs. Solid-state NMR spectroscopy (ssNMR) provides a method by which to examine MPs in different membrane systems, ranging from synthetic bilayers to whole cells (Baker & Baldus, 2014). This flexibility enables ssNMR experiments to be directly correlated with complementary approaches probing structure on different length scales or function via functional assays or methods such as electrophysiology.

As with other spectroscopic methods, sensitivity is a critical factor for NMR studies on MPs. Unless specific experimental conditions are established (such as low temperature; Rocchigiani, Ciancaleoni, Zuccaccia, & Macchioni, 2011) and/or hyperpolarization methods are used, it is not uncommon for a single ssNMR sample to contain 5–10 mg of the MP of interest as well as environmental molecules such as water and lipids. At the same time, "NMR-active" nuclei such as (^{13}C or ^{15}N, respectively) must be incorporated into the molecule, usually by adding isotopically labeled molecules to the growth medium. Separate from the issue of sensitivity, spectral resolution in ssNMR studies on MPs is most easily established by conducting experiments under magic-angle-spinning (MAS) ssNMR. Under MAS, a randomly packed sample is rotated about an axis at an angle of 54.7° relative to the magnetic field (the "magic angle"), at speeds ranging from 1 to 100 kHz, to average out some of the interactions and mimic the effects of molecular tumbling. The rest of this chapter will assume the use of MAS ssNMR, as it is applicable to a wide range of specimens.

Figure 1 summarizes the overall workflow for MAS ssNMR experiments on MPs. Although each step will be discussed in detail below, the choices made at each step have repercussions throughout, so it is worthwhile to

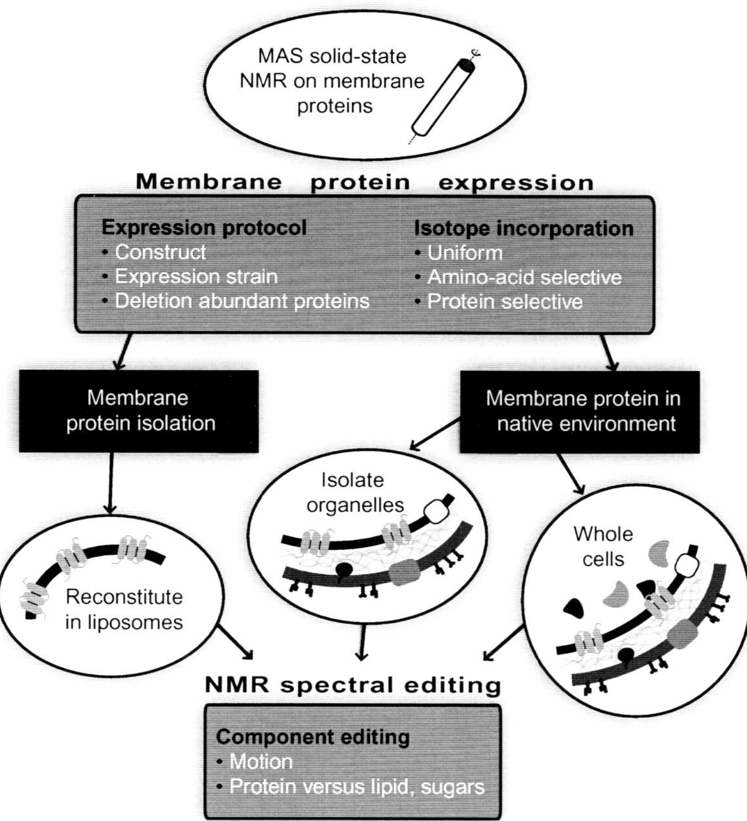

Figure 1 Experimental pathways for magic-angle-spinning solid-state NMR of membrane proteins. In general, the first step is to establish expression conditions for the protein of choice. Since NMR requires significant quantities of isotope-labeled protein, this step will determine the outcome of subsequent steps. After suitable expression has been obtained, a variety of samples can be prepared, using synthetic or native membranes. Specific NMR experiments can be used to target specific types of molecules in heterogeneous membrane samples.

consider the experimental path as a whole. In particular, the choice of membrane system for the final sample will depend on the achieved levels of expression, the stability of the protein during purification, and the need for refolding (such as for the purification of overexpressed β-barrels from inclusion bodies discussed below). These considerations underline the fact that there is no single general approach that works best for every MP.

With sufficient expression levels of folded protein, correctly inserted into the membrane, native membrane samples, such as isolated cellular membranes, can remove many of the biochemical challenges associated with purifying large quantities of MPs. This approach also has the potential to reduce the volume of cell culture required to produce samples (Baker, Daniëls, van der Cruijsen, Folkers, & Baldus, in preparation), significantly reducing costs. In addition, the choice of membrane system will influence the type of NMR experiments that are best suited to the system—for example, spectroscopic editing of lipid signals from fully isotopically labeled native membrane samples.

2. PRODUCTION OF RECOMBINANT PROTEINS IN *ESCHERICHIA COLI*

Production of recombinant proteins in *E. coli* is a well-established (Studier, Rosenberg, Dunn, & Dubendorff, 1990) and frequently summarized method for preparation of protein samples for structural and functional studies (Acton et al., 2011; Kubicek, Block, Maertens, Spriestersbach, & Labahn, 2014). We will therefore focus on techniques specifically useful for the preparation of MP samples for ssNMR.

2.1 Optimization of expression conditions using small-scale cultures

The choice of expression conditions can have a significant impact on the yield of MPs, either correctly folded in the native membrane environment or in inclusion bodies. In a benchmark study for a set of soluble proteins, it was observed that relatively subtle changes in expression protocols can easily change expression levels by an order of magnitude (Berrow et al., 2006). Since expression parameters are not independent, multidimensional optimization of expression conditions is required, and small-scale cultures serve as a convenient way to test many dependent conditions. In our experience, small-scale expression screening experiments serve as a good predictor for

Table 1 Optimization parameters for expression of MPs in *E. coli*

	Starting condition	**Possible variations**
Bacterial strain	BL21 Rosetta2	BL21, Lemo21, C41/C43, Origami, KRS
Culture medium	M9	LB, SB/TB, autoinduction media
Induction (OD_{600})	0.6–0.8	0.25–2.0
Temperature (°C)	18	16, 20, 28, 37
Induction time (h)	12	1–24

large-scale sample production, provided that the conditions used for screening closely mimic the production process (Folkers, van Buuren, & Kaptein, 2004). In Table 1 and below, we summarize some of the culture conditions that may have a large effect on recombinant protein expression. For a more detailed protocol for isotope labeling and bacterial growth, see (Studier, 2005, 2014).

2.1.1 Bacterial strain

The choice of bacterial strain can critically affect both expression and solubility. For the expression of eukaryotic proteins in *E. coli*, it is crucial to use strains that overexpress tRNAs that are frequently used in eukaryotes but not in *E. coli* (e.g., Rosetta, EMD Millipore). The so-called Walker strains C41 (DE3) and C43(DE3), derived from BL21 cells, often permit expression of MPs in high amounts with lower toxicity (Miroux & Walker, 1996). A study by Wagner et al. (2008) suggested that mutations in these cells affect the expression of the T7 polymerase, leading the authors to engineer a BL21-derived strain called Lemo21 (New England Biolabs), which allows for regulation of T7 polymerase activity via controlled expression of T7 lysozyme, its natural inhibitor. However, the gene for T7 lysozyme is under the control of a rhamnose-inducible promotor that is inhibited by glucose, complicating its use when growing in M9 media. We have seen improved expression in Lemo21 cells in M9 cultures for a MP, but only under specific culture conditions.

For cellular preparations, a BL21 strain that is deficient in the genes for the highly abundant outer MPs OmpA and OmpF can be used to reduce background signals (Renault, Tommassen-van Boxtel, et al., 2012). However, the use of this strain is not needed when specific labeling is achieved by treatment with rifampicin (as described in Section 2.2).

2.1.2 Preculture conditions

Since MPs are often toxic for the host strain, it is important to prevent expression of the protein prior to induction (see also Studier, 2014). In most cases, both T7 polymerase and recombinant protein expression are suppressed by the lac repressor. To keep the lac repressor in its active state, 0.5–1% (w/v) glucose can be added during transformation and culturing. Similarly, we found that for toxic proteins, bacteria kept in the exponential growth phase at low densities throughout all precultures exhibit better expression levels during induction (Romanuka, van den Bulke, Kaptein, Boelens, & Folkers, 2009).

2.1.3 Culture medium

As described above, isotope labeling is required for sample preparation for NMR and expression is normally done in a defined minimal medium. As such, we generally perform expression screening in minimal medium. Although the overall yield per liter in richer media (Luria Broth (LB), Super Broth, Terrific Broth) is often better, the amount of correctly folded protein per cell is not necessarily so. It is possible that the slower growth rate in minimal media enables more effective protein folding, increasing the expression yield.

For MPs, uniformly labeled samples often are too complex and crowded for resonance assignment. More sophisticated labeling schemes have been described previously (Filipp, Sinha, Jairam, Bradley, & Opella, 2009; Verardi, Traaseth, Masterson, Vostrikov, & Veglia, 2012), but general strategies include amino acid specific labels and specific ^{13}C incorporation via metabolic precursors. When adding labeled amino acids to the growth media, it is important to consider the metabolic pathways associated with each amino acid. For example, threonine can be converted to isoleucine via α-ketobutyrate; to label threonine but not isoleucine, $^{12}C^{14}N$ isoleucine may be added to the media in addition to $^{13}C^{15}N$ threonine. For a summary of the metabolic interconversion of amino acids, please see (e.g., fig. S5, Sinnige et al., 2014). In our experience, addition of amino acids to M9 cultures can change growth patterns and expression levels, and expression conditions might need to be reoptimized. Amino acid labeling schemes also may use combinations of singly labeled amino acids (i.e., only ^{13}C or ^{15}N labeled) for assignment of specific sequential correlations (i.e., between ^{13}C in residue (i) and ^{15}N in residue ($i+1$)). Spectra can also be simplified using specifically labeled metabolic precursors as a carbon source, such as glycerol labeled at either the C2 or the C1/3 positions. These strategies results in each amino acid ^{13}C labeled at specific positions (Filipp et al., 2009).

Lastly, introducing deuterons via D_2O and/or appropriate precursors provides additional routes in ssNMR spectroscopy, from the level of spectral editing to ^1H detection schemes to the analysis of water molecules (see Section 5). In general, the doubling time of *E. coli* in D_2O is significantly slower than in H_2O, and an additional preculture in D_2O should be included to give the bacteria time to adapt before induction of protein expression. D_2O can be introduced between the LB and M9 cultures as deuterated LB or subsequent to the M9 H_2O culture (see below).

2.1.4 Expression conditions

Our experience with both soluble (Berrow et al., 2006) and MPs (Fig. 2) reveals that both expression levels and amount of folded protein can be optimized by changing the timing of induction according to the growth phase of the bacteria. Although induction at an OD_{600} of 0.6–0.8 serves

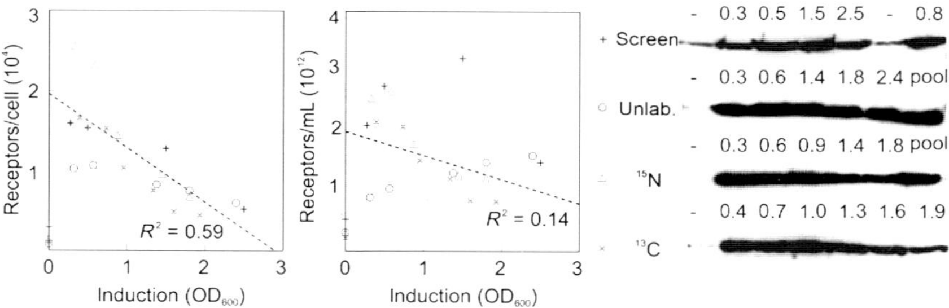

Figure 2 Effect of growth phase on the expression of membrane protein expression. The GPCR NTR1 (UniProt ID: P20789) was expressed in *E. coli* DH5α and induced at various densities (OD_{600}) and various media: small-scale expression screening in unlabeled M9 medium (+), or in 2-L flasks containing 500 mL unlabeled M9 medium (○). ^{15}N-enriched medium ( ) or ^{13}C 15N-enriched medium (×). The left two panels show the number of receptors as quantified by a radioligand-binding assay. The left panel shows the number of receptors per cell, while the middle panel shows the number of receptors per mL of culture. Both the total amount of protein/ml cell culture as well as the amount of protein per cell is negatively affected by the OD_{600} of induction. The right panel shows a Western blot of the total expression of an equivalent volume of cell culture for each condition, as a function of cell density at the time of induction (OD_{600}) as labeled above each row. The label "pool" represents a combined "average" sample from all expression conditions.

as a good starting point, induction earlier or later in the exponential phase, and occasionally in the stationary growth phase, can result in better expression and folding. Optimization of this condition is particularly important for NMR of native membranes, since the amount of folded protein per cell, not the total amount of protein in a culture, is critical. The length of time between induction and harvesting also needs to be optimized, especially for toxic proteins where cultures can stop growing almost immediately after induction. Lastly, growth temperature at induction, as well as the concentration of the induction agent (such as IPTG), should be optimized. Slower growth due to lower temperature can be beneficial for MP folding.

2.1.5 Evaluation of results

To evaluate small-scale expression screening, SDS-PAGE can be used to compare the total and membrane fractions of a lysed cell culture in a detergent-containing lysis buffer. If expression levels are low, or the MP of interest stains poorly, Western blotting can be used. Ideally, a functional assay or ligand-binding assay would be used to quantify the amount of properly folded MP produced, providing a quantitative optimization target.

2.1.6 Protocol: Generalized small-scale expression tests

(1) Transformation: plate on LB + antibiotics + glucose—overnight

(2) Preculture: a few colonies, each in 2 mL LB + antibiotics + glucose—4–6 h

(3) Media change: once cells have reached OD_{600} of 0.5–1.0, centrifuge for 3 min at 6000 rpm in a benchtop minicentrifuge and remove supernatant.

(4) Preculture: resuspend cells in M9 + antibiotics, and split into 3 × 25 mL cultures. Grow one culture at 20, 30, and 37 °C overnight.

(5) For subsequent experiments, the precultures with an OD_{600} of ~1.0 can be used to inoculate 50 mL freshly prepared M9 at OD_{600} 0.05–0.1.

(6) This preculture is then split evenly into 10 × 5.0 mL in 50-mL Grenier tubes, each with a different expression condition.

(7) Approximately 30 min prior to induction, cultures are transferred to the temperature used for induction, and when the required OD_{600} is reached, expression is started by adding inducers such as IPTG.

(8) Harvest cells at different time points after induction by centrifuging at 4000 × *g* for 15 min.

(9) Resuspend 100 μL of cell culture in 100-μL SDS sample buffer per OD_{600} 1.0 culture density, for SDS-PAGE analysis. The remainder of the culture can be used for small-scale cell envelope preparation, lysis in a suitable buffer or functional characterization.

2.2 Selective isotope incorporation

For preparation of NMR samples of MPs in their native cellular environments, it is desirable that only the protein of interest be isotopically labeled, and the cellular background be NMR silent. Rifampicin, an antibiotic that targets the native *E. coli* RNA polymerase (Di Mauro et al., 1969; Wehrli, Knüsel, Schmid, & Staehelin, 1968), and the single-protein production (SPP) system (Suzuki, Zhang, Liu, Woychik, & Inouye, 2005), which makes use of the highly specific endoribonuclease MazF, both inhibit background protein production prior to translation and have been used for NMR in both the solution (Almeida et al., 2001; Mao et al., 2009) and solid states (Baker et al., in preparation; Mao et al., 2011). The SPP method requires replacement of any ACA sequences in the mRNA of the target protein to avoid degradation, while the rifampicin method requires use of a nonbacterial RNA polymerase (such as the T7 system) for expression. A detailed protocol for the use of the SPP system exists (Suzuki, Mao, & Inouye, 2007). Since the rifampicin procedure is applicable for the generally used T7 expression system, we will focus here on this method. This protocol has been used to produce cellular membrane samples of YidC from *E. coli* (UniProt ID: P25714) and KcsA from *S. lividans* (UniProt ID: P0A334) for ssNMR (Baker et al., in preparation).

2.2.1 Protocol: Selective isotope incorporation for native membrane samples

(1) Preculture: 5-mL LB + antibiotics—8 h at 37 °C

(2) Preculture: 50-mL M9 + antibiotics—overnight at 37 °C

(3) (If deuteration desired, 50–100 mL preculture in D_2O M9 + antibiotics) *optional*

(4) Culture: 50–500 mL M9 + antibiotics—^{12}C, ^{14}N, $^{1/2}H$

  a. Inoculate at ~OD 0.1 and grow to OD ~1.8

  b. Pellet cells (4000 × *g*, 10–15 m, 20 °C)

  c. Replace media with appropriately isotopically labeled M9 + antibiotics

d. Add IPTG to 1 m*M* final concentration and grow 15–30 min at 25–28 °C

e. Add rifampicin to 100 μg/mL final concentration and grow at 25–28 °C overnight, shielded from light[1]

For solid-state NMR of MPs, an additional aspect that needs to be taken into account arises for native membrane samples prepared with either rifampicin or the SPP method. Labeling with ^{13}C-glucose results in labeled lipids in the cellular membranes, resulting in high background levels in experiments not involving ^{15}N. The high background levels can be overcome spectroscopically (using magnetization transfer via ^{15}N—see Section 5), or by labeling via amino acids instead of glucose and ammonium chloride, or by the use of cerulenin, which inhibits phospholipid biosynthesis (Mao et al., 2011).

3. ISOLATION OF CELLS AND CELLULAR MEMBRANES

3.1 From cells to ssNMR samples of MPs

All ssNMR samples begin with a similar isolation protocol. The differences for whole cell, native membrane, purified membranes, and purified and reconstituted samples are indicated in the general protocol below.

3.1.1 Protocol: Isolation of cells and membranes

(1) Harvest cells by centrifugation at 4000 × *g* for 15 min at 4 °C.

(2) Resuspend cells in ~5–10 mL (depending on cell volume) of an ice-cold buffer of choice (e.g., 50 m*M* Tris pH 7.4, 100 m*M* NaCl, or phosphate-buffered saline (PBS)). Cells may be frozen at −80 °C at this point for several months.

Note: Whole cells may be washed with PBS by repeating the centrifugation (step 1) and packed into a ssNMR rotor.

(3) Lyse cells with a chilled (4 °C minimum 1 h prior to lysis) French press (8000 psi)—it normally takes about four cycles for complete lysis (can be monitored by measuring OD_{600})

(4) Pellet cell debris at 8000 × *g* for 15 min.

a. Optional step: Isolation of inclusion bodies for further purification by centrifugation at 25,000–100,000 × *g* for 15 min.

(5) Pellet membranes at ~100,000 × *g* for 1 h

[1] *Note*: rifampicin is light sensitive—keep stock solutions and bacterial cultures covered in aluminum foil.

Note: At this stage, membranes are ready for protein purification, if desired (see Section 4 for a discussion of next steps for purified protein samples).

(6) Wash membranes in ~20-mL 10 m*M* phosphate buffer (e.g., pH 6.8)

(7) Pellet membranes at ~100,000 × *g* for 1 h.

Note: At this stage, membranes are ready for separation of inner and outer membranes, if desired (see below for a discussion of next steps for membrane separation).

(8) Resuspend in 1-mL 10 m*M* phosphate buffer (e.g., pH 6.8).

(9) Pellet membranes at ~125,000 × *g* for 2–3 h.

(10) Remove supernatant and pack into a ssNMR MAS rotor for native membrane experiments.

3.2 Purification of specific membranes

If expression levels in native membrane samples are limiting, further purification of the membranes containing the MP of interest can improve the sensitivity of ssNMR experiments. Inner and outer *E. coli* membranes can be separated by the following protocol.

3.2.1 Protocol: Separation of inner and outer bacterial membranes

(1) Harvest membranes as described above (steps 1–7).

(2) Prepare a sucrose gradient in 50 m*M* Tris pH 8.0 buffer. For a 27-mL gradient, layer:

- **a.** 2 mL of 55% (w/v) sucrose
- **b.** 8 mL of 51% (w/v) sucrose
- **c.** 8 mL of 45% (w/v) sucrose
- **d.** 5 mL of 36% (w/v) sucrose
- **e.** Membranes resuspended in 4 mL of 20% (w/v) sucrose

(3) Centrifuge overnight at 100,000 × *g*, preferably in a swinging bucket rotor (e.g., SW32-Ti (Beckman)). A fixed angle rotor will result in angled layers that will need to be kept flat.

(4) Harvest with a syringe:

- **a.** Outer membranes at the interface of the 55% and 51% sucrose steps.
- **b.** Inner membranes at the interface of the 45% and 36% sucrose steps.

(5) Wash membranes in ~20-mL 10 m*M* phosphate buffer (pH 7).

(6) Pellet membranes at ~100,000 × *g* for 1 h and discard supernatant.

(7) Repeat steps 5 and 6 to remove any sucrose and Tris from the samples.

(8) Resuspend in 1-mL 10 m*M* phosphate buffer (pH 6.8)
(9) Pellet membranes at ~125,000 × *g* for 2–3 h
(10) Remove supernatant and pack into a ssNMR MAS rotor for native membrane experiments.

In our experience, after running sucrose gradients with membrane preparations from *E. coli* BL21-derived cell lines grown in M9, an additional band is observed at the interface between the 51% and 45% sucrose steps. This band seems to be composed of a mixture of inner and outer membranes and reduces the amount of purified components that can be obtained. This band is not observed from membrane preparations of these cell lines grown in LB.

4. PURIFICATION AND RECONSTITUTION OF MPs FOR ssNMR

Although unnecessary for native membrane samples, purification and reconstitution of MPs can provide valuable insight into the stability and environmental sensitivity of the MP of interest. As each protein requires a different treatment during purification, we only provide guidelines for parts of the process that are consistent for two broad classes of MPs: those that are produced folded in membranes, and those produced in inclusion bodies. Detailed discussions of chromatography of MPs, for example, can be found elsewhere (Asenjo & Andrews, 2009; Crowe et al., 1994; Saraswat et al., 2013).

4.1 Detergent solubilization of folded MPs

Once membranes have been harvested (see Section 3.1.1), they are solubilized with detergent to produce soluble protein-detergent complexes that can then be purified by traditional chromatography. In general, solubilization is done with high detergent concentrations, which can be an important cost factor, and it is worth optimizing solubilization conditions to improve protein yield and efficiency. Table 2 below summarizes conditions that can be varied during solubilization and gives starting conditions to try for new target MPs. As for optimization of expression conditions, the best conditions are most easily determined by comparison of yields by SDS-PAGE or functional assays. Many of the parameters that affect solubilization will also affect the stability of the protein during purification; however, the same conditions might not be best for both steps. If protein stability is an issue, consider a second optimization for the subsequent purification process.

Table 2 Common parameters for optimization of detergent solubilization of folded MPs

	Starting condition	**Possible variations**
Type of detergent	Dodecyl maltoside	CHAPS, Triton, Digitonin, decyl maltoside
Detergent concentration	10 × critical micelle concentration (CMC)	5–20 × CMC
Solubilization time	2 h	30 min—overnight
Solubilization temperature	4 °C	4–25 °C
Salt concentrations	250 m*M* NaCl	0–500 m*M* NaCl, KCl, $MgSO_4$
Glycerol concentration	10% (v/v)	0–30% (v/v)

4.2 β-Barrel purification and refolding

Overexpression as cytoplasmic inclusion bodies in *E. coli* can lead to high-recombinant protein yields. Commonly, β-barrel proteins can be refolded from a denatured state into detergent micelles or preformed lipid bilayers. In general, inclusion bodies are solubilized in a denaturing agent and diluted into a refolding buffer-containing detergent. However, buffer conditions and refolding protocols need to be established to obtain a sufficiently good yield of refolded protein. Table 3 shows common parameters to optimize during denaturation and refolding. In the choice of detergent, the critical micelle concentration (CMC) should be taken into account—the CMC should be sufficiently high to enable efficient removal of the detergent during the subsequent reconstitution in proteoliposomes (see Section 4.3). In addition, the amount of detergent required for optimal refolding should be cost efficient. Varying the concentration of detergent in the refolding buffer as well as the dilution factor of the unfolded protein into this buffer can help mitigate potential expenses.

β-Barrel proteins have the additional advantage that folded and unfolded protein can be discriminated on (regular) SDS-PAGE or in some cases on seminative SDS-PAGE, a native gel that is run on ice in the presence of 0.2% SDS in the sample buffer and 0.075% SDS in the running buffer. The folded protein migrates faster than heat-denatured species (see Dekker, Merck, Tommassen, & Verheij, 1995; Heller, 1978; Nakamura & Mizushima, 1976; Robert et al., 2006), quantification of the band intensities leads to estimates of the refolding yield (Fig. 3A).

Table 3 Common parameters for optimization of β-barrel refolding

	Starting condition	Possible variations
Denaturant for inclusion body solubilization	8 *M* urea	6 *M* guanidine hydrochloride
Denatured protein concentration	100 μ*M*	20–500 μ*M*
Type of detergent	*N*-dodecyl-*N*,*N*-dimethylamine-*N*-oxide (LDAO)	Octyl glucoside, tetraethylene glycol monooctyl ether (C8E4), diheptanoyl phosphocholine (DHPC), dodecylphosphocholine (DPC), sulfobetaine 12 (SB12)
Detergent concentration	10 × critical micelle concentration (CMC)	5–20 × CMC
Dilution speed	Fast (rapid addition with stirring)	Slow (drop-wise addition with stirring)
Dilution factor (protein solution:refolding buffer)	1:10	1:2–1:100

4.3 Reconstitution of detergent solubilized MPs into lipid bilayers

Protein can be reconstituted in proteoliposomes by addition of lipids and removal of the detergent. For the latter, biobeads that absorb the detergent or dialysis are most commonly used. In some cases, biobeads do not result in the appearance of proteoliposomes that are sufficiently large for ssNMR sample preparation, possibly due to the quick removal of the detergent. Dialysis has the disadvantage that it takes longer to remove the detergent compared to biobeads, but depending on the MP this may be the preferential method.

4.3.1 Protocol: Reconstitution of detergent-solubilized MPs into lipid bilayers

(1) Remove any aggregates by centrifugation for 20 min at 4000 × *g* and 4 °C.

(2) If necessary, concentrate the protein with a centrifuge concentrator (Millipore) and remove aggregates by centrifugation for 20 min at 4000 × *g* and 4 °C.

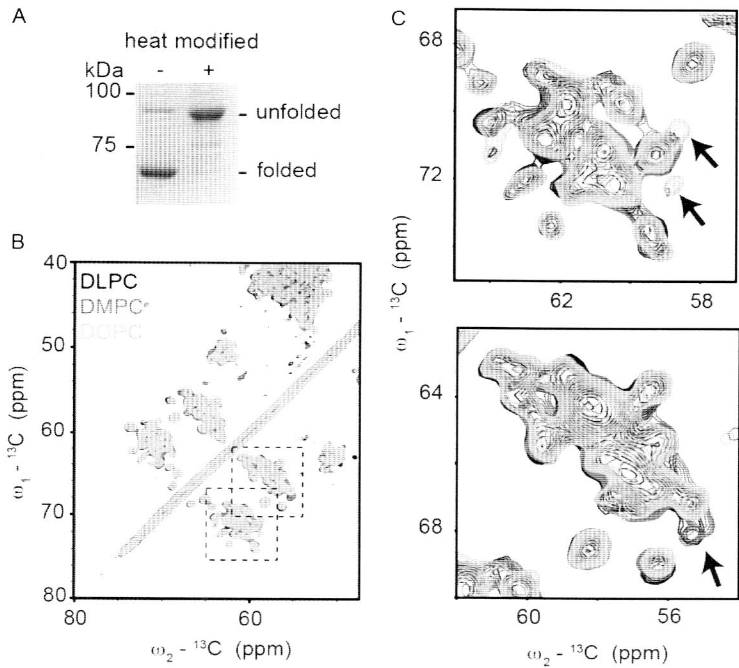

Figure 3 Proteins with β-barrel folds such as BamA (β-barrel assembly machinery A, UniProt ID: P0A940) from *E. coli* can be studied by ssNMR. (A) Folding of BamA analyzed on seminative SDS-PAGE. BamA reconstituted in DOPC vesicles at a molar lipid: protein ratio of 25:1 is shown. Native, folded BamA shows a characteristic shift of the electrophoretic mobility from 88 to 70 kDa, compared to heat-denatured protein. (B) and (C) Comparison of BamA reconstituted in different lipid bilayers. Shown in (B) is a section of a 2D ^{13}C,^{13}C correlation experiment with 30 ms PARIS mixing of BamA in DLPC (black), DMPC (red), and DOPC (blue) at a molar lipid: protein ratios of 25:1. (C) Close-ups of the boxed regions in (B) reveal small differences in certain regions of the spectra. (See the color plate.)

(3) Dry the lipids (such as *E. coli* lipids) in chloroform under a stream of nitrogen, followed by >1 h vacuum drying.

(4) Take up the lipid film in 1-mL reconstitution buffer and incubate the lipids for 5 min at 37 °C (necessary for certain lipids to reach the liquid crystalline phase).

(5) Resuspend the lipids thoroughly by vortexing to form multilamellar vesicles.

(6) Add the lipids to the detergent-solubilized protein (the suspension should become clear as the detergent micelles solubilize the lipids).

(7) **(a)** If removing detergent with dialysis, add reconstitution buffer to dilute below the CMC of the detergent and incubate 30 min at 37 °C.

(b) If using biobeads, wash the correct amount (according to manufacturer's recommendations for the amount of detergent in the sample) of beads 3 × with water and then 3 × with reconstitution buffer and add the beads to the protein and lipid solution.

(8) **(a)** Dialyze against reconstitution buffer (optional if using biobeads) at room temperature or 4 °C for 1–7 days, changing the dialysis buffer once or twice a day, until no more detergent (bubbles or foam) is observed when replacing the buffer. The protein–lipid solution has become turbid.

(b) If using biobeads, separate the biobeads from the protein–lipid suspension on a gravity flow column.

(9) Harvest the proteoliposomes by ultracentrifugation for 1–2 h at 100,000–125,000 × *g* and 4 °C and pack into a MAS rotor.

For reconstitution, it is useful to test a variety of pure lipids and lipid mixtures (see Fig. 3B and C, such as *E. coli* polar lipid extract). The lipid-to-protein ratio, as well as the composition of the reconstitution buffer (pH, addition of salts) should also be screened to give optimal results. Reconstitution yields can be checked on SDS-PAGE by comparing the amount of protein in the proteoliposome pellet and that in the supernatant. For β-barrels, the amounts of folded protein can be estimated on (seminative) SDS-PAGE as described above. Ultimately, however, ssNMR experiments such as a 2D ^{13}C,^{13}C correlation experiment are most suited (see Section 5) to judge the success of sample preparation. It is not unusual for the best samples to yield ^{13}C line widths of 0.7–1 ppm. An example is shown in Fig. 4 for the bacterial KcsA channel that was reconstituted in Asolectin. The (^{13}C,^{13}C) correlation spectrum was obtained using proton-driven spin diffusion with a mixing time of 30 ms (Fig. 4A and 4B) shows results of an NCA experiment (see, e.g., Ref. Weingarth et al. (2014) and references therein).

5. DEDICATED ssNMR EXPERIMENTS

In the last decade, a series of multidimensional ssNMR correlation experiments have been developed to obtain resonance assignments and structurally characterize solid-phase proteins. In principle, all these

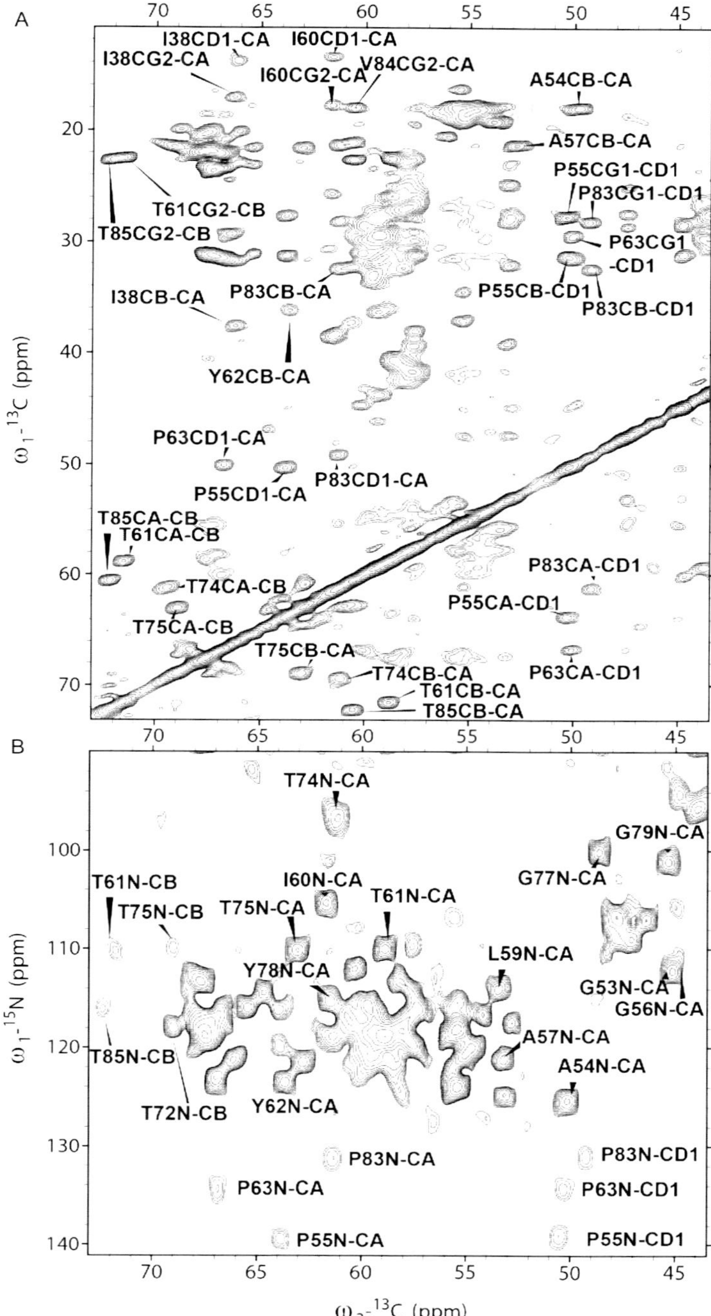

Figure 4 High resolution spectra of U[^{13}C,^{15}N] labeled KcsA reconstituted in asolectin lipids. (A) 2D ^{13}C,^{13}C correlation experiment recorded with 600 μs CP time, 30 ms proton-driven spin diffusion, and 10.921 KHz MAS. (B) 2D NCA spectrum recorded with 600 μs CP time, 3.2 ms SPECIFIC CP time. Both spectra were recorded at 700 MHz at 273 K.

experiments are applicable to MPs embedded in synthetic membranes or when associated with cellular membranes/organelles or cells as described above.

To this end, most ssNMR-studies have used experiments that correlate N–C moieties or C–C spin networks in two, three, or even four spectral dimensions. With such techniques several 3D structures have been obtained and functional aspects of MPs have been elucidated (see, e.g., (Baker & Baldus, 2014) for a recent review). More recently, proton detected experiments have been used for MPs embedded in synthetic bilayers (Linser et al., 2011; Weingarth et al., 2014), and already helped to understand structural aspects related to their mode of action.

With increasing protein size or the presence of endogenous proteinaceous or nonproteinaceous components, dedicated signal filtering and spectral simplification methods become an important aspect in ssNMR. As discussed in Section 2, signal overlap or unwanted molecular contributions may be suppressed by dedicated preparation procedures. Alternatively, the use of ^{15}N-edited dimensions greatly reduces the spectral components in nonproteinaceous cellular samples, for example, leaving ^{13}C spectra largely containing protein signals or, depending on the experimental conditions and labeling procedures, containing labeled lipids (Cukkemane & Baldus, 2013) or nucleotides (Renault, Pawsey, et al., 2012). Another powerful means to deduce structural information in large integral MPs or to zoom on specific molecular components in native preparations relates to solid-state NMR filtering approaches. Different motional scales often characterize signals from water-exposed and transmembrane protein segments. Accordingly, mobile protein loops and rigid transmembrane segments might be detected using through-bond and through-space transfer pathways, respectively (Andronesi et al., 2005). Finally, spectral complexity can also be reduced by "divide-and-conquer" approaches. Here, NMR data are obtained on reference samples such as in synthetic bilayers (Renault, Tommassen-van Boxtel, et al., 2012) or in solution to aid the analysis of membrane-embedded MPs by ssNMR. We have used such methods in the case of membrane-embedded histidine kinases (Etzkorn et al., 2008), retinal complexes (Etzkorn et al., 2010) or, more recently for the outer MP BamA (Sinnige et al., 2015).

On the other hand, correlating signals for the MP of interest to molecular interaction partners, or the surrounding cellular setting offers a unique opportunity to study the supramolecular structure of MPs or the functional roles. As we have described recently, there are a variety of experimental approaches that have been successfully applied to, mostly, synthetic bilayer preparations. Combining such experimental data with information from computational resources such as docking or molecular dynamics simulations further enhances the possibility to study MPs and their functional aspects (Weingarth & Baldus, 2013).

6. CONCLUSIONS

The protocols and tips provided in this chapter are only a starting point for studying MPs by ssNMR. One of the strongest advantages of solid-state NMR is its flexibility, and experiments can be designed to investigate highly specific questions unique to each biological system. In addition to this experimental flexibility, there is also the potential to compare a reconstituted system, where tight control over the molecular constituents is possible, to a native system such as a cellular membrane, where biological relevance is ensured but heterogeneity must be addressed. Although we restricted our discussion here to bacterial MP preparations, reports in the literature (Goncalves et al., 2013) as well as ongoing work in our own laboratory suggest that future ssNMR studies of MP will also be in many cases possible for eukaryotic systems.

As with other spectroscopic methods, sensitivity and spectral resolution are a critical factor. However, the development of hyperpolarization methods that were already demonstrated in the context of cellular ssNMR studies (Renault, Pawsey, et al., 2012) and the further development of ultra-high field NMR systems are likely to further improve the prospects of studying complex MP systems by ssNMR under *in situ* conditions.

ACKNOWLEDGMENTS

Work done in the authors' laboratories and described in this review was funded in part by the Max-Planck society, the DFG, NIH, and NWO (grant numbers 700.11.344 and 700.58.102). L. B. thanks the International Human Frontier Science Program for a Long Term Fellowship. We thank Mark Daniëls for his expertise in the wetlab.

REFERENCES

Acton, T. B., Xiao, R., Anderson, S., Aramini, J., Buchwald, W. A., Ciccosanti, C., et al. (2011). Preparation of protein samples for NMR structure, function, and small-molecule screening studies. *Methods in Enzymology*, *493*, 21–60. http://dx.doi.org/10.1016/B978-0-12-381274-2.00002-9.

Almeida, F. C., Amorim, G. C., Moreau, V. H., Sousa, V. O., Creazola, A. T., Américo, T. A., et al. (2001). Selectively labeling the heterologous protein in Escherichia coli for NMR studies: A strategy to speed up NMR spectroscopy. *Journal of Magnetic Resonance, 148*(1), 142–146.

Andronesi, O. C., Becker, S., Seidel, K., Heise, H., Young, H. S., & Baldus, M. (2005). Determination of membrane protein structure and dynamics by magic-angle-spinning solid-state NMR spectroscopy. *Journal of the American Chemical Society, 127*, 12965–12974. http://dx.doi.org/10.1021/ja0530164.

Asenjo, J. A., & Andrews, B. A. (2009). Protein purification using chromatography: Selection of type, modelling and optimization of operating conditions. *Journal of Molecular Recognition, 22*(2), 65–76. http://dx.doi.org/10.1002/jmr.898.

Baker, L. A., & Baldus, M. (2014). Characterization of membrane protein function by solid-state NMR spectroscopy. *Current Opinion in Structural Biology, 27*, 48–55.

Baker, L. A.; Daniëls, M.; van der Cruijsen, E. A. W.; Folkers, G. E.; & Baldus, M. (in preparation). Improved cellular solid-state NMR of membrane proteins with specific target labeling.

Berrow, N. S., Büssow, K., Coutard, B., Diprose, J., Ekberg, M., Folkers, G. E., et al. (2006). Recombinant protein expression and solubility screening in Escherichia coli: A comparative study. *Acta Crystallographica. Section D Biological Crystallography, 62*(Pt 10), 1218–1226. http://dx.doi.org/10.1107/S0907444906031337.

Crowe, J., Döbeli, H., Gentz, R., Hochuli, E., Stüber, D., & Henco, K. (1994). 6xHis-Ni-NTA chromatography as a superior technique in recombinant protein expression/purification. *Methods in Molecular Biology (Clifton, N.J.), 31*, 371–387. http://dx.doi.org/10.1385/0-89603-258-2:371.

Cukkemane, A., & Baldus, M. (2013). Characterization of a cyclic nucleotide-activated K(+) channel and its lipid environment by using solid-state NMR spectroscopy. *Chembiochem: A European Journal of Chemical Biology, 14*(14), 1789–1798. http://dx.doi.org/10.1002/cbic.201300182.

Dekker, N., Merck, K., Tommassen, J., & Verheij, H. M. (1995). In vitro folding of Escherichia coli outer-membrane phospholipase A. *European Journal of Biochemistry/FEBS, 232*(1), 214–219.

Di Mauro, E., Synder, L., Marino, P., Lamberti, A., Coppo, A., & Tocchini-Valentini, G. P. (1969). Rifampicin sensitivity of the components of DNA-dependent RNA polymerase. *Nature, 222*(5193), 533–537.

Etzkorn, M., Kneuper, H., Dünnwald, P., Vijayan, V., Krämer, J., Griesinger, C., et al. (2008). Plasticity of the PAS domain and a potential role for signal transduction in the histidine kinase DcuS. *Nature Structural & Molecular Biology, 15*(10), 1031–1039. http://dx.doi.org/10.1038/nsmb.1493.

Etzkorn, M., Seidel, K., Li, L., Martell, S., Geyer, M., Engelhard, M., et al. (2010). Complex formation and light activation in membrane-embedded sensory rhodopsin II as seen by solid-state NMR spectroscopy. *Structure (London, England: 1993), 18*(3), 293–300. http://dx.doi.org/10.1016/j.str.2010.01.011.

Filipp, F. V., Sinha, N., Jairam, L., Bradley, J., & Opella, S. J. (2009). Labeling strategies for 13C-detected aligned-sample solid-state NMR of proteins. *Journal of Magnetic Resonance (San Diego, Calif: 1997), 201*(2), 121–130. http://dx.doi.org/10.1016/j.jmr.2009.08.012.

Folkers, G. E., van Buuren, B. N. M., & Kaptein, R. (2004). Expression screening, protein purification and NMR analysis of human protein domains for structural genomics. *Journal of Structural and Functional Genomics, 5*(1–2), 119–131. http://dx.doi.org/10.1023/B:JSFG.0000029200.66197.0c.

Goncalves, J., Eilers, M., South, K., Opefi, C. A., Laissue, P., Reeves, P. J., et al. (2013). Magic angle spinning nuclear magnetic resonance spectroscopy of G protein-coupled receptors. *Methods in Enzymology, 522*, 365–389. http://dx.doi.org/10.1016/B978-0-12-407865-9.00017-0.

Heller, K. B. (1978). Apparent molecular weights of a heat-modifiable protein from the outer membrane of Escherichia coli in gels with different acrylamide concentrations. *Journal of Bacteriology, 134*(3), 1181–1183.

Kubicek, J., Block, H., Maertens, B., Spriestersbach, A., & Labahn, J. (2014). Expression and purification of membrane proteins. *Methods in Enzymology, 541*, 117–140. http://dx.doi.org/10.1016/B978-0-12-420119-4.00010-0.

Linser, R., Dasari, M., Hiller, M., Higman, V., Fink, U., Lopez del Amo, J.-M., et al. (2011). Proton-detected solid-state NMR spectroscopy of fibrillar and membrane proteins. *Angewandte Chemie (International Ed. in English), 50*(19), 4508–4512. http://dx.doi.org/10.1002/anie.201008244.

Mao, L., Inoue, K., Tao, Y., Montelione, G. T., McDermott, A. E., & Inouye, M. (2011). Suppression of phospholipid biosynthesis by cerulenin in the condensed Single-Protein-Production (cSPP) system. *Journal of Biomolecular NMR, 49*(2), 131–137.

Mao, L., Tang, Y., Vaiphei, S. T., Shimazu, T., Kim, S.-G., Mani, R., et al. (2009). Production of membrane proteins for NMR studies using the condensed single protein (cSPP) production system. *Journal of Structural and Functional Genomics, 10*(4), 281–289. http://dx.doi.org/10.1007/s10969-009-9072-0.

Miroux, B., & Walker, J. E. (1996). Over-production of proteins in Escherichia coli: Mutant hosts that allow synthesis of some membrane proteins and globular proteins at high levels. *Journal of Molecular Biology, 260*(3), 289–298. http://dx.doi.org/10.1006/jmbi.1996.0399.

Nakamura, K., & Mizushima, S. (1976). Effects of heating in dodecyl sulfate solution on the conformation and electrophoretic mobility of isolated major outer membrane proteins from Escherichia coli K-12. *Journal of Biochemistry, 80*(6), 1411–1422.

Renault, M., Pawsey, S., Bos, M. P., Koers, E. J., Nand, D., Tommassen-van Boxtel, R., et al. (2012). Solid-state NMR spectroscopy on cellular preparations enhanced by dynamic nuclear polarization. *Angewandte Chemie (International Ed. in English), 51*(12), 2998–3001.

Renault, M., Tommassen-van Boxtel, R., Bos, M. P., Post, J. A., Tommassen, J., & Baldus, M. (2012). Cellular solid-state nuclear magnetic resonance spectroscopy. *Proceedings of the National Academy of Sciences of the United States of America, 109*, 4863–4868.

Robert, V., Volokhina, E. B., Senf, F., Bos, M. P., Van Gelder, P., & Tommassen, J. (2006). Assembly factor Omp85 recognizes its outer membrane protein substrates by a species-specific C-terminal motif. *PLoS Biology, 4*(11), e377. http://dx.doi.org/10.1371/journal.pbio.0040377.

Rocchigiani, L., Ciancaleoni, G., Zuccaccia, C., & Macchioni, A. (2011). Low-temperature kinetic NMR studies on the insertion of a single olefin molecule into a Zr-C bond: Assessing the counterion–solvent interplay. *Angewandte Chemie, International Edition, 50*(49), 11752–11755. http://dx.doi.org/10.1002/anie.201105122.

Romanuka, J., van den Bulke, H., Kaptein, R., Boelens, R., & Folkers, G. E. (2009). Novel strategies to overcome expression problems encountered with toxic proteins: Application to the production of Lac repressor proteins for NMR studies. *Protein Expression and Purification, 67*(2), 104–112. http://dx.doi.org/10.1016/j.pep.2009.05.008.

Saraswat, M., Musante, L., Ravidá, A., Shortt, B., Byrne, B., & Holthofer, H. (2013). Preparative purification of recombinant proteins: Current status and future trends. *BioMed Research International, 2013*, 312709. http://dx.doi.org/10.1155/2013/312709.

Sinnige, T., Weingarth, M., Renault, M., Baker, L., Tommassen, J., & Baldus, M. (2014). Solid-state NMR studies of full-length BamA in lipid bilayers suggest limited overall POTRA mobility. *Journal of Molecular Biology, 426*(9), 2009–2021. http://dx.doi.org/10.1016/j.jmb.2014.02.007.

Sinnige, T., Houben, K., Pritisanac, I., Renault, M., Boelens, R., & Baldus, M. (2015). Insight into the conformational stability of membrane-embedded BamA using a combined solution and solid-state approach. *Journal of Biomolecular NMR*. http://dx.doi.org/10.1007/s10858-014-9891-6, in press.

Studier, F. W. (2005). Protein production by auto-induction in high density shaking cultures. *Protein Expression and Purification*, *41*(1), 207–234.

Studier, F. W. (2014). Stable expression clones and auto-induction for protein production in E. coli. *Methods in Molecular Biology (Clifton, N.J.)*, *1091*, 17–32. http://dx.doi.org/10.1007/978-1-62703-691-7_2.

Studier, F., Rosenberg, A., Dunn, J., & Dubendorff, J. (1990). Use of T7 RNA polymerase to direct expression of cloned genes. *Methods in Enzymology*, *185*(1986), 60–89.

Suzuki, M., Mao, L., & Inouye, M. (2007). Single protein production (SPP) system in Escherichia coli. *Nature Protocols*, *2*(7), 1802–1810. http://dx.doi.org/10.1038/nprot.2007.252.

Suzuki, M., Zhang, J., Liu, M., Woychik, N. A., & Inouye, M. (2005). Single protein production in living cells facilitated by an mRNA interferase. *Molecular Cell*, *18*(2), 253–261. http://dx.doi.org/10.1016/j.molcel.2005.03.011.

Verardi, R., Traaseth, N. J., Masterson, L. R., Vostrikov, V. V., & Veglia, G. (2012). Isotope labeling for solution and solid-state NMR spectroscopy of membrane proteins. *Advances in Experimental Medicine and Biology*, *992*, 35–62. http://dx.doi.org/10.1007/978-94-007-4954-2_3.

Wagner, S., Klepsch, M. M., Schlegel, S., Appel, A., Draheim, R., Tarry, M., et al. (2008). Tuning Escherichia coli for membrane protein overexpression. *Proceedings of the National Academy of Sciences of the United States of America*, *105*(38), 14371–14376. http://dx.doi.org/10.1073/pnas.0804090105.

Wehrli, W., Knüsel, F., Schmid, K., & Staehelin, M. (1968). Interaction of rifamycin with bacterial RNA polymerase. *Proceedings of the National Academy of Sciences of the United States of America*, *61*(2), 667–673.

Weingarth, M., & Baldus, M. (2013). Solid-state NMR-based approaches for supramolecular structure elucidation. *Accounts of Chemical Research*, *46*, 2037–2046. http://dx.doi.org/10.1021/ar300316e.

Weingarth, M., Van der Cruijsen, E. A., Ostmeyer, J., Lievestro, S., Roux, B., et al. (2014). Quantitative analysis of the water occupancy around the selectivity filter of a K + channel in different gating modes. *Journal of the American Chemical Society*, *136*, 2000–2007. http://dx.doi.org/10.1021/ja411450y.

CHAPTER SIXTEEN

Solution NMR Structure Determination of Polytopic α-Helical Membrane Proteins: A Guide to Spin Label Paramagnetic Relaxation Enhancement Restraints

Linda Columbus*,[1], Brett Kroncke[†]

*Department of Chemistry, University of Virginia, Charlottesville, Virginia, USA
[†]Department of Biochemistry and Center for Structural Biology, Vanderbilt University School of Medicine, Nashville, Tennessee, USA
[1]Corresponding author: e-mail address: lcolumbus@gmail.com

Contents

Methods in Enzymology, Volume 557
ISSN 0076-6879
http://dx.doi.org/10.1016/bs.mie.2014.12.005

Abstract

Solution nuclear magnetic resonance structures of polytopic α-helical membrane proteins require additional restraints beyond the traditional Nuclear Overhauser Effect (NOE) restraints. Several methods have been developed and this review focuses on paramagnetic relaxation enhancement (PRE). Important aspects of spin labeling, PRE measurements, structure calculations, and structural quality are discussed.

Polytopic α-helical membrane proteins are challenging systems for solution nuclear magnetic resonance (NMR) approaches. Four major bottlenecks are (i) membrane protein expression in sufficient yields required for *in vitro* investigations; (ii) extraction, solubilization, and stabilization of folded and functional membrane proteins in membrane mimics such as detergents; (iii) structural heterogeneity and dynamics that are integral to membrane protein function, but interfere with *in vitro* structural investigations; and (iv) Nuclear Overhauser Effect restraints (NOEs)—the workhorse of NMR structure calculations—do not provide tertiary structural information for deuterated α-helical membrane proteins. For the most recent reviews of challenges i–iii, readers should seek these references for expression (Guerfal et al., 2013; Gupta, Kueppers, & Schmitt, 2014; Kimura-Soyema, Shirouzu, & Yokoyama, 2014; Parker & Newstead, 2014; Su, Si, Baker, & Berger, 2013), solubilization (Columbus et al., 2009; Duquesne & Sturgis, 2010; Schimerlik, 2001), and structural heterogeneity (Cafiso, 2014; Pan, Piyadasa, O'Neil, & Konermann, 2012). There are several methods (e.g., residual dipolar couplings (RDCs) and methyl NOEs) developed to tackle the lack of NOE-based structural restraints and this review will focus on nitroxide spin label paramagnetic relaxation enhancement (PRE)-derived restraints as applied to polytopic α-helical membrane proteins. PRE measurements rely on the incorporation of a paramagnetic species site specifically (e.g., a nitroxide spin label), which enhances the relaxation of neighboring NMR active nuclei in a distance-dependent manner. PRE-proton-derived distances are long range (15–25 Å) compared to the standard ^1H–^1H NOE (5–6 Å) and were used in the solution NMR structure determination of eight (Berardi, Shih, Harrison, & Chou, 2011; Eichmann et al., 2014; Maslennikov et al., 2010; Reckel et al., 2011; Van Horn et al., 2009; Zhou et al., 2008) of the nine polytopic α-helical membrane proteins determined to date (Table 1; with sensory rhodopsin the exception (Gautier & Nietlispach, 2012)). The general workflow for utilizing PRE restraints for the NMR structure determination of α-helical membrane proteins is shown in Fig. 1 and is divided into

Table 1 Solution NMR structures of polytopic integral membrane proteins

		Restraints in addition to backbone NOEs[a]			# of TM[a]	Oligomeric state			
	PDB IDs	SL-PRE	RDCs	Inter-NOEs[b]			Year	X-ray structure	rmsd (Å)
DsbB	2k73, 2k74	1144	337	39	4	Monomer	2008	2zuq	2.4
DAGK	2kdc	208	67		3	Trimer	2009	3ze3	[d]
Sensory rhodopsin	2ksy			1536	7	Monomer	2010	1h68	1.06
KdpD	2ksf	845			4	Monomer	2010	None	
ArcB	2ksd	291			2	Monomer	2010	None	
QseC	2kse	295			2	Monomer	2010	None	
Proteorhodopsin	2l6x	1006	81	87	7	Monomer	2011	4jq6[c]	3.71
UCP2	2lck	452	470		6	Monomer	2011	None	
YgaP	2mpn	8		216	2	Dimer	2014	None	

[a]Per monomer.
[b]Nonsequential and long range ($|1-j| \geq 5$).
[c]Protein with 57% identity.
[d]The NMR structure has domain swapping and the crystal structure does not.

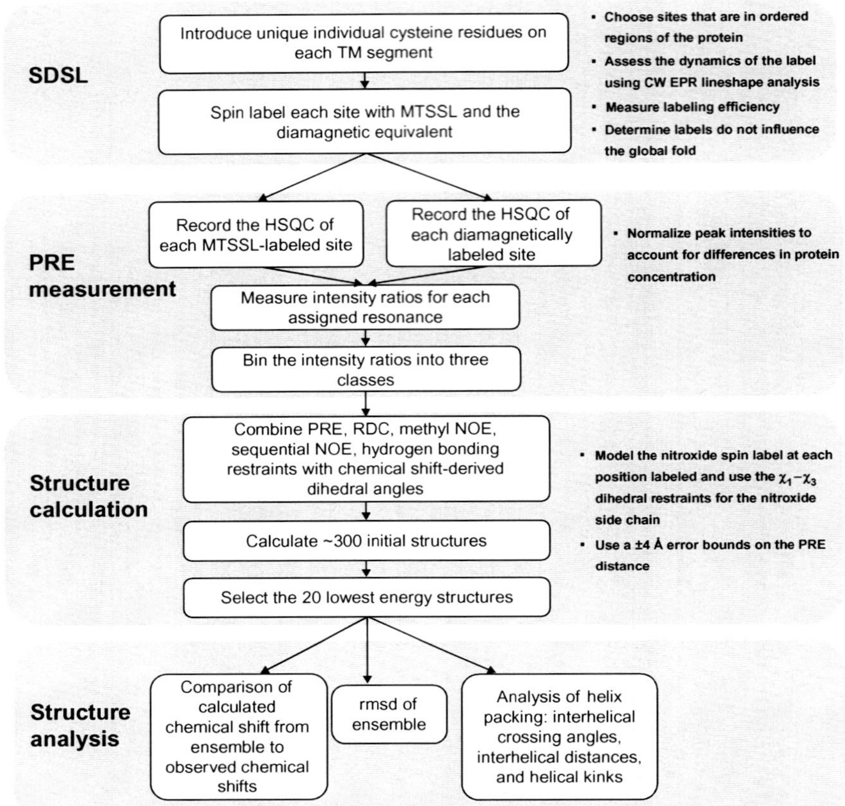

Figure 1 Workflow diagram for utilizing PRE restraints for the NMR structure determination of α-helical membrane proteins. There are four steps: (i) site-directed spin labeling (SDSL), (ii) the PRE measurement, (iii) the structure calculation, and (iv) the evaluation of the accuracy and precision of the structure. Tips are also provided in the text on the right.

four stages: (i) site-directed spin labeling (SDSL), (ii) PRE measurement, (iii) structure calculation, and (iv) accuracy and precision evaluation of the structure.

1. SITE-DIRECTED SPIN LABELING

Typically, SDSL is used to introduce a nitroxide spin label at a unique cysteine residue. Mutagenesis introduces alanine or serine in place of native cysteine residues such that a single cysteine can be introduced throughout the α-helical membrane protein. Selection of where to place the cysteine residue, and subsequently the spin label, requires a few important considerations.

1.1 Nitroxide dynamics

The dynamics of the spin label, which is determined by internal side-chain motions, backbone fluctuations, and rigid body motions, can contribute to large ranges of distances between the label and the relaxed nuclei. However, the PRE measurement is most influenced by the closer distances; the distance range interpreted may only capture a subpopulation of states sampled. Therefore, placing the spin label in dynamic regions may complicate the interpretation of the structure calculated and is best avoided by placing the spin label in structured regions. The dynamics of the spin label, which reflect both the internal flexibility of the spin label and the flexibility of the protein backbone, should always be evaluated using CW EPR (Columbus, Kalai, Jeko, Hideg, & Hubbell, 2001; Hubbell, McHaourab, Altenbach, & Lietzow, 1996; Kroncke, Horanyi, & Columbus, 2010; McHaourab, Lietzow, Hideg, & Hubbell, 1996).

For α-helical membrane proteins, the spin label should be placed on the detergent-exposed helical sites, which will have an ordered backbone, but will likely not disrupt the structure, though there are many instances of the spin label introduced at tertiary contact sites in α-helical membrane proteins without disrupting fold or function (Claxton et al., 2010; Columbus et al., 2009; Gross, Columbus, Hideg, Altenbach, & Hubbell, 1999; Lo, Kroncke, Solomon, & Columbus, 2014; Perozo, Kloda, Cortes, & Martinac, 2001; Wegener, Tebbe, Steinhoff, & Jung, 2000). The CW EPR lineshapes at detergent-exposed α-helical sites are distinct and are slightly less mobile compared to their soluble protein counterparts (Columbus & Hubbell, 2002; Kroncke et al., 2010). Crystal structures of two spin-labeled sites on detergent-exposed sites of LeuT reveal that the spin label interacts with the surface of the membrane protein and restricts the motion of the nitroxide label (Kroncke et al., 2010). In the soluble protein counterparts, density for the spin label is often not observed due to a lack of interaction with the protein and oscillations about the two terminal bonds of the spin label (Guo, Cascio, Hideg, & Hubbell, 2008; Langen, Oh, Cascio, & Hubbell, 2000).

1.2 Choice of spin label

The most well-studied spin label is MTSSL (Fig. 2A; referred to as *R*1 once incorporated into a protein). The internal dynamics, dihedral preferences,

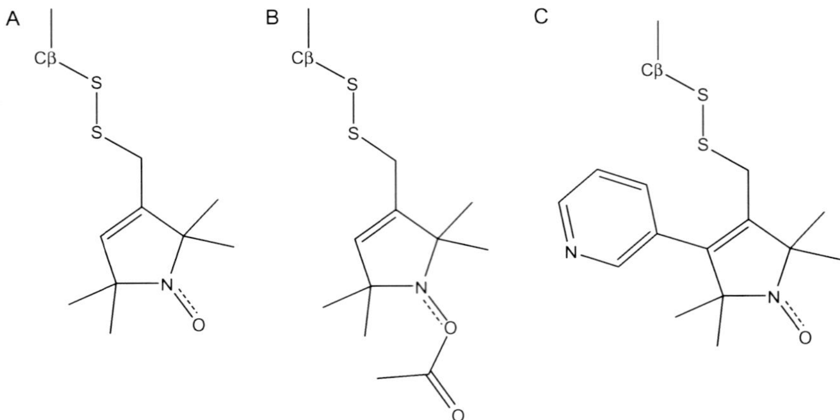

Figure 2 Structures of spin labels. The most common spin labels for PRE measurements MTSSL and the diamagnetic equivalent are shown in (A) and (B), respectively. An example of an alternative label with restricted motion is shown in (C).

and structure on helical sites of soluble and membrane proteins were determined and, thus, information crucial to the interpretation of PRE measurements is known for this spin label. However, there are some drawbacks in using this label: (i) the disulfide formed between the protein and label is labile resulting in increased "free" spin label over time, (ii) a contaminant is often observed in protein–detergent complexes and needs to be recognized and removed (Kroncke & Columbus, 2012), and (iii) oscillations about the internal side-chain bonds can contribute to a broad spatial distribution. To address the spatial distribution of the label, there are additional spin labels that have restricted motion compared to *R*1 (such as that shown in Fig. 2C) (Fawzi et al., 2011; Hubbell, Lopez, Altenbach, & Yang, 2013); however, the labels are larger, have not been investigated on membrane proteins, and may increase the nonspecifically bound contaminants observed in protein–detergent complexes (described below; Kroncke & Columbus, observation).

1.3 Protein topology and PRE restraint coverage

In addition to the dynamic considerations of placement of the spin label, the PRE restraint coverage (the number and topological impact of each PRE restraint) required for an accurate structure calculation should be considered. Two research groups have investigated the influence of topological position, as well as the number of spin-labeled sites, required in order to obtain an accurate overall topology (Chen et al., 2011) and structure (Gottstein, Reckel, Dotsch, & Guntert, 2012). For overall topology, Chen et al. suggest that for α-helical membrane proteins with five to seven transmembrane helices two to three labels are sufficient. This minimal set of long-range restraints when combined with a sufficient number of RDC restraints and medium-range intraresidue NOEs can lead to a well-determined solution NMR structure. Gottstein et al. determined that one spin label per helix was the minimum number of restraints to obtain correct structures. The spin label is best incorporated into the center of the helix (Gottstein et al., 2012) and facing the detergent (Chen et al., 2011). If labeling is only efficient toward the termini of the helices, then you will need to have two PRE restraint sets per helix. The main differences between these two studies are resolution and assumptions utilized in the structure calculation. Chen et al. were developing computational methods for using a minimal set of PREs to determine topology and Gottstein et al. aimed to evaluate PREs as sufficient restraints for high-resolution structure determination. Therefore, in the context of this review, a minimum of one label per helix is recommended.

1.4 Structure and function perturbation

The *R*1 spin label is well tolerated in many regions of α-helical membrane proteins (Claxton et al., 2010; Gross et al., 1999; Lo et al., 2014; Perozo et al., 2001; Wegener et al., 2000). However, all spin-labeled mutants should be evaluated for disruption of fold and function. A major challenge is that functional assays are often hindered in detergent systems. The fold is best analyzed by recording the Heteronuclear Single Quantum Coherence (HSQC) spectrum of the protein labeled with the diamagnetically equivalent spin label (Fig. 2B) (Eichmann et al., 2014; Lo et al., 2014; Zhou et al., 2008); this spectrum needs to be recorded for the PRE measurement regardless and should also be evaluated for the probe's influence on the global fold.

1.5 Labeling efficiency

The labeling efficiency varies between sites on the protein and to some extent correlates with the accessibility of the reactive sulfhydryl. Typically, excess MTSSL is needed to drive the reaction to completion. Even at a molar ratio of 1:6 (protein:spin label), labeling overnight, the labeling efficiency is highly variable between different sites, with more occluded sites labeling less efficiently than exposed sites (Kelly & Gross, 2003;

Kroncke & Columbus, 2012). The high labeling efficiency required for PRE measurements cannot be achieved unless high concentrations (relative to protein) of spin label are used.

1.6 Spin label contaminant

In micellar systems, unreacted, nonspecifically associated spin label contaminant is prevalent and complicates the interpretation of relaxation data. To circumvent this challenge, a strategy of separation that includes affinity chromatography and incubation times is proposed by Kroncke & Columbus (2012). The contaminant is easily identified by a characteristic Electron Paramagnetic Resonance (EPR) lineshape and the EPR spectrum can be used to easily quantify the amount of contaminant (Kroncke & Columbus, 2012).

2. PRE MEASUREMENTS

The nitroxide spin label enhances both longitudinal ($R1$) and transverse ($R2$) relaxation rates, and both measurements have been used for calculating PRE restraints for soluble proteins; however, only $R2$ measurements have been utilized for polytopic α-helical membrane proteins. In terms of nitroxide-derived PREs, $R2$ measurements are preferred because they are less sensitive to the internal motions of the nitroxide side chain (Iwahara, Schwieters, & Clore, 2004). For the PRE measurement, there are specific considerations at the acquisition and processing steps.

2.1 The reference spectrum

Two TROSY-based ^{15}N-HSQC spectra are acquired—one with the nitroxide present and the other with either the reduced nitroxide or the diamagnetically equivalent spin label. Ascorbic acid can be used to reduce the nitroxide; however, a nitroxide buried in a micelle is often not fully reduced with this method (Liang, Bushweller, & Tamm, 2006; Lo et al., 2014). The partial reduction could lead to an underestimate of distances and introduces variability between samples because of the dependence on the accessibility of the label to the reductant. A suggested alternative is the use of the diamagnetic spin label (Fig. 2B), although this approach doubles the number of samples needed. For the eight proteins listed in Table 1, two used the diamagnetic label (YgaP and DsbB), five used ascorbic acid (DAGK, ArcB, KdpD, QseC, and UCP2), and one used the unlabeled native protein as the reference spectrum (proteorhodopsin).

2.2 Calculation of distances from intensity ratios

Once the paramagnetic and diamagnetic spectra are acquired, Lorentzian peak models, which are fit to both dimensions, and intensity ratios ($I_{\text{para}}/I_{\text{dia}}$) are calculated for each assigned resonance (Battiste & Wagner, 2000). The intensities should be normalized using the average of a set of unperturbed resonance peaks (6 or more) to account for differences in protein concentrations. For resonance peaks that broadened beyond detection in the paramagnetic spectrum, the intensity can be estimated from the noise (Battiste & Wagner, 2000). Using the intensity ratio, the relaxation rate enhancements ($R2^{\text{sp}}$) can be quantified by the following relationship:

$$\frac{I_{\text{para}}}{I_{\text{dia}}} = \frac{R2\ \mathrm{e}^{-R2^{\text{sp}}t}}{R2 + R2^{\text{sp}}} \tag{1}$$

where $R2$ is the intrinsic relaxation rate for each amide and is estimated from the half-height Lorentzian linewidths in the diamagnetic spectrum, t is the total INEPT evolution time (~9–12 ms), I_{para} is the resonance peak intensity in the paramagnetic spectrum, and I_{dia} is the resonance peak intensity of the diamagnetic spectrum (Battiste & Wagner, 2000; Gillespie & Shortle, 1997a, 1997b). Solving for $R2^{\text{sp}}$ can be done numerically or the exponential can be well approximated with the first-order term in a Taylor series expansion (i.e., $\mathrm{e}^{R2^{\text{sp}}t} \approx 1 - R2^{\text{sp}}t$) resulting in the following relationship:

$$R2^{\text{sp}} \approx \frac{R2\left(1 - \dfrac{I_{\text{para}}}{I_{\text{dia}}}\right)}{\dfrac{I_{\text{para}}}{I_{\text{dia}}} + R2t}. \tag{2}$$

The modified Solomon–Bloembergen equation (Krugh, 1976; Solomon & Bloembergen, 1956) for transverse relaxation can be used to convert the relaxation enhancement rates into distances:

$$r = \left[\frac{K}{R2^{\text{sp}}}\left(4\tau_{\text{c}} + \frac{3\tau_{\text{c}}}{1 + \omega_{\text{h}}^{2}\tau_{\text{c}}^{2}}\right)\right]^{1/6} \tag{3}$$

where r is the electron–nuclear distance, ω_{h} is the Larmor frequency of the nuclear spin, τ_{c} is the correlation time for the electron–nuclear interaction and is estimated from the global protein correlation time from ^{15}N relaxation measurements, and $K=\frac{1}{15}\left(S(S+1)\gamma^2 g^2\beta^2\right)=1.23\times10^{-32}\text{cm}^6\text{s}^{-2}$, where g is the electron g-factor, γ is the nuclear gyromagnetic ratio, and β is the Bohr magneton.

2.3 Distance classifications

The distance restraints are typically classified into three categories based on the intensity ratio (Battiste & Wagner, 2000; Liang et al., 2006), although the intensity ratio cutoff values and bounds have varied in different structure calculations. Resonances that are not significantly broadened ($I_{para}/I_{dia}>0.85$) can be treated as unrestrained or only restrained with a lower bound distance calculated for the 0.85 intensity ratio. For resonances with intensity ratios between 0.15 and 0.85, the intensity ratio versus distance is relatively linear; thus, the calculated distance from Eq. (3) is used. For peaks broadened beyond detection or with an intensity ratio <0.15, an upper distance of 15 Å and a lower distance limit are not used. If a crystal structure is available of the protein, additional optimization of the distances can be calculated; however, this review focuses on *de novo* structure calculation in systems of which limited structural data is available.

3. STRUCTURE CALCULATION WITH PRE RESTRAINTS

Although NMR structure calculation typically depends on thousands of NOE restraints (on average at least four per amino acid), not all restraints are equal in terms of their impact on the accuracy and precision of the structure calculation. PRE restraints are similar, yet with each spin-labeled mutant you obtain one restraint per assigned amide proton (e.g., one restraint per residue except proline). So, to obtain thousands of PRE restraints, 10 or more mutants are needed (although this depends on the size of the protein). In most cases, a large number of PRE restraints are not obtained and the treatment of the nitroxide and distance restraints in the structure calculation is extremely important.

3.1 The nitroxide spin label

The treatment of the spin label in *de novo* membrane protein structure calculations has varied; three described approaches thus far are not modeling the nitroxide label (e.g., the native residue Cβ was used) (Eichmann et al., 2014; Gottstein et al., 2012), representing the nitroxide as a floating point (artificial atom with a radius of zero) (Zhou et al., 2008), and modeling the nitroxide side chain, but without a rotamer library (which normally generates an all *trans* conformer of the label) (Chen et al., 2011; Van Horn et al., 2009). However, with many crystal structures of the spin label on soluble proteins and two structures of the label on α-helical membrane proteins, a preferred rotameric state of the spin label can be used in the structure calculation (Fleissner, Cascio, & Hubbell, 2009; Kroncke et al., 2010). The dihedral that defines the disulfide bond is always ±90° and the preferred χ_1 and χ_2 rotamers are both −60°. The terminal two dihedrals should not be restricted. The suggested error bounds for the dihedrals are ±5–10°.

The necessity for modeling the spin label side-chain structure is demonstrated for a model two-transmembrane α-helical membrane protein (TM0026) (Columbus et al., 2009; Kroncke & Columbus, 2013; Lo et al., 2014). For the represented structure calculations (Fig. 3 and Table 2), only 35 hydrogen bond restraints and either 87 (A13R1 and V15R1) or 55 (just A13R1) PRE restraints were used. Six structure calculations were performed with three nitroxide side-chain dihedral restraints for each PRE restraint set (A13R1 and V15R1 or just A13R1). In addition to rmsd, the structural ensembles were evaluated with solvent accessible surface area (SASA) which should assess helical packing (larger values of SASA indicate poor helical packing) (Table 2). In the case of 55 PRE restraints (Fig. 3, bottom three structures; 15 of which are across the helical interface), the helices are too far apart from each other and the rmsd for both helices 1 and 2 are greater than 4 Å. When the additional PRE restraints are included from V15R1 (which increases the interhelical restraints to 28), the helical packing and rmsd values begin to improve, but the resulting structures are vastly different depending on the nitroxide side-chain rotameric restraints. With only the disulfide dihedral restraint, the SASA and the rmsd values are larger than when additional side-chain restraints are incorporated. However, if all the dihedrals of the nitroxide are restrained, the SASA is the lowest of all the calculations, but the rmsd values are higher than with just χ_1–χ_3 restricted. Closer inspection of this structure indicates that the second helix is bent and shorter than the first helix and, thus, potentially inaccurate. The best structure in terms of both SASA and rmsd values is with the χ_1–χ_3 dihedrals of −60°, −60°, and 90°. These calculations demonstrate the need for the proper treatment of the nitroxide spin label side chain in the structure calculations.

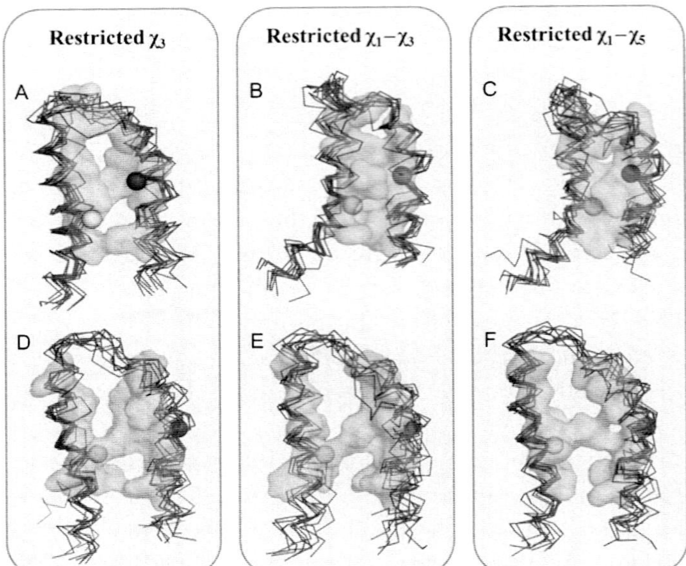

Figure 3 Lowest energy structures (10 lowest) of TM0026. Structures were calculated with 35 hydrogen bond (identified from amide proton NOEs) and PRE restraints. PRE restraints were obtained from A13R1 (a tertiary contact site) and V15R1 (a detergent-exposed site) (A–C; 87 PRE restraints) or from only A13R1 (D–F; 55 PRE restraints). Each set of structures (A and D, B and E, and C and F) was calculated using a different restriction on the conformation of the spin label as labeled. The rmsds for each set of structures, calculated from the most average structure, are shown in Table 2. Light green and blue spheres show the beta carbon of A13 and V43, respectively, to aid in helical twist assessment. A transparent gray surface shows the interface between the two helices (and correlates with the solvent accessible surface area). (See the color plate.)

Table 2 Solvent accesible surface area (SASA) and rmsd values for the structures in Fig. 2

	PREs from A13R1 and V15R1 (87 PRE restraints: 28 of which are interhelical)				PREs from A13R1 only (55 PRE restraints: 15 of which are interhelical)			
	rmsd (Å)				rmsd (Å)			
Label restriction	**Helix 1**	**Helix 2**	**Helices 1 and 2**	**SASA (Å^2)**	**Helix 1**	**Helix 2**	**Helices 1 and 2**	**SASA (Å^2)**
Restricted χ_3	1.28	1.32	3.89	1610	1.36	1.03	4.95	1980
Restricted χ_1–χ_3	1.47	1.04	2.99	1260	1.04	1.39	4.53	2010
Restricted χ_1–χ_5	1.59	1.29	3.14	1130	1.08	1.49	4.06	1970

3.2 Error bounds for PRE distances

For soluble proteins, the error bound value (ranges from ±4 to ±7 Å) did not significantly affect the rmsd of the structural ensemble (Battiste & Wagner, 2000; Gillespie & Shortle, 1997a, 1997b); thus, the error bound typically is estimated as ±4 Å. For α-helical membrane proteins, Gottstein et al. demonstrated that with a sufficient number of restraints, the error bound value weakly influences the accuracy of the structure calculated. These results are consistent with the various error bounds (±2–4 Å) used in the different structure calculations thus far.

3.3 Planar bilayer restraints

Planar bilayer restraints (treated as square potentials) can be used in the structure calculations. The planes are defined at the headgroup regions at a set distance from each other and given a ±4 Å error bound. These restraints can be used to prevent soluble regions of the protein from sampling conformers within the space occupied by the detergent or native lipid bilayer and can prevent the transmembrane helices from sampling unrealistic packing angles. The bilayer restraint distance (~30–40 Å) can be determined on the basis of (1) the hydrophobic thickness of structures of membrane proteins from the same native bilayer (which can be found in the Orientations of Proteins in Membranes database (Lomize, Lomize, Pogozheva, & Mosberg, 2006)) and (2) the residues that are observed to have NOEs with the detergent headgroup protons. Although this approach has only been used with a β-barrel membrane protein NMR structure (Fox et al., 2014), bilayer restraints hold promise for α-helical membrane protein structure calculations.

4. ASSESSMENT OF STRUCTURE QUALITY

Currently, only the backbone rmsd values of the calculated ensemble are used to assess the accuracy of a solution NMR α-helical membrane protein structure. In the cases where a large number of restraints are used, rmsd is likely a good measure of the accuracy of the structure. However, when the number of restraints is limited, ensembles can be calculated with low rmsd values that have questionable helical packing. Chemical shift predictions (Shen & Bax, 2010) calculated from the structural ensemble can be compared to the observed chemical shift values to identify questionable regions of the structures. Additional "model versus data" metrics of structure quality should be used as appropriate (e.g., the RDC Q-factor (Clore & Schwieters, 2004)). Finally, the discriminating power which can be calculated with the RPF program (Huang, Powers, & Montelione, 2005) is a strong assessment of the structural accuracy of soluble proteins (Bagaria, Jaravine, Huang, Montelione, & Guntert, 2012; Huang, Rosato, Singh, & Montelione, 2012) and will likely perform similarly on α-helical membrane proteins. Readers interested in the quality assessment of NMR structures should refer to a recent comprehensive review (Rosato, Tejero, & Montelione, 2013).

As more is learned about α-helical membrane protein fold, stability, and structure, knowledge-based metrics of structure quality need to be considered. Helical packing motifs have been analyzed from known structures (Bowie, 1997; Dong, Sharma, Zhou, & Cross, 2012; Gimpelev, Forrest, Murray, & Honig, 2004; Lo et al., 2009; Walters & DeGrado, 2006) and a database (TMPad) (Lo, Cheng, Chiu, Sung, & Hsu, 2011) has been created cataloguing these motifs. Recently, Maslennikov et al. used interhelical crossing angles, interhelical distances, and helical kinks to assess the structures of KpdD, ArcB, and QseC using Helix Packing Pair program (Dalton, Michalopoulos, & Westhead, 2003). Thus, as these databases grow, standards of evaluation for helical packing angles, packing surface area, likely residues at these interfaces, and much more can be developed.

5. FUTURE DEVELOPMENTS

5.1 Molecular dynamics, explicit solvent, and the native bilayer

The current methods for calculating NMR structures depend on a large number of restraints between side chain and backbone atoms. As larger complexes, such as α-helical membrane proteins in detergent micelles, are being investigated, the number of restraints is decreasing. Therefore, there is a need to improve the computational approaches used during structure calculations and refinement. Currently, all-atom molecular dynamics (MD) simulations of membrane proteins in lipid bilayers are possible and tools to simplify the execution of these simulations are being developed (Wu et al., 2014). All NMR restraints can be easily incorporated into the molecular simulations and MD refinement has already been used for a β-barrel membrane protein NMR structure (Fox et al., 2014).

5.2 Carbon PREs

In the past decade, deuteration and the increased sensitivity of cryoprobes have allowed ^{13}C detection experiments to be utilized for the investigation of large biomolecular systems (Bermel, Bertini, Felli, Kummerle, & Pierattelli, 2006; Bermel, Bertini, Felli, & Pierattelli, 2009; Bertini, Jimenez, & Piccioli, 2005; Bertini, Jimenez, Pierattelli, Wedd, & Xiao, 2008; Bertini, Luchinat, Parigi, & Pierattelli, 2005; Machonkin, Westler, & Markley, 2002; Shimba et al., 2004; Takeuchi, Sun, & Wagner, 2008). Although paramagnetic effects on ^{13}C nuclei have been investigated with metalloproteins (Bermel, Bertini, Felli, Kummerle, & Pierattelli, 2003; Bertini, Felli, Luchinat, Parigi, & Pierattelli, 2007; Bertini, Jimenez, et al., 2005; Bertini, Luchinat, et al., 2005; Kostic, Pochapsky, & Pochapsky, 2002; Machonkin et al., 2002; Machonkin, Westler, & Markley, 2004), only a few studies have investigated ^{13}C-detected PREs with spin labels on soluble globular proteins (Madl, Felli, Bertini, & Sattler, 2010). An added advantage of using ^{13}C PRE restraints is that they are less sensitive to the internal dynamics of the nitroxide side chain (Madl et al., 2010). Utilizing nitroxide-derived PREs, ^{13}C methyl labeling (Goto, Gardner, Mueller, Willis, & Kay, 1999; Religa & Kay, 2010; Shih, Stoica, & Goto, 2008), and ^{13}C direct detection would provide extremely valuable distance restraints (~3–13 Å) (Madl et al., 2010) for α-helical membrane proteins.

6. SUMMARY

The use of nitroxide-derived PRE restraints is extremely valuable to solution NMR structure determination of α-helical membrane proteins when executed properly. Investigators need to be cognizant of the challenges and the solution to these challenges. Specifically, the common nitroxide contaminant in micellar systems, the dynamics of the label, labeling efficiency, the effects the probe has on the global structure, and the treatment of the label in the structure calculation should all be determined.

REFERENCES

Bagaria, A., Jaravine, V., Huang, Y. J., Montelione, G. T., & Guntert, P. (2012). Protein structure validation by generalized linear model root-mean-square deviation prediction. *Protein Science*, *21*, 229–238.

Battiste, J. L., & Wagner, G. (2000). Utilization of site-directed spin labeling and high-resolution heteronuclear nuclear magnetic resonance for global fold determination of large proteins with limited nuclear overhauser effect data. *Biochemistry*, *39*, 5355–5365.

Berardi, M. J., Shih, W. M., Harrison, S. C., & Chou, J. J. (2011). Mitochondrial uncoupling protein 2 structure determined by NMR molecular fragment searching. *Nature*, *476*, 109–113.

Bermel, W., Bertini, I., Felli, I. C., Kummerle, R., & Pierattelli, R. (2003). ^{13}C direct detection experiments on the paramagnetic oxidized monomeric copper, zinc superoxide dismutase. *Journal of the American Chemical Society*, *125*, 16423–16429.

Bermel, W., Bertini, I., Felli, I. C., Kummerle, R., & Pierattelli, R. (2006). Novel ^{13}C direct detection experiments, including extension to the third dimension, to perform the complete assignment of proteins. *Journal of Magnetic Resonance*, *178*, 56–64.

Bermel, W., Bertini, I., Felli, I. C., & Pierattelli, R. (2009). Speeding up (13)C direct detection biomolecular NMR spectroscopy. *Journal of the American Chemical Society*, *131*, 15339–15345.

Bertini, I., Felli, I. C., Luchinat, C., Parigi, G., & Pierattelli, R. (2007). Towards a protocol for solution structure determination of copper(II) proteins: The case of Cu(II)Zn(II) superoxide dismutase. *ChemBioChem*, *8*, 1422–1429.

Bertini, I., Jimenez, B., & Piccioli, M. (2005). ^{13}C direct detected experiments: Optimization for paramagnetic signals. *Journal of Magnetic Resonance*, *174*, 125–132.

Bertini, I., Jimenez, B., Pierattelli, R., Wedd, A. G., & Xiao, Z. (2008). Protonless ^{13}C direct detection NMR: Characterization of the 37 kDa trimeric protein CutA1. *Proteins*, *70*, 1196–1205.

Bertini, I., Luchinat, C., Parigi, G., & Pierattelli, R. (2005). NMR spectroscopy of paramagnetic metalloproteins. *ChemBioChem*, *6*, 1536–1549.

Bowie, J. U. (1997). Helix packing in membrane proteins. *Journal of Molecular Biology*, *272*, 780–789.

Cafiso, D. S. (2014). Identifying and quantitating conformational exchange in membrane proteins using site-directed spin labeling. *Accounts of Chemical Research*, *47*(10), 3102–3109.

Chen, H., Ji, F., Olman, V., Mobley, C. K., Liu, Y., Zhou, Y., et al. (2011). Optimal mutation sites for PRE data collection and membrane protein structure prediction. *Structure*, *19*, 484–495.

Claxton, D. P., Quick, M., Shi, L., de Carvalho, F. D., Weinstein, H., Javitch, J. A., et al. (2010). Ion/substrate-dependent conformational dynamics of a bacterial homolog of neurotransmitter: Sodium symporters. *Nature Structural & Molecular Biology*, *17*, 822–829.

Clore, G. M., & Schwieters, C. D. (2004). How much backbone motion in ubiquitin is required to account for dipolar coupling data measured in multiple alignment media as assessed by independent cross-validation? *Journal of the American Chemical Society*, *126*, 2923–2938.

Columbus, L., & Hubbell, W. L. (2002). A new spin on protein dynamics. *Trends in Biochemical Sciences*, *27*, 288–295.

Columbus, L., Kalai, T., Jeko, J., Hideg, K., & Hubbell, W. L. (2001). Molecular motion of spin labeled side chains in alpha-helices: Analysis by variation of side chain structure. *Biochemistry*, *40*, 3828–3846.

Columbus, L., Lipfert, J., Jambunathan, K., Fox, D. A., Sim, A. Y., Doniach, S., et al. (2009). Mixing and matching detergents for membrane protein NMR structure determination. *Journal of the American Chemical Society*, *131*, 7320–7326.

Dalton, J. A., Michalopoulos, I., & Westhead, D. R. (2003). Calculation of helix packing angles in protein structures. *Bioinformatics*, *19*, 1298–1299.

Dong, H., Sharma, M., Zhou, H. X., & Cross, T. A. (2012). Glycines: Role in alpha-helical membrane protein structures and a potential indicator of native conformation. *Biochemistry*, *51*, 4779–4789.

Duquesne, K., & Sturgis, J. N. (2010). Membrane protein solubilization. *Methods in Molecular Biology*, *601*, 205–217.

Eichmann, C., Tzitzilonis, C., Bordignon, E., Maslennikov, I., Choe, S., & Riek, R. (2014). Solution NMR structure and functional analysis of the integral membrane protein YgaP from *Escherichia coli*. *The Journal of Biological Chemistry*, *289*, 23482–23503.

Fawzi, N. L., Fleissner, M. R., Anthis, N. J., Kalai, T., Hideg, K., Hubbell, W. L., et al. (2011). A rigid disulfide-linked nitroxide side chain simplifies the quantitative analysis of PRE data. *Journal of Biomolecular NMR*, *51*, 105–114.

Fleissner, M. R., Cascio, D., & Hubbell, W. L. (2009). Structural origin of weakly ordered nitroxide motion in spin-labeled proteins. *Protein Science*, *18*, 893–908.

Fox, D. A., Larsson, P., Lo, R. H., Kroncke, B. M., Kasson, P. M., & Columbus, L. (2014). Structure of the Neisserial outer membrane protein Opa(6)(0): Loop flexibility essential to receptor recognition and bacterial engulfment. *Journal of the American Chemical Society*, *136*, 9938–9946.

Gautier, A., & Nietlispach, D. (2012). Solution NMR studies of integral polytopic alpha-helical membrane proteins: The structure determination of the seven-helix transmembrane receptor sensory rhodopsin II, pSRII. *Methods in Molecular Biology*, *914*, 25–45.

Gillespie, J. R., & Shortle, D. (1997a). Characterization of long-range structure in the denatured state of staphylococcal nuclease. I. Paramagnetic relaxation enhancement by nitroxide spin labels. *Journal of Molecular Biology*, *268*, 158–169.

Gillespie, J. R., & Shortle, D. (1997b). Characterization of long-range structure in the denatured state of staphylococcal nuclease. II. Distance restraints from paramagnetic relaxation and calculation of an ensemble of structures. *Journal of Molecular Biology*, *268*, 170–184.

Gimpelev, M., Forrest, L. R., Murray, D., & Honig, B. (2004). Helical packing patterns in membrane and soluble proteins. *Biophysical Journal*, *87*, 4075–4086.

Goto, N. K., Gardner, K. H., Mueller, G. A., Willis, R. C., & Kay, L. E. (1999). A robust and cost-effective method for the production of Val, Leu, Ile (delta 1) methyl-protonated ^{15}N-, ^{13}C-, ^{2}H-labeled proteins. *Journal of Biomolecular NMR*, *13*, 369–374.

Gottstein, D., Reckel, S., Dotsch, V., & Guntert, P. (2012). Requirements on paramagnetic relaxation enhancement data for membrane protein structure determination by NMR. *Structure*, *20*, 1019–1027.

Gross, A., Columbus, L., Hideg, K., Altenbach, C., & Hubbell, W. L. (1999). Structure of the KcsA potassium channel from Streptomyces lividans: A site-directed spin labeling study of the second transmembrane segment. *Biochemistry*, *38*, 10324–10335.

Guerfal, M., Claes, K., Knittelfelder, O., De Rycke, R., Kohlwein, S. D., & Callewaert, N. (2013). Enhanced membrane protein expression by engineering increased intracellular membrane production. *Microbial Cell Factories*, *12*, 122.

Guo, Z., Cascio, D., Hideg, K., & Hubbell, W. L. (2008). Structural determinants of nitroxide motion in spin-labeled proteins: Solvent-exposed sites in helix B of T4 lysozyme. *Protein Science*, *17*, 228–239.

Gupta, R. P., Kueppers, P., & Schmitt, L. (2014). New examples of membrane protein expression and purification using the yeast based Pdr1-3 expression strategy. *Journal of Biotechnology*, *191*, 158–164.

Huang, Y. J., Powers, R., & Montelione, G. T. (2005). Protein NMR recall, precision, and F-measure scores (RPF scores): Structure quality assessment measures based on information retrieval statistics. *Journal of the American Chemical Society*, *127*, 1665–1674.

Huang, Y. J., Rosato, A., Singh, G., & Montelione, G. T. (2012). RPF: A quality assessment tool for protein NMR structures. *Nucleic Acids Research*, *40*, W542–W546.

Hubbell, W. L., Lopez, C. J., Altenbach, C., & Yang, Z. Y. (2013). Technological advances in site-directed spin labeling of proteins. *Current Opinion in Structural Biology*, *23*, 725–733.

Hubbell, W. L., McHaourab, H. S., Altenbach, C., & Lietzow, M. A. (1996). Watching proteins move using site-directed spin labeling. *Structure*, *4*, 779–783.

Iwahara, J., Schwieters, C. D., & Clore, G. M. (2004). Ensemble approach for NMR structure refinement against (1)H paramagnetic relaxation enhancement data arising from a flexible paramagnetic group attached to a macromolecule. *Journal of the American Chemical Society*, *126*, 5879–5896.

Kelly, B. L., & Gross, A. (2003). Potassium channel gating observed with site-directed mass tagging. *Nature Structural Biology*, *10*, 280–284.

Kimura-Soyema, T., Shirouzu, M., & Yokoyama, S. (2014). Cell-free membrane protein expression. *Methods in Molecular Biology*, *1118*, 267–273.

Kostic, M., Pochapsky, S. S., & Pochapsky, T. C. (2002). Rapid recycle (13)C′, (15)N and (13)C, (13)C′ heteronuclear and homonuclear multiple quantum coherence detection for resonance assignments in paramagnetic proteins: Example of Ni(2+)-containing acireductone dioxygenase. *Journal of the American Chemical Society*, *124*, 9054–9055.

Kroncke, B. M., & Columbus, L. (2012). Identification and removal of nitroxide spin label contaminant: Impact on PRE studies of alpha-helical membrane proteins in detergent. *Protein Science*, *21*, 589–595.

Kroncke, B. M., & Columbus, L. (2013). Backbone (1)H, (1)(3)C and (1)(5)N resonance assignments of the alpha-helical membrane protein TM0026 from Thermotoga maritima. *Biomolecular NMR Assignments*, 7, 203–206.

Kroncke, B. M., Horanyi, P. S., & Columbus, L. (2010). Structural origins of nitroxide side chain dynamics on membrane protein alpha-helical sites. *Biochemistry*, *49*, 10045–10060.

Krugh, T. R. (1976). Spin-Label-Induced Nuclear Magnetic Resonance Relaxation Studies of Enzymes. In L. J. Berliner (Ed.), *Spin labeling: Theory and applications* (pp. 339–372). New York: Academic Press.

Langen, R., Oh, K. J., Cascio, D., & Hubbell, W. L. (2000). Crystal structures of spin labeled T4 lysozyme mutants: Implications for the interpretation of EPR spectra in terms of structure. *Biochemistry*, *39*, 8396–8405.

Liang, B., Bushweller, J. H., & Tamm, L. K. (2006). Site-directed parallel spin-labeling and paramagnetic relaxation enhancement in structure determination of membrane proteins by solution NMR spectroscopy. *Journal of the American Chemical Society*, *128*, 4389–4397.

Lo, A., Cheng, C. W., Chiu, Y. Y., Sung, T. Y., & Hsu, W. L. (2011). TMPad: An integrated structural database for helix-packing folds in transmembrane proteins. *Nucleic Acids Research*, *39*, D347–D355.

Lo, A., Chiu, Y. Y., Rodland, E. A., Lyu, P. C., Sung, T. Y., & Hsu, W. L. (2009). Predicting helix-helix interactions from residue contacts in membrane proteins. *Bioinformatics*, *25*, 996–1003.

Lo, R. H., Kroncke, B. M., Solomon, T. L., & Columbus, L. (2014). Mapping membrane protein backbone dynamics: A comparison of site directed spin labeling to NMR ^{15}N-relaxation measurements. *Biophysical Journal*, *107*(7), 1697–1702.

Lomize, M. A., Lomize, A. L., Pogozheva, I. D., & Mosberg, H. I. (2006). OPM: Orientations of proteins in membranes database. *Bioinformatics*, *22*, 623–625.

Machonkin, T. E., Westler, W. M., & Markley, J. L. (2002). (13)C[(13)C] 2D NMR: A novel strategy for the study of paramagnetic proteins with slow electronic relaxation rates. *Journal of the American Chemical Society*, *124*, 3204–3205.

Machonkin, T. E., Westler, W. M., & Markley, J. L. (2004). Strategy for the study of paramagnetic proteins with slow electronic relaxation rates by nmr spectroscopy: Application to oxidized human [2Fe-2S] ferredoxin. *Journal of the American Chemical Society*, *126*, 5413–5426.

Madl, T., Felli, I. C., Bertini, I., & Sattler, M. (2010). Structural analysis of protein interfaces from ^{13}C direct-detected paramagnetic relaxation enhancements. *Journal of the American Chemical Society*, *132*, 7285–7287.

Maslennikov, I., Klammt, C., Hwang, E., Kefala, G., Okamura, M., Esquivies, L., et al. (2010). Membrane domain structures of three classes of histidine kinase receptors by cell-free expression and rapid NMR analysis. *Proceedings of the National Academy of Sciences of the United States of America*, *107*, 10902–10907.

McHaourab, H. S., Lietzow, M. A., Hideg, K., & Hubbell, W. L. (1996). Motion of spin-labeled side chains in T4 lysozyme. Correlation with protein structure and dynamics. *Biochemistry*, *35*, 7692–7704.

Pan, Y., Piyadasa, H., O'Neil, J. D., & Konermann, L. (2012). Conformational dynamics of a membrane transport protein probed by H/D exchange and covalent labeling: The glycerol facilitator. *Journal of Molecular Biology*, *416*, 400–413.

Parker, J. L., & Newstead, S. (2014). Method to increase the yield of eukaryotic membrane protein expression in *Saccharomyces cerevisiae* for structural and functional studies. *Protein Science*, *23*, 1309–1314.

Perozo, E., Kloda, A., Cortes, D. M., & Martinac, B. (2001). Site-directed spin-labeling analysis of reconstituted MscL in the closed state. *The Journal of General Physiology*, *118*, 193–206.

Reckel, S., Gottstein, D., Stehle, J., Lohr, F., Verhoefen, M. K., Takeda, M., et al. (2011). Solution NMR structure of proteorhodopsin. *Angewandte Chemie (International Ed. in English)*, *50*, 11942–11946.

Religa, T. L., & Kay, L. E. (2010). Optimal methyl labeling for studies of supra-molecular systems. *Journal of Biomolecular NMR*, *47*, 163–169.

Rosato, A., Tejero, R., & Montelione, G. T. (2013). Quality assessment of protein NMR structures. *Current Opinion in Structural Biology*, *23*, 715–724.

Schimerlik, M. I. (2001). Overview of membrane protein solubilization. *Current Protocols in Neuroscience*, Chapter 5, Unit 5.9.

Shen, Y., & Bax, A. (2010). SPARTA+: A modest improvement in empirical NMR chemical shift prediction by means of an artificial neural network. *Journal of Biomolecular NMR*, *48*, 13–22.

Shih, S. C., Stoica, I., & Goto, N. K. (2008). Investigation of the utility of selective methyl protonation for determination of membrane protein structures. *Journal of Biomolecular NMR*, *42*, 49–58.

Shimba, N., Kovacs, H., Stern, A. S., Nomura, A. M., Shimada, I., Hoch, J. C., et al. (2004). Optimization of ^{13}C direct detection NMR methods. *Journal of Biomolecular NMR*, *30*, 175–179.

Solomon, I., & Bloembergen, N. (1956). Nuclear magnetic interactions in the HF molecule. *Journal of Chemical Physics*, *25*, 261–266.

Su, P. C., Si, W., Baker, D. L., & Berger, B. W. (2013). High-yield membrane protein expression from *E. coli* using an engineered outer membrane protein F fusion. *Protein Science*, *22*, 434–443.

Takeuchi, K., Sun, Z. Y., & Wagner, G. (2008). Alternate ^{13}C-^{12}C labeling for complete mainchain resonance assignments using C alpha direct-detection with applicability toward fast relaxing protein systems. *Journal of the American Chemical Society*, *130*, 17210–17211.

Van Horn, W. D., Kim, H. J., Ellis, C. D., Hadziselimovic, A., Sulistijo, E. S., Karra, M. D., et al. (2009). Solution nuclear magnetic resonance structure of membrane-integral diacylglycerol kinase. *Science*, *324*, 1726–1729.

Walters, R. F., & DeGrado, W. F. (2006). Helix-packing motifs in membrane proteins. *Proceedings of the National Academy of Sciences of the United States of America*, *103*, 13658–13663.

Wegener, C., Tebbe, S., Steinhoff, H. J., & Jung, H. (2000). Spin labeling analysis of structure and dynamics of the Na(+)/proline transporter of *Escherichia coli*. *Biochemistry*, *39*, 4831–4837.

Wu, E. L., Cheng, X., Jo, S., Rui, H., Song, K. C., Davila-Contreras, E. M., et al. (2014). CHARMM-GUI membrane builder toward realistic biological membrane simulations. *Journal of Computational Chemistry*, *35*(27), 1997–2004.

Zhou, Y., Cierpicki, T., Jimenez, R. H., Lukasik, S. M., Ellena, J. F., Cafiso, D. S., et al. (2008). NMR solution structure of the integral membrane enzyme DsbB: Functional insights into DsbB-catalyzed disulfide bond formation. *Molecular Cell*, *31*, 896–908.

SECTION IV

Crystallization of Membrane Proteins

CHAPTER SEVENTEEN

Inducing Two-Dimensional Crystallization of Membrane Proteins by Dialysis for Electron Crystallography

Yusuf M. Uddin*, Ingeborg Schmidt-Krey*,†,1

*School of Biology, Georgia Institute of Technology, Atlanta, Georgia, USA

†School of Chemistry and Biochemistry, Georgia Institute of Technology, Atlanta, Georgia, USA

[1]Corresponding author: e-mail address: ingeborg.schmidt-krey@biology.gatech.edu

Contents

Abstract

Electron crystallography is an electron cryo-microscopy (cryo-EM) method that is particularly suitable for structure–function studies of small membrane proteins, which are crystallized in two-dimensional (2D) arrays for subsequent cryo-EM data collection and image processing. This approach allows for structural analysis of membrane proteins in a close-to-native, phospholipid bilayer environment. The process of growing 2D crystals from purified membrane proteins by dialysis detergent removal is described in this chapter. A short section covers screening for and identifying 2D crystals by transmission electron microscopy, and in the last section, optimization of the purification to obtain crystals of higher quality is discussed.

1. INTRODUCTION

Membrane proteins play critical roles in the form of receptors, transporters, enzymes, and anchors as well as in cell identification and adhesion. As with any protein, one of the important approaches to understanding function is to solve the structure of the membrane protein. Major techniques

Methods in Enzymology, Volume 557
ISSN 0076-6879
http://dx.doi.org/10.1016/bs.mie.2014.12.022

in the field of structural biology include X-ray diffraction, nuclear magnetic resonance spectroscopy, and electron cryo-microscopy (cryo-EM). Via the cryo-EM method electron crystallography, it is possible to study the structure and function by crystallizing the membrane protein of interest within a two-dimensional (2D) phospholipid bilayer (Glaeser, 1985; Kühlbrandt, 1992; Kühlbrandt, Schägger, & Hunte, 2003). This is a particularly useful technique for membrane proteins, as the membrane bilayer constitutes an environment that closely resembles physiological conditions and can greatly stabilize these often-fragile proteins. Purification of membrane proteins requires detergents to remove them from their native membrane (Vinothkumar, Edwards, & Standfuss, 2013). This may lead to inactivation of the protein, aggregation, and even unfolding, especially over time. 2D crystallization by dialysis reconstitutes membrane proteins into a lipid bilayer and frequently confers stability even in the early stages of dialysis due to the presence of phospholipids (Dreaden et al., 2013; Schmidt-Krey, 2007). The crystal symmetry depends on crystallization factors and the type of protein being crystallized. Well-characterized examples of successful 2D crystallization and structural studies of membrane proteins include Aquaporin-0, bacteriorhodopsin, the nicotinic acetylcholine receptor, light harvesting complex II, and several ATPases as well as transporters (Coudray et al., 2013; Gonen, Sliz, Kistler, Cheng, & Walz, 2004; Grigorieff, Ceska, Downing, Baldwin, & Henderson, 1996; Henderson & Unwin, 1975; Kühlbrandt, Wang, & Fujiyoshi, 1994; Miyazawa, Fujiyoshi, & Unwin, 2003; Pomfret, Rice, & Stokes, 2007; Subramaniam & Henderson, 1999; Unwin, 2005; Unwin & Henderson, 1975).

Due to the recent development of direct electron detectors for cryo-EM data collection, single-particle cryo-EM is rapidly developing for the study of structures of both soluble and membrane proteins at increasingly high resolutions, and in the range of resolutions that could previously only be attained by crystallographic methods (Allegretti, Mills, McMullan, Kühlbrandt, & Vonck, 2014; Li et al., 2013; Liao, Cao, Julius, & Cheng, 2014; McMullan, Chen, Henderson, & Faruqi, 2009). One major limitation to imaging individual proteins is that they must be of a size sufficient for identification and alignment (Ludtke, Baldwin, & Chiu, 1999), even though the use of direct electron detectors now allows for analysis of increasingly smaller particles than was previously possible with film and charge-coupled device cameras (Liao et al., 2014). In many instances, membrane proteins of interest are still too small for single-particle cryo-EM. Therefore, proteins of this nature can be crystallized into a 2D lattice for electron crystallographic analysis.

Though the formation of ordered arrays can be induced by different methods, dialysis is the most commonly used approach and will be the focus of this chapter. Dialysis is so frequently employed to generate 2D crystals due to its past success in producing highly ordered crystals (Abeyrathne, Chami, Pantelic, Goldie, & Stahlberg, 2010; Hasler, Heymann, Engel, Kistler, & Walz, 1998; Kühlbrandt, 1992; Vink, Derr, Love, Stokes, & Ubarretxena-Belandia, 2007), as well as the relatively large quantities of highly stable crystals that can be produced (Johnson et al., 2013). A membrane protein that has been purified and solubilized in detergent is mixed with lipids, and the detergent is removed by dialysis against a detergent-free buffer, resulting in reconstitution of the protein into a phospholipid bilayer and ordering under optimal conditions. Data is then collected by cryo-EM, and image processing allows for the calculation of projection maps and the three-dimensional (3D) structure by electron crystallographic methods (Crowther, Henderson, & Smith, 1996; Gipson, Zeng, Zhang, & Stahlberg, 2007; Wisedchaisri & Gonen, 2013; Wisedchaisri, Reichow, & Gonen, 2011). Occasionally, 2D crystals will stack, which precluded structure determination until the very recent development of microED (Iadanza & Gonen, 2014; Nannenga & Gonen, 2014; Nannenga, Shi, Leslie, & Gonen, 2014; Shi, Nannenga, Iadanza, & Gonen, 2013). MicroED allows for electron diffraction data collection and analysis of small 3D crystals grown by either 2D or 3D crystallization approaches. Thus, while primarily aimed at the formation of 2D ordered arrays and structural analysis by electron crystallography, the protocols described here may also lead to structure determination by microED in the case of 2D crystal stacking.

2. 2D CRYSTALLIZATION BY DIALYSIS

2D crystallization trials are conducted once the membrane protein of interest can be purified reproducibly and in sufficient quantities. At this point, the protein is solubilized in detergent, and the overall goal is to remove the detergent and replace it with a phospholipid bilayer under conditions that favor 2D crystallization of the protein within the newly formed membrane. Conditions such as the amount and type of reconstituting lipid as well as dialysis buffer components, such as glycerol and salt, can have an impact on whether the membrane protein will crystallize (Hasler et al., 1998; Johnson et al., 2013; Kühlbrandt et al., 2003; Schmidt-Krey, 2007).

Prior to and in preparation for dialysis, the purified protein is mixed with a precise amount of lipid and placed into dialysis tubing with a pore size large enough to remove detergent monomers, but small enough to retain the protein and most of the added lipid (Johnson et al., 2013; Kühlbrandt, 1992; Kühlbrandt et al., 2003; Schmidt-Krey, 2007). The amount of lipid to be added is defined by the lipid-to-protein ratio (LPR). This unit can be expressed in molar units or by weight. The LPR is an important factor when determining optimal conditions for crystallization (Kühlbrandt, 1992).

There are several ways to reconstitute a protein within a bilayer while removing the detergent. Dialysis "machines" typically allow for parallel testing of multiple potential crystallization conditions (Jap et al., 1992; Renault et al., 2006; Stokes, Rice, Hu, Kim, & Ubarretxena-Belandia, 2010). Bio-Beads rapidly remove detergent and have proven very effective for some membrane proteins that were only available in small amounts (Lévy, Bluzat, Seigneuret, & Rigaud, 1990; Rigaud et al., 1997). Dialysis buttons, such as the kinds also used for 3D crystallization, allow for trials of a range of volumes, depending on the sample (Wisedchaisri et al., 2011). Cyclodextrins can be added to the mixture to improve detergent removal (Signorell, Kaufmann, Kukulski, Engel, & Rémigy, 2007). In this chapter, we will discuss the use of a simple, and in many cases, equally effective method of detergent removal using dialysis membrane tubing.

The first step is to set aside all glassware that will be used for the experiment. One 400 ml beaker is used for each condition. The glassware should be washed, rinsed, and dried thoroughly, making sure not to introduce any oils, detergents, and other contaminants before the experiment. Next, cut 8 cm or more of membrane tubing with a molecular weight cutoff below the size of the protein. Soak the tubing in deionized water to both hydrate it and to remove any possible chemicals used for storage. Prepare 250 ml of detergent-free dialysis buffer for each 2D crystallization condition to be tested. As a starting point, the detergent-free dialysis buffer typically contains the same components that were used during the last purification step and for storage of the protein. With the protein sample set aside on ice, place a series of Eppendorf tubes on ice (room temperature for membrane proteins that are not temperature sensitive), and add the detergent-solubilized membrane protein sample, in the range of 50–100 μl and at a concentration around 1 mg/ml (Johnson et al., 2013; Kühlbrandt, 1992; Schmidt-Krey, 2007), into each Eppendorf tube. Volumes as low as 50–75 μl are mostly chosen for proteins that are difficult and/or expensive to purify. Next, add the appropriate amount of lipid to each Eppendorf tube. Generally, a good starting point is an LPR of 0 and a very high LPR of 40 or more, depending on the size of the membrane protein. At LPR 0, an abundance of copurified lipid will result in the formation of proteoliposomes, and likely will require additional purification steps (see below). Testing a very high and vastly excessive LPR, such as 40 or above, provides information on the suitability of conditions for reconstitution. Any LPR range in between will serve the purpose of identifying the most optimal condition for the membrane protein 2D crystallization. A commonly used lipid for 2D crystallization is 1,2-dimyristoyl-*sn*-glycero-3-phosphocholine due to its size and transition temperature, thus it is useful to start with this phospholipid if there is no previous crystallization or reconstitution data (Kühlbrandt et al., 2003). It is not recommended to use a diluted stock of lipid; rather, 10 mg/ml or 1 mg/ml, depending on the LPR, is common to avoid altering the conditions of the protein sample. In addition, consider creating your stock lipids in a solution containing the detergent that your protein is suspended in. Note that it is helpful to add protein before adding lipid because the lipid amounts are often very small (when using 1 or 10 mg/ml), and it will be easier to remove them from the pipette tip completely by pipetting up and down several times in the protein solutions. Vortex the mixture for several seconds and allow the protein to incubate with the lipid in the Eppendorf tubes for 10 min before you vortex the sample again (Johnson et al., 2013).

At this point, you will want to disinfect your gloves and be sure any lipids or oils from the outside environment do not come in contact during the next few steps. Use 70% ethanol to wipe down the bench and clean the gloves. Attach one dialysis membrane clip to the bottom of the 8 cm piece of dialysis tubing, with roughly 5 mm of tubing exposed above the clip (Fig. 1). Flick the tubing vigorously by holding on to the clip to remove water from the tubing, before pipetting the protein–lipid mixture into the dialysis tubing. The second clip is then attached at the top of the tubing perpendicular to the first clip. The angle ensures that the sample will stay immersed in the dialysis buffer and place the sample inside. Fill a beaker with 250 ml of dialysis buffer. Cover each beaker with foil and label it. Incubate the beakers at a temperature above the phase transition temperature of the lipid, even though lower temperatures may also be considered for highly fragile membrane proteins (Schmidt-Krey, Haase, Mutucumarana, Stafford, & Kühlbrandt, 2007). The temperature and incubation length are usually determined empirically; however, a good starting point is room temperature incubations for 3–7 days, depending on the type and concentration of the detergent (Johnson et al., 2013; Johnson & Schmidt-Krey, 2013; Schmidt-Krey, 2007).

Figure 1 Membrane tubing containing 100 μl of sample to be dialyzed against 250 ml of dialysis buffer. The plastic clip on the top should be positioned perpendicular to the opposite clip in order to keep the sample submerged during dialysis. Foil or a large Petri dish is placed over the beaker to prevent contamination and evaporation.

The critical micelle concentration (CMC) determines whether or not a detergent can be easily removed (Hasler et al., 1998). Detergents with low CMCs can often take more than 1 week to remove. Sometimes it may be helpful to dilute the sample at a 1:1 or 1:2 ratio with water or MES/HEPES buffer to reduce some of the detergent-protein binding before incubation with lipids and dialysis (Rémigy et al., 2003; Wang & Kühlbrandt, 1991). At the end of the incubation, the dialysis tubing is cut with a razor blade, and the sample is stored in an Eppendorf tube. Electron microscope (EM) grids are immediately prepared by negative staining with 1–2% uranyl acetate.

3. EM SCREENING OF 2D CRYSTALLIZATION CONDITIONS

Screening a batch of purified and solubilized membrane protein before dialysis may provide information on either protein aggregation or dispersed particles if the detergent is still present, and is thus often used to prescreen a sample before 2D crystallization is attempted. This will give important guidance for further optimization of the purification.

While automated screening is being developed and offers promise for 2D crystallization trials (Stokes, Ubarretxena-Belandia, Gonen, & Engel, 2013; Suloway et al., 2005), visual screening at the EM is common and, at least currently, much more inexpensive. Whether automated or not, screening for optimal 2D crystallization conditions involves several steps. First, the dialysis sample is evaluated for proper removal of the detergent. High detergent concentrations will already become visible when the grid is negatively stained since the detergent will break the surface tension of the sample droplet on the grid (Fig. 2). Even low concentrations of detergent may lead to membrane deterioration after just days of storage, which is an added incentive to immediately prepare EM grids after sampling of the dialysis. Next, the negative control with an LPR of 0 will be evaluated. The grid is expected to only contain protein aggregates. For a detailed analysis of successful lipid removal during purification, thin layer chromatography and/or mass spectrometry are used after the purification and before dialysis (Petković et al., 2001; Skipski, Peterson, & Barclay, 1964). However, a fast and easy way to find out whether substantial amount of lipids have been copurified, is to image the sample at the electron microscope. Samples with excessive amount of copurified lipid will often contain vesicles or other membrane structures at LPRs of 0. Higher LPRs as well as samples under different buffer conditions and dialysis parameters, such as temperature and time, are then screened. At low magnification, note is taken of membrane morphology and size. The morphologies of membranes within which the crystals are commonly found are in the form of vesicles, tubes, or sheets (Fig. 3). Wider tubular, or rather planar–tubular, crystals look similar to vesicles but

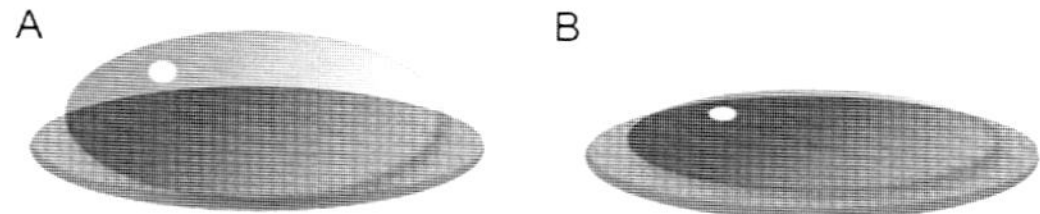

Figure 2 First information on detergent removal after dialysis may be gained during EM grid preparation of the protein sample. Approximately 2 μl of postdialysis sample is added onto a carbon-coated copper grid. The sample in (A) depicts better detergent removal due to the higher surface tension compared to (B), which illustrates the appearance of a sample droplet that still contains detergent. The surface tension is broken, and the sample rapidly spreads across the grid. Crystals that have formed in a sample with detergent still present may be subject to deterioration over time if stored at room temperature, or potentially even lower crystal quality.

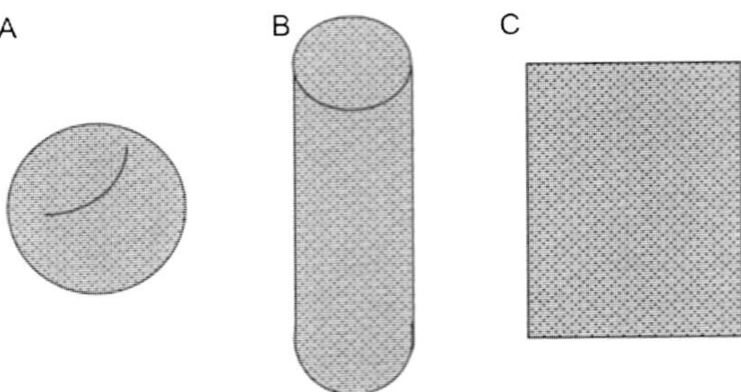

Figure 3 Proteoliposomes and 2D crystals can vary in size and shape; however, they will exist as one or more of the following morphologies: vesicles, planar tubes, and sheets. (A) Vesicles and (B) planar–tubular crystals are similar in structure, but tubes are much more elongated. (C) Lipid sheets are flat, single-layer lipid bilayers.

have become elongated during crystal growth. Tubes with very narrow diameters can be helical crystals and do not collapse into two bilayers on the EM grid, such as vesicles and wider tubes would. Crystal sheets differ from the other morphologies as one bilayer is seen on the EM grid as opposed to the two of vesicles and planar–tubular crystals.

Different membrane morphologies are then evaluated at high magnification. Depending on the membrane size, high magnification is frequently in the range of 30–60 K magnification. Images are then evaluated by online fast Fourier transform (FFTs) for order. Live FFTs of smaller image areas have the advantage that even smaller 2D crystals within larger membranes can be identified due to a better signal-to-noise ratio. Identification of even small ordered arrays is likely to rapidly lead to identification of conditions for growing larger 2D crystals.

4. OPTIMIZATION OF 2D CRYSTALLIZATION AND PROTEIN PURIFICATION

In the 2D crystallization of membrane proteins, the perfect condition for crystals is rarely discovered in the first trial. Before or after screening different 2D crystallization parameters and identifying promising conditions, one may consider various steps towards crystal formation as well as optimizing quality and size. Parameters, such as salt type and concentration, glycerol, type of reconstituting lipid, or a mixture of types, are tested and screened (Johnson et al., 2013). One additional important step may involve the optimization of the protein purification as well as the removal of copurified lipids. To generate high quality crystals, it is important that the protein be as pure as possible. A monodisperse sample contains protein of identical structure and dimensions. To achieve monodispersity, one usually considers a "polishing" step. Gel filtration chromatography can be used to separate particles by size.

A minimal amount of copurified lipid may be permissible for 2D crystallization and even necessary to prevent instability (Nußberger, Dörr, Wang, & Kühlbrandt, 1993). Some proteins may form 2D crystals even if there is a limited amount of copurified lipids. However, if one needed to remove more lipids due to lack of control over the crystallization, one or more of the following steps could be applied.

Hydrophobic interaction columns (HIC) are one of many chromatography methods of purification. The lipid interaction with the protein is weakened under high salt conditions. Thus, as the sample is added into the HIC column, the lipids are exchanged into the bulk solution and bound tightly to the surface of the ligands, while the protein is washed away with lower salt concentrations (Velkov, Lim, Capuano, & Prankerd, 2008).

Another way to optimize the purification of membrane proteins is to resuspend the protein in buffer with detergent to further remove copurified lipids. As detergent levels increases, the protein–lipid belt will begin to be exchanged with detergent to separate individual particles of membrane protein. Then the excess, nonbound lipids can be washed away. If the protein is purified using an affinity purification, such as Ni-NTA chromatography, it is recommended to incorporate additional washing steps to elute contaminants before the actual protein of interest is eluted. Other "washing" methods include spinning the sample down into a pellet and resuspension in fresh buffer several times. Phospholipase A_2 enzyme treatment is another option for lipid removal as it is able to break down the lipids and help disassociate them from the protein (Nußberger et al., 1993). In addition, a way to "polish" the protein purification and improve crystal quality is to apply a size-exclusion chromatography step to separate the protein from large and small impurities.

Increasing purity by removing lipids might seem like an obvious optimization step, but it is sometimes necessary to understand the lipid composition of the protein of interest. Some membrane proteins require the presence of their natural lipids in order to maintain structural and functional integrity, such as the light harvesting complex II and photosystem II (Cevc, 1987; Nußberger et al., 1993).

5. CONCLUSION

Membrane proteins are ubiquitous in organisms and their functions are vital for life. Thus, understanding the structure of these proteins is relevant for a wide range of studies of membrane proteins. This chapter outlines the critical steps and parameters in inducing and identifying 2D crystals for the subsequent structure determination by electron crystallography via cryo-EM data collection and image processing. Optimization of the protein purification and removal of copurified lipids will be a critical step before or after the first crystallization trials, and the best conditions will usually be determined after screening for and identifying parameters by EM to achieve optimal crystal quality and size.

The major bottlenecks of electron crystallography are constituted by the purification and 2D crystallization, yet progress in identifying key parameters for 2D crystallization and automation will contribute to the future success of this method (Abeyrathne et al., 2010; Johnson et al., 2013; Stokes et al., 2013; Wisedchaisri & Gonen, 2013).

REFERENCES

Abeyrathne, P. D., Chami, M., Pantelic, R. S., Goldie, K. N., & Stahlberg, H. (2010). Chapter one—Preparation of 2D crystals of membrane proteins for high-resolution electron crystallography data collection. In J. J. Grant (Ed.), *Methods in Enzymology: Vol. 481* (pp. 25–43). Amsterdam, Netherlands: Academic Press.

Allegretti, M., Mills, D. J., McMullan, G., Kühlbrandt, W., & Vonck, J. (2014). Atomic model of the F420-reducing [NiFe] hydrogenase by electron cryo-microscopy using a direct electron detector. *Elife*, *3*, e01963.

Cevc, G. (1987). How membrane chain melting properties are regulated by the polar surface of the lipid bilayer. *Biochemistry*, *26*(20), 6305–6310.

Coudray, N., Valvo, S., Hu, M., Lasala, R., Kim, C., Vink, M., et al. (2013). Inward-facing conformation of the zinc transporter YiiP revealed by cryoelectron microscopy. *Proceedings of the National Academy of Sciences*, *110*(6), 2140–2145.

Crowther, R. A., Henderson, R., & Smith, J. M. (1996). MRC image processing programs. *Journal of Structural Biology*, *116*(1), 9–16.

Dreaden, T. M., Metcalfe, M., Kim, L. Y., Johnson, M. C., Barry, B. A., & Schmidt-Krey, I. (2013). Screening for two-dimensional crystals by transmission electron microscopy of negatively stained samples. *Methods in Molecular Biology*, *955*, 73–101.

Gipson, B., Zeng, X., Zhang, Z. Y., & Stahlberg, H. (2007). 2dx—User-friendly image processing for 2D crystals. *Journal of Structural Biology*, *157*(1), 64–72.

Glaeser, R. M. (1985). Electron crystallography of biological macromolecules. *Annual Review of Physical Chemistry*, *36*(1), 243–275.

Gonen, T., Sliz, P., Kistler, J., Cheng, Y., & Walz, T. (2004). Aquaporin-0 membrane junctions reveal the structure of a closed water pore. *Nature*, *429*(6988), 193–197.

Grigorieff, N., Ceska, T. A., Downing, K. H., Baldwin, J. M., & Henderson, R. (1996). Electron-crystallographic refinement of the structure of bacteriorhodopsin. *Journal of Molecular Biology*, *259*(3), 393–421.

Hasler, L., Heymann, J. B., Engel, A., Kistler, J., & Walz, T. (1998). 2D crystallization of membrane proteins: Rationales and examples. *Journal of Structural Biology, 121*(2), 162–171.

Henderson, R., & Unwin, P. N. T. (1975). Three-dimensional model of purple membrane obtained by electron microscopy. *Nature, 257*(5521), 28–32.

Iadanza, M. G., & Gonen, T. (2014). A suite of software for processing MicroED data of extremely small protein crystals. *Journal of Applied Crystallography, 47*(3), 1140–1145.

Jap, B. K., Zulauf, M., Scheybani, T., Hefti, A., Baumeister, W., Aebi, U., et al. (1992). 2D crystallization: From art to science. *Ultramicroscopy, 46*(1–4), 45–84.

Johnson, M. C., Dreaden, T. M., Kim, L. Y., Rudolph, F., Barry, B. A., & Schmidt-Krey, I. (2013). Two-dimensional crystallization of membrane proteins by reconstitution through dialysis. *Methods in Molecular Biology, 955*, 31–58.

Johnson, M. C., & Schmidt-Krey, I. (2013). Two-dimensional crystallization by dialysis for structural studies of membrane proteins by the cryo-EM method electron crystallography. *Methods in Cell Biology, 113*, 325–337.

Kühlbrandt, W. (1992). Two-dimensional crystallization of membrane proteins. *Quarterly Reviews of Biophysics, 25*(01), 1–49.

Kühlbrandt, W., Schägger, H., & Hunte, C. (2003). *Membrane protein purification and crystallization: A practical approach*. San Diego, CA: Academic Press.

Kühlbrandt, W., Wang, D., & Fujiyoshi, Y. (1994). Atomic model of plant light-harvesting complex by electron crystallography. *Nature, 367*(6464), 614–621.

Lévy, D., Bluzat, A., Seigneuret, M., & Rigaud, J.-L. (1990). A systematic study of liposome and proteoliposome reconstitution involving Bio-bead-mediated triton X-100 removal. *Biochimica et Biophysica Acta (BBA)-Biomembranes, 1025*(2), 179–190.

Li, X., Mooney, P., Zheng, S., Booth, C. R., Braunfeld, M. B., Gubbens, S., et al. (2013). Electron counting and beam-induced motion correction enable near-atomic-resolution single-particle cryo-EM. *Nature Methods, 10*(6), 584–590.

Liao, M., Cao, E., Julius, D., & Cheng, Y. (2014). Single particle electron cryo-microscopy of a mammalian ion channel. *Current Opinion in Structural Biology, 27*, 1–7.

Ludtke, S. J., Baldwin, P. R., & Chiu, W. (1999). EMAN: Semiautomated software for high-resolution single-particle reconstructions. *Journal of Structural Biology, 128*(1), 82–97.

McMullan, G., Chen, S., Henderson, R., & Faruqi, A. (2009). Detective quantum efficiency of electron area detectors in electron microscopy. *Ultramicroscopy, 109*(9), 1126–1143.

Miyazawa, A., Fujiyoshi, Y., & Unwin, N. (2003). Structure and gating mechanism of the acetylcholine receptor pore. *Nature, 423*(6943), 949–955.

Nannenga, B. L., & Gonen, T. (2014). Protein structure determination by MicroED. *Current Opinion in Structural Biology, 27*, 24–31.

Nannenga, B. L., Shi, D., Leslie, A. G. W., & Gonen, T. (2014). High-resolution structure determination by continuous-rotation data collection in MicroED. *Nature Methods, 11*(9), 927–930.

Nußberger, S., Dörr, K., Wang, D. N., & Kühlbrandt, W. (1993). Lipid-protein interactions in crystals of plant light-harvesting complex. *Journal of Molecular Biology, 234*(2), 347–356.

Petković, M., Schiller, J., Müller, M., Benard, S., Reichl, S., Arnold, K., et al. (2001). Detection of individual phospholipids in lipid mixtures by matrix-assisted laser desorption/ionization time-of-flight mass spectrometry: Phosphatidylcholine prevents the detection of further species. *Analytical Biochemistry, 289*(2), 202–216.

Pomfret, A. J., Rice, W. J., & Stokes, D. L. (2007). Application of the iterative helical real-space reconstruction method to large membranous tubular crystals of P-type ATPases. *Journal of Structural Biology, 157*(1), 106–116.

Rémigy, H. W., Caujolle-Bert, D., Suda, K., Schenk, A., Chami, M., & Engel, A. (2003). Membrane protein reconstitution and crystallization by controlled dilution. *FEBS Letters, 555*(1), 160–169.

Renault, L., Chou, H.-T., Chiu, P.-L., Hill, R. M., Zeng, X., Gipson, B., et al. (2006). Milestones in electron crystallography. *Journal of Computer-Aided Molecular Design, 20*(7–8), 519–527.

Rigaud, J.-L., Mosser, G., Lacapere, J.-J., Olofsson, A., Levy, D., & Ranck, J.-L. (1997). Bio-beads: An efficient strategy for two-dimensional crystallization of membrane proteins. *Journal of Structural Biology, 118*(3), 226–235.

Schmidt-Krey, I. (2007). Electron crystallography of membrane proteins: Two-dimensional crystallization and screening by electron microscopy. *Methods, 41*(4), 417–426.

Schmidt-Krey, I., Haase, W., Mutucumarana, V., Stafford, D. W., & Kühlbrandt, W. (2007). Two-dimensional crystallization of human vitamin K-dependent γ-glutamyl carboxylase. *Journal of Structural Biology, 157*(2), 437–442.

Shi, D., Nannenga, B. L., Iadanza, M. G., & Gonen, T. (2013). Three-dimensional electron crystallography of protein microcrystals. *Elife, 2*.

Signorell, G. A., Kaufmann, T. C., Kukulski, W., Engel, A., & Rémigy, H.-W. (2007). Controlled 2D crystallization of membrane proteins using methyl-β-cyclodextrin. *Journal of Structural Biology, 157*(2), 321–328.

Skipski, V., Peterson, R., & Barclay, M. (1964). Quantitative analysis of phospholipids by thin-layer chromatography. *Biochemical Journal, 90*(2), 374.

Stokes, D. L., Rice, W. J., Hu, M., Kim, C., & Ubarretxena-Belandia, I. (2010). Two-dimensional crystallization of integral membrane proteins for electron crystallography. *Methods in Molecular Biology, 654*, 187–205, Springer.

Stokes, D. L., Ubarretxena-Belandia, I., Gonen, T., & Engel, A. (2013). High-throughput methods for electron crystallography. *Methods in Molecular Biology, 955*, 273–296.

Subramaniam, S., & Henderson, R. (1999). Electron crystallography of bacteriorhodopsin with millisecond time resolution. *Journal of Structural Biology, 128*(1), 19–25.

Suloway, C., Pulokas, J., Fellmann, D., Cheng, A., Guerra, F., Quispe, J., et al. (2005). Automated molecular microscopy: The new leginon system. *Journal of Structural Biology, 151*(1), 41–60.

Unwin, N. (2005). Refined structure of the nicotinic acetylcholine receptor at 4 Å resolution. *Journal of Molecular Biology, 346*(4), 967–989.

Unwin, N., & Henderson, R. (1975). Molecular structure determination by electron microscopy of unstained crystalline specimens. *Journal of Molecular Biology, 94*(3), 425–440.

Velkov, T., Lim, M. L., Capuano, B., & Prankerd, R. (2008). A protocol for the combined sub-fractionation and delipidation of lipid binding proteins using hydrophobic interaction chromatography. *Journal of Chromatography B, 867*(2), 238–246.

Vink, M., Derr, K. D., Love, J., Stokes, D. L., & Ubarretxena-Belandia, I. (2007). A high-throughput strategy to screen 2D crystallization trials of membrane proteins. *Journal of Structural Biology, 160*(3), 295–304.

Vinothkumar, K. R., Edwards, P. C., & Standfuss, J. (2013). Practical aspects in expression and purification of membrane proteins for structural analysis. *Methods in Molecular Biology, 955*, 17–30, Springer.

Wang, D. N., & Kühlbrandt, W. (1991). High-resolution electron crystallography of light-harvesting chlorophyll ab-protein complex in three different media. *Journal of Molecular Biology, 217*(4), 691–699.

Wisedchaisri, G., & Gonen, T. (2013). Phasing electron diffraction data by molecular replacement: Strategy for structure determination and refinement. *Methods in Molecular Biology, 955*, 243–272.

Wisedchaisri, G., Reichow, S., & Gonen, T. (2011). Advances in structural and functional analysis of membrane proteins by electron crystallography. *Structure, 19*(10), 1381–1393.

CHAPTER EIGHTEEN

Crystallization of Membrane Proteins by Vapor Diffusion

Jared A. Delmar*, Jani Reddy Bolla†, Chih-Chia Su*, Edward W. Yu*,†,1
*Department of Physics and Astronomy, Iowa State University, Ames, Iowa, USA
†Department of Chemistry, Iowa State University, Ames, Iowa, USA
[1]Corresponding author: e-mail address: ewyu@iastate.edu

Contents

Abstract

X-ray crystallography remains the most robust method to determine protein structure at the atomic level. However, the bottlenecks of protein expression and purification often discourage further study. In this chapter, we address the most common problems encountered at these stages. Based on our experiences in expressing and purifying antimicrobial efflux proteins, we explain how a pure and homogenous protein sample can be successfully crystallized by the vapor diffusion method. We present our current

Methods in Enzymology, Volume 557
ISSN 0076-6879
http://dx.doi.org/10.1016/bs.mie.2014.12.018

protocols and methodologies for this technique. Case studies show step-by-step how we have overcome problems related to expression and diffraction, eventually producing high-quality membrane protein crystals for structural determinations. It is our hope that a rational approach can be made of the often anecdotal process of membrane protein crystallization.

1. INTRODUCTION

As of this writing, nearly 100,000 protein structures have been deposited to the Protein Data Bank (PDB). Approximately 90% were solved by X-ray crystallography—by far the most successful technique used to study protein structure. However, of more than 30,000 unique X-ray structures, only approximately 1% represent membrane proteins (http://blanco.biomol.uci.edu/mpstruc/). In contrast to their share of the known structures, membrane proteins are estimated to comprise approximately 30% of genes (Krogh, Larsson, von Heijne, & Sonnhammer, 2001; Wallin & von Heijne, 1998). They play the critical roles of gatekeepers, mediating messages and materials moving into and out of the cell, for example, G-protein-coupled receptors, which are responsible for vision, taste, smell, and immune response in eukaryotes (Lohse, Benovic, Codina, Caron, & Lefkowitz, 1990; Palczewski et al., 2000); multidrug efflux transporters, which confer antibiotic resistance to pathogenic bacteria (Saier, Tam, Reizer, & Reizer, 1994); voltage-gated ion channels, such as those responsible for neuron firing and muscle contraction (Lee, Lee, Chen, & MacKinnon, 2005; Long, Tao, Campbell, & MacKinnon, 2007); and porins, through which bacteria, mitochondria, and chloroplasts exchange material, such as nutrients, with the extracellular environment (Cowan et al., 1992; Weiss et al., 1991). These proteins comprise at least 60% of all drug targets (Drew, Fröderberg, Baars, & de Gier, 2003; Overington, Al-Lazikani, & Hopkins, 2006).

Unfortunately, understanding the structures and action mechanisms of these important integral membrane proteins is often hampered by the difficulties associated with expression, purification, and crystallization. There are at least two nontrivial tasks in the way of straightforward application of X-ray crystallography to membrane proteins: (i) obtaining a sufficient quantity of purified protein and (ii) producing high-quality crystals (Bernaudat et al., 2011). The latter will be the focus of this chapter. However, as any practicing membrane protein crystallographer can attest to, obtaining diffraction-quality crystals can be a formidable task, which requires a pure and homogeneous protein sample. In this respect, sample preparation and crystallization are inseparable. Thus, a significant portion of this review will be devoted to the topics of membrane protein expression and purification. In addition, extensive examples and detailed discussions, based on our experiences in expressing, purifying, and crystallizing antimicrobial efflux proteins, will be provided. Hopefully, by introducing the available methodologies and protocols, a rational approach can be made to the often anecdotal process of membrane protein crystallization.

2. MEMBRANE PROTEIN EXPRESSION

With few exceptions, membrane proteins are not abundant in their native environment. For example, *Escherichia coli* cells were found to contain only eight transcripts per cell corresponding to the inner membrane copper transporter CusA, even after induction (Thieme, Neubauer, Nies, & Grass, 2008). Those few exceptions include the rhodopsins, porins, ATPases, photosynthetic reaction centers, and other light-harvesting complexes (Bill et al., 2011). Coincidentally, the first membrane protein structure determined by X-ray crystallography was the photosynthetic reaction center of *Rhodopseudomonas viridis* (Deisenhofer, Epp, Miki, Huber, & Michel, 1985). It was not until 1998, more than a decade later, that a membrane protein structure was obtained from a non-native, i.e., recombinant, source (Doyle et al., 1998; Wiener, 2004). To date, recombinant overexpression is the most common approach to obtain proteins of interest. Often, this step is the first bottleneck that must be overcome in order to produce the milligram quantities of pure protein necessary for a single crystallization experiment (Bernaudat et al., 2011; Bill et al., 2011).

2.1 Host

Thus far, *E. coli* remains the favorite host for heterologous protein expression. It grows quickly, tolerates a high cell density, and does so on a relatively lean diet (Sahdev, Khattar, & Saini, 2008). The protein yield from *E. coli* cells is also particularly high, compared to other organisms. In one experiment, the expression of 20 membrane proteins in six host organisms (three prokaryotic, including *E. coli*, and three eukaryotic) was measured (Bernaudat et al., 2011). Overall, the highest yields were achieved with *E. coli* cells. Further, only two membrane proteins tested were not successfully overexpressed in *E. coli*.

Bacterial proteins do tend to express better in *E. coli*, in general, but it is not a cure-all for heterologous membrane protein expression. Especially for eukaryotic membrane proteins, other host organisms must be considered (Bernaudat et al., 2011; Pryor & Leiting, 1997; Schertler, 1992). Often, eukaryotic proteins contain posttranslational modifications, such as phosphorylation, glycosylation, methylation, or acetylation, which a non-native host may be incapable of reproducing (Gabrielsen, Gardiner, Fromme, & Cogdell, 2009). *E. coli* and other bacterial hosts, in particular, tend to lack these posttranslational capabilities (Sahdev et al., 2008).

Translational machinery will also vary between the host and native organisms. *E. coli* cells exhibit a clear codon bias, with the population of tRNA approximately proportional to the chromosomal frequency (Kane, 1995). Occurrence of rare codons or many minor codons in the target protein sequence has a marked effect on expression levels. The reason for this is apparent when considering that an overexpressed membrane protein may represent a significant fraction of the total cellular protein. In some cases, even a single rare codon can be responsible for reduced expression levels (Kane, 1995). Accordingly, optimal expression of a protein might be achieved by optimizing the codons for the particular host. This can be a time-consuming process. For example, the *Neisseria gonorrhoeae* gene *norM*, encoding the multidrug efflux pump NorM, contains nine rare codons for *E. coli*. To overexpress the *N. gonorrhoeae* NorM membrane protein in *E. coli* TOP10 cells, our lab had to spend a few months fixing each of these codons (Long, Rouquette-Loughlin, Shafer, & Yu, 2008). Even the *E. coli* gene *acrD*, which produces the aminoglycoside efflux pump AcrD, has 15 rare codons. Thus, it is incompatible with *E. coli* host cells. We had to correct each of these 15 codons in order to overexpress this membrane protein in *E. coli* BL21(DE3) cells. Fortunately, several companies, such as Integrated DNA Technologies (IDT), offer protein codon optimization tools to correct for the codon bias encountered in many host cells.

Upon translation, membrane proteins must then be chaperoned from the cytoplasm and properly inserted into the corresponding membrane. This is of particular importance to outer membrane proteins of Gram-negative bacteria. A proper signaling peptide is essential to guide the expressed protein across the inner membrane and anchor to the outer membrane. Differences in the lipid bilayer structure and composition, as well as differences in insertion machinery, of the native organism compared to the host determine whether this process results in an active membrane protein or an intractable aggregate (Caffrey, 2003; Loll, 2003).

Finally, properly overexpressed, folded, and active membrane proteins tend to be toxic to their host (Bernaudat et al., 2011). The reason for this is obvious in the case of membrane channels or transporters, which directly control the permeability and delicate intracellular chemistry of their host, and can have direct roles in metabolism. In an experiment conducted with *Rhodopseudomonas blastica*, the overexpression of porins resulted directly in lysis of the host cell (Bannwarth & Schulz, 2003; Schmid, Krömer, & Schulz, 1996). Thus, the amount of inducer, such as isopropyl β-D-1-thiogalactopyranoside or arabinose, as well as the expression time and temperature, are often found to be critical variables for optimal protein expression.

Various strains of *E. coli* have been commercially developed to address the challenges encountered in membrane protein overexpression. Among those used in our lab, the BL21(DE3) strain (Novagen) is deficient in OmpT and lon proteases, which allows for increased protein stability (Sahdev et al., 2008); mutations in the *lac* promoter render C41(DE3) and C43(DE3) strains far less sensitive to the toxicity of membrane protein overexpression (Molina et al., 2008; Wagner et al., 2008); and the BL21-CodonPlus strain (Stratagene) contains additional copies of rare *E. coli* tRNAs.

In addition to using these specialized strains of *E. coli*, our lab also generates Δ*acrB* knockout strains by removing the *acrB* gene, which encodes the multidrug efflux pump AcrB. Deletions of abundant native membrane proteins have been shown to facilitate the purification process and produce a more homogenous protein (Bannwarth & Schulz, 2003). In many cases, without this specific deletion, our attempts at crystallizing membrane proteins would result in crystals of AcrB, instead.

2.2 Vector

In general, it is not easy to predict whether an expression vector is suitable for the protein of interest. Choosing a vector that is compatible with both the membrane protein target and host organism, and one that is tailored to the desired expression and purification systems, is typically a trial-and-error process. Luckily, many powerful expression vectors are commercially available. In particular, our lab likes the pET-type vectors (Novagen) for the expression of α-helical transmembrane proteins. For β-barrel transmembrane proteins, we have inordinate success with the pBAD-type vectors (Life Technologies).

If a particular protein cannot be expressed with any combination of vectors and host cells, one may want to think about expressing the protein fused with green fluorescent protein or maltose-binding protein. In many cases, the expression levels of these fusion proteins are much higher than the parental ones (Pryor & Leiting, 1997). It has also been reported that protein expression can be improved simply by single-point mutagenesis. To this end, Molina et al. subjected eight bacterial and one human membrane proteins to cycles of random mutagenesis (Molina et al., 2008). In five of these proteins, expression levels could be increased, as much as 40-fold, by one round of mutagenesis corresponding to approximately 6 mutations per 1000 residues. Further, the structure of the first β_1-adrenergic receptor was solved with one mutation, cysteine to lysine, which reportedly improved expression (Warne et al., 2008).

3. MEMBRANE PROTEIN PURIFICATION

Successfully overexpressed membrane proteins must be extracted and purified from the host's lipid membrane before crystallization, with the intention of providing as similar an environment as possible to the native bilayer (Fig. 1). At the same time, we wish to leave out all the properties

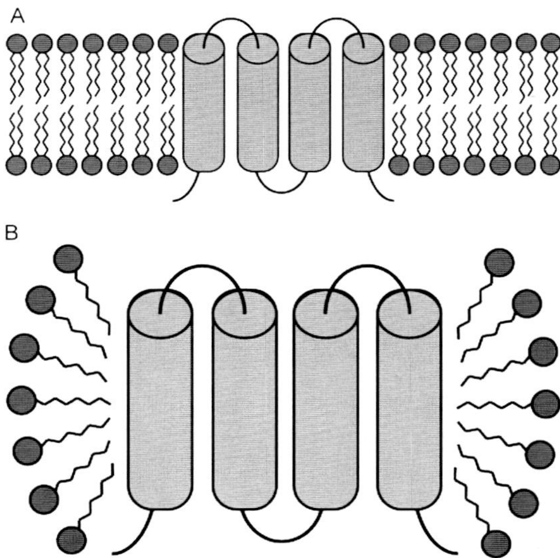

Figure 1 Cartoon representation cross-section of a membrane protein solubilized in (A) lipid bilayer of the host cell and (B) detergent introduced during the purification process. Upon removal of the protein from the native lipid (blue (light gray in the print version)) environment, the protein–detergent complex consists of a uniform disordered ring of detergent monomers (red (dark gray in the print version)), making hydrophobic contacts with the surface of the protein (green (light gray in the print version)).

that make the bilayer unsuitable for crystallization, such as heterogeneity, polarity, and charge (Caffrey, 2003). Ideally, these should be replaced by the most important indicators of crystallizability, monodispersity, and stability (Mancusso, Karpowich, Czyzewski, & Wang, 2011). This is a difficult task, indeed. As stated by Bowie, membrane proteins are "delicate weaklings, unable to withstand the rigors of life outside the safety of the bilayer" (Bowie, 2001). While detergent-based solubilization is the most popular and effective technique to replace the bilayer, most membrane proteins are only marginally stable even in the best available detergent (Zhou & Bowie, 2000).

The reason for their weakness is twofold: (i) detergent, often smaller than the lipid environment it is replacing, may not completely cover the largely hydrophobic surface of the membrane protein and (ii) membrane proteins, of diverse function, tend to exhibit much conformational flexibility, with some conformations more stable than others. Without the ordering imposed by the bilayer, they are free to adopt those conformations that may favor aggregation (Zhou & Bowie, 2000).

3.1 Detergent

Beginning with the first membrane protein crystal structure in 1980, detergents have been rationally designed especially for the purpose of facilitating protein purification and crystallization. The number of these detergents is now estimated in the 100s (Wiener, 2004). Nonetheless, only a small subset have been successful. The search for a suitable detergent benefits greatly from Internet-based databases, which compile the crystallization conditions of all membrane proteins to date. For example, "Membrane Proteins of Known Structure" (http://blanco.biomol.uci.edu/mpstruc/) is particularly helpful.

At least structurally, all detergents are similar—a hydrophilic headgroup joined to a hydrophobic tail. This amphipathic character allows the detergent monomers to interact with both the hydrophobic surfaces of membrane proteins and solvents, as well as each other (Garavito, Picot, & Loll, 1996). All detergents used for membrane protein crystallization are alkyl chain, between 7 and 14 carbons in length, with varying headgroups (Wiener, 2004). A detergent falls into one of three categories based on the properties of the headgroup: ionic, nonionic, or zwitterionic (Fig. 2). Importantly, the most effective families of mild detergents for purification and crystallization have nonionic or, less commonly, zwitterionic headgroups. Besides solubilization, other detergent effects on membrane proteins can include inactivation, denaturation, or aggregation (De Grip, 1982). In these cases, harsh

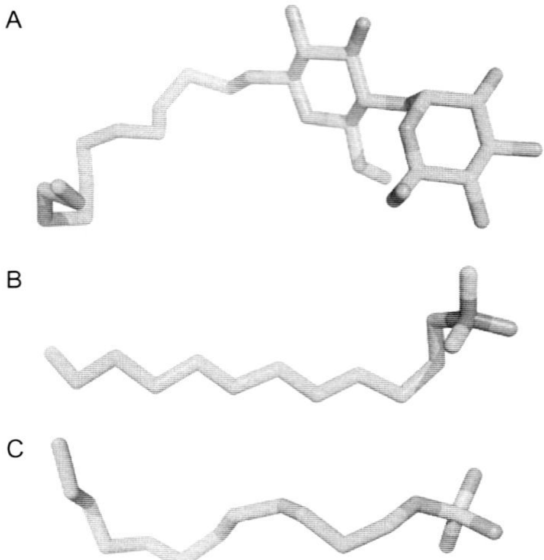

Figure 2 Representative structures of commonly used detergents of three classes. (A) *n*-Dodecyl-β-D-maltoside (DDM)—nonionic. (B) Lauryldimethylamine oxide (LDAO)—zwitterionic. (C) Sodium dodecyl sulfate (SDS)—anionic. Atoms are colored: C, green (gray in the print version); O, red (dark gray in the print version); N, blue (black in the print version); and S, yellow (light gray in the print version).

ionic detergents are often responsible. For example, the ionic detergent sodium dodecyl sulfate is used in gel electrophoresis for exactly the purpose of protein denaturation. As of the publication of their paper in 2008, Newstead et al. reported that no outer membrane proteins to date had been crystallized with ionic detergent (Newstead, Hobbs, Jordan, Carpenter, & Iwata, 2008).

To determine the effect of nonionic and zwitterionic detergents on stability, Sonoda et al. used six common detergents to solubilize 24 prokaryotic and eukaryotic membrane transporters (Sonoda et al., 2011). The average stability measured was greatest for the two longest chain lengths of detergent, *n*-dodecyl-β-D-maltoside (DDM) and dodecyl nonaethylene glycol ether ($C_{12}E_9$), both nonionic. In a similar study, Mancusso et al. examined 20 homologous lipid phosphatases (Mancusso et al., 2011). All 20 purified membrane proteins tended to be unstable in DDM solution, and all precipitated when concentrated over 1 mg/ml. When DDM was exchanged for other detergents, both the stability and monodispersity of the sample varied. In one of the tested samples, LP-1, both these traits improved when solubilized with 6-cyclohexyl-1-hexyl-β-D-maltoside (CYMAL-6). The size-exclusion chromatography peaks were sharper and the protein could be concentrated to 4 mg/ml. Taken together, these results indicate the preference of membrane proteins for long-chain, nonionic detergents. For instance, DDM, with its 12 alkyl carbons, is well known to stabilize membrane proteins and is often used for both solubilization and crystallization (Kang, Lee, & Drew, 2013).

3.2 Protein–detergent complex

After purification in the best available detergent, the membrane protein carries between 40% and 200% of its own weight in detergent (Garavito et al., 1996; Møller & Le Maire, 1993). Save one or two molecules, all of these detergent coats are believed to be unstructured (Deisenhofer & Michel, 1989). Their shapes are also known, based on the low-resolution neutron structures of two reaction centers, from *R. viridis* and *Rhodobacter sphaeroides* (Roth, Arnoux, Ducruix, & Reiss-Husson, 1991; Roth et al., 1989). Remarkably, the shape of detergent surrounding each protein molecule is somewhat similar and is independent of the detergent molecules (*n*-dodecyl-*N*,*N*-dimethylamine-*N*-oxide (LDAO) and *n*-octyl-β-D-glucopyranoside (β-OG), respectively). In each case, the hydrophobic surface of the protein was covered in detergent molecules, wrapping the protein in a toroidal ring. Approximately 200 molecules cover 40% of the surface area of each reaction center. The shear amount of detergent in each protein–detergent complex, the building blocks of the crystal lattice, demonstrates that the detergent is at least as important as the protein itself in crystallization.

Within the solvent surrounding the protein–detergent complex, detergent monomers exist which self-associate and bind to the membrane protein surface in a concentration-dependent manner, with a minimum concentration equal to the critical micelle concentration (CMC). For example, at 0.3% (w/v) β-OG, each monomer of human prostaglandin synthase was found to bind at least 40 molecules of detergent. At 0.7% (w/v) β-OG, the number of bound detergent molecules roughly doubled (Garavito et al., 1996). Crystallization is mostly done at detergent concentrations greater than the CMC, where equilibrium exists between detergent monomers, micelles, and detergent involved with the protein–detergent complex, with the concentration of monomers approximately equal to the CMC (Wiener, 2004). As a higher CMC detergent will have more monomers in solution, it is expected that those monomers will exchange more readily between detergent micelles and the protein molecule surface (Garavito et al., 1996).

4. MEMBRANE PROTEIN CRYSTALLIZATION VIA VAPOR DIFFUSION

Many techniques exist to crystallize membrane proteins, including the recently developed lipidic cubic phase (Landau & Rosenbusch, 1996; Nollert, Navarro, & Landau, 2002), bicelle (Faham & Bowie, 2002), and vesicle fusion (Takeda et al., 1998) methods. This chapter will focus on the vapor diffusion approach. The most common and successful method of membrane protein crystallization, to date, this approach has allowed researchers in our lab to determine the crystal structures of a number of membrane proteins. These include the inner membrane efflux pumps CusA (Long et al., 2010), AcrB (Su et al., 2006), and MtrD (Bolla et al., 2014); the outer membrane channels CusC (Lei, Bolla, Bishop, Su, & Yu, 2014), MtrE (Lei, Chou, et al., 2014), and CmeC (Su et al., 2014); as well as the CusBA adaptor–transporter efflux complex (Su et al., 2011).

If not for the necessary presence of detergent, the crystallization of membrane proteins via vapor diffusion would be exactly similar to that of soluble proteins. While protein–protein interactions dominate the crystallization of soluble proteins, both detergent–detergent and protein–detergent interactions are additional considerations for the membrane protein crystallographer. Unlike crystallization of water-soluble proteins, choosing the right detergent has been the key success for our membrane protein crystallization efforts. In our opinion, this is the most important factor for obtaining high-quality membrane protein crystals suitable for X-ray diffraction. Thus, the rule of thumb is that more efforts should be made to screen a variety of mild detergents than to optimize other parameters when crystallizing a new membrane protein.

In a simple crystallization experiment, a precipitant solution, containing at least buffer, precipitant, and salt components, is mixed with a protein solution, containing the purified protein–detergent complexes. This forces the protein solution into a state of supersaturation. To escape, the protein–detergent complex has two routes: (i) aggregation and (ii) crystallization. If the crystallization solution is not too harsh and the protein–detergent complex has enough time to make specific interactions with its neighbors, the latter results. In general, a slow introduction of the crystallization solution to the protein–detergent complex facilitates this, resulting in larger crystals with fewer imperfections (Benvenuti & Mangani, 2007).

Compared to other techniques, the conditions for crystallizing membrane protein via vapor diffusion tend to be very mild, involving a mixture of relatively low concentrations of precipitant, salt, and buffer. In short, a small quantity of the purified protein solution is mixed, usually in equal ratio, with the crystallization solution contained in the well (precipitant solution). This drop, now half precipitant solution and half protein solution, is either mounted in the well itself (sitting drop) (Fig. 3A) or inverted above the well (hanging drop) (Fig. 3B), and the chamber is made airtight. Vapor diffusion describes the evaporation of volatile species in the drop into the chamber, followed by diffusion across the chamber into the well. The end result of this process is chemical equilibrium between the drop and well. Because the well is much bigger, the equilibrium concentrations of drop components are approximated by the concentrations in the well.

For simplicity, it is generally assumed that only water is exchanged between the drop and well (Fowlis et al., 1988). However, any volatile species can evaporate into the chamber and exchange between the drop and well. When it reaches equilibrium, or close to it, the hanging or sitting drop will manifest different types and degrees of phase behavior. In addition to the protein, the detergent and the interactions between the two must be considered. The phase transition of detergent is relatively complex. It can

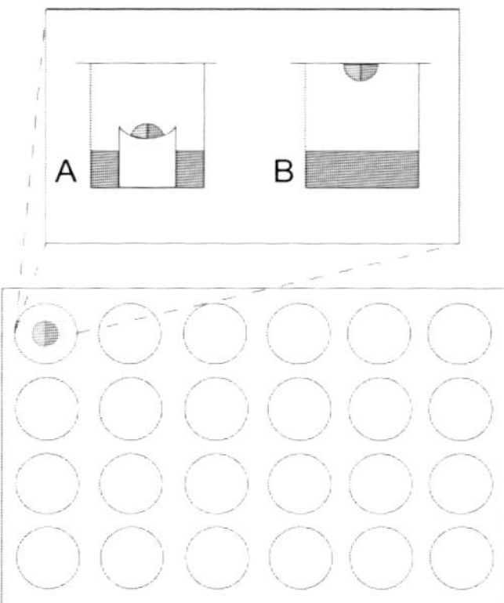

Figure 3 A 24-well plate in which either (A) sitting-drop or (B) hanging-drop vapor diffusion is done. The well solution (colored red (dark gray in the print version)) is mixed in equal proportion with protein solution (colored blue (light gray in the print version)). In the sitting-drop technique, the resulting mixture sits on a pedestal above the well solution and the top of the chamber is sealed with grease and glass. In the hanging-drop technique, the mixed drop is included on the glass slide itself, inverted, and sealed with grease atop the chamber.

crystallize, aggregate into micelles, or separate into detergent-rich and detergent-poor phases (Loll, 2003; Newby et al., 2009). Thus, searching for the right crystallization condition involves exploring point by point a complicated many-dimensional phase diagram for both protein and detergent.

The dimensions of this phase space include, at least, both protein and detergent concentrations, detergent type, precipitant concentration and type, buffer pH and type, salt concentration and type, and additive concentration and type. Further, the optimal crystallization conditions occupy only a very small volume of this phase space. Finally, the contribution of these factors to crystallization is nonlinear. In general, the effect of one component on the solubility of the protein is coupled to the effect of another (McPherson, 2004). Even with the advent of robotic screening technology, capable of testing up to 40,000 conditions per day (Stevens, 2000), crystallization cannot be approached naively.

There are two general approaches to navigating the phase space of a protein: (i) a "grid search," by systematic screening of what are believed to be the important variables (McPherson, 2009), or (ii) using a "coarse matrix screen," heavily biased on known crystallization conditions of other proteins (Jancarik & Kim, 1991). To date, many coarse matrix screens designed specifically for membrane proteins, including MemGold (Molecular Dimensions) and MembFac (Hampton Research), have been made commercially available. In many cases, these screening kits are instrumental in the search for initial crystallization conditions. The individual components of these conditions and their roles in the crystallization process will be discussed in the following sections, with an emphasis on the most common and successful ones.

4.1 Detergent

Some say that the biggest hurdle in membrane protein crystallization is not obtaining the crystals, but optimizing them for high-resolution X-ray diffraction (Sonoda et al., 2011). In this respect, no other variable is more influential in the success of membrane protein crystallization than detergent.

Although no single detergent category can be generally applied to membrane protein crystallization, statistical evidence suggests that some are more useful than others. Almost all successfully used detergents are nonionic alkyl sugars with C7–C14 alkyl chains. By far, the most crystal structures have been solved using the alkylmaltoside family of detergents—nearly one-half of all α-helical proteins (Newstead, Ferrandon, & Iwata, 2008). This family includes the single most popular detergent, DDM, which contains 12 carbons in the main alkyl chain. By contrast, the alkylglucoside family of detergents are most commonly used to crystallize outer membrane proteins (Sonoda et al., 2011). They include the second most popular detergent, β-OG, which is an 8-carbon alkyl-chained detergent.

Recently, the development of new detergents has emerged as a tool for handling difficult membrane proteins (Bolla, Su, & Yu, 2012). Non-conventional approaches, such as amphipols (Popot, 2010), nanodiscs (Popot, 2010), hemifluorinated surfactants (Breyton et al., 2009; Popot, 2010), cholate acid-based amphiphiles (Zhang et al., 2007), and the new family of maltose–neopentyl glycol amphiphiles (Chae et al., 2010), have been proven to aid in the solubilization of membrane proteins, while retaining their structure and function. This new generation of amphiphiles is particularly promising and may eventually have a large role in the solubilization, stabilization, and crystallization of membrane proteins for high-resolution structural analysis.

4.2 Precipitant

Based on their mechanism of action, precipitants can be divided into three categories: (i) salts, (ii) organic solvents, and (iii) long-chain polymers. As will be discussed in the following section, salt acts as a precipitant by either dehydrating the protein molecules or altering the ionic strength.

Organic solvents, such as ethanol or isopropanol, generally act to reduce the dielectric constant of the crystallization solution. This makes electrostatic interactions weaker and reduces the solubility of ionic compounds slightly. Interestingly, one of the first membrane proteins to be crystallized, the hydrophobic seed protein crambin, was initially solubilized in ethanol and crystallized by slowly adding water (Garavito et al., 1996; Teeter, 1984). At high concentrations, most of these compounds are denaturing. However, 2-methyl-2,4-pentanediol has been remarkably successful for outer membrane protein crystallization (Newstead, Hobbs, et al., 2008).

The third and most successful category of precipitants comprises the long-chain polymers, including polyethylene glycols (PEGs). More than 80% of all membrane proteins have been successfully crystallized with various molecular weights of PEGs (Newstead, Ferrandon, et al., 2008; Newstead, Hobbs, et al., 2008) and 100% of all membrane proteins in our lab. The mechanism of action of these polymers is attributed to volume exclusion effects (McPherson, 2004). Unlike other components of the crystallization solution, PEGs have no consistent conformation, resembling a flail. PEGs with molecular weights between 400 and 20,000 Da have been effective for membrane protein crystallization, with the larger ones obviously being more forceful.

4.3 Salt

Salt ions, both cationic and anionic, can alter the solubility of proteins. This effect was first discovered by Hofmeister (1888). The action mechanism is not entirely clear. However, it is thought to be due to specific interactions between ions, protein macromolecules, and water molecules in the solvent layer surrounding the protein. The effect of anions is usually ordered:

$$SO_4^{2-} > HPO_4^{2-} > CH_3CO_2^- > Cl^- > NO_3^- > Br^- > ClO_3^- > I^- > ClO_4^- > SCN^-$$

The order of cations is usually given as:

$$NH_4^+ > K^+ > Na^+ > Li^+ > Mg^{2+} > Ca^{2+}$$

Early members of the series tend to increase solvent surface tension and decrease the solubility of nonpolar molecules. They strengthen the hydrophobic interaction and this effect is often referred to "salting out." In this way, ammonium sulfate precipitation is commonly used in protein purification. Additionally, more than 10% of membrane protein crystals reported in the PDB are crystallized with the help of this salt (Newstead, Ferrandon, et al., 2008). By contrast, later ions in the series tend to increase the solubility of nonpolar molecules and decrease the order of water molecules. They weaken the hydrophobic interaction and this effect is called "salting in." Typically, ions that have a strong "salting in" effect are strong denaturants, because they interact much more strongly with the unfolded protein than with its native folded form.

4.4 Buffer

The actual net charge of the protein is determined not only by its composite amino acids but also by the pH of the solvent (Rosenberger, 1996). For each amino acid, its charge is determined by its pK_a; at a pH below the pK_a, an amino acid is positively charged and vice versa. Therefore, altering the pH of the crystallization solution will alter the degree of electrostatic interaction, mainly, between the protein–detergent complexes.

At a pH equal to the p*I*, which is an average of the pK_a of the surface residues of the protein, the net charge is neutralized and solubility drops dramatically. While this is usually too strong a condition for crystallization, it is common to find the optimal crystallization pH nearby (Rosenberger, 1996). The pH determines the charge of the zwitterionic detergents in a similar way.

Although no direct correlation appears to exist between predicted p*I* and the pH at which crystallization was reported, there is a good correlation between their difference, pH–p*I*, and p*I*. Specifically, acidic proteins tend to crystallize between 0 and 2.5 pH units above their p*I*, while basic proteins crystallize between 0.5 and 3 pH units below their p*I* (Kantardjieff & Rupp, 2004).

5. CASE STUDIES

5.1 Crystallization of the *E. coli* CusA heavy-metal efflux pump

Initially, we cloned the *E. coli cusA* gene into pET15b to form the expression vector pET15bΩ*cusA*. This expression vector was then used to transform *E. coli* BL21(DE3) cells and overexpress the full-length CusA membrane protein, containing a 6xHis tag at the N-terminus. The expressed protein was purified using a Ni^{2+}-affinity column to >95% purity. The final purified protein was then dialyzed and concentrated to 20 mg/ml in buffer containing 20 m*M* Na-HEPES (pH 7.5) and 0.05% DDM for crystallization trials. The original protein crystals crystallized fairly quickly, and these crystals diffracted X-rays to 4 Å resolution. However, mass spectrometry identified these as crystals of the *E. coli* AcrB protein, which was probably copurified with the CusA protein as a contaminant.

To avoid further contamination with the AcrB protein, we made an *E. coli* knockout strain BL21(DE3)Δ*acrB*, which harbors a deletion in the chromosomal *acrB* gene. The CusA transporter was then overexpressed using the transformed cells BL21(DE3)Δ*acrB*/pET15bΩ*cusA* and purified using a Ni^{2+}-affinity column. The purified protein was then dialyzed, concentrated to 20 mg/ml in buffer containing 20 m*M* Na-HEPES (pH 7.5) and 0.05% DDM, and subjected to crystallization trials using sitting-drop vapor diffusion. This time, mass spectrometry confirmed that the crystals were composed of CusA molecules. However, these crystals diffracted X-rays poorly, to ~9 Å resolution. After extensive screening, with as many salts, PEGs, pHs, temperatures, additives, and detergents available to us, we finally obtained plate-shaped crystals by adding 0.05% CYMAL-6 to the initial crystallization condition. Although these plate-shaped crystals were much better in quality, they diffracted X-rays anisotropically to a resolution of ~5 Å in one dimension and ~8 Å in the other.

As the addition of CYMAL-6 significantly improved the crystal quality, we decided to use CYMAL-6 to purify and crystallize the CusA protein. During this process, we also found that the additives Jeffamine M-600 (JM-600) and glycerol further improved their quality. Thus, the final crystallization condition was 2 μl protein solution (containing 20 mg/ml CusA protein in 20 m*M* Na-HEPES (pH 7.5) and 0.05% (w/v) CYMAL-6) mixed with 2 μl of reservoir solution containing 10% PEG 3350, 0.1 *M* Na-MES (pH 6.5), 0.4 *M* $(NH_4)_2SO_4$, 1% JM-600, and 10% glycerol. The resulting mixture was equilibrated against 500 μl of the reservoir solution. Typically, the dimensions of the crystals were 0.2 mm × 0.2 mm × 0.2 mm (Fig. 4A). Crystals of CusA were found to be trigonal with space group *R*32 ($a=b=178.4$ Å, $c=285.8$ Å) (Fig. 4B) (Long et al., 2010). Based on the molecular weight of CusA (115.72 kDa), it was found that a single molecule occupied the asymmetric unit with a solvent content of 67.5%. The final crystal structure of the CusA transporter was determined to a resolution of 3.5 Å.

5.2 Crystallization of the *E. coli* CusBA adaptor–transporter complex

As of this article's publication, CusBA remains the only adaptor–transporter complex for which a high-resolution X-ray structure is available. For purification of this important complex, the individual proteins of the CusA transporter and CusB (Su et al., 2009) adaptor were individually overexpressed in *E. coli* cells BL21(DE3)*ΔacrB*/pET15bΩ*cusA* and BL21 (DE3)*ΔacrB*/pET15bΩ*cusB*, respectively. The CusA protein was solubilized with 2% CYMAL-6 and purified using Ni^{2+}-affinity chromatography, as mentioned previously. The CusB protein was also purified using a Ni^{2+}-affinity column, without the addition of any detergent. Cocrystals of the CusBA complex were obtained using sitting-drop vapor diffusion. After extensive optimization, the CusBA crystals were grown at room temperature in 24-well plates with a final crystallization condition of 2 μl protein solution (containing 0.1 m*M* CusA and 0.1 m*M* CusB in 20 m*M* Na-HEPES (pH 7.5) and 0.05% (w/v) CYMAL-6) mixed with 2 μl of reservoir solution containing 10% PEG 6000, 0.1 *M* Na-HEPES (pH 7.5), 0.1 *M* ammonium acetate, and 20% glycerol. The resulting mixture was

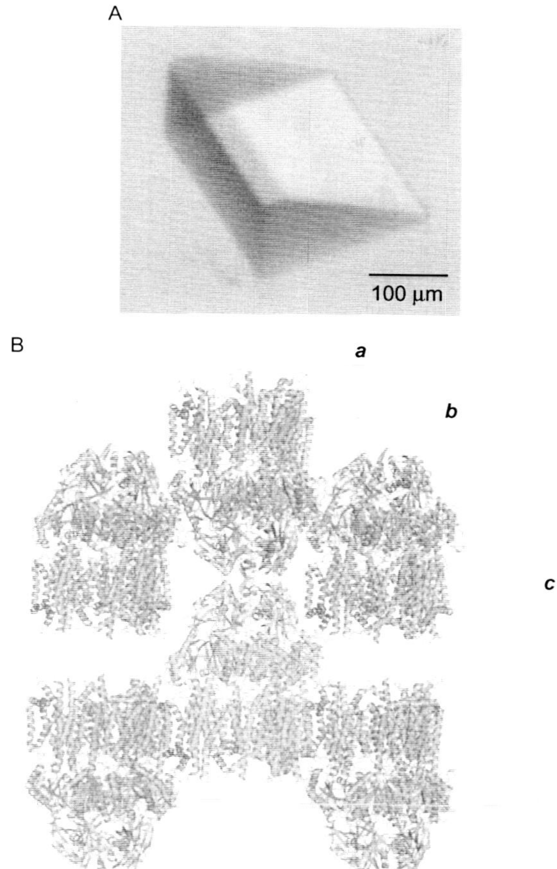

Figure 4 Crystal structure of the *E. coli* inner membrane heavy-metal efflux pump CusA. (A) A single crystal of CusA. (B) Packing diagram of the CusA crystal viewed orthogonal to the long axis of the unit cell. (See the color plate.)

equilibrated against 500 μl of the reservoir solution. Cocrystals of CusBA grew to a full size in the drops within 2 months. Typically, the dimensions of the crystals were 0.1 mm × 0.1 mm × 0.1 mm (Fig. 5A).

The cocrystals of the CusBA adaptor–transporter complex took a trigonal space group *R*32 with unit-cell parameters $a=b=160.2$ Å, $c=682.7$ Å (Fig. 5B) (Su et al., 2011). Based on the molecular weights of CusA (115.7 kDa) and CusB (42.3 kDa), the asymmetric unit contained one CusA and two CusB molecules with a solvent content of 70.8%. As it turns out, the presence of CusB drastically improved crystal quality in comparison with the

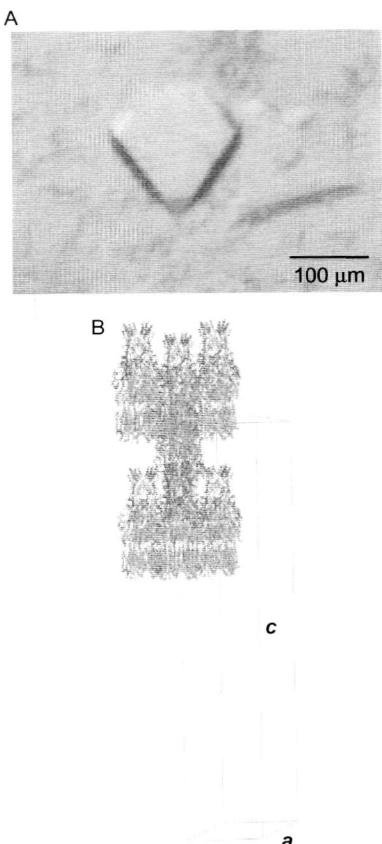

Figure 5 Crystal structure of the *E. coli* inner membrane heavy-metal efflux pump CusA in complex with the periplasmic membrane fusion protein CusB. (A) A single crystal of the CusBA complex. (B) Packing diagram of the CusBA cocrystal viewed orthogonal to the long axis of the unit cell.

crystals of CusA alone. The reason for this may be that CusB enlarges that hydrophilic surface of the membrane protein, thereby providing additional surface for crystal contacts. The final structure of the CusBA adaptor–transporter complex was determined to a resolution of 2.9 Å.

5.3 Crystallization of the *N. gonorrhoeae* MtrD multidrug efflux pump

It took us a few years of effort to crystallize the MtrD membrane protein. First, two different constructs, based on the protein sequence of MtrD in

N. gonorrhoeae strain FA19, were used to express the MtrD protein, containing a 6xHis tag at the N- or C-terminus, respectively. The open reading frame (ORF) of *mtrD* from *N. gonorrhoeae* FA19 was cloned into pET15b to form each expression vectors. After screening of expression conditions, only the C-terminal 6xHis-tagged MtrD protein could be expressed in *E. coli* C43(DE3)Δ*acrB* cells, which harbor a deletion in the chromosomal *acrB* gene. Using these conditions, we then purified the MtrD protein using a Ni^{2+} column. Optimization of the initial crystals with respect to primary and secondary detergent, salt, precipitant, pH, and temperature, yielded only showers of microcrystals. Unfortunately, these microcrystals did not even diffract X-rays to low resolution.

Based on protein sequence alignments of different *N. gonorrhoeae* strains, we decided to focus on the homologous MtrD efflux pump from *N. gonorrhoeae* strain PID332. The alignment indicated that MtrD of strain PID332 is 11 amino acids shorter at the C-terminus in comparison with FA19 MtrD. We then cloned the *N. gonorrhoeae* PID332 *mtrD* gene into pET15b to generate the pET15bΩ*mtrD* expression vector. This recombinant plasmid encoding the *N. gonorrhoeae* PID332 MtrD protein, with a 6xHis at the C-terminus, was then transformed into C43(DE3)Δ*acrB* cells to overexpress this protein. Upon expression, MtrD was solubilized in 2% CYMAL-6 and subjected to purification using Ni^{2+}-affinity chromatography. Initial crystallization trials, using 0.05% CYMAL-6 as a primary detergent, did not yield any crystals. However, small cubic-shaped crystals appeared in the drops after adding sucrose monododecanoate (SM) as a secondary detergent. Subsequent crystallization trials were carried out using both CYMAL-6 and SM detergents. After extensive optimization, the final MtrD crystals were grown at room temperature using sitting-drop vapor diffusion with the following procedures. A 2 μl protein solution containing 0.2 m*M* MtrD in buffer solution (20 m*M* Na-HEPES (pH 7.5), 0.05% (w/v) CYMAL-6, and 0.5% (w/v) SM) was mixed with 2 μl of reservoir solution, containing 30% PEG 400, 0.1 *M* Na-Bicene (pH 8.5), 0.1 *M* NH_4SO_4, 0.05 *M* $BaCl_2$, and 9% glycerol. The resulting mixture was equilibrated against 500 μl of the reservoir solution. Typically, the dimensions of the crystals were 0.2 mm × 0.2 mm × 0.2 mm (Fig. 6A). Crystals of MtrD belonged to space group *R*32 ($a=b=153.0$ Å, $c=360.7$ Å) (Fig. 6B) (Bolla et al., 2014). Analysis of the Matthews coefficient indicated the presence of one MtrD molecule (113.69 kDa) per asymmetric unit, with a solvent content of 66.7%. The final crystal structure of the MtrD transporter was determined to a resolution of 3.5 Å.

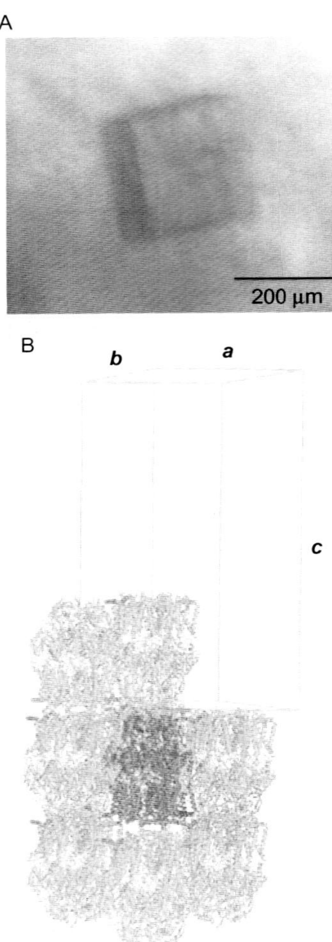

Figure 6 Crystal structure of the *N. gonorrhoeae* inner membrane multidrug efflux pump MtrD. (A) A single crystal of MtrD. (B) Packing diagram of the MtrD crystal viewed orthogonal to the long axis of the unit cell.

5.4 Crystallization of the *E. coli* CusC heavy-metal efflux channel

The CusC outer membrane channel is one of the most well-studied proteins in our laboratory. Its structures have allowed us to unmask the sequential transition of conformations leading to the folding and membrane insertion of this channel. To express CusC, the ORF of *cusC* from *E. coli* K12 genomic DNA was cloned into pBAD22 to produce the expression vector pBAD22Ω*cusC*. This expression system included a CusC signaling peptide at the N-terminus as well as a 6xHis tag at the C-terminal end. The N-terminal signaling peptide was needed to guide the channel protein to express at the outer membrane. It should be noted that this signaling peptide would be removed automatically when the CusC protein was matured in the cell. This recombinant plasmid was then used to transform the C43 (DE3)*ΔacrB* cells to overexpress the channel protein. To ensure that the harvested CusC protein was attached to or anchored the *E. coli* outer membrane, and not the inner membrane, we performed a pre-extraction procedure using 0.5% sodium lauroyl sarcosinate to selectively dissolve and remove proteins of the inner membrane. The CusC outer membrane protein was then solubilized in 2% (w/v) DDM. The extracted protein was purified with a Ni^{2+}-affinity column to >95% purity. The purified protein was then dialyzed and concentrated to 15 mg/ml in buffer containing 20 m*M* Na-HEPES (pH 7.5), 200 m*M* NaCl, and 0.05% (w/v) DDM for crystallization trials.

Sitting-drop vapor diffusion was employed for initial crystal screening. During this screening, we found that the full-length CusC channel could be crystallized only in the presence of 2% (w/v) β-OG, which served as a secondary detergent to DDM. It was also found that the addition of a small amount of JM 600 could improve the quality of the crystals. The best crystals of CusC were grown at room temperature with a reservoir solution containing 8% PEG 3350, 0.05 *M* sodium acetate (pH 4.0), 0.2 *M* $(NH_4)_2SO_4$, 1% JM 600, and 2% β-OG. Typically, the dimensions of the crystals were 0.2 mm × 0.2 mm × 0.2 mm (Fig. 7A).

Crystals of the full-length CusC channel took the space group *R*32 with the unit-cell parameters $a=b=88.5$ Å, $c=474.42$ Å (Fig. 7B) (Lei, Bolla, et al., 2014). Based on the molecular weight of CusC (49.30 kDa), only one CusC molecule occupied the asymmetric unit with a solvent content of 67.8%. The final model was refined to a resolution of 2.1 Å.

It is interesting to note that a single-point mutation in the first cysteine residue of CusC resulted in a dramatically different conformation. Instead of the four-stranded beta-sheet present in the wild-type CusC monomer, the corresponding residues in this mutant form two independent random loops. By mutating or deleting this key residue, we obtained a different structure entirely, which corresponds to the state immediately before membrane insertion (Lei, Bolla, et al., 2014). In this membrane protein, even the modification of one residue can prevent anchoring and insertion into the bilayer. Therefore, one should be cautious when crystallizing membrane proteins with any mutations or truncated sequences.

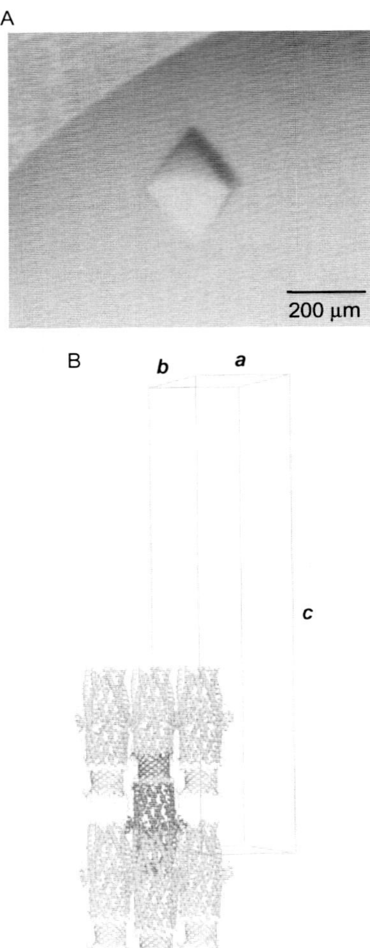

Figure 7 Crystal structure of the *E. coli* outer membrane channel CusC. (A) A single crystal of CusC. (B) Packing diagram of the CusC crystal viewed orthogonal to the long axis of the unit cell.

5.5 Crystallization of the *N. gonorrhoeae* MtrE multidrug efflux channel

The construct used to express the *N. gonorrhoeae* MtrE outer membrane channel was similar to that of *E. coli* CusC. However, after several trials, we found that the signaling peptide sequence of *Pseudomonas aeruginosa* OprM was the best peptide for expressing the *N. gonorrhoeae* MtrE heterogeneously in *E. coli*. In short, the ORF of *mtrE* from *N. gonorrhoeae* FA19 was cloned into pBAD22 to form the expression vector pBAD22Ω*mtrE*, which included the *P. aeruginosa* OprM signaling peptide at the N-terminus and a 6xHis tag at the C-terminus. The full-length MtrE protein was then expressed in *E. coli* C43(DE3) cells possessing this vector. The procedures for protein extraction and purification were the same as those for the CusC membrane protein. The purified protein was dialyzed and concentrated to 15 mg/ml in buffer containing 20 m*M* Na-HEPES (pH 7.5), 200 m*M* NaCl, and 0.05% (w/v) DDM for crystallization trials.

Crystals of MtrE were obtained using sitting-drop vapor diffusion. Based on crystal screening, the following procedure was adopted. A 2 μl protein solution containing 15 mg/ml MtrE protein in 20 m*M* Na-HEPES (pH 7.5), 200 m*M* NaCl, and 0.05% (w/v) DDM was mixed with a 2 μl of reservoir solution containing 20% PEG 400, 0.2 *M* sodium acetate (pH 4.6), 0.25 *M* $MgSO_4$, and 2% (w/v) β-OG. The resultant mixture was equilibrated against 500 μl of the reservoir solution at room temperature. Crystals of MtrE grew to a full size in the drops within 2 weeks. Typically, the dimensions of the crystals were 0.1 mm × 0.1 mm × 0.2 mm (Fig. 8A).

Crystals of the MtrE channel protein belonged to the space group $P6_322$ ($a=b=93.9$ Å, $c=391.5$ Å) (Fig. 8B) (Lei, Chou, et al., 2014). Analysis of the Matthews coefficient indicated the presence of one MtrE protomer (49.29 kDa) per asymmetric unit, with a solvent content of 75.8%. The final structural model was resolved to a resolution of 3.3 Å.

5.6 Crystallization of the *Campylobacter jejuni* CmeC multidrug efflux channel

The construct used to express the *C. jejuni* CmeC outer membrane channel was similar to that of *N. gonorrhoeae* MtrE. Briefly, the full-length CmeC membrane protein containing a 6xHis tag at the C-terminus was overproduced in *E. coli* C43(DE3)/pBAD22bΩ*cmeC* cells. This expression system included an OprM signaling peptide at the N-terminus and a 6xHis tag at the C-terminus. The purified protein was dialyzed and concentrated to 15 mg/ml in buffer containing 20 m*M* Na-HEPES (pH 7.5), 200 m*M* NaCl, and 0.05% DDM.

Initial crystallization trials were not successful and did not yield even poor-quality crystals. These trials were followed by extensive screening of secondary detergents. Fortunately, hexagonal-shaped CmeC crystals were obtained, but only in drops containing the detergent C_8E_4. Further optimization of the crystallization conditions eventually produced well-diffracting crystals.

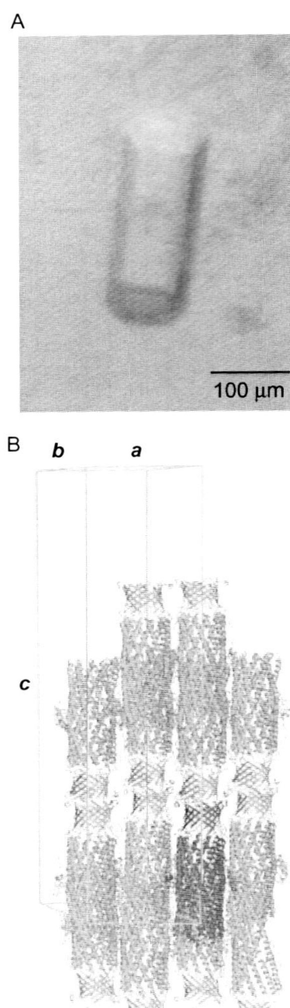

Figure 8 Crystal structure of the *N. gonorrhoeae* outer membrane channel MtrE. (A) A single crystal of MtrE. (B) Packing diagram of the MtrE crystal viewed orthogonal to the long axis of the unit cell.

Crystals of CmeC were obtained using sitting-drop vapor diffusion at room temperature in 24-well plates with the following procedures. A 2 µl protein solution containing 15 mg/ml CmeC protein in 20 m*M* Na-HEPES (pH 7.5), 200 m*M* NaCl, and 0.05% (w/v) DDM was mixed with 2 µl of reservoir solution containing 18% PEG 400, 0.1 *M* sodium

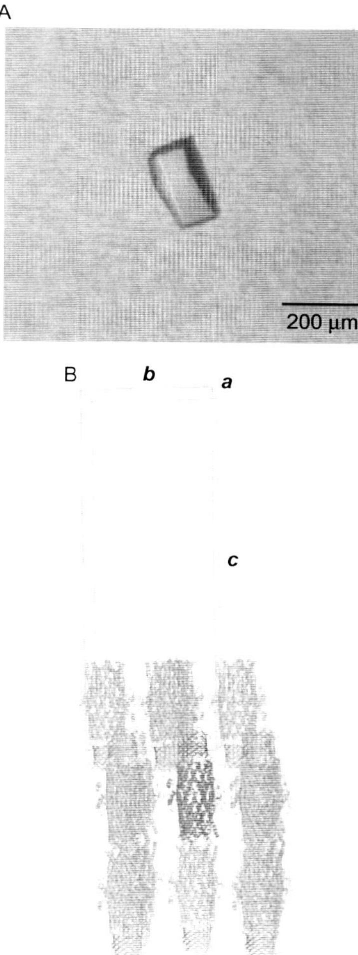

Figure 9 Crystal structure of the *C. jejuni* outer membrane channel CmeC. (A) A single crystal of CmeC. (B) Packing diagram of the CmeC crystal viewed orthogonal to the long axis of the unit cell.

acetate (pH 4.0), 0.3 *M* $(NH_4)_2SO_4$, and 2% C_8E_4. The resulting mixture was equilibrated against 500 μl of the reservoir solution. Crystals of CmeC grew to a full size in the drops within 2 weeks. Typically, the dimensions of the crystals were 0.1 mm × 0.2 mm × 0.2 mm (Fig. 9A).

Crystals of the CmeC outer membrane channel belonged to the space group $C222_1$ with unit-cell parameters: $a = 92.38$ Å, $b = 147.35$ Å, $c = 420.43$ Å (Fig. 9B) (Su et al., 2014). Based on the molecular weight of

CmeC (54.26 kDa), three molecules were found in the asymmetric unit with a solvent content of 80.6%. These three molecules assembled to form a trimeric channel within the unit cell. The final crystal structure was determined to a resolution of 2.4 Å.

6. CONCLUDING REMARKS

For the routine availability of three-dimensional membrane protein structures to become possible, there is still a mountain to climb. Like the state of soluble protein crystallization 20 years ago, initial data are still being acquired to facilitate the large-scale investigations demanded today. Only then can more lofty goals of crystallization be achieved. For example, no atomic-resolution model of a drug efflux complex, spanning both the inner and outer bilayers, has been reported to date. In Gram-negative bacteria, such complexes are responsible for conferring resistance to commonly used antibiotics, a problem that is exacerbated by the increased use of these drugs. While bacterial infections remain a leading cause of death worldwide, invaluable structural information leading to antibiotic resistance mechanisms would provide a platform to produce new drugs and inhibitors that increase the efficacy of our weakened therapies.

ACKNOWLEDGMENT

This work is supported by an NIH Grant R01GM086431 (E. W. Y.).

REFERENCES

Bannwarth, M., & Schulz, G. E. (2003). The expression of outer membrane proteins for crystallization. *Biochimica et Biophysica Acta*, *1610*(1), 37–45.

Benvenuti, M., & Mangani, S. (2007). Crystallization of soluble proteins in vapor diffusion for X-ray crystallography. *Nature Protocols*, *2*(7), 1633–1651.

Bernaudat, F., Frelet-Barrand, A., Pochon, N., Dementin, S., Hivin, P., Boutigny, S., et al. (2011). Heterologous expression of membrane proteins: Choosing the appropriate host. *PLoS One*, *6*, e29191.

Bill, R. M., Henderson, P. J. F., Iwata, S., Kunji, E. R. S., Michel, H., Neutze, R., et al. (2011). Overcoming barriers to membrane protein structure determination. *Nature Biotechnology*, *29*(4), 335–340.

Bolla, J. R., Su, C., Do, S. V., Radhakrishnan, A., Kumar, N., Long, F., et al. (2014). Crystal structure of the *Neisseria gonorrhoeae* MtrD inner membrane multidrug efflux pump. *PLoS One*, *9*(6), e97903.

Bolla, J. R., Su, C., & Yu, E. W. (2012). Biomolecular membrane protein crystallization. *Philosophical Magazine (Abingdon)*, *92*(19), 2648–2661.

Bowie, J. U. (2001). Stabilizing membrane proteins. *Current Opinion in Structural Biology*, *11*(4), 397–402.

Breyton, C., Gabel, F., Abla, M., Pierre, Y., Lebaupain, F., Durand, G., et al. (2009). Micellar and biochemical properties of (hemi)fluorinated surfactants are controlled by the size of the polar head. *Biophysical Journal, 97*(4), 1077–1086.

Caffrey, M. (2003). Membrane protein crystallization. *Journal of Structural Biology, 142*(1), 108–132.

Chae, P. S., Rasmussen, S. G. F., Rana, R. R., Gotfryd, K., Chandra, R., Goren, M. A., et al. (2010). Maltose-neopentyl glycol (MNG) amphiphiles for solubilization, stabilization, and crystallization of membrane proteins. *Nature Methods, 7*(12), 1003–1008.

Cowan, S. W., Schirmer, T., Rummel, G., Steiert, M., Ghosh, R., Pauptit, R. A., et al. (1992). Crystal structures explain functional properties of two *E. coli* porins. *Nature, 358*(6389), 727–733.

De Grip, W. J. (1982). Thermal stability of rhodopsin and opsin in some novel detergents. *Methods in Enzymology, 81*, 256–265.

Deisenhofer, J., Epp, O., Miki, K., Huber, R., & Michel, H. (1985). Structure of the protein subunits in the photosynthetic reaction centre of *Rhodopseudomonas viridis* at 3 Å resolution. *Nature, 318*(6047), 618–624.

Deisenhofer, J., & Michel, H. (1989). The photosynthetic reaction center from the purple bacterium *Rhodopseudomonas viridis*. *Science, 245*(4925), 1463–1473.

Doyle, D. A., Cabral, J. M., Pfuetzner, R. A., Kuo, A., Gulbis, J. M., Cohen, S. L., et al. (1998). The structure of the potassium channel: Molecular basis of K^+ conductivity and selectivity. *Science, 280*(5360), 69–77.

Drew, D., Fröderberg, L., Baars, L., & de Gier, J. L. (2003). Assembly and overexpression of membrane proteins in *Escherichia coli*. *Biochimica et Biophysica Acta, 1610*(1), 3–10.

Faham, S., & Bowie, J. U. (2002). Bicelle crystallization: A new method for crystallizing membrane proteins yields a monomeric bacteriorhodopsin structure. *Journal of Molecular Biology, 316*(1), 1–6.

Fowlis, W. W., DeLucas, L. J., Twigg, P. J., Howard, S. B., Meehan, E. J., Jr., & Baird, J. K. (1988). Experimental and theoretical analysis of the rate of solvent equilibration in the hanging drop method of protein crystal growth. *Journal of Crystal Growth, 90*, 117–129.

Gabrielsen, M., Gardiner, A. T., Fromme, P., & Cogdell, R. J. (2009). Membrane protein crystallization: Approaching the problem and understanding the solutions. *Current Topics in Membranes, 63*, 127–149.

Garavito, R. M., Picot, D., & Loll, P. J. (1996). Strategies for crystallizing membrane proteins. *Journal of Bioenergetics and Biomembranes, 28*(1), 13–27.

Hofmeister, F. (1888). Zur lehre von der wirkung der saize. *Archiv für Experimentelle Pathologie und Pharmakologie, 24*(4), 247–260.

Jancarik, J., & Kim, S. H. (1991). Sparse matrix sampling: A screening method for the crystallization of proteins. *Journal of Applied Crystallography, 24*, 409–411.

Kane, J. F. (1995). Effects of rare codon clusters on high-level expression of heterologous proteins in *Escherichia coli*. *Current Opinion in Biotechnology, 6*(8), 494–500.

Kang, H. J., Lee, C., & Drew, D. (2013). Breaking the barriers in membrane protein crystallography. *The International Journal of Biochemistry and Cell Biology, 45*(3), 636–644.

Kantardjieff, K. A., & Rupp, B. (2004). Protein isoelectric point as a predictor for increased crystallization screening efficiency. *Bioinformatics, 20*(14), 2162–2168.

Krogh, A., Larsson, B., von Heijne, G., & Sonnhammer, E. L. L. (2001). Predicting transmembrane topology with a hidden Markov model: Application to complete genomes. *Journal of Molecular Biology, 305*, 567–580.

Landau, E. M., & Rosenbusch, J. P. (1996). Lipidic cubic phases: A novel concept for the crystallization of membrane proteins. *Proceedings of the National Academy of Sciences of the United States of America, 93*(25), 14532–14535.

Lee, S., Lee, A., Chen, J., & MacKinnon, R. (2005). Structure of the KvAP voltage-dependent K^+ channel and its dependence on the lipid membrane. *Proceedings of the National Academy of Sciences of the United States of America, 102*(43), 15441–15446.

Lei, H., Bolla, J. R., Bishop, N. R., Su, C., & Yu, E. W. (2014). Crystal structures of CusC review conformational changes accompanying folding and transmembrane channel formation. *Journal of Molecular Biology*, *426*(2), 403–411.

Lei, H., Chou, T., Su, C., Bolla, J. R., Kumar, N., Radhakrishnan, A., et al. (2014). Crystal structure of the open state of the *Neisseria gonorrhoeae* MtrE outer membrane channel. *PLoS One*, *9*, e97475.

Lohse, M. J., Benovic, J. L., Codina, J., Caron, M. G., & Lefkowitz, R. J. (1990). Beta-arrestin: A protein that regulates beta-adrenergic receptor function. *Science*, *248*(4962), 1547–1550.

Loll, P. J. (2003). Membrane protein structural biology: The high throughput challenge. *Journal of Structural Biology*, *142*(1), 144–153.

Long, F., Rouquette-Loughlin, C., Shafer, W. M., & Yu, E. W. (2008). Functional cloning and characterization of the multidrug efflux pumps NorM from *Neisseria gonorrhoeae* and YdhE from *Escherichia coli*. *Antimicrobial Agents and Chemotherapy*, *52*(9), 3052–3060.

Long, F., Su, C., Zimmermann, M. T., Boyken, S. E., Rajashankar, K. R., Jernigan, R. L., et al. (2010). Crystal structures of the CusA efflux pump suggest methionine-mediated metal transport. *Nature*, *467*(7314), 484–488.

Long, S. B., Tao, X., Campbell, E. B., & MacKinnon, R. (2007). Atomic structure of a voltage-dependent K^+ channel in a lipid membrane-like environment. *Nature*, *450*(7168), 376–382.

Mancusso, R., Karpowich, N., Czyzewski, B., & Wang, D. (2011). Simple screening method for improving membrane protein thermostability. *Methods*, *55*(4), 324–329.

McPherson, A. (2004). Protein crystallization in the structural genomics era. *Journal of Structural and Functional Genomics*, *5*(1), 3–12.

McPherson, A. (2009). Introduction to the crystallization of biological macromolecules. *Current Topics in Membranes*, *63*, 5–23.

Molina, D. M., Cornvik, T., Eshaghi, S., Haeggström, J. Z., Nordlund, P., & Sabet, M. I. (2008). Engineering membrane protein overproduction in *Escherichia coli*. *Protein Science*, *17*, 673–680.

Møller, J., & Le Maire, M. (1993). Detergent binding as a measure of hydrophobic surface area of integral membrane proteins. *Journal of Biological Chemistry*, *268*(25), 18659–18672.

Newby, Z. E. R., O'Connell, J. D., III, Gruswitz, F., Hays, F. A., Harries, W. E. C., Harwood, I. M., et al. (2009). A general protocol for the crystallization of membrane proteins for X-ray structural determination. *Nature Protocols*, *4*(5), 619–637.

Newstead, S., Ferrandon, S., & Iwata, S. (2008). Rationalizing α-helical membrane protein crystallization. *Protein Science*, *17*(1), 466–472.

Newstead, S., Hobbs, J., Jordan, D., Carpenter, E. P., & Iwata, S. (2008). Insights into outer membrane protein crystallisation. *Molecular Membrane Biology*, *25*(8), 631–638.

Nollert, P., Navarro, J., & Landau, E. M. (2002). Crystallization of membrane proteins in cubo. *Methods in Enzymology*, *343*, 183–199.

Overington, J. P., Al-Lazikani, B., & Hopkins, A. L. (2006). How many drug targets are there? *Nature Reviews Drug Discovery*, *5*(12), 993–996.

Palczewski, K., Kumasaka, T., Hori, T., Behnke, C. A., Motoshima, H., Fox, B. A., et al. (2000). Crystal structure of rhodopsin: A G protein-couple receptor. *Science*, *289*(5480), 739–745.

Popot, J. L. (2010). Amphipols, nanodiscs, and fluorinated surfactants: Three non-conventional approaches to studying membrane proteins in aqueous solutions. *Annual Review of Biochemistry*, *79*, 737–775.

Pryor, K. D., & Leiting, B. (1997). High-level expression of soluble protein in *Escherichia coli* using a His6-tag and maltose-binding-protein double-affinity fusion system. *Protein Expression and Purification*, *10*(3), 309–319.

Rosenberger, F. (1996). Protein crystallization. *Journal of Crystal Growth*, *166*(1), 40–54.

Roth, M., Arnoux, B., Ducruix, A., & Reiss-Husson, F. (1991). Structure of the detergent phase and protein-detergent interactions in crystals of the wild-type (strain Y) *Rhodobacter sphaeroides* photochemical reaction center. *Biochemistry*, *30*(39), 9403–9413.

Roth, M., Lewit-Bentley, A., Michel, H., Deisenhofer, J., Huber, R., & Oesterhelt, D. (1989). Detergent structure in crystals of a bacterial photosynthetic reaction centre. *Nature*, *340*, 659–662.

Sahdev, S., Khattar, S. K., & Saini, K. S. (2008). Production of active eukaryotic proteins through bacterial expression systems: A review of the existing biotechnology strategies. *Molecular and Cellular Biochemistry*, *307*(1), 249–264.

Saier, M. H., Jr., Tam, R., Reizer, A., & Reizer, J. (1994). Two novel families of bacterial membrane proteins concerned with nodulation, cell division and transport. *Molecular Microbiology*, *11*(5), 841–847.

Schertler, G. F. X. (1992). Overproduction of membrane proteins. *Current Opinion in Structural Biology*, *2*(4), 534–544.

Schmid, B., Krömer, M., & Schulz, G. E. (1996). Expression of porin from *Rhodopseudomonas blastica* in *Escherichia coli* inclusion bodies and folding into exact native structure. *Federation of European Biochemical Societies Letters*, *381*(1), 111–114.

Sonoda, Y., Newstead, S., Hu, N., Alguel, Y., Niji, E., Beis, K., et al. (2011). Benchmarking membrane protein detergent stability for improving throughput of high-resolution X-ray structures. *Structure*, *19*(1), 17–25.

Stevens, R. C. (2000). High-throughput protein crystallization. *Current Opinion in Structural Biology*, *10*, 558–563.

Su, C., Li, M., Gu, R., Takatsuka, Y., McDermott, G., Nikaido, H., et al. (2006). Conformation of the AcrB multidrug efflux pump in mutants of the putative proton relay pathway. *Journal of Bacteriology*, *188*(20), 7290–7296.

Su, C., Long, F., Zimmermann, M. T., Rajashankar, K. R., Jernigan, R. L., & Yu, E. W. (2011). Crystal structure of the CusBA heavy-metal efflux complex of *Escherichia coli*. *Nature*, *470*(7335), 558–562.

Su, C., Radhakrishnan, A., Kumar, N., Long, F., Bolla, J. R., Lei, H., et al. (2014). Crystal structure of the *Campylobacter jejuni* CmeC outer membrane channel. *Protein Science*, *23*(7), 954–961.

Su, C., Yang, F., Long, F., Reyon, D., Routh, M. D., Kuo, D. W., et al. (2009). Crystal structure of the membrane fusion protein CusB from *Escherichia coli*. *Journal of Molecular Biology*, *393*(2), 342–355.

Takeda, K., Sato, H., Hino, T., Kono, M., Fukuda, K., Sakurai, I., et al. (1998). A novel three-dimensional crystal of bacteriorhodopsin obtained by successive fusion of the vesicular assemblies. *Journal of Molecular Biology*, *283*(2), 463–474.

Teeter, M. (1984). Water structure of a hydrophobic protein at atomic resolution: Pentagon rings of water molecules in crystals of crambin. *Proceedings of the National Academy of Sciences of the United States of America*, *81*(19), 6014–6018.

Thieme, D., Neubauer, P., Nies, D. H., & Grass, G. (2008). Sandwich hybridization assay for sensitive detection of dynamic changes in mRNA transcript levels in crude *Escherichia coli* cell extracts in response to copper ions. *Applied and Environmental Microbiology*, *74*(24), 7463–7470.

Wagner, S., Klepsch, M. M., Schlegel, S., Appel, A., Draheim, R., Tarry, M., et al. (2008). Tuning Escherichia coli for membrane protein overexpression. *Proceedings of the National Academy of Sciences of the United States of America*, *105*(38), 14371–14376.

Wallin, E., & von Heijne, G. (1998). Genome-wide analysis of integral membrane proteins from eubacterial, archaean, and eukaryotic organisms. *Protein Science*, 7(4), 1029–1038.

Warne, T., Serrano-Vega, M. J., Baker, J. G., Moukhametzianov, R., Edwards, P. C., Henderson, R., et al. (2008). Structure of a β1-adrenergic G-protein-coupled receptor. *Nature*, *454*(7203), 486–491.

Weiss, M. S., Abele, U., Weckesser, J., Welte, W., Schiltz, E., & Schulz, G. E. (1991). Molecular architecture and electrostatic properties of a bacterial porin. *Science*, *254*(5038), 1627–1630.

Wiener, M. C. (2004). A pedestrian guide to membrane protein crystallization. *Methods*, *34*(3), 364–372.

Zhang, Q., Ma, X., Ward, A., Hong, W. X., Jaakola, V. P., Stevens, R. C., et al. (2007). Designing facial amphiphiles for the stabilization of integral membrane proteins. *Angewandte Chemie International Edition*, *46*(37), 7023–7025.

Zhou, Y., & Bowie, J. U. (2000). Building a thermostable membrane protein. *Journal of Biological Chemistry*, *275*(10), 6975–6979.

CHAPTER NINETEEN

Bicelles Coming of Age: An Empirical Approach to Bicelle Crystallization

Sandra Poulos, Jacob L.W. Morgan, Jochen Zimmer, Salem Faham[1]
Center for Membrane Biology, Department of Molecular Physiology and Biological Physics, University of Virginia, Charlottesville, Virginia, USA
[1]Corresponding author: e-mail address: sf3bb@cms.mail.virginia.edu

Contents

Abstract

Biological membranes represent a unique environment in which integral membrane proteins (MPs) fold to perform diverse biological functions. In many cases, lipids support the native conformation or mediate important interactions between MPs. It is therefore imperative to develop methods that maintain this support for the structural and

Methods in Enzymology, Volume 557
ISSN 0076-6879
http://dx.doi.org/10.1016/bs.mie.2014.12.024

functional analyses of an exceedingly important class of biological macromolecules. Bicelles are detergent-stabilized phospholipid bilayer discs into which MPs can be reconstituted for biophysical studies. Here, we review recent advances and emerging concepts in employing bicelles for the crystallization and structure determination of MPs. We discuss variations of established procedures as well as alternative approaches, and we present a summary and analysis of the conditions used for bicelle-mediated MP crystallization.

1. INTRODUCTION

Membrane proteins (MPs) are critical for the survival of all living organisms. They reside in lipid bilayers that offer a unique and intricate environment. The bilayer milieu consists of a hydrophobic core, the phospholipid head group region that is followed by the water–lipid interface, and the hydrophilic intra- or extracellular space. Purification of MPs requires detergents, which only partially mimic the amphipathic environment of biological membranes. Analyzing MPs by biochemical and biophysical techniques presents numerous challenges to structural biologists. However, due to their profound biological significance, structures of MPs are highly sought after. To date, X-ray crystallography is the most successful method to obtain MP structures, and several methods and techniques (Cherezov et al., 2007; Landau & Rosenbusch, 1996; Ostermeier, Iwata, Ludwig, & Michel, 1995; Prive et al., 1994) have been developed to coax MP into high-quality crystals. Traditionally, crystallization trials with MPs were carried out in detergent micelles, which coat the hydrophobic transmembrane region of the protein, thereby reducing the surface area available to mediate crystal contacts (Newstead, Ferrandon, & Iwata, 2008). The lipidic cubic phase (LCP) method was the first to demonstrate that MPs can produce three-dimensional crystals from a lipidic medium (Landau & Rosenbusch, 1996). The LCP method utilizes the lipid monoolein and derivatives thereof, which form a continuous interconnected bilayer system (Caffrey, 2000). The bicelle method falls somewhere in between the purely detergent method and the purely lipidic method. Bicelles are formed by mixtures of a phospholipid and a detergent (Sanders & Prestegard, 1990; Sanders & Schwonek, 1992). Previously, this space was explored by using lipids as additives to detergent-purified MPs. More recently, a high-lipid high-detergent approach has been exploring this space further (as discussed below) (Gourdon et al., 2011). The composition used to form bicelles is variable within a semidefined set of parameters. The lipid is typically 1,2-dimyristoyl-*sn*-glycero-3-phosphocholine (DMPC) and the most frequently used detergent is 3-([3-cholamidopropyl]dimethylammonio)-2-hydroxy-1-propanesulfonate (CHAPSO), or a CHAPSO analog. Bicelles are bilayer micelles consisting of a disc-like phospholipid membrane whose edges are stabilized by the detergent (Fig. 1A). However, this composition undergoes several phase transitions depending on the temperature, the phospholipid:detergent ratio (q), and the lipid concentration (Li et al., 2013).

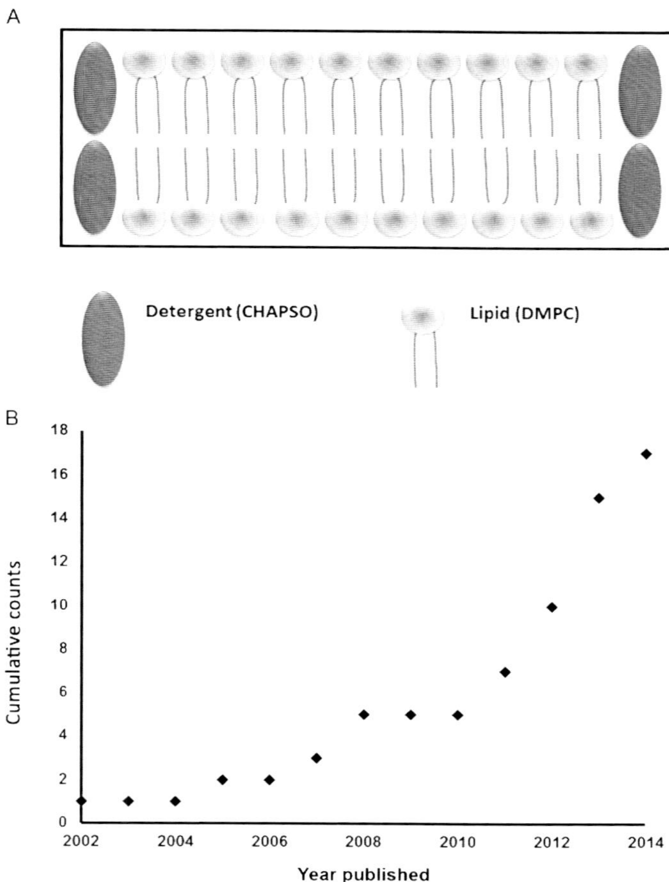

Figure 1 (A) Bicelle bilayers are edge stabilized by detergent to form bilayer discs. (B) Cumulative number of membrane protein structures solved using the bicelle method. A sharp increase is evident in the last few years.

These phases include a perforated lamellar phase, multilamellar vesicles, ribbons or worm-like micelles, and the bicellar phase. Notably, the bicelle-forming mixture is a liquid at lower temperatures while in bicellar phase and forms a gel at higher temperatures in the perforated lamellar phase, a high-viscosity gel-like phase (Nieh et al., 2011; Triba, Warschawski, & Devaux, 2005). This feature is significant for crystallization since access to the liquid phase makes crystallization setups easy using liquid-handling robotics and access to the gel phase allows for exploration of crystallization conditions in an alternate environment that may be more favorable for crystal growth.

2. PROTEIN FUNCTION AND FOLDING IN LIPID BICELLES

Bicelles represent a versatile tool for MP biochemistry since they consist of phospholipid bilayers that can accommodate MPs, thus depending on the lipid composition used, bicelles can provide a native-like environment for MPs. DMPC, the most frequently used bicelle lipid, forms an unnaturally thin bilayer with a hydrophobic core of only about 26 Å (Kucerka, Nieh, & Katsaras, 2011). However, several MPs have been shown to retain native-like properties in DMPC bilayers (Claridge, Aittoniemi, Cooper, & Schnell, 2013; De Angelis & Opella, 2007; Sharonov et al., 2014), perhaps due to the modulation of the bilayer thickness by the hydrophobic region of the integrated MP (Lee, 2011; Nielsen & Andersen, 2000). It is therefore likely that bicelles form a suitable environment for biochemical analyses on MPs in general. However, only a limited number of functional studies employing bicelle-embedded MP have been published to date.

The first detailed analysis of a bicelle-reconstituted MP was reported by the Sanders laboratory in 2000, describing the reconstitution of functional diacylglycerol kinase (DAGK) into DMPC/CHAPSO or 1,2-dipalmitoyl-*sn*-glycero-3-phosphocholine/CHAPSO bicelles (Czerski & Sanders, 2000). DAGK enzymatic activity was observed by coupling the kinase activity with the activities of pyruvate kinase (PK) and lactate dehydrogenase (LDH), the latter oxidizes NADH to NAD^+ upon reduction of pyruvate to lactate. DAGK exhibited comparable activity levels in bicelles, detergent micelles, and lipid vesicles (Czerski & Sanders, 2000). Several other studies with proteoliposome- and even lipidic cubic phase-reconstituted enzymes demonstrate that PK and LDH retain their activities in a lipid-rich environment, attesting to the suitability of this assay to analyze membrane-integrated enzymes that generate nucleoside-diphosphate products (primarily adenosine diphosphate (ADP), and uridine diphosphate (UDP)) (Belli, Elsener, Wunderli-Allenspach, & Kramer, 2009; Hubbard, McNamara, Azumaya, Patel, & Zimmer, 2012; Li & Caffrey, 2011; Omadjela et al., 2013). In another study, the Henzler-Wildman group reconstituted the small multidrug resistance transporter EmrE into lipid bicelles of different compositions and demonstrated the functionality of the transporter in drug-binding assays (Morrison & Henzler-Wildman, 2012; Morrison et al., 2012).

Recent observations indicate the suitability of bicelles as a membrane surrogate in cell-free MP expression systems. *In vitro* biosynthesis of MPs in the absence of a membrane-integrated translocation machinery either results in aggregation and precipitation of the synthesized proteins or requires additional surfactants to stabilize the protein (Harbers, 2014; Park et al., 2011). Recently, Uhlemann and colleagues reported the efficient *in vitro* expression of subunits *a* and *c* of the *Escherichia coli* F_oF_1 ATP synthase in the presence of 1,2-diheptanoyl-*sn*-glycero-3-phosphocholine (DHPC)/DMPC bicelles (Uhlemann, Pierson, Fillingame, & Dmitriev, 2012). While DHPC alone was insufficient to stabilize the synthesized subunits, the presence of bicelles allowed efficient protein recovery. Similarly, Shimono and coworkers reported cell-free expression and folding of bacteriorhodopsin (BR) in bicelle mixtures containing 3-[(3-cholamidopropyl) dimethylammonio]-1-propanesulfonate hydrate (CHAPS) or sodium cholate together with egg phosphotidylcholine lipids (Shimono et al., 2009). Interestingly, the folding efficiency of BR markedly improved when sterol detergents were used in the bicelle preparations.

In addition to the increasing number of biochemical studies on membrane-integrated enzymes and transporters in bicelles, isotropic bicelles have long been a valuable model system for analyzing lipid-binding properties and conformational changes of peripheral MPs and peptides. These studies include lipases; pore-forming peptides; and membrane fusion proteins analyzed by NMR, EPR, and CD spectroscopy as well as various fluorescence-based techniques (Andersson & Maler, 2005; Durr, Yamamoto, Im, Waskell, & Ramamoorthy, 2007; Khatun & Mukhopadhyay, 2013; Nusair et al., 2012; Shi et al., 2013; Whiles, Deems, Vold, & Dennis, 2002). Similar to bicelles, nanodiscs also form bilayer discs (Borch & Hamann, 2009). In the case of the nanodiscs, membrane scaffold proteins wrap around the lipids covering the hydrophobic surface of the bilayer (Inagaki, Ghirlando, & Grisshammer, 2013). Unlike lipid/detergent bicelles, nanodiscs are less dynamic and can be essentially diluted indefinitely without destabilizing the lipid bilayer. Therefore, nanodiscs have proven to be a very useful tool for many biochemical studies of MPs (Alami, Dalal, Lelj-Garolla, Sligar, & Duong, 2007; Baas, Denisov, & Sligar, 2004; Boldog, Grimme, Li, Sligar, & Hazelbauer, 2006; Whorton et al., 2007).

3. BICELLE CRYSTALLIZATION OF MPs

BR was the first MP crystallized in DMPC/CHAPSO bicelles (Faham & Bowie, 2002). It was crystallized both at a higher temperature (37 °C) condition that promotes gel formation and at room temperature (RT) favoring the liquid bicelle phase (Faham et al., 2005). Studies have demonstrated that the bicelle composition can be altered in order to lower the gelling temperature, thereby allowing access to the gel phase at RT (Ottiger & Bax, 1998). The next structure to be determined from bicelles was the β-adrenergic G protein-coupled receptor (GPCR) (Rasmussen et al., 2007), which was an impressive accomplishment since GPCRs have eluded crystallographers for a long time. Since then, the bicelle method has been used successfully on a variety of MPs such as β-barrels (Gruss et al., 2013; Noinaj et al., 2013; Ujwal et al., 2008), membrane enzymes (Huang et al., 2012; Morgan, McNamara, & Zimmer, 2014; Vinothkumar, 2011), channels (Payandeh, Scheuer, Zheng, & Catterall, 2011; Tang et al., 2014), transporters (Chen, Oldham, Davidson, & Chen, 2013; Symersky et al., 2012; Wang, Elferich, & Gouaux, 2012), and additional seven-transmembrane helical MPs (Lee et al., 2013; Luecke et al., 2008; Ran et al., 2013; Table 1). They also include an example of a single-pass integral MP (Huang et al., 2012), demonstrating that the bicelle method is compatible with MPs having only one-transmembrane helix. The increase in the number of successful cases has become apparent in the last few years (Fig. 1B). Many of these structures have been determined to high resolution, with the average resolution equaling 2.5 Å (Table 1). Since the number of structures determined using the bicelle method has significantly increased, it may be now possible to recognize the presence of common elements that lead to successful crystallization.

4. USER GUIDE TO BICELLE CRYSTALLIZATION

The following describes a general bicelle preparation protocol. A detailed case study for the crystallization of an MP complex using the bicelle method is discussed below.

Table 1 Membrane Proteins Crystallized Using the Bicelle Method

Protein Name	PDB Code	Protein (mg/mL)	Bicelle (%)	Bicelle Composition	Lipid Additives	Cryo	Resolution (Å)
7 TMs							
Bacteriorhodopsin (BR)	1KME	8	8	DMPC/CHAPSO (3:1)			2.0
BR—room temperature	1XJI	8	8	DTPC/CHAPSO (3:1)		Reservoir + 35% PEG 2K	2.2
β2 Adrenergic GPCR	2R4R	8–12	8.3	DMPC/CHAPSO (3:1)		Reservoir + 20% glycerol	3.4
Xanthorhodopsin	3DDL	3.75	9.2	DMPC/NM (0.6:1)		15% Ethylene glycol	1.9
BR	4HYX, 4HWL		9	DMPC/FA-4 or FA-7 (10:1)			2.0
Proteorhodopsin	4JQ6					Paratone-N	2.30
Enzymes							
Rhomboid protease	2XTV	9	2	DMPC/CHAPSO (2.6:1)		Reservoir + 25% glycerol	1.7
Transglycosylase	3VMT	15	3	DMPC/CHAPS (3:1)	Lipid II analog	36% PEG 400	2.3
Cellulose synthase	4P02	8	5.7	DMPC/CHAPS (2.34:1)	POPE	Reservoir + 20% Glycerol	2.65
Transporters							
LeuT-Leu	3USG	8–10	7	DMPC/CHAPSO (2.8:1)			2.5

Continued

Table 1 Membrane Proteins Crystallized Using the Bicelle Method—cont'd

Protein Name	PDB Code	Protein (mg/mL)	Bicelle (%)	Bicelle Composition	Lipid Additives	Cryo	Resolution (Å)
LeuT-SeMet	3USL	8–10	7	DMPC/CHAPSO (2.52:1)			2.7
c_{10} ring ATP synthase	3U2F	6		DMPC/CHAPSO (2:1)	DMPE	28% MPD	2.0
Maltose transporter	4JBW	12.5	7	DMPC/CHAPSO (2.7:1)	Cardiolipin	35% PEG MME 2000	3.9
Channels							
Voltage-gated Na channel	3RVY			DMPC/CHAPSO		Reservoir + 28% glucose	2.7
Voltage-gated Ca channel	4MS2			DMPC/CHAPSO		Reservoir + 26% glucose	2.75
β-Barrels							
VDAC	3EMN	12	7	DMPC/CHAPSO (2.8:1)		Reservoir + 10% PEG 400	2.3
TamA	4C00	10	8	DMPC/CHAPSO (2.43:1)		Perfluoropolyether oil	2.3
*Hd*BamAΔ3	4K3C	8	7	DMPC/CHAPSO (2.8:1)			2.9
*Ng*BamA	4K3B	8	7	DMPC/CHAPSO (2.8:1)			3.2

5. BICELLE CRYSTALLIZATION GENERAL PROTOCOL

5.1 Preparation of the Bicelles

- To prepare 250 μL 30% bicelles, 175 μL H_2O and 75 mg lipid/detergent mixture will be needed. Using a nitrogen stream and a glass vial, dry the appropriate amount of chloroform-dissolved DMPC. Ensure that the chloroform is completely evaporated by placing the lipid film under vacuum for 2 h at RT.
- Dissolve the appropriate amount of detergent in H_2O. Common initial detergents to use are DHPC, CHAPS, or CHAPSO. The amount of detergent used is based on the desired lipid:detergent molar ratio. Molar ratios in the range of 2.2–2.8 are the most common, and we suggest a molar ratio of 2.6 as a good starting point. Add the detergent solution to the dried lipid film.
- Overlay the bicelle stock solution with nitrogen, seal the vial, and vortex at 4 °C for $\geq$5 h. Alternatively, a few sonication intervals with a microtip sonicator can be used to achieve proper mixing. The bicelle stock solution should be a clear liquid at lower temperatures with a profound phase transition to a gel-like phase at elevated temperatures (usually $>$25 °C, depending on the lipid/detergent composition used). Occasionally, repeated cooling–warming cycles are necessary to completely dissolve the lipids.

5.2 Reconstitution of the Protein into Bicelles

- Concentrate the purified protein to ~7–12 mg/mL using the largest possible molecular weight cutoff concentrator to avoid excessive detergent concentrations.
- Add one volume of bicelle stock solution to four volumes of protein solution on ice. Keep the bicelle solution cold in order to remain in the liquid phase for easier pipetting.
- Mix by pipetting the solution up and down and incubate on ice for $\geq$30 min.

5.3 Crystal Trial Setup

- Crystallization trials can be set up manually or by standard liquid-handling robotics. The protein/bicelle mixture should be kept cold during crystallization setup in order for it to remain in the liquid phase for easy pipetting.

- Crystal trays can be incubated at various temperatures.

5.4 Visualization

One frequent concern with the bicelle method is the appearance of non-protein crystals or crystal-like features in the crystallization drops. One approach to deal with this problem is to use a UV microscope (Judge, Swift, & Gonzalez, 2005). Tryptophan fluorescence can help identify protein crystals (Fig. 2). Another approach is to set up crystallization trials in the absence of protein and determine if the bicelle solution alone can produce crystals under the crystallization condition of interest. Figure 3 shows an example of crystal-like features grown at 17 °C from a (7%, 2.8:1, DMPC:CHAPSO) bicelle-only solution in condition 88 from crystal screen JCSG+ (Qiagen), containing 0.2 *M* $CaCl_2$, 0.1 *M* Bis–Tris, pH 5.5, and

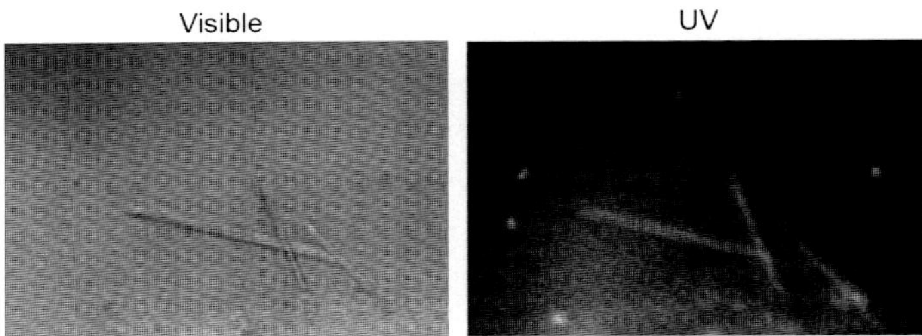

Figure 2 *E. coli* BcsA crystals grown in bicelles (JZ, unpublished). UV microscope reporting on the protein Trp fluorescence aids in identifying protein crystals.

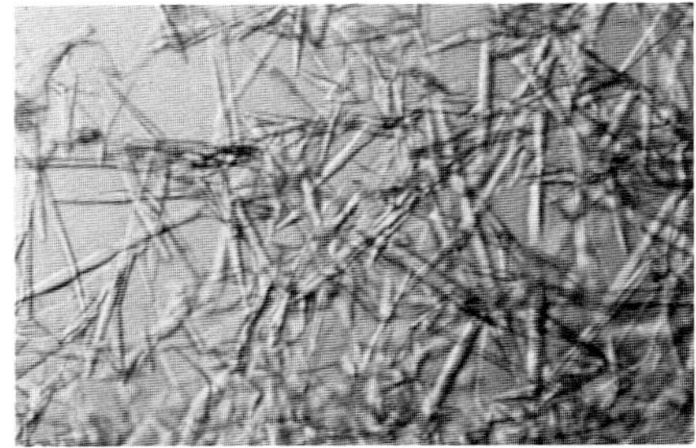

Figure 3 An example of the appearance of nonprotein crystal-like objects observed in bicelles at 17 °C, but not at 21 °C. Crystal-like features grew in condition 88 from crystal screen JCSG+ (Qiagen), containing 0.2 *M* $CaCl_2$, 0.1 *M* Bis–Tris, pH 5.5, and 45% MPD.

45% 2-methyl-2,4-pentanediol (MPD). However, a condition producing such "false positives" should not be dismissed automatically. A very similar condition (20 m*M* $CaCl_2$, 32% MPD, and 100 m*M* Na acetate, pH 4.5) was reported to produce crystals of proteorhodopsin in similar bicelles (Ran et al., 2013). Additionally, a number of the conditions that have been used successfully to produce crystals in bicelles can also generate the appearance of crystalline-like objects in the absence of protein at 17 °C, but not at 21 °C.

5.5 Optimization

- Once protein crystals are identified, it may be necessary to optimize them. In addition to the traditional approach of fine-tuning the crystallization condition, bicelles offer additional parameters to be adjusted. These include the protein:bicelle ratio, the lipid:detergent ratio, and bicelle composition, including the addition of native or nonbicelle-forming lipids.

5.6 Cryo Protection

- The crystallization examples listed in Table 1 document numerous procedures for flash-cooling protein crystals grown using the bicelle method. Sufficient cryo protection has been achieved in the presence of 20–25% glycerol, 26–28% glucose, 15% ethylene glycol, 10–36% polyethylene glycol (PEG) 400, 35% PEG MME 2000, 28% MPD, or oil-based cryoprotectants, such as Paratone-N.

6. SUCCESSFUL CRYSTALLIZATION CONDITIONS

Selection of crystallization conditions can often be difficult and time consuming. In addition, bicelles can be formed by a variety of compositions, thus the number of parameters that can be screened may seem daunting. Over the last 10 years, an increasing number of MPs have been crystallized in bicelles (Fig. 1B), as the method has been more frequently explored. Based on those successes, it is possible to identify reliable patterns used in crystallization, which should be useful in designing initial crystallization screens when using the bicelle method. Here, we review the available data and examine the patterns of conditions and bicelle compositions used. Based on the known crystallization procedures, we have generated a collection of conditions that have produced diffraction quality MP crystals (Table 2). This empirically determined screen contains 22 conditions and can be easily

Table 2 Empirically Derived Bicelle Crystal Screen[a]

	Additives	Buffer	Precipitant
1	0.18 *M* Hexanediol, 3.5% (v/v) triethylene glycol		2.4 *M* NaH_2PO_4
2	0.18 *M* Hexanediol, 3.5% (v/v) triethylene glycol		3.6 *M* NaH_2PO_4
3			3.2 *M* NaH_2PO_4
4			3.0 *M* Sodium phosphate, pH 5.6
5	0.18 *M* Sodium acetate, 0.005 *M* EDTA	0.1 *M* HEPES, pH 7.0	1.8 *M* Ammonium sulfate
6		0.1 *M* Sodium citrate, pH 5.0	2.0 *M* Ammonium sulfate
7		0.1 *M* Sodium citrate, pH 3.3	1.8 *M* Sodium acetate
8		0.1 *M* Imidazole, pH 6.0	0.9 *M* Sodium acetate
9		0.1 *M* Bis–Tris, pH 7.0	1.5 *M* Sodium chloride
10		0.1 *M* MES, pH 6.5	1.0 *M* LiCl, 10% PEG 400
11	0.002 *M* Magnesium chloride, 7% (v/v) propylene glycol		64% (v/v) MPD
12	0.02 *M* Calcium chloride	0.1 *M* Sodium acetate, pH 4.5	32% (v/v) MPD
13	0.3 *M* NaCl	0.1 *M* Tris–HCl, pH 7.5	15% Ethanol
14	0.3 *M* NaCl	0.1 *M* Tris–HCl, pH 7.5	9% Isopropanol
15		0.1 *M* Sodium acetate, pH 4.5	25% (v/v) MPD, 5% (v/v) PEG 400
16	0.05 *M* Magnesium chloride, 0.1 *M* urea	0.1 *M* Sodium acetate, pH 5.0	35% (v/v) MPD, 10% (v/v) PEG 400
17	0.1 *M* Sodium citrate	0.1 *M* HEPES, pH 7.5	12% (v/v) MPD
18		0.1 *M* Tris–HCl, pH 8.5	18% (v/v) MPD, 10% (v/v) PEG 400
19	0.1 *M* Sodium chloride, 0.2 *M* sodium malonate	0.1 *M* Potassium phosphate, pH 7.0	32% (v/v) PEG 300

Table 2 Empirically Derived Bicelle Crystal Screen—cont'd

	Additives	Buffer	Precipitant
20	0.1 *M* Magnesium chloride	0.1 *M* HEPES, pH 8.0	25% (w/v) PEG 400
21	0.18 *M* Hexanediol, 0.28 *M* ammonium sulfate	0.1 *M* Sodium formate, pH 4.3	28.5% (w/v) PEG 2000
22	0.1 *M* NDSB-256	0.1 *M* Sodium cacodylate, pH 5.6	13% (w/v) PEG MME 2000

[a]Images of these conditions with bicelles in the absence of protein can be found at http://people.virginia.edu/~sf3bb/Bicelles/Bicelle-Screen1-Images.html.

expanded into 96 conditions or more by simply varying the pH, the concentration of the main precipitant in each condition, or the additive.

A summary of the observations derived from these conditions are discussed here.

6.1 Precipitants

Although salts initially appeared as the most common category of crystallization reagents, organic solvents and PEG conditions have emerged as particularly useful and are now almost as prevalent (Fig. 4A). Sodium phosphate and ammonium sulfate are the most common salts. Organics, specifically MPD, are also frequently used and are often combined with lesser amounts of low-molecular-weight PEGs. Low-MW (PEG 300–400) and medium-MW PEGs (PEG 2000 and PEG 2000 MME) have been used as the main precipitant in about one-third of the cases. Precipitants can dramatically alter the behavior of lipid bilayers (Kimble-Hill, 2013). For example, PEGs have been reported to increase rigidity of phospholipids and to facilitate fusion events (Boni et al., 1984).

6.2 Buffer pH

The buffer pH can be an important consideration when attempting to crystallize proteins in bicelles. The buffers used in the successful crystallization conditions fall within a range of pH 3.0–8.5 (Fig. 4B). The most apparent feature of the pH distribution is that conditions with acidic pH (<7) far outnumber conditions with basic pH (>7).

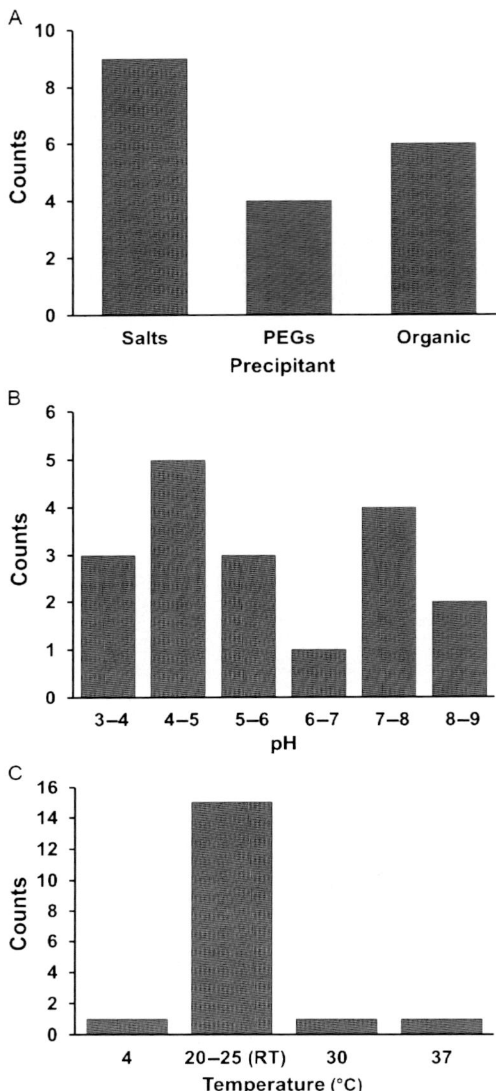

Figure 4 Analysis of the crystallization conditions used successfully for bicelle crystallization. Comparisons of (A) the main precipitants, (B) pHs, and (C) temperature.

6.3 Protein Concentration

The final protein concentration used, after mixing with bicelles, is most commonly around 8–10 mg/mL, with examples as low as 4 mg/mL and as high as 15 mg/mL.

6.4 Temperature

Crystallization of MPs by the bicelle method has been the most successful in the temperature range of 20–25 °C, though there has been some success at 4, 30, and 37 °C as well (Fig. 4C). The choice of the temperature is expected to be critical as the bicellar mixture can undergo significant phase transitions, for example, from the liquid bicelle phase to the gel-like perforated phase. Additionally, low temperatures (4 °C) can result in lower solubility and precipitation of the lipids, while higher temperatures (>30 °C) may lead to instability of the MP of interest.

6.5 Bicelle Composition

Bicelle composition and lipid/detergent ratios are important factors in obtaining and improving crystals. The most frequently used bicelles are composed of a mixture of DMPC/CHAPSO (Fig. 5A). Alternative compositions made by replacing CHAPSO with CHAPS, nonylmaltoside, or facial amphiphiles (FAs) have also been successful in producing good quality crystals (Lee et al., 2013). Of note, DMPC/CHAPS bicelles have been shown to be more stable than DMPC/DHPC bicelles (McKibbin, Farmer, Edwards, Villa, & Booth, 2009), making it an interesting alternative worthy of further exploration. It has also been reported that steroid-based FAs stabilize MPs and are also capable of replacing CHAPSO in the formation of bicelles (Lee et al., 2013). This can be particularly useful in case a MP is destabilized by CHAPSO or DHPC. The FAs are CHAPSO analogs that include a maltoside group (FA-4) or a phosphocholine group (FA-7) as the hydrophilic part of the molecule.

During optimization, small amounts of other lipids may be used in combination with DMPC, such as 1,2-dimyristoyl-*sn*-glycero-3-phosphoethanolamine (DMPE) or 1-palmitoyl-2-oleoyl-*sn*-glycero-3-phosphoethanolamine (POPE). For example, high-quality crystals of the cellulose synthase BcsA–B complex were obtained by using a modified bicelle composition consisting of DMPC/POPE/CHAPS (2.34:0.05:1) (Morgan et al., 2014) as described in detail later. In another case, LeuT-SeMet crystals were obtained by using bicelles in which 10% of the DMPC was replaced with DMPE (Wang et al., 2012). The c_{10} ring of the yeast mitochondrial ATP synthase was crystallized with DMPC/CHAPSO (2:1) bicelles with 3% (w/w) 1,1′,2,2′-tetramyristoyl cardiolipin (Symersky et al., 2012). Finally, the *Staphylococcus aureus* membrane-bound transglycosylase was crystallized with a substrate analog, a lipid II derivative, in addition to DMPC/CHAPS

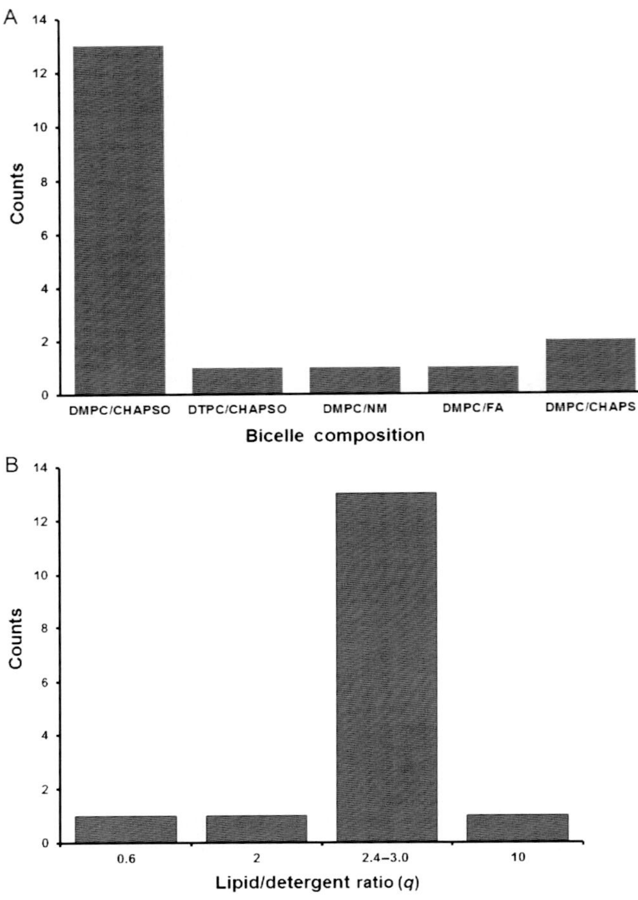

Figure 5 Comparison of bicelle compositions (A) and q-ratios (B) used successfully for crystallization.

(Huang et al., 2012). These examples show that addition of lipids can be an important factor when optimizing crystal growth and diffraction quality.

The majority of crystals were produced when working within q-ratios of 2.4:1–3:1 (Fig. 5B). However, a wide range of q-ratios can be used to form bicelles, and examples of successful crystallization at the far ends of this spectrum (0.6 and 10) have been reported (Lee et al., 2013; Luecke et al., 2008).

6.6 Detergent Used for Purification

A frequent question is whether the detergent used in the purification is compatible with the bicelle method. There has not been a systematic investigation on how the various detergents influence the behavior of the bicellar mixture or how they may alter the bicelle phase diagram (which is also likely influenced by many crystallization reagents). The successful examples listed in Table 1 show that the bicelle method is compatible with several commonly used detergents (Table 3). The most frequently used detergent is dodecylmaltoside (DDM), though many other common detergents can be used effectively. Regarding the choice of detergent, priority should be assigned first to those detergents that are compatible with the protein.

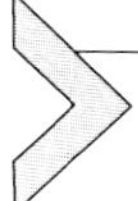

7. CASE STUDY OF CELLULOSE SYNTHASE CRYSTALLIZATION

In the case of the *Rhodobacter sphaeroides* bacterial cellulose synthase BcsA–B complex (Morgan et al., 2014), initial crystallization trials were carried out using bicelle mixtures containing 100 mg total of DMPC:detergent at a 2.6:1 molar ratio dissolved in 250 μL of water (28.5% bicelles). Initial bicelle mixtures tested included DMPC/DHPC, DMPC/CHAPS, and DMPC/CHAPSO. The BcsA–B complex (purified in 5 m*M* N,N-dimethyldodecylamine N-oxide (LDAO)) was concentrated to ~10 mg/mL in a 100 kDa cutoff Amicon spin concentrator, and bicelle stock was added to the concentrated protein at a volume/volume ratio of four parts protein to one part bicelle stock, resulting in a bicelle concentration of ~5.7%. The mixture was incubated on ice for ≥1 h, after which the allosteric activator

Table 3 Detergents that Have Been Demonstrated to Be Compatible with the Bicelle Method

Detergents
DDM
OG
C8E4
LDAO
DM
Digitonin
NG

Detergents used for purification before the sample was added to the bicelle mixture.

cyclic-di-GMP was added to 1 mM final concentration, diluting the protein/bicelle mixture by an additional 10%. Initial crystallization trials were performed using a mosquito crystallization robot in INTELLI-PLATE 96-3 by combining 100 nL protein/bicelle and 100 nL well solution. Qiagen Anions Suite, Molecular Dimensions MemGold 2, and the Qiagen JCSG + Suite screens were set up at 4, 17, 23, and 30 °C using each of the bicelle compositions. Initial hits were obtained in 1.25 M sodium acetate (NaOAc) and 100 mM NaOAc, pH 4.6 (Qiagen Anion Suite solution #2) with DMPC/CHAPS bicelles at 30 °C. The crystals were reproduced in 24-well crystallization plates by screening pH against NaOAc concentrations, with best results at pH 4.5 and ~1.7 M NaOAc.

In order to increase the crystal size, additional low pH buffers were screened and the bicelle mixtures were modified by replacing 10% of DMPC with lipids carrying different head groups and/or longer chain fatty acids, such as oleic and palmitic acids, expected to alter the bilayer properties. DMPC/DMPE/CHAPS, DMPC/POPE/CHAPS, and DMPC/DOPC/CHAPS mixtures were all set up with the refined NaOAc versus pH screen. A significant improvement of the crystal size was observed in 1.55–1.8 M NaOAc, 50 mM sodium citrate, pH 2.5–3.5 (final pH ~5.4–5.6), and DMPC/POPE/CHAPS or DMPC/DOPC/CHAPS bicelles. Notably, these bicelle mixtures appeared much more fluid in that they did not form a white solid phase on ice and formed a less viscous gel phase at 37 °C compared to the DMPC/CHAPS and the DMPC/DMPE/CHAPS mixtures. Therefore, a bicelle mixture containing additional CHAPS (resulting in a final ratio of 2.34:0.05:1, DMPC:POPE:CHAPS), which appeared to further increase bicelle fluidity, was also screened and further improved crystal size. Finally, incubating the protein/bicelle/c-di-GMP mixture on ice overnight before setting up crystal trays resulted in highly reproducible, large crystals that grew to full size of ~$0.5 \times 0.1 \times 0.1$ mm in 1.5–2 weeks.

The crystals were cryoprotected by the gradual replacement of mother solution with a solution containing 1.92 M NaOAc, 50 mM Na citrate, pH 3.5, 2% DMPC/POPE/CHAPS bicelles, and 20% glycerol and flash cooled in liquid nitrogen. The crystals belonged to spacegroup $P2_12_12_1$ with a solvent content of approximately 80% and diffracted X-rays isotropically to 2.65 Å at beamline 22ID-D, APS, Argonne National Laboratory. Although typically crystals from bicelles have type I packing (Agah & Faham, 2012), in this case, we find the packing of the molecules in the crystal is inconsistent with a stacking of two-dimensional crystals, suggesting that the molecules are surrounded by discontinuous lipid discs rather than being embedded in a planar lipid bilayer. Crystal contacts are primarily mediated by hydrophilic intra- and extracellular regions, yet electron density for five ordered lipids is visible in the final map including two near the interface between adjacent molecules.

8. ALTERNATIVE APPROACHES

Reports from several laboratories document improved purification and crystallization successes when supplementing MPs with either native or synthetic phospholipids, as for example in the case of the mammalian voltage-gated K^+ channel (Long, Campbell, & Mackinnon, 2005). A number of cases even state the dependence of crystal growth on supplemented lipids, exemplified by several P-type ATPases (Jidenko et al., 2005), lactose permease (Guan, Smirnova, Verner, Nagamori, & Kaback, 2006), or the cytochrome *b6f* complex of oxygenic photosynthesis (Zhang, Kurisu, Smith, & Cramer, 2003).

Accordingly, Poul Nissen's laboratory developed a MP crystallization protocol that resembles the bicelle crystallization approach in many aspects, yet it is potentially more versatile and applicable to essentially any type of lipid required. The "high lipid and detergent" (HiLiDe) method systematically relipidates purified MPs at various detergent–lipid ratios prior to crystallization, resulting in detergent/lipid compositions that resemble q-ratios frequently used for bicelle preparations, Table 1 (Gourdon et al., 2011). As such, the method is not limited to "bicelle-forming" detergent/lipid combinations and yields low-viscosity samples that can be dispensed by crystallization robotics. Furthermore, it facilitates the systematic variation of lipid(s) and detergent(s) to optimize crystal growth and quality. Recent MP structures resulting from HiLiDe crystallization methods include the NMDA receptor (Lee et al., 2014), copper and zinc P-type ATPases (Andersson et al., 2014; Wang et al., 2014), and bacterial vitamin K epoxide reductase (Liu, Cheng, Fowle Grider, Shen, & Li, 2014), among others (Raman, Cherezov, & Caffrey, 2006).

9. SUMMARY

There has been a significant increase in the number of structures determined from bicelle crystallized MPs. The lack of a bicelle-specific crystallization screen has been an obstacle when using this method, but with the larger number of examples available now, it has become possible to compare crystallization conditions and attempt to identify patterns. Indeed, we have been able to assemble a collection of bona fide conditions that can be used as a core for a larger crystallization screen customized for the bicelle method.

ACKNOWLEDGMENTS

J.L.W.M. is supported by a National Science Foundation Graduate Research Fellowship, Grant DGE-1315231. J.Z. is supported by the National Institutes of Health, Grant 1R01GM101001, and the Center for LignoCellulose Structure and Formation, Energy Frontier Research Center, US Department of Energy, Office of Science, Grant DE-SC0001090.

REFERENCES

Agah, S., & Faham, S. (2012). Crystallization of membrane proteins in bicelles. *Methods in Molecular Biology*, *914*, 3–16.

Alami, M., Dalal, K., Lelj-Garolla, B., Sligar, S. G., & Duong, F. (2007). Nanodiscs unravel the interaction between the SecYEG channel and its cytosolic partner SecA. *The EMBO Journal*, *26*(8), 1995–2004.

Andersson, A., & Maler, L. (2005). Magnetic resonance investigations of lipid motion in isotropic bicelles. *Langmuir*, *21*(17), 7702–7709.

Andersson, M., Mattle, D., Sitsel, O., Klymchuk, T., Nielsen, A. M., Moller, L. B., et al. (2014). Copper-transporting P-type ATPases use a unique ion-release pathway. *Nature Structural & Molecular Biology*, *21*(1), 43–48.

Baas, B. J., Denisov, I. G., & Sligar, S. G. (2004). Homotropic cooperativity of monomeric cytochrome P450 3A4 in a nanoscale native bilayer environment. *Archives of Biochemistry and Biophysics*, *430*(2), 218–228.

Belli, S., Elsener, P. M., Wunderli-Allenspach, H., & Kramer, S. D. (2009). Cholesterol-mediated activation of P-glycoprotein: Distinct effects on basal and drug-induced ATPase activities. *Journal of Pharmaceutical Sciences*, *98*(5), 1905–1918.

Boldog, T., Grimme, S., Li, M., Sligar, S. G., & Hazelbauer, G. L. (2006). Nanodiscs separate chemoreceptor oligomeric states and reveal their signaling properties. *Proceedings of the National Academy of Sciences of the United States of America*, *103*(31), 11509–11514.

Boni, L. T., Hah, J. S., Hui, S. W., Mukherjee, P., Ho, J. T., & Jung, C. Y. (1984). Aggregation and fusion of unilamellar vesicles by poly(ethylene glycol). *Biochimica et Biophysica Acta*, *775*(3), 409–418.

Borch, J., & Hamann, T. (2009). The nanodisc: A novel tool for membrane protein studies. *Biological Chemistry*, *390*(8), 805–814.

Caffrey, M. (2000). A lipid's eye view of membrane protein crystallization in mesophases. *Current Opinion in Structural Biology*, *10*(4), 486–497.

Chen, S., Oldham, M. L., Davidson, A. L., & Chen, J. (2013). Carbon catabolite repression of the maltose transporter revealed by X-ray crystallography. *Nature*, *499*(7458), 364–368.

Cherezov, V., Rosenbaum, D. M., Hanson, M. A., Rasmussen, S. G., Thian, F. S., Kobilka, T. S., et al. (2007). High-resolution crystal structure of an engineered human beta2-adrenergic G protein-coupled receptor. *Science*, *318*(5854), 1258–1265.

Claridge, J. K., Aittoniemi, J., Cooper, D. M., & Schnell, J. R. (2013). Isotropic bicelles stabilize the juxtamembrane region of the influenza M2 protein for solution NMR studies. *Biochemistry*, *52*(47), 8420–8429.

Czerski, L., & Sanders, C. R. (2000). Functionality of a membrane protein in bicelles. *Analytical Biochemistry*, *284*(2), 327–333.

De Angelis, A. A., & Opella, S. J. (2007). Bicelle samples for solid-state NMR of membrane proteins. *Nature Protocols*, *2*(10), 2332–2338.

Durr, U. H., Yamamoto, K., Im, S. C., Waskell, L., & Ramamoorthy, A. (2007). Solid-state NMR reveals structural and dynamical properties of a membrane-anchored electron-carrier protein, cytochrome b5. *Journal of the American Chemical Society*, *129*(21), 6670–6671.

Faham, S., Boulting, G. L., Massey, E. A., Yohannan, S., Yang, D., & Bowie, J. U. (2005). Crystallization of bacteriorhodopsin from bicelle formulations at room temperature. *Protein Science*, *14*(3), 836–840.

Faham, S., & Bowie, J. U. (2002). Bicelle crystallization: A new method for crystallizing membrane proteins yields a monomeric bacteriorhodopsin structure. *Journal of Molecular Biology*, *316*(1), 1–6.

Gourdon, P., Andersen, J. L., Hein, K. L., Bublitz, M., Pedersen, B. P., Liu, X.-Y., et al. (2011). HiLiDe—Systematic approach to membrane protein crystallization in lipid and detergent. *Crystal Growth & Design*, *11*(6), 2098–2106.

Gruss, F., Zahringer, F., Jakob, R. P., Burmann, B. M., Hiller, S., & Maier, T. (2013). The structural basis of autotransporter translocation by TamA. *Nature Structural & Molecular Biology*, *20*(11), 1318–1320.

Guan, L., Smirnova, I. N., Verner, G., Nagamori, S., & Kaback, H. R. (2006). Manipulating phospholipids for crystallization of a membrane transport protein. *Proceedings of the National Academy of Sciences of the United States of America*, *103*(6), 1723–1726.

Harbers, M. (2014). Wheat germ systems for cell-free protein expression. *FEBS Letters*, *588*, 2762–2773.

Huang, C. Y., Shih, H. W., Lin, L. Y., Tien, Y. W., Cheng, T. J., Cheng, W. C., et al. (2012). Crystal structure of *Staphylococcus aureus* transglycosylase in complex with a lipid II analog and elucidation of peptidoglycan synthesis mechanism. *Proceedings of the National Academy of Sciences of the United States of America*, *109*(17), 6496–6501.

Hubbard, C., McNamara, J. T., Azumaya, C., Patel, M. S., & Zimmer, J. (2012). The hyaluronan synthase catalyzes the synthesis and membrane translocation of hyaluronan. *Journal of Molecular Biology*, *418*(1–2), 21–31.

Inagaki, S., Ghirlando, R., & Grisshammer, R. (2013). Biophysical characterization of membrane proteins in nanodiscs. *Methods*, *59*(3), 287–300.

Jidenko, M., Nielsen, R. C., Sorensen, T. L., Moller, J. V., le Maire, M., Nissen, P., et al. (2005). Crystallization of a mammalian membrane protein overexpressed in *Saccharomyces cerevisiae*. *Proceedings of the National Academy of Sciences of the United States of America*, *102*(33), 11687–11691.

Judge, R. A., Swift, K., & Gonzalez, C. (2005). An ultraviolet fluorescence-based method for identifying and distinguishing protein crystals. *Acta Crystallographica. Section D, Biological Crystallography*, *61*(Pt. 1), 60–66.

Khatun, U. L., & Mukhopadhyay, C. (2013). Interaction of bee venom toxin melittin with ganglioside GM1 bicelle. *Biophysical Chemistry*, *180–181*, 66–75.

Kimble-Hill, A. (2013). A review of factors affecting the success of membrane protein crystallization using bicelles. *Frontiers in Biology*, *8*(3), 261–272.

Kucerka, N., Nieh, M. P., & Katsaras, J. (2011). Fluid phase lipid areas and bilayer thicknesses of commonly used phosphatidylcholines as a function of temperature. *Biochimica et Biophysica Acta*, *1808*(11), 2761–2771.

Landau, E. M., & Rosenbusch, J. P. (1996). Lipidic cubic phases: A novel concept for the crystallization of membrane proteins. *Proceedings of the National Academy of Sciences of the United States of America*, *93*(25), 14532–14535.

Lee, A. G. (2011). Lipid-protein interactions. *Biochemical Society Transactions*, *39*(3), 761–766.

Lee, S. C., Bennett, B. C., Hong, W. X., Fu, Y., Baker, K. A., Marcoux, J., et al. (2013). Steroid-based facial amphiphiles for stabilization and crystallization of membrane proteins. *Proceedings of the National Academy of Sciences of the United States of America*, *110*(13), E1203–E1211.

Lee, C. H., Lu, W., Michel, J. C., Goehring, A., Du, J., Song, X., et al. (2014). NMDA receptor structures reveal subunit arrangement and pore architecture. *Nature*, *511*(7508), 191–197.

Li, D., & Caffrey, M. (2011). Lipid cubic phase as a membrane mimetic for integral membrane protein enzymes. *Proceedings of the National Academy of Sciences of the United States of America*, *108*(21), 8639–8644.

Li, M., Morales, H. H., Katsaras, J., Kucerka, N., Yang, Y., Macdonald, P. M., et al. (2013). Morphological characterization of DMPC/CHAPSO bicellar mixtures: A combined SANS and NMR study. *Langmuir*, *29*(51), 15943–15957.

Liu, S., Cheng, W., Fowle Grider, R., Shen, G., & Li, W. (2014). Structures of an intramembrane vitamin K epoxide reductase homolog reveal control mechanisms for electron transfer. *Nature Communications*, *5*, 3110.

Long, S. B., Campbell, E. B., & Mackinnon, R. (2005). Crystal structure of a mammalian voltage-dependent Shaker family K^+ channel. *Science*, *309*(5736), 897–903.

Luecke, H., Schobert, B., Stagno, J., Imasheva, E. S., Wang, J. M., Balashov, S. P., et al. (2008). Crystallographic structure of xanthorhodopsin, the light-driven proton pump with a dual chromophore. *Proceedings of the National Academy of Sciences of the United States of America*, *105*(43), 16561–16565.

McKibbin, C., Farmer, N. A., Edwards, P. C., Villa, C., & Booth, P. J. (2009). Urea unfolding of opsin in phospholipid bicelles. *Photochemistry and Photobiology*, *85*(2), 494–500.

Morgan, J. L., McNamara, J. T., & Zimmer, J. (2014). Mechanism of activation of bacterial cellulose synthase by cyclic di-GMP. *Nature Structural & Molecular Biology*, *21*(5), 489–496.

Morrison, E. A., DeKoster, G. T., Dutta, S., Vafabakhsh, R., Clarkson, M. W., Bahl, A., et al. (2012). Antiparallel EmrE exports drugs by exchanging between asymmetric structures. *Nature*, *481*(7379), 45–50.

Morrison, E. A., & Henzler-Wildman, K. A. (2012). Reconstitution of integral membrane proteins into isotropic bicelles with improved sample stability and expanded lipid composition profile. *Biochimica et Biophysica Acta*, *1818*(3), 814–820.

Newstead, S., Ferrandon, S., & Iwata, S. (2008). Rationalizing alpha-helical membrane protein crystallization. *Protein Science*, *17*(3), 466–472.

Nieh, M. P., Raghunathan, V. A., Pabst, G., Harroun, T., Nagashima, K., Morales, H., et al. (2011). Temperature driven annealing of perforations in bicellar model membranes. *Langmuir*, *27*(8), 4838–4847.

Nielsen, C., & Andersen, O. S. (2000). Inclusion-induced bilayer deformations: Effects of monolayer equilibrium curvature. *Biophysical Journal*, *79*(5), 2583–2604.

Noinaj, N., Kuszak, A. J., Gumbart, J. C., Lukacik, P., Chang, H., Easley, N. C., et al. (2013). Structural insight into the biogenesis of beta-barrel membrane proteins. *Nature*, *501*(7467), 385–390.

Nusair, N. A., Mayo, D. J., Dorozenski, T. D., Cardon, T. B., Inbaraj, J. J., Karp, E. S., et al. (2012). Time-resolved EPR immersion depth studies of a transmembrane peptide incorporated into bicelles. *Biochimica et Biophysica Acta*, *1818*(3), 821–828.

Omadjela, O., Narahari, A., Strumillo, J., Melida, H., Mazur, O., Bulone, V., et al. (2013). BcsA and BcsB form the catalytically active core of bacterial cellulose synthase sufficient for in vitro cellulose synthesis. *Proceedings of the National Academy of Sciences of the United States of America*, *110*(44), 17856–17861.

Ostermeier, C., Iwata, S., Ludwig, B., & Michel, H. (1995). Fv fragment-mediated crystallization of the membrane protein bacterial cytochrome c oxidase. *Nature Structural Biology*, *2*(10), 842–846.

Ottiger, M., & Bax, A. (1998). Characterization of magnetically oriented phospholipid micelles for measurement of dipolar couplings in macromolecules. *Journal of Biomolecular NMR*, *12*(3), 361–372.

Park, K. H., Billon-Denis, E., Dahmane, T., Lebaupain, F., Pucci, B., Breyton, C., et al. (2011). In the cauldron of cell-free synthesis of membrane proteins: Playing with new surfactants. *New Biotechnology*, *28*(3), 255–261.

Payandeh, J., Scheuer, T., Zheng, N., & Catterall, W. A. (2011). The crystal structure of a voltage-gated sodium channel. *Nature*, *475*(7356), 353–358.

Prive, G. G., Verner, G. E., Weitzman, C., Zen, K. H., Eisenberg, D., & Kaback, H. R. (1994). Fusion proteins as tools for crystallization: The lactose permease from *Escherichia coli*. *Acta Crystallographica. Section D, Biological Crystallography*, *50*(Pt. 4), 375–379.

Raman, P., Cherezov, V., & Caffrey, M. (2006). The membrane protein data bank. *Cellular and Molecular Life Sciences*, *63*(1), 36–51.

Ran, T., Ozorowski, G., Gao, Y., Sineshchekov, O. A., Wang, W., Spudich, J. L., et al. (2013). Cross-protomer interaction with the photoactive site in oligomeric proteorhodopsin complexes. *Acta Crystallographica. Section D, Biological Crystallography*, *69*(Pt. 10), 1965–1980.

Rasmussen, S. G., Choi, H. J., Rosenbaum, D. M., Kobilka, T. S., Thian, F. S., Edwards, P. C., et al. (2007). Crystal structure of the human beta2 adrenergic G-protein-coupled receptor. *Nature*, *450*(7168), 383–387.

Sanders, C. R., 2nd., & Prestegard, J. H. (1990). Magnetically orientable phospholipid bilayers containing small amounts of a bile salt analogue, CHAPSO. *Biophysical Journal*, *58*(2), 447–460.

Sanders, C. R., 2nd., & Schwonek, J. P. (1992). Characterization of magnetically orientable bilayers in mixtures of dihexanoylphosphatidylcholine and dimyristoylphosphatidylcholine by solid-state NMR. *Biochemistry*, *31*(37), 8898–8905.

Sharonov, G. V., Bocharov, E. V., Kolosov, P. M., Astapova, M. V., Arseniev, A. S., & Feofanov, A. V. (2014). Point mutations in dimerization motifs of the transmembrane domain stabilize active or inactive state of the EphA2 receptor tyrosine kinase. *The Journal of Biological Chemistry*, *289*(21), 14955–14964.

Shi, X., Bi, Y., Yang, W., Guo, X., Jiang, Y., Wan, C., et al. (2013). Ca^{2+} regulates T-cell receptor activation by modulating the charge property of lipids. *Nature*, *493*(7430), 111–115.

Shimono, K., Goto, M., Kikukawa, T., Miyauchi, S., Shirouzu, M., Kamo, N., et al. (2009). Production of functional bacteriorhodopsin by an *Escherichia coli* cell-free protein synthesis system supplemented with steroid detergent and lipid. *Protein Science*, *18*(10), 2160–2171.

Symersky, J., Pagadala, V., Osowski, D., Krah, A., Meier, T., Faraldo-Gomez, J. D., et al. (2012). Structure of the c(10) ring of the yeast mitochondrial ATP synthase in the open conformation. *Nature Structural & Molecular Biology*, *19*(5), 485–491, S481.

Tang, L., Gamal El-Din, T. M., Payandeh, J., Martinez, G. Q., Heard, T. M., Scheuer, T., et al. (2014). Structural basis for Ca^{2+} selectivity of a voltage-gated calcium channel. *Nature*, *505*(7481), 56–61.

Triba, M. N., Warschawski, D. E., & Devaux, P. F. (2005). Reinvestigation by phosphorus NMR of lipid distribution in bicelles. *Biophysical Journal*, *88*(3), 1887–1901.

Uhlemann, E. M., Pierson, H. E., Fillingame, R. H., & Dmitriev, O. Y. (2012). Cell-free synthesis of membrane subunits of ATP synthase in phospholipid bicelles: NMR shows subunit a fold similar to the protein in the cell membrane. *Protein Science*, *21*(2), 279–288.

Ujwal, R., Cascio, D., Colletier, J. P., Faham, S., Zhang, J., Toro, L., et al. (2008). The crystal structure of mouse VDAC1 at 2.3 Å resolution reveals mechanistic insights into metabolite gating. *Proceedings of the National Academy of Sciences of the United States of America*, *105*, 17742–17747.

Vinothkumar, K. R. (2011). Structure of rhomboid protease in a lipid environment. *Journal of Molecular Biology*, *407*(2), 232–247.

Wang, H., Elferich, J., & Gouaux, E. (2012). Structures of LeuT in bicelles define conformation and substrate binding in a membrane-like context. *Nature Structural & Molecular Biology*, *19*(2), 212–219.

Wang, K., Sitsel, O., Meloni, G., Autzen, H. E., Andersson, M., Klymchuk, T., et al. (2014). Structure and mechanism of Zn^{2+}-transporting P-type ATPases (Letter). *Nature, Advance Online Publication*.

Whiles, J. A., Deems, R., Vold, R. R., & Dennis, E. A. (2002). Bicelles in structure-function studies of membrane-associated proteins. *Bioorganic Chemistry*, *30*(6), 431–442.

Whorton, M. R., Bokoch, M. P., Rasmussen, S. G., Huang, B., Zare, R. N., Kobilka, B., et al. (2007). A monomeric G protein-coupled receptor isolated in a high-density lipoprotein particle efficiently activates its G protein. *Proceedings of the National Academy of Sciences of the United States of America*, *104*(18), 7682–7687.

Zhang, H., Kurisu, G., Smith, J. L., & Cramer, W. A. (2003). A defined protein-detergent-lipid complex for crystallization of integral membrane proteins: The cytochrome b6f complex of oxygenic photosynthesis. *Proceedings of the National Academy of Sciences of the United States of America*, *100*(9), 5160–5163.

CHAPTER TWENTY

Fluorescence Recovery After Photobleaching in Lipidic Cubic Phase (LCP-FRAP): A Precrystallization Assay for Membrane Proteins

Gustavo Fenalti[1], Enrique E. Abola[2], Chong Wang[3], Beili Wu[4], Vadim Cherezov[5]

Department of Integrative Structural and Computational Biology, The Scripps Research Institute, La Jolla, California, USA

[5]Corresponding author: e-mail address: cherezov@usc.edu

Contents

[1] Current address: Celgene Corporation, 10300 Campus Point Drive, San Diego, CA 92121, USA.

[2,5] Current address: The Bridge@USC, University of Southern California, Los Angeles, CA 90089, USA.

[3] Current address: Department of Chemistry and Chemical Biology, Harvard University, Cambridge, MA 02138, USA.

[4] Current address: CAS Key Laboratory of Receptor Research, Shanghai Institute of Materia Medica, Chinese Academy of Sciences, 555 Zuchongzhi Road, Pudong, Shanghai 201203, PR China.

Methods in Enzymology, Volume 557
ISSN 0076-6879
http://dx.doi.org/10.1016/bs.mie.2014.12.008

Abstract

Crystallization of integral membrane proteins (MPs) is notoriously difficult, given their poor stability outside native membrane environment and due to the interference of detergent micelles with crystallization process. MP crystallization in a membrane mimetic matrix, known as lipidic cubic phase (LCP), has recently started to gain popularity, following successes in structure determination of G protein-coupled receptors (GPCRs), transporters, and enzymes. Unlike crystallization trials in aqueous solutions where protein molecules are free to move, diffusion of MPs in LCP is restricted, and, thus, a high level of protein mobility can serve as an early indication for subsequent crystallization success. Prompted by our initial observations that precipitant conditions can dramatically affect diffusion of GPCRs in LCP, we have developed a simple pre-crystallization assay, based on measuring protein diffusion at a number of different conditions by fluorescence recovery after photobleaching (LCP-FRAP). Over the last few years, the LCP-FRAP assay was incorporated in our GPCR structure determination pipeline and proved as a powerful technique allowing for a faster identification of crystallization conditions for many different receptors. The assay is used to screen for the best protein constructs, ligands, LCP host lipids, precipitants, and additives, thereby focusing subsequent crystallization trials on the most promising parts of the multi-dimensional crystallization phase diagram, substantially increasing the likelihood of finding the right crystallization condition. Here, we describe our LCP-FRAP protocols for guiding GPCR crystallization, which can be adapted to any other MP, and discuss some of the critical considerations related to application of this assay.

1. INTRODUCTION

Membrane proteins (MPs) represent an important class of proteins involved in essential cellular and physiological processes, including signal and energy transduction, transport of ions and nutrients, and catalysis of chemical reactions and other functions. Developing an understanding of their mechanisms of action requires access to their high-resolution structures as well as that of their complexes.

The most successful technique for elucidating three-dimensional structures of macromolecules is X-ray crystallography, which calls for growing well-ordered protein crystals. However, MPs are notoriously difficult to crystallize. Traditional approach includes extracting MPs with detergents and using vapor diffusion to crystallize them in the form of protein/detergent complexes. Many MPs, however, are not stable in detergent solutions, do not crystallize or form only poorly ordered crystals. An alternative approach, introduced almost 20 years ago by Landau and Rosenbusch (1996), utilizes lipidic cubic phase (LCP) for crystallizing MPs. This technique takes advantage of a membrane-like environment to stabilize proteins and typically results in crystals with better packing and lower solvent content compared to the traditional vapor diffusion techniques. Initial great success of the LCP (also known as *in meso*) technique was in elucidating details of the bacteriorhodopsin photocycle by providing high-resolution structures of photocycle intermediates and various bacteriorhodopsin mutants (Katona, Andreasson, Landau, Andreasson, & Neutze, 2003; Landau, Pebay-Peyroula, & Neutze, 2003). After many years of LCP technology development, including miniaturization and automation of crystallization trials as well as improvements in crystal detection and crystallographic data collection and processing (Cherezov, 2011), another grand success came with the publication of the high-resolution structure of β_2-adrenergic receptor (β_2AR; Cherezov et al., 2007; Rosenbaum et al., 2007), a G protein-coupled receptor (GPCR). Since then, the structures of more than 25 novel GPCRs have been solved using this approach (Katritch et al., 2014), including those of GPCR–protein complexes (Rasmussen et al., 2011). By the end of 2014, there are a total of 196 structures for 65 unique MPs in the Protein Data Bank, for which successful structure determination is attributed to the *in meso* technique.

Despite the successful crystallization of many novel MPs using the LCP method, obtaining crystals for a given target remains highly challenging and a number of assays and technologies have been developed to characterize the stability and physicochemical properties of proteins before and after reconstitution in LCP, prior to entering into extensive crystallization trials (Cherezov, 2011; Xu, Liu, Hanson, Stevens, & Cherezov, 2011). Here, we present general considerations and protocols for using fluorescence recovery after photobleaching in LCP (LCP-FRAP) as an efficient precrystallization screening assay for profiling diffusion of fluorescently labeled MPs in a variety of conditions. LCP-FRAP can serve to select the most promising protein constructs, ligands, LCP host lipids, and precipitants for subsequent crystallization trials, thus, considerably reducing time and effort for obtaining initial crystal hits.

2. EXPERIMENTAL COMPONENTS AND CONSIDERATIONS FOR LCP-FRAP ASSAYS

2.1 LCP structure and properties

Lipids are amphiphilic molecules which upon mixing with water form a variety of liquid crystalline phases determined by their chemical structure, temperature, and hydration level (Briggs & Caffrey, 1994; Qiu & Caffrey, 2000). The most common class of lipids used for LCP crystallization is monoacylglycerides (MAGs; Caffrey, Lyons, Smyth, & Hart, 2009). They contain a hydrophilic glycerol head group attached to a hydrophobic monounsaturated fatty acid chain through an ester bond. At high hydration levels, many MAGs form LCP in a wide range of temperatures. Topologically, LCP is composed of a single lipid bilayer that separates the space into two interpenetrating nonintersecting networks of water channels. Both lipid bilayer and water channels are continuous in all three dimensions, and therefore, this type of LCP is often referred to as a bicontinuous phase. Such continuity allows MPs to diffuse within the bilayer and for precipitants to penetrate inside the LCP and induce crystallization. The lipid bilayer of LCP follows an infinite periodic minimal surface (IPMS) with a cubic symmetry (Hyde, Andersson, Ericsson, & Larsson, 1984). Each point on the IPMS corresponds to a saddle point with the two principal orthogonal curvatures of the same absolute values but opposite directions resulting in a mean curvature of zero (Rummel et al., 1998). LCP lattices with three different space groups have been observed: *Ia3d* (gyroid), *Pn3m* (diamond), and *Im3m* (primitive). The spatial structure of LCP imposes certain limits on the size of MPs, within which the proteins are not affected by the curvature and can pass through the narrow channels. Typical lattice parameter of a monoolein-based LCP with *Pn3m* space group is 110 Å, with a water channel diameter of 50 Å. We have previously shown that one needs to swell the LCP transforming it into a more disordered liquid-like sponge phase in order to crystallize the light-harvesting complex 2 (LH2), a 123-kDa protein (Cherezov, Clogston, Papiz, & Caffrey, 2006). We hypothesized that the swelling relaxes spatial constraints and likely increases mobility of large proteins. Indeed, for example, photosynthetic reaction center from *Rhodobacter sphaeroides* (101 kDa; Katona et al., 2003) and *Blastochloris viridis* (144 kDa; Wohri et al., 2009), ba3 cytochrome *c* oxidase (94 kDa; Tiefenbrunn et al., 2011), and β_2AR–T4L/Gs/nanobody complex (~165 kDa, including T4L and nanobody; Rasmussen et al., 2011), were crystallized from sponge mesophases. Macroscopically, LCP appears as a sticky, viscous, transparent, and optically isotropic gel. This material does not flow on its own or under gravity and cannot be pipetted, but it can be extruded through a needle using a positive displacement syringe, requiring development of special tools and protocols for LCP applications (Cherezov, 2011).

2.2 Development of an LCP-FRAP assay

An important consideration for successful crystallization is the ability of the target MP to diffuse within the LCP matrix leading to nucleation and crystal growth. Diffusion in LCP is generally dependent on the protein size and shape relative to the narrowest parts of the water channels (~50 Å in the case of monoolein; Grabe, Neu, Oster, & Nollert, 2003). However, since lipidic mesophases are relatively soft and dynamic materials, it is difficult to predict just from their structural parameters whether a particular protein will or will not diffuse inside them and what effect the spatial constraints will have on the diffusion rates of the protein. Therefore, we started to use the FRAP technique to systematically study diffusion within the LCP matrix of amphiphilic and soluble macromolecules of a variety of size and architecture, including lipids, proteins, and polymers. We examined the effects of LCP shrinking and swelling by common precipitant agents on the diffusion rates of these macromolecules. For LCP-FRAP, the protein is labeled by a fluorescent tag and inserted in LCP under different conditions. Fluorescence is bleached by a laser in a spot of a few microns diameter generally defined by the microscope objective; the recovery of fluorescence inside the spot is then followed over time. These experiments led to an unexpected observation that even relatively small proteins, such as some members of the GPCR family, do not always diffuse in LCP, in spite of being similar by architecture and size to bacteriorhodopsin, a protein that freely diffuses and easily crystallizes in LCP (Cherezov, Liu, Griffith, Hanson, & Stevens, 2008). Further experiments demonstrated that the composition of precipitant solutions strongly affects mobility of receptors in LCP. Therefore, screening of numerous salts and their concentrations is required to identify conditions in which proteins diffuse in LCP, with the higher protein mobile fractions correlating strongly with the conditions that induce crystallization. These studies motivated the development of an LCP-FRAP assay for guiding crystallization trials (Cherezov et al., 2008; Xu et al., 2011). This assay substantially reduces the time spent on trial and error crystallization steps for MPs as one can rapidly identify and focus on only those conditions in which protein can diffuse in LCP, and thus has a chance to crystallize. Moreover, protein stability and its ability to diffuse in LCP are also often interrelated as unstable protein samples usually unfold or aggregate, resulting in loss of their mobility and inhibition of crystal formation. On the other hand, stable MP samples that maintain their integrity when incorporated into the LCP typically diffuse well in the LCP matrix. It should be noted that while LCP-FRAP assay is applicable to any MP, the initial development of this method and our own experience are mainly related to its application to GPCRs, hence we provide here protocols that we routinely use in our GPCR studies. These protocols and principles, however, can easily be adapted to other families of MPs.

Apart from the composition of precipitant solutions, a number of other factors can affect the ability of a given MP to diffuse in LCP. In order to obtain crystals suitable for X-ray diffraction in LCP, target proteins need to be extensively tested for conditions supporting a more stable conformation. In the case of our studies with GPCRs, these include screening for different LCP host lipids, ligands, fusion protein partners (T4-lysozyme, cytochrome *b*562 RIL (BRIL), rubredoxin; Chun et al., 2012), and receptor mutations capable of increasing receptor's stability and the likelihood of crystallization. The choice of ligands can dramatically affect receptor diffusion rates and it is common to observe no diffusion with one type of ligand and high protein mobile fractions with a different ligand. A similar situation is also observed in respect to different fusion partners, and even with the same fusion partner but with different insertion sites.

Another important consideration is the introduction of a fluorescent tag, which can be coupled to the protein via free amine groups or cysteine residues. Protein samples are labeled with 5,5′-disulfato-1′-ethyl-3,3,3′,3′-tetramethylindocarbocyanine (Cy3) dye, which was selected because of its relatively small size, hydrophilic properties, low sensitivity to environment, and favorable properties for FRAP applications. Two commercially available Cy3 protein conjugation carriers can be used: Cy3 mono-maleimide reacting with free sulfhydryl groups of cysteine residues and *N*-hydroxylsuccinimidyl (NHS) ester reacting with free amino groups (e.g., N-terminus, lysine residues). The most common practice, and thus the first method to be tried, is to use Cy3 NHS at pH 7.0–7.4 to predominately label the N-terminus, as it generally does not affect receptor conformation due to the introduction of a fluorescent tag, thus maintaining the receptor stability required for LCP diffusion. It is important to mention, however, that this method can label free amines of lipid head groups, copurified with the protein, which can lead to an increase in the background and fluorescence recovery in the assay. Although this increase in the background fluorescence is easily distinguishable from receptor recovery (lipids have faster diffusion and fluorescence recovery then proteins), in more difficult cases, when receptor diffusion is marginal, it can be difficult to differentiate between receptor and lipid recovery. Aminolabeling is usually performed at a low dye to protein ratio to avoid attachments of two or more dyes to one protein molecule, which could complicate data interpretation. Labeling cysteine residues, in our experience with GPCRs, often affects receptor stability, and therefore this type of labeling is used as a second choice. Nonetheless, while some receptors seem prone to destabilization and aggregation upon cysteine labeling, others are mostly unaffected. In any case, it is important to conduct tests to ensure that protein behavior and function are not altered by the dye.

2.3 High-throughput LCP-FRAP assay

Routine laboratory use of the LCP-FRAP assay is greatly facilitated by the use of an high-throughput (HT) approach. Protein diffusion in LCP is relatively slow, typically requiring tens to hundreds minutes to record a complete fluorescence recovery curve; therefore, a faster HT method was designed in which all samples in a 96-well glass sandwich plate are bleached sequentially and then probed at the end of a specified delay time (typically 30 min), which allows for measuring mobile fractions of protein in 96 samples within 2 h (Xu et al., 2011). We developed a 96-well glass sandwich plate for LCP studies (Cherezov & Caffrey, 2003; Cherezov, Peddi, Muthusubramaniam, Zheng, & Caffrey, 2004) to overcome problems that we had previously encountered, mainly associated with difficulties in locating small colorless protein crystals growing in LCP. The base of the plate is made of a 1-mm thick hydrophobically coated flat glass slide. The 96 wells of 5-mm diameter are defined by a perforated double sticky tape with the thickness of about 150 μm attached to the base glass slide. After samples are delivered, the wells are sealed by a 0.2-mm thick glass cover slide. LCP therefore is sandwiched between two flat glass slides providing excellent optical properties for crystal detection by different methods, including UV-fluorescence, and for other applications, such as the LCP-FRAP assay. The plate has an SBS-compatible footprint and is suitable for robotic crystallization and automated imaging. For LCP-FRAP, the use of a thinner spacer of ~60 μm is preferred to allow for a more uniform bleaching across the sample thickness. Glass sandwich plates with different spacer thicknesses are available commercially from Marienfeld-Superior, Hampton Research, and Molecular Dimensions.

While FRAP can be performed on any confocal microscope or a fluorescent microscope equipped with a laser, the HT screening approach requires special considerations. A brief description of our prototype instrument consisting of a fluorescent microscope, a dye-cell laser, an automatic

XYZ stage with an SBS plate holder, and a sensitive cooled CCD camera has been published (Cherezov et al., 2008). This prototype was used to build a commercial LCP-FRAP instrument, available from Formulatrix, which can be used for both a high-throughput LCP-FRAP (HT LCP-FRAP) analysis of 96-well conditions in glass sandwich plates, which measures the mobile fractions, and a subsequent full FRAP, allowing for an in-depth analysis of the protein diffusion in LCP in selected individual conditions.

The Formulatrix FRAP instrument has a user-friendly interface and has been designed to perform the HT LCP-FRAP in a fully automatic mode providing results in a color-coded 96-well grid (Fig. 1). Alternatively, results can be exported and plotted for better visualization of diffusion trends (Fig. 2). Setting up an HT LCP-FRAP experiment is facilitated with the instrument presets, but it also allows custom settings including parameters for fluorescent imaging and location of the bleaching spot (center of the drop or close to periphery).

Since the efficiency of protein labeling varies between samples, the exposure time for reading the fluorescence signal is one of the parameters that most often needs adjustment, as poorly labeled samples with low

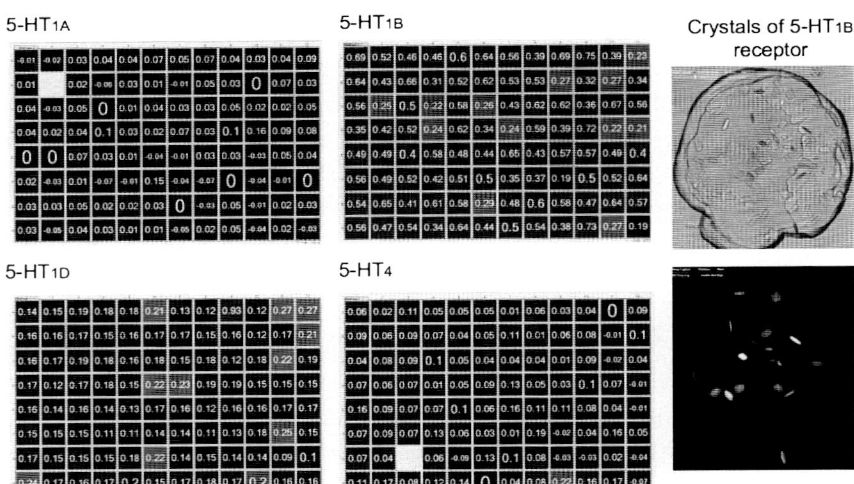

Figure 1 HT LCP-FRAP assays are used to define the best serotonin (5-HT) receptor subtype for subsequent crystallization trials. Typical results of the HT LCP-FRAP assays for four serotonin receptor subtypes, 5-HT$_{1A}$, 5-HT$_{1B}$, 5-HT$_{1D}$, and 5-HT$_4$ are shown in the format representing mobile fractions of the receptor in 96 different screening conditions. Only 5-HT$_{1B}$ receptor showed substantial diffusion in LCP and yielded crystals used for the structure determination.

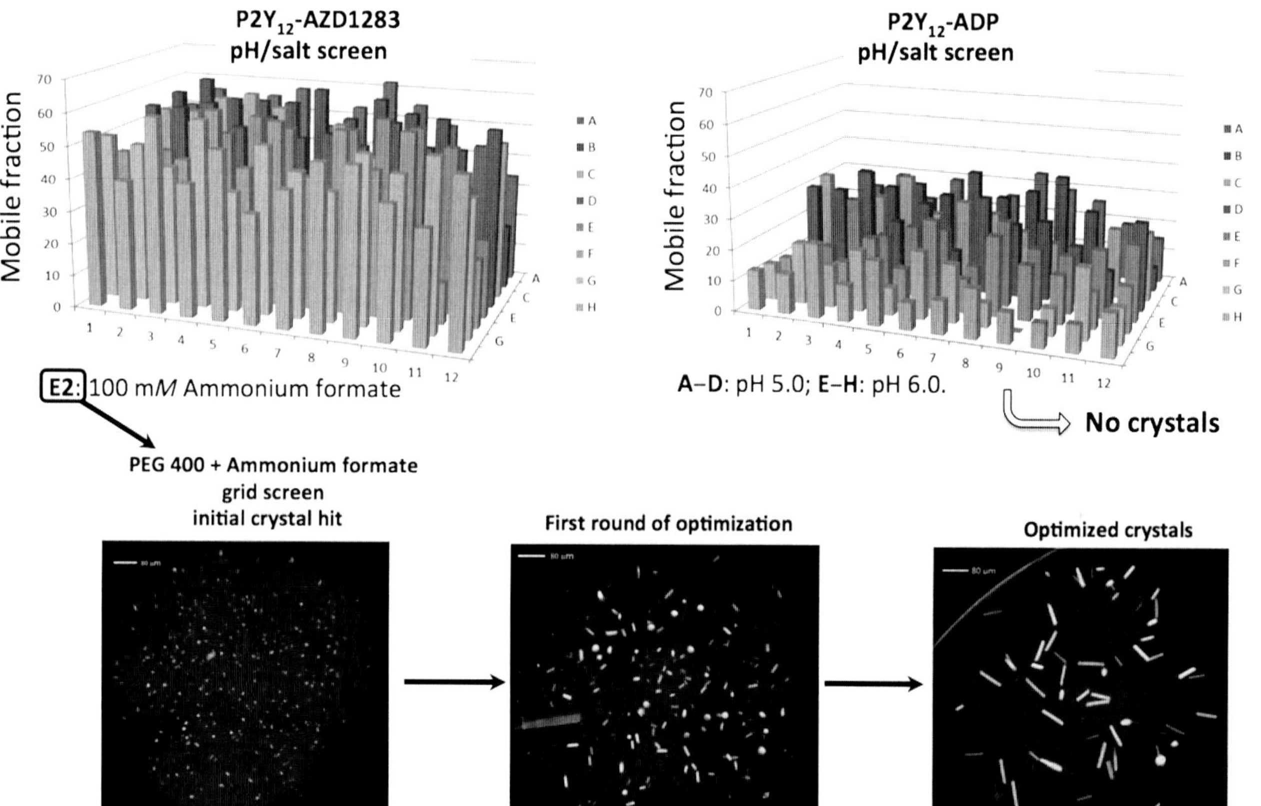

Figure 2 HT LCP-FRAP assays are used to select the best ligand for subsequent crystallization trials. Typical results of the HT LCP-FRAP assays for purinergic $P2Y_{12}$ receptor in complex with different ligands. Receptor complex with AZD1283 showed much higher mobility in LCP and yielded crystals after optimization of precipitant conditions, while complex with ADP did not produce crystals even after extensive screening.

fluorescence can lead to unreliable results. Exposure times over 1 s, however, in general, should be avoided because of a high noise level, and instead the protein should be concentrated further or relabeled.

HT LCP-FRAP assays can be conducted either at a relatively low protein concentration of about 0.5–1 mg/ml to conserve valuable protein or at a concentration of 10–50 mg/ml, at which crystallization trials in LCP are typically performed. While absolute results may vary slightly, both types of experiments should reveal similar trends. At low protein concentration, one may need to optimize labeling efficiency to have sufficiently strong fluorescence signal without overlabeling the protein, i.e., attaching more than one fluorophore per protein. The advantage of performing LCP-FRAP assays at high protein concentrations is the possibility to detect early microcrystal hits by fluorescence, which are often too small for detection by conventional bright-field or cross-polarized microscopy.

2.4 Full LCP-FRAP assay

While HT LCP-FRAP is convenient and fast, it can only characterize the mobile fractions of the proteins. In order to determine the diffusion rate, data collection of a full recovery curve is required, which includes photobleaching of a single spot on the LCP drop and measuring fluorescence recovery of the same spot in defined intervals during 20–30 min (Fig. 3). In a typical LCP-FRAP experiment, the user runs an HT FRAP first on a 96-well plate, following by a full FRAP analysis of selected wells that show recovery in the HT experiment. Analysis of the full FRAP recovery curves can provide clues about the lipid diffusion, and about the heterogeneity of the sample, such as when the recovery curve cannot be described by a single component (see, for example, experiments with bacteriorhodopsin in Cherezov et al., 2008).

2.5 Design of screens for HT LCP-FRAP

The HT LCP-FRAP assay provides important initial information about the protein mobility in LCP. The assay gives a full spectrum of values describing the protein mobile fraction in different conditions as well as how the protein responds to the changing environment presented by additives and screening conditions. HT LCP-FRAP screens are therefore typically designed to test a variety of precipitants, salts, and additives at different concentrations, thus getting an early indication of optimal ranges for subsequent crystallization trials.

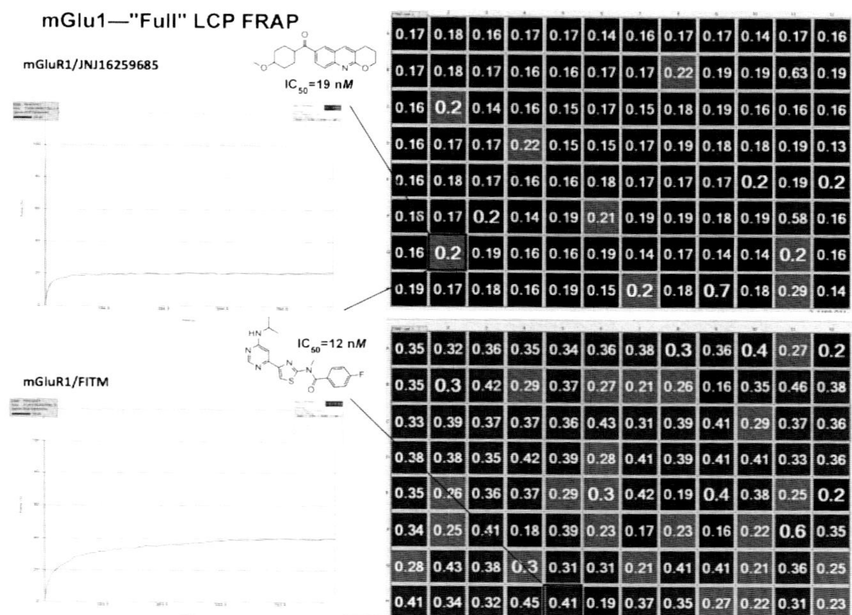

Figure 3 HT LCP-FRAP and full LCP-FRAP assays are used to select the best ligand and initial crystallization screening conditions for glutamate receptor mGlu1. HT LCP-FRAP assays showed better diffusion for mGlu1 in complex with FITM as compared to JNJ16259685, despite the similar binding affinities for these ligands. Full LCP-FRAP assay confirmed that only lipid diffusion was present in case of mGlu1 receptor bound to JNJ16259685, while both lipid and protein diffusion was detected in case of mGlu1 receptor bound to FITM. Crystallization trials were focused on mGlu1/FITM complex and produced crystals suitable for structure determination.

For assaying GPCRs, we designed a series of 96-well screens using 48 different salts (commercially available as StockOption Salt kit from Hampton Research) at two concentrations 100 and 400 m*M* (Xu et al., 2011). A constant precipitant concentration of PEG 400 of 30% (v/v) is used and buffered at a pH ranging from 5 to 8. Typically, three 96-well screens are prepared at pH 6.0, 7.0, and 8.0, covering the best initial pH range for receptor diffusion. These screening conditions can be further expanded by using additional salt concentrations (for example, 200 and 300 m*M*), PEG 400 concentrations (for example, 25% and 35% v/v), and pH (for example, 5.0, 5.5, 6.5, 7.5, and 8.5).

It is important to note that PEG 400 has a major effect on LCP properties that users need to be aware (Joseph et al., 2011). The increase in concentration of PEG 400 swells the LCP matrix, which could allow for better diffusion of the more bulky MPs, for example, receptors that are fused to protein partners, such as T4L and BRIL (Chun et al., 2012). PEG 400 concentrations above 35%, however, can drive monoolein-based LCP into a liquid-like sponge phase. While such transition could be desired when working with large MPs, the predictive power of the LCP-FRAP assays is diminished in this case, because even very large protein aggregates would move relatively fast in the sponge phase. The LCP–sponge transitions for shorter chain MAGs occur at lower PEG 400 concentrations, thus requiring adjustments of screens.

Optimization of LCP-FRAP diffusion is often obtained by fine-tuning salt and PEG concentrations as well as pH values and type of buffer. The use of additives is also helpful in increasing protein diffusion and often facilitates crystal growth and improves crystal quality.

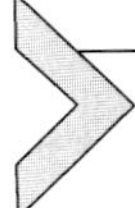

3. EXAMPLES OF USING LCP-FRAP TO GUIDE GPCR CRYSTALLIZATION

LCP-FRAP assays have been designed and validated using β_2 adrenergic and adenosine A_{2A} receptors as test samples (Xu et al., 2011), and then successfully used to guide crystallization trials and obtain initial crystal hits for dopamine D3 (Chien et al., 2010) and chemokine CXCR4 (Wu et al., 2010) receptors. Subsequently, these assays were included in the standard GPCR structure determination pipeline used by GPCR Network Center (Stevens et al., 2013) and applied to numerous receptors from different GPCR classes and families, including the serotonin (5-HT_{1B} and 5-HT_{2B}) receptors (Wacker et al., 2013; Wang et al., 2013), glucagon receptor (Siu et al., 2013), opioid receptors (Fenalti et al., 2014; Wu et al., 2012), chemokine CCR5 receptor (Tan et al., 2013), glutamate receptor mGluR1 (Wu et al., 2014), and purinergic $P2Y_{12}$ receptor (Zhang et al., 2014). Application of HT LCP-FRAP led to rapid identification of initial diffusing conditions and subsequent crystallization hits for these receptors. As an example, Fig. 2 demonstrates the results of a typical HT LCP-FRAP experiment for the $P2Y_{12}$ receptor that has been purified with different small molecule ligands. $P2Y_{12}$ receptor in complex with AZD1283 showed a good diffusion under many different conditions and was easily crystallized, while the same receptor construct bound to ADP diffused poorly and no crystals could be obtained after extensive trials. A growing number of cases like this have now demonstrated agreements between conditions maintaining protein mobility and conditions inducing crystal nucleation and growth for GPCRs. Another relevant example of the agreement between conditions maintaining protein diffusion and induction of crystal seeds is given by work on the 5-HT$_{1B}$ receptor (Fig. 1; Wang et al., 2013). Serotonin receptors, 5-HT$_{1A}$, 5-HT$_{1B}$, 5-HT$_{1D}$, and 5-HT$_4$, were screened in HT LCP-FRAP assay. Each receptor was purified with the ligand that provided an optimal stabilizing effect in detergent-solubilized state. The results showed that only the 5-HT$_{1B}$ receptor diffused in LCP, while the other receptors diffused poorly or had no diffusion. Initial crystal hits for the 5-HT$_{1B}$ receptor were detected during LCP-FRAP assays and could be further optimized for high-resolution structure determination, while the other receptor subtypes could not be crystallized even after extensive screening. This study demonstrates that the LCP-FRAP assay can facilitate the identification of targets with good crystallogenic properties within a receptor subfamily. A great variability between different receptors and even between different constructs of the same receptor (see section on construct design) is often observed, and while some receptors require little protein engineering and diffuse in a large number of conditions, more challenging proteins require extensive engineering and a careful choice of ligands to obtain initial diffusion in LCP. The successful use of this assay to guide crystallization trials for challenging novel targets has therefore established HT LCP-FRAP as a key assay to obtain critical information on a given receptor.

It is well accepted that in order to obtain stable GPCR samples, an extensive screening of different constructs is needed. The crystallization of GPCRs in LCP has initially relied on a fusion of the T4L lysozyme into the receptor's intracellular loop 3 (ICL3; Rosenbaum et al., 2007), but, subsequently, a number of different fusion partners have been identified as suitable for crystallization when inserted in the ICL3, or fused to the N-terminal of receptors (Chun et al., 2012). The particular positions of the fusion protein junction sites in the ICL3 also require optimization and may affect receptor diffusion in LCP. The choice of ligand used for GPCR purifications have a profound effect on receptor stability and therefore on its behavior in the LCP matrix. Many receptors show no diffusion when in complex with certain small-molecule ligands but display excellent diffusion when in complex with other ligands. As an example, full LCP-FRAP assay results were used to illustrate the effect of different ligands on mGlu1 diffusion. Although two ligands JNJ16259685 and FITM have similar high affinities, mGlu1/FITM complex showed substantial protein diffusion, while mGlu1/JNJ16259685 revealed only diffusion of copurified with the receptor and labeled lipids (Fig. 3). Consistent with these distinct diffusion properties, mGlu1–FITM complex could be crystallized, while the mGlu1–JNJ16259685 did not yield any crystals. This example reveals the power of LCP-FRAP in guiding ligand selection during receptor crystallization trials.

4. PROTOCOLS FOR LCP-FRAP ASSAYS

A flowchart for a typical LCP-FRAP experiment is shown in Fig. 4. It starts with protein expression, purification, and labeling. The labeled protein is then incorporated in LCP, and several 96-well assay plates are set up using an LCP crystallization robot, in which small-volume LCP boluses are overlaid with different screening solutions. Furthermore, the plates are incubated for 12–24 h at 20 °C allowing for equilibration, after which the plates are run through an HT LCP-FRAP measurements sequence. The results of the HT LCP-FRAP run are then analyzed and selected wells are subject to a full LCP-FRAP protocol for obtaining additional information on protein diffusion rates. After the performed experiments are fully analyzed, a decision is made based on the following considerations. If no appreciable protein diffusion is detected in any of the screened conditions, then LCP-FRAP assays are repeated using a different ligand, LCP host lipid, or protein construct. If several screening conditions support protein diffusion, then they are used as a basis for formulation of new expanded grid screens, which are further used for new LCP-FRAP assays or directly for crystallization trials.

4.1 GPCR expression, purification, and labeling

GPCR constructs are generally expressed in *Spodoptera frugiperda* (Sf9) insect cells using a modified Bac-to-Bac expression system (Invitrogen). Successful GPCR purification protocols follow similar steps, as have been described previously (Chien et al., 2010; Fenalti et al., 2014; Tan et al., 2013; Wacker et al., 2013; Wu et al., 2010). Protein is typically labeled during purification, while bound to a metal affinity resin. All steps are performed either on ice or in a cold room at 4 °C.

1. Lyse cells in a hypotonic buffer.
2. Wash membranes extensively in a low-salt and high-salt buffers.
3. Incubate the purified membranes with 2 mg/ml iodoacetamide (Sigma) to block solvent-exposed cysteine residues (this treatment is omitted in case of cysteine labeling), and a ligand of choice (50–100 μ*M*) in the presence of protease inhibitors.

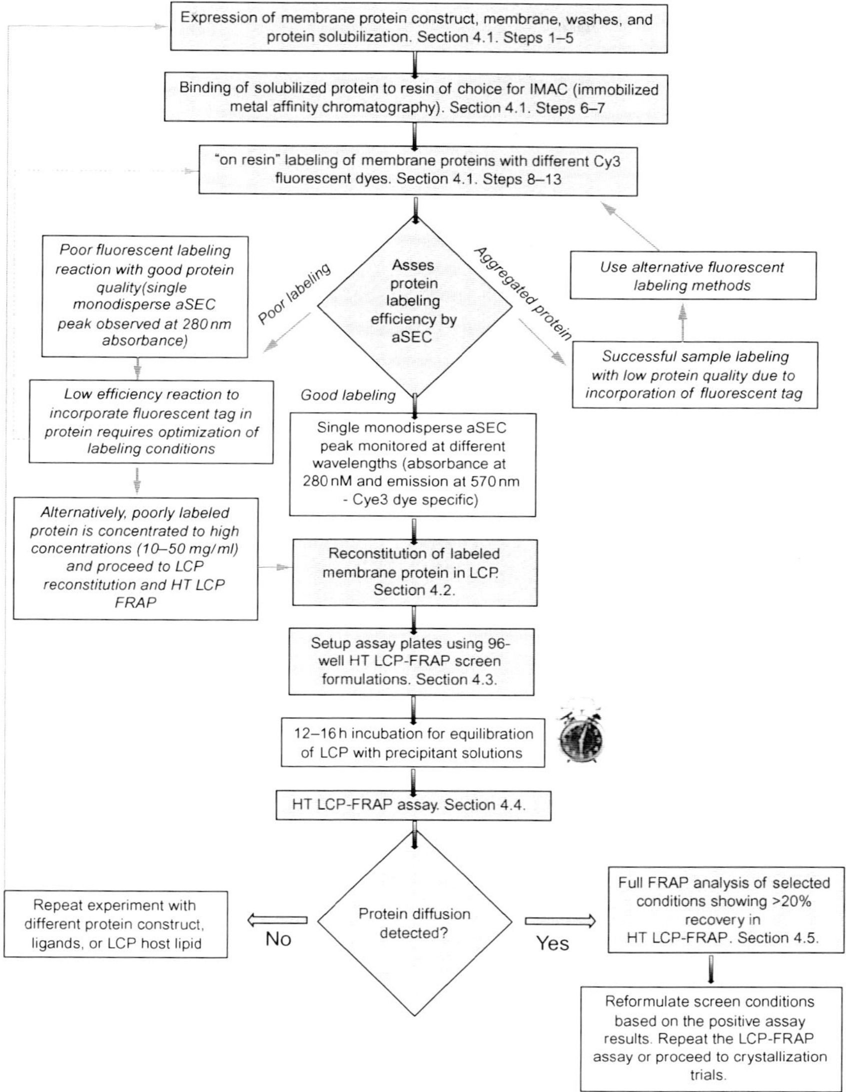

Figure 4 A flowchart summarizing the steps involved in setting up, executing, and analyzing LCP-FRAP assays.

4. Solubilize the target protein with 0.5–1% (w/v) *n*-dodecyl-*β*-D-maltopyranoside (DDM, Anatrace), 0.1–0.2% (w/v) cholesteryl hemisuccinate (CHS, Sigma), 50 m*M* HEPES, pH 7.5, and 150–500 m*M* of NaCl.

5. Remove the insolubilized membranes by centrifugation at 150,000 × *g* for 45 min.
6. Collect the supernatant and incubate it with TALON IMAC resin (Clontech) for at least 5–6 h, or overnight.
7. Wash out impurities with a wash buffer (0.1% w/v DDM; 0.02% w/v CHS; 50 m*M* HEPES, pH 7.5; 150–500 m*M* NaCl; 10% v/v glycerol; 5 m*M* ATP (Sigma); 10 m*M* $MgCl_2$, and the ligand of choice).
8. Label the protein with Cy3-mono NHS ester (GE healthcare) while it is bound on the TALON resin. For protein labeling, 5–10 μl of 5 mg/ml stock solution of Cy3-mono NHS ester in DMF (Sigma) to 500 μg of protein bound to 1 ml of TALON resin in 5 ml of buffer at pH 7.2 (0.1% w/v DDM; 0.02% w/v CHS; 50 m*M* HEPES, pH 7.2; 150–500 m*M* NaCl; 10% v/v glycerol).
9. Incubate for 3 h in the dark.
10. Wash out the unreacted dye with wash buffer using a gravity column (Bio-Rad).
11. Elute the labeled protein using five column volumes of wash buffer with an increased imidazole concentration (250–300 m*M*).
12. Remove imidazole by running the eluate through a desalting column.
13. Concentrate the protein to ∼20 mg/ml (or higher concentrations) using a 100-kDa cutoff concentrator (Vivascience).

If receptor is glycosylated it can be optionally deglycosilated by the treatment with PNGase F, after which the enzyme is removed by using another IMAC purification step (either TALON or Nickel-NTA resin), and protein is concentrated.

Concentrated protein is immediately used for LCP-FRAP sample preparation. It is important to note that the final receptor concentration could range from ∼0.5 mg/ml and up to 50 mg/ml. The lower protein concentrations could be used when the amount of protein is limited. The advantage of using 20 mg/ml and higher concentrations is that LCP-FRAP assays can lead to formation of crystals, which can be easily identified by fluorescent imaging even when they are very small in size (Fig. 2).

All protein samples are then evaluated for purity, monodispersity, and labeling efficiency by analytical size-exclusion chromatography (aSEC) prior to LCP-FRAP sample preparation. Labeled GPCR samples should display a single monomer peak on aSEC, when monitoring the fluorescence emission of the fluorescent dye used, with a typical protein labeling ratio of 0.1–5%.

4.2 Reconstitution of labeled receptor in LCP

Once receptor has been successfully purified, labeled, and characterized by analytical SEC, it is reconstituted in the lipid bilayer of LCP for pre-crystallization assays. The reconstitution is achieved spontaneously upon mechanical mixing of a purified protein in detergent solution with LCP host lipids or lipid mixtures using a lipid syringe mixer (Caffrey & Cherezov, 2009). Lipid additives can modulate LCP properties or can preferentially interact with proteins reconstituted in LCP and therefore are useful for optimization of crystallization conditions. Cholesterol was found to be the best lipid additive significantly improving the growth of β_2AR-T4L crystals (Cherezov et al., 2007), and since that it has been successfully used to facilitate crystallization of most GPCRs.

1. Transfer a necessary amount (15–50 mg) of an LCP host lipid (e.g., monoolein) or a lipid mixture (e.g., monoolein/cholesterol) into a small 0.5 ml plastic vial.
2. Melt lipid at ~40 °C.
3. Attach a 100 μl Hamilton gas-tight syringe to a syringe coupler (Formulatrix). Remove the plunger and transfer molten lipid into the syringe barrel through the plunger end using an adjustable volume pipette.
4. Insert the plunger back and slowly move the lipid up in the syringe to remove any trapped air bubbles. Stop when the lipid reaches the end of the coupler needle.
5. Use a 25 or 50-μl syringe with a flat-tipped 26-s gauge needle to transfer an appropriate amount of the protein solution in the second 100 μl gas-tight syringe to achieve 40 wt% of protein solution in the final mixture (for example, for 30 mg of lipid use 20 μl of protein solution). Avoid trapping air bubbles during the syringe loading.
6. Connect the syringe containing protein solution to the open end of the coupler attached to the lipid syringe.
7. Move the protein solution and the lipid through the coupler inner needle back and forth from one syringe to another by pushing alternatively on the corresponding plungers until the lipid mesophase becomes homogeneous and transparent. This sequence of motions mechanically mixes the lipid with the protein solution through the action of shearing forces forming inside the narrow coupler needle. Upon mixing, an LCP forms spontaneously and the protein becomes inserted into the lipid bilayer of the LCP. Complete mixing requires about 100 passages and typically takes less than 5 min.

4.3 LCP-FRAP sample preparation

Labeled protein reconstituted in LCP is then dispensed in 96-well glass sandwich plates (Marienfeld-Superior; Hampton Research; Molecular Dimensions) using an LCP crystallization robot (Formulatrix NT8 LCP; TTP Labtech Mosquito LCP; Art Robbins Griphon LCP) by delivering 40–50 nl boluses of the protein-laden LCP and overlaying them with 0.8 μl of screening solutions. After setting up and sealing wells with a glass coverslip, plates are wrapped with foil and incubated at 20 °C overnight (12–16 h) to equilibrate LCP drops with the screening solutions before conducting FRAP measurements.

4.4 HT LCP-FRAP data collection and analysis

HT LCP-FRAP data acquisition and image analysis are fully automated using the Formulatrix FRAP instrument, and ideally data are collected within 12–24 h after setting up plates. The data acquisition sequence starts from a drop location step, in which *X* and *Y* coordinates of all LCP drops in 96 wells are defined. It then proceeds to a focusing step for each well. If the fluorescence signal is not sufficient to reliably locate all LCP drops, then the exposure time can be adjusted, or alternatively the protein should be further concentrated or relabeled. When all the drops are located and focused, each sample sequentially is subject to the following procedure: recording a fluorescence image of the initial prebleached state, bleaching at a user-defined location (typically at the center or about 100 μm from the LCP drop edge), and immediate recording an image of the postbleached state. Then after a user-defined incubation time (an adjustable parameter, typically 30 min after bleaching), the plate is scanned to record the end-state fluorescence recovery images for each sample in the 96-well plate.

Recorded images are processed automatically to locate the bleached spot and to integrate the intensity within the spot before bleaching, immediately after bleaching, and after recovery. The results are then expressed as a percentage of the recovery in a color-coded table representing the 96-well plate. The thresholds depend on the percentage of the labeled lipid in the sample and are user-adjustable. As an example, in Figures 1 and 3, values below 20% are shown as blue, indicating the absence of protein diffusion, whereas values of 21–30% are shown as green, indicating potential protein diffusion, and above 30% are shown as red, indicating good protein diffusion. At 21–30% recovery samples need to be carefully analyzed by full FRAP to confirm receptor diffusion. Users should be aware of occasional false positives with recoveries above 30%, which are often related to changes in the lipid phase state (e.g., transition to a sponge phase). Such false positives can easily be identified by checking the actual recorded prebleached, post-bleached, and recovered images. Phase change is often accompanied by an oversaturated fluorescence signal. Transition into a sponge phase can be spotted by the motion of large fluorescent specks, which otherwise remain stationary in a gel-like LCP.

4.5 Full LCP-FRAP data collection and analysis

Full time-course fluorescence recovery data are collected on selected samples to characterize the protein diffusion profile in more detail. Data acquisition and analysis of the full LCP-FRAP curves are performed as previously described (Cherezov et al., 2008). Fluorescence recovery curves after correction and normalization are fit with one- or two-component diffusion equation to obtain the mobile fraction, M, and the characteristic diffusion time, T, of the labeled molecules. The two-component diffusion equation can be used to obtain diffusion parameters for the slow moving proteins in the presence of fast diffusing lipids. The one-component equation is used when no lipid diffusion is detected. The radius of the bleached spot, R, is determined by fitting the radially integrated intensity of the bleached spot by a Gaussian. The diffusion coefficient, D, is calculated as $D=R^2/4T$.

ACKNOWLEDGMENTS

We thank members of the group, past and present, who participated in the development of LCP-FRAP protocols, in particular, Jeffrey Liu, Fei Xu, Wei Liu, Joshua Kunken, and Michael Hanson. The work was supported in parts by the National Institutes of Health Grants R21 GM103972 and P50 GM073197.

REFERENCES

Briggs, J., & Caffrey, M. (1994). The temperature–composition phase diagram and mesophase structure characterization of monopentadecenoin in water. *Biophysical Journal*, *67*(4), 1594–1602. http://dx.doi.org/10.1016/S0006-3495(94)80632-0.

Caffrey, M., & Cherezov, V. (2009). Crystallizing membrane proteins using lipidic mesophases. *Nature Protocols*, *4*, 706–731. http://dx.doi.org/10.1038/nprot.2009.31.

Caffrey, M., Lyons, J., Smyth, T., & Hart, D. J. (2009). Monoacylglycerols: The workhorse lipids for crystallizing membrane proteins in mesophases. *Current Topics in Membranes*, *63*, 83–108.

Cherezov, V. (2011). Lipidic cubic phase technologies for membrane protein structural studies. *Current Opinion in Structural Biology*, *21*(4), 559–566. http://dx.doi.org/10.1016/j.sbi.2011.06.007.

Cherezov, V., & Caffrey, M. (2003). Nano-volume plates with excellent optical properties for fast, inexpensive crystallization screening of membrane proteins. *Journal of Applied Crystallography*. *36*, 1372–1377.

Cherezov, V., Clogston, J., Papiz, M. Z., & Caffrey, M. (2006). Room to move: Crystallizing membrane proteins in swollen lipidic mesophases. *Journal of Molecular Biology, 357*(5), 1605–1618. http://dx.doi.org/10.1016/j.jmb.2006.01.049.

Cherezov, V., Liu, J., Griffith, M., Hanson, M. A., & Stevens, R. C. (2008). LCP-FRAP assay for pre-screening membrane proteins for in meso crystallization. *Crystal Growth & Design, 8*(12), 4307–4315. http://dx.doi.org/10.1021/cg800778j.

Cherezov, V., Peddi, A., Muthusubramaniam, L., Zheng, Y. F., & Caffrey, M. (2004). A robotic system for crystallizing membrane and soluble proteins in lipidic mesophases. *Acta Crystallographica. Section D, Biological Crystallography, 60*(Pt. 10), 1795–1807. http://dx.doi.org/10.1107/S0907444904019109.

Cherezov, V., Rosenbaum, D. M., Hanson, M. A., Rasmussen, S. G., Thian, F. S., Kobilka, T. S., et al. (2007). High-resolution crystal structure of an engineered human beta2-adrenergic G protein-coupled receptor. *Science, 318*(5854), 1258–1265. http://dx.doi.org/10.1126/science.1150577.

Chien, E. Y., Liu, W., Zhao, Q., Katritch, V., Han, G. W., Hanson, M. A., et al. (2010). Structure of the human dopamine D3 receptor in complex with a D2/D3 selective antagonist. *Science, 330*(6007), 1091–1095. http://dx.doi.org/10.1126/science.1197410.

Chun, E., Thompson, A. A., Liu, W., Roth, C. B., Griffith, M. T., Katritch, V., et al. (2012). Fusion partner toolchest for the stabilization and crystallization of G protein-coupled receptors. *Structure, 20*(6), 967–976. http://dx.doi.org/10.1016/j.str.2012.04.010.

Fenalti, G., Giguere, P. M., Katritch, V., Huang, X. P., Thompson, A. A., Cherezov, V., et al. (2014). Molecular control of delta-opioid receptor signalling. *Nature, 506*(7487), 191–196. http://dx.doi.org/10.1038/nature12944.

Grabe, M., Neu, J., Oster, G., & Nollert, P. (2003). Protein interactions and membrane geometry. *Biophysical Journal, 84*(2 Pt. 1), 854–868. http://dx.doi.org/10.1016/S0006-3495(03)74904-2.

Hyde, S. T., Andersson, S., Ericsson, B., & Larsson, K. (1984). A cubic structure consisting of a lipid bilayer forming an infinite periodic minimum surface of the gyroid type in the glycerolmonooleat-water system. *Zeitschrift für Kristallographie, 168*, 213–219. http://dx.doi.org/10.1524/zkri.1984.168.1-4.213.

Joseph, J. S., Liu, W., Kunken, J., Weiss, T. M., Tsuruta, H., & Cherezov, V. (2011). Characterization of lipid matrices for membrane protein crystallization by high-throughput small angle X-ray scattering. *Methods, 55*(4), 342–349. http://dx.doi.org/10.1016/j.ymeth.2011.08.013.

Katona, G., Andreasson, U., Landau, E. M., Andreasson, L. E., & Neutze, R. (2003). Lipidic cubic phase crystal structure of the photosynthetic reaction centre from Rhodobacter sphaeroides at 2.35A resolution. *Journal of Molecular Biology, 331*(3), 681–692.

Katritch, V., Fenalti, G., Abola, E. E., Roth, B. L., Cherezov, V., & Stevens, R. C. (2014). Allosteric sodium in class A GPCR signaling. *Trends in Biochemical Sciences, 39*(5), 233–244. http://dx.doi.org/10.1016/j.tibs.2014.03.002.

Landau, E. M., Pebay-Peyroula, E., & Neutze, R. (2003). Structural and mechanistic insight from high resolution structures of archaeal rhodopsins. *FEBS Letters, 555*(1), 51–56.

Landau, E. M., & Rosenbusch, J. P. (1996). Lipidic cubic phases: A novel concept for the crystallization of membrane proteins. *Proceedings of the National Academy of Sciences of the United States of America, 93*(25), 14532–14535.

Qiu, H., & Caffrey, M. (2000). The phase diagram of the monoolein/water system: Metastability and equilibrium aspects. *Biomaterials, 21*(3), 223–234.

Rasmussen, S. G., DeVree, B. T., Zou, Y., Kruse, A. C., Chung, K. Y., Kobilka, T. S., et al. (2011). Crystal structure of the beta2 adrenergic receptor-Gs protein complex. *Nature, 477*(7366), 549–555. http://dx.doi.org/10.1038/nature10361.

Rosenbaum, D. M., Cherezov, V., Hanson, M. A., Rasmussen, S. G., Thian, F. S., Kobilka, T. S., et al. (2007). GPCR engineering yields high-resolution structural insights into beta2-adrenergic receptor function. *Science*, *318*(5854), 1266–1273. http://dx.doi.org/10.1126/science.1150609.

Rummel, G., Hardmeyer, A., Widmer, C., Chiu, M. L., Nollert, P., Locher, K. P., et al. (1998). Lipidic cubic phases: New matrices for the three-dimensional crystallization of membrane proteins. *Journal of Structural Biology*, *121*(2), 82–91. http://dx.doi.org/10.1006/jsbi.1997.3952.

Siu, F. Y., He, M., de Graaf, C., Han, G. W., Yang, D., Zhang, Z., et al. (2013). Structure of the human glucagon class B G-protein-coupled receptor. *Nature*, *499*(7459), 444–449. http://dx.doi.org/10.1038/nature12393.

Stevens, R. C., Cherezov, V., Katritch, V., Abagyan, R., Kuhn, P., Rosen, H., et al. (2013). The GPCR network: A large-scale collaboration to determine human GPCR structure and function. *Nature Reviews. Drug Discovery*, *12*(1), 25–34. http://dx.doi.org/10.1038/nrd3859.

Tan, Q., Zhu, Y., Li, J., Chen, Z., Han, G. W., Kufareva, I., et al. (2013). Structure of the CCR5 chemokine receptor-HIV entry inhibitor maraviroc complex. *Science*, *341*(6152), 1387–1390. http://dx.doi.org/10.1126/science.1241475.

Tiefenbrunn, T., Liu, W., Chen, Y., Katritch, V., Stout, C. D., Fee, J. A., et al. (2011). High resolution structure of the ba3 cytochrome c oxidase from Thermus thermophilus in a lipidic environment. *PLoS One*, *6*(7), e22348. http://dx.doi.org/10.1371/journal.pone.0022348.

Wacker, D., Wang, C., Katritch, V., Han, G. W., Huang, X. P., Vardy, E., et al. (2013). Structural features for functional selectivity at serotonin receptors. *Science*, *340*(6132), 615–619. http://dx.doi.org/10.1126/science.1232808.

Wang, C., Jiang, Y., Ma, J., Wu, H., Wacker, D., Katritch, V., et al. (2013). Structural basis for molecular recognition at serotonin receptors. *Science*, *340*(6132), 610–614. http://dx.doi.org/10.1126/science.1232807.

Wohri, A. B., Wahlgren, W. Y., Malmerberg, E., Johansson, L. C., Neutze, R., & Katona, G. (2009). Lipidic sponge phase crystal structure of a photosynthetic reaction center reveals lipids on the protein surface. *Biochemistry*, *48*(41), 9831–9838. http://dx.doi.org/10.1021/bi900545e.

Wu, B., Chien, E. Y., Mol, C. D., Fenalti, G., Liu, W., Katritch, V., et al. (2010). Structures of the CXCR4 chemokine GPCR with small-molecule and cyclic peptide antagonists. *Science*, *330*(6007), 1066–1071. http://dx.doi.org/10.1126/science.1194396.

Wu, H., Wacker, D., Mileni, M., Katritch, V., Han, G. W., Vardy, E., et al. (2012). Structure of the human kappa-opioid receptor in complex with JDTic. *Nature*, *485*(7398), 327–332. http://dx.doi.org/10.1038/nature10939.

Wu, H., Wang, C., Gregory, K. J., Han, G. W., Cho, H. P., Xia, Y., et al. (2014). Structure of a class C GPCR metabotropic glutamate receptor 1 bound to an allosteric modulator. *Science*, *344*(6179), 58–64. http://dx.doi.org/10.1126/science.1249489.

Xu, F., Liu, W., Hanson, M. A., Stevens, R. C., & Cherezov, V. (2011). Development of an automated high throughput LCP-FRAP assay to guide membrane protein crystallization in lipid mesophases. *Crystal Growth & Design*, *11*(4), 1193–1201. http://dx.doi.org/10.1021/cg101385e.

Zhang, K., Zhang, J., Gao, Z. G., Zhang, D., Zhu, L., Han, G. W., et al. (2014). Structure of the human P2Y12 receptor in complex with an antithrombotic drug. *Nature*, *509*(7498), 115–118. http://dx.doi.org/10.1038/nature13083.

CHAPTER TWENTY-ONE

Crystallization of Proteins from Crude Bovine Rod Outer Segments☆

Bo Y. Baker*,†,1, Sahil Gulati*,†, Wuxian Shi‡, Benlian Wang‡, Phoebe L. Stewart*,†, Krzysztof Palczewski*,†,1

*Department of Pharmacology, Case Western Reserve University, Cleveland, Ohio, USA
†Cleveland Center for Membrane and Structural Biology, School of Medicine, Case Western Reserve University, Cleveland, Ohio, USA
‡Center for Proteomics and Bioinformatics, Center for Synchrotron Biosciences, School of Medicine, Case Western Reserve University, Cleveland, Ohio, USA
[1]Corresponding authors: e-mail address: byb@case.edu; kxp65@case.edu

Contents

Abstract

Obtaining protein crystals suitable for X-ray diffraction studies comprises the greatest challenge in the determination of protein crystal structures, especially for membrane proteins and protein complexes. Although high purity has been broadly accepted as one of the most significant requirements for protein crystallization, a recent study of the *Escherichia coli* proteome showed that many proteins have an inherent propensity to crystallize and do not require a highly homogeneous sample (Totir et al., 2012). As

☆The structures reported in this study have been deposited in the Protein Data Bank (PDB) under accession codes 4O63 (bovine GAPDH_3NAD) and 4Q2R (bovine CK-B).

Methods in Enzymology, Volume 557
ISSN 0076-6879
http://dx.doi.org/10.1016/bs.mie.2014.11.045

exemplified by RPE65 (Kiser, Golczak, Lodowski, Chance, & Palczewski, 2009), there also are cases of mammalian proteins crystallized from less purified samples. To test whether this phenomenon can be applied more broadly to the study of proteins from higher organisms, we investigated the protein crystallization profile of bovine rod outer segment (ROS) crude extracts. Interestingly, multiple protein crystals readily formed from such extracts, some of them diffracting to high resolution that allowed structural determination. A total of seven proteins were crystallized, one of which was a membrane protein. Successful crystallization of proteins from heterogeneous ROS extracts demonstrates that many mammalian proteins also have an intrinsic propensity to crystallize from complex biological mixtures. By providing an alternative approach to heterologous expression to achieve crystallization, this strategy could be useful for proteins and complexes that are difficult to purify or obtain by recombinant techniques.

ABBREVIATIONS

CK-B creatine kinase-B type
DEAE diethylaminoethyl
GAPDH glyceraldehyde 3-phosphate dehydrogenase
G_t transducin (rod-specific G protein)
HEPES 2-[4-(2-hydroxyethyl)piperazin-1-yl]ethanesulfonic acid
LDH L-lactate dehydrogenase
MS mass spectroscopy
NAD nicotinamide adenine dinucleotide
PDE6 phosphodiesterase 6
PEG polyethylene glycol
ROS rod outer segment(s)

1. INTRODUCTION

While structural studies provide valuable information needed to understand physiological processes at the molecular level, X-ray crystallography has been the leading approach to determine protein structures at the atomic level. As of March 2014, the number of protein structures solved by X-ray diffraction methods reached 87,945, accounting for ~88.9% of the total protein structures in the Protein Data Bank (PDB). Growing protein crystals suitable for X-ray diffraction studies comprises the major bottleneck in the process of solving a protein structure, especially for membrane proteins and protein complexes because variations in the molecular properties of each protein require specific crystallization conditions.

High purity of a protein sample has been broadly accepted as one of the most important requirements for crystallization. However, it is also well documented that protein crystallization can be used effectively for protein purification. Examples of proteins for which crystallization was used to obtain the highest level of purity include glycogen phosphorylase (Fischer & Krebs, 1958), cytochrome c2 (Pettigrew, Bartsch, Meyer, & Kamen, 1978), and lysozyme (Judge, Forsythe, & Pusey, 1998). Several unsuccessful protein complex crystallization attempts have still resulted in the crystallization of a single-protein component. Quite recently, the structure of bovine arrestin-1 splice variant (p^{44}) was determined from a mixture of arrestin with opsin (Kim et al., 2013). Though this co-crystallization failed, investigators were able to obtain crystals of the arrestin variant.

Several groups have studied the effect of impurities on protein crystallization. In 1998, Judge et al. investigated the effect of protein impurities on the crystallization of lysozyme from fresh chicken egg whites. Other proteins did not affect lysozyme crystal face growth even at concentrations of up to 50% of total protein in the sample (Judge et al., 1998). In 2006, Dong et al. reported an increase in the number of proteins forming diffraction-quality crystals upon addition of trace amounts of protease (Dong et al., 2007). The authors found that despite the presence of protease as well as cleaved peptides and protein fragments in these samples, protein crystals still formed readily, suggesting a certain tolerance for sample impurities during crystallization. Indeed, the presence of protease helped to remove flexible regions of target proteins and generated protein domains that improved crystallization.

Membrane proteins, particularly those derived from a vertebrate source/genome, pose a more difficult problem (Muller, Wu, & Palczewski, 2008). Certain properties of membrane proteins, including their hydrophobic nature, propensity to bind other hydrophobic partners in heterogeneous complexes, instability in detergents required for purification, conformational heterogeneity, and flexibility often constitute major obstacles for accurate structural and functional studies. In studying the effect of sample purity on membrane protein crystallization, Kors et al. (2009) discovered that a membrane protein, photosynthetic reaction center from *Rhodobacter sphaeroides*, produced crystals in the presence of substantial amounts of impurities (Kors et al., 2009). Unexpectedly, crystals were obtained with a protein/contaminant ratio as high as 50%, including lipid material and membrane fragments. In 2012, Alber's group investigated a native source of proteins for high-throughput protein crystallography applications (Totir et al., 2012). The authors separated 408 unique fractions from the *E. coli* proteome with protein purities ranging from 95% to less than 5%. Despite different levels of sample complexity, the authors were able to obtain crystals from 295 of these fractions, representing 73% of the total protein (Totir et al., 2012). This suggests that many proteins, if not all, have an inherent propensity to crystallize, even from samples of varying heterogeneity.

Mammalian proteins differ from bacterial proteins in several respects. During biosynthesis, mammalian proteins undergo various posttranslational modifications, such as phosphorylation for signal transduction and glycosylation needed for protein folding, distribution, and activity. Whether the findings discussed above apply more generally to proteins from higher organisms has not been established. Here, we used crude extracts from bovine rod outer segment (ROS) to investigate the potential of mammalian proteins from a native source to form crystals suitable for structural determination. ROSs are compartments of vertebrate rod photoreceptor cells in the eye's retina that are essential for visual phototransduction. The proteome of ROS contains about 516 proteins involved in transduction of visual signals, maintenance of retinal structure, and metabolic pathways (Kwok, Holopainen, Molday, Foster, & Molday, 2008), whereas the internal disc membranes are likely to have less than a dozen proteins (Skiba et al., 2013). Some proteins, such as the phosphodiesterase 6 (PDE6) heterotetramer and the transducin (G_t) heterotrimer complex are difficult to obtain by recombinant expression methods. Because of their high abundance in ROS, PDE6, and G_t are routinely purified from these compartments for biochemical and structural studies (Baker & Palczewski, 2011; Goc et al., 2008).

Here, we investigated the crystallization profiles of seven proteins from bovine ROS using a general screening approach. This study is the first attempt to crystallize mammalian proteins from a native source under highly heterogeneous conditions. The results provide an alternative approach to achieving crystallization of proteins that resist expression and purification or display conformational heterogeneity due to diverse folding. This study provides a powerful addition to current structural genomics initiatives by circumventing the need to attain homogeneity for protein crystallization.

2. EXPERIMENTAL PROCEDURES

2.1. Protein sample preparation

ROS membranes were isolated from 300 frozen bovine retinas (W. L. Lawson Co., Lincoln, NE) under dim red light as previously described (Baker & Palczewski, 2011). Soluble proteins were extracted from isolated ROS membranes three times with isotonic buffer containing 20 m*M* 2-[4-(2-hydroxyethyl)piperazin-1-yl]ethanesulfonic acid (HEPES), pH

7.5, 0.1 *M* NaCl, and 5 m*M* $MgCl_2$. Soluble protein extracts were combined and centrifuged at 30,940 × *g* for 30 min at 4 °C. The supernatant was further centrifuged at 30,940 × *g* at 4 °C for 50 min to remove any traces of membranes and dialyzed overnight at 4 °C against buffer containing 20 m*M* HEPES, pH 7.5. The dialyzed sample (named F_A) was used directly for crystallization screening or subjected to weak ion exchange chromatography on a 5 ml HiTrap Fast Flow DEAE Sepharose column (GE Healthcare Bio-Sciences Corp, Piscataway, NJ) at a flow rate of 1 ml/min. The column was washed with 200 ml of buffer containing 20 m*M* HEPES, pH 7.5, and bound proteins were eluted by a linear NaCl gradient from 0 to 1 *M* over 120 ml at a flow rate of 1 ml/min. Fractions were collected for each of the eluted peaks and retained for further processing.

To separate glyceraldehyde 3-phosphate dehydrogenase (GAPDH), the flow-through (named F_B) from the diethylaminoethyl (DEAE)-Sepharose column was loaded onto a nicotinamide adenine dinucleotide (NAD)-agarose (Bio-WORLD, Dublin, OH) column. In preparation, lyophilized NAD-agarose (1 g) was solubilized in water and packed into an Econo-Column (0.5 × 5 cm, BioRad, Hercules, CA). Before protein loading, the column was equilibrated with binding buffer (20 m*M* HEPES, pH 7.3). Protein samples were loaded onto the column at flow rate of 1 ml/min. The column was washed with 200 ml of binding buffer followed by elution of GAPDH with the elution buffer (20 m*M* HEPES, pH 7.3, 5 m*M* NAD). GAPDH was concentrated to 2–3 mg/ml prior to crystallization trials.

To separate arrestin-1, a method described previously (Buczylko & Palczewski, 1993) was used with a few alterations. Combined fractions of elution peak 1 (P1, named F_C) from the DEAE-cellulose column were dialyzed against buffer containing 20 m*M* HEPES, pH 7.3, and 0.1 *M* NaCl at 4 °C. The dialyzed sample was incubated with 4 ml (dry volume) of Heparin-Sepharose (GE Healthcare) for 3 h. The formed slurry was loaded onto a 1 × 7.5 cm Econo-Column (BioRad, Hercules, CA) and washed with 100 ml of loading buffer (20 m*M* HEPES, pH 7.3, 0.1 *M* NaCl). Arrestin-1 was eluted with elution buffer (20 m*M* HEPES, pH 7.3, 0.4 *M* NaCl), concentrated to ~5 mg/ml, and used for crystallization trials. Arrestin-1 was exclusively crystallized from F_C. The flow-through from Heparin-Sepharose (named F_G) was collected and dialyzed against 20 m*M* HEPES, pH 7.3, 0.1 *M* NaCl. The F_G was concentrated to ~10 mg/ml prior to crystallization trials. LDH crystals were grown from F_G.

Opsin was obtained from bovine ROS as previously described (Sachs, Maretzki, & Hofmann, 2000). Briefly, a ROS pellet was resuspended in buffer containing 10 m*M* sodium phosphate, pH 7.0, and 50 m*M* hydroxylamine. The ROS suspension was kept on ice and photobleached for 30 min under a 60-W incandescent white light bulb. The photobleached suspension then was centrifuged at 16,110 × *g* for 10 min at 4 °C, and the pellet was resuspended in 1 ml of solution containing 10 m*M* sodium phosphate, pH 6.5, and 2% bovine serum albumin (BSA). To remove hydroxylamine, the suspension was centrifuged again at 16,110 × *g* for 10 min at 4 °C, and the pellet was washed five times with BSA solution. In the last step, BSA was removed by washing the pellet six times with 1 ml of 10 m*M* sodium phosphate, pH 6.5. The resulting pellet containing opsin membranes was solubilized with 1 ml of solution containing 20 m*M* Bis-Tris-propane, pH 7.5, 130 m*M* NaCl, 1 m*M* $MgCl_2$, 10% sucrose, and 1% β-D-octylglucopyranoside for 1 h at room temperature. Insoluble material was removed by centrifugation at 16,110 × *g* for 5 min at 4 °C. Solubilized opsin then was concentrated to ~5 mg/ml with a 50 kDa cutoff Amicon Ultra-0.5 concentrator (Millipore, Billerica, MA) and used for crystallization.

2.2. Protein crystallization and validation

Crystallization trials were performed by the sitting-drop vapor-diffusion method. All screens were set up in 96-well plates with a Phoenix crystallization robot (Art Robbins Instruments, Sunnyvale, CA) (Mcpherson, 1991). Initial screenings were carried out with commercially available kits, included polyethylene glycol (PEG)/Ion, PEG/Ion2, and PEG/Rx Index from Hampton Research (Aliso Viejo, CA); AmSO4 Suite, JCSG+ Suite, and Protein Complex Suite from Qiagen (Valencia, CA); and Wizard from Emerald Bio (Bainbridge Island, WA). Plates were incubated at 4 °C or RT, and checked on a regular basis at least once every week for the first month and once per month thereafter. Once crystals (or "hits") were observed, experiments were repeated with minor optimizations such as varying the type and concentration of precipitant, pH, and temperature along with the mode of vapor diffusion (sitting or hanging drops).

Crystals were obtained by mixing equal volumes (150 nl) of protein and reservoir solution in 96-well plates (Intelli-Plate 96-2, Art Robbins, Sunnyvale, CA). GAPDH crystals formed in over 10 conditions from PEG/Ion screen kits, including conditions No. 2, 6, 21, 22, 23, and 27 from PEG/Ion and conditions No. 6, 14, 20, 40, and 46 from PEG/Ion 2. LDH crystals formed in 10% PEG 3.5 K, 0.1 *M* $MgSO_4$ from a custom-designed screen kit. Arrestin-1 crystals were found in over five conditions from PEG/Ion screen kits, including conditions No. 3, 5, 13, 25, and 29 from PEG/Ion 2. The majority of protein crystals in this study grew after incubation for 1–2 weeks at 4 °C with protein sample buffer containing 20 m*M* HEPES, pH 7.3, and 0.1 *M* NaCl. GAPDH crystals grew at 4 °C with a reservoir solution containing 20% PEG 3.5 K, 0.2 *M* succinic acid, pH 7.0. Diffraction-suitable LDH crystals grew at room temperature with a reservoir solution containing 10% PEG 3.5 K, 0.2 *M* sodium acetate. Arrestin-1 crystals grew at 4 °C with a reservoir solution containing 7.5% PEG 3.5 K, 0.1 *M* sodium malonate, pH 5.0, and 4% PEG 400.

Opsin crystals were obtained by hanging drop vapor diffusion after mixing equal volumes of protein at 5 mg/ml and the reservoir solution described above. Crystals reached their full size within 25 days at 4 °C. Prior to freezing in liquid nitrogen, crystals were soaked in their respective reservoir solutions supplemented with 10% glycerol. After X-ray diffraction experiments, single crystals were dissolved, and the proteins were separated with SDS-PAGE followed by silver staining.

2.3. Mass spectrometry analysis

Crystals (5–10 for each protein) were washed and analyzed by SDS-PAGE. Coomassie-stained SDS-PAGE bands were excised and bleached in a solution of 50 m*M* ammonium bicarbonate in 50% acetonitrile. Bands were dehydrated in 100% acetonitrile and dried in a Speedvac centrifuge. Prior to overnight in-gel trypsin digestion, thiol groups were reduced with 20 m*M* dithiothreitol at room temperature for 1 h, followed by alkylation with 50 m*M* iodoacetamide in 50 m*M* ammonium bicarbonate for 30 min in the dark. Proteolytic peptides were extracted from gels with 50% acetonitrile in 5% formic acid and then resuspended in 0.1% formic acid after being completely dried under vacuum. Analyses of the resulting peptides were performed with an Orbitrap Elite Hybrid Mass Spectrometer (Thermo Electron, San Jose, CA, USA) equipped with a Waters nanoAcquity UPLC system (Waters, Taunton, MA, USA). Spectra were recorded by data-dependent methods with an alternating full scan followed by 20 MS/MS scans. Obtained data were analyzed using Mascot Daemon (Matrix Science, Boston, MA), with the settings of: 10 ppm for parent ions and 0.8 Da for product ions. Carbamidomethylation of Cys residues was set as a fixed modification, and oxidation of Met residues as a variable modification.

2.4. Data collection and structural determination

Preliminary diffraction was tested with a Rigaku MicroMax-007HF in-house X-ray source. Higher resolution data were collected with synchrotron sources at beamline X29 of Brookhaven National Laboratory (BNL) and NETCAT-24-ID-C at the Advanced Photon Source (APS), Argonne National Laboratory. Diffraction datasets were processed and scaled with HKL2000 (Otwinowski & Minor, 1997) and XDS (Kabsch, 2010a, 2010b). Protein structures were determined with the CCP4 suite version 6.4.0 (Winn et al., 2011) by molecular replacement with closely related homolog structures used as search models. Structures were refined by alternate rounds of manual rebuilding with COOT version 0.7.2 (Emsley, Lohkamp, Scott, & Cowtan, 2010) and automated refinement by REFMAC (Vagin et al., 2004). Figures were prepared with Chimera-1.8.1-win64 (Goddard, Huang, & Ferrin, 2007; Pettersen et al., 2004).

3. PILOT EXPERIMENTAL RESULTS

Isolation of ROS from bovine retinas was followed by isotonic extraction of proteins and DEAE-cellulose chromatography to enrich some proteins and prevent disruption of protein complexes, especially those with transient interactions. Unlike other proteins, opsin preparation from bovine ROS did not require further downstream treatment of ROS membranes. The overall experimental design is illustrated in Fig. 1.

Isotonic extracts of ROS (named F_A) obtained from 300 bovine retinas yielded ~20 mg of total protein. Fractionation of F_A extracts by DEAE-cellulose chromatography yielded three major peaks (Fig. 2A). Fractions from these peaks were collected along with the flow-through and analyzed by SDS-PAGE (Fig. 2B). All fractions were subsequently buffer exchanged by dialysis and concentrated prior to crystallization screening.

3.1. Protein crystals readily formed from ROS extracts

Initial screening of F_A extracts yielded more than 10 different types of crystals with various morphologies, including needles and thin rods. Subsequent SDS-PAGE and diffraction analysis of these crystals revealed that they all belonged to a single protein. Mass spectroscopy (MS) analysis identified this protein as GAPDH, a key enzyme involved in glycolysis. Because its high propensity to crystallize complicated our screening results and could have prevented crystallization of other proteins, the F_A fraction was subjected to DEAE-ion-exchange chromatography.

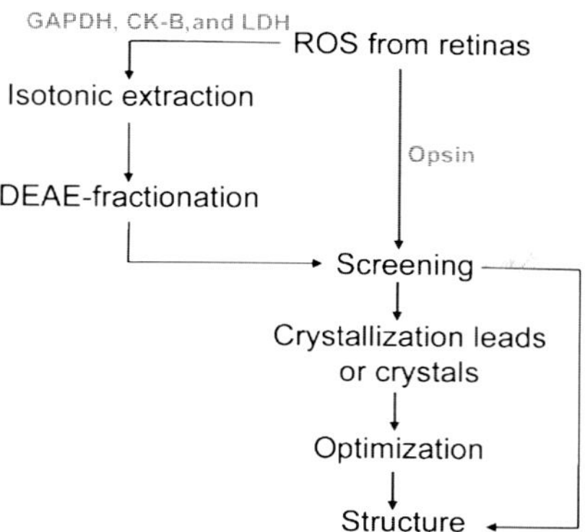

Figure 1 Flowchart illustrating the protocol used for crystallizing proteins from bovine ROS. Isotonic extraction and DEAE-ion-exchange chromatography were employed for fractionating all proteins from bovine ROS except opsin.

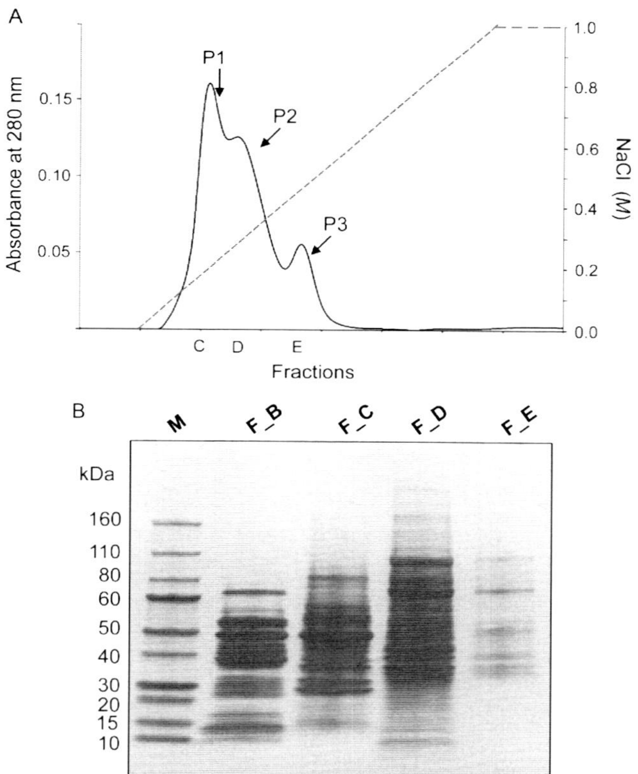

Figure 2 Protein sample preparation from bovine ROS. (A) Proteins from ROS isotonic extracts were fractionated by DEAE-ion-exchange chromatography. In a typical preparation, proteins eluted from the column in three major peaks. (B) Fractions corresponding to the three major elution peaks (P1, P2, and P3) were collected along with the flow-through (F). The protein composition of each peak was checked by SDS-PAGE with M indicating the molecular marker lane. F_B: flow-through; F_C: elution peak 1; F_D: elution peak 2; F_E: elution peak 3.

The DEAE-cellulose chromatography step added two benefits to sample preparation; it removed GAPDH from the F_A extract and increased the yield of less abundant proteins. However, using DEAE-cellulose chromatography has some drawbacks. This step could disturb interactions of weakly bound protein complexes in the F_A extract. GAPDH did not bind to the DEAE column and remained in the flow-through, from which it was recovered by affinity binding to a NAD-agarose column. The flow-through from NAD-agarose (named F_F) was collected, dialyzed against 20 m*M* HEPES, pH 7.3, 0.1 *M* NaCl, and concentrated to ~10 mg/ml for crystallization. Two proteins were crystallized from F_F, namely enolase and malate dehydrogenase. Enolase crystals grew in a reservoir containing 20% PEG 3.5 K and 0.2 *M* sodium thiocyanate. Maltase crystals were grown in 25% PEG 1 K, 0.1 *M* Na/K-phosphate, pH 6.5, and 0.2 *M* NaCl.

Bovine GAPDH crystals diffracted to ~2 Å as illustrated in Fig. 3A. The structure was determined by molecular replacement using the rabbit GAPDH structure (PDB ID: 1J0X) as the search model (Fig. 4A). The R_{work} and R_{free} of the final structure were 19% and 24%, respectively. Unit cell dimensions were a=79.8 Å, b=126.5 Å, c=83.8 Å; α=90°, β=118°, γ =90°. Although bovine GAPDH existed as a homotetramer identical to known structures in this protein family, it had three NAD molecules bound instead of either two or four found in other mammalian GAPDHs, a feature that distinguishes it from others PDB entries, 1J0X (Cowan-Jacob, Kaufmann, Anselmo, Stark, & Grutter, 2003), 1U8F (Jenkins & Tanner, 2006), and 1ZNQ (Baker, Shi, Wang, & Palczewski, 2014).

Sparse-matrix commercial screening kits are designed to accommodate a broad range of conditions for crystallization of various proteins. However, these kits may not suffice for each individual protein. To establish crystallization conditions for interesting proteins in crude ROS extracts, we referred to the ROS proteome database (Kwok et al., 2008) and selected several groups of proteins as test targets. These were L-lactate dehydrogenase (LDH), macrophage migration inhibitory factor, creatine kinase-B type (CK-B), dynamin, pyrophosphatase, phosphofructokinase, T-complex protein, and elongation factor 2. Screening conditions were designed based on published conditions for these protein structures or their closely related homologs. From these targeted screenings, we obtained crystals for two proteins, CK-B (Fig. 3B) and LDH that had not emerged from earlier screenings with commercial kits. Both proteins formed three-dimensional crystals from ROS isotonic fractions. Crystals of CK-B diffracted to 1.6 Å, a better

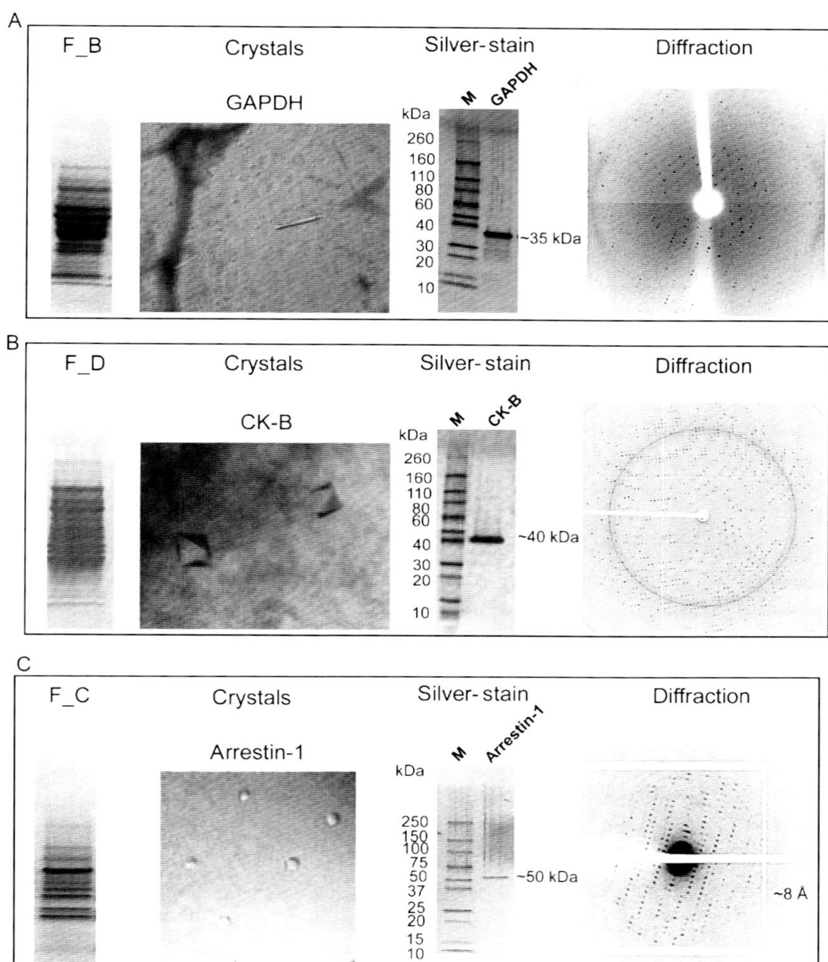

Figure 3 Three examples of crystallization of proteins from complex samples. Single crystals were examined by a silver-stained gel readout and a diffraction test. (A) Study of GAPDH showing rod-shaped crystals. (B) Study of CK-B exhibiting triclinic crystals. (C) Study of arrestin-1 revealing cubic crystals. F_B indicates the DEAE-cellulose flow-through fraction, F_C designates the DEAE-cellulose peak 1 fraction and F_D corresponds to the DEAE-cellulose peak 2 fraction.

resolution than the previously reported structure for bovine creatine kinase (2.3 Å, PDB code: 1G0W) (Tisi, Bax, & Loew, 2001). The structure of CK-B was solved with a human brain-type creatine kinase structure (PDB code: 3B6R) used as the search model (Bong et al., 2008; Fig. 4B);

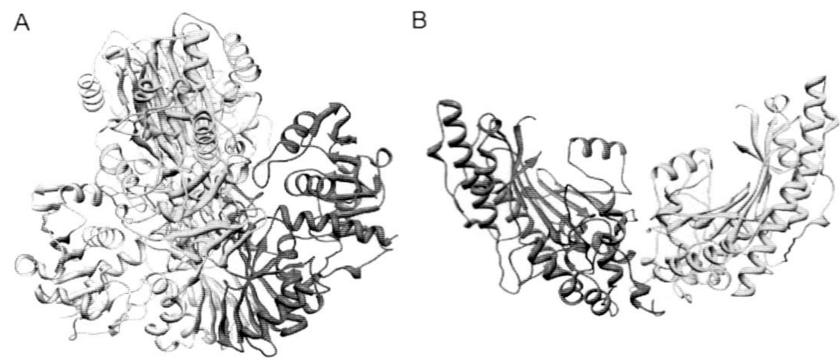

Figure 4 Ribbon representations of bovine GAPDH (left panel) solved to 1.93 Å and bovine CK-B (right panel) solved to 1.65 Å. Individual subunits are presented in different colors. (See the color plate.)

the sequence identity between bovine CK-B and human CK-B was 97%. The structure of CK-B was refined to R_{work} of 22% and R_{free} of 25%. The CK-B crystals belonged to the $P4_3$ space group with unit cells of $a = 96$ Å, $b = 96$ Å, $c = 107$ Å, $\alpha = 90°$, $\beta = 90°$, $\gamma = 90°$. Interestingly, proteins extracted from LDH crystals migrated as two close bands on SDS-PAGE gels (Fig. 5). MS analysis indicated that these crystals were composed of LDH chains A and B. Functional LDHs are either homo- (LDHA or LDHB) or heterotetramers (LDH A/B) comprised of either four chains of LDHA or two chains each of LDHA and LDHB, respectively. LDH A/B participates in adjusting LDH levels at different stages of cellular development (Ahmad & Hasnain, 2005). Apparently, LDH crystals obtained from F_G fractions had a heterotetrameric composition, whereas all other available crystal structures of the LDH protein family are in the homotetrameric form. Hence, we were able to capture a unique state of LDH that has never been reported. Figure 5 illustrates the optimization of LDH A/B crystallization from ROS extracts. Initially, LDH A/B crystallized as needles at 4 °C. Although lowering the concentration of polyethylene glycol (PEG 3.5 K) resulted in fewer but longer rod-shaped crystals at RT, these crystals were unstable and degraded within a week. Conditions then were further modified by introducing various salts into the mother liquor. Replacement of $MgSO_4$ with Na-acetate yielded stable pencil-shaped crystals of LDH A/B suitable for diffraction studies. But the diffraction quality for LDH was insufficient to determine the protein's structure.

Crystallization of several proteins with this screening strategy suggests that it can be successfully applied to other designated proteins. Our initial

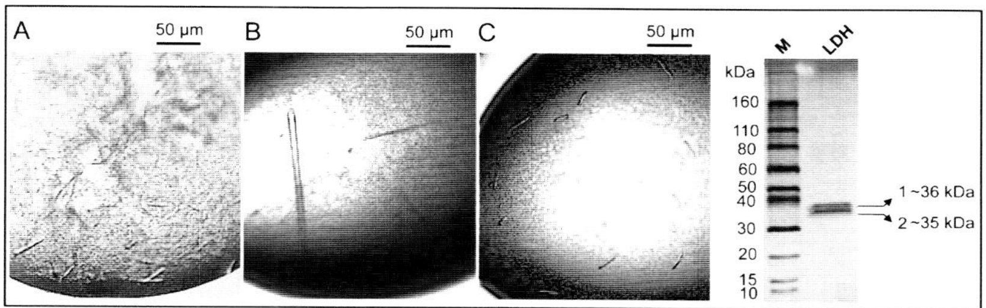

Figure 5 Optimization of LDH crystallization conditions. (A) LDH initially formed crystalline needles at 4 °C. (B) By lowering the PEG concentration, fewer and longer rod crystals were obtained at room temperature. (C) Replacement of $MgSO_4$ with 0.2 *M* Na-acetate yielded stable pencil-shaped crystals suitable for diffraction studies. Size bars of 50 µm are shown at the top of the crystal images. (D) Silver-staining analysis of a single LDH crystal revealed two distinguishable bands on a SDS-PAGE gel.

screenings employed commercial protein crystallization kits. Then to obtain protein crystals of specifically targeted proteins, screening conditions were further refined.

3.2. A membrane protein crystallized from ROS membranes

To determine the possibility of crystallizing membrane proteins from ROS membranes, we used opsin as a test model that is expressed at very high levels in these compartments (Palczewski, 2006, 2012). ROS extracts were solubilized with detergent-containing buffer, concentrated, and used directly for protein crystallization. Conditions for crystallization were based on those previously reported for opsin structures (Park, Scheerer, Hofmann, Choe, & Ernst, 2008; Scheerer et al., 2008). Crystals appeared after 1 week of incubation at 4 °C and took about 25 days to achieve their full sizes. X-ray diffraction analysis of these crystals revealed a symmetry consistent with the H3 space group with unit cell dimensions of $a = 241.86$ Å, $b = 241.85$ Å, $c = 110.45$ Å, $\alpha = 90°$, $\beta = 90°$, and $\gamma = 120°$. These diffraction data matched well with those obtained from highly purified opsin crystals (PDB ID: 3CAP). Furthermore, silver staining indicated a molecular weight of ~35 kDa (Fig. 6), which matches the molecular weight of purified opsin and confirms the identity of the crystallized protein as opsin.

3.3. Summary of protein crystallization

Proteins crystallized from ROS had molecular weights ranging from 30 to 50 kDa. Their functions can be classified as phototransduction (arrestin-1 and opsin), glycolysis (GAPDH, enolase, and LDH A/B), and metabolism

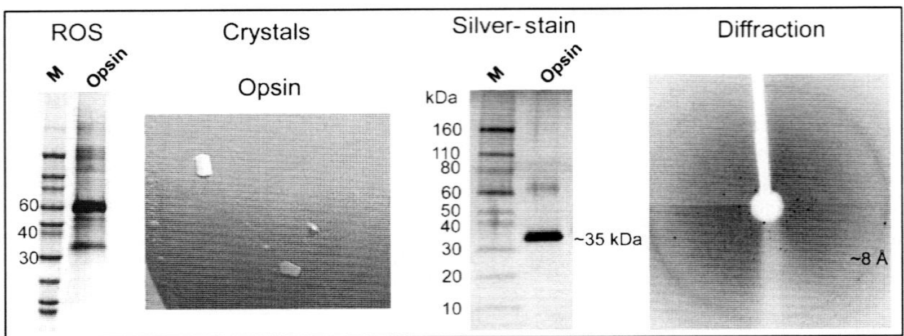

Figure 6 Opsin crystallization from bovine ROS. Hexagonal crystals (~90 μm in their longest dimension) were obtained from photobleached ROS membranes. Crystals were characterized by silver staining and X-ray diffraction. The outermost reflections correspond to 8 Å.

(CK-B and malate dehydrogenase). Structurally, CK-B existed as a dimer and GAPDH as a tetramer. The other four crystals did not diffract well enough for structural determination or are under optimization. Based on the structures of their related homologs, these proteins are expected to form tetramers in crystals (arrestin-1: PDB code, 1CF1; LDH: PDB code, 1I0Z; enolase: PDB code, 3B97; maltase: PDB code, 1MLD). Therefore, all crystals obtained from ROS were likely oligomeric, a phenomenon similar to that observed for proteins crystallized from *E. coli* (Totir et al., 2012). Table 1 summarizes the crystallization conditions for all proteins crystallized from ROS extracts.

PEGs were the most successful precipitants for crystallization of proteins from ROS extracts, contributing to five out of seven conditions that produced crystals. Most protein crystals were obtained with PEG concentrations between 10% and 20%. For crystals obtained with PEG, the medium molecular weight PEG 3–4 K was the most common, representing 80% of cases. Other conditions with PEGs as major precipitants included 0.1–0.2 *M* salts as coprecipitants. Ammonium sulfate was the second most successful precipitant in these trials. The concentration of ammonium sulfate used for obtaining all crystals ranged between 2 and 3.6 *M*. No crystals were obtained with organic solvents. The most common buffers used were HEPES, Tris, and Na-acetate with a pH range from 5.0 to 8.0.

Protein crystals were obtained at all concentrations above 5 mg/ml for opsin and above 10 mg/ml for all other proteins. Protein crystallization is also affected by the abundance of a particular protein in heterogeneous samples like ROS extracts. High-abundance proteins crystallize more readily

Table 1 Protein crystals from bovine ROS

Crystal number	ROS fraction[a]	MW (kDa)	Protein identification	Crystallization conditions	Diffraction (Å)	Structure type
#1	F_A and F_B	36	Glyceraldehyde-3-phosphate dehydrogenase	20% PEG 3.5 K; 0.2 *M* succinic acid, pH 7.0	1.9	Homotetramer
#2	F_D	42	Creatine kinase-B type	1.8 *M* ammonium sulfate; 0.1 *M* HEPES, pH 7.5; 3% PEG 1 K	1.6	Dimer
#3	F_G	36	L-Lactate dehydrogenase A and B chain	10% PEG 3.5 K; 0.2 *M* Na-acetate	10	Under optimization
#4	F_C	45	S-Antigen (arrestin-1)	7.5% PEG 3.5 K; 0.1 *M* Na-malonate, pH 5.0; 4% PEG 400	6	Under optimization
#5	F_F	36	Malate dehydrogenase	25% PEG 1 K; 0.1 *M* Na/K-phosphate, pH 6.5; 0.2 *M* NaCl	8	Under optimization
#6	F_F	47	α-Enolase	20% PEG 3.5 K; 0.2 *M* Na-thiocyanate	20	Under optimization
#7	ROS	35	Opsin	3.0–3.6 *M* ammonium sulfate; 0.1 *M* Na-acetate, pH 5.6	8	Under optimization

[a]F_A: ROS isotonic extraction. F_B: ROS isotonic extraction/DEAE-cellulose flow-through. F_C: ROS isotonic extraction/DEAE-cellulose, peak 1. F_D: ROS isotonic extraction/DEAE-cellulose, peak 2. F_F: ROS isotonic extraction/DEAE-cellulose, flow/NAD-flow. F_G: ROS isotonic extraction/DEAE-cellulose, peak 1/Heparin-flow.

than proteins at low abundance. Hence, crystallization of less abundant proteins (CK-B, LDH A/B, enolase, and maltase) was further enhanced by removing the most abundant proteins from ROS extracts. Thus, removal of GAPDH from the F_B fraction led to the formation of enolase and maltase crystals much more readily during a subsequent screen. Likewise, LDH crystals formed following the removal of arrestin-1 from the F_C fraction.

In total, crystals of seven different proteins were obtained from ROS isotonic fractions with maximum resolutions ranging from 1.6 to 20 Å (data not shown). Some crystals were detected in about one week, whereas others took longer. Among these proteins, structures were previously reported for CK-B (PDB code: 1G0W) [26], opsin (PDB code: 3CAP) [29], and arrestin-1 (PDB code: 1CF1) (Fig. 3C) [30]. Structures of GAPDH, LDH, enolase, and maltase from a bovine source have yet to be published.

3.4. Reproducibility

Crystallization occurs as a result of random association events influenced by several factors. ROS extracts were comprised of a mixture of various proteins, making it essential to assess the reproducibility of crystallization. For hits obtained from initial screens, the basic conditions were repeated and optimized to obtain crystals for further studies. Even though the morphology and size of the crystals responded to the changes in type and concentration of precipitants employed, crystallization of all proteins discussed in this study appeared to be reproducible.

4. CONCLUSIONS

Here, we investigated the crystallization profile of crude extracts from bovine ROS. In total, seven proteins were crystallized among which three formed diffraction-quality crystals after a series of crystallization condition optimizations. Structures are reported here for two of these proteins (bovine GAPDH and CK-B).

Successful crystallization of multiple proteins from complex ROS extracts indicates that many proteins from mammals also have high propensity to crystallize. Among the studied proteins, GAPDH ranked first with respect to its tendency to crystallize from ROS extracts. Arrestin was next, crystallizing under many different conditions. Although some proteins formed poorly diffracting crystals, LDH crystals featured a heterologous tetrameric composition, whereas all the other available structures for this protein family were homotetramers. Thus, we captured an unreported, naturally occurring oligomeric state of LDH. Furthermore, crystallization of opsin from ROS extracts validates the use of crude ROS membranes or heterogeneous sample preparations for initial crystallization screening of highly abundant membrane proteins. This would facilitate the identification of crystallization hits for such proteins and markedly reduce the time and expense involved in testing multiple crystallization conditions.

The tendency of mammalian proteins to crystallize has other implications. Traditionally, proteins are separated from their native environment and purified to homogeneity before crystallization. Our findings regarding the intrinsic crystallization propensity of mammalian proteins open a new avenue to conduct structural studies on native-source proteins, circumventing the need for exhausting purification and further downstream processing. Methodology used in this study is easily scalable for structural studies of proteomes from other tissues and organisms. It could also be especially helpful for protein complexes that are not well expressed in heterologous systems. Another major advantage of using complex rather than purified protein samples is that the former retain proteins and their complexes in their native states, and hence increase their likelihood of crystallizing in an unperturbed manner.

Even though the methods used in this study are biased toward proteins with a high tendency to crystallize, such proteins could serve as crystallization-inducers by cocrystallizing along with a targeted binding partner. An interesting avenue for future studies would be to explore the feasibility of using native-source proteins for structural studies of protein–protein interactions and other complex biological events.

ACKNOWLEDGMENTS

We thank Drs. Leslie T. Webster, Jr., Marcin Golczak, and Yaroslav Tsybovsky (Case Western Reserve University, CWRU) for their comments on the manuscript, Dr. Narayanasami Sukumar helped with the data collection on the native form of GAPDH at NE-CAT/ Cornell University. We thank Dr. Philip Kiser for helping with synchrotron data collection, Dr. Beata Jastrzebska for her long-time support on PDE6 preparation, Susan Farr and Satsumi Roos for helping with ROS preparations. We also thank The Protein Expression Purification Crystallization Core at Case Western Reserve University for use of their equipment and Summer J. Watterson for her help. This study was inspired by the RPE65 crystallization experiment, Kiser, Golczak, Lodowski, Chance, & Palczewski. (2009). The work was supported by funding from the National Eye Institute, National Institutes of Health Grants R24EY021126 (to K. P.), R01EY008061 (to K. P.). K. P. is John H. Hord Professor of Pharmacology.

This publication was made possible by the Center for Synchrotron Biosciences Grant, P30-EB-00998, in support of beamline X29 from the National Institute of Biomedical

Imaging and Bioengineering. The work is also based upon research conducted at the Advanced Photon Source on the Northeastern Collaborative Access Team beamlines, which are supported by a grant from the National Institute of General Medical Sciences (P41 GM103403) from the National Institutes of Health. Use of the Advanced Photon Source, an Office of Science User Facility operated for the U.S. Department of Energy (DOE) Office of Science by Argonne National Laboratory, was supported by the U.S. DOE under Contract No. DE-AC02-06CH11357.

REFERENCES

Ahmad, R., & Hasnain, A. U. (2005). Ontogenetic changes and developmental adjustments in lactate dehydrogenase isozymes of an obligate air-breathing fish Channa punctatus during deprivation of air access. *Comparative Biochemistry and Physiology. Part B, Biochemistry & Molecular Biology, 140*(2), 271–278. http://dx.doi.org/10.1016/j.cbpc.2004.10.012.

Baker, B. Y., & Palczewski, K. (2011). Detergents stabilize the conformation of phosphodiesterase 6. *Biochemistry, 50*(44), 9520–9531. http://dx.doi.org/10.1021/bi2014695.

Baker, B. Y., Shi, W., Wang, B., & Palczewski, K. (2014). High-resolution crystal structures of the photoreceptor glyceraldehyde 3-phosphate dehydrogenase (GAPDH) with three and four-bound NAD molecules. *Protein Science, 23*(11), 1629–1639. doi.10.1002/pro.2543. Epub 2014 Sep 25.

Bong, S. M., Moon, J. H., Nam, K. H., Lee, K. S., Chi, Y. M., & Hwang, K. Y. (2008). Structural studies of human brain-type creatine kinase complexed with the ADP-Mg^{2+}-NO_3^--creatine transition-state analogue complex. *FEBS Letters, 582*(28), 3959–3965. http://dx.doi.org/10.1016/j.febslet.2008.10.039.

Buczylko, J., & Palczewski, K. (1993). Purification of arrestin from bovine retinas. *Methods in Neurosciences, 15*, 226–236.

Cowan-Jacob, S. W., Kaufmann, M., Anselmo, A. N., Stark, W., & Grutter, M. G. (2003). Structure of rabbit-muscle glyceraldehyde-3-phosphate dehydrogenase. *Acta Crystallographica. Section D, Biological Crystallography, 59*(Pt. 12), 2218–2227.

Dong, A., Xu, X., Edwards, A. M., Midwest Center for Structural Genomics, Structural Genomics Consortium, Chang, C., et al. (2007). In situ proteolysis for protein crystallization and structure determination. *Nature Methods, 4*(12), 1019–1021. http://dx.doi.org/10.1038/nmeth1118.

Emsley, P., Lohkamp, B., Scott, W. G., & Cowtan, K. (2010). Features and development of Coot. *Acta Crystallographica. Section D, Biological Crystallography, 66*(Pt. 4), 486–501. http://dx.doi.org/10.1107/S0907444910007493.

Fischer, E. H., & Krebs, E. G. (1958). The isolation and crystallization of rabbit skeletal muscle phosphorylase b. *The Journal of Biological Chemistry, 231*(1), 65–71.

Goc, A., Angel, T. E., Jastrzebska, B., Wang, B., Wintrode, P. L., & Palczewski, K. (2008). Different properties of the native and reconstituted heterotrimeric G protein transducin. *Biochemistry, 47*(47), 12409–12419. http://dx.doi.org/10.1021/bi8015444.

Goddard, T. D., Huang, C. C., & Ferrin, T. E. (2007). Visualizing density maps with UCSF Chimera. *Journal of Structural Biology, 157*(1), 281–287. http://dx.doi.org/10.1016/j.jsb.2006.06.010.

Jenkins, J. L., & Tanner, J. J. (2006). High-resolution structure of human D-glyceraldehyde-3-phosphate dehydrogenase. *Acta Crystallographica. Section D, Biological Crystallography, 62*(Pt. 3), 290–301. http://dx.doi.org/10.1107/S0907444905042289.

Judge, R. A., Forsythe, E. L., & Pusey, M. L. (1998). The effect of protein impurities on lysozyme crystal growth. *Biotechnology and Bioengineering, 59*(6), 776–785.

Kabsch, W. (2010a). Integration, scaling, space-group assignment and post-refinement. *Acta Crystallographica. Section D, Biological Crystallography, 66*(Pt. 2), 133–144. http://dx.doi.org/10.1107/S0907444909047374.

Kabsch, W. (2010b). XDS. *Acta Crystallographica. Section D, Biological Crystallography, 66*(Pt. 2), 125–132. http://dx.doi.org/10.1107/S0907444909047337.

Kim, Y. J., Hofmann, K. P., Ernst, O. P., Scheerer, P., Choe, H. W., & Sommer, M. E. (2013). Crystal structure of pre-activated arrestin p44. *Nature, 497*(7447), 142–146. http://dx.doi.org/10.1038/nature12133.

Kiser, P. D., Golczak, M., Lodowski, D. T., Chance, M. R., & Palczewski, K. (2009). Crystal structure of native RPE65, the retinoid isomerase of the visual cycle. *Proceedings of the National Academy of Sciences of the United States of America, 106*(41), 17325–17330. http://dx.doi.org/10.1073/pnas.0906600106, 0906600106 [pii].

Kors, C. A., Wallace, E., Davies, D. R., Li, L., Laible, P. D., & Nollert, P. (2009). Effects of impurities on membrane-protein crystallization in different systems. *Acta Crystallographica. Section D, Biological Crystallography, 65*(Pt. 10), 1062–1073. http://dx.doi.org/10.1107/S0907444909029163, S0907444909029163 [pii].

Kwok, M. C., Holopainen, J. M., Molday, L. L., Foster, L. J., & Molday, R. S. (2008). Proteomics of photoreceptor outer segments identifies a subset of SNARE and Rab proteins implicated in membrane vesicle trafficking and fusion. *Molecular & Cellular Proteomics*, 7(6), 1053–1066. http://dx.doi.org/10.1074/mcp.M700571-MCP200.

Mcpherson, A. (1991). A brief-history of protein crystal-growth. *Journal of Crystal Growth, 110*(1–2), 1–10. http://dx.doi.org/10.1016/0022-0248(91)90859-4.

Muller, D. J., Wu, N., & Palczewski, K. (2008). Vertebrate membrane proteins: Structure, function, and insights from biophysical approaches. *Pharmacological Reviews, 60*(1), 43–78. http://dx.doi.org/10.1124/pr.107.07111, pr.107.07111 [pii].

Otwinowski, Z., & Minor, W. (1997). Processing of X-ray diffraction data collected in oscillation mode. *Macromolecular Crystallography, Part A, 276*, 307–326. http://dx.doi.org/10.1016/S0076-6879(97)76066-X.

Palczewski, K. (2006). G protein-coupled receptor rhodopsin. *Annual Review of Biochemistry, 75*, 743–767. http://dx.doi.org/10.1146/annurev.biochem.75.103004.142743.

Palczewski, K. (2012). Chemistry and biology of vision. *The Journal of Biological Chemistry, 287*(3), 1612–1619. http://dx.doi.org/10.1074/jbc.R111.301150.

Park, J. H., Scheerer, P., Hofmann, K. P., Choe, H. W., & Ernst, O. P. (2008). Crystal structure of the ligand-free G-protein-coupled receptor opsin. *Nature, 454*(7201), 183–187. http://dx.doi.org/10.1038/nature07063, nature07063 [pii].

Pettersen, E. F., Goddard, T. D., Huang, C. C., Couch, G. S., Greenblatt, D. M., Meng, E. C., et al. (2004). UCSF Chimera–a visualization system for exploratory research and analysis. *Journal of Computational Chemistry, 25*(13), 1605–1612. http://dx.doi.org/10.1002/jcc.20084.

Pettigrew, G. W., Bartsch, R. G., Meyer, T. E., & Kamen, M. D. (1978). Redox potentials of photosynthetic bacterial cytochromes-C2 and structural bases for variability. *Biochimica et Biophysica Acta, 503*(3), 509–523. http://dx.doi.org/10.1016/0005-2728(78)90150-0.

Sachs, K., Maretzki, D., & Hofmann, K. P. (2000). Assays for activation of opsin by all-trans-retinal. *Methods in Enzymology, 315*, 238–251.

Scheerer, P., Park, J. H., Hildebrand, P. W., Kim, Y. J., Krauss, N., Choe, H. W., et al. (2008). Crystal structure of opsin in its G-protein-interacting conformation. *Nature, 455*(7212), 497–502. http://dx.doi.org/10.1038/nature07330, nature07330 [pii].

Skiba, N. P., Spencer, W. J., Salinas, R. Y., Lieu, E. C., Thompson, J. W., & Arshavsky, V. Y. (2013). Proteomic identification of unique photoreceptor disc components reveals the presence of PRCD, a protein linked to retinal degeneration. *Journal of Proteome Research, 12*(6), 3010–3018. http://dx.doi.org/10.1021/pr4003678.

Tisi, D., Bax, B., & Loew, A. (2001). The three-dimensional structure of cytosolic bovine retinal creatine kinase. *Acta Crystallographica. Section D, Biological Crystallography*, *57*(Pt. 2), 187–193.

Totir, M., Echols, N., Nanao, M., Gee, C. L., Moskaleva, A., Gradia, S., et al. (2012). Macro-to-micro structural proteomics: Native source proteins for high-throughput crystallization. *PLoS One*, 7(2), e32498. http://dx.doi.org/10.1371/journal.pone.0032498.

Vagin, A. A., Steiner, R. A., Lebedev, A. A., Potterton, L., McNicholas, S., Long, F., et al. (2004). REFMAC5 dictionary: Organization of prior chemical knowledge and guidelines for its use. *Acta Crystallographica. Section D, Biological Crystallography*, *60*(Pt. 12 Pt. 1), 2184–2195. http://dx.doi.org/10.1107/S0907444904023510.

Winn, M. D., Ballard, C. C., Cowtan, K. D., Dodson, E. J., Emsley, P., Evans, P. R., et al. (2011). Overview of the CCP4 suite and current developments. *Acta Crystallographica. Section D, Biological Crystallography*, *67*(Pt. 4), 235–242. http://dx.doi.org/10.1107/S0907444910045749.

CHAPTER TWENTY-TWO

Crystallization of Photosystem II for Time-Resolved Structural Studies Using an X-ray Free Electron Laser

Jesse Coe, Christopher Kupitz, Shibom Basu, Chelsie E. Conrad, Shatabdi Roy-Chowdhury, Raimund Fromme, Petra Fromme[1]

Department of Chemistry and Biochemistry, Arizona State University, Tempe, Arizona, USA

[1]Corresponding author: e-mail address: petra.fromme@asu.edu

Contents

Abstract

Photosystem II (PSII) is a membrane protein supercomplex that executes the initial reaction of photosynthesis in higher plants, algae, and cyanobacteria. It captures the light from the sun to catalyze a transmembrane charge separation. In a series of four charge separation events, utilizing the energy from four photons, PSII oxidizes two water

Methods in Enzymology, Volume 557
ISSN 0076-6879
http://dx.doi.org/10.1016/bs.mie.2015.01.011

molecules to obtain dioxygen, four protons, and four electrons. The light reactions of photosystems I and II (PSI and PSII) result in the formation of an electrochemical transmembrane proton gradient that is used for the production of ATP. Electrons that are subsequently transferred from PSI via the soluble protein ferredoxin to ferredoxin-$NADP^+$ reductase that reduces $NADP^+$ to NADPH. The products of photosynthesis and the elemental oxygen evolved sustain all higher life on Earth. All oxygen in the atmosphere is produced by the oxygen-evolving complex in PSII, a process that changed our planet from an anoxygenic to an oxygenic atmosphere 2.5 billion years ago. In this chapter, we provide recent insight into the mechanisms of this process and methods used in probing this question.

1. INTRODUCTION

Two and a half billion years ago oxygenic photosynthesis evolved resulting in the drastic oxygenation of Earth's atmosphere which led to an explosion of biodiversity and, eventually, the evolution of higher organisms. Photosystem II (PSII) is crucial to oxygenic photosynthesis. As indicated by the close homology of the oxygen-evolving complex (OEC) across many species, it has only substantially evolved once and the core functions and structure have been maintained through billions of years of evolution. PSII is large membrane protein complex made up of 19 protein subunits and over 50 noncovalent cofactors. During photoactivation, the OEC of this massive complex proceeds through a five state photochemical reaction, the Kok cycle, over the course of which four charge separations occur (Fig. 1). One electron and one proton are extracted in each of the charge separation events, leading to two water molecules being deconstructed until the formation of dioxygen each cycle. PSII is able to oxidize water, driven by visible light and catalyzed by earth abundant metals at a low overpotential of +1.1 V (Wydrzynski & Satoh, 2006). However, with such a high redox potential, it operates at the limit of the stability of biomolecules. The unraveling of the mechanism of water splitting in PSII is one of the major goals in bioenergetics as it would open the door to the development of synthetic oxygen evolving systems that combine the major catalytic features of PSII with the stability of artificial systems.

The first static structure of PSII was determined to 3.8 Å in 2001 (Zouni). The resolution was increased subsequently (Ferreira, Iverson, Maghlaoui, Barber, & Iwata, 2004; Loll, Kern, Saenger, Zouni, & Biesiadka, 2005; Umena, Kawakami, Shen, & Kamiya, 2011) and the highest resolution structures are now available at the near atomic resolution

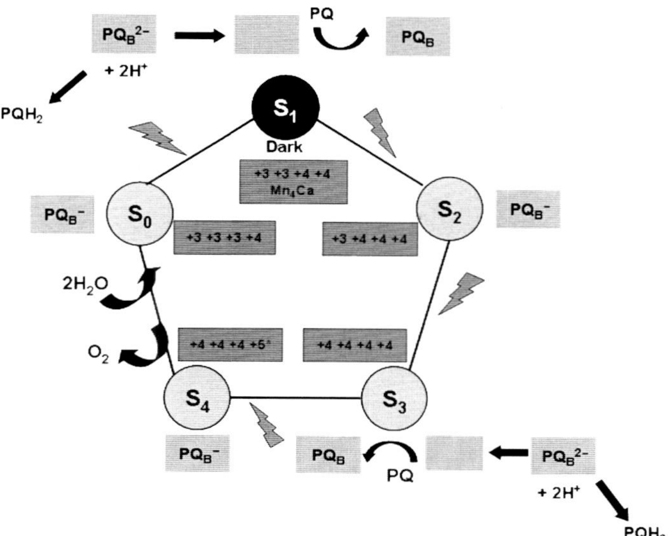

Figure 1 Schematic of the Kok cycle showing photoinduced progression from the dark S_1 state to the S_4 state, where oxygen evolution occurs followed by a final photoexcitation returning the PSII to the dark S_1 state. PQ_B is reduced at each excitation, leaving, and being replaced by a fresh PQ_B upon arrival at the S_3 and S_1 states. *Figure originally published in Kupitz, Basu, et al. (2014).*

of 1.9 Å (Suga et al., 2014). In order to understand the catalytic mechanism at work, time-resolved studies with microsecond time resolution are needed to explore the multiple photoexcited states in the Kok cycle (Renger, 2012). Traditional macrocrystallography is unable to unravel the structure of the OEC in its different oxidation states due to site-specific radiation damage of the OEC by X-ray photoreduction. X-ray absorption fine structure spectroscopic studies have indicated that the manganese ions in the OEC possess a high propensity to site-specific X-ray-induced reduction (Allakhverdiev et al., 2005). A bias may also arise with cryo-cooled crystals as data are collected far removed from biologically relevant temperatures. The recently developed method of serial femtosecond X-ray crystallography (SFX) has been used to investigate the structure of selected photoexcited states of PSII (Kern et al., 2013, 2014; Kupitz, Basu, et al., 2014). Very recently, X-ray free electron lasers (XFELs) have also been used to determine the first high resolution undamaged dark structure of PSII based on serial data collection on large cryo-cooled crystals (Suga et al., 2014).

Fundamentally, SFX is based upon using highly coherent, extremely brilliant, femtosecond hard X-ray pulses to collect thousands of X-ray diffraction snapshots on a hydrated stream of small crystals at room temperature (Chapman et al., 2011), enabling diffraction information to be obtained before the ensuing Coulomb explosion destroys the molecules including each measured crystal (Neutze et al., 2000). In this way, radiation damage is "outrun" in the data collection and samples can be probed at more biologically mimetic conditions compared to traditional protein macrocrystallography. Furthermore, using a fresh crystal to produce each diffraction pattern, time-resolved photoexcitation processes can be probed by coupling an optical pump probe to the sample prior to interaction with the X-ray beam (Aquila et al., 2012; Neutze & Moffat, 2012; Spence, Weierstall, & Chapman, 2012). For the first time, irreversible processes are now able to be probed in a time-resolved experiment due to the use of discrete crystals for each diffraction pattern. A delay time between the optical pumping and beam interaction as well as the number of times the sample is flashed with light can be modulated to explore different transitory states. Recent work has advanced the knowledge of the structure and further advancements will ultimately be able to resolve the mechanism of photosynthetic water splitting and oxygen evolution. The method of time-resolved SFX (TR-SFX) has the potential to lead to a "molecular movie" of photosynthetic water splitting in the future. In this chapter, recent advancements of methods developed for TR-SFX of PSII will be summarized and compared.

2. ISOLATION OF PHOTOSYSTEM II

All structures solved so far from PSII are based on PSII isolated from the thermophilic cyanobacteria *Thermosynechococcus elongatus* (*T. elongatus*) (Ferreira et al., 2004; Kern et al., 2013; Kupitz, Basu, et al., 2014; Loll et al., 2005; Zouni et al., 2001) and *Thermosynechococcus vulcanus* (*T. vulcanus*) (Kamiya & Shen, 2003; Suga et al., 2014). As PSII undergoes the process of photodamage and repair (Aro, Virgin, & Andersson, 1993), the reproducible growth of large qualities of the cyanobacteria is an essential prerequisite of all functional studies. Our group uses a large 122L photobioreactor that has been developed together with the Pulse Institute that controls temperature, pH, CO_2, and air flux and measures cell density which can be coupled to the light intensity for growth of *T. elongatus* for reproducible growth of the cells at low light conditions to minimize photodamage. The complete isolation procedure from cell harvest to growth of crystals is performed in one setting within 48 h under dim green light which involves three recrystallization steps. The methods have recently been published in Kupitz, Grotjohann, et al. (2014).

3. CRYSTALLIZATION FOR STUDIES WITH FELs

Due to X-ray damage, macromolecular crystallography requires very large single crystals of PSII on the order of millimeters in size to allow for a shift of the crystals after each image during data collection. However, even with an extremely careful shift strategy such as applied in Umena et al. (2011), photoreduction is difficult to minimize, leading to a photoreduced structure of the OEC.

SFX studies depend on a continuously fresh stream of small crystals to obtain diffraction before destruction. This leads to the need for size homogeneous nano- or microcrystals (on the order of 500 nm to 5 μm) created using novel crystallization techniques. New methods have been developed by our team using free interface crystallization to allow for reproducible growth of large quantities of microcrystals of PSII with a very narrow size distribution of centered around 1 μm^3 as described in Kupitz, Grotjohann, et al. (2014). We will discuss the different methods used for the growth of PSII microcrystals below in more detail.

3.1 Batch methods and establishing the phase diagram

The batch method describes crystallization through a homogenized solution containing the protein and precipitant. Initial trials to establish knowledge of phase space are easily accessible using this method, owing to the highly controllable and quantifiable solution environment. Seed crystals can be added at different points in phase space and monitored to map out the phase diagram. Seed crystals will dissolve in the nonsaturated zone, grow without additional nucleation in the metastable zone, and induce nucleation of additional crystals in the nucleation zone. Particularly important is that the amount of induced nucleation scales with distance from the border between the metastable and nucleation zones, permitting optimization of conditions with respect to yield, crystal size, and homogeneity.

3.2 Free interface diffusion

Figure 1 shows a schematic of the free interface diffusion (FID) method developed for the growth of PSII nanocrystals that have been used for our TR-SFX studies (Kupitz, Grotjohann, et al., 2014). The method is based on the idea that in order to develop a so-called "shower" of small crystals, one must rapidly access a point in the nucleation zone of phase space so that a high nucleation rate is achieved. This provides an alternative to vapor diffusion experiments where it is often difficult to control crystal growth at high supersaturation.

The following procedure describes a way to access the phase space. This construction also results in the deposition of crystals into a pellet, which can be easily collected and allows for further quenching and control of crystal density. FID has been used for growth of larger crystals in a traditional setup, where crystallization occurs inside a capillary (Ng, Gavira, & García-Ruíz, 2003). By the use of thin capillaries, diffusion is slowed so larger crystals are grown along a concentration gradient (McPherson, 1991). In contrast, for growth of nanocrystals rapid nucleation is desired. The precipitant, which has a higher density than the protein solution, is slowly added dropwise to the protein solution (see Fig. 2). The drops of precipitant passing through the protein solution cause the protein to interact with a high concentration of precipitant. This continues as the dense precipitant forms a layer underneath the solubilized protein. The interface between the two solutions allows for rapid nucleation. Once the crystals reach a certain size, they sediment into the precipitant where their growth is quenched. Centrifugation immediately after the formation of the interface not only results in crystals that can be seen within 30 min but also results in a more homogenous size distribution of crystals due to a rapid density separation. Thereby, the FID method provides an intrinsic quenching of crystal growth and a uniform size distribution for PSII crystals of 1 μm ± 500 nm (Kupitz, Grotjohann, et al., 2014).

3.3 Quenching

Once a yield of crystals has been determined to be at an acceptable size and distribution, it is desirable to quench any further growth prior to diffraction or during sample delivery. While the FID setup has a temporary quenching mechanism, it is not sustainable due to the slow onset of equilibrium and complete mixing of the layers. It is important to remove the free protein that could lead to growth of existing crystals, further leading to clogging of injectors and a broader size distribution of crystals. Once the crystals form a pellet, the supernatant is removed and replaced by stabilization buffer that contains 1.25 times of all solutes in the crystallizing precipitant solution. This ensures

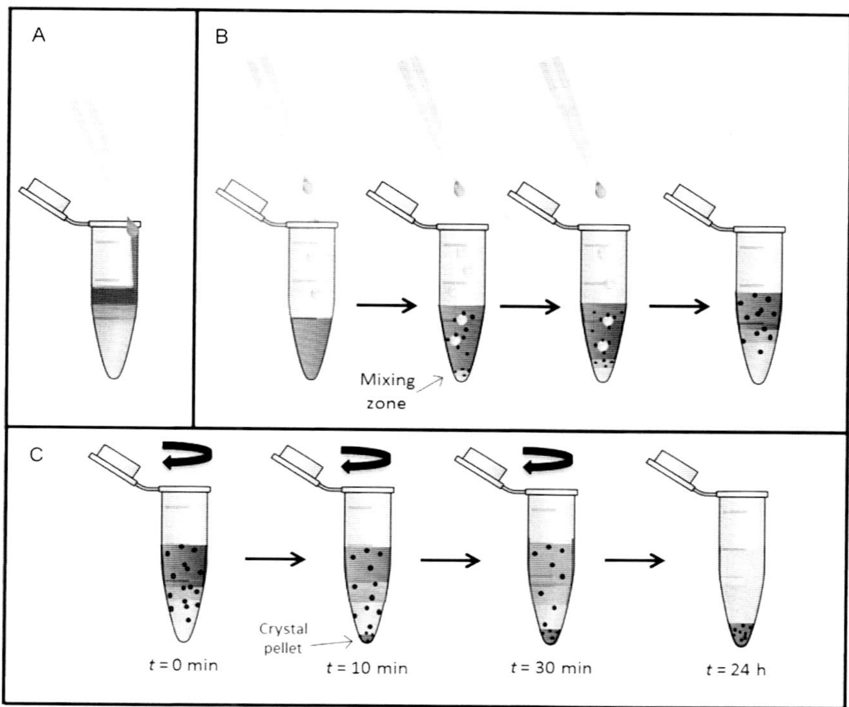

Figure 2 Stepwise procedure depicting FID and FID centrifugation nanocrystallization techniques. (A) Shows a minimal mixing layering scheme between the protein (green; dark gray in the print version) and the precipitant (white). This results in few small crystals originating at the interface. (B) Shows the precipitant being dropped through the protein, creating a slightly larger mixing zone than (A), resulting in a large shower of microcrystals. (C) Shows the time progression when the experimental setup depicted in (B) is subjected to centrifugation, expediting the formation of crystals, and causing a pellet of size homogeneous crystals to form. *Figure originally published in Kupitz, Grotjohann, et al. (2014).*

the stabilization of existing crystals, both in the absence of free protein and through the presence of a more thermodynamically unfavorable solute environment, enabling a strong preference for PSII molecules to remain in the solid crystal phase.

3.4 Quantification of natural plastoquinone and addition of PQ_{decyl}

In order to investigate conformational changes at both the acceptor and donor site of PSII and to allow multiple laser excitation steps, it is important to verify the quinone content of the PSII in the crystals. PQ_B is a mobile electron carrier and light exposure must therefore be avoided during all preparation steps to ensure high occupancy of the QB-binding site with PQ. We have determined the PQ content of our crystals using high-pressure liquid chromatography (HPLC) with a (C-18) column after each PEG 2000 precipitation step. The protein was then subjected to a pigment extraction using acetone according to the protocol from Patzlaff and Barry (1996). The presumed ratio of chlorophyll *a* to PQ is 76:4 at full quinone occupancy in *T. elongatus*. The area under each peak was integrated to obtain a ratio between chlorophyll *a* and PQ in the extracted sample. From this ratio, a percentage of PQ_B occupancy was calculated from an average value taken from three HPLC runs. This resulted in occupancies of 91.8% before initial crystallization, 88.4 after the first recrystallization, 86.4 after the second recrystallization, and 81.1% after the third recrystallization step.

After two optical events, Q_B becomes doubly reduced to PQ^{2-} and leaves the binding site as PQH_2, and thus needs to be replaced for further oxidation to occur in order to reach the S_4 state. However, PQ is extremely difficult to obtain by synthesis or purification due to poor solubility caused by the long isoprene tail. As a substitute, a derivative of PQ with the same head group but an *N*-decyl chain instead of the isoprene tail, referred to as PQ_{decyl}, was synthesized. This PQ_{decyl} was added to the crystals so that it could repopulate the binding site after the departure of the native PQH_2. Thus, the addition of the PQ_{decyl} allows for the S_4 state to be reached even when the protein is not in its' natural membrane environment.

4. DETECTION AND CHARACTERIZATION OF NANO- AND MICROCRYSTALS

4.1 Optical microscopy

During all crystallization trials, drops taken from an individual experiment can be imaged using optical light microscopy, by which crystals >1 μm can be detected. However, one must be cautious drawing conclusions from optical microscopy (OM) alone since the size of the crystals are on or past the edge of what is visible and reliance on OM alone cannot identify nanocrystals <1 μm. One of the most useful methods in OM is the use of polarized light to check for birefringence since protein crystals often possess refractive anisotropy. This can help to distinguish between crystalline protein and amorphous precipitate, provided crystals are >1 μm. Another caution should be mentioned that many salts are also birefringent and, since the birefringence signal will scale with the size of the crystal, crystals that are on the order of 1 μm or less will not be recognizable as such by birefringence.

4.2 Ultraviolet fluorescence microscopy

The use of ultraviolet fluorescence microscopy (UVM) allows the confirmation of protein in the crystals via tryptophan fluorescence. This is complimentary to OM when trying to determine whether or not a birefringent signal comes from a salt or protein crystal. Diffuse signal can also indicate undersaturated conditions or the presence of free protein amongst crystals. In the general case of non-SONICC active protein crystals (see Section 4.3) that may have low birefringence or have a birefringent salt as a precipitant, discrete spots in a UVM image can provide an alternative indication of crystals.

4.3 Second-order nonlinear imaging of chiral crystals

The second-order nonlinear imaging of chiral crystals (SONICC) technique is an ideal method for detecting crystals of a chiral molecule with non-centrosymmetric crystals, which is common among protein crystals (Wampler, Begue, & Simpson, 2008). When a substantially intense electric field is produced by a laser pulse, molecular dipoles are induced. In the case of a chiral crystal, these induced dipoles are anisotropic on their potential energy surface and allow the sampling of nonlinear, even numbered higher order polarizability terms such as the second-generation harmonic (frequency doubling) (Haupert & Simpson, 2011). Enhanced signal can be measured at half the wavelength of the incident pulse, indicative of chiral crystals due to constructive interference provided by the ordered lattice. PSII is crystallized in space group of $P2_12_12_1$ which is SONICC active and provides positive confirmation of crystals too small to image optically, distinguishing them from amorphous precipitate or identifying them in a visibly clear drop. Protein crystals as small as 100 nm in size can be detected with SONICC. Second harmonic-generated signal was measured at 532 nm, indicative of a two photon process that is only enhanced by the presence of crystals with anisotropic unit cells and is negligible otherwise.

4.4 Dynamic light scattering

Dynamic light scattering (DLS) allows the calculation of particle size and size distribution using a temporal autocorrelation function of the scattered light signal over time in tandem with the Stokes–Einstein equation for particle radius. This comes from the stochastic Brownian motion of particles resulting in a time-dependent scattering intensity caused by interference with the surrounding particles. The first-order autocorrelation function is given by Eq. (1) as a function of the scattering radius q with delay time τ being parameterized. The diffusion coefficient, D_t, can then be identified and further used to calculate the hydrodynamic radius, r, assuming a sphere with known viscosity, η, at a temperature T (Eq. 2). Through computational Fourier decomposition, multiple signals can be identified in the raw data leading to measurement of size dispersion. Thus, once other methods have been employed to confirm the existence of crystals, DLS provides the ability to monitor size and homogeneity at various time intervals during crystallization:

$$g(q;\tau) = e^{-q^2 D_t \tau} \tag{1}$$

$$D = \frac{k_B T}{6\pi\eta r} \tag{2}$$

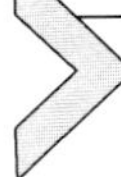

5. TIME-RESOLVED CRYSTALLOGRAPHY OF PSII USING FELs

The TR-SFX approach using FELs allows for the determination of the structure of undamaged biomolecules at room temperature, as diffraction occurs before destruction takes place (Barty et al., 2012). Furthermore, time-resolved studies can be performed where a reaction is initiated by light or rapid mixing, even on irreversible processes due to the serial delivery of the single crystals in a liquid jet where each femtosecond X-ray pulse hits a new crystal.

Time-resolved studies in crystallography were pioneered with the Laue method which uses a relatively polychromatic "pink" beam and large crystals to study light-induced reversible reactions. Pioneering work has been done with the p21–GTP complex, myoglobin, and PYP (Rajagopal et al., 2005; Schlichting et al., 1990; Šrajer et al., 2001). However, Laue crystallography cannot be used for the study of the S-state cycle of PSII due to the X-ray-induced reduction of the metal cluster of the OEC. Further limitations are the limited light penetration of the large crystals, prohibiting a uniform population of transitory states. SFX overcomes these problems and opens a new window of opportunity for time-resolved studies toward molecular movies of biomolecules at work.

5.1 Considerations for PSII in TR-SFX

For time-resolved studies, it is important that a sufficient population of the protein molecules must be in the same state when the crystal is probed by the X-rays in order to elucidate changes in the electron density of conformationally active localities. Thus, for light-induced time-resolved experiments, it is imperative that crystals be small enough so that a high majority of the molecules in the crystal are excited by a saturating laser flash. Furthermore, when processing serial data, it is important that Bragg diffraction intensities be comparable between. This leads to the need for size homogeneous nano- or microcrystals (on the order of 500 nm to 5 μm) to allow maximal uniform excitation from the optical pump laser.

Recently, the dark state of PSII was investigated at the FEL at SACLA by a group led by Jian-Ren Shen. In this study, they collected FEL data on PSII crystals using an alternate fixed target approach whereby they solved the first high resolution undamaged dark structure of PSII based on data collection of very large single crystals at cryogenic temperatures. The dark structure was determined at 1.9 Å resolution and showed smaller distances between Mn atoms of the metal cluster of the OEC compared to the first high resolution structure of PSII based on data collection at synchrotrons (Umena et al., 2011). In this incredibly vast experiment, data were collected on 336 individual large crystals of millimeter size with femtosecond X-ray pulses at the FEL in SACLA. To minimize progression of X-ray damage, data were collected under cryogenic conditions with a defocused beam (1 μm). With the FEL's relatively large beam focus, the crystals had to be translated 50 μm between each shot.

While this breakthrough work led to the first undamaged high resolution structure of the dark state of PSII, time-resolved studies, however, will be very difficult using this setup as uniform light excitation cannot be achieved with large single crystals. Additionally, data collection at cryogenic temperatures would not allow progression of the S-states beyond the S_2 state which is reached after a single laser excitation.

When the PSII is excited, light is captured by a large antenna system and the excitation energy is transferred into the center of the complex, where charge separation takes catalyzed by the primary donor P680. The charge is passed through an electron transfer chain from P680 through chlorophyll *a*, a pheophytin, the plastoquinone PQ_A and finally to the terminal acceptor plastoquinone PQ_B. After two charge separation events, PQ_B is doubly reduced to PQ^{2-}, picks up two protons, and leaves the binding site as plastoquinol PQH_2.

Once PQH_2 departs, it is subsequently replaced by another PQ_B from the PQ pool located in the photosynthetic membrane. $P680^+$ is concurrently reduced by extracting one electron at a time from two substrate water molecules bound at the OEC via the redox active tyrosine. After four electrons have been extracted in subsequent charge separation events, oxygen is evolved. The OEC consists of a cubanoid Mn_4O_5Ca cluster which undergoes four corresponding oxidation events and cycles through the Kok cycle. In the absence of light, PSII's ground state is S_1 state where one positive charge has already been accumulated in PQ_B. The oxidation states of the four manganese atoms in the OEC is currently under debate but one likely scenario consists of oxidation states of (+3+3+3+4) for the S_0 state, (+3+3+4+4) for the S_1 ground state, (+3+4+4+4) for S_2, (+4+4+4+4) for S_3, and (+4+4+4+5) for S_4. In order to ensure that the desired excited states (S_2 and S_3) are being reached, it is necessary to verify the enzymatic activity (oxygen evolution) and quinone exchange.

5.2 The pump-probe experiment

In the time-resolved SFX experiments, an optical pump laser is used to excite the crystals preceding interaction with the probe XFEL beam. This technique necessitates a laser excitation scheme with the goal to achieve a maximum excited population amongst PSII molecules in the crystals in each pump pulse and allowing for uniform evolution of transition states between each pump. Two different setups have been developed for TR-SFX studies on PSII.

In the experimental scheme described in Kupitz, Basu, et al. (2014), the crystals are delivered to the FEL beam in a fast running jet (10 m/s) at ambient temperature and hydrated in their mother liquor. The experimental setup is depicted in Fig. 3. The optical laser pulses were triggered by the linac coherent light source (LCLS), meaning that the time delay between the flashes is known exactly and is independent of the flow rate of the liquid jet. The fast running jet uses high amounts of sample at a flow rate of 10 μL/min. However, it ensures that the sample is fully replenished before the next FEL pulse arrives, eliminating the possibility of upstream excitation or damage. Data are collected simultaneously in pulse by pulse alternating light and dark sets, where the pump lasers are triggered with a frequency of 60 Hz synched to the 120 Hz LCLS FEL pulses. To reiterate, this manifests as one snapshot being a "dark" snapshot with the next one being "light" snapshot and so on, resulting in 60 dark images and 60 light images collected per second. This ensures that all variables during data collection are identical for the light

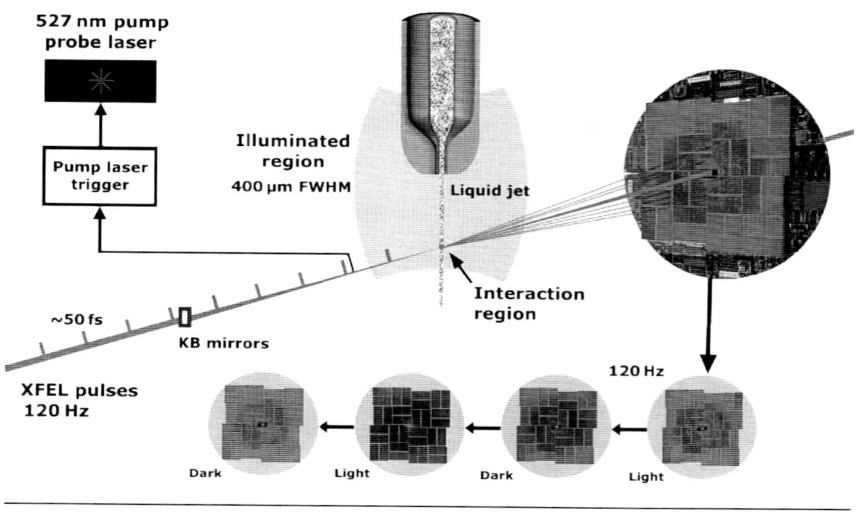

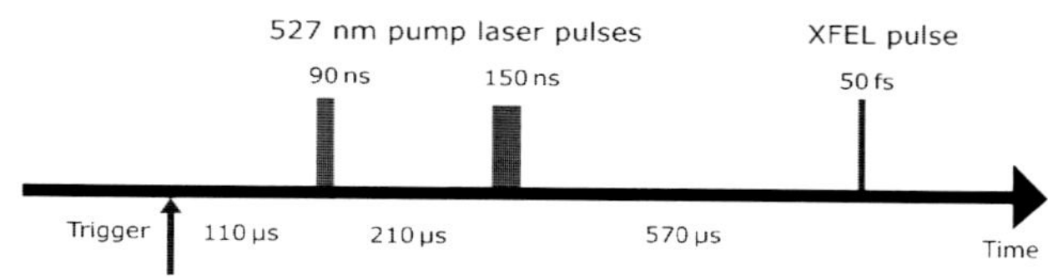

Figure 3 Diagram of the experimental setup for the time-resolved SFX experiment in which the S_1 and S_3 states are probed alternately at 60 Hz each. The optical pump laser scheme is shown at the bottom, indicating delay times of 210 µs between the first and second pump, and 570 µs between the second pump and interaction with the FEL beam, allowing population of the S_2 and S_3 states, respectively. *Figure originally published in Kupitz, Basu, et al. (2014).* (See the color plate.)

and dark datasets. Data were also collected without any laser excitation and comparison shows that the data of the purely dark runs and the alternating dark/light runs are identical. The laser excitation scheme is shown in the bottom of Fig. 3. Delay times of 210 µs between flashes 1 and 2 and 560 µs between flash 2 and the "probing" with the FEL pulse were used, corresponding to three times the measured time constants of the OEC progression from the S_1 state to the S_2 state and from the S_2 state to the S_3 state (Dekker & Van Grondelle, 2000). Data in the literature for the electron transfer between PQ_A and PQ_B greatly vary and are in the range of 200–800 µs. Future planned studies will extend the time points to up to 2 ms.

Ideal time delays between flashes should be in the range between 200 µs and 2 ms to study the conformational changes associated with electron transfer at the acceptor site and oxidation of the Mn_4CaO_x cluster at the donor site. A variation of time delays, including longer time delays up to 2 ms and an improvement in resolution of the structural model, are important to the resolution of conformational changes at atomic detail in any future work.

An alternative approach used by Kern et al. (2013, 2014) for TR-SFX studies. In this approach, the crystals are delivered to the FEL beam in a slow-running jet where the sample is exposed to light by flowing across multiple windows in the nozzle. In this setup, the time delays between the first sets of laser excitations are dictated by the flow rate of the sample in the nozzle and only the last flash is triggered by the incoming FEL pulse. While this sample delivery method has the advantage of low sample consumption, it is limited in that the time delay for the first set of flashes is determined by the flow rate, which is often not constant and also varies within the nozzle as there is an order of magnitude difference in the flow rates of the center and sides of the nozzle. This limits the setup to long time delays between the flashes with an average time delay between the first flashes on the order of 500 ms. This number is only a rough estimate as the flow rates have not been directly determined and were instead estimated from volume filled into the reservoirs and the time the sample ran out (see Kern et al., 2014). Regardless, these long time delays inhibit realizing the conformational changes at the acceptor site as PQ_A^- is oxidized by side reaction with oxygen in 2–3 ms before arrival of the second electron (de Wijn & van Gorkom, 2001).

In summary, the most commonly successful sample delivery method for TR-SFX to date is the gas dynamic virtual nozzle which uses amplified liquid pressure and gas focusing to deliver the sample. It has not only be used for the TR-SFX studies on PSII but also formed the basis for the first TR-SFX study that reached atomic resolution, using the photoactive yellow protein as a model system (Tenboer et al., 2014). Using the lipidic cubic phase (LCP) as a delivery media allows for minimal sample consumption with flow rates on the order of nL/min (Weierstall et al., 2014). This media has also been shown to successfully support membrane protein crystals with datasets solved with <0.5 mg. However, the crystals must be grown in the LCP, necessitating possibly new conditions, and the optical density of the material and slow flow rate is prohibitive toward pump-probe studies. Another recent approach to SFX sample delivery is through the use of a nanoflow electrospinning microjet as described in Sierra et al. (2012). This method conserves sample by use of an applied electrical current to induce an electrospinning jet, as opposed to gas focusing, and the use of a viscous media such as glycerol to control droplet formation which allows a flow rate for PSII crystals around 3 μL/min. While this balances the need for sample conservation with a jet that is able to be optically pumped, the nature of the jet only allows for delay times on the order of seconds and concern arises with regards to structural artifacts generated by the applied voltage on the protein. A primary drawback is sample consumption with typical flow rates of 10–15 μL/min but this allows access to the microsecond time range for pump-probe experiments and avoids any possible upstream scattering excitation.

5.3 Evaluating PSII SFX data

For Kupitz, Basu, et al. (2014), the program Cheetah (Barty et al., 2014), developed specifically for SFX data, was used for background correction and hit-finding. The patterns were then indexed and merged using the *CrystFEL* software suite (White et al., 2012) and refined with the *Phenix* software suite (Adams et al., 2002). In SFX each diffraction snapshot represents a thin slice through reciprocal space, resulting in each measured reflection representing only a partial measure of scattering factor. Furthermore, the intensity between individual X-ray shots varies by more than 200%. In light of this, the determination of accurate structure factors requires a high multiplicity of Bragg intensities with a recommended minimum of 50, a sharp contrast to traditional synchrotron crystallography. A multiplicity of >600 for the dark dataset and >300 for the double flash datasets (putative S_3 state) have been achieved in the Kern et al. experiment (2014). Kern et al. (2014) applied a program described for data evaluation. A key difference between this program and Cheetah is that it uses resolution to select images, leading to low multiplicity of the data in the higher resolution shells. By contrast, the hits are selected in Cheetah according to a threshold of spots at a selected signal-to-noise ratio.

5.4 Structural changes of PSII in the Kok cycle

In Kupitz et al., the structure of PSII was solved at 5.0 and 5.5 Å for dark and double-excited datasets, respectively. Large changes were detected in the unit cell constants between the dark S_1 state and the double flash putative S_3 state, which are reversed in triple flash experiments.

Despite the large changes of the unit cell constants, the overall dimensions of PSII do not increase as shown in the overlay of the transmembrane helices in Fig. 4D. However, larger differences are detected in the acceptor side loop regions and the nonheme iron coordinated thereby (Fig. 5). After PQH_2 is formed and leaves the Q_B-binding site, an empty binding site ensues which could trigger the changes of the loops structures and the expansion of the unit cell constants. The fact that all crystals undergo this change in unit cell constant is an independent indication for the

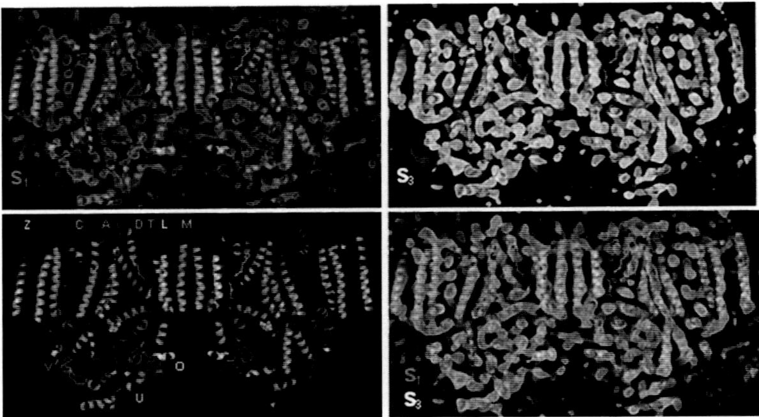

Figure 4 Electron density omit maps at 1.5σ of the PSII homodimer. (A) The dark (S_1 ground) state, (B) the doubly excited (putative S_3 state), (C) ribbon and loop model of PSII with labeled subunits, and (D) overlay of S_1 and putative S_3 states, revealing conformational changes evolved in progression through the Kok cycle. *Figure originally published in Kupitz, Basu, et al. (2014).*

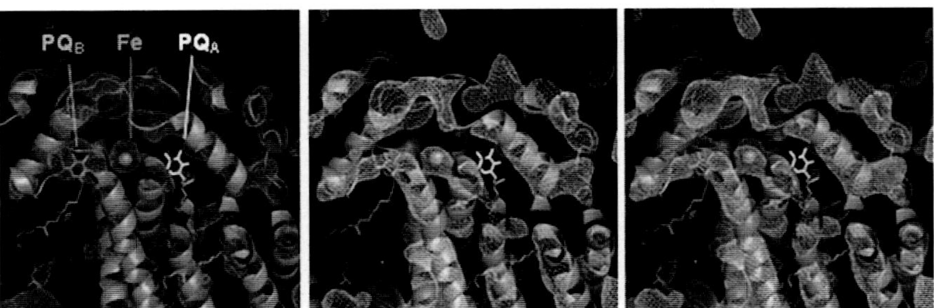

Figure 5 Overlay of omit maps for the dark S_1 (green; dark gray in the print version) and double flash (putative S_3 (white) state). (A) View of the transmembrane region of photosystem II along the membrane plane (B) more detailed view of the acceptor site which contains the binding sites for PQA and PQB. Note the nonheme iron (red (gray in the print version) sphere) and the loop regions exhibiting significant conformational changes between the two states. *Figure originally published in Kupitz, Basu, et al. (2014).*

homogenous progression of PSII in our crystals through the Kok cycle. This change is reversible as it is reversed in a three-flash experiment where PQ_{decyl} is incorporated to allow for the binding site to be filled before the third flash arrives.

In order to detect conformational changes of the OEC and avoid phase bias simulated annealed omit maps were calculated of the OEC and its protein environment (see Fig. 6A and B). These omit maps tentatively show changes in the OEC and surrounding environment. From the omit map shown in Fig. 6B, an elongation of the Mn_4CaO_x portion of the OEC can be seen in the S_3 state with respect to the S_1 dark state. This may allow for the second substrate water molecule to bind between the "dangler" Mn_4 and the Mn_3CaO_4 cubane-like structure during the S-state transition. This is in agreement with hybrid density functional theoretical modeling (Isobe et al., 2012), which has shown that the substrate water has a probable minimum on its potential energy surface when coordinated by the Mn_4 and modulated by the O_5 (Fig. 6C). Furthermore, a decrease in $Mn–Ca^{2+}$

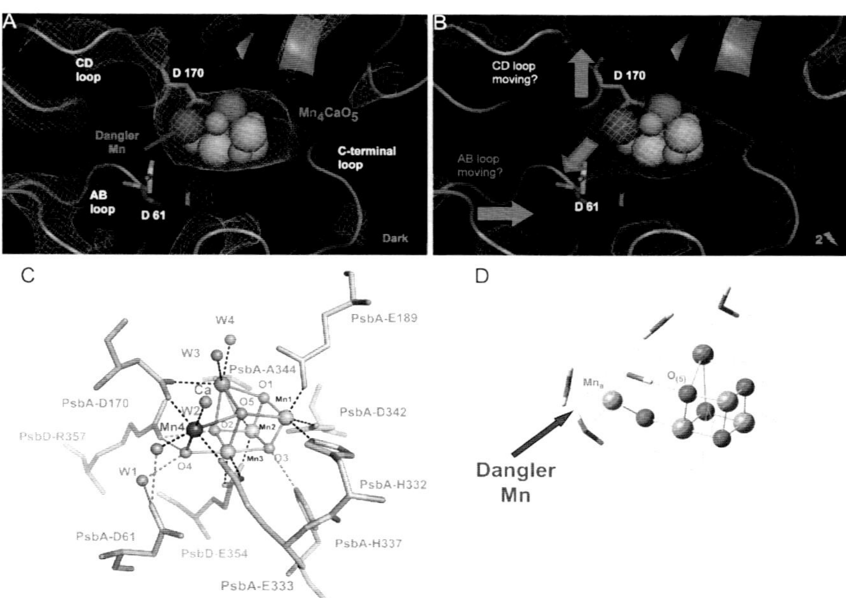

Figure 6 Simulated annealing omit maps of the OEC. (A) and (B) Exhibit the S_1 and S_3 states, respectively, at 1.5σ. In (B), the blue represents the S_1 state, whereas the yellow represents the S_3 state for comparison. (C) Shows the crystal structure from the 1.9 Å structure of the OEC from Umena et al. (2011). (D) Shows the proposed S_3 state derived from DFT calculations performed in Isobe et al. (2012). *Figure originally published in Kupitz, Basu, et al. (2014).* (See the color plate.)

distances seen through extended X-ray absorption fine structure (EXAFS) spectroscopy, pointing to a change in the character of the Mn_4–O_5 bond that would occur during elongation.

The observed electron density changes shown in Fig. 6 agree with the recent theoretical studies of Isobe et al. (2012), who predicted a "breakage" of the dangler Mn from the cubane cluster in the S_3 state. In addition to the elongation of the electron density in the direction of the dangler Mn, the overall dimensions of the Mn_4CaO_5 cluster appears to shrink in the S_3 state, and the distance between the Ca and the 3 Mn in the cluster decreases, as part of the Ca sticks out of the electron density map in the S_3 state. EXAFS studies on PSII where the Ca was substituted with Sr showed very similar spectra in S_1 and S_2, indicating that no significant changes occur in the Mn–Mn or Mn–Ca distances in this S_2-state transition, while significant changes in EXAFS spectra were observed in the S_3 state (Pushkar, Yano, Sauer, Boussac, & Yachandra, 2008), which included the prediction that the distances between Mn and Ca would shrink in the S_3 state. Experimental findings in Kupitz et al. (2014) support a shrinking of the Mn_4CaO_5 cluster in S_3 which would support the hypothesis of a condensation of the Mn_4CaO_5 cluster in S_3 based on the Jahn–Teller (JT) effect which has also been studied in several model Mn compounds (for more details on studies on Mn model compounds, see Yamaguchi et al., 2013 and references therein). Mn–O distances derived from recently published model Mn–O and Mn_3Ca–O cubane structures (Kanady, Tsui, Day, & Agapie, 2011; Mukherjee et al., 2012) indicate that Mn–O distances depend on the oxidation states of the Mn ions. The average Mn(II)–O distance is 2.2 Å, the average Mn(III)–O distance is 2.0 Å and shrinks to 1.8 Å for the Mn(IV)–O distance. Based on X-ray absorption and emission spectroscopy, two models exist for the oxidation states of the Mn_4CaO_5 in S_3, which is either described as $Mn(III)(IV)_3$ or $Mn(IV)_4$ (Dau, Zaharieva, & Haumann, 2012; Yano & Yachandra, 2007). In the model of S_3 where all Mn ions have reached the Mn(IV) oxidation state, a significant shrinking of the dimension of the cluster is expected due to the JT distortion with the average Mn–O distance being reduced to 1.8 Å (Yamaguchi et al., 2013). The shrinking of the overall dimensions of the metal cluster, which is supported by our maps of the putative S_3 state, appears to be the first experimental indication of the role that the JT distortion plays in the mechanism of water splitting (Kanady et al., 2011; Mukherjee et al., 2012).

Changes are also visible in the protein environment of the metal cluster. These are much more difficult to interpret and validate at low resolution than the changes of the metal cluster and have to be confirmed at higher resolution. While the dark state SA-omit map (Figs. 4A and 5A) matches the structural model of dark state (Umena et al., 2011), there are significant changes visible on the SA-omit map of the double flash putative S_3 state. The SA-omit map of the putative S_3 state is suggestive of conformational changes which may indicate movement of the CD loop (including the ligand D170) away from the cluster. If this could be confirmed at higher resolution, it would indicate that ASP170, which provides ligands to both Ca and the danger Mn in the dark state, may not be a ligand in the higher S-states. The loop

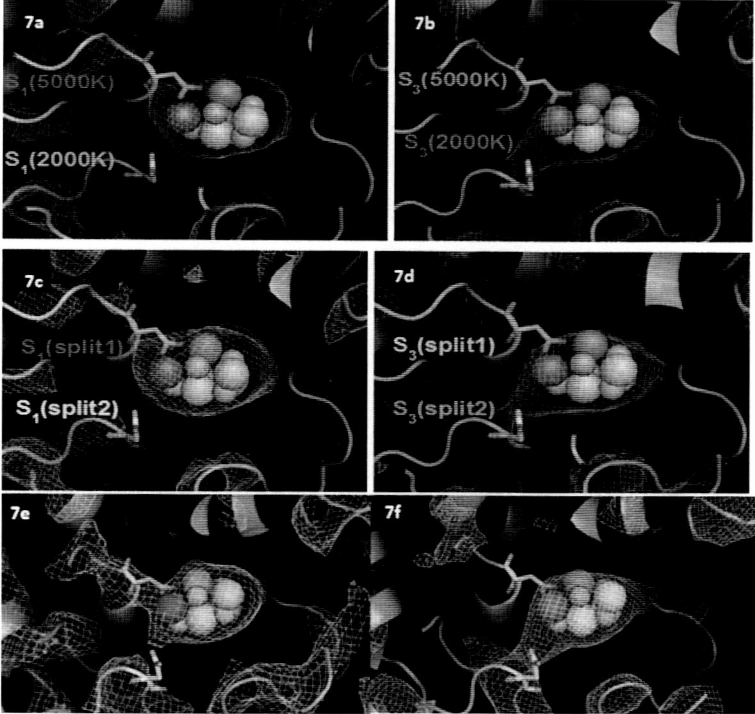

Figure 7 (A and B) Comparison between SA-omit maps of the two halves of the randomly split dark dataset (S_1) at the contour level of 1.5σ (A) and comparison between SA-omit maps of the two halves of the double-excited dataset (S_3) at a contour level of 1.5σ (B). (C and D) Comparison between two SA-omit maps made at two different start temperatures (2000 and 5000 K) for S_1 state at 1.5σ level (C) and the comparison between the two SA-omit maps made at two different start temperatures (2000 and 5000 K) for S_3 state at 1.5σ contour level (D). (E and F) SA-omit maps calculated at 500 K with the same *hkl* values at 1.5 contour level for the dark dataset (E) and the double flash dataset (F).

between the transmembrane helices A and B (AB loop) may change its confirmation so that it moves closer to the metal cluster. A density feature connects this loop at the position of Asp61 to the dangler Mn of the metal cluster in the putative S_3 state. Mutagenesis studies and recent spectroscopic evidence also support the current interpretation. Mutagenesis experiments have questioned Asp170 as a ligand in the higher S-states, as mutants still show 80% oxygen evolving activity and the FTIR spectra are not significantly altered in mutants of Asp170 (Debus, Strickler, Walker, & Hillier, 2005). While Asp61 only serves as a second sphere ligand in the 1.9 Å crystal structure (Umena et al., 2011) mutagenesis studies indicated an important role in the water oxidation process as the S_2 to S_3 transition is blocked in Asp61 mutants (Debus, 2014; Dilbeck, Bao, Neveu, & Burnap, 2013; Pokhrel & Brudvig, 2014; Service, Hillier, & Debus, 2010).

At low resolution an interpretation of changes in the protein environment is challenging and the question may arise: how robust are these changes in the SA-omit map?

Three validation tests have been performed, comprised of the annealing temperature of the SA-omit map (Fig. 7A), splitting of the data for both the dark and light datasets in half and calculation of the SA-omit maps with the split datasets (Fig. 7B), and calculation of the SA-omit maps for the light and dark states using exactly the same structure factors (Fig. 7C). The results in Fig. 7 and the comparison with the SA-omit map in Fig. 6 show that the changes of the metal cluster and its protein environment between the dark and putative S_3 state are visible in all cases.

6. SUMMARY

The structures of PSII in the S_1 and S_3 states were determined at 5.0 and 5.5 Å, respectively, by TR-SFX. The experiments required new developments in crystal growth and crystal characterization so that a high density of microcrystals of PSII of very homogenous size could be achieved. Uniform growth of nano- and microcrystals requires knowledge of the phase diagram and analytical techniques such as SONICC and DLS are crucial to detection and characterization. Using the pump-probe TR-SFX setup, radiation free snapshots of the S-state cycle can be recorded at biological temperatures, allowing for deeper understanding of the enzymatic mechanism behind PSII. Future developments may eventually lead to an evolution from snapshots of transitory states to molecular movies, enabling thorough understanding of enzymatic mechanisms such as PSII. Elucidating the mechanism of photosynthetic water oxidation and oxygen evolution is critical to the understanding the underlying photosynthetic mechanism and holds high potential for applications such as renewable energy and membrane protein bioengineering.

ACKNOWLEDGMENTS

This work was supported by the following agencies: the Center for Bio-Inspired Solar Fuel Production; an Energy Frontier Research Center funded by the DOE; Office of Basic Energy Sciences (award DE-SC0001016); the National Institutes of Health (award 1R01GM095583); the US National Science Foundation (award MCB-1021557 and MCB-1120997); the DFG Clusters of Excellence "Inflammation at Interfaces" (EXC 306) and the "Center for Ultrafast Imaging"; the Deutsche Forschungsgemeinschaft (DFG); the Max Planck Society, the Atomic, Molecular and Optical Sciences Program; Chemical Sciences Geosciences and Biosciences Division, DOE OBES (M.J.B.) and the SLAC LDRD program (M.J.B., H.L.); the US DOE through Lawrence Livermore National Laboratory under the contract DE-AC52-07NA27344 and supported by the UCOP Lab Fee Program (award 118036) and the LLNL LDRD program (12-ERD-031); the Hamburg Ministry of Science and Research and Joachim Herz Stiftung as part of the Hamburg Initiative for Excellence in Research. The National Science Foundation through the BioXFEL Science Technology Center (award 1231306).

REFERENCES

Adams, P. D., Grosse-Kunstleve, R. W., Hung, L. W., Ioerger, T. R., McCoy, A. J., Moriarty, N. W., et al. (2002). PHENIX: Building new software for automated crystallographic structure determination. *Acta Crystallographica. Section D, Biological Crystallography*, *58*(11), 1948–1954.

Allakhverdiev, S. I., Nishiyama, Y., Takahashi, S., Miyairi, S., Suzuki, I., & Murata, N. (2005). Systematic analysis of the relation of electron transport and ATP synthesis to the photodamage and repair of photosystem II in Synechocystis. *Plant Physiology*, *137*(1), 263–273.

Aquila, A., Hunter, M. S., Doak, R. B., Kirian, R. A., Fromme, P., White, T. A., et al. (2012). Time-resolved protein nanocrystallography using an X-ray free-electron laser. *Optics Express*, *20*(3), 2706–2716.

Aro, E. M., Virgin, I., & Andersson, B. (1993). Photoinhibition of photosystem II. Inactivation, protein damage and turnover. *Biochimica et Biophysica Acta. Bioenergetics*, *1143*(2), 113–134.

Barty, A., Caleman, C., Aquila, A., Timneanu, N., Lomb, L., White, T. A., et al. (2012). Self-terminating diffraction gates femtosecond X-ray nanocrystallography measurements. *Nature Photonics*, *6*(1), 35–40.

Barty, A., Kirian, R. A., Maia, F. R., Hantke, M., Yoon, C. H., White, T. A., et al. (2014). Cheetah: Software for high-throughput reduction and analysis of serial femtosecond X-ray diffraction data. *Journal of Applied Crystallography*, *47*(3), 1118–1131.

Chapman, H. N., Fromme, P., Barty, A., White, T. A., Kirian, R. A., Aquila, A., et al. (2011). Femtosecond X-ray protein nanocrystallography. *Nature*, *470*(7332), 73–77.

Dau, H., Zaharieva, I., & Haumann, M. (2012). Recent developments in research on water oxidation by photosystem II. *Current Opinion in Chemical Biology*, *16*(1), 3–10.

de Wijn, R., & van Gorkom, H. J. (2001). Kinetics of electron transfer from QA to QB in photosystem II. *Biochemistry*, *40*(39), 11912–11922.

Debus, R. J. (2014). Evidence from FTIR difference spectroscopy that D1-Asp61 influences the water reactions of the oxygen-evolving Mn_4CaO_5 cluster of photosystem II. *Biochemistry, 53*, 2941–2955.

Debus, R. J., Strickler, M. A., Walker, L. M., & Hillier, W. (2005). No evidence from FTIR difference spectroscopy that aspartate-170 of the D1 polypeptide ligates a manganese ion that undergoes oxidation during the S0 to S1, S1 to S2, or S2 to S3 transitions in photosystem II. *Biochemistry, 44*(5), 1367–1374.

Dekker, J. P., & Van Grondelle, R. (2000). Primary charge separation in photosystem II. *Photosynthesis Research, 63*(3), 195–208.

Dilbeck, P. L., Bao, H., Neveu, C. L., & Burnap, R. L. (2013). Perturbing the water cavity surrounding the manganese cluster by mutating the residue D1-valine 185 has a strong effect on the water oxidation mechanism of photosystem II. *Biochemistry, 52*(39), 6824–6833.

Ferreira, K. N., Iverson, T. M., Maghlaoui, K., Barber, J., & Iwata, S. (2004). Architecture of the photosynthetic oxygen-evolving center. *Science, 303*(5665), 1831–1838.

Haupert, L. M., & Simpson, G. J. (2011). Screening of protein crystallization trials by second order nonlinear optical imaging of chiral crystals (SONICC). *Methods, 55*(4), 379–386.

Isobe, H., Shoji, M., Yamanaka, S., Umena, Y., Kawakami, K., Kamiya, N., et al. (2012). Theoretical illumination of water-inserted structures of the $CaMn_4O_5$ cluster in the S2 and S3 states of oxygen-evolving complex of photosystem II: Full geometry optimizations by B3LYP hybrid density functional. *Dalton Transactions, 41*(44), 13727–13740.

Kamiya, N., & Shen, J. R. (2003). Crystal structure of oxygen-evolving photosystem II from Thermosynechococcus vulcanus at 3.7-Å resolution. *Proceedings of the National Academy of Sciences, 100*(1), 98–103.

Kanady, J. S., Tsui, E. Y., Day, M. W., & Agapie, T. (2011). A synthetic model of the Mn_3Ca subsite of the oxygen-evolving complex in photosystem II. *Science, 333*(6043), 733–736.

Kern, J., Alonso-Mori, R., Tran, R., Hattne, J., Gildea, R. J., Echols, N., et al. (2013). Simultaneous femtosecond X-ray spectroscopy and diffraction of photosystem II at room temperature. *Science, 340*(6131), 491–495.

Kern, J., Tran, R., Alonso-Mori, R., Koroidov, S., Echols, N., Hattne, J., et al. (2014). Taking snapshots of photosynthetic water oxidation using femtosecond X-ray diffraction and spectroscopy. *Nature Communications, 5*, 4371.

Kupitz, C., Basu, S., Grotjohann, I., Fromme, R., Zatsepin, N. A., Rendek, K. N., et al. (2014). Serial time-resolved crystallography of photosystem II using a femtosecond X-ray laser. *Nature, 513*, 261–265.

Kupitz, C., Grotjohann, I., Conrad, C. E., Roy-Chowdhury, S., Fromme, R., & Fromme, P. (2014). Microcrystallization techniques for serial femtosecond crystallography using photosystem II from Thermosynechococcus elongatus as a model system. *Philosophical Transactions of the Royal Society, B: Biological Sciences, 369*(1647), 20130316.

Loll, B., Kern, J., Saenger, W., Zouni, A., & Biesiadka, J. (2005). Towards complete cofactor arrangement in the 3.0 Å resolution structure of photosystem II. *Nature, 438*(7070), 1040–1044.

McPherson, A. (1991). Review. Current approaches to macromolecular crystallization. In *EJB reviews 1990* (pp. 49–71). Berlin, Heidelberg: Springer.

Mukherjee, S., Stull, J. A., Yano, J., Stamatatos, T. C., Pringouri, K., Stich, T. A., et al. (2012). Synthetic model of the asymmetric [Mn3CaO4] cubane core of the oxygen-evolving complex of photosystem II. *Proceedings of the National Academy of Sciences, 109*(7), 2257–2262.

Neutze, R., & Moffat, K. (2012). Time-resolved structural studies at synchrotrons and X-ray free electron lasers: Opportunities and challenges. *Current Opinion in Structural Biology, 22*(5), 651–659.

Neutze, R., Wouts, R., van der Spoel, D., Weckert, E., & Hajdu, J. (2000). Potential for biomolecular imaging with femtosecond X-ray pulses. *Nature*, *406*(6797), 752–757.

Ng, J. D., Gavira, J. A., & García-Ruíz, J. M. (2003). Protein crystallization by capillary counterdiffusion for applied crystallographic structure determination. *Journal of Structural Biology*, *142*(1), 218–231.

Patzlaff, J. S., & Barry, B. A. (1996). Pigment quantitation and analysis by HPLC reverse phase chromatography: A characterization of antenna size in oxygen-evolving photosystem II preparations from cyanobacteria and plants. *Biochemistry*, *35*(24), 7802–7811.

Pokhrel, R., & Brudvig, G. W. (2014). Oxygen-evolving complex of photosystem II: Correlating structure with spectroscopy. *Physical Chemistry Chemical Physics*, *16*(24), 11812–11821.

Pushkar, Y., Yano, J., Sauer, K., Boussac, A., & Yachandra, V. K. (2008). Structural changes in the Mn_4Ca cluster and the mechanism of photosynthetic water splitting. *Proceedings of the National Academy of Sciences*, *105*(6), 1879–1884.

Rajagopal, S., Anderson, S., Srajer, V., Schmidt, M., Pahl, R., & Moffat, K. (2005). A structural pathway for signaling in the E46Q mutant of photoactive yellow protein. *Structure*, *13*(1), 55–63.

Renger, G. (2012). Mechanism of light induced water splitting in photosystem II of oxygen evolving photosynthetic organisms. *Biochimica et Biophysica Acta. Bioenergetics*, *1817*(8), 1164–1176.

Schlichting, I., Almo, S. C., Rapp, G., Wilson, K., Petratos, K., Lentfer, A., et al. (1990). Time-resolved X-ray crystallographic study of the conformational change in Ha-Ras p21 protein on GTP hydrolysis. *Nature*, *345*, 309–315.

Service, R. J., Hillier, W., & Debus, R. J. (2010). Evidence from FTIR difference spectroscopy of an extensive network of hydrogen bonds near the oxygen-evolving Mn4Ca cluster of photosystem II Involving D1-Glu65, D2-Glu312, and D1-Glu329. *Biochemistry*, *49*(31), 6655–6669.

Sierra, R. G., Laksmono, H., Kern, J., Tran, R., Hattne, J., Alonso-Mori, R., et al. (2012). Nanoflow electrospinning serial femtosecond crystallography. *Acta Crystallographica. Section D, Biological Crystallography*, *68*(11), 1584–1587.

Spence, J. C. H., Weierstall, U., & Chapman, H. N. (2012). X-ray lasers for structural and dynamic biology. *Reports on Progress in Physics*, *75*(10), 102601.

Šrajer, V., Ren, Z., Teng, T. Y., Schmidt, M., Ursby, T., Bourgeois, D., et al. (2001). Protein conformational relaxation and ligand migration in myoglobin: A nanosecond to millisecond molecular movie from time-resolved Laue X-ray diffraction. *Biochemistry*, *40*(46), 13802–13815.

Suga, M., Akita, F., Hirata, K., Ueno, G., Murakami, H., Nakajima, Y., et al. (2014). Native structure of photosystem II at 1.95 A resolution viewed by femtosecond X-ray pulses. *Nature*, *517*, 99–103.

Tenboer, J., Basu, S., Zatsepin, N., Pande, K., Milathianaki, D., Frank, M., et al. (2014). Time-resolved serial crystallography captures high-resolution intermediates of photoactive yellow protein. *Science*, *346*(6214), 1242–1246.

Umena, Y., Kawakami, K., Shen, J. R., & Kamiya, N. (2011). Crystal structure of oxygen-evolving photosystem II at a resolution of 1.9 Å. *Nature*, *473*(7345), 55–60.

Wampler, R. D., Begue, N. J., & Simpson, G. J. (2008). Molecular design strategies for optimizing the nonlinear optical properties of chiral crystals. *Crystal Growth and Design*, *8*(8), 2589–2594.

Weierstall, U., James, D., Wang, C., White, T. A., Wang, D., Liu, W., et al. (2014). Lipidic cubic phase injector facilitates membrane protein serial femtosecond crystallography. *Nature Communications*, *5*, 3309.

White, T. A., Kirian, R. A., Martin, A. V., Aquila, A., Nass, K., Barty, A., et al. (2012). CrystFEL: A software suite for snapshot serial crystallography. *Journal of Applied Crystallography*, *45*(2), 335–341.

Wydrzynski, T., & Satoh, K. (Eds.). (2006). *Advances in photosynthesis and respiration: Vol. 22. Photosystem II: The light-driven water: Plastoquinone oxidoreductase:* Springer.

Yamaguchi, K., Isobe, H., Yamanaka, S., Saito, T., Kanda, K., Shoji, M., et al. (2013). Full geometry optimizations of the mixed-valence $CaMn_4O_4X$ $(H_2O)_4$ (X = OH or O) cluster in OEC of PS II: Degree of symmetry breaking of the labile Mn-X-Mn bond revealed by several hybrid DFT calculations. *International Journal of Quantum Chemistry, 113*(4), 525–541.

Yano, J., & Yachandra, V. K. (2007). Oxidation state changes of the Mn_4Ca cluster in photosystem II. *Photosynthesis Research, 92*(3), 289–303.

Zouni, A., Witt, H. T., Kern, J., Fromme, P., Krauss, N., Saenger, W., et al. (2001). Crystal structure of photosystem II from Synechococcus elongatus at 3.8 Å resolution. *Nature, 409*(6821), 739–743.

Computational Approaches to Understand Membrane Proteins

CHAPTER TWENTY-THREE

Major Intrinsic Protein Superfamily: Channels with Unique Structural Features and Diverse Selectivity Filters

Ravi Kumar Verma*,2, Anjali Bansal Gupta*,2,3, Ramasubbu Sankararamakrishnan*,†,1

*Department of Biological Sciences and Bioengineering, Indian Institute of Technology Kanpur, Kanpur, India
†Center of Excellence for Chemical Biology, Indian Institute of Technology Kanpur, Kanpur, India
[1]Corresponding author: e-mail address: rsankar@iitk.ac.in

Contents

Abstract

Members of the superfamily of major intrinsic proteins (MIPs) facilitate water and solute permeability across cell membranes and are found in sources ranging from bacteria to humans. Aquaporin and aquaglyceroporin channels are the prominent members of the MIP superfamily. Experimental studies show that MIPs are involved in important physiological processes in mammals and plants. They are implicated in several human

[2] Both authors contributed equally.
[3] Present Address: Mechanobiology Institute, National University of Singapore, Singapore.

Methods in Enzymology, Volume 557
ISSN 0076-6879
http://dx.doi.org/10.1016/bs.mie.2014.12.006

diseases and are considered to be attractive drug targets for a wide range of diseases such as cancer, brain edema, epilepsy, glaucoma, and congestive heart failure. Three-dimensional structures of MIP channels from diverse sources reveal that MIPs adopt a unique conserved hourglass helical fold consisting of six transmembrane helices (TM1–TM6) and two half-helices (LB and LE). Conserved NPA motifs near the center and the aromatic/arginine selectivity filter (Ar/R SF) toward the extracellular side constitute two narrow constriction regions within the channel. Structural knowledge combined with simulation studies have helped to investigate the role of these two constriction regions in the transport and selectivity of the solutes. With the availability of many genome sequences from diverse species, a large number of MIP genes have been identified. Homology models of 1500 MIP channels have been used to derive structure-based sequence alignment of TM1–TM6 helices and the two half-helices LB and LE. Thirteen residues are highly conserved in different transmembrane helices and half-helices. High group conservation of small and weakly polar residues is observed in 27 positions at the interface of two interacting helices. Thus, although the MIP sequences are diverse, the hourglass helical fold is maintained during evolution with the conservation of these 40 positions within the transmembrane region. We have proposed a generic structure-based numbering scheme for the MIP channels that will facilitate easier comparison of the MIP sequences. Analysis of Ar/R SF in all 1500 MIPs indicates the extent of diversity in the four residues that form this narrow region. Certain residues are completely avoided in the SF, even if they have the same chemical nature as that of the most frequently observed residues. For example, arginine is the most preferred residue in a specific position of Ar/R SF, whereas lysine is almost always avoided in any of the four positions. MIP channels with highly hydrophobic or hydrophilic Ar/R SF have been identified. Similarly, there are examples of MIP channels in which all four residues of Ar/R SF are bulky, thus almost occluding the pore. Many plant MIPs possess small residues at all SF positions, resulting in a larger pore diameter. A majority of MIP channels are yet to be functionally characterized, and their *in vivo* substrates are not yet identified. A complete understanding of the relationship between the nature of Ar/R SF and the solutes that are transported is required to exploit MIP channels as potential drug targets.

1. INTRODUCTION

Channels constitute an important group of integral membrane proteins that facilitate the transport of molecules and ions across the membranes. They are highly selective and are involved in efficient transport of selected molecules or ions. Examples of ion channels belonging to voltage-gated (Catterall, 2000; Yellen, 2002), ligand-gated (Hilf & Dutzler, 2008; Reeves & Lummis, 2002), mechano-gated (Patel, Lazdunski, & Honore, 2001), or temperature-dependent gating channels (Voets et al., 2004) have been recognized and well characterized. Three-dimensional structures of several of these ion channels have been determined, and the structural knowledge together with computer simulations helped to understand or speculate how the channel function is regulated (Amaral, Carnevale, Klein, & Treptow, 2012; Berneche & Roux, 2001; Jentsch, Stein, Weinreich, & Zdebik, 2002; Jiang et al., 2003 Law, Henchman, & McCammon, 2005; Payandeh, Scheuer, Zheng, & Catterall, 2011; Shrivastava & Sansom, 2000). Apart from the ion channels, there is plethora of other channels that transport a wide variety of solutes. This includes gap junctions that form intercellular channels between adjacent cells (Maeda et al., 2009; Sohi & Willecke, 2004), bacterial outer membrane porin channels that transport hydrophilic solutes (Koebnik, Locher, & van Gelder, 2000; Nikaido, 2003), protein-translocating prokaryotic SecY or eukaryotic Sec61 channels (Hessa et al., 2007; Osborne, Rapoport, & van den Berg, 2005), mitochondrial outer membrane VDAC channels that control the exchange of ions and metabolites between cytosol and mitochondria (Colombini, 2004; Lemasters & Holmuhamedov, 2006), and the huge nuclear pore complexes that are involved in the exchange of macromolecules between the nucleus and cytoplasm (Rout et al., 2000; Suntharalingam & Wente, 2003). These disparate molecules are involved in the transport of diverse entities ranging from small ions to biomacromolecules like proteins.

The first channel protein to transport water was discovered in early 1990s (Preston, Carrol, Guggino, & Agre, 1992; Zeidel, Ambudkar, Smith, & Agre, 1992), and since then, a large number of proteins belonging to this family have been identified in all kingdoms of life, from bacteria to humans (Calamita, Bishai, Preston, Guggino, & Agre, 1995; Daniels, Mirkov, & Chrispeels, 1994; Ishibashi, Hara, & Kondo, 2009; Kozono et al., 2003; Li, Santoni, & Maurel, 2014; Pettersson, Filipsson, Becit, Brive, & Hohmann, 2005). Although the water-transporting aquaporins (AQPs) were first recognized, later studies showed that channels of a similar type are involved in transporting other solutes (Borgnia, Nielsen, Engel, & Agre, 1999; Liu et al., 2002; Rojek, Praetorius, Frokiaer, Nielsen, & Fenton, 2008; Wu & Beitz, 2007). They are broadly referred as aquaglyceroporins (AQGPs) because glycerol was initially identified as the solute that is transported. Both classes of channels constitute one of the largest families of channel proteins known today, and they form the superfamily of major intrinsic proteins (MIPs; Gupta et al., 2012; Zardoya, 2005). MIP channels have been identified in microbes including bacteria and archaea (Calamita et al., 1995; Kozono et al., 2003; Tanghe, van Dijck, &

Thevelein, 2006), insects (Spring, Robichaux, & Hamlin, 2009), parasites (Hansen, Kun, Schultz, & Beitz, 2002), algae (Anderberg, Danielson, & Johanson, 2011), fungi (Verma, Prabh, & Sankararamakrishnan, 2014b; Xu, Cooke, & Zwiazek, 2013), plants (Gupta & Sankararamakrishnan, 2009; Johanson et al., 2001), teleosts (Tingaud-Sequeira et al., 2010), and mammals (Ishibashi et al., 2009). In organisms, such as plants and mammals, MIPs are present in large numbers (Gupta & Sankararamakrishnan, 2009; Park, Scheffler, Bauer, & Cambell, 2010). MIP sequences exhibit internal homology with two highly conserved NPA (Asn-Pro-Ala) sequence motifs. Its presence in almost every organism makes it an ideal candidate to investigate the molecular evolution of this channel from bacteria to humans. Phylogenetic analyses of MIPs have been performed across different species groups or within the same organism or species group (Abascal, Irisarri, & Zardoya, 2014; Verma et al., 2014b). In addition to water and glycerol, reports indicate that the repertoire of solutes that are being transported by MIP channels keep growing. This includes neutral solutes (urea, methylamine, lactic acid), gases (CO_2, ammonia, nitric oxide), reactive oxygen species (hydrogen peroxide), and metalloids (arsenite, antimonite, boric acid, silicic acid; Bienert, Schussler, & Jahn, 2008; Choi & Roberts, 2007; Ma et al., 2006; Maurel, Verdoucq, Luu, & Santoni, 2008; Soria, Marrone, Calamita, & Marinelli, 2013; Uehlein, Lovisolo, Siefritz, & Kaldenhoff, 2003; Wu & Beitz, 2007). There is also an exception to the transport properties of MIP channels, and at least one MIP, aquaporin 6, has been reported to be an anion channel (Ikeda et al., 2002). The solutes that are transported across a large number of MIPs are yet to be identified.

With a large number of diverse solutes that are transported across the MIP channels, the physiological significance of MIPs in different species is being studied. In mammals, MIPs have a significant role in urinary concentration in the kidneys (Bockenhauer & Bichet, 2014; Nielsen et al., 2002), regulating the water fluxes into and out of brain (Badaut, Fukuda, Jullienne, & Petry, 2014; Papadopoulos & Verkman, 2007), follicle development in reproductive systems (Huang et al., 2006), skin hydration and elasticity (Hara, Ma, & Verkman, 2002; Lee et al., 2012), adipose metabolism (Kuriyama et al., 2002), maintaining corneal and lens transparency in the eye (Schey, Wang, Wenke, & Qi, 2014; Verkman, Ruiz-Ederra, & Levin, 2008), and lung edema (Jin, Yu, Peng, Zhang, & Xin, 2013). In plants, MIPs seem to have diverse roles including leaf movement (Uehlein & Kaldenhoff, 2008), nutrient transport (Li, Santoni, et al., 2014), seed germination (Liu et al., 2013), hydraulic conductivity

(Chaumont & Tyerman, 2014; Murai-Hatano et al., 2008), plant growth, stress responses (Vandeleur et al., 2009), and regulation of plant–water relations (Maurel et al., 2008). They are also important in symbiotic relationship such as plant–fungi interactions (Aroca, Porcel, & Ruiz-Lozano, 2007; Dietz, von Bulow, Baker, & Nehls, 2011). The functions of MIP channels are regulated in various ways. Phosphorylation, ubiquitination, interaction with metal ions, and other types of posttranslational modifications have been reported as regulation mechanisms in some plant and mammalian MIPs (Chaumont, Moshelion, & Daniels, 2005; Gunnarson, Zelenina, & Aperia, 2004; Moeller, Aroankins, Slengerik-Hansen, Pisitkun, & Fenton, 2014; Yukutake, Hirano, Suematsu, & Yasui, 2009; Yukutake & Yasui, 2010). With vital physiological role in the functioning of many organs in humans, MIP channels have been implicated in a variety of diseases and disorders (King, Kozono, & Agre, 2004; Misu et al., 2007; Verkman, 2009). They are considered as promising drug targets to treat diseases like cancer, glaucoma, obesity, epilepsy, and wound healing (Binder, Nagelhus & Ottersen, 2012; Fruhbeck, Catalan, Gomez-Ambrosi, & Rodriguez, 2006; Hara-Chikuma & Verkman, 2008; Verkman, 2009; Verkman, Hara-Chikuma, & Papadopoulos, 2008). Several excellent review articles have been published recently related to aquaporin function, regulation, its relation to diseases, and its potential as candidate drug targets (Ahmadpour, Geijer, Tamas, Lindkvist-Petersson, & Hohmann, 2014; Fenton, Pedersen, & Moeller, 2013; Huber, Tsujita, & Nakada, 2012; Kun & de Carvalho, 2009; Verkman, Anderson, & Papadopoulos, 2014; Wudick, Luu, & Maurel, 2009).

2. STRUCTURAL BIOLOGY OF MIP CHANNELS

Many channel proteins are generally known to be strongly selective in transporting specific solutes/ions, and they are also efficient in transporting the substrates. Gating is another important feature in channels that depends on several factors. Mutation studies help to some extent in identifying important residues used to determine the selectivity and efficiency in the transport function. However, the mechanism of selectivity and transport can be best understood if the structures of the channel proteins are known at atomic resolution. High-resolution structures of several members of MIP family have been determined from different sources (mammals, plant, archaea, yeast, *Escherichia coli*, and *Plasmodium falciparum* [PfAQP]). To date, 24 MIP structures are available (see http://blanco.biomol.uci.edu/mpstruc/), and 11 are unique structures (AQP0, AQP1, AQP2, AQP4, AQP5, AqpZ, GlpF, SoPIP2;1, Aqy1, AqpM, and PfAQP). The yeast aquaporin, Aqy1, is the only membrane protein to date whose structure has been determined at subangstrom resolution (PDB ID: 3ZOJ; resolution: 0.88 Å; Eriksson et al., 2013). All MIP channels adopt the unique hourglass helical fold with six TM helices (TM1–TM6) and two half-helices (LB and LE; Fig. 1A). MIP channels form tetramers with each monomer as a functional channel. The review

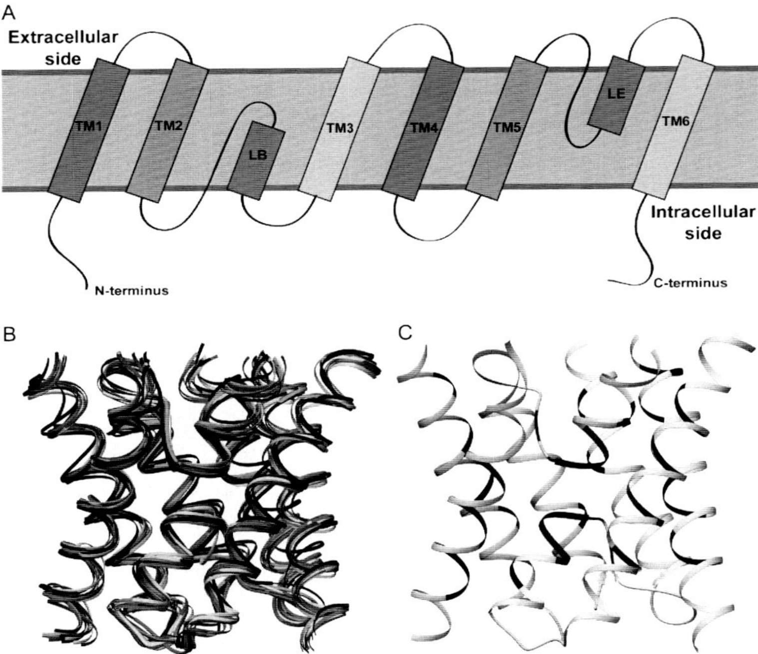

Figure 1 (A) Membrane topology diagram of an MIP channel. Each MIP channel has six transmembrane helical segments (TM1–TM6) and two half-helices formed by the loops LB and LE. The transmembrane region is shown in dark gray color. (B) Superposition of 11 unique MIP structures determined to date. These are AQP0, AQP1, AQP2, AQP4, AQP5, AqpZ, GlpF, SoPIP2;1, Aqy1, AqpM, and PfAQP. Their respective PDB IDs are 2B6O, 1J4N, 4NEF, 3GD8, 3D9S, 1RC2, 1FX8, 1Z98, 2W1P, 2F2B, and 3C02. Only the transmembrane region (TM1–TM6 helices and LB and LE loop regions) was considered for the superposition. All MIP structures from diverse sources adopt the unique hourglass helical fold. (C) Positions of 13 highly conserved residues (orange) and those 27 positions in which small and weakly polar residues exhibit a high level of group conservation (blue) are shown in the ribbon diagram of the hourglass fold. The plant MIP structure SoPIP2;1 (PDB ID: 1Z98) was used as a reference structure to illustrate the highly conserved positions within the transmembrane region. (See the color plate.)

article by Gonen and Waltz on AQP structures has discussed 7 of 11 unique MIP structures in detail (Gonen & Walz, 2006). Since then, the additional four unique structures of MIP channels available recently have further improved our understanding of MIP function and its interactions with lipids and other proteins. Hence, the next sections are restricted to discussing these MIP channels only.

2.1 AQP5 and AQP2

The high-resolution structures of human AQP5 (Horsefield et al., 2008) and AQP2 (Frick et al., 2014) have been determined recently. AQP5 and AQP2 are the closest paralogs among the 13 human MIPs, and they share 63% sequence identity. Both AQPs are targeted from intracellular vesicles to the plasma membrane, and phosphorylation plays an important role in trafficking (Hasegawa et al., 2011; Lu et al., 2008). Structures of AQP2 and AQP5 exhibit the canonical hourglass helical fold, which is also adopted by other mammalian aquaporins AQP0 (Hite, Li, & Walz, 2010), AQP1 (Sui, Han, Lee, Walian, & Jap, 2001), and AQP4 (Ho et al., 2009; and also by MIP channels from *E. coli* [Fu et al., 2000; Savage, Egea, Robles-Colmenares, O'Connell & Stroud, 2003], plant [Tornroth-Horsefield et al., 2006], yeast [Eriksson et al., 2013], archaea [Lee et al., 2005], and *Plasmodium* [Newby et al., 2008]; see later discussion). Both AQPs assemble as tetramers, and the calculation of channel radius profiles revealed that the narrowest point within the channel is at the aromatic/arginine (Ar/R) selectivity filter (SF). It is interesting to note that in both AQPs, the small C-terminal helix seems to show conformational variations between the protomers. In AQP5, phosphorylation and conformational changes of C-terminal helix are suggested to be potentially important events for trafficking (PDB ID: 3D9S). In AQP2, conformation of C-terminal helix is different in each protomer (PDB ID: 4NEF). The flexibility in C-terminus is attributed to two consecutive proline residues present in this region. It is also speculated that the side of the C-terminal helix rich in leucine may be involved in interactions with the protein LIP5 (lysosomal trafficking regulator interacting protein 5), a protein known to interact with leucine-rich regions (van Balkom et al., 2009). The N-terminus of AQP2 also displays variation in its conformation, and two distinct conformations have been observed in the crystal structure. Hydrogen-bond interactions between a Glu residue from N-terminus and Ser and Arg from loop D give rise to a structural triad in one of the conformations. Such conserved interactions are also observed in AQP5 (Horsefield et al., 2008) and SoPIP2;1 (Tornroth-Horsefield et al., 2006). In another conformation, an additional helical turn extending to cytoplasm is observed, also seen in the structure of AQP1 (Sui et al., 2001). It is suggested that these distinct N-termini conformations have an important role in trafficking and gating the channels. Two Cd^{2+} ion binding sites have also been identified in AQP2 structure, and radioactive binding studies indicate that these could be Ca^{2+} binding sites *in vivo*. Binding of Ca^{2+} could influence the conformation of loop D, which in turn could be significant in gating the channel as observed in SoPIP2;1 (Tornroth-Horsefield et al., 2006).

Mutations in the AQP2 gene have been linked to nephrogenic diabetes insipidus (NDI), a disorder related to maintaining water balance (Mulders et al., 1997; Sasaki, Chiga, Kikuchi, Rai, & Uchida, 2013). NDI patients suffer severe dehydration due to their inability to concentrate urine (Morello & Bichet, 2001). With the available AQP2 structure, most of the NDI-causing mutations have been located at helix–helix interfaces within the TM region, indicating that such mutations are likely to disturb the helix packing and folding of the protein. A few mutations are also observed in loop C, in loop D, and near the Ca^{2+}-binding sites. Substitutions in these sites could be related to structural changes resulting in ER retention of AQP2 in NDI patients.

2.2 *Plasmodium* AQGPs

The structure of the only MIP channel from the malaria parasite PfAQP has been determined by Stroud and coworkers (Newby et al., 2008). PfAQP conducts both water and glycerol efficiently and has an important role in malarial biology (Hansen et al., 2002). The availability of this structure may be important to consider PfAQP as an attractive target for developing antimalarial drugs (Kun & de Carvalho, 2009). PfAQP also adopts remarkably similar right-handed helical bundle structures as other MIP channels (PDB ID: 3C02). The 2.05 Å resolution structure reveals glycerol and water molecules in the extracellular vestibule, within the conduction pore, and near the Ar/R SF region. The long loop C connecting TM3 and TM4 defines the characteristics of the extracellular vestibule. Structural knowledge divulges specific interactions between the glycerol molecules and channel residues in different parts of the channel. This indicates the possible role of these residues in the selectivity of the solutes to be transported. This is also the only structure available in which substitutions are observed in both the conserved NPA motifs (NLA and NPS at LB and LE half-helices, respectively). The substitutions in the NPA motifs are compensated by interactions from other covariant changes that occur throughout the channel. As a result, interactions between the N-terminal regions of the two half-helices and the orientation of asparagines side chains toward the channel interior are preserved. The Ar/R SF is identical to that of glycerol-transporting GlpF, and this could be the reason for the high conductance of glycerol by PfAQP. Then the intriguing question is, what makes PfAQP an equally efficient water channel (Hansen et al., 2002)? Unlike GlpF, the guanidinium group of highly conserved Arg in the Ar/R SF is involved in three hydrogen-bond interactions within the channel. This reduces the cost of desolvation of the Arg side chain in the narrow SF region. This is very similar to what is observed in the structures of water-transporting MIP channels. This might explain PfAQP's ability to transport both water and glycerol efficiently.

2.3 Yeast aquaporin Aqy1

The yeast aquaporin Aqy1 plays an essential physiological role in the survival of the host during rapid freezing and thawing (Tanghe et al., 2002). Structure of Aqy1 has been determined by Fischer et al. in two different pH conditions (Fischer et al., 2009). An ultra-high-resolution structure of Aqy1 is also available from Neutze, Tajkhorshid, and coworkers (Eriksson et al., 2013). The high (pH 8.0) and low (pH 3.5) pH structures of Aqy1 are identical (PDB IDs: 2W1P and 2W2E) and represent the closed conformation of the channel. The longer N-terminus in Aqy1, a common feature observed in yeast MIP channels, seems to have a specific functional role. The N-terminal regions of each protomer participate in extensive interactions with the same region from the neighboring protomers. A tyrosine residue before the beginning of the first TM helix inserts inside the channel at the cytoplasmic side of the channel. This is the narrowest point in the channel (0.8 Å diameter), and this residue is involved in a hydrogen-bond network with water molecules and backbone oxygen atoms of glycine residues. Experiments measuring the water transport activity with wild-type Aqy1 and N-terminus truncated Aqy1 suggest a gating role for the N-terminal region of Aqy1 (Fischer et al., 2009). This is in contrast to plant aquaporin SoPIP2;1 in which the loop D connecting TM4 and TM5 undergoes conformational changes to block the channel from the cytoplasmic side (Tornroth-Horsefield et al., 2006). Equilibrium and nonequilibrium molecular dynamics simulations, site-directed mutagenesis data, and functional studies support the conclusion that gating of orthodox yeast aquaporins involve the N-terminal region and is regulated by two putative modes, phosphorylation and mechanosensitivity.

The subangstrom resolution structure of Aqy1 (resolution: 0.88 Å; Eriksson et al., 2013) enabled analyzing the electron densities associated with atoms of important residues such as Asn of NPA motifs. Hydrogen-bond interactions involving side-chain nitrogen atoms of Asn residues with passing water molecules in the channel interior are confirmed from this analysis. This structure (PDB ID: 3ZOJ) renders support to the bipolar distribution of water–water hydrogen bonds inside the channel. Together with a strongly correlated movement of water molecules, it provides an explanation of the mechanism of proton exclusion. The two SF residues in the Ar/R constriction, Arg and His, can adopt different tautomeric states, which is usually difficult to deduce from the crystal structure. However, this ultra-high-resolution structure reveals the exact tautomeric states of the two SF residues based on the electron clouds associated with the side-chain atoms of these residues. Occupancy of water molecules and the geometry of water selective sites in the SF region supported by molecular dynamics simulations explain how the Ar/R tetrad discriminates water from other solutes.

2.4 Protein–lipid interactions in MIP channels

It has been shown that the structure and function of membrane proteins are influenced by the lipid environment in which they are present (Li, Shi, Guo, Li, & Xu, 2014; Mondal, Khelashvili, & Weinstein, 2014; Poveda et al., 2014). Knowledge of interactions between specific protein residues and different components of lipids in the dynamic membrane environment is vital in understanding the role of lipids in the structure–function relationship of specific membrane proteins. Solid-state nuclear magnetic resonance (Huster, 2014), electron spin resonance (Marsh, 2008), mass spectrometry (Barrera, Zhou, & Robinson, 2013), electron crystallography (Reichow & Gonen, 2009), time-resolved fluorescence microscopy (de Almeida, Loura, & Prieto, 2009), two-dimensional infrared IR spectroscopy (Smith, 2012), and computer simulations (Lindahl & Sansom, 2008) are some of the techniques used to investigate lipid–protein interactions. Experimental studies have indicated that the structure and function of G-protein coupled receptors (GPCRs; Gawrisch & Soubias, 2008), potassium channels (Poveda et al., 2014), virus ion channels (Zhou & Cross, 2013), and lactose permease (Dowhan & Bogdanov, 2009) are modulated by the surrounding lipid environment. Hence, it would be of great interest to learn the role of lipids in the structure and function of MIP channels.

Structures of AQP0 (Gonen et al., 2005; Hite et al., 2010), AQP4 (Ho et al., 2009), and AQP5 (Horsefield et al., 2008) reveal bound lipid molecules (Fig. 2). The structure of AQP0 has been determined in the presence of two different lipids—DMPC (PDB ID: 2B6O) and *E. coli* polar lipids (EPL) (PDB ID: 3M9I). DMPC and EPL differ in the chemical nature of headgroup, length of hydrophobic tails, and saturation of acyl chains. There is no difference in the AQP0 structures determined in the presence of these two different lipid environments. In both AQP0 structures, seven annular lipids were identified surrounding each AQP0 monomer, and they are found in very similar positions on the protein surface. Although the chain lengths are different, the bilayer thickness for DMPC and EPL was found to be the same. This is due to the fact that the longer acyl chains of EPL lipids interdigitate at the center of the bilayer. Differences between EPL and DMPC lipids are found in the positions of acyl chains when extracellular and cytoplasmic leaflets are compared. Compared to the lipid headgroups, acyl chains seem to define the interactions between the annular lipids and

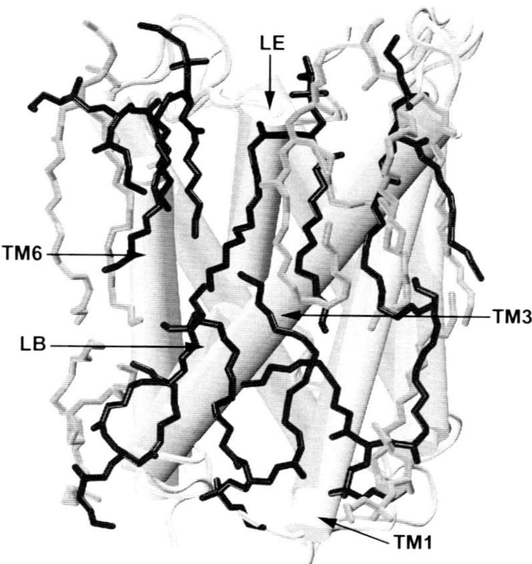

Figure 2 Protein–lipid interactions in MIP channels. AQP0 (PDB ID: 3M9I) and AQP4 (PDB ID: 2ZZ9) structures are superposed, and their bound lipids are displayed. Lipids associated with AQP0 and AQP4 are displayed in green (black shade in the print version) and violet (gray shade in the print version) colors, respectively. Lipid-interacting transmembrane helices and half-helices in MIP channels are labeled.

the AQP0 channel. These interactions are defined by the acyl chains filling in the gaps on the AQP0 surface. With these two very different lipids and the AQP0 structure being identical, it appears that the hydrophobic mismatch did not have a major role in influencing the AQP0 structures. Instead, the lipids have shown flexibility by bending and interdigitating to adapt to AQP0. It would be interesting to see if this conclusion is valid for lipids that have very different hydrophobic thickness and in that case whether the AQP0 structure adjusts to the bilayer thickness. Determination of AQP0 structure by growing 2-D crystals with diverse lipids will answer some of the questions mentioned earlier including the role of lipid headgroups.

E. coli lipids were also used to obtain 2-D crystals of AQP4 (Ho et al., 2009), and five lipid molecules were identified in the 2.8 Å resolution structure (PDB ID: 2ZZ9; Fig. 2). Three lipids in the extracellular leaflet interact with specific residues forming a negatively charged surface area of AQP4 in the adjoining bilayer. These interactions could contribute to the stability of double-layered 2-D crystals. Two lipids in the cytosolic leaflet are positioned between the AQP4 tetramers. In AQP5 structure (Horsefield et al., 2008), a lipid has been located in the central pore formed by the four monomers (PDB ID: 3D9S). A network of hydrogen bonds is observed between the phosphatidylserine lipid head group and backbone oxygen atoms of specific residues at the cytosolic surface. Although the central pore of the structures of AqpZ (Jiang, Daniels, & Fu, 2006) and PfAQP (Newby et al., 2008) contains one or two phospholipids, bulky hydrophobic side chains occlude the central pore in other MIP channels like AQP0 (Gonen et al., 2005) and AQP1 (Sui et al., 2001). The functional significance of the lipid insertion at the central pore of these AQP structures could be to block the transport of gases or ions. Although some AQPs accommodate lipid molecule(s) in the central pore, others seem to use hydrophobic side chains to block the central pore. In these AQPs if there is enough room for conformational changes of these hydrophobic side chains, then the intriguing question is whether these side chains will be part of the gating of the central tetrameric pore. If so, the central pore is likely to be a conduction pathway, albeit a restricted one. Further mutation studies and simulations are likely to provide an explanation for a possible functional role for the central pore.

3. MIPModDB DATABASE: A DATABASE OF MIP MODELS

The structures of MIP channels determined experimentally belong to different organisms, and their amino acid sequences are diverse. They all adopted a conserved helical fold (Fig. 1B). MIP channels have certain unique characteristics. They have six transmembrane segments and two half-helices. They have two highly conserved NPA motifs. They exhibit internal homology and quasi twofold symmetry at the sequence and structure levels, respectively. With more and more genome sequences available, these unique features of MIP channels were exploited to identify new MIP genes. Expression and functional studies provided support to the identified genes in some studies. Genome sequences of plants, fungi, mammals, and microbial organisms available in NCBI database (http://www.ncbi.nlm.nih.gov) were searched, and MIP genes were identified at the genome level from very diverse species, from microbes to mammals (Bansal & Sankararamakrishnan, 2007; Gupta & Sankararamakrishnan, 2009; Gupta et al., 2012; Verma et al., 2014b; Xu et al., 2013). With diverse MIP sequences adopting the unique hourglass fold, homology modeling studies were carried out on newly identified MIP channel proteins with experimentally determined structures from diverse sources as templates (Bansal & Sankararamakrishnan, 2007; Wallace & Roberts, 2004).

The protocol followed to model MIP channels is described in our earlier studies on plant and fungal MIPs (Bansal & Sankararamakrishnan, 2007; Gupta & Sankararamakrishnan, 2009; Gupta et al., 2012; Verma et al., 2014b). Briefly, three experimentally determined MIP structures from bacteria (PDB ID: 1FX8), archaea (PDB ID: 2F2B), and bovine (PDB ID: 1J4N) were used as templates, and the software MODELLER (ver. 7v7 or higher; Eswar, Eramian, Webb, Shen, & Sali, 2008) was used to model each MIP channel. The important step in the modeling procedure is to get alignment of the target MIP sequence with the template sequences. We manually examined to make sure that no gaps were present in the TM helical segments or the functionally important half-helical regions when the target sequence was aligned with the template sequences. In our earlier study on plant MIPs (Bansal & Sankararamakrishnan, 2007), we also showed that small and weakly polar residues at the helix–helix interface are group conserved, and we identified 17 such positions. The presence of six TM helical regions, two highly conserved NPA or NPA-like motifs, and group conservation at 17 positions are the criteria used to confirm that the protein identified indeed belongs to MIP superfamily. For each MIP sequence, 10 models were built, and the 1 with the optimal MODELLER objective function was selected for further refinement. We used MODELLER's loop optimization protocol and the graph-theory-based SCWRL3 program (Canutescu, Shelenkov, & Dunbrack, 2003) for refining the loop and side-chain conformation, respectively. The model was minimized using the GROMACS program (Hess, Kutzner, van der Spoel, & Lindahl, 2008) to ensure that the predicted structure is free of steric clashes, and the program PROCHECK (Laskowski, MacArthur, Moss, & Thornton, 1993) was used to examine the quality of models. MIP structural models were built for a large number of MIP channels from diverse organisms. We have developed a database called MIPModDB to store all MIP-related sequence and structural information.

MIPModDB is currently available at http://bioinfo.iitk.ac.in/MIPModDB and contains sequences and structures of 1500 MIP sequences (Gupta et al., 2012). This database allows the user to download modeled MIP structures. It also provides information about substitution(s) in the conserved NPA motifs, residues forming the Ar/R SF, gene structure, and pairwise sequence similarity between the target and template sequences. One can also perform phylogenetic analysis and structure-based sequence alignment for a selected set of MIP sequences. MIPModDB also provides facility to search the database based on the Ar/R SF, NPA substitution, accession ID, or organism.

4. UNIQUE STRUCTURAL FEATURES WITHIN THE TM REGION

Conventional pairwise alignment tools have been used to compare MIP sequences within the same species and across different species. It has been reported that MIP sequences from different subfamilies exhibit sequence identity below 20% (Chaumont, Barrieu, Wojcik, Chrispeels, & Jung, 2001). Previous studies on plant and fungal MIP sequences have shown that as many as 17 positions within the TM hourglass fold show a high degree of group conservation (Bansal & Sankararamakrishnan, 2007; Gupta & Sankararamakrishnan, 2009; Verma et al., 2014b). Small and weakly polar residues (Ser, Thr, Cys, Ala, and Gly) have been shown to be group conserved mostly in positions that occur at the helix–helix interface. This was found by doing structure-based sequence alignment of MIP sequences. In this procedure, sequences corresponding to the respective TM segments are compared instead of the full sequences. With 1500 sequences from all three kingdoms, we have performed structure-based sequence alignment for all six transmembrane segments (TM1–TM6) and two half-helices. Our results show that very high group conservation is observed in 16 of 17 previously identified positions (Table 1). In addition to these

Table 1 Group conservation of small and weakly polar residues in the transmembrane region of MIP channels

Residue[a]	Occurrence at the interface of two helical segments[b]	Conservation (%)[c]
A43 (TM1)	TM1–TM2	94
A47 (TM1)	TM1–TM2	97
T48 (TM1)	TM1–TM3	89
T55 (TM1)	TM1–TM2	93
A57 (TM1)	TM1–TM3	79
T58 (TM1)	TM1–TM2	80
A78 (TM2)	TM2–TM5	88
A80 (TM2)	TM1–TM2	87
G82 (TM2)	TM2–TM5	95
A(103) (LB)	LB–TM4, LB–LE, LB–TM6	95
T105 (LB)	LB–TM3	98
G107 (LB)	LB–TM4, LB–TM6	95
A125 (TM3)	TM1–TM3	86
G129 (TM3)	TM1–TM3	100
A130 (TM3)	TM1–TM3, TM3–LE	99
G133 (TM3)	TM1–TM3	99
V134 (TM3)	TM1–TM3, TM3–LE	89
A164 (TM4)	TM4–TM5	83
T172 (TM4)	LB–TM4, TM4–TM6	96
G203 (TM5)	TM2–TM5	92
G220 (LE)	LE–TM6	95
A224 (LE)	LB–LE, TM3–LE	97
G228 (LE)	TM3–LE	99
G248 (TM6)	TM4–TM6	96
G252 (TM6)	TM4–TM6	99

Continued

Table 1 Group conservation of small and weakly polar residues in the transmembrane region of MIP channels—cont'd

Residue	Occurrence at the interface of two helical segments	Conservation (%)
A253 (TM6)	LB–TM6, TM4–TM6	97
A256 (TM6)	TM4–TM6	95

[a]Residue numbers correspond to the spinach MIP channel SoPIP2;1. The PDB ID of this structure is 1Z98.
[b]High group conservation of small and weakly polar residues (Ala, Gly, Ser, Cys, and Thr) occurs between two TM helical segments or between the half-helix (LB or LE) and a TM helix or between two half-helices.
[c]Structure-based sequence alignments of 1500 MIP channels for each TM helix and the two half-helices are used to find the group conservation of small and weakly polar residues. These MIP channels belong to seven organism groups, namely, animals, plants, fungi, heterokonta, bacteria, archaea, and protozoa. Only those residues that show at least 75% group conservation are listed in this table.

positions, we also identified another 11 positions revealing high group-based conservation of small and weakly polar residues (Figs. 1C and 3). In addition, we find a very high level of conservation of specific residues in TM1, TM3, TM4, and TM6 (Table 2). These include a Glu in TM1 and TM4, Gln in TM3, and Pro in TM6. If we include the highly conserved residues from the half-helices, there are 13 such residues. In total, 40 positions are either group conserved or display conservation of specific residues. This clearly demonstrates the limitations of the existing sequence alignment tools. Although the MIP sequences are diverse, almost one-quarter of the positions within the TM region exhibit an extremely high level of conservation either at the residue level or at the group level. Each TM helix, with the exception of TM5, has three to seven group-conserved residues. TM5 has only one such position (Table 1). A similar exercise using structure-based sequence alignment for other helical membrane proteins within TM regions is likely to reveal high group-based conservation in addition to the conservation of specific residues. In addition to plant and fungal MIPs, group conservation of small and weakly polar residues has been reported earlier in GPCRs by Smith and coworkers (Liu, Eilers, Patel, & Smith, 2004). In the majority of cases, these residues occur at the helix–helix interface. In both MIP channels and GPCRs, it has been speculated that these small and weakly polar residues facilitate the tight helix packing and bring two interacting TM helices closer together. In most of the cases, this is the crossing point of the two interacting helices, and such an arrangement will strengthen other favorable interactions between residues of the interacting TM helices.

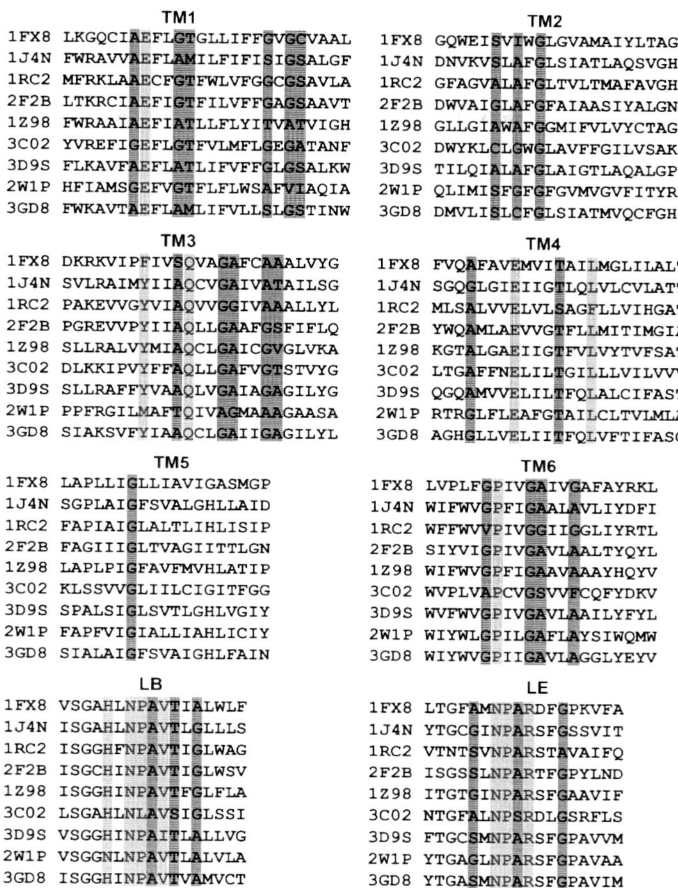

Figure 3 Structure-based sequence alignments of all six TM helical and the two half-helical segments are shown for the experimentally determined MIP structures. Positions that are highly group conserved are shown in blue (dark gray shade in the print version) background. Highly conserved bulky residues are displayed in orange (light gray shade in the print version) background. Alignment of any pair of sequences in this diagram show less than 50% sequence identity. For each sequence, their unique PDB IDs are shown in the left. Because AQP0 and AQP2 exhibit more than 50% sequence identity with AQP5, only AQP5 is selected as a representative of these three sequences.

4.1 Generic numbering scheme for TM residues of MIP channels

With MIP channels present in almost every organism and the MIP sequences showing high diversity, we are proposing a new and generic numbering scheme that will assist in comparing residue positions from two different

Table 2 Highly conserved bulky residues in the transmembrane region of MIP channels

Residue[a]	Occurrence at the TM helical bundle	Conservation (%)[b]	Remarks
E44 (TM1)	Occurs at the TM1–TM3 interface; interacts with LB loop residues before the beginning of half-helix	98.0	Conserved in all seven organism groups
H99 (LB)	Side chain facing the channel; occurs before the beginning of the LB half-helix; can interact with residues from TM1 and TM3	83.8	Not conserved in fungi
N101 (LB)	Side chain facing the channel; occurs at the interface of LB–TM3 and LB–LE helical segments	97.9	Part of the first NPA motif; conserved in all seven organism groups
P102 (LB)	Occurs at the interface of LB–TM3 and LB–LE helical segments	98.3	Part of the first NPA motif; conserved in all seven organism groups
V104 (LB)	Side chain facing the channel; occurs at the LB–TM4 interface	81.2	Not conserved in fungi
Y122 (TM3)	Occurs at the interface of TM1–TM3 and LB–TM3 helical segments	84.0	Not conserved in fungi
Q126 (TM3)	Occurs at the interface of TM1–TM3, LB–TM3, and TM3–LE helical segments	98.2	Conserved in all seven organism groups
E168 (TM4)	Occurs the interface of TM4–TM5 and TM4–TM6 helical segments; interact with LE loop residues before the beginning of half-helix	95.0	Conserved in all seven organism groups
L175 (TM4)	Side chain facing the channel; occurs at the interface of LB–TM4 and TM4–TM5 helical segments	81.3	Not conserved in bacteria

Table 2 Highly conserved bulky residues in the transmembrane region of MIP channels—cont'd

Residue	Occurrence at the TM helical bundle	Conservation (%)	Remarks
N222 (LE)	Side chain facing the channel; occurs the interface of TM1–LE and LB–LE helical segments	99.1	Part of the second NPA motif; conserved in all seven organism groups
P223 (LE)	Occurs at the interface of LB–LE and LE–TM6 helical segments	89.9	Part of the second NPA motif; conserved in all seven organism groups
R225 (LE)	Side chain facing the channel	89.4	Ar/R SF residue; not conserved in protozoa
P249 (TM6)	Occurs at the interface of LB–TM6, TM4–TM6, and LE–TM6 helical segments	88.1	Not conserved in heterokonta

[a]See footnote *a* of Table 1.
[b]Conservation is based on the structure-based sequence alignment of 1500 MIP channels for each TM helix and the two half-helical segments. These MIP channels belong to seven organism groups, namely, animals, plants, fungi, heterokonta, bacteria, archaea, and protozoa.

MIP channels. A similar scheme was first recommended for the superfamily of GPCRs (Ballesteros & Weinstein, 1995) and later extended to neurotransmitter/sodium symporter proteins (Beuming, Shi, Javitch, & Weinstein, 2006). This scheme is structure-based, and the most conserved position in the structure-based sequence alignment is chosen for each of the six TM helices and the two half-helices as a reference position. This position is given the number 50, and all other positions within the same TM helix are relative to this position. For example, the preceding and succeeding residues in the primary sequence are assigned 49 and 51, respectively. With the structure-based sequence alignment, we have identified the most conserved residues in each TM helix and the two half-helices. The highly conserved residue in TM1 is Glu (E44 in SoPIP2;1), and it is referred as $E_{44}1.50$ in the generic numbering scheme. The most conserved residues in TM3, TM4, and TM6 are Q126, E168, and P249, respectively (the residue numbers correspond to that in SoPIP2;1), and their generic numbers are $Q_{126}3.50$, $E_{168}4.50$, and $P_{249}6.50$. For the two half-helices, the highly conserved Asn residues from the NPA motifs will serve as reference positions, and their generic numbers are $N_{101}LB.50$ and $N_{222}LE.50$. Only for TM2 and TM5 helices, the positions that are highly group conserved are chosen as reference

Table 3 Structure-based generic numbering scheme for MIP channels

TM helix/half-helix[a]	Residue[b]	Generic number[c]
TM1	E44	1.50
TM2	G82	2.50
TM3	Q126	3.50
LB	N101	LB.50
TM4	E168	4.50
TM5	G203	5.50
LE	N222	LE.50
TM6	P249	6.50

[a]See the topology diagram in Fig. 1A for the assignment of different secondary structural elements.
[b]Residue numbers correspond to that of spinach MIP channel SoPIP2;1.
[c]Highly conserved residues in each TM helix or half-helix is assigned the number 50. E4.50 indicates the most conserved Glu residue from TM4 helix, and all other residues in this helix are numbered relative to this position. See text for details.

positions. G82 in TM2 and G203 in TM5 will be designated as $G_{82}2.50$ and $G_{203}5.50$, respectively, in the structure-based generic numbering scheme. The amino acids used as reference positions for each TM helix and the two half-helices and the corresponding structure-based generic numbers are summarized in Table 3. The four residues that form the Ar/R SF come from TM2 and TM5 helices and the loop LE region. In the generic numbering schemes, these positions in spinach SoPIP2;1 will be $F_{81}2.49$, $H_{210}5.57$, $T_{219}LE.47$, and $R_{225}LE.53$, respectively. Such a numbering scheme is expected to make it easier to compare two distinct MIP sequences.

5. RESIDUES FORMING Ar/R SF ARE DIVERSE

Four residues from TM2 and TM5 helices and the loop LE form one of the constriction sites in the MIP channel. As the name suggests, this narrow region consists of aromatic and arginine residues and contributes to the selectivity of the solutes to be transported across the channels. Knowledge of Ar/R SF residues derived from the homology models has been used to derive structural subclasses of MIP channels (Bansal & Sankararamakrishnan, 2007; Gupta & Sankararamakrishnan, 2009; Verma et al., 2014b; Wallace & Roberts, 2004). Arginine at LE.53 position is mostly conserved in MIP channels. Mutation at this position in the tonoplast intrinsic protein subfamily of plant MIPs demonstrated the critical role played by the residue at LE.53 position (Azad, Yoshikawa, Ishikawa, Sawa, & Shibata, 2012). In another study, it has been shown that site-directed mutagenesis at the 5.57 position in *Arabidopsis* and rice MIP channels belonging to the nodulin-26-like intrinsic proteins (NIPs) family revealed the key role played by this residue from TM5 helix (Mitani-Ueno, Yamaji, Zhao, & Ma, 2011). Both mutation and simulation studies have highlighted the Ar/R SF region responsible for impeding proton permeation in AQP1 channel (Li et al., 2011). Molecular dynamics simulation studies of AQP1 and AQP4 channels showed that arginine at the LE.53 position could adopt two different conformational states depending on the electrostatic potential and thus could regulate the water flux across the membrane (Hub, Aponte-Santamaria, Grubmuller, & de Groot, 2010). Beitz, Wu, Holm, Schultz, and Zeuthen (2006) constructed a series of single and double mutants of the AQP1 channel and studied the transport properties. They demonstrated the transport of glycerol, urea, and ammonia in this water-specific channel in most of the mutants when bulky aromatic and arginine residues were replaced by Ala/Val. When the positively charged Arg at LE.53 was substituted by the hydrophobic Val, even proton conduction was observed. Hence, size, volume, and the chemical nature of the side chains of individual residues in the Ar/R SF will have enormous influence on the type of solutes that are transported across the MIP channels.

Although a large number of MIP genes have been identified from sequence search, Ar/R SFs of very few MIP channels have been investigated experimentally and computationally. The homology models of 1500 MIPs have provided a wealth of information regarding the Ar/R SFs. They are from seven different organism groups. Among the 1500 MIPs, there are 34, 252, 79, 17, 395, 537, and 186 MIPs from protozoa, bacteria, archaea, heterokonta, fungi, plants, and animals, respectively. To further enhance our understanding on the nature of residues preferred or avoided in the Ar/R SFs, we have analyzed several properties of the four residues that define this narrow region from all seven organism groups.

5.1 Some residues are totally avoided in the Ar/R SF

We first calculated the frequencies of all 20 amino acids that form the SF in 1500 MIP channels for which homology models are available in the MIPModDB database. Because Ar/R SF is formed by four residues, we considered all 6000 residues of the entire set of MIP channels. A histogram of amino acid distribution in the narrow SF region is plotted in Fig. 4A. Our

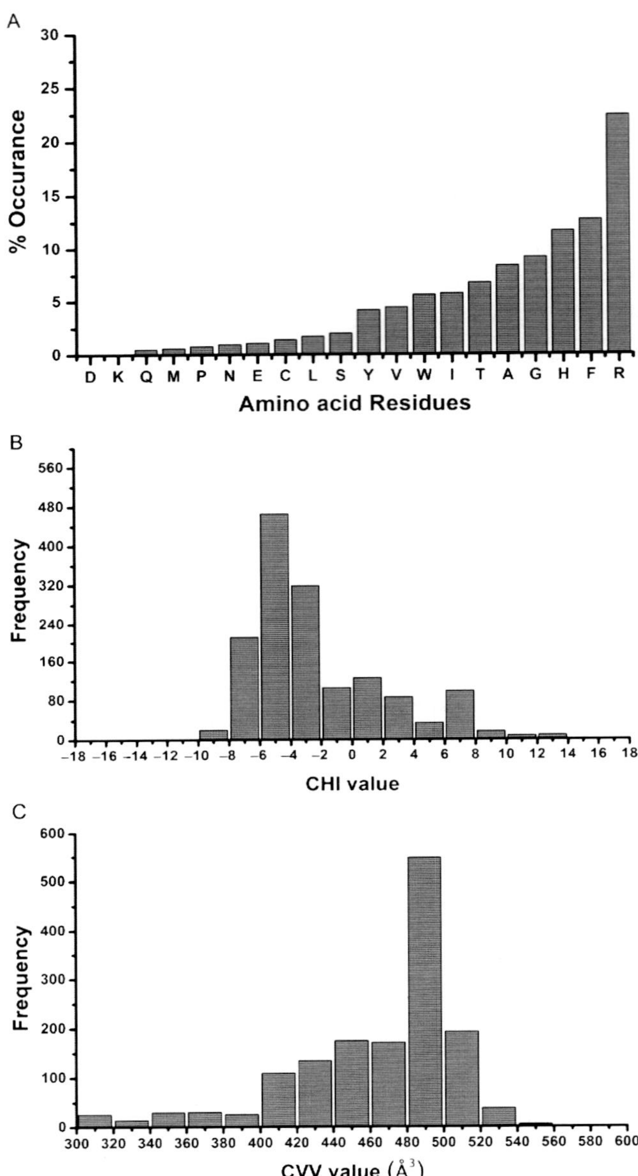

Figure 4 (A) Percentage occurrence of amino acid residues in the aromatic/arginine selectivity filter of MIP channels. All four residues of Ar/R SF in each MIP channel have been considered for this purpose (6000 Ar/R residues from 1500 MIP channels). (B) Distribution of cumulative hydropathy index (CHI) value calculated for the Ar/R SF residues. For each MIP channel, CHI value was calculated by summing up the hydropathy values of the respective residues that form the SF region. Hydropathy scale proposed by Kyte and Doolittle (1982) was considered for this purpose. (C) Distribution of cumulative vdw volume (CVV) calculated in $Å^3$ for the Ar/R SF residues of each MIP channel. CVV was calculated by adding the vdw volume (Creighton, 1997; Richards, 1974) of each of the four residues of the Ar/R SF region.

analysis has shown some surprising insights about the preferences of amino acids in this region. Some amino acids are almost totally avoided in the SF in any of the four positions. The amino acids Asp and Lys are rarely found in MIP channels, and they occur only 0.02% and 0.08%, respectively. Similarly, Gln, Glu, Asn, Pro, and Met are also found only in a handful of MIP channels. Arginine is the most frequently found amino acid in the SF, and it occurs 22.4% of the time. This is anticipated because arginine is predominantly found in the LE.53 position in most of the MIP channels. Although this positively charged amino acid has been shown to be very important in proton exclusion, it is intriguing to find that Lys, another basic residue, is almost totally avoided. Arg is followed by Phe (12.7%) and His (11.62%) as the next prominently observed amino acids in the SF, confirming the role of aromatic residues in selecting the solutes. The next most frequently observed amino acids are small in size (Gly, Ala, and Thr), and they each occur in between 7% and 9% of 6000 residues. The mostly hydrophobic Ile, Trp, Val, and Tyr are also present in significant number (4–6%). We have also looked at all MIP channels that are devoid of Arg as well as aromatic residues. We found that only 4.4% of all MIP channels have neither Arg nor aromatic residues in the narrow Ar/R constriction region.

5.2 Chemical nature of Ar/R SF

Mutation studies demonstrated that substitution of highly conserved basic residue Arg by a hydrophobic residue can alter the transport properties of MIP channels (Beitz et al., 2006). We have examined the hydrophobic or hydrophilic nature of all MIP channels in our dataset. We used the hydropathy scale proposed by Kyte and Doolittle (1982). For each MIP channel, the cumulative hydropathy value was found by summing the hydropathy values of the four individual residues that constitute the Ar/R SF. The most hydrophobic and hydrophilic residues according to the Kyte and Doolittle scale are Ile and Arg, respectively, and their hydropathy values are +4.5 and −4.5. If all four SF residues are Ile, then the cumulative hydropathy index (CHI) of SF will be 18.0. Similarly, SF with Arg in all four positions will have a CHI value of −18.0. We have plotted the distribution of CHI values in Fig. 4B. It is clear that the majority of MIPs have values between −8 and +4, indicating that Ar/R constriction of MIP channels in general is more hydrophilic. This is also probably due to the presence of highly conserved Arg in the LE.53 position in almost all the MIP channels. CHI values of only 18 MIP channels are above +10, indicating that

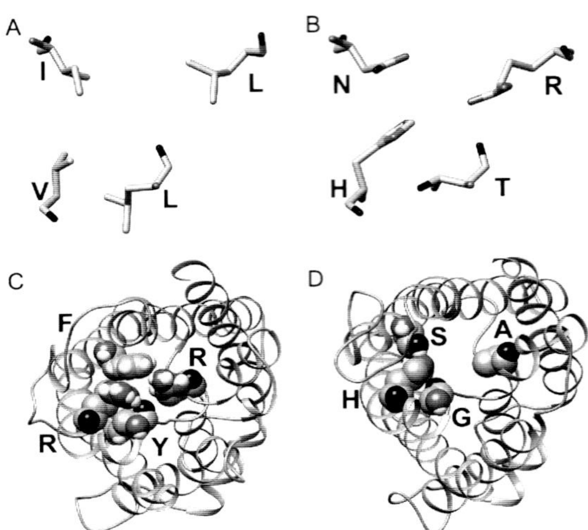

Figure 5 Ar/R selectivity filters of the (A) most hydrophobic and (B) most hydrophilic MIP channels. The most hydrophobic SF is from *Trypanosoma brucei* and has all bulky hydrophobic residues. The MIP channel from the model tree plant *Populus trichocarpa* has the most hydrophilic SF, and this channel belongs to the PIP subfamily. MIP channels with the highest and the lowest CVV values are shown in (C) and (D), respectively. The Ar/R SF residues are shown in space-filling representation. The channel is viewed down the channel axis from the extracellular side. The MIP channel similar to AQP7 from human and the plant SIP channel from *Arabidopsis thaliana* have the highest and the lowest CVV values, respectively. The residues forming the SF are labeled with one-letter amino acid code in all four examples.

they have the most hydrophobic Ar/R constriction. The maximum CHI value of 16.3 is observed for the *Trypanosoma brucei* MIP channel with SF residues IVLL (Fig. 5A). With all large hydrophobic residues in the Ar/R SF, it has been suggested that this channel is likely to facilitate the transport of large drug molecules (Baker et al., 2012). At the other extreme, there is just one MIP channel with a CHI value less than −10 (Fig. 5B), and 22 MIPs have CHI values below −9. These are the MIP channels with the most hydrophilic SF. MIP channels with the most hydrophobic and hydrophilic Ar/R constriction regions are listed in Table 4.

5.3 Amino acid size preference for the Ar/R SF region

With the Ar/R SF region forming one of the two narrow constriction regions in the MIP channels, the size and volume of the four amino acids can decide the pore diameter of the narrow filter. This, in turn, can be a major factor whether bigger or smaller solutes can pass through the channel.

Table 4 Top five MIP channels with the most hydrophobic and the most hydrophilic Ar/R SF residues

Ar/R SF residues[a]	CHI value[b]	Organism	Phylogenetic group[c]	MIPModDB accession ID[d]
Hydrophobic				
IVLL	16.3	*Trypanosoma brucei*	Protozoa	TRBRUC0986
LVAL	13.6	*Bos taurus*, *Canis familiaris*, *Macaca mulatta*, and two others	Mammalian AQP11–12	BOTAUR0488, CAFAMI0355, MAMULA0437, and two others
VIGI	12.8	*Deinococcus geothermalis* DSM 11300	Bacterial AQP	DEGEOT0194
VCMV	12.8	*Caenorhabditis elegans*	Nonmammalian AQP11-12	CAELEG0337
IAAV	12.3	*Salinispora tropica* CNB-440	Bacterial AQP	SATROP0144
Hydrophilic				
NHTR	−11.9	*Populus trichocarpa*	Plant PIPs	POTRIC0512
WNSR	−9.7	*Candida glabrata*, *Zygosaccharomyces rouxii*, *Torulaspora delbrueckii*	Fungi Fps01-like MIPs	CAGLAB0342, ZYROUX0847, TODELB1403
NGYR	−9.7	*Paracoccidioides brasiliensis*	Fungi alpha	PABRAS0901
WNTR	−9.6	*Candida glabrata*, *Saccharomyces cerevisiae*, *Naumovozyma dairenensis*, and three others	Fungi Fps01-like MIPs	CAGLAB0343, SACERE0481, NADAIR1397, and three others
HHAR	−9.1	*Cenarchaeum symbiosum*, *Nitrosopumilus maritimus*, and nine others	Archaeal AQPs	CESYMB1556, NIMARI1564

[a]The four residues correspond to those from TM2, TM5 helices, and LE region, and their positions in the generic numbering scheme are 2.49, 5.57, LE.47, and LE.53.
[b]CHI value is the cumulative hydropathy index and is calculated by summing up the respective hydropathy values of all four residues that form the Ar/R SF. The hydropathy scale proposed by Kyte and Doolittle (1982) has been used for calculating CHI values.
[c]For different phylogenetic group for each organism, see Verma, Prabh, and Sankararamakrishnan (2014a) and Verma et al. (2014b).
[d]The unique accession IDs for each MIP channel as given in the MIPModDB database (http://bioinfo.iitk.ac.in/MIPModDB).

We used the van der Waals (vdw) volume of amino acids as given in Creighton (Creighton, 1997; Richards, 1974). While Trp has the largest vdw volume of 163 Å^3, Gly has the smallest vdw volume (48 Å^3). Hypothetically, if all four residues of the SF are Trp residues, then the total volume occupied by these residues will be 652 Å^3. On the other hand, the smallest possible volume in the narrow SF region could be from four Gly residues (192 Å^3). We have calculated the cumulative vdw volume (CVV) of all four SF residues by summing up the volumes of the four individual SF residues. The distribution of CVVs for all the MIP channels present in our dataset is plotted in Fig. 4C. The majority of MIP channels have CVV values between 400 and 520 Å^3. The MIP channel with the largest CVV value (572 Å^3) belongs to humans and is annotated as "similar to AQP7 channel." It has SF residues FRYR (Table 5 and Fig. 5C). The next-largest CVV value (559 Å^3) belongs to a group of fungal MIP channels with WMLR in the SF. The MIP channel with the SF residues SHGA has the smallest CVV value (306 Å^3), and it belongs to a plant MIP from *Arabidopsis* (Table 5 and Fig. 5D). Other MIP channels with smaller CVV values have SF residues ISGA (312 Å^3), SHGS (312 Å^3), GQGV (315 Å^3), and GSGR (317 Å^3). These channels are found in protozoa or plant SIPs (small and basic intrinsic proteins), a subfamily of plant MIPs. In these channels with small CVV values, at least three positions are occupied by small residues such as Gly, Ala, and Ser.

Thus, our analysis of Ar/R SF regions of MIP channels from diverse organisms has revealed the extent of diversity in the residues forming this narrow region. Some residues are rarely found in the SF, and this includes acidic residues Asp and Glu. What is most surprising is the near total absence of Lys in the SF, whereas the other basic residue Arg is the most preferred among the 20 amino acids. In terms of chemical nature, we see examples of MIP channels from both extremes. In some MIP channels, all four SF residues are hydrophobic, and even the most conserved Arg residue is absent in these channels. At the other extreme, MIP channels with SF are formed by all hydrophilic residues. Similarly, there is significant variation in the size of the residues occupying the positions that contribute to the formation of Ar/R SF. Although bulky residues can result in narrow Ar/R constriction, the smaller residues can keep the channel radius larger, thus allowing the passage of bigger solutes. The extent of diversity in the chemical nature and the size of Ar/R SF residues indicate that MIP channels are probably involved in transporting a larger number of diverse solutes than what we know today.

Table 5 Top five MIP channels with Ar/R SF residues having largest and smallest cumulative volumes

Ar/R SF residues[a]	CVV (in Å^3)[b]	Organism	Phylogenetic group[c]	MIPModDB accession ID[d]
Largest cumulative volume				
FRYR	572	*Homo sapiens*	Mammalian AQGPs	HOSAPI0722
WMLR	559	*Leptosphaeria maculans*, *Phaeosphaeria nodorum*, *Pyrenophora teres*, and two others	Fungi Yfl054-like MIPs	LEMACU1025, PHNODO0856, PYTERE1363, and two others
WTFR	539	*Schizophyllum commune*	Fungi beta	SCCOMM1010
YILR	537	*Fibroporia radiculosa*, *Postia placenta*, *Wolfiporia cocos*, and three others	Fungi delta	FIRADI1432, POPLAC1527, WOCOCO1543, and three others
YIIR	537	*Laccaria bicolor*, *Mixia osmundae* IAM 14324	Fungi delta	LABICO0950, MIOSMU1446
Smallest cumulative volume				
SHGA	306	*Arabidopsis thaliana*	Plant SIPs	ARTHAL0025
ISGA	312	*Leishmania braziliensis*, *Leishmania infantum JPCM5*, *Leishmania major*, *Trypanosoma cruzi*	Protozoa	LEBRAZ0415, LEINFA0420, LEMAJO0425, TRCRUZ0993
SHGS	312	*Oryza sativa* subsp. *japonica*, *Populus trichocarpa*, *Zea mays*	Plant SIPs	ORSATI0250, POTRIC0525, ZEMAYS0281
GQGV	315	*Leishmania braziliensis*	Protozoa	LEBRAZ0416
GSGR	317	*Populus trichocarpa*, *Zea mays*, *Hordeum vulgare*, and 14 others	Plant NIPs	POTRAC0500, ZEMAYS0263, HOVULG1049, and 14 others

[a]See footnote *a* of Table 4.
[b]CVV is the cumulative vdw volume calculated by adding the vdw volumes (Creighton, 1997; Richards, 1974) of the four individual residues that constitute the Ar/R SF.
[c]See footnote *c* of Table 4.
[d]See footnote *d* of Table 4.

6. CONCLUSION

Since the discovery of the first AQP channel, many MIP channels have been investigated and functionally characterized. However, identification of a large number of MIP genes has given rise to several questions. For a majority of MIPs, little is known regarding several aspects of these channels such as cellular localization, expression pattern, transport properties, possible posttranslational modifications, gating, interaction with other proteins and involvement in important physiological processes. Modeling studies have helped to identify the residues that form the Ar/R SF region. However, this knowledge can only help to speculate the kind of solutes that can pass through. The large gap between the available MIP sequences and their functional significance has to be filled by a concerted approach between different experimental and computational techniques. It is especially intriguing to find a large number of MIP paralogs in higher organisms like plants and mammals. What is the need for so many MIPs in a single species, and what kind of solutes do they transport *in vivo*? Although a large number of diverse solutes have been already identified for a limited set of MIPs, are there many more different kinds of solutes that are yet to be identified? Beyond the two constriction regions, how do other regions within the channel influence in selecting the solutes? What is the exact role of lipids in an MIP structure–function relationship? MIP channels being possible drug targets to treat different diseases, these questions are expected to occupy researchers for many years in the field of MIP biology.

ACKNOWLEDGMENTS

Funding for this research came from the National Bioscience Award for Career Development to R. S. by the Department of Biotechnology (BT/HRD/34/17/2008). R. S. is a USV Chair Professor at IIT-Kanpur. MHRD, Government of India is gratefully acknowledged for establishing the Center of Excellence for Chemical Biology at IIT-Kanpur. R. K. V. acknowledges CSIR for the Senior Research Fellowship. We thank all our lab members for the discussion.

REFERENCES

Abascal, F., Irisarri, I., & Zardoya, R. (2014). Diversity and evolution of membrane intrinsic proteins. *Biochimica et Biophysica Acta*, *1840*, 1468–1481.

Ahmadpour, D., Geijer, C., Tamas, M. J., Lindkvist-Petersson, K., & Hohmann, S. (2014). Yeast reveals unexpected roles and regulatory features of aquaporins and aquaglyceroporins. *Biochimica et Biophysica Acta*, *1840*, 1482–1491.

Amaral, C., Carnevale, V., Klein, M. L., & Treptow, W. (2012). Exploring conformational states of the bacterial voltage-gated sodium channel NavAb via molecular dynamics simulations. *Proceedings of the National academy of Sciences of the United States of America, 109*, 21336–21341.

Anderberg, H. I., Danielson, J. A. H., & Johanson, U. (2011). Algal MIPs, high diversity and conserved motifs. *BMC Evolutionary Biology, 11*, 110.

Aroca, R., Porcel, R., & Ruiz-Lozano, J. M. (2007). How does arbuscular mycorrhizal symbiosis regulate root hydraulic properties and plasma membrane aquaporins in Phaseolus vulgaris under drought, cold or salinity stresses? *New Phytologist, 173*, 808–816.

Azad, A. K., Yoshikawa, N., Ishikawa, T., Sawa, Y., & Shibata, H. (2012). Substitution of a single amino acid residue in the aromatic/arginine selectivity filter alters the transport profiles of tonoplast aquaporin homologs. *Biochimica et Biophysica Acta, 1818*, 1–11.

Badaut, J., Fukuda, A. M., Jullienne, A., & Petry, K. G. (2014). Aquaporin and brain diseases. *Biochimica et Biophysica Acta, 1840*, 1554–1565.

Baker, N., Glover, L., Munday, J. C., Andres, D. A., Barrett, M. P., de Koning, H. P., et al. (2012). Aquaglyceroporin 2 controls susceptibility to melarsoprol and pentamidine in African trypanosomes. *Proceedings of the National academy of Sciences of the United States of America, 109*, 10996–11001.

Ballesteros, J. A., & Weinstein, H. (1995). Integrated methods for the construction of three-dimensional models and computational probing of structure-function relations in G protein-coupled receptors. *Methods in Neuroscience, 25*, 366–428.

Bansal, A., & Sankararamakrishnan, R. (2007). Homology modeling of major intrinsic proteins in rice, maize and Arabidopsis: Comparative analysis of transmembrane helix association and aromatic/arginine selectivity filters. *BMC Structural Biology*, 7, 27.

Barrera, N. P., Zhou, M., & Robinson, C. V. (2013). The role of lipids in defining membrane protein interactions: Insights from mass spectrometry. *Trends in Cell Biology, 23*, 1–8.

Beitz, E., Wu, B., Holm, L. M., Schultz, J. E., & Zeuthen, T. (2006). Point mutations in the aromatic/arginine region in aquaporin 1 allow passage of urea, glycerol, ammonia, and protons. *Proceedings of the National academy of Sciences of the United States of America, 103*, 269–274.

Berneche, S., & Roux, B. (2001). Energetics of ion conduction through the K+ channel. *Nature, 414*, 73–77.

Beuming, T., Shi, L., Javitch, J. A., & Weinstein, H. (2006). A comprehensive structure-based alignment of prokaryotic and eukaryotic neurotransmitter/Na^+ symporters (NSS) aids in the use of the LeuT structure to probe NSS structure and function. *Molecular Pharmacology, 70*, 1630–1642.

Bienert, G. P., Schussler, M. D., & Jahn, T. P. (2008). Metalloids: Essential, beneficial or toxic? Major intrinsic proteins sort it out. *Trends in Biochemical Sciences, 33*, 20–26.

Binder, D. K., Nagelhus, E. A., & Ottersen, O. P. (2012). Aquaporin-4 and epilepsy. *Glia, 60*, 1203–1214.

Bockenhauer, D., & Bichet, D. G. (2014). Urinary concentration: Different ways to open and close the tap. *Pediatric Nephrology, 29*, 1297–1303.

Borgnia, M., Nielsen, S., Engel, A., & Agre, P. (1999). Cellular and molecular biology of the aquaporin water channels. *Annual Review of Biochemistry, 68*, 425–458.

Calamita, G., Bishai, W. R., Preston, G. M., Guggino, W. B., & Agre, P. (1995). Molecular cloning and characterization of AqpZ, a water channel from Escherichia coli. *Journal of Biological Chemistry, 270*, 29063–29066.

Canutescu, A. A., Shelenkov, A. A., & Dunbrack, R. L. (2003). A graph-theory algorithm for rapid protein side-chain prediction. *Protein Science, 12*, 2001–2014.

Catterall, W. A. (2000). Structure and regulation of voltage-gated Ca2+ channels. *Annual Review of Cell and Developmental Biology, 16*, 521–555.

Chaumont, F., Barrieu, F., Wojcik, E., Chrispeels, M. J., & Jung, R. (2001). Aquaporins constitute a large and highly divergent protein family in maize. *Plant Physiology*, *125*, 1206–1215.

Chaumont, F., Moshelion, M., & Daniels, M. J. (2005). Regulation of plant aquaporin activity. *Biology of the Cell*, *97*, 749–764.

Chaumont, F., & Tyerman, S. D. (2014). Aquaporins: Highly regulated channels controlling plant water relations. *Plant Physiology*, *164*, 1600–1618.

Choi, W. G., & Roberts, D. M. (2007). Arabidopsis NIP2;1, a major intrinsic protein transporter of lactic acid induced by anoxic stress. *Journal of Biological Chemistry*, *282*, 24209–24218.

Colombini, M. (2004). VDAC: The channel at the interface between mitochondria and the cytosol. *Molecular and Cellular Biochemistry*, *256*, 107–115.

Creighton, T. E. (1997). *Proteins: Structures and molecular properties*. New York: W. H. Freeman and Company.

Daniels, M. J., Mirkov, T. E., & Chrispeels, M. J. (1994). The plasma membrane of Arabidopsis thaliana contains a mercury-insensitive aquaporin that is a homolog of the tonoplast water channel protein TIP. *Plant Physiology*, *106*, 1325–1333.

de Almeida, R. F. M., Loura, L. M. S., & Prieto, M. (2009). Membrane lipid domains and rafts: Current applications of fluorescence lifetime spectroscopy and imaging. *Chemistry and Physics of Lipids*, *157*, 61–77.

Dietz, S., von Bulow, J., Baker, N., & Nehls, U. (2011). The aquaporin gene family of the ectomycorrhizal fungus *Laccaria bicolor*: Lessons for symbiotic functions. *New Phytologist*, *190*, 927–940.

Dowhan, W., & Bogdanov, M. (2009). Lipid-dependent membrane protein topogenesis. *Annual Review of Biochemistry*, *78*, 515–540.

Eriksson, U. K., Fischer, G., Friemann, R., Enkavi, G., Tajkhorshid, E., & Neutze, R. (2013). Subangstrom resolution X-ray structure details aquaporin-water interactions. *Science*, *340*, 1346–1349.

Eswar, N., Eramian, D., Webb, B., Shen, M.-Y., & Sali, A. (2008). Protein structure modeling with MODELLER. *Methods in Molecular Biology*, *426*, 145–159.

Fenton, R. A., Pedersen, C. N., & Moeller, H. B. (2013). New insights into regulated aquaporin-2 function. *Current Opinion in Nephrology and Hypertension*, *22*, 551–558.

Fischer, G., Kosinska-Eriksson, U., Aponte-Santamaria, C., Palmgren, M., Geijer, C., Hedfalk, K., et al. (2009). Crystal structure of a yeast aquaporin at 1.15 angstrom reveals a novel gating mechanism. *PLoS Biology*, *7*, e1000130.

Frick, A., Eriksson, U. K., de Mattia, F., Oberg, F., Hedfalk, K., Neutze, R., et al. (2014). X-ray structure of human aquaporin 2 and its implications for nephrogenic diabetes insipidus and trafficking. *Proceedings of the National academy of Sciences of the United States of America*, *111*, 6305–6310.

Fruhbeck, G., Catalan, V., Gomez-Ambrosi, J., & Rodriguez, A. (2006). Aquaporin-7 and glycerol permeability as novel obesity drug-target pathways. *Trends in Pharmacological Sciences*, *27*, 345–347.

Fu, D., Libson, A., Miercke, L. J., Weitzman, C., Nollert, P., Krucinski, J., et al. (2000). Structure of a glycerol-conducting channel and the basis for its selectivity. *Science*, *290*, 481–486.

Gawrisch, K., & Soubias, O. (2008). Structure and dynamics of polyunsaturated hydrocarbon chains in lipid bilayers—Significance of GPCR function. *Chemistry and Physics of Lipids*, *153*, 64–75.

Gonen, T., Cheng, Y. F., Sliz, P., Hiroaki, Y., Fujiyoshi, Y., Harrison, S. C., et al. (2005). Lipid-protein interactions in double-layered two-dimensional AQP0 crystals. *Nature*, *438*, 633–638.

Gonen, T., & Walz, T. (2006). The structure of aquaporins. *Quarterly Reviews of Biophysics, 39*, 361–396.

Gunnarson, E., Zelenina, M., & Aperia, A. (2004). Regulation of brain aquaporins. *Neuroscience, 129*, 947–955.

Gupta, A. B., & Sankararamakrishnan, R. (2009). Genome-wide analysis of major intrinsic proteins in the tree plant Populus trichocarpa: Characterization of XIP subfamily of aquaporins from evolutionary perspective. *BMC Plant Biology, 9*, 134.

Gupta, A. B., Verma, R. K., Agarwal, V., Vajpai, M., Bansal, V., & Sankararamakrishnan, R. (2012). MIPModDB: A central resource for the superfamily of major intrinsic proteins. *Nucleic Acids Research, 40*, D362–D369.

Hansen, M., Kun, J. F. J., Schultz, J. E., & Beitz, E. (2002). A single, bi-functional aquaglyceroporin in blood-stage *Plasmodium falciparum* malaria parasites. *Journal of Biological Chemistry, 277*, 4874–4882.

Hara, M., Ma, T. H., & Verkman, A. S. (2002). Selectively reduced glycerol in skin of aquaporin-3-deficient mice may account for impaired skin hydration, elasticity, and barrier recovery. *Journal of Biological Chemistry, 277*, 46616–46621.

Hara-Chikuma, M., & Verkman, A. S. (2008). Aquaporin-3 facilitates epidermal cell migration and proliferation during wound healing. *Journal of Molecular Medicine, 86*, 221–231.

Hasegawa, T., Azlina, A., Javkhlan, P., Yao, C. J., Akamatsu, T., & Hosoi, K. (2011). Novel phosphorylation of aquaporin 5 at its threonine 259 through cAMP signaling in salivary gland cells. *American Journal of Physiology. Cell Physiology, 301*, C667–C678.

Hess, B., Kutzner, C., van der Spoel, D., & Lindahl, E. (2008). GROMACS 4: Algorithms for highly efficient, load-balanced, and scalable molecular simulation. *Journal of Chemical Theory and Computation, 4*, 435–447.

Hessa, T., Meindl-Beinker, N. M., Bernsel, A., Kim, H., Sato, Y., Lerch-Bader, M., et al. (2007). Molecular code for transmembrane-helix recognition by the Sec61 translocon. *Nature, 450*, 1026–1030.

Hilf, R. J. C., & Dutzler, R. (2008). X-ray structure of a prokaryotic pentameric ligand-gated ion channel. *Nature, 452*, 375–379.

Hite, R. K., Li, Z., & Walz, T. (2010). Principles of membrane protein interactions with annular lipids deduced from aquaporin-0 2D crystals. *The EMBO Journal, 29*, 1652–1658.

Ho, J. D., Yeh, R., Sandstrom, A., Chorny, I., Harries, W. E. C., Robbins, R. A., et al. (2009). Crystal structure of human aquaporin 4 at 1.8 angstrom and its mechanism of conductance. *Proceedings of the National academy of Sciences of the United States of America, 106*, 7437–7442.

Horsefield, R., Norden, K., Feller, M., Backmark, A., Tornroth-Horsefield, S., van Scheltinga, A. C. T., et al. (2008). High-resolution X-ray structure of human aquaporin 5. *Proceedings of the National academy of Sciences of the United States of America, 105*, 13327–13332.

Huang, H. F., He, R. H., Sun, C. C., Zhang, Y., Meng, Q. X., & Ma, Y. Y. (2006). Function of aquaporins in female and male reproductive systems. *Human Reproduction Update, 12*, 785–795.

Hub, J. S., Aponte-Santamaria, C., Grubmuller, H., & de Groot, B. L. (2010). Voltage-regulated water flux through aquaporin channels in silico. *Biophysical Journal, 99*, L97–L99.

Huber, V. J., Tsujita, M., & Nakada, T. (2012). Aquaporins in drug discovery and pharmacotherapy. *Molecular Aspects of Medicine, 33*, 691–703.

Huster, D. (2014). Solid-state NMR spectroscopy to study protein lipid interactions. *Biochimica et Biophysica Acta, 1841*, 1146–1160.

Ikeda, M., Beitz, E., Kozono, D., Guggino, W. B., Agre, P., & Yasui, M. (2002). Characterization of aquaporin-6 as a nitrate channel in mammalian cells—Requirement of pore-lining residue threonine 63. *Journal of Biological Chemistry*, *277*, 39873–39879.

Ishibashi, K., Hara, S., & Kondo, S. (2009). Aquaporin water channels in mammals. *Clinical and Experimental Nephrology*, *13*, 107–117.

Jentsch, T. J., Stein, V., Weinreich, F., & Zdebik, A. A. (2002). Molecular structure and physiological function of chloride channels. *Physiological Reviews*, *82*, 503–568.

Jiang, J., Daniels, B. V., & Fu, D. (2006). Crystal structure of AqpZ tetramer reveals two distinct Arg-189 conformations associated with water permeation through the narrowest constriction of the water-conducting channel. *Journal of Biological Chemistry*, *281*, 454–460.

Jiang, Y. X., Lee, A., Chen, J. Y., Ruta, V., Cadene, M., Chait, B. T., et al. (2003). X-ray structure of a voltage-dependent K+ channel. *Nature*, *423*, 33–41.

Jin, Y. Y., Yu, G. Z., Peng, P., Zhang, Y. F., & Xin, X. M. (2013). Down-regulated expression of AQP5 on lung in rat DIC model induced by LPS and its effect on the development of pulmonary edema. *Pulmonary Pharmacology*, *26*, 661–665.

Johanson, U., Karlsson, M., Johansson, I., Gustavsson, S., Sjovall, S., Fraysse, L., et al. (2001). The complete set of genes encoding major intrinsic proteins in Arabidopsis provides a framework for a new nomenclature for major intrinsic proteins in plants. *Plant Physiology*, *126*, 1358–1369.

King, L. S., Kozono, D., & Agre, P. (2004). From structure to disease: The evolving tale of aquaporin biology. *Nature Reviews. Molecular Cell Biology*, *5*, 687–698.

Koebnik, R., Locher, K. P., & van Gelder, P. (2000). Structure and function of bacterial outer membrane proteins: Barrels in a nutshell. *Molecular Microbiology*, *37*, 239–253.

Kozono, D., Ding, X. D., Iwasaki, I., Meng, X. Y., Kamagata, Y., Agre, P., et al. (2003). Functional expression and characterization of an archaeal aquaporin—AqpM from *Methanothermobacter marburgensis*. *Journal of Biological Chemistry*, *278*, 10649–10656.

Kun, J. F., & de Carvalho, E. G. (2009). Novel therapeutic targets in *Plasmodium falciparum*: Aquaglyceroporins. *Expert Opinion on Therapeutic Targets*, *13*, 385–394.

Kuriyama, H., Shimomura, I., Kishida, K., Kondo, H., Furuyama, N., Nishizawa, H., et al. (2002). Coordinated regulation of fat-specific and liver-specific glycerol channels, aquaporin adipose and aquaporin 9. *Diabetes*, *51*, 2915–2921.

Kyte, J., & Doolittle, R. F. (1982). A simple method for displaying the hydropathic character of a protein. *Journal of Molecular Biology*, *157*, 105–132.

Laskowski, R. A., MacArthur, M. W., Moss, D. S., & Thornton, J. M. (1993). PROCHECK: A program to check the stereochemical quality of protein structures. *Journal of Applied Crystallography*, *26*, 283–291.

Law, R. J., Henchman, R. H., & McCammon, J. A. (2005). A gating mechanism proposed from a simulation of a human alpha 7 nicotinic acetylcholine receptor. *Proceedings of the National academy of Sciences of the United States of America*, *102*, 6813–6818.

Lee, Y., Je, Y. J., Lee, S. S., Li, Z. J., Choi, D. K., Kwon, Y. B., et al. (2012). Changes in transepidermal water loss and skin hydration according to expression of aquaporin-3 in psoriasis. *Annals of Dermatology*, *24*, 168–174.

Lee, J. K., Kozono, D., Remis, J., Kitagawa, Y., Agre, P., & Stroud, R. M. (2005). Structural basis for conductance by the archaeal aquaporin AqpM at 1.68 A. *102*, 18932–18937.

Lemasters, J. J., & Holmuhamedov, E. (2006). Voltage-dependent anion channel (VDAC) as mitochondrial governator—Thinking outside the box. *Biochimica et Biophysica Acta*, *1762*, 181–190.

Li, H., Chen, H. N., Steinbronn, C., Wu, B. H., Beitz, E., Zeuthen, T., et al. (2011). Enhancement of proton conductance by mutations of the selectivity filter of aquaporin-1. *Journal of Molecular Biology*, *407*, 607–620.

Li, G. W., Santoni, V., & Maurel, C. (2014). Plant aquaporins: Roles in plant physiology. *Biochimica et Biophysica Acta, 1840*, 1574–1582.

Li, L. Y., Shi, X. S., Guo, X. D., Li, H., & Xu, C. Q. (2014). Ionic protein lipid interaction at the plasma membrane: What can the charge do? *Trends in Biochemical Sciences, 39*, 130–140.

Lindahl, E., & Sansom, M. S. P. (2008). Membrane proteins: Molecular dynamics simulations. *Current Opinion in Structural Biology, 18*, 425–431.

Liu, W., Eilers, M., Patel, A. B., & Smith, S. O. (2004). Helix packing moments reveal diversity and conservation in membrane protein structure. *Journal of Molecular Biology, 337*, 713–729.

Liu, C. W., Fukumoto, T., Matsumoto, T., Gena, P., Frascaria, D., Kaneko, T., et al. (2013). Aquaporin OsPIP1;1 promotes rice salt resistance and seed germination. *Plant Physiology and Biochemistry, 63*, 151–158.

Liu, Z. J., Shen, J., Carbrey, J. M., Mukhopadhyay, R., Agre, P., & Rosen, B. P. (2002). Arsenite transport by mammalian aquaglyceroporins AQP7 and AQP9. *Proceedings of the National academy of Sciences of the United States of America, 99*, 6053–6058.

Lu, H. J., Matsuzaki, T., Bouley, R., Hasler, U., Qin, Q. H., & Brown, D. (2008). The phosphorylation state of serine 256 is dominant over that of serine 261 in the regulation of AQP2 trafficking in renal epithelial cells. *The American Journal of Physiology. Renal Physiology, 295*, F290–F294.

Ma, J. F., Tamai, K., Yamaji, N., Mitani, N., Konishi, S., Katsuhara, M., et al. (2006). A silicon transporter in rice. *Nature, 440*, 688–691.

Maeda, S., Nakagawa, S., Suga, M., Yamashita, E., Oshima, A., Fujiyoshi, Y., et al. (2009). Structure of the connexin 26 gap junction channel at 3.5 angstrom resolution. *Nature, 458*, 597–602.

Marsh, D. (2008). Electron spin resonance in membrane research: Protein-lipid interactions. *Methods, 46*, 83–96.

Maurel, C., Verdoucq, L., Luu, D. T., & Santoni, V. (2008). Plant aquaporins: Membrane channels with multiple integrated functions. *Annual Review of Plant Biology, 59*, 595–624.

Misu, T., Fujihara, K., Kakita, A., Konno, H., Nakamura, M., Watanabe, S., et al. (2007). Loss of aquaporin 4 in lesions of neuromyelitis optica: Distinction from multiple sclerosis. *Brain, 130*, 1224–1234.

Mitani-Ueno, N., Yamaji, N., Zhao, F. J., & Ma, J. F. (2011). The aromatic/arginine selectivity filter of NIP aquaporins plays a critical role in substrate selectivity for silicon, boron, and arsenic. *Journal of Experimental Botany, 62*, 4391–4398.

Moeller, H. B., Aroankins, T. S., Slengerik-Hansen, J., Pisitkun, T., & Fenton, R. A. (2014). Phosphorylation and ubiquitylation are opposing processes that regulate endocytosis of the water channel aquaporin-2. *Journal of Cell Science, 127*, 3174–3183.

Mondal, S., Khelashvili, G., & Weinstein, H. (2014). Not just an oil slick: How the energetics of protein-membrane interactions impacts the function and organization of transmembrane proteins. *Biophysical Journal, 106*, 2305–2316.

Morello, J. P., & Bichet, D. G. (2001). Nephrogenic diabetes insipidus. *Annual Review of Physiology, 63*, 607–630.

Mulders, S. M., Knoers, N. V. A. M., vanLieburg, A. F., Monnens, L. A. H., Leumann, E., Wuhl, E., et al. (1997). New mutations in the AQP2 gene in nephrogenic diabetes insipidus resulting in functional but misrouted water channels. *Journal of the American Society of Nephrology, 8*, 242–248.

Murai-Hatano, M., Kuwagata, T., Sakurai, J., Nonami, H., Ahamed, A., Nagasuga, K., et al. (2008). Effect of low root temperature on hydraulic conductivity of rice plants and the possible role of aquaporins. *Plant and Cell Physiology, 49*, 1294–1305.

Newby, Z. E. R., O'Connell, J., III, Robles-Colmenares, Y., Khademi, S., Miercke, L. J., & Stroud, R. M. (2008). Crystal structure of the aquaglyceroporin PfAQP from the malarial parasite *Plasmodium falciparum*. *Nature Structural and Molecular Biology*, *15*, 619–625.

Nielsen, S., Frokiaer, J., Marples, D., Kwon, T. H., Agre, P., & Knepper, M. A. (2002). Aquaporins in kidney: From molecules to medicine. *Physiological Reviews*, *82*, 205–244.

Nikaido, H. (2003). Molecular basis of bacterial outer membrane permeability revisited. *Microbiology and Molecular Biology Reviews*, *67*, 593–656.

Osborne, A. R., Rapoport, T. A., & van den Berg, B. (2005). Protein translocation by the Sec61/SecY channel. *Annual Review of Cell and Developmental Biology*, *21*, 529–550.

Papadopoulos, M. C., & Verkman, A. S. (2007). Aquaporin-4 and brain edema. *Pediatric Nephrology*, *22*, 778–784.

Park, W., Scheffler, B. E., Bauer, P. J., & Cambell, B. T. (2010). Identification of the family of aquaporin genes and their expression in upland cotton (*Gossypium hirsutum* L.). *BMC Plant Biology*, *10*, 142.

Patel, A. J., Lazdunski, M., & Honore, E. (2001). Lipid and mechano-gated 2P domain K+ channels. *Current Opinion in Cell Biology*, *13*, 422–427.

Payandeh, J., Scheuer, T., Zheng, N., & Catterall, W. A. (2011). The crystal structure of a voltage-gated sodium channel. *Nature*, *475*, 353–358.

Pettersson, N., Filipsson, C., Becit, E., Brive, L., & Hohmann, S. (2005). Aquaporins in yeasts and filamentous fungi. *Biology of the Cell*, *97*, 487–500.

Poveda, J. A., Giudici, A. M., Renart, M. L., Molina, M. L., Montoya, E., Fernandez-Carvajal, A., et al. (2014). Lipid modulation of ion channels through specific binding sites. *Biochimica et Biophysica Acta*, *1838*, 1560–1567.

Preston, G. M., Carrol, T. P., Guggino, W. B., & Agre, P. (1992). Appearance of water channels in Xenopus oocytes expressing red cell CHIP28 protein. *Science*, *256*, 385–387.

Reeves, D. C., & Lummis, S. C. R. (2002). The molecular basis of the structure and function of the 5-HT3 receptor: A model ligand-gated ion channel (Review). *Molecular and Membrane Biology*, *19*, 11–26.

Reichow, S. L., & Gonen, T. (2009). Lipid-protein interactions probed by electron crystallography. *Current Opinion in Structural Biology*, *19*, 560–565.

Richards, F. M. (1974). Interpretation of protein structures—total volume, group volume distributions and packing density. *Journal of Molecular Biology*, *82*, 1–14.

Rojek, A., Praetorius, J., Frokiaer, J., Nielsen, S., & Fenton, R. A. (2008). A current view of the mammalian aquaglyceroporins. *Annual Review of Physiology*, *70*, 301–327.

Rout, M. P., Aitchison, J. D., Suprapto, A., Hjertaas, K., Zhao, Y. M., & Chait, B. T. (2000). The yeast nuclear pore complex: Composition, architecture, and transport mechanism. *Journal of Cell Biology*, *148*, 635–651.

Sasaki, S., Chiga, M., Kikuchi, E., Rai, T., & Uchida, S. (2013). Hereditary nephrogenic diabetes insipidus in Japanese patients: Analysis of 78 families and report of 22 new mutations in AVPR2 and AQP2. *Clinical and Experimental Nephrology*, *17*, 338–344.

Savage, D. F., Egea, P. F., Robles-Colmenares, Y., O'Connell, J. D., III, & Stroud, R. M. (2003). Architecture and selectivity in aquaporins 2.5 A X-ray structure of aquaporin Z. *PLoS Biology*, *1*, 334–340.

Schey, K. L., Wang, Z., Wenke, J. L., & Qi, Y. (2014). Aquaporins in the eye: Expression, function, and roles in ocular disease. *Biochimica et Biophysica Acta*, *1840*, 1513–1523.

Shrivastava, I. H., & Sansom, M. S. P. (2000). Simulations of ion permeation through a potassium channel: Molecular dynamics of KcsA in a phospholipid bilayer. *Biophysical Journal*, *78*, 557–570.

Smith, A. W. (2012). Lipid-protein interactions in biological membranes: A dynamic perspective. *Biochimica et Biophysica Acta*, *1818*, 172–177.

Sohi, G., & Willecke, K. (2004). Gap junctions and the connexin protein family. *Cardiovascular Research*, *62*, 228–232.

Soria, L. R., Marrone, J., Calamita, G., & Marinelli, R. A. (2013). Ammonia detoxification via ureagenesis in rat hepatocytes involves mitochondrial aquaporin-8 channels. *Hapatology*, *57*, 2061–2071.

Spring, J. H., Robichaux, S. R., & Hamlin, J. A. (2009). The role of aquaporins in excretion in insects. *Journal of Experimental Biology*, *212*, 358–362.

Sui, H., Han, B. G., Lee, J. K., Walian, P., & Jap, B. K. (2001). Structural basis of water-specific transport through the AQP1 water channel. *Nature*, *414*, 872–878.

Suntharalingam, M., & Wente, S. R. (2003). Peering through the pore: Nucear pore complex structure, assembly, and function. *Developmental Cell*, *4*, 775–789.

Tanghe, A., Van Dijck, P., Dumortier, F., Teunissen, A., Hohmann, S., & Thevelein, J. A. (2002). Aquaporin expression correlates with freeze tolerance in baker's yeast and over-expression improves freeze tolerance in industrial strains. *Applied and Environmental Microbiology*, *68*, 5981–5989.

Tanghe, A., van Dijck, P., & Thevelein, J. M. (2006). Why do microorganisms have aquaporins. *Trends in Microbiology*, *14*, 78–85.

Tingaud-Sequeira, A., Calusinska, M., Finn, R. N., Chauvigne, F., Lozano, J., & Cerda, J. (2010). The zebrafish genome encodes the largest vertebrate repertoire of functional aquaporins with dual paralogy and substrate specificities similar to mammals. *BMC Evolutionary Biology*, *10*, 38.

Tornroth-Horsefield, S., Wang, Y., Hedfalk, K., Johanson, U., Karlsson, M., Tajkhorshid, E., et al. (2006). Structural mechanism of plant aquaporin gating. *Nature*, *439*, 688–694.

Uehlein, N., & Kaldenhoff, R. (2008). Aquaporins and plant leaf movement. *Annals of Botany*, *101*, 1–4.

Uehlein, N., Lovisolo, C., Siefritz, F., & Kaldenhoff, R. (2003). The tobacco aquaporin NtAQP1 is a membrane CO_2 pore with physiological functions. *Nature*, *425*, 734–737.

van Balkom, B. W. M., Boone, M., Hendriks, G., Kamsteeg, E. J., Robben, J. H., Stronks, H. C., et al. (2009). LIP5 interacts with aquaporin 2 and facilitates its lysosomal degradation. *Journal of the American Society of Nephrology*, *20*, 990–1001.

Vandeleur, R. K., Mayo, G., Shelden, M. C., Gilliham, M., Kaiser, B. N., & Tyerman, S. D. (2009). The role of plasma membrane intrinsic protein aquaporins in water transport through roots: Diurnal and drought stress responses reveal different strategies between isohydric and anisohydric cultivars of grapevine. *Plant Physiology*, *149*, 445–460.

Verkman, A. S. (2009). Aquaporins: Translating bench research to human disease. *Journal of Experimental Biology*, *212*, 1707–1715.

Verkman, A. S., Anderson, M. O., & Papadopoulos, M. C. (2014). Aquaporins: Important but elusive drug targets. *Nature Reviews. Drug Discovery*, *13*, 259–277.

Verkman, A. S., Hara-Chikuma, M., & Papadopoulos, M. C. (2008). Aquaporins—New players in cancer biology. *Journal of Molecular Medicine*, *86*, 523–529.

Verkman, A. S., Ruiz-Ederra, J., & Levin, M. H. (2008). Functions of aquaporins in the eye. *Progress in Retinal and Eye Research*, *27*, 420–433.

Verma, R. K., Prabh, N. D., & Sankararamakrishnan, R. (2014a). *Intra-helical salt-bridge and helix destabilizing residues within the same helical turn: Role of functionally important loop E half-helix in channel regulation of major intrinsic proteins*. Manuscript submitted.

Verma, R. K., Prabh, N. D., & Sankararamakrishnan, R. (2014b). New subfamilies of major intrinsic proteins in fungi suggest novel transport properties in fungal channels: Implications for the host-fungal interactions. *BMC Evolutionary Biology*, *14*, 173.

Voets, T., Droogmans, G., Wissenbach, U., Janssens, A., Flockerzi, V., & Nilius, B. (2004). The principle of temperature-dependent gating in cold- and heat-sensitive TRP channels. *Nature*, *430*, 748–754.

Wallace, I. S., & Roberts, D. M. (2004). Homology modeling of representative subfamilies of Arabidopsis major intrinsic proteins. Classification based on the aromatic/arginine selectivity filter. *Plant Physiology*, *135*, 1059–1068.

Wu, B., & Beitz, E. (2007). Aquaporins with selectivity for unconventional permeants. *Cellular and Molecular Life Sciences*, *64*, 2413–2421.

Wudick, M. M., Luu, D. T., & Maurel, C. (2009). A look inside: Localization patterns and functions of intracellular plant aquaporins. *New Phytologist*, *184*, 289–302.

Xu, H., Cooke, J. E. K., & Zwiazek, J. J. (2013). Phylogenetic analysis of fungal aquaporins provide insight into their possible role in water transport of mycorrhizal associations. *Botany-Botanique*, *91*, 495–504.

Yellen, G. (2002). The voltage-gated potassium channels and their relatives. *Nature*, *419*, 35–42.

Yukutake, Y., Hirano, Y., Suematsu, M., & Yasui, M. (2009). Rapid and reversible inhibition of aquaporin-4 by zinc. *Biochemistry*, *48*, 12059–12061.

Yukutake, Y., & Yasui, M. (2010). Regulation of water permeability through aquaporin-4. *Neuroscience*, *168*, 885–891.

Zardoya, R. (2005). Phylogeny and evolution of the major intrinsic protein family. *Biology of the Cell*, *97*, 397–414.

Zeidel, M. L., Ambudkar, S. V., Smith, B. L., & Agre, P. (1992). Reconstitution of functional water channels in liposomes containing purified red-cell CHIP28 protein. *Biochemistry*, *31*, 7436–7440.

Zhou, H. X., & Cross, T. A. (2013). Modeling the membrane environment has implications for membrane protein structure and function: Influenza A M2 protein. *Protein Science*, *22*, 381–394.

CHAPTER TWENTY-FOUR

Comparative Sequence–Function Analysis of the Major Facilitator Superfamily: The "Mix-and-Match" Method

M. Gregor Madej[1]

Department of Physiology, David Geffen School of Medicine, University of California Los Angeles, Los Angeles, California, USA

[1]Corresponding author: e-mail address: gregor.madej@gmail.com

Contents

Abstract

The major facilitator superfamily (MFS) is a diverse group of secondary transporters with members found in all kingdoms of life. The paradigm for MFS is the lactose permease (LacY) of *Escherichia coli*, which has been the test bed for the development of many methods applied for the analysis of transport proteins. X-ray structures of an inward-facing conformation and the most recent structure of an almost occluded conformation confirm many conclusions from previous studies. One fundamentally important problem for understanding the mechanism of secondary active transport is the identification and physical localization of residues involved in substrate and H^+ binding. This information is exceptionally difficult to obtain with the MFS because of the broad sequence diversity among the members. The increasing number of solved MFS structures has led to the recognition of a common feature: inverted structure-repeat, formed by fused

Methods in Enzymology, Volume 557
ISSN 0076-6879
http://dx.doi.org/10.1016/bs.mie.2014.12.015

triple-helix domains with opposite orientation in the membrane. The presented method here exploits this feature to predict functionally homologous positions of known relevant positions in LacY. The triple-helix motifs are aligned in combinatorial fashion so as to detect substrate and H^+-binding sites in symporters that transport substrates, ranging from simple ions like phosphate to more complex disaccharides.

1. INTRODUCTION

Regions of functional or structural similarity in proteins are often identified by arranging the sequences of amino acids to align conserved side chains. Proteins that are related by homologous sequences are expected to exhibit functional, structural, or evolutionary relationships. This principle provides the motivation to define protein phylogenetic relationships to correlate specific residues with function. However, this is particularly difficult for the major facilitator superfamily (MFS), one of the largest groups of secondary active transporters conserved from archaea, bacteria, and plants to humans (Pao, Paulsen, & Saier, 1998; Reddy, Shlykov, Castillo, Sun, & Saier, 2012; Saier et al., 1999). The sequence identity ranges around only 12–18% and regions of functional similarity (e.g., substrate- or H^+-binding sites) align for only very closely related MFS transporters (Kasho, Smirnova, & Kaback, 2006; Sugihara, Smirnova, Kasho, & Kaback, 2011). A hydrophobic amino acid content of 60–70% of most MFS members, high α-helix content and the inherent symmetry of the proteins with regard to helix kinks and bends provide extended nonspecific overlapping of residues and probably account for the reported similarities (Madej, Dang, Yan, & Kaback, 2013). MFS proteins catalyze the transport of a wide range of substrates including ions, carbohydrates, lipids, amino acids and peptides, nucleosides, and other small molecules (Reddy et al., 2012; Saier, 2000; Saier et al., 1999); therefore, the side chains involved in substrate binding vary among the MFS members, further concealing functional homology. Moreover, in many instances, MFS transporters catalyze active transport by transducing the energy stored in an H^+ or Na^+ electrochemical gradient into a concentration gradient of the substrates (symport); however, some members in this superfamily catalyze transport by antiport of cytosolic and cellar substrates or by facilitated diffusion (uniport), possibly increasing the sequence diversity. Thus, it seems not surprising that in less closely related transporters, residues with similar functional roles are often difficult to discern by sequence alignments.

1.1 Structural organization of MFS

In the past decade, a clear picture of about a dozen MFS structures has been unveiled by X-ray crystallography (Abramson et al., 2003; Chaptal et al., 2011; Dang et al., 2010; Deng et al., 2014; Doki et al., 2013; Ethayathulla et al., 2014; Guan, Mirza, Verner, Iwata, & Kaback, 2007; Guettou et al., 2013; Huang, Lemieux, Song, Auer, & Wang, 2003; Iancu, Zamoon, Woo, Aleshin, & Choe, 2013; Kumar et al., 2014; Mirza, Guan, Verner, Iwata, & Kaback, 2006; Newstead et al., 2011; Parker & Newstead, 2014; Pedersen et al., 2013; Quistgaard, Low, Moberg, Tresaugues, & Nordlund, 2013; Solcan et al., 2012; Sun et al., 2012, 2014; Yan et al., 2013; Yin, He, Szewczyk, Nguyen, & Chang, 2006; Zheng, Wisedchaisri, & Gonen, 2013). Despite the low sequence similarity and distinct substrate specificities, the structures exhibit a common structural fold—the MFS fold (Abramson, Kaback, & Iwata, 2004; Abramson et al., 2003; Huang et al., 2003). In most MFS transporters, the 400–600 amino acyl residues comprising the proteins are arranged into two pseudo-symmetrical six-helix bundles (Pao et al., 1998; Reddy et al., 2012; Saier et al., 1999) and the two bundles contain two inverted-topology repeats of three-helix bundles each (Radestock & Forrest, 2011). The two six-helix bundles form a deep hydrophilic central cavity and many residues postulated to play an important role in catalysis are located at the apex of this cavity (Yan, 2013).

1.2 Functionally relevant positions

Only for the lactose permease (LacY) from *Escherichia coli*, a galactoside/H^+ symporter, the complete Cys-scanning mutagenesis has been accomplished to determine which of the 417 residues play an obligatory role in the mechanism. A library of mutants with a single-Cys residue at each position of the molecule was created for structure–function studies; each residue was replaced individually with Cys in a functional mutant devoid of all eight native Cys residues (Frillingos, Sahin-Toth, Wu, & Kaback, 1998). The great majority of the single-Cys mutants are expressed normally in the membrane and catalyze accumulation of lactose against a significant concentration gradient, thereby demonstrating that Cys replacement at most positions does not induce severe perturbations in the structure of LacY or in the symport mechanism. It is striking that fewer than 10 residues are irreplaceable for active lactose transport: Glu126 (helix IV) and Arg144 (helix V), are critical for substrate binding, as well as Trp151 (helix V) where an aromatic side chain is essential; Glu269 (helix VIII), which cannot even be replaced with Asp without markedly decreasing affinity (Smirnova, Kasho, Sugihara, Choe, & Kaback, 2009) is the acceptor of hydrogen bonds from the galactopyranosyl ring, making it a likely candidate for a primary role in specificity (Kumar et al., 2014). Furthermore, the irreplaceable residues, His322 (helix X) and Asn272 (helix VIII) are also ligands to the sugar (Jiang, Villafuerte, Andersson, White, & Kaback, 2014; Kumar et al., 2014). Tyr236 (helix VII) acts as a hydrogen-bond donor to His322 stabilizing its conformation (Kumar et al., 2014). All replacements for these residues result in a protein with markedly decreased binding affinity and little or no transport activity (Jiang et al., 2014; Smirnova et al., 2009), thereby demonstrating that they are essential constituents of the sugar-binding site in LacY. Arg302 (helix IX) and Glu325 (helix X) are likely exclusively involved in H^+ translocation—neutral replacements for Glu325 yield mutants that are defective in all transport reactions that involve lactose/H^+ symport but catalyze equilibrium exchange and/or counterflow as well as or better than WT (Carrasco, Antes, Poonian, & Kaback, 1986; Carrasco et al., 1989; Sahin-Tóth & Kaback, 2001), and replacement of Arg302 with Ser or Ala results in a similar phenotype (Sahin-Tóth & Kaback, 2001). Other important positions, although not directly associated with sugar or H^+ binding/transport, are positions promoting structural integrity (i.e., charged pairs). The Asp237 (helix VII)/Lys358 (helix XI) charge pair, which is important for membrane insertion and stability of LacY, is a genuine salt bridge. Thus, Asp237 can be replaced with Glu, carboxymethyl-Cys, or sulfonylethylthio-Cys with Lys or Arg at position 358, and good activity is retained. Similarly, LacY tolerates replacement of Lys358 with Arg or ammoniumethylthio-Cys with Asp or Glu at position 237. Remarkably, LacY with Lys, Arg, or ammoniumethylthio-Cys in place of Asp237 is highly active when Lys358 is replaced with Asp or Glu, thereby demonstrating that the polarity of the charge interaction can be reversed without loss of activity (Dunten, Sahin-Toth, & Kaback, 1993; Sahin-Toth & Kaback, 1993). In contrast, the Asp68 (helix II)/Lys131 (helix IV) charge pair, which is involved in the coordinated opening and closing of the cytoplasmic and periplasmic cavities, exhibits marked sensitivity to changes in the nature or polarity of the side chains. Even a short extension of Asp residue by replacement with Glu inactivates transport and activity is not rescued by replacing Lys131 with Arg, nor will this pair tolerate interchange (Liu, Madej, & Kaback, 2010).

1.3 Helix interactions

The overall structures of MFS members show that the transporters consist of irregular, highly kinked helices, and residues that introduce structural irregularity (i.e., Pro, Gly) aligned in many kinked regions of the helices (Madej, Soro, & Kaback, 2012; Radestock & Forrest, 2011). However, most replacements of the 12 Pro residues in LacY with Gly, Ala, or Leu and deletion of Pro403 and Pro405 in the C-terminal tail are well tolerated (Consler, Tsolas, & Kaback, 1991). With the exception of Pro28, it is found that primarily the hydrophobicity and/or size of the side chain is significant for efficient transport function. Similar studies have been performed on the Gly residues in LacY (Jung, Jung, Colacurcio, & Kaback, 1995). The results indicate that although none of the Gly residues in LacY is mandatory for activity, the bulk of the side chain, rather than conformational flexibility is particularly important.

The tight interactions between N- and C-terminal six-helix bundles are facilitated by two pairs of Gly residues on the periplasmic side of LacY, located at the ends of helices II and XI (Gly46 and Gly370, respectively) and helices V and VIII (Gly159 and Gly262, respectively). When Gly46 and Gly262 are replaced with bulky Trp residues, transport activity is abrogated with little or no effect on galactoside binding (Kasho et al., 2006). Moreover, site-directed alkylation and stopped-flow binding kinetics indicate that the G46W/G262W mutant is open on the periplasmic side (Smirnova, Kasho, Sugihara, & Kaback, 2013). However, the recently solved crystallographic structure of LacY double-Trp mutant (G46W/G262W) with a bound, high-affinity lactose analog shows the protein in an almost occluded, outward-open conformation (Kumar et al., 2014). The presence of Trp residues at these positions causes the periplasmic cavity, which is closed in absence of substrate, to open. But when galactoside binding occurs and the mutant attempts to transition into an occluded state, it cannot do so completely because the bulky Trp residues at positions 46 and 262 block complete closure. These results indicate that the tight interactions between N- and C-terminal six-helix bundles are critical for the turnover, and indeed the Gly-Gly pairs are conserved in closely related MFS transporters (Kasho et al., 2006).

Although a substantial progress has been made in determining three-dimensional structures of MFS members, the mechanism of transport remains unclear in many cases. One fundamentally important problem for understanding the mechanism of active or passive transporters is the identification and physical localization of residues involved in substrate and ion binding. A conventional, linear sequence alignment is very successful for detecting functionally homologous positions in MFS transporters, which are closely related (Kasho et al., 2006; Pedersen et al., 2013; Sugihara et al., 2011; Sun et al., 2012). However, for a growing number of crystallographic structures with only distant evolutionary relationship the identification of functionally related residues by a conventional sequence alignment fails—residues with similar functional roles are often at disparate locations in the proteins (Madej et al., 2013; Madej & Kaback, 2013; Madej, Sun, Yan, & Kaback, 2014). Developing a concept that relates these positions would enable the systematic prediction of functionally homologous residues, which can be tested experimentally. The purpose of this chapter is to describe a method where a combinatorial sequence alignment of pseudo-symmetrical structure motifs (helix-triplets) leads to an improved alignment of functionally related side chains.

2. HELIX-TRIPLETS

The concept of combinatorial sequence alignments was inspired by a series of observations regarding structure–function relationships and the evolution of proteins. Initial experiments indicated that Asp46 mutants of L-fucose permease (L-fucose/H^+ symporter; FucP) from *E. coli*, a distantly related MFS member (~10% sequence identity to LacY), do not catalyze L-fucose/H^+ symport, but the D46A and D46N mutants catalyze counterflow (Dang et al., 2010), a reaction that does not involve H^+ translocation. Furthermore, Asp46 mutants exhibit L-fucose-induced Trp fluorescence quenching, and the mutants bind with essentially WT affinity (Sugihara, Sun, Yan, & Kaback, 2012); a conclusion consistent with isothermal calorimetry measurements (Dang et al., 2010). In this respect, Asp46 mutants exhibit the main characteristics of neutral replacement mutants of Glu325 in LacY, which are also unable to catalyze lactose/H^+ symport, but catalyze counterflow, as well as equilibrium exchange (reviewed in Madej, 2014; Madej & Kaback, 2014). However, Glu325 is located in the C-terminal six-helix bundle of LacY, at a different position from Asp46 in FucP, which is in the N-terminal six-helix bundle (Dang et al., 2010). Moreover, the recent crystallographic model of FucP was obtained with the central cavity open to the periplasm, an outward-facing conformation opposite to that of LacY (Dang et al., 2010). It was demonstrated that the inverted-topology repeats (Fig. 1; repeat-A, H1–H3; repeat-B, H4–H6; repeat-C, H7–H9; and repeat-D, H10–H12) encode all the structural information required

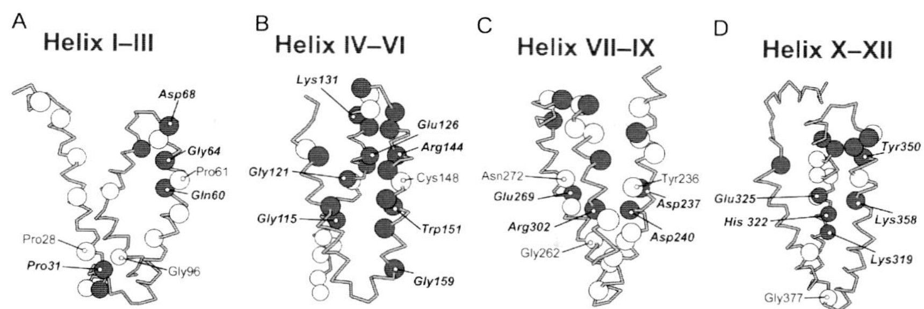

Figure 1 Location of functionally relevant side chains in LacY helix-triplets. Cys-scanning mutagenesis has revealed functionally relevant positions (Table 2; shown as spheres); positions of Cys-mutants with 0–25% residual activity of WT are shown as dark gray spheres, and positions with 25–50% residual activity are shown as white spheres. Some specific positions of relevant positions are indicated for orientation.

for the generation of the inward- and outward-facing conformers. Using a modeling strategy that involves swapping the three-dimensional structures of the inverted-topology repeats, a transformation between the inward- and outward-facing conformations for LacY was obtained, and the resulting model agrees well with the comparative models of outward-open LacY build based on the crystallographic coordinates of FucP (Radestock & Forrest, 2011). This postulate is in line with the idea, based on sequence analysis, that each of the pseudo-symmetrical structure motifs may have originated from a gene duplication event and a subsequent fusion of the genes (Hvorup & Saier, 2002). Remarkably, the topology of the inverted-topology repeats (triple-helix bundles) is similar to that of the monosaccharide and disaccharide transporters, the SemiSWEETs and SWEETs (Chen et al., 2010; Xuan et al., 2013). The recently solved X-ray structures show that the two triple-helix bundles in SemiSWEETs and SWEETs are arranged in tandems with parallel configuration to produce the respective 6- and 6+1-membrane-helix pores (Xu et al., 2014). The four triple-helix bundles in the MFS transporters are arranged in two tandems with an alternating antiparallel configuration. This observation points to two possibilities: that SWEETs and MFS transporters evolved from an ancestral triple-helix bundle or that the triple-helix bundle represents convergent evolution (Xu et al., 2014).

Sequence analysis of the MFS predicts that the majority of the members comprise 12 transmembrane helices, with some containing more (Jin, Guffanti, Beck, & Krulwich, 2001; Reddy et al., 2012), and the crystallographic structures identified the triple-helix motif in the two pseudo-symmetrical six-helix bundles as the most common topological feature of

MFS transporters. Thus, suggesting that the MFS members likely evolved from primordial helix-triplets that formed functional transporters (Hvorup & Saier, 2002); however, the functional segments assembled in a different consecutive order (Madej et al., 2013). This idea suggests a simple, parsimonious chain of events that may have led to the enormous sequence diversity within the MFS.

3. DETECTION OF FUNCTIONALLY HOMOLOGOUS POSITIONS

As a result of the strong symmetry within LacY and FucP, all superimpositions of the inverted-topology repeats (Fig. 2A) display similar qualities with root mean square deviations (rmsds) ranging from ~1.8 to ~2.6 Å (Table 1). Clearly, a comparison of the tertiary structures does not discriminate defective alignments since the triple-helix motif is universal to MFS members. Although the use of sequence alignments to identify conserved residues has provided many important clues about their function, this goal is difficult to achieve for MFS. In MFS, the sequence alignments are biased by the high content of hydrophobic residues that are in most cases not directly involved in the mechanism, but overlap nonspecifically in the alignments (Madej & Kaback, 2013). In contrast, polar side chains are rare within the α-helical core of membrane proteins because many polar groups that are removed from the contact with water require stabilization by interactions with other polar groups and contacts with the hydrophobic tails of lipids disturb the overall stability (Beuming & Weinstein, 2004; Isom, Cannon, Castaneda, Robinson, & Garcia-Moreno, 2008; White & Wimley, 1999).

3.1 Sequence simplification

Very few residues are absolutely irreplaceable in LacY (Guan & Kaback, 2006), but Cys substitution at in total 92 positions has a significant effect on activity (Table 2), inhibiting the steady-state level of accumulation by >50% (Frillingos et al., 1998). Although for many of the 92 side chains the specific function has yet to be established, the spatial arrangement of these positions represents the signature mold of functionally relevant positions (Fig. 1). The recently solved structures for xylose, phosphate, and peptide/H^+ symporters (XylE, PiPT, and $PepT_{Gk}$, respectively) in the substrate-bound state (Doki et al., 2013; Pedersen et al., 2013; Sun et al., 2012) allow the identification of residues involved in substrate binding. A combinatorial supposition of the helix-triplets on LacY exhibits that polar residues involved in substrate binding of these transporters are located

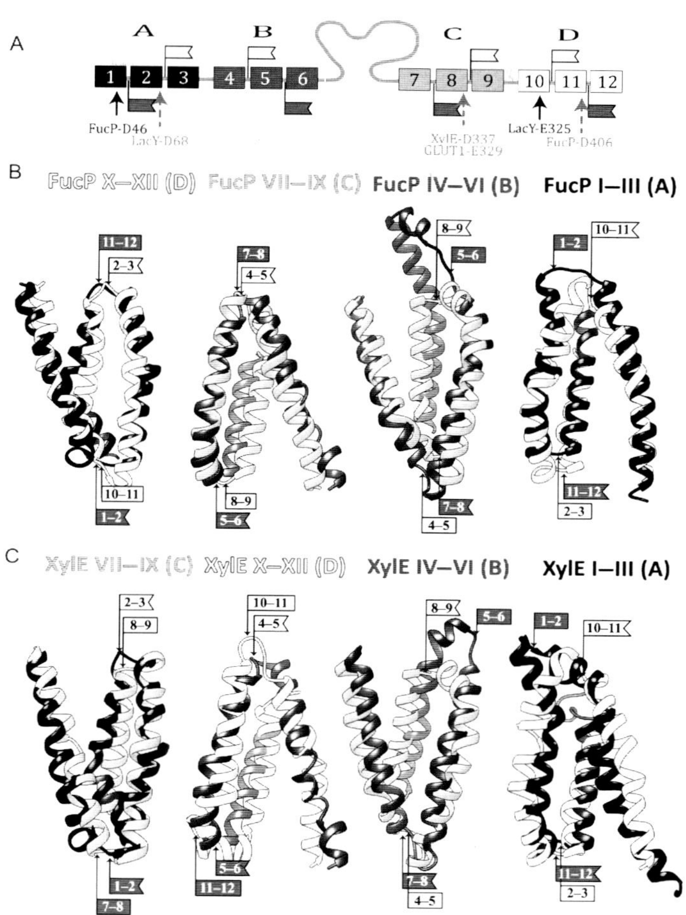

Figure 2 Alignment. (A) Helices represented by boxes are shown in consecutive order in LacY, and are colored according to helix-triplets (helices 1–3, black; helices 4–6, dark gray; helices 7–9, light gray; helices 10–12, white). Arrows of the same color indicate the positions of correlating residues. (B) Helix-triplets from FucP and LacY are aligned (helices 1–3, black; helices 4–6, dark gray; helices 7–9, light gray; helices 10–12, white). (C) Helix-triplets from XylE and LacY are aligned (color coding same as in A and B). Helix-triplets from LacY and XylE are colored as in A. The alignments are oriented with the LacY cytoplasmic side at the top. The flags indicate loops within triple-helix motifs (white, cytoplasmic loop; gray, periplasmic loop; swallowtail flags indicate LacY loops and rectangular flags indicate the aligned loops of FucP or XylE). The numbers on the flags indicate the two helices that are connected by the respective loop.

Table 1 Alternative suppositions of FucP

	FucP			
LacY	**A**	**B**	**C**	**D**
A	2.5	2.5	2.2	2.5
B	2.3	2.1	1.8	2.2
C	2.5	1.8	1.9	2.1
D	2.4	2.2	2.2	2.4

Structure symmetry motifs on LacY structure symmetry motifs (A, helix-triplet H1–H3; B, helix-triplet H4–H6; C, helix-triplet H7–H9; D, helix-triplet H10–H12). Respective rmsds of the Cα atoms are expressed in Å.

at different positions in the protein but at overlapping positions in the structural topology repeats (helix-triplets) with respect to LacY (Madej & Kaback, 2013). However, as a consequence of the very disparate substrate specificities, the side chains themselves vary as expected in these transporters.

Similarity of proteins can be assessed using correlation tables, where the 20 canonical amino acids are categorized according to a quality or a combination of qualities. Generally, the simplest similarity categorization of side chains is performed based on hydrophobicity: hydrophobic residues or polar residues (Chan, 1999). One of the advantages of such simplification matrices is the reduction of the protein sequence complexity (simplification of the sequence) with the objective to retain most information for the simplified protein sequence compared with the parent sequence using global sequence alignment (Cieplak, Koplik, & Banavar, 2001; Murphy, Wallqvist, & Levy, 2000; Solis & Rackovsky, 2000).

In order to quantify position similarities of 92 functionally relevant positions in LacY, in different alignments to other MFS proteins, side chains were sorted into similarity groups (reduction clades, Table 3); reasonable sequence simplification is achieved using the reduction alphabets according to Li et al. (2003) modified with regard to the relative occurrence in α-helical membrane proteins, the packing value in α-helical membrane proteins and the hydrophobicity (Eilers, Patel, Liu, & Smith, 2002)—clustering aminoacid types with similar parameters are classified as similar (Madej & Kaback, 2013). The main difference to Li et al. (2003) is that His is grouped with positively charged residues Arg and Lys instead of with Asn in the first reduction level. Additionally, Ala and Cys are grouped with Ser and Thr based on similar helix packing and hydrophobicity and with small side chains (Gly, Pro) in the last reduction level (Table 3). The scoring matrix (Table 4) is generated by assigning score values depending on the affiliation of a

Table 2 Functionally relevant and critical positions in LacY

Res.	Triplet	Res.	Triplet	Res.	Triplet	Res.	Triplet
Y3	*n*	I103	*n*	F185	*n*	**K319**	**D**
F9	**A**	S107	**B**	A213	*n*	**<u>H322</u>**	
F15		I108		L225	**C**	F324	
F20		G111		**Y228**		**<u>E325</u>**	
P28		**G115**		V229		P327	
P31		A120		G231		F328	
H35		**G121**		Y236		G332	
D36		**P123**		**<u>D237</u>**		**F334**	
I52		**<u>E126</u>**		**D240**		**K335**	
L57		**E130**		F243		A347	
Q60		**K131**		A244		**T348**	
P61		S133		F247		**I349**	
G64		R134		F251		**Y350**	
S67		**S136**		T258		F354	
D68		**E139**		<u>G262</u>		**<u>K358</u>**	
L76		**F140**		G268		G377	
L84		**G141**		**<u>E269</u>**		L385	
G96		**<u>R144</u>**		<u>N272</u>		L400	
N102		**G147**		**P280**			
		C148		N284			
		G150		**G287**			
		<u>W151</u>		G288			
		C154		**K289**			
		<u>G159</u>		A291			
		A177		L292			
		L184		**G296**			
				<u>R302</u>			
				I304			

Functionally relevant positions, where mutations to Cys cause >50% inhibition of the transport rate with Cys-less LacY are shown and critical positions where mutations to Cys cause >75% inhibition are shown in bold font. Residues for which specific functions are suggested are underlined. Residues in loop-regions are marked with "*n*."

Table 3 Reduction matrix for sequence simplification

Reduction levels	Reduction clades																			
	L	I	V	F	Y	W	M	G	P	A	T	S	C	N	Q	E	D	H	R	K
	LIV			F	YW		M	GP		ATS			C	NQ		ED		H	RK	
	LIVF				YW		M	GP		ATSC				NQED				HRK		
	LIVF				YWM			GPATSC						NQEDHRK						

Grouping of residues based on reasonable simplification of protein sequence with no interlacing for different levels of reduction. Letter grouped without spaces represent a reduction clade and are treated as similar in scoring of sequence alignments (Table 4).

Modified from Li, Fan, Wang, and Wang (2003).

Table 4 Scoring matrix

	K	R	H	D	E	Q	N	C	S	T	A	P	G	M	W	Y	F	V	I	L
L	0	0	0	0	0	0	0	0	0	0	0	0	0	0	0	0	0	0	0	0
I	0	0	0	0	0	0	0	0	0	0	0	0	0	0	0	0	0	0	0	
V	0	0	0	0	0	0	0	0	0	0	0	0	0	0	0	0	0	0		
F	0	0	0	0	0	0	0	0	0	0	0	0	0	0	0	0	0			
Y	0	0	0	0	0	0	0	0	0	0	0	0	0	6.6	10	20				
W	0	0	0	0	0	0	0	0	0	0	0	0	0	6.6	20					
M	0	0	0	0	0	0	0	0	0	0	0	0	0	20						
G	0	0	0	0	0	0	0	3.3	3.3	3.3	3.3	10	20							
P	0	0	0	0	0	0	0	3.3	3.3	3.3	3.3	20								
A	0	0	0	0	0	0	0	5	6.6	6.6	20									
T	0	0	0	0	0	0	0	5	6.6	20										
S	0	0	0	0	0	0	0	5	20											
C	0	0	0	0	0	0	0	20												
N	2.8	2.8	2.8	5	5	10	20													
Q	2.8	2.8	2.8	5	5	20														
E	2.8	2.8	2.8	10	20															
D	2.8	2.8	2.8	20																
H	6.6	6.6	20																	
R	10	20																		
K	20																			

The score is calculated as the reciprocal of noninterlacing pairing probability in the reduction clade. The number of events is the number of objects in the reduction clade (Table 3), and the number of outcomes is the total number of possible residue type. Residues L, I, V, and F are excluded and are scored 0.

residue pair to the same reduction clade with the lowest reduction level (Table 3). The score values are the reciprocals of noninterlacing pairing probability (quotient of pairing events and number of outcomes) in the reduction clade. For example, the alignment pair Ala-Ala returns a score of 20 (1 out of 20) but the pair Ala-Cys returns a score of 5 (4 out of 20). The hydrophobic residues Leu, Ile, Val, and Phe are ignored and receive always the score 0. The similarity score (Q) is calculated as the quotient of the sum of the individual scoring values and the theoretical score for identical sequences (Eq. 1).

3.2 Functional similarity

Q is the similarity score, V_{all} is the score assigned from the scoring matrix (Table 4), AA_{all} is the number of functionally relevant positions in the alignment (Table 2), and AA_{LIVF} is the number of not compared side chains (L, I, V, and F). For 100% conservation, Q should equal to 1.

$$Q = \frac{\Sigma V_{\text{all}}}{(AA_{\text{all}} - AA_{\text{LIVF}})20}. \quad (1)$$

The combinatorial alignment of the symmetry motifs relies on mapping known functional markers (Fig. 1) in the helix-triplets of LacY followed by inferential mapping of unknown functional positions in other MFS members. The triple-helix units are aligned individually by pairwise sequence alignments of the protein fragments using the blocks substitution matrix 62 (BLOSUM-62, or lower BLOSOM matrices for evolutionary more distant proteins) and the Needleman–Wunsch algorithm (Needleman & Wunsch, 1970); in addition, the coordinates of the protein fragments (if available) are aligned using a secondary structure term, analogous to residue similarity, but the values depend on the type of secondary structure where the residues are found (helix, strand, or other) weighted relative to residue similarity term [total score = 0.30 (residue similarity score from BLOSUM-62) + 0.70 (secondary structure score) − gap penalties] (Pettersen et al., 2004). Subsequently, the alignment combinations are tested for functional similarity according to Eq. (1). All alignments are inspected visually and compared to experimental data for consistence.

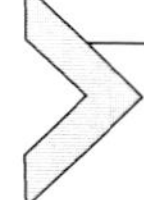

4. EXAMPLES

4.1 FucP

Although the structure suppositions of FucP and LacY helix-triplets do not display any significant differences (Table 1), the analysis of the respective sequence alignments reveal for particular helix-triplet pairs a significantly higher similarity score (Table 5). The helix-triplet A of LacY shows highest scoring value in combination with FucP triplet D, FucP triplets B and C reveal highest similarity values with LacY-C and -B, respectively, and LacY-D scores highest value with triplet A among the FucP triplets. Thus, a meaningful sequence alignment between LacY and FucP is obtained if the order of the triple-helix motifs is altered (Fig. 2A). This alignment was tested based on the phenotype of 34 mutants in FucP (Fig. 3); the transport activity of

Table 5 Combinatorial alignment of LacY and FucP symmetry motifs

	LacY-A	**LacY-B**	**LacY-C**	**LacY-D**
FucP-A	0.45	0.24	0.19	**0.38**
FucP-B	0.26	0.25	**0.38**	0.16
FucP-C	0.15	**0.40**	0.14	0.20
FucP-D	**0.51**	0.24	0.19	0.30

The values represent the similarity score (Q) according to Eq. (1). The comparison indicates the D–C–B–A sequence order with respect to LacY (indicated in bold font).

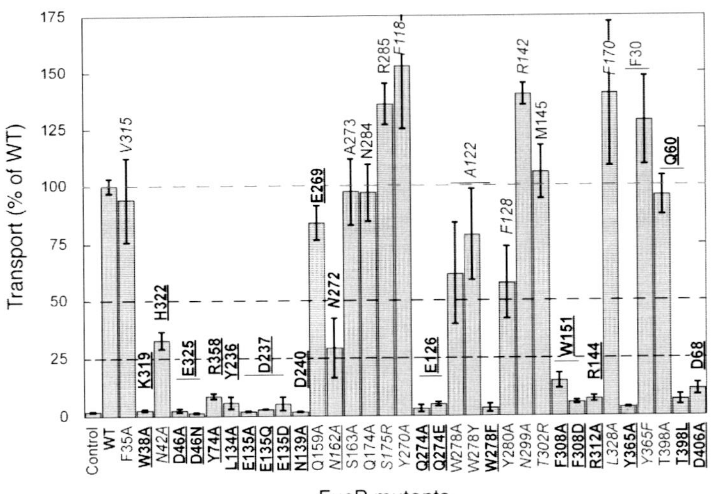

Figure 3 Transport activity of FucP mutants. Results of a cell-based *in vivo* transport, active transport assay, with different FucP mutants are shown and compared to results obtained with Cys-mutants of Cys-less LacY (Madej et al., 2013). Transport activities of the FucP mutants are means of four measurements and are normalized to WT. Control refers to transport by FucP-deficient *E. coli* cells transformed with empty vector. Error bars represent stabdatr deiation (SDs). Residual activity of WT is indicated: 0–25%, bold, underlined; 25–50%, underlined, italics; 50–100%, regular; >100%, italics. On top of given histograms, the activities of the corresponding Cys-mutants of LacY are indicated for comparison.

the corresponding positions in LacY and FucP reveals equivalent susceptibility to mutations (Madej et al., 2013).

The superimposition of helix-triplet LacY-D on FucP-A, where Asp46 of FucP (helix I) is located, reveals the conservation of many functional residues. In addition to Asp46 in FucP, which is located in the same position as Glu325 (helix X) in LacY, Asn42 (helix I) is in the same position as His322 (helix X, LacY), and Trp38 (helix I) aligns with Lys319 (helix X, LacY). As in LacY, mutation of these residues in FucP severely cripple transport (Fig. 3), thereby implicating the residues as participants in substrate binding or other functionally relevant processes (Madej & Kaback, 2014). In the superposed helix-triplets FucP-D and LacY-A, Asp406 (helix XI) of FucP aligns with Asp68 (helix II) in LacY. Asp68 interacts with Lys131 (helix IV) in LacY, forming an essential charge pair between helices II and IV. Replacement of Asp68 blocks a conformational transition involving the two helices, which inactivates transport (Liu et al., 2010). In FucP, Asp406 is charge paired with Arg283 (helix VII) in helix-triplet FucP-C, and transport is severely inhibited in FucP mutant D406A (Fig. 3; Madej et al., 2013). Thus, the Arg283/Asp406 pair corresponds structurally and probably functionally to the charge pair between helix-triplets A and B in LacY (Madej et al., 2013); however, in FucP, it is helix-triplets C and D that interact. The recently presented X-ray structure of LacY with bound galactopyranoside provided structural evidence for residues involved in sugar recognition and binding (Kumar et al., 2014). The indole ring of Trp151 (helix V) in helix-triplet B and the galactopyranosyl ring provide the primary hydrophobic interaction between LacY and substrate (Guan, Hu, & Kaback, 2003). Superimposition of FucP triplet C and LacY triplet B reveals that Phe308 (helix VIII) in FucP is at exactly the same position as Trp151 in LacY (Fig. 4). An aromatic side chain at position 151 in LacY, ideally Trp, is an absolute requirement for binding and transport (Guan et al., 2003). FucP mutants F308A and F308D transport very poorly, consistent with the possible functional equivalence of Phe308 to Trp151 in LacY (Fig. 3). Remarkably, the altered consecutive order results in inverted orientation of the helix-triplets in LacY and FucP with respect to each other—the cytoplasmic loops superpose on the periplasmic loops (Fig. 2A). Arg144 (helix V), which is positioned two helix turns from Trp151 toward the cytoplasm, is another essential sugar-binding residue in LacY (Sahin-Toth et al., 1999). In FucP, Phe308 is flanked by Arg312 (helix VIII) on the cytoplasmic side (Fig. 4), and, as in LacY, mutation of Arg312 drastically impairs transport (Fig. 3). Neutral replacement of

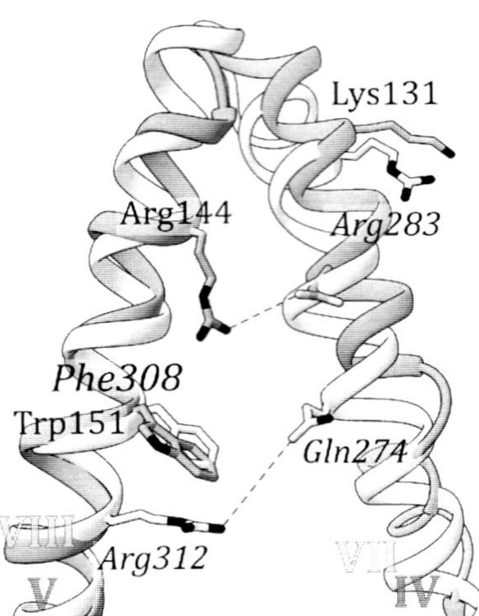

Figure 4 Superposition of triple-helix motifs. Superposition of FucP-C (light gray) on LacY-B (darker gray). Significantly equivalent residue pairs are shown as sticks (labels of LacY residues are shown in regular type and for FucP in italic type). A dashed line indicates interactions between residues Arg144, Glu126 in LacY and Arg312, and Gln274 in FucP.

Glu126 (helix IV) in LacY, also an irreplaceable residue in close proximity to Arg144, completely abolishes binding and transport of substrate (Sahin-Toth et al., 1999). In FucP, mutation of Gln274 (helix VII), which is close to Arg312, markedly impairs transport (Madej et al., 2013). Thus, it appears that the helix-triplets LacY-B and FucP-C contain many residues that are essential for sugar recognition and binding. However, the putative sugar-binding site in FucP seems to be arranged upside down within the helix-triplets, but with the same orientation with respect to the membrane relative to LacY (Madej et al., 2013).

The Gly-Gly motifs in LacY (Gly46/Gly370 and Gly159/Gly262), mentioned afore, promote tight interactions between helix-triplets A and D, as well as, B and C, respectively, on the periplasmic side of the molecule (Kumar et al., 2014; Smirnova et al., 2013). The Gly-Gly motifs are conserved in FucP and other MFS members (Kasho et al., 2006). However, in FucP, the Gly-Gly pairs seem to be located on the opposite side of the sugar-binding site but on the same site with respect to the membrane relative to LacY (Fig. 5). The detected similarities support the hypothesis that the helix-triplets may share common evolutionary origins. LacY and FucP might have evolved from primordial noncovalently fused helix-triplets that formed symporters. Modern symporters formed as functional mutations

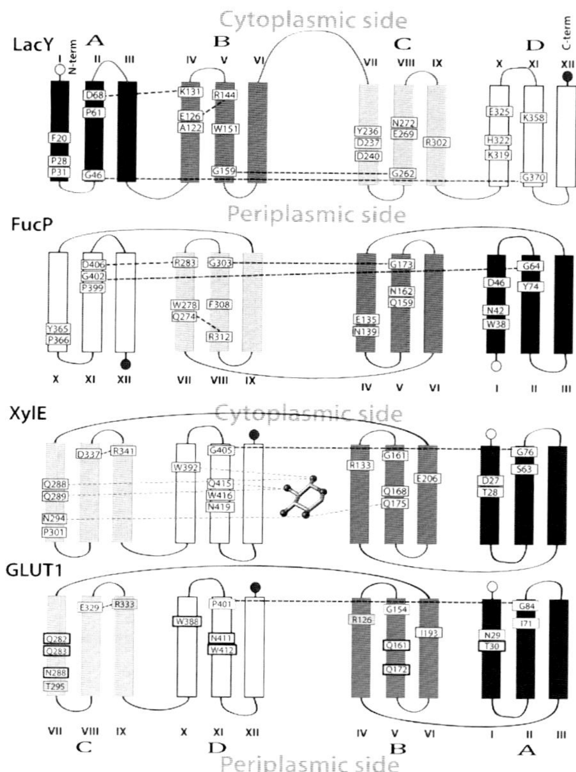

Figure 5 Schematic representation of the superimposition of the triple-helix motifs. Helix-triplets are numbered from A through D according to the sequence order and are colored as in Fig. 2. Overlapping side chain positions are aligned. Prominent contacts with the substrate or between side chains are indicated as broken lines. Bold boxes around residues in GLUT1 indicate tested positions, where a mutation to Cys significantly impairs the transport activity; gray background indicates positions implicated in a medical condition.

were being introduced and the functional segments assembled in a varying order resulting in altered consecutive orders of homologous helix-triplets (LacY: A–B–C–D, FucP: D–C–B–A).

4.2 XylE, PiPT, and GLUT1

The recently presented X-ray structures for XylE (xylose/H^+ symporter from *E. coli*) in the substrate-bound state allow the identification of residues involved in substrate binding (Sun et al., 2012). The overall architecture reviles a noticeable similarity of the substrate-binding residues located in

Table 6 Combinatorial alignment of LacY and XylE symmetry motifs

	LacY-A	LacY-B	LacY-C	LacY-D
XylE-A	0.28	0.28	0.26	**0.28**
XylE-B	0.07	0.13	**0.35**	0.13
XylE-C	**0.36**	0.20	0.20	0.14
XylE-D	0.22	**0.37**	0.09	0.18

The values represent the similarity score (Q) according to Eq. (1). The comparison indicates the C–D–B–A sequence order with respect to LacY (indicated in bold font).

the C-terminal six-helix bundle of XylE to the N-terminal portion of the substrate-binding site in LacY (Madej & Kaback, 2013). The examination of the symmetrically disposed triple-helix units of XylE and LacY for similarity conservation (Eq. 1) indicates positional shifts of the XylE helix-triplets in the sequence with respect to LacY. The superimposition of helix-triplets XylE-C and -D on LacY-A and -B, respectively, prompts the highest similarity scores (Table 6). In LacY, Glu269 (helix VIII) is the acceptor of hydrogen bonds from the C4–OH group of the galactopyranosyl ring, making it a likely candidate for a primary role in specificity (Kumar et al., 2014). Together with Asn272 (helix VIII), another essential residue for sugar binding (Jiang et al., 2014), the residues form a distinctive binding motif, which is likely represented in FucP by Gln159 (helix V) and Asn162 (helix V) (Madej et al., 2013). The corresponding residues Glu168 (helix V) and Gln175 (helix V) in XylE, which appear to be involved in substrate binding (Sun et al., 2012), are observed at the same positions in homologous symmetry motifs: LacY-C, FucP-B, and XylE-B (Madej & Kaback, 2013). In addition, aromatic side chains also play an important role in substrate binding in XylE. For example, in XylE, Trp416 and Asn415 (helix XI) align with Trp151 and Arg144 (helix V) in LacY. Asn415 of XylE is involved in a complex H-bound network with Gln289 (helix VII) and Asn294 (helix VII) that interacts with xylose (Sun et al., 2012). In LacY, Glu126 (helix IV) acts as H^+-bond acceptor from the hydroxyl group of the galactopyranoside and Arg144 (helix V) (Kumar et al., 2014), explaining why conservative replacement with Asp for Glu126 causes markedly diminished binding affinity and removal of the carboxyl group abolishes binding and transport (Sahin-Toth et al., 1999; Smirnova et al., 2009); as well as in FucP, mutation at the homolog position, Gln274, which is close to Arg312, impairs transport (Madej et al., 2013). Remarkably, this position aligns with Trp392 (helix X) in XylE, which coordinates xylose by providing H^+-bond of εN to C1–OH (Sun et al., 2012), rendering it functionally equivalent to Glu126 in LacY

(Madej & Kaback, 2013). It appears that the positions of most residue types interactions with the substrate are conserved in XylE, FucP, and LacY, although the residue types themselves vary. This principle is even more evident if the structure of XylE is compared with the recently presented structure of phosphate/H^+ symporter (PiPT) from the fungus *Piriformospora indica* with bound phosphate (Pedersen et al., 2013). PiPT shares the motif sequence order with XylE (C–D–B–A relative to LacY) and the superposition of the structures reveals a precise overlap of the substrate-binding sites (Fig. 6); however, the individual residue types are different (Madej & Kaback, 2013).

Remarkably, the sequence order of the motifs in LacY (A–B–C–D), FucP (D–C–B–A) and XylE or PiPT (C–D–B–A) is different and consequently the orientation of the helix-triplets differs with respect to the membrane. Unlike in FucP where the orientation of all four motifs is inverted with respect to LacY (Fig. 2A), in XylE and PiPT, only motifs A and B are inverted with respect to motifs C and D in LacY, thereby placing the C terminus of the first six-helix bundle (motifs A and B) and the

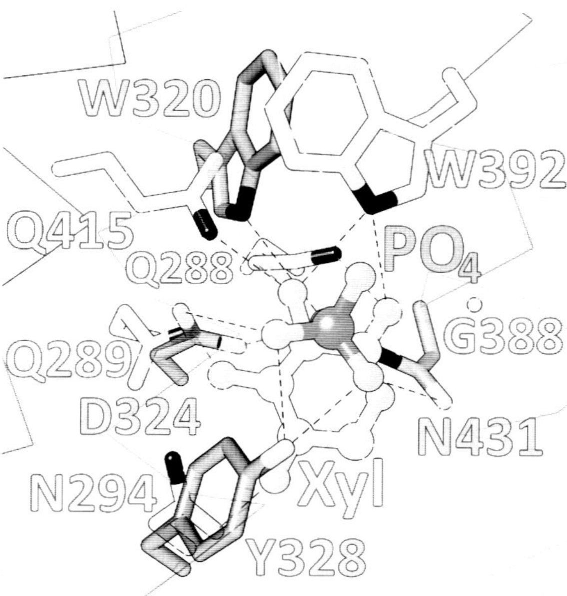

Figure 6 Overall architecture of the substrate-binding sites in XylE and PiPT. View on the C-terminal six-helix bundle of XylE (PDB ID: 4GBY; colored in light gray) is superposed on the C-terminal six-helix bundle of PiPT (PDB ID: 4J05; side chains shown colored in darker gray). A satisfying structure alignment is achieved (rmsd achieved: 2.6 Å, number of residues reference (4GBY): 475, number of residues moving (4J05): 422, number of aligned residues: 346, sequence identity: 19.9%)—the putative H^+-binding sites (not shown) and the substrate-binding residues overlap.

N terminus of the following six-helix bundle (motifs C and D) on opposite sides of the protein in the alignment of the structure motifs (Fig. 2B). This is not the case in the native XylE or PiPT; rather, in these proteins, the N-terminal six-helix bundle is rotated 180° with respect to the membrane plane relative to LacY. With a single substrate-binding site (Guan & Kaback, 2006), the primordial MFS symporters may have exhibited substantial flexibility with regard to the orientation of the helix bundles to each other. This is not surprising considering that substrate-binding sites are generally located at the approximate middle of these symporters. Therefore, inversion of triple-helix motifs may be well tolerated. In this context, although the antiporter EmrE is not an MFS protein, it is a dimer with the binding site placed at the dimer interface in the middle of the complex. Moreover, it has been demonstrated that the EmrE dimer is functional in both parallel and antiparallel configurations (Nasie, Steiner-Mordoch, Gold, & Schuldiner, 2010).

One objective of determining crystallographic models of the bacterial representatives is identification and physical localization of residues important for catalysis in transporters with medical relevance. XylE from *E. coli* is recognized as a homolog of human D-glucose transporters, the GLUTs (SLC2) (Maiden, Davis, Baldwin, Moore, & Henderson, 1987), although XylE catalyzes xylose/H^+ symport, and the active transport is driven by the H^+ electrochemical gradient ($\Delta\tilde{\mu}_{H^+}$); most of the GLUTs are uniporters (Mueckler & Thorens, 2013). H^+ symport in LacY is critically dependent on the carboxyl side chain of Glu325 (Carrasco et al., 1989), and like Glu325 mutants, mutants of Asp46 in FucP are specifically blocked in reactions that do involve H^+ translocation (Dang et al., 2010; Sugihara et al., 2012), indicating that these positions may share the same function (Madej et al., 2013). Combinatorial alignment of the triple-helix units D of LacY and A of XylE reveals that Glu325 (helix X) superposes with Asp27 (helix I) in XylE, suggesting functional homology (Madej & Kaback, 2013). Similar to the findings obtained with mutants of Glu325 in LacY or Asp46 in FucP, replacement of Asp27 with Ala or Asn in XylE results in mutants that do not catalyze xylose/H^+ symport, but binding, counterflow, and equilibrium exchange are totally unaffected (Madej et al., 2014). This supports the conclusion that Asp27 in XylE is functionally homologous to Glu325 in LacY possibly being the primary side chain involved in H^+ translocation. To draw functional conclusions with respect to the GLUTs, a homology model of GLUT1 was generated based on the crystallographic coordinates of XylE (Madej et al., 2014). The recent structural resolution of the human glucose transporter GLUT1 (Deng et al., 2014) verified the GLUT1 theoretical model (rmsd of the X-ray structure and the theoretical model:

N-terminal domains: 1.105 Å over 157 C atoms, C-terminal domains: 1.037 Å over 133 C atoms, and ICH domain in the middle loop: 0.959 Å over 28 C atoms). This indicates that the functional side chain correlations in GLUT1 concluded from positional similarity of corresponding residues in XylE, FucP, and LacY (Madej et al., 2014) are likely correct (Table 7).

Table 7 Functional side chain correlations in GLUT1 concluded from positional homology of corresponding residues in XylE, FucP, and LacY

Function	GLUT1	XylE	FucP	LacY
Substrate ligand	F26	F24	N42	H322
	Q161	Q168	Q159	E269
	I168	Q175	N162	N272
	Q282	Q288	–	F20
	Q283	Q299	–	–
	N288	N294	Y365	–
	W388 (A396[a])	W392	Q274	E126
	N411 (H419[a])	Q415	R312	R144
	W412	W416	F308	W151
H^+ site	N29 (D79[a])	D27	D46	E325
Structural	G31	T28	P50	P327
	G75/G76	G67/C68	G73/–	A361/M362
	G79/G134	G71/G141	E135/Y74	D237/K358
	S80	G72	I79	S366
	G84/P401	G76/G405	G402/G64	G370/G159
	G286/S414	G292/A418	A359/C307	G24/C154
	S294	P301	P366	P31
	G384	G388	G272	G121
	P385	P389	A276	P123
Salt-bridged	E329/R400	D337/R404	D406/R283	D68/K131
	R126/–/N29	R133/E206/D27	Y101/Y212/D45[#]	K319/D240/R302

[a]GLUT13.
[#]polar interaction; –, homologous residue not found.

As expected from the high sequence similarity, GLUT1 shares the motif sequence order with XylE (C-D-B-A relative to LacY); the sugar-binding site and other functionally relevant positions are well conserved (Deng et al., 2014). Moreover, the positions and in certain instances the function of the residues that cause greater than 50% inhibition of D-glucose transport activity in GLUT1 upon replacement with Cys (Alisio & Mueckler, 2004; Hruz & Mueckler, 1999, 2000; Mueckler & Makepeace, 2002, 2004, 2005, 2006, 2009; Mueckler, Roach, & Makepeace, 2004) are consistent with the results obtained by Cys-scanning mutagenesis of LacY (Fig. 5; Frillingos et al., 1998).

In most GLUTs, a highly conserved Asn (Asn29, helix I in GLUT1) or an Ala in GLUT6 and GLUT8 aligns with Asp27 of XylE or Glu325 in LacY. In GLUT13 (HMIT), which is unlike most other GLUTs $\Delta\widetilde{\mu}_{H^+}$ dependent, this position corresponds to Asp79 (helix I), indicating that active transport utilizing $\Delta\widetilde{\mu}_{H^+}$ might be critically dependent on protonation and deprotonation of a carboxyl side chain in this location. Moreover, the substrate affinity (K_D) of LacY is unaffected by the imposition of a $\Delta\widetilde{\mu}_{H^+}$ and K_D is virtually the same from both sides of the membrane (Guan & Kaback, 2004). Thus, it is suggested that the sugar binding involves induced fit, as the N- and C-terminal bundles converge when side chains from both helix bundles ligate the substrate (Madej et al., 2014). By this means, substrate binding may accounts for the energetic cost of the conformational transition, and the energy is regained upon dissociation. Thus, MFS transporters, including the GLUT uniporters, operate in an analogous fashion to enzymes, and the difference being that the protein rather than the substrate forms the transition state.

The crystallographic model of GLUT1 was derived from a double-mutant (Deng et al., 2014), where the glycosylation site Asn45 was eliminated to avoid heterogeneity, and the inherent conformational flexibility was lowered by introducing the mutation E329Q (Schürmann et al., 1997). Remarkably, Glu329 (helix VIII) in GLUT1 corresponds to Asp68 (helix II) in LacY, which interacts with Lys131 (helic IV) in LacY, forming an essential charge pair between helices II and IV as pointed out earlier (Liu et al., 2010). Replacement of Asp68 blocks a conformational transition involving the two helices in LacY. Site-directed alkylation of residues in the cytoplasmic and periplasmic cavity of LacY mutants D68E and D68N reveals that sugar binding to the Asp68 mutant causes closure of the cytoplasmic cavity, similar to WT LacY; however, strikingly different from WT, opening of the periplasmic cavity upon sugar binding is paralyzed in Asp68 mutants (Liu et al., 2010). Consistent with the predicted state for the mutant E329Q, GLUT1 was found in an inward-open conformation in the X-ray structure (Deng et al., 2014).

5. CONCLUSION

The paradigm of the MFS, lactose permease from *E. coli*, is arguably the most intensively studied secondary transporter at present (Guan & Kaback, 2006; Madej, 2014; Madej & Kaback, 2014; Smirnova et al., 2011). Each of the 417 aminoacyl side chains in LacY has been mutated (Frillingos et al., 1998), and only fewer than 10 residues are irreplaceable for active lactose transport. Moreover, an abundance of biochemical and spectroscopic data demonstrates that substrate binding causes the molecule to open reciprocally on either side of the membrane, thereby providing almost unequivocal evidence for an alternating-access model (Smirnova et al., 2011). As a result of intensive study over three decades, an overall mechanism for coupling in LacY has been worked out (Madej & Kaback, 2014):

- **(i)** Active lactose/H^+ symport, in the presence of the H^+ electrochemical gradient ($\Delta\widetilde{\mu}_{H^+}$), or passive equilibration, in the absence of $\Delta\widetilde{\mu}_{H^+}$, is the same reaction, but the rate-limiting steps differ; the limiting step in the absence of $\Delta\widetilde{\mu}_{H^+}$ is deprotonation, but the limiting step in the presence of $\Delta\widetilde{\mu}_{H^+}$ is likely the return of apo LacY to the other side of the membrane.
- **(ii)** Galactoside binding and dissociation, not $\Delta\widetilde{\mu}_{H^+}$, are the driving force for alternating access.
- **(iii)** LacY must be protonated to bind galactoside (the pK_a for binding is ~10.5).
- **(iv)** Binding of galactoside involves induced fit that prompts the transition to an occluded intermediate that undergoes alternating access.
- **(v)** Galactoside dissociates, releasing the energy of binding.
- **(vi)** Arg302 comes into proximity with protonated Glu325 causing deprotonation.

Structure models are critical but not sufficient to explain the catalysis of transport. Secondary transport is a dynamic process that can be described only partially by the static snapshots provided by X-ray crystallography. The clues to understanding transport mechanisms are based on the principles of enzyme kinetics—the functional characterization of strategically placed mutants is crucial in this respect. However, without structural information, it is virtually impossible to deduce the chemistry underlying ion-coupled transport. Although crystallographic coordinates are becoming increasingly available, functional data lags behind. The triple-helix motifs of LacY, FucP, PiPT (homolog of human organic ion transporters OCT1 and OAT3), and XylE (homolog of the GLUTs) align in a combinatorial fashion according to the "mix-and-match" method, thereby suggesting functionally homologous positions. Remarkably, substrate and H^+-binding sites in the corresponding helix-triplets of these symporters are found to be in similar locations, and in contrast to the conventional, linear sequence alignment, conserved motifs are observed (Madej & Kaback, 2013). It appears likely that the overall mechanism of transport, in particular, the ordered kinetic mechanism homologous to LacY might be operative in other H^+-coupled MFS symporters. These are certainly novel observations, and they must be tested by functional characterization of the mutants.

It seems apparent that a meaningful sequence alignment between MFS members can be obtained only if the triple-helix motifs are compared independently. The evolutionary relationship suggested by the structural similarity of the domains (Radestock & Forrest, 2011) is consistent with the presence of evolutionarily conserved sequence motifs. It appears that the evolutionary origins of these transporters are triple-helix bundles, which correspond to the inverted structural symmetry motifs, fused in tandem by intragenic duplication. An alternative although less dynamic scenario is that a multispanning protein evolved first, possibly for stability reasons, and functional mutations were introduced later (Vastermark, Lunt, & Saier, 2014; Vastermark & Saier, 2014). However, this scenario remains without experimental support. In any case, using the notion of a "mix-and-match" evolution involving the triple-helix motifs in MFS, a more useful annotation of sequence motifs than a conventional, linear sequence alignment may be appropriate.

ACKNOWLEDGMENTS

M. Gregor Madej is grateful to Avner Schlessinger (Icahn School of Medicine at Mount Sinai) for reading the manuscript and for helpful comments on the paper. M.G.M. is also grateful to H. Ronald Kaback (UC, Los Angeles) for helpful and critical discussions. This work was supported by National Institutes of Health Grants DK51131, DK069463, and GM073210; National Science Foundation Grant MCB-1129551 (to H. R. K.).

REFERENCES

Abramson, J., Kaback, H. R., & Iwata, S. (2004). Structural comparison of lactose permease and the glycerol-3-phosphate antiporter: Members of the major facilitator superfamily. *Current Opinion in Structural Biology*, *14*(4), 413–419.

Abramson, J., Smirnova, I., Kasho, V., Verner, G., Kaback, H. R., & Iwata, S. (2003). Structure and mechanism of the lactose permease of Escherichia coli. *Science*, *301*(5633), 610–615.

Alisio, A., & Mueckler, M. (2004). Relative proximity and orientation of helices 4 and 8 of the GLUT1 glucose transporter. *The Journal of Biological Chemistry*, *279*(25), 26540–26545.

Beuming, T., & Weinstein, H. (2004). A knowledge-based scale for the analysis and prediction of buried and exposed faces of transmembrane domain proteins. *Bioinformatics*, *20*(12), 1822–1835.

Carrasco, N., Antes, L. M., Poonian, M. S., & Kaback, H. R. (1986). *Lac* permease of *Escherichia coli*: Histidine-322 and glutamic acid-325 may be components of a charge-relay system. *Biochemistry*, *25*(16), 4486–4488.

Carrasco, N., Puttner, I. B., Antes, L. M., Lee, J. A., Larigan, J. D., Lolkema, J. S., et al. (1989). Characterization of site-directed mutants in the lac permease of Escherichia coli. 2. Glutamate-325 replacements. *Biochemistry*, *28*(6), 2533–2539.

Chan, H. S. (1999). Folding alphabets. *Nature Structural Biology*, *6*(11), 994–996.

Chaptal, V., Kwon, S., Sawaya, M. R., Guan, L., Kaback, H. R., & Abramson, J. (2011). Crystal structure of lactose permease in complex with an affinity in activator yields unique insight into sugar recognition. *Proceedings of the National Academy of Sciences of the United States of America*, *108*, 9361–9366.

Chen, L. Q., Hou, B. H., Lalonde, S., Takanaga, H., Hartung, M. L., Qu, X. Q., et al. (2010). Sugar transporters for intercellular exchange and nutrition of pathogens. *Nature*, *468*(7323), 527–532.

Cieplak, M., Koplik, J., & Banavar, J. R. (2001). Boundary conditions at a fluid–solid interface. *Physical Review Letters*, *86*(5), 803–806.

Consler, T. G., Tsolas, O., & Kaback, H. R. (1991). Role of proline residues in the structure and function of a membrane transport protein. *Biochemistry*, *30*(5), 1291–1298.

Dang, S., Sun, L., Huang, Y., Lu, F., Liu, Y., Gong, H., et al. (2010). Structure of a fucose transporter in an outward-open conformation. *Nature*, *467*(7316), 734–738.

Deng, D., Xu, C., Sun, P., Wu, J., Yan, C., Hu, M., et al. (2014). Crystal structure of the human glucose transporter GLUT1. *Nature*, *510*, 121–125.

Doki, S., Kato, H. E., Solcan, N., Iwaki, M., Koyama, M., Hattori, M., et al. (2013). Structural basis for dynamic mechanism of proton-coupled symport by the peptide transporter POT. *Proceedings of the National Academy of Sciences of the United States of America*, *110*(28), 11343–11348.

Dunten, R. L., Sahin-Toth, M., & Kaback, H. R. (1993). Role of the charge pair aspartic acid-237-lysine-358 in the lactose permease of Escherichia coli. *Biochemistry*, *32*(12), 3139–3145.

Eilers, M., Patel, A. B., Liu, W., & Smith, S. O. (2002). Comparison of helix interactions in membrane and soluble alpha-bundle proteins. *Biophysical Journal*, *82*(5), 2720–2736.

Ethayathulla, A. S., Yousef, M. S., Amin, A., Leblanc, G., Kaback, H. R., & Guan, L. (2014). Structure-based mechanism for Na(+)/melibiose symport by MelB. *Nature Communications*, *5*, 3009.

Frillingos, S., Sahin-Toth, M., Wu, J., & Kaback, H. R. (1998). Cys-scanning mutagenesis: A novel approach to structure function relationships in polytopic membrane proteins. *The FASEB Journal*, *12*(13), 1281–1299.

Guan, L., Hu, Y., & Kaback, H. R. (2003). Aromatic stacking in the sugar binding site of the lactose permease. *Biochemistry*, *42*(6), 1377–1382.

Guan, L., & Kaback, H. R. (2004). Binding affinity of lactose permease is not altered by the H+ electrochemical gradient. *Proceedings of the National Academy of Sciences of the United States of America*, *101*(33), 12148–12152.

Guan, L., & Kaback, H. R. (2006). Lessons from lactose permease. *Annual Review of Biophysics and Biomolecular Structure*, *35*, 67–91.

Guan, L., Mirza, O., Verner, G., Iwata, S., & Kaback, H. R. (2007). Structural determination of wild-type lactose permease. *Proceedings of the National Academy of Sciences of the United States of America*, *104*(39), 15294–15298.

Guettou, F., Quistgaard, E. M., Tresaugues, L., Moberg, P., Jegerschold, C., Zhu, L., et al. (2013). Structural insights into substrate recognition in proton-dependent oligopeptide transporters. *EMBO Reports*, *14*(9), 804–810.

Hruz, P. W., & Mueckler, M. M. (1999). Cysteine-scanning mutagenesis of transmembrane segment 7 of the GLUT1 glucose transporter. *The Journal of Biological Chemistry*, *274*, 36176–36180.

Hruz, P. W., & Mueckler, M. M. (2000). Cysteine-scanning mutagenesis of transmembrane segment 11 of the GLUT1 facilitative glucose transporter. *Biochemistry*, *39*(31), 9367–9372.

Huang, Y., Lemieux, M. J., Song, J., Auer, M., & Wang, D. N. (2003). Structure and mechanism of the glycerol-3-phosphate transporter from Escherichia coli. *Science*, *301*(5633), 616–620.

Hvorup, R. N., & Saier, M. H., Jr. (2002). Sequence similarity between the channel-forming domains of voltage-gated ion channel proteins and the C-terminal domains of secondary carriers of the major facilitator superfamily. *Microbiology*, *148*(Pt 12), 3760–3762.

Iancu, C. V., Zamoon, J., Woo, S. B., Aleshin, A., & Choe, J. Y. (2013). Crystal structure of a glucose/H + symporter and its mechanism of action. *Proceedings of the National Academy of Sciences of the United States of America*, *110*(44), 17862–17867.

Isom, D. G., Cannon, B. R., Castaneda, C. A., Robinson, A., & Garcia-Moreno, B. (2008). High tolerance for ionizable residues in the hydrophobic interior of proteins. *Proceedings of the National Academy of Sciences of the United States of America*, *105*(46), 17784–17788.

Jiang, X., Villafuerte, M. K., Andersson, M., White, S. H., & Kaback, H. R. (2014). Galactoside-binding site in LacY. *Biochemistry*, *53*(9), 1536–1543.

Jin, J., Guffanti, A. A., Beck, C., & Krulwich, T. A. (2001). Twelve-transmembrane-segment (TMS) version (DeltaTMS VII-VIII) of the 14-TMS Tet(L) antibiotic resistance protein retains monovalent cation transport modes but lacks tetracycline efflux capacity. *Journal of Bacteriology*, *183*(8), 2667–2671.

Jung, K., Jung, H., Colacurcio, P., & Kaback, H. R. (1995). Role of glycine residues in the structure and function of lactose permease, an Escherichia coli membrane transport protein. *Biochemistry*, *34*(3), 1030–1039.

Kasho, V. N., Smirnova, I. N., & Kaback, H. R. (2006). Sequence alignment and homology threading reveals prokaryotic and eukaryotic proteins similar to lactose permease. *Journal of Molecular Biology*, *358*(4), 1060–1070.

Kumar, H., Kasho, V., Smirnova, I., Finer-Moore, J. S., Kaback, H. R., & Stroud, R. M. (2014). Structure of sugar-bound LacY. *Proceedings of the National Academy of Sciences of the United States of America*, *111*(5), 1784–1788.

Li, T., Fan, K., Wang, J., & Wang, W. (2003). Reduction of protein sequence complexity by residue grouping. *Protein Engineering*, *16*(5), 323–330.

Liu, Z., Madej, M. G., & Kaback, H. R. (2010). Helix dynamics in LacY: Helices II and IV. *Journal of Molecular Biology*, *396*(3), 617–626.

Madej, M. G. (2014). Function, structure, and evolution of the major facilitator superfamily: The LacY manifesto. *Advances in Biology, 2014*. Article ID: 523591.

Madej, M. G., Dang, S., Yan, N., & Kaback, H. R. (2013). Evolutionary mix-and-match with MFS transporters. *Proceedings of the National Academy of Sciences of the United States of America*, *110*(15), 5870–5874.

Madej, M. G., & Kaback, H. R. (2013). Evolutionary mix-and-match with MFS transporters II. *Proceedings of the National Academy of Sciences of the United States of America*, *110*, E4831–E4838.

Madej, M.G. Kaback, H.R. (2014). The life and times of lac permease: Crystals ain't enough, but they certainly do help. In C. Ziegler, & R. Kraemer (Eds.), *Springer series in biophysics:*

Transporters: Vol. 17. Membrane transporter function: To structure and beyond (pp. 121–158). Springer-Verlag: Berlin Heidelberg.

Madej, M. G., Soro, S. N., & Kaback, H. R. (2012). Apo-intermediate in the transport cycle of lactose permease (LacY). *Proceedings of the National Academy of Sciences of the United States of America, 109*(44), E2970–E2978.

Madej, M. G., Sun, L., Yan, N., & Kaback, H. R. (2014). Functional architecture of MFS D-glucose transporters. *Proceedings of the National Academy of Sciences of the United States of America, 111*(7), E719–E727.

Maiden, M. C., Davis, E. O., Baldwin, S. A., Moore, D. C., & Henderson, P. J. (1987). Mammalian and bacterial sugar transport proteins are homologous. *Nature, 325*(6105), 641–643.

Mirza, O., Guan, L., Verner, G., Iwata, S., & Kaback, H. R. (2006). Structural evidence for induced fit and a mechanism for sugar/H(+) symport in LacY. *The EMBO Journal, 25*, 1177–1183.

Mueckler, M., & Makepeace, C. (2002). Analysis of transmembrane segment 10 of the Glut1 glucose transporter by cysteine-scanning mutagenesis and substituted cysteine accessibility. *The Journal of Biological Chemistry, 277*(5), 3498–3503.

Mueckler, M., & Makepeace, C. (2004). Analysis of transmembrane segment 8 of the GLUT1 glucose transporter by cysteine-scanning mutagenesis and substituted cysteine accessibility. *The Journal of Biological Chemistry, 279*(11), 10494–10499.

Mueckler, M., & Makepeace, C. (2005). Cysteine-scanning mutagenesis and substituted cysteine accessibility analysis of transmembrane segment 4 of the Glut1 glucose transporter. *The Journal of Biological Chemistry, 280*(47), 39562–39568.

Mueckler, M., & Makepeace, C. (2006). Transmembrane segment 12 of the Glut1 glucose transporter is an outer helix and is not directly involved in the transport mechanism. *The Journal of Biological Chemistry, 281*(48), 36993–36998.

Mueckler, M., & Makepeace, C. (2009). Model of the exofacial substrate-binding site and helical folding of the human Glut1 glucose transporter based on scanning mutagenesis. *Biochemistry, 48*(25), 5934–5942.

Mueckler, M., Roach, W., & Makepeace, C. (2004). Transmembrane segment 3 of the Glut1 glucose transporter is an outer helix. *The Journal of Biological Chemistry, 279*(45), 46876–46881.

Mueckler, M., & Thorens, B. (2013). The SLC2 (GLUT) family of membrane transporters. *Molecular Aspects of Medicine, 34*(2–3), 121–138.

Murphy, L. R., Wallqvist, A., & Levy, R. M. (2000). Simplified amino acid alphabets for protein fold recognition and implications for folding. *Protein Engineering, 13*(3), 149–152.

Nasie, I., Steiner-Mordoch, S., Gold, A., & Schuldiner, S. (2010). Topologically random insertion of EmrE supports a pathway for evolution of inverted repeats in ion-coupled transporters. *The Journal of Biological Chemistry, 285*(20), 15234–15244.

Needleman, S. B., & Wunsch, C. D. (1970). A general method applicable to the search for similarities in the amino acid sequence of two proteins. *Journal of Molecular Biology, 48*(3), 443–453.

Newstead, S., Drew, D., Cameron, A. D., Postis, V. L., Xia, X., Fowler, P. W., et al. (2011). Crystal structure of a prokaryotic homologue of the mammalian oligopeptide-proton symporters, PepT1 and PepT2. *The EMBO Journal, 30*(2), 417–426.

Pao, S. S., Paulsen, I. T., & Saier, M. H., Jr. (1998). Major facilitator superfamily. *Microbiology and Molecular Biology Reviews, 62*(1), 1–32.

Parker, J. L., & Newstead, S. (2014). Molecular basis of nitrate uptake by the plant nitrate transporter NRT1.1. *Nature, 507*(7490), 68–72.

Pedersen, B. P., Kumar, H., Waight, A. B., Risenmay, A. J., Roe-Zurz, Z., Chau, B. H., et al. (2013). Crystal structure of a eukaryotic phosphate transporter. *Nature, 496*(7446), 533–536.

Pettersen, E. F., Goddard, T. D., Huang, C. C., Couch, G. S., Greenblatt, D. M., Meng, E. C., et al. (2004). UCSF Chimera—A visualization system for exploratory research and analysis. *Journal of Computational Chemistry*, *25*(13), 1605–1612.

Quistgaard, E. M., Low, C., Moberg, P., Tresaugues, L., & Nordlund, P. (2013). Structural basis for substrate transport in the GLUT-homology family of monosaccharide transporters. *Nature Structural & Molecular Biology*, *20*(6), 766–768.

Radestock, S., & Forrest, L. R. (2011). The alternating-access mechanism of MFS transporters arises from inverted-topology repeats. *Journal of Molecular Biology*, *407*(5), 698–715.

Reddy, V. S., Shlykov, M. A., Castillo, R., Sun, E. I., & Saier, M. H., Jr. (2012). The major facilitator superfamily (MFS) revisited. *The FEBS Journal*, *279*(11), 2022–2035.

Sahin-Toth, M., & Kaback, H. R. (1993). Properties of interacting aspartic acid and lysine residues in the lactose permease of Escherichia coli. *Biochemistry*, *32*(38), 10027–10035.

Sahin-Tóth, M., & Kaback, H. R. (2001). Arg-302 facilitates deprotonation of Glu-325 in the transport mechanism of the lactose permease from *Escherichia coli*. *Proceedings of the National Academy of Sciences of the United States of America*, *98*(11), 6068–6073.

Sahin-Toth, M., le Coutre, J., Kharabi, D., le Maire, G., Lee, J. C., & Kaback, H. R. (1999). Characterization of Glu126 and Arg144, two residues that are indispensable for substrate binding in the lactose permease of Escherichia coli. *Biochemistry*, *38*(2), 813–819.

Saier, M. H., Jr. (2000). Families of transmembrane sugar transport proteins. *Molecular Microbiology*, *35*, 699–710.

Saier, M. H., Jr., Beatty, J. T., Goffeau, A., Harley, K. T., Heijne, W. H., Huang, S. C., et al. (1999). The major facilitator superfamily. *Journal of Molecular Microbiology and Biotechnology*, *1*(2), 257–279.

Schürmann, A., Doege, H., Ohnimus, H., Monser, V., Buchs, A., & Joos, H.-G. (1997). Role of conserved arginine and glutamate residues on the cytosolic surface of the glucose transporters for transporter function. *Biochemistry*, *36*, 12897–12902.

Smirnova, I., Kasho, V., & Kaback, H. R. (2011). Lactose permease and the alternating access mechanism. *Biochemistry*, *50*(45), 9684–9693.

Smirnova, I., Kasho, V., Sugihara, J., Choe, J. Y., & Kaback, H. R. (2009). Residues in the H+ translocation site define the pKa for sugar binding to LacY. *Biochemistry*, *48*(37), 8852–8860.

Smirnova, I., Kasho, V., Sugihara, J., & Kaback, H. R. (2013). Trp replacements for tightly interacting Gly-Gly pairs in LacY stabilize an outward-facing conformation. *Proceedings of the National Academy of Sciences of the United States of America*, *110*(22), 8876–8881.

Solcan, N., Kwok, J., Fowler, P. W., Cameron, A. D., Drew, D., Iwata, S., et al. (2012). Alternating access mechanism in the POT family of oligopeptide transporters. *The EMBO Journal*, *31*(16), 3411–3421.

Solis, A. D., & Rackovsky, S. (2000). Optimized representations and maximal information in proteins. *Proteins*, *38*(2), 149–164.

Sugihara, J., Smirnova, I., Kasho, V., & Kaback, H. R. (2011). Sugar recognition by CscB and LacY. *Biochemistry*, *50*(51), 11009–11014.

Sugihara, J., Sun, L., Yan, N., & Kaback, H. R. (2012). Dynamics of the L-fucose/H+ symporter revealed by fluorescence spectroscopy. *Proceedings of the National Academy of Sciences of the United States of America*, *109*(37), 14847–14851.

Sun, J., Bankston, J. R., Payandeh, J., Hinds, T. R., Zagotta, W. N., & Zheng, N. (2014). Crystal structure of the plant dual-affinity nitrate transporter NRT1.1. *Nature*, *507*(7490), 73–77.

Sun, L., Zeng, X., Yan, C., Sun, X., Gong, X., Rao, Y., et al. (2012). Crystal structure of a bacterial homologue of glucose transporters GLUT1-4. *Nature*, *490*(7420), 361–366.

Vastermark, A., Lunt, B., & Saier, M. (2014). Major facilitator superfamily porters, LacY, FucP and XylE of Escherichia coli appear to have evolved positionally dissimilar catalytic residues without rearrangement of 3-TMS repeat units. *Journal of Molecular Microbiology and Biotechnology*, *24*(2), 82–90.

Vastermark, A., & Saier, M. H. (2014). Major facilitator superfamily (MFS) evolved without 3-transmembrane segment unit rearrangements. *Proceedings of the National Academy of Sciences of the United States of America*, *111*(13), E1162–E1163.

White, S. H., & Wimley, W. C. (1999). Membrane protein folding and stability: Physical principles. *Annual Review of Biophysics and Biomolecular Structure*, *28*, 319–365.

Xu, Y., Tao, Y., Cheung, L. S., Fan, C., Chen, L. Q., Xu, S., et al. (2014). Structures of bacterial homologues of SWEET transporters in two distinct conformations. *Nature*, *515*, 448–452.

Xuan, Y. H., Hu, Y. B., Chen, L. Q., Sosso, D., Ducat, D. C., Hou, B. H., et al. (2013). Functional role of oligomerization for bacterial and plant SWEET sugar transporter family. *Proceedings of the National Academy of Sciences of the United States of America*, *110*(39), E3685–E3694.

Yan, N. (2013). Structural advances for the major facilitator superfamily (MFS) transporters. *Trends in Biochemical Sciences*, *38*(3), 151–159.

Yan, H., Huang, W., Yan, C., Gong, X., Jiang, S., Zhao, Y., et al. (2013). Structure and mechanism of a nitrate transporter. *Cell Reports*, *3*(3), 716–723.

Yin, Y., He, X., Szewczyk, P., Nguyen, T., & Chang, G. (2006). Structure of the multidrug transporter EmrD from Escherichia coli. *Science*, *312*(5774), 741–744.

Zheng, H., Wisedchaisri, G., & Gonen, T. (2013). Crystal structure of a nitrate/nitrite exchanger. *Nature*, *497*(7451), 647–651.

CHAPTER TWENTY-FIVE

Elucidating Ligand-Modulated Conformational Landscape of GPCRs Using Cloud-Computing Approaches

Diwakar Shukla*,†,1, Morgan Lawrenz*, Vijay S. Pande*,†

*Department of Chemistry, Stanford University, Stanford, California, USA
†SIMBIOS NIH Center for Biomedical Computation, Stanford University, Stanford, California, USA
[1]Corresponding author: e-mail address: shukla@stantford.edu

Contents

Abstract

G-protein-coupled receptors (GPCRs) are a versatile family of membrane-bound signaling proteins. Despite the recent successes in obtaining crystal structures of GPCRs, much needs to be learned about the conformational changes associated with their activation. Furthermore, the mechanism by which ligands modulate the activation of GPCRs has remained elusive. Molecular simulations provide a way of obtaining detailed an atomistic description of GPCR activation dynamics. However, simulating GPCR activation is

Methods in Enzymology, Volume 557
ISSN 0076-6879
http://dx.doi.org/10.1016/bs.mie.2014.12.007

challenging due to the long timescales involved and the associated challenge of gaining insights from the "Big" simulation datasets. Here, we demonstrate how cloud-computing approaches have been used to tackle these challenges and obtain insights into the activation mechanism of GPCRs. In particular, we review the use of Markov state model (MSM)-based sampling algorithms for sampling milliseconds of dynamics of a major drug target, the G-protein-coupled receptor β_2-AR. MSMs of agonist and inverse agonist-bound β_2-AR reveal multiple activation pathways and how ligands function via modulation of the ensemble of activation pathways. We target this ensemble of conformations with computer-aided drug design approaches, with the goal of designing drugs that interact more closely with diverse receptor states, for overall increased efficacy and specificity. We conclude by discussing how cloud-based approaches present a powerful and broadly available tool for studying the complex biological systems routinely.

1. INTRODUCTION

G-protein-coupled receptors (GPCRs) represent the largest family of transmembrane proteins involved in cellular signaling and mediate a variety of physiological processes involving hormones, neurotransmitters, and drugs (Cherezov et al., 2007; Gether & Kobilka, 1998; Shukla, Sun, & Lefkowitz, 2008). The structural and functional understanding of GPCRs, like many other membrane proteins, has lagged behind soluble drug targets such as kinases mainly due to the challenges associated with membrane protein crystallization. With recent technological advancements, crystal structures have been obtained for key drug targets such as adenosine A_{2a} receptor, the β_1 and β_2 adrenergic receptors (Rosenbaum et al., 2011). The PDB database now (as of August 2014) contains 115 crystal structures of GPCRs in complex with a variety of ligands and intracellular binding partners. These structures have provided the much-needed molecular basis for the mechanism of drug action (Shoichet & Kobilka, 2012). However, most of these structures correspond to the inactive form of the receptors, which is the most predominant structure in the conformational ensemble of GPCRs and selected for by crystallography, which captures the energetic minima.

The signaling process involves a conformational change from the inactive to a functional or active state of the GPCR. These conformational changes affect the receptor's functional interactions with intracellular signaling molecules in a manner that is selected for by the ligand molecule—a property termed functional selectivity (Reiter, Ahn, Shukla, & Lefkowitz, 2012). There is now growing experimental and computational evidence, which suggests that transient GPCR intermediate states mediate the process of GPCR activation (Dror et al., 2011; Nygaard et al., 2013). However, the structures and the exact role of these intermediate states have remained elusive. Therefore, understanding the functional dynamics of GPCRs will remain an important direction for molecular biophysics in the years to come.

Given the wealth of information from crystal structures, simulations can now play a key role in predicting the chemically detailed nature of *how* these proteins function, especially in terms of functional dynamics and conformational change. A natural example of this paradigm is using molecular dynamics simulations to gain new insight into the function of GPCRs and the mechanisms of drug action. In principle, long-timescale molecular simulations can provide all possible conformations of the receptor in atomistic detail and the rates of transitions between them. However, long activation timescales (milliseconds) for GPCRs push such problems out of the realm of high performance computing and even dedicated hardware (Lane, Shukla, Beauchamp, & Pande, 2013). In a recent study, we have shown how massively distributed simulations on cloud-computing platforms and Markov state model (MSM)-based sampling protocols could be used for studying such long-timescale phenomena (Kohlhoff et al., 2014). In particular, we ran tens of thousands of independent short simulations of apo, agonist, and inverse agonist-bound β_2 adrenergic receptor (β_2-AR) on Google Exacycle, a cloud-computing initiative that provides an interface for running compute jobs directly on Google's production infrastructure. MSMs were then used to obtain long-timescale behavior ($\sim$100 μs) from the individual short trajectories ($\sim$10 ns). Our results provide a human-readable view of how different ligands bias the conformational landscape of GPCRs by altering the conformational preferences of key structural elements within the protein. We find that activation and deactivation proceed through multiple pathways and typically visit metastable intermediate states, which open up a new set of opportunities for drug screening and development.

In this chapter, we provide a brief description of our approach for studying long-timescale phenomena in protein dynamics and summarize the basic findings of how ligands modulate conformational preferences of GPCRs. It is our intent to provide a general guide to construction of the MSMs of the functional dynamics of proteins using our recent massively distributed simulations of GPCRs in the cloud as an example. This chapter is not a thorough review of techniques used for MSM construction and is written for experimentalist or nonexperts working on membrane proteins.

2. WHAT ARE MARKOV STATE MODELS?

MSMs are kinetic models that describe the conformational dynamics of proteins in terms of jumps between the conformational states of the protein. These models are similar to kinetic models of chemical reaction networks, where the rate of formation and disappearance of different chemical species in the network could be used to describe their time evolution. In the MSMs, the distinct protein conformational states represent the different chemical species in the reaction network. In a nutshell, MSMs are computationally parameterized discrete time master equations: computational models parameterized typically from molecular dynamics to simulate long-timescale processes. Recently, discrete-state master equation or MSMs have had success at modeling long-time statistical dynamics (Chodera & Noé, 2014). In these models, metastable conformational states are identified such that the intrastate dynamics are much faster than interstate dynamics (states visited over time from a discrete Markov chain). Transition rates between the states are estimated from molecular dynamics simulations. If the model is shown to self-consistently recapitulate the statistical dynamics of the trajectories it was constructed from, and if it obeys the Markov property, i.e., the next state visited by the system can be predicted using only the information about the current state of the system, it can be used to simulate the statistical evolution of a noninteracting ensemble of molecules over much longer times than the lengths of the individual trajectories from which it is constructed. Experimental observables, such as distance between a particular residue pair and chemical shifts, can be computed directly from MSM probability-weighted linear combinations of the observables of each state, and direct comparison with experimental data can be computed. The key steps in the MSM building process involve (a) identification of distinct kinetically relevant conformational states of protein by clustering of the simulation data and (b) obtaining the probabilities of transition between states from the kinetic information embedded in the simulated trajectories.

2.1 Generation of the initial dataset

For identification of kinetically relevant states and the transition probability among those states, one must start with some initial simulation data. Many different types of simulation methods can be used for obtaining the simulation datasets and molecular dynamics is just one of the popular methods used for simulating biomolecular dynamics. The initial simulations are typically started from the available crystal structures of the protein. In the case of β_2-AR, the inactive (PDB: 2RH1; Cherezov et al., 2007; ligand: inverse agonist, carazolol) and active (PDB: 3P0G; Rasmussen et al., 2011; ligand: agonist BI-167107) crystal structures were chosen as the starting structures. The agonist-bound inactive structure and the inverse agonist-bound active structures were generated by docking the respective ligands in the *orthosteric* binding site of β_2-AR. This protocol could be used for any GPCR for which multiple crystal structures in the presence of different ligands are available. Simulations from these initial structures were started with an aim to sample a well-connected conformational landscape of β_2-AR using adaptive sampling protocols described below. In the adaptive sampling scheme, MSMs are built after an initial dataset is obtained to identify new conformational states of protein, which are then used as the starting points for future simulations (Weber & Pande, 2011). Since one must build an MSM for adaptive sampling, we have first described the procedure for building MSMs below.

2.2 Clustering the simulation data

In order to define states, we need metrics to distinguish among different conformational states of a protein. Structural metrics such as root mean square deviation (RMSD) of the protein or subregions within the protein are typically used as a metric. Recently, several new methods have been developed which use a more agnostic approach to clustering by using the set of all protein dihedrals or inter-residue distances to identify the most kinetically relevant metrics (Pérez-Hernández, Paul, Giorgino, Fabritiis, & Noé, 2013; Schwantes & Pande, 2013). In the case of β_2-AR, such structural metrics are already available from the decades of mutagenesis and structural studies. Specifically, these structural metrics were identified using the crystal structures of active and inactive forms of the β_2-AR (Fig. 1). These metrics represent the four key regions of β_2-AR: the RMSD of the ligand binding pocket residues within 4 Å of both carazolol and BI-16707, the RMSD of the connector region (1121 and F282), the RMSD of the NPxxY region (N322-C327), and the distance between Cα atoms of R131 on Helix 3 and L272 on Helix 6. We used a hybrid metric that inversely weights the four criteria according to their relative magnitudes of change between active and inactive states, as the latter movement is significantly larger (6 Å) than the others (1–2 Å). The hybrid metric is used to partition the conformational space into 1000s of conformational states using clustering algorithms such as *k*-centers or *k*-medoids.

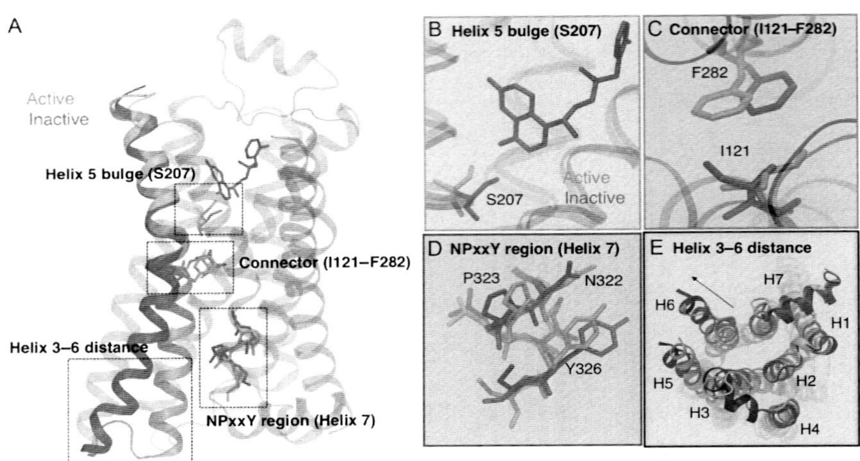

Figure 1 (A) Structural metrics used for the construction of MSMs of B2-AR dynamics. The structures shown in green (light gray in the print version) and magenta (dark gray in the print version) correspond to inactive (2RH1) and active (3P0G) crystal structures of B2-AR. The four key structural metrics involved in activation are (B) Helix 5 Bulge (S207), (C) flipping of connector residues (I121-F282), (D) twisting of NPxxY region (N322-327) on Helix 7, and (E) outward movement of Helix 6.

2.3 Estimating the transition probability matrix

Each snapshot from the simulation data is assigned to the closest conformational state (distance is estimated using the hybrid metric for β_2-AR). The result is a translation of molecular dynamics (MD) simulation snapshots along a trajectory into a series of conformational states over time. These modified trajectories are then used to estimate the transition count matrix C_{ij}, which denotes the number of observed transitions from state i at time t to state j at time $t+\tau$, where τ is the lag time of the model. Finally, the transition probability matrix $T(\tau)$ includes the probability of transitions from state i to state j in a certain time interval τ is obtained by normalizing C_{ij} with the sum of all transitions from state i. To ensure that the total population of all the states is conserved, a maximum likelihood estimate of the transition probability matrix that obeys the detailed balance is obtained. The eigenvalues/eigenvectors of the transition matrix gives information about the aggregate transitions between subsets of states in the model and what timescales these transitions occur on. The equilibrium populations of the individual states are estimated from the first eigenvector of the transition probability matrix. The timescales of the dynamical processes occurring on the conformational landscape of the protein can be obtained by estimating the eigenvalues of this matrix.

2.4 Improving the MSM using adaptive sampling

With all of its promise, the key challenge in MSM construction is its computational demands. The most straightforward way of building MSMs involves running long trajectories that sample the deactivation and activation processes of GPCRs. However, due to the presence of stable inactive and intermediate states of GPCRs, the simulation trajectories get stuck in these stable states and thus further sampling these states does not provide additional information about the conformational landscape. Adaptive sampling procedures seek to identify the starting configurations for future simulations such that the conformational landscapes can be sampled efficiently. As detailed below, adaptive approaches can dramatically aid in overcoming the sampling challenge: i.e., one runs simulations only where needed (e.g., as dictated by link between the uncertainty in observables of interest and sampling from specific states). The application of adaptive sampling, combined with GPUs, and massively parallel computing is particularly powerful, enabling simulations on millisecond to seconds timescales (Kohlhoff et al., 2014; Shukla, Meng, Roux, & Pande, 2014). The simplest and the most efficient adaptive sampling process is the least-count-based sampling, where the future simulations are started from the least populated MSM states (Weber & Pande, 2011). This procedure ensures that the future simulations are started from the edges of the conformational landscape, thereby avoiding the already well-sampled stable states. A schematic representing the least-count-based sampling on a model conformational landscape (Mueller potential) from the three rounds of adaptive sampling is shown in Figs. 2 and 3. Several other adaptive sampling procedures have been reported such as eigenvalue-based sampling, which aims to reduce the error in the estimate of the kinetics of the slowest dynamical process (Bowman, Ensign, & Pande, 2010). In this sampling procedure, the states that introduce the largest error in the second eigenvalue of the transition probability matrix (which represents the slowest dynamical process of the system) are used as the starting configurations for future simulations.

2.5 Validating the MSM

There are various ways to test whether an MSM is self-consistent (agrees with the data used for MSM construction) such as Swope–Pitera eigenvalue test, Chapman–Kolmogorov test, and Bayesian model selection approaches. In this section, we describe the most commonly used MSM validation protocols. The eigenvalues μ of the transition probability matrix are related to

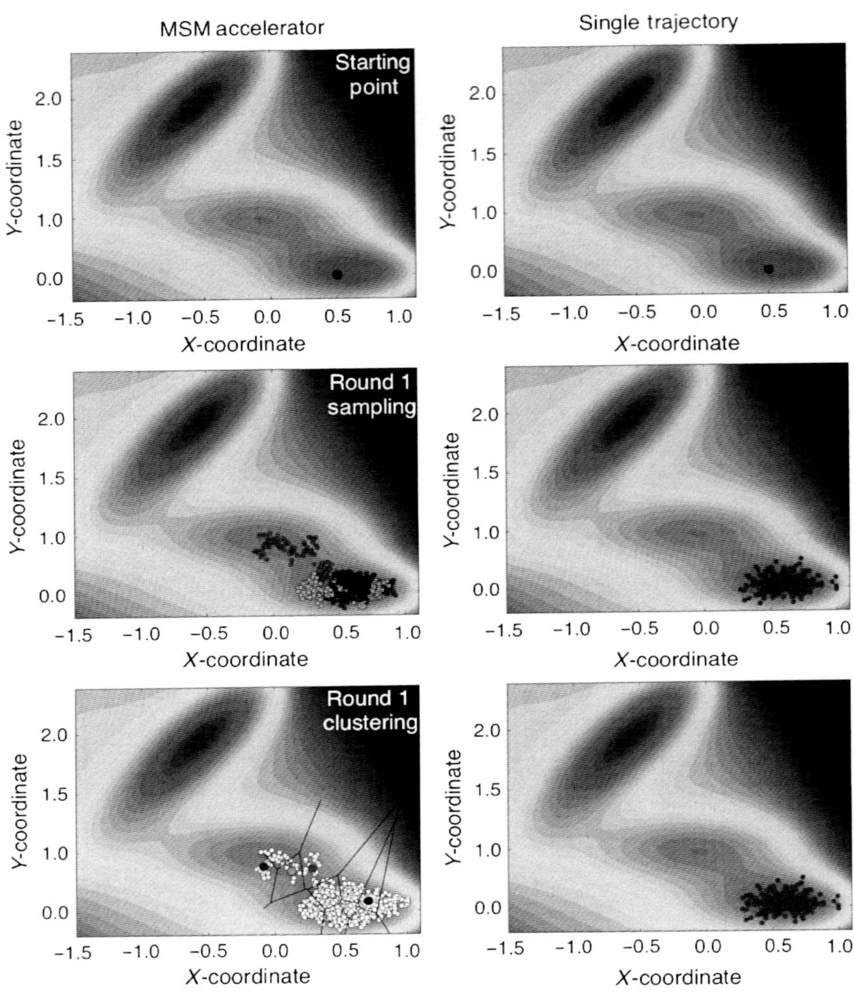

Figure 2 Two-dimensional demo of MSM accelerator, running an adaptive sampling algorithm to efficiently sample space. MSM adaptive sampling (AS) (left) versus a fast, single long trajectory (right), with equal total sampling at each step. MSM AS proceeds by running many trajectories, building an MSM, then shooting more trajectories. Due to its ability to drive sampling, AS is very efficient. See http://www.youtube.com/watch?v=oh-xqR79QM0 for more details. Here, we show the first three rounds. Figure 3 shows the final rounds.

the implied timescales $1/k$ of the transitions and the lag time via the following equation $k = -\tau/\ln(\mu)$. The lag time is the time difference between trajectory snapshots used for construction of MSM. These rates, or implied time scales, should remain unchanged (with respect to lag time) when a

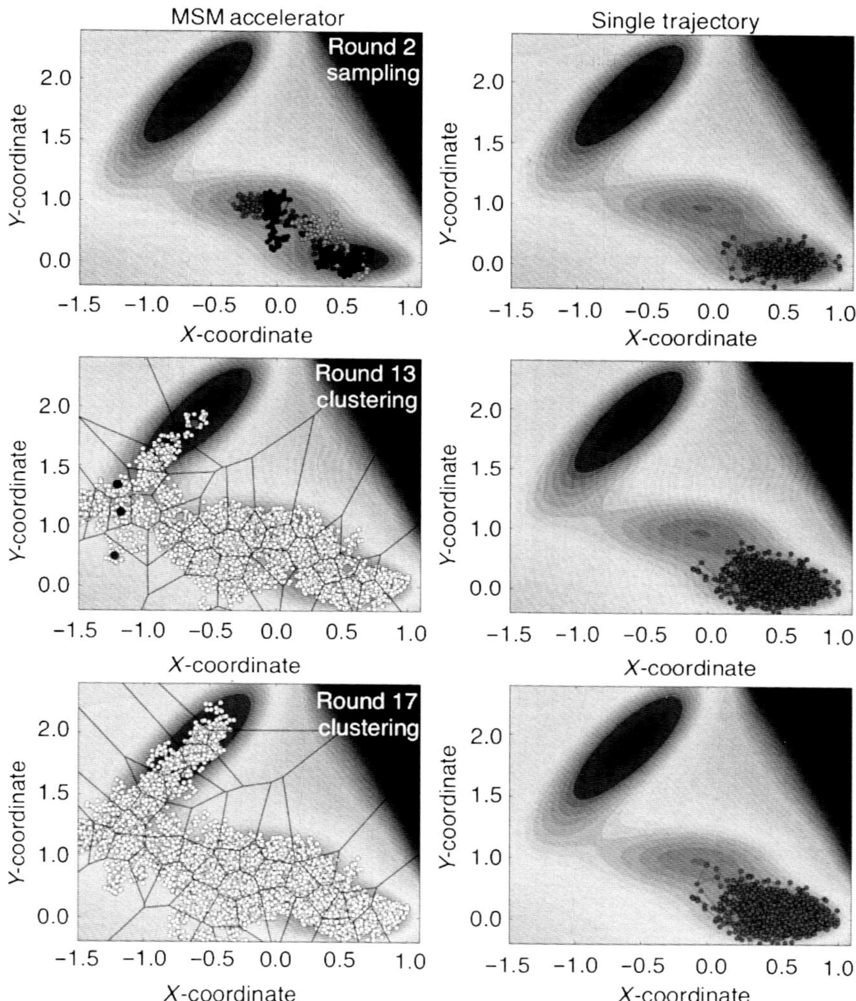

Figure 3 Two-dimensional demo of MSM accelerator, running an adaptive sampling algorithm to efficiently sample space. Here, we show the final rounds, where the MSM has fully converged with much less computer power needed to simulate the event using traditional molecular dynamics (right pane). (See the color plate.)

system is Markovian. Therefore, convergence of implied timescales is monitored to assess the quality of the MSM. The implied timescales of 3000-state models of the MSMs of β_2-AR show a converged spectrum of timescales after a lag time of 7.5 ns. This timescale convergence criterion is obtained from the Chapman–Kolmogorov test of the Markov model. This test ensures that the MSM kinetics are robust. The transition probability matrix can also be used to obtain the population of the system at any time using the following relationship: $p(t+\tau) = p(t)T(\tau)$, where $p(t)$ is the vector of state population at time t. We used this relationship to estimate the relaxation of key structural metrics as GPCR activates or deactivates. Dror et al. have recently reported the long deactivation simulations of β_2-AR (Dror et al., 2011). The comparison between the autocorrelations of the individual long deactivation trajectories and the MSM-propagated dynamics of Helix 3–6 distance and NPxxY RMSD from the inactive structure of β_2-AR shows that the long-timescale behavior of MSMs obtained using short (tens of nanoseconds) trajectories is similar to those obtained from individual trajectories that are 100–1000 times longer. Furthermore, the cloud-based massively distributed simulations also represent a much more efficient use of simulation time than simply running a long brute-force simulation of the same length.

3. HOW DO MSMs ENABLE NOVEL INSIGHT?

In the above section, we reviewed how one builds an MSM and how MSM-based adaptive sampling schemes can be used to efficiently sample the conformational landscape of proteins. However, the main goal of the MSM construction process is to obtain novel insights into the mechanism at hand. The natural question to ask is how such mechanistic insights could be gleaned in an unbiased manner from MSMs of protein conformational change? MSMs allow for a direct calculation of any sort of property that could be in principle calculated from a very long trajectory. This includes structure prediction, free energies, mechanistic properties, rates, etc. However, beyond these quantitative features, MSMs allow one to gain new insight by systematic coarse graining (often called "lumping") of the high-resolution MSM to yield a simplified, few state model which is still faithful to the original data (or rather as quantitatively faithful a simplified model can be), by elucidating the most probable pathways for conformational change in atomistic description, and providing a long-timescale behavior of protein which is not immediately accessible from short simulation trajectories.

3.1 Coarse-graining MSMs to identify key conformational states

From a structural viewpoint, the key questions about protein function involve identifying the functionally important conformational states of protein. High-resolution MSMs (often comprising of thousands of states) provide a detailed atomistic view of conformational landscape, which is useful for making quantitative comparison with the experiments. However, few states models provide a more human-comprehensible picture of protein function. MSMs and the free energy landscapes of proteins are fundamentally hierarchical in nature where fast dynamics occurs between shallow local free energy minima, and slow dynamics is observed between deep free energy minima. Therefore, coarse graining of MSMs is performed to obtain few state models by lumping together states rapidly interconverting states into macrostates.

The 3000-state microstate model of β_2-AR was used to build a 10-state macromodel (Fig. 4) using Perron-Cluster Cluster Analysis (PCCA+) method (Deuflhard & Weber, 2005). This method uses the eigenvalues of the transition probability matrix to find the few state partition that best captures the slowest transitions. Other advanced methods have been reported recently, which take into account the statistical uncertainty in the transition

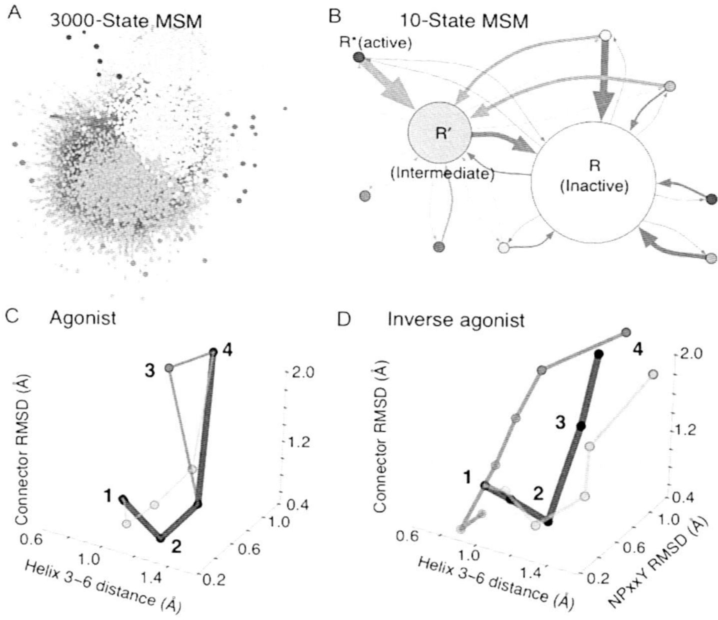

Figure 4 (A) Network representation of the 3000-state MSM built from the simulations of agonist-bound GPCR with each circle representing an individual conformational state. (B) Ten-state MSM built from the 3000-state MSMs using spectral clustering methods to identify kinetically relevant states. The circles in the 3000-state MSM are colored according to their membership in the coarse-grained 10-state MSM. The weight of arrow indicates the transition probability between states. Three distinct activation pathways adopted in the presence of an (C) agonist and (D) inverse agonist.

probabilities to obtain an optimal coarse-grained model (Bowman, 2012). The few state macromodel for an agonist-bound β_2-AR reveals two dominant conformational states of β_2-AR, the inactive and intermediate states representing 93.8% and 5.7% of the total population, respectively. The simple model also shows that the active state population for an agonist-bound GPCR in the absence of G-protein is negligible. For an agonist-bound receptor, the key structural difference between the inactive and intermediate states is the flipping of the I121 and F282 residues in the connector region whereas for the inverse agonist, the twisting of the Helix 7 NPxxY region residues distinguishes the two states. The intermediate state for the apo receptor is the same as the intermediate state in the MSM of the agonist-bound receptor.

3.2 Finding activation pathways using transition path theory

The mechanism by which ligands modulate the allosteric communication between extracellular ligand binding site and intracellular G-protein binding site lies at the heart of GPCR Pharmacology. MSMs could be used along with advanced theoretical frameworks such as transition path theory (TPT) (E & Vanden-Eijnden, 2010) to identify the dominant activation pathways involved in allosteric communication. The transition pathways in an MSM are a chain of states connecting the two-end points such as active and inactive states of β_2-AR. The relative probability of each activation pathway is given by the magnitude of the flux from the initial (inactive) to final (active) states. Recall that we know the equilibrium populations and the probability of transition between states from the MSM transition probability matrix. The probability of observing a particular chain of states connecting the active and inactive states can be calculated using the population of the states along the chain and the transition probabilities between them. One can also compute the mean first passage time between the defined active and inactive states to approximate the rate of the transition. A detailed description of TPT and its application to discrete Markov jump processes are included in several reviews by Vanden-Eijnden and coworkers.

TPT was applied to the MSMs of agonist-bound, inverse agonist-bound, and apo simulations. The identification of activation pathways requires the definition of states that belong to the active and inactive states of β_2-AR. For this analysis, active and inactive states are defined according to the order parameters used to construct the MSMs. In particular, the order parameters from crystal structures of inactive (2RH1) and active (3SN6) β_2-AR were used to define the cutoffs for active and inactive states, respectively. The

TPT analysis indicates a multitude of pathways with similar flux contribute to the overall activation mechanism of β_2-AR (Fig. 4). Projection of the pathways with highest flux along the three key structural metrics shows that there are distinct activation pathways for agonist and inverse agonist-bound β_2-AR. These results suggest a model of GPCR allostery, which is distinct from the widely used Monod–Wyman–Changeux (MWC) and the Koshland–Némethy–Filmer (KNF) models of protein allostery (Koshland, Némethy, & Filmer, 1966; Monod, Wyman, & Changeux, 1965). These models would describe allosteric communication in GPCRs as a binding of ligand on the extracellular side leading to a conformational change on the intracellular side. However, our results suggest that there are multiple preexisting activation pathways in the conformational ensemble of β_2-AR and different ligands modulate the probability of observing a particular path by stabilizing or destabilizing conformational states along the pathway. The shifts in the relative stability of intermediate states can lead to differences in observed functional effects.

3.3 Generating long-timescale trajectories using MSMs

In the above sections, we discussed how low-resolution models are required for obtaining a human-comprehensible view of the stable basins in the conformational landscape. MSMs can also provide a human-comprehensible view of the kinetic processes occurring on the conformational landscape by generating long-timescale dynamics between the MSM states. A long-timescale trajectory is simply a series of states visited over time starting from a particular MSM state. A kinetic Monte Carlo scheme is used for generating long trajectories by selecting the next state in the series using the probability of transition from the current state to all other states. The use of transition probabilities between states ensures that the long trajectory is consistent the microscopic rates between individual states. Each jump process in the time series corresponds to an increment of τ (lag time of the model) in real time.

The long (150 μs) trajectories of β_2-AR were generated using the kinetic Monte Carlo scheme. These trajectories are shown as projections along the four described structural criteria for the receptor with agonist BI-167107 and inverse agonist carazolol bound (Fig. 5). The agonist-bound simulations show a stable active state but the antagonist-bound simulations only show infrequent jumps to the active state values in some of the projections. These differences in activation dynamics are consistent with the biological functions of the ligands.

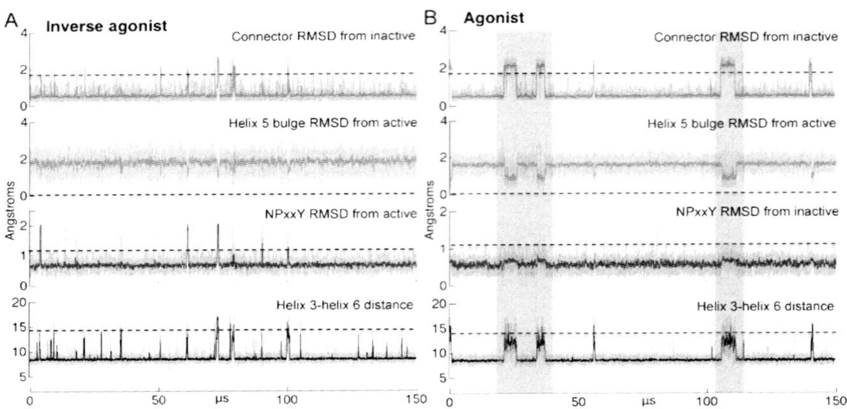

Figure 5 MSM trajectory generated using kinetic Monte Carlo scheme shows the toggling of molecular switches during activation of β_2-AR in the presence of (A) inverse agonist carazolol and (B) agonist BI-167107.

4. HOW CAN MSMs BE USED FOR IMPROVED DRUG DESIGN?

We want to further mine the rich structural dataset generated here for useful information that can guide drug design for these key drug targets. We describe a small molecule docking screen that uses MSMs for choosing functionally important protein states that can enrich diverse ligand chemotypes, a key advantage of virtual screens over empirical screens alone (Shoichet, 2004). We validate our approach with a retrospective screen of known GPCR ligands, but application in a prospective screen could discover new GPCR ligands that interact more closely with diverse receptor states, for increased efficacy and specificity.

4.1 Preparing protein states and ligands for small molecule docking

We want to select protein structures for docking that cover the full range of the conformational change during the activation process. To select these structures, we choose MSM states along the TPT pathways with flux greater than 30% of the maximum. Unique selected states totaled 20 for the agonist-bound TPT pathways, 43 for the inverse agonist-bound TPT pathways, and 102 for the apo TPT pathways. For comparison to our MSM docking methodology, we also dock to the available experimental crystal structures for both the active (3P0G) and inactive (2RH1) GPCR, as well as to 20 randomly selected MD snapshots from agonist-bound GPCR deactivation simulations in a previous study (Dror et al., 2011). All protein conformations were aligned to the active crystal structure (3P0G), for easy comparison of results.

Two hundred known β_2-AR agonists and antagonists, along with ~8000 structurally similar decoys were taken from the GPCR Decoy and Ligand Database (Gatica & Cavasotto, 2012). The decoys were used for docking validation, described below. Ligand protonation states were set at pH 7 using the protonate command in the docking program Surflex (Spitzer & Jain, 2012). OpenEye Omega (Hawkins, Skillman, Warren, Ellingson, & Stahl, 2010) was used to enumerate stereoisomers up to four chiral centers for each ligands. Ligands with experimentally known stereochemical activities were annotated, such as timolol. For scoring, the stereoisomer with the highest docking score was selected.

4.2 Evaluating docking performance

We used the docking program Surflex to dock all compounds. Inverse 3D representations of the binding site, called protomols in Surflex, were generated by identifying the active site with the agonist BI-16707 in the crystal structure 3P0G. These protomols serve as the sampling region for the docked molecules. The program samples ligand conformations within the protomol region and computes an approximate binding affinity based on a linear combination of steric, polar, entropic, and solvation terms. To further utilize our rich structural dataset, we used a protein ensemble docking approach, in which Surflex docks and scores each ligand to 20 protein snapshots provided for each MSM state, and reports the best-scoring complex.

To validate our approach, we compare the performance of docking to the MSM states and to the available crystal structure, as is standard for early-stage virtual drug screens. We also compare to performance of docking to random MD snapshots. We evaluate docking performance with receiving operator curves (ROC), which reflects the enrichment, or ranking, of true ligands over decoys. These curves plot the true-positive rate with the false-positive rate for a range of docking score cutoffs. Then, the area under the curve (AUC) for the ROC plot is used as a metric of performance, where AUC = 1.0 is accurate ranking of true ligands over decoys, and AUC = 0.5 reflects random performance. Error bars can be computed for the AUC values as the 95% confidence interval (CI) and statistical significance was evaluated by a test of proportions for AUC values from the MSM states that reflect better accuracy than the reference crystal structure docking AUC metric. We computed a one-sided Z-test statistic, with a p-value requirement of 0.05 for significant improvement.

4.3 Clustering top-scoring ligands by chemotype

The rich structural diversity available from the simulations could be further utilized to identify the chemotypes that selectively bind to a particular MSM state. For each MSM state, the top 10% of scored true ligands, excluding decoys, were selected. This resulted in a total of 3300 compounds for each agonist and antagonist docking set. These compounds were clustered using a k-centers algorithm and a Tanimoto metric that evaluates 3D shape, based on steric overlap, and chemistry, such as complement of proton donor and acceptor groups and π-stacking of rings. The Tanimoto computes overlap O between ligands A and B as the ratio of O_{AB} and $(O_{AA} + O_{BB} - O_{AB})$. The ligands were not realigned to each other before computing the Tanimoto overlap, in order to preserve the docked conformation, which reflects the binding site shape in a given MSM state. The OpenEye program ROCS (ROCS version 3.1.2, n.d.) was used to compute all Tanimoto scores, and the "Combination Score" was used, which combined shape and chemistry overlap, so that the perfect overlap of ligands is 2.0. A cutoff for the clustering was selected as the value separating the top 5% of all Tanimoto scores computed for all compound pair overlaps, which was 0.4 for agonists and 0.5 for inverse agonists. The low Tanimoto scores reflect the diversity of poses in the binding site. The selected cutoff gave 935 clusters for antagonist chemotypes and 497 agonist chemotypes, including different stereoisomers.

We next perform an analysis that maps the MSM states, and its preferred ligand chemotypes, onto a 1D representation of the activation pathway, in order to find a correspondence between the inactive/active state of the receptor and ligand affinity. The ligand chemotype clusters were assigned a progress variable ξ based on the MSM state that selected (highly ranked) the chemotype cluster center. This progress variable is a score for the MSM protein conformation that is a linear combination of the previously described structural metrics used to build the MSM and describe GPCR activation. The H5 bulge, connector, and NPxxY RMSD from the active state are scored a point in the range of [0, 2.0, 0.5] Å, and the H3-6 distance in the range of [8, 12, 1] Å. The scores are then normalized by the maximum and adjusted such that the end states correspond to the inactive ($\xi = 0.0$) and active ($\xi = 1.0$) crystal structures. Chemotypes were given binary counts across all states within each progress variable ξ, with a tally of 1 if a member from the cluster was found at a given ξ.

4.4 Docking to MSMs perform best at selecting true ligands over decoys

Figure 6 compares the AUC values for docking to states selected from the MSMs, active and inactive crystal structures, and to random MD protein snapshots. We see statistically significant improvement for both agonist and antagonist docking when docking to MSM states. In Fig. 6, the CI for agonist docking performance (0.81, 0.84) for active crystal is improved to (0.86, 0.88) in the agonist-bound MSM docking and is compared to a (0.77, 0.82) CI for random snapshot docking. Also, a substantial improvement for the antagonist docking is seen from the (0.75, 0.79) CI for the best crystal AUC to the (0.83, 0.84) CI for antagonist-bound MSM docking, compared to the random snapshot docking CI (0.74, 0.78). Thus, in summary, the agonist-bound MSM docking performs best for discriminating agonists over agonist decoys, and inverse agonist-bound MSM docking performs best for discriminating antagonists over antagonist decoys.

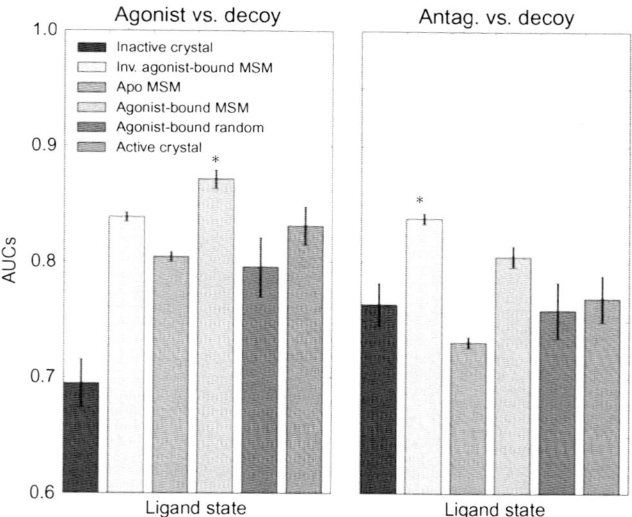

Figure 6 AUCs for evaluating docking performance of the MSM states. AUC values and 95% confidence intervals from the docking results of the GPCR Decoy and Ligand Database. AUCs for docking to MSM states with different ligand conditions, to random agonist-bound MD snapshots, and to inactive (2RH1) and active (3P0G) crystal structures are shown.

Furthermore, we observe that docking to MSM intermediate states enrich particular ligand chemotypes. The clustered chemotypes indicate a correspondence between ligand types and kinetically stable intermediate receptor conformations. Figure 7 shows two examples of agonist and antagonist chemotypes that are enriched by select MSM states along the activation pathways, plotted as the percentage of total ligands at a given ξ that are represented by the chemotype. For example, the agonist chemotype 2 and antagonist chemotype 3 are only enriched at intermediate states and would have been undiscovered by virtual screen docking to the crystal structures alone or without knowledge of the full activation pathway. These results reveal the power of MSMs for picking functional intermediate protein states that have different estimated affinities for known ligand chemotypes and can give testable predictions for ligands that may isolate rare intermediate conformations of receptors.

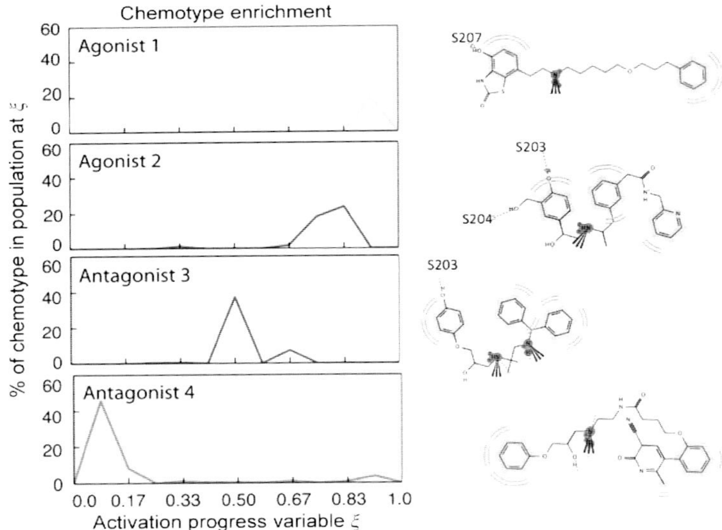

Figure 7 Examples of GPCR ligand chemotypes enriched at MSM states along activation pathways. Four examples of chemotype clusters enriched by select MSM states along the activation pathways are shown, with the percentage of a chemotype represented in the total ligands enriched at a given ξ. The chemical structures and pharmacophore is shown for each panel. Example agonist chemotypes are catecholamine and ethanolamine derivatives that make hydrogen bonds to key activating serine residues. Example antagonist chemotypes share a 2-hydroxy propyl amino core and exchange serine hydrogen bonds for hydrophobic interactions in the active site.

5. CONCLUSIONS

We have provided the readers with a detailed view of the MSM construction and analysis process with an emphasis on building models of GPCR dynamics. We have presented the algorithms used for building MSMs in an intuitive manner while avoiding the mathematical formalism. For a more detailed mathematical description of algorithms and the recent advances in MSM construction, we recommend recent reviews and research papers cited in this chapter. There are two publically available software that can be used for building MSMs: MSMBuilder (Beauchamp et al., 2011) and EMMA (Senne, Trendelkamp-Schroer, Mey, Schütte, & Noé, 2012). MSMBuilder was used for the analysis of β_2-AR simulations. At the time of this writing, public releases of MSMBuilder and EMMA are available online (http://msmbuilder.org; http://simtk.org/home/emma).

The take-home message of this chapter is that MSMs represent a natural statistical framework for understanding and sampling the complex dynamics of biological systems. MSM analysis stands in stark contrast to the standard methods used for brute-force sampling and single reaction coordinate-based analysis of protein dynamics. The presence of multiple preexisting pathways and intermediate states in conformational landscape of GPCRs reemphasizes the need to move from a two-state description of GPCRs to a conformational ensemble view of GPCR function. In particular, screening of drug candidates using few crystal structures of GPCRs have failed to provide leads with diverse efficacy profiles beyond antagonists and inverse agonists. Using MSM-based ensemble screening protocols creates an opportunity to develop drugs that interact more closely with diverse receptor states, for overall increased efficacy and specificity.

MSMs have been used widely now for investigating biological systems. However, there are caveats to consider while analyzing simulation data using MSMs. MSMs present a statistical description of underlying dynamics using the force field used in MD simulations. Therefore, MSMs are only as accurate as the underlying force fields used for describing the system. Molecular mechanics force fields for proteins have shown consistent improvement over the last decade but kinetic predictions of these models have not been accurate (Beauchamp, Lin, Das, & Pande, 2012). Recently, several groups have reported improved comparison of MSM and experimental kinetics (Lapidus et al., 2014; Noé et al., 2011) and also more systematic and reproducible approaches for building force fields (Paulechka, Kroenlein, Kazakov, &

Frenkel, 2012; Wang, Martinez, & Pande, 2014). MSMs can also be used as a framework for testing these new force field development protocols on complex systems.

Finally, the application of cloud-computing approaches in medicine would open up new avenues for tacking challenging problems in an efficient and easily accessible manner. Existing cloud-computing platforms such as Amazon Web Services, Google Exacycle (Hellerstein, Kohlhoff, & Konerding, 2012), Folding@home (Shirts & Pande, 2000), and even standard GPU clusters running MD packages optimized for GPUs (Eastman et al., 2013) would enable researchers to perform large-scale computations on demand, while keeping the complexity of the underlying infrastructure hidden. The impact of the commodity-computing approach is further enhanced by integration of novel distributed sampling algorithms such as MSMs for investigating the biological systems and extracting the knowledge from the vast information contained in "Big" datasets.

ACKNOWLEDGMENTS

D. S. acknowledges support by the Biomedical Data Science Initiative Postdoc Scholars Program of the Stanford School of Medicine. M. L. was supported by the Ruth Kirschstein fellowship for postdoctoral researchers. D. S. and V. S. P. acknowledge support from the Simbios, NIH Center for Biomedical Computation (NIH U54 Roadmap GM072970). V. S. P. acknowledges NIH (R01GM62828).

REFERENCES

Beauchamp, K. A., Bowman, G. R., Lane, T. J., Maibaum, L., Haque, I. S., & Pande, V. S. (2011). MSMBuilder2: Modeling conformational dynamics on the picosecond to millisecond scale. *Journal of Chemical Theory and Computation*, 7(10), 3412–3419.

Beauchamp, K. A., Lin, Y.-S., Das, R., & Pande, V. S. (2012). Are protein force fields getting better? A systematic benchmark on 524 diverse NMR measurements. *Journal of Chemical Theory and Computation*, *8*(4), 1409–1414.

Bowman, G. R. (2012). Improved coarse-graining of Markov state models via explicit consideration of statistical uncertainty. *The Journal of Chemical Physics*, *137*(13), 134111.

Bowman, G. R., Ensign, D. L., & Pande, V. S. (2010). Enhanced modeling via network theory: Adaptive sampling of Markov state models. *Journal of Chemical Theory and Computation*, *6*(3), 787–794.

Cherezov, V., Rosenbaum, D. M., Hanson, M. A., Rasmussen, S. G. F., Thian, F. S., Kobilka, T. S., et al. (2007). High-resolution crystal structure of an engineered human β_2-adrenergic G protein-coupled receptor. *Science*, *318*(5854), 1258–1265.

Chodera, J. D., & Noé, F. (2014). Markov state models of biomolecular conformational dynamics. *Current Opinion in Structural Biology*, *25*, 135–144.

Deuflhard, P., & Weber, M. (2005). Robust Perron cluster analysis in conformation dynamics. *Linear Algebra and its Applications*, *398*, 161–184.

Dror, R. O., Arlow, D. H., Maragakis, P., Mildorf, T. J., Pan, A. C., Xu, H., et al. (2011). Activation mechanism of the β_2-adrenergic receptor. *Proceedings of the National Academy of Sciences of the United States of America*, *108*(46), 18684–18689.

Eastman, P., Friedrichs, M. S., Chodera, J. D., Radmer, R. J., Bruns, C. M., Ku, J. P., et al. (2013). OpenMM 4: A reusable, extensible, hardware independent library for high performance molecular simulation. *Journal of Chemical Theory and Computation*, *9*(1), 461–469.

Gatica, E. A., & Cavasotto, C. N. (2012). Ligand and decoy sets for docking to G protein-coupled receptors. *Journal of Chemical Information and Modeling*, *52*(1), 1–6.

Gether, U., & Kobilka, B. K. (1998). G protein-coupled receptors II. Mechanism of agonist activation. *The Journal of Biological Chemistry*, *273*(29), 17979–17982.

Hawkins, P. C. D., Skillman, A. G., Warren, G. L., Ellingson, B. A., & Stahl, M. T. (2010). Conformer generation with OMEGA: Algorithm and validation using high quality structures from the Protein Databank and Cambridge Structural Database. *Journal of Chemical Information and Modeling*, *50*(4), 572–584.

Hellerstein, J. L., Kohlhoff, K. J., & Konerding, D. E. (2012). Science in the cloud: Accelerating discovery in the 21st century. *IEEE Internet Computing*, *16*(4), 64–68.

Kohlhoff, K. J., Shukla, D., Lawrenz, M., Bowman, G. R., Konerding, D. E., Belov, D., et al. (2014). Cloud-based simulations on Google exacycle reveal ligand modulation of GPCR activation pathways. *Nature Chemistry*, *6*(1), 15–21. http://dx.doi.org/10.1038/nchem.1821.

Koshland, D. E., Némethy, G., & Filmer, D. (1966). Comparison of experimental binding data and theoretical models in proteins containing subunits*. *Biochemistry*, *5*(1), 365–385.

Lane, T. J., Shukla, D., Beauchamp, K. A., & Pande, V. S. (2013). To milliseconds and beyond: Challenges in the simulation of protein folding. *Current Opinion in Structural Biology*, *23*(1), 58–65.

Lapidus, L. J., Acharya, S., Schwantes, C. R., Wu, L., Shukla, D., King, M., et al. (2014). Complex pathways in folding of protein G explored by simulation and experiment. *Biophysical Journal*, *107*(4), 947–955.

Monod, J., Wyman, J., & Changeux, J.-P. (1965). On the nature of allosteric transitions: A plausible model. *Journal of Molecular Biology*, *12*(1), 88–118.

Noé, F., Doose, S., Daidone, I., Löllmann, M., Sauer, M., Chodera, J. D., et al. (2011). Dynamical fingerprints for probing individual relaxation processes in biomolecular dynamics with simulations and kinetic experiments. *Proceedings of the National Academy of Sciences of the United States of America*, *108*(12), 4822–4827.

Nygaard, R., Zou, Y., Dror, R. O., Mildorf, T. J., Arlow, D. H., Manglik, A., et al. (2013). The dynamic process of β(2)-adrenergic receptor activation. *Cell*, *152*(3), 532–542.

Paulechka, E., Kroenlein, K., Kazakov, A., & Frenkel, M. (2012). A systematic approach for development of an OPLS-like force field and its application to hydrofluorocarbons. *The Journal of Physical Chemistry. B*, *116*(49), 14389–14397.

Pérez-Hernández, G., Paul, F., Giorgino, T., Fabritiis, G. D., & Noé, F. (2013). Identification of slow molecular order parameters for Markov model construction. *The Journal of Chemical Physics*, *139*(1), 015102.

Rasmussen, S. G. F., Choi, H.-J., Fung, J. J., Pardon, E., Casarosa, P., Chae, P. S., et al. (2011). Structure of a nanobody-stabilized active state of the β_2 adrenoceptor. *Nature*, *469*(7329), 175–180.

Reiter, E., Ahn, S., Shukla, A. K., & Lefkowitz, R. J. (2012). Molecular mechanism of β-arrestin-biased agonism at seven-transmembrane receptors. *Annual Review of Pharmacology and Toxicology*, *52*, 179–197.

ROCS, version 3.1.2. (n.d.). OpenEye Scientific Software, Inc., Santa Fe, NM, 2011.

Rosenbaum, D. M., Zhang, C., Lyons, J. A., Holl, R., Aragao, D., Arlow, D. H., et al. (2011). Structure and function of an irreversible agonist-β_2 adrenoceptor complex. *Nature*, *469*(7329), 236–240.

Schwantes, C. R., & Pande, V. S. (2013). Improvements in Markov state model construction reveal many non-native interactions in the folding of NTL9. *Journal of Chemical Theory and Computation*, *9*(4), 2000–2009.

Senne, M., Trendelkamp-Schroer, B., Mey, A. S. J. S., Schütte, C., & Noé, F. (2012). EMMA: A software package for Markov model building and analysis. *Journal of Chemical Theory and Computation*, *8*(7), 2223–2238.

Shirts, M., & Pande, V. S. (2000). Screen savers of the world unite! *Science*, *290*(5498), 1903–1904.

Shoichet, B. K. (2004). Virtual screening of chemical libraries. *Nature*, *432*(7019), 862–865.

Shoichet, B. K., & Kobilka, B. K. (2012). Structure-based drug screening for G-protein-coupled receptors. *Trends in Pharmacological Sciences*, *33*(5), 268–272.

Shukla, D., Meng, Y., Roux, B., & Pande, V. S. (2014). Activation pathway of Src kinase reveals intermediate states as targets for drug design. *Nature Communications*, *5*, 3397.

Shukla, A. K., Sun, J.-P., & Lefkowitz, R. J. (2008). Crystallizing thinking about the β_2-adrenergic receptor. *Molecular Pharmacology*, *73*(5), 1333–1338.

Spitzer, R., & Jain, A. N. (2012). Surflex-Dock: Docking benchmarks and real-world application. *Journal of Computer-Aided Molecular Design*, *26*(6), 687–699.

Wang, L.-P., Martinez, T. J., & Pande, V. S. (2014). Building force fields: An automatic, systematic, and reproducible approach. *The Journal of Physical Chemistry Letters*, *5*(11), 1885–1891.

Weber, J. K., & Pande, V. S. (2011). Characterization and rapid sampling of protein folding Markov state model topologies. *Journal of Chemical Theory and Computation*, 7(10), 3405–3411.

Weinan, E., & Vanden-Eijnden, E. (2010). Transition-path theory and path-finding algorithms for the study of rare events. *Annual Review of Physical Chemistry*, *61*(1), 391–420.

AUTHOR INDEX

Note: Page numbers followed by "*f*" indicate figures, and "*t*" indicate tables and "*np*" indicate footnotes.

C

D

E

F

G

H

I

J

L

M

N

O

P

Q

R

S

T

W

X

Y

Z

SUBJECT INDEX

Note: Page numbers followed by "*f*" indicate figures and "*t*" indicate tables.

C

D

E

J

K

L

M

N

O

T

U

V

X

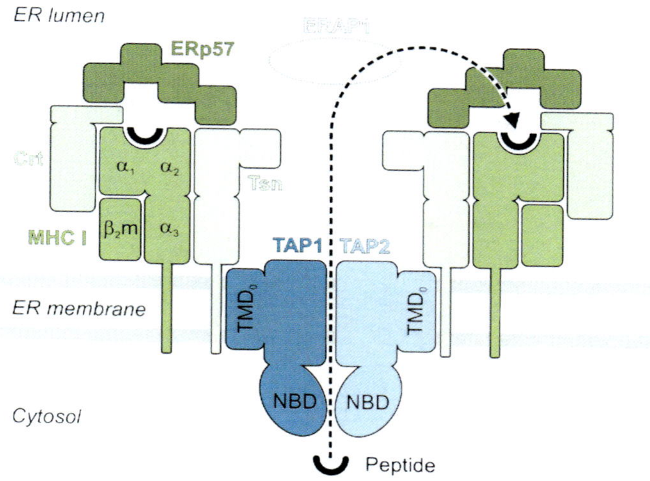

Simon Trowitzsch and Robert Tampé, Figure 3 Structural organization of the PLC. Subunits are schematically depicted according to their domain organization. Transmembrane domains (TMD$_0$) and the nucleotide-binding domains (NBDs) of TAP1 and TAP2 are shown. The cartoon highlights multivalent intersubunit interactions, which stabilize the PLC (see text for details). Peptides are translocated from the cytosol to the ER lumen *via* TAP (dashed arrow), where they are further processed *via* the aminopeptidase ERAP1, prior to loading onto MHC I molecules. ER, endoplasmic reticulum; PLC, peptide-loading complex; Tsn, tapasin; Crt, calreticulin; MHC I, major histocompatibility complex class I; β_2m, β_2-microglobulin; TAP, transporter associated with antigen processing; ERAP1, endoplasmic reticulum aminopeptidase 1.

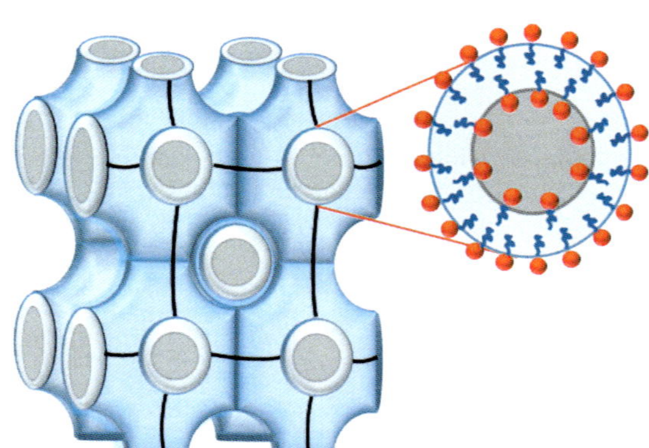

Aiman Sadaf *et al.*, Figure 15 Schematic representation of lipidic cubic phase (LCP). In this architecture, the three-dimensional structure of a single lipid bilayer is associated with an interconnected aqueous channel. Monoolein with a single double bond within the alkyl chain is the most popular lipid component in LCP system.

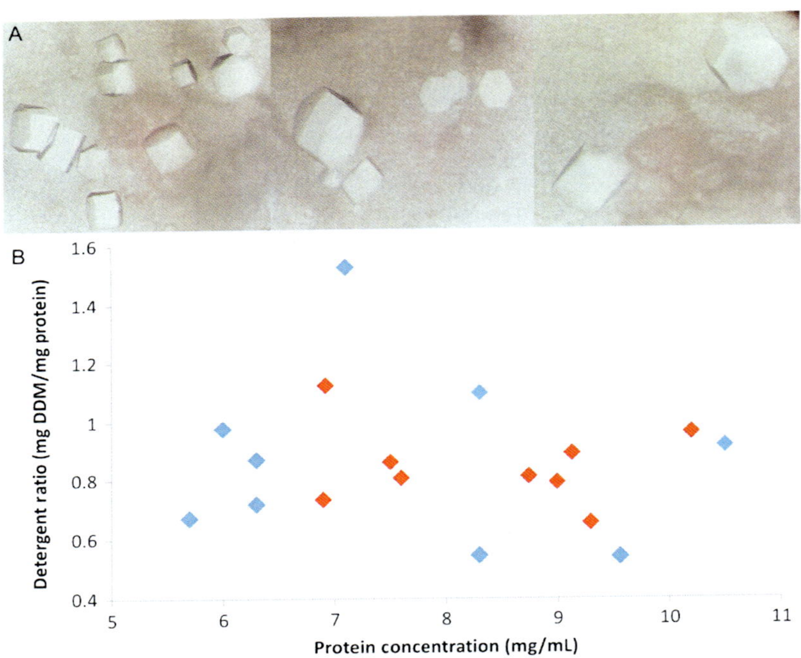

Chelsy Prince and Zongchao Jia, Figure 12 (A) Etk NM2 crystals grow as small to medium hexagonal prisms. (B) Plot of protein samples that successfully crystallized (red) and those that did not (blue) as a function of both the protein and detergent concentration in the sample.

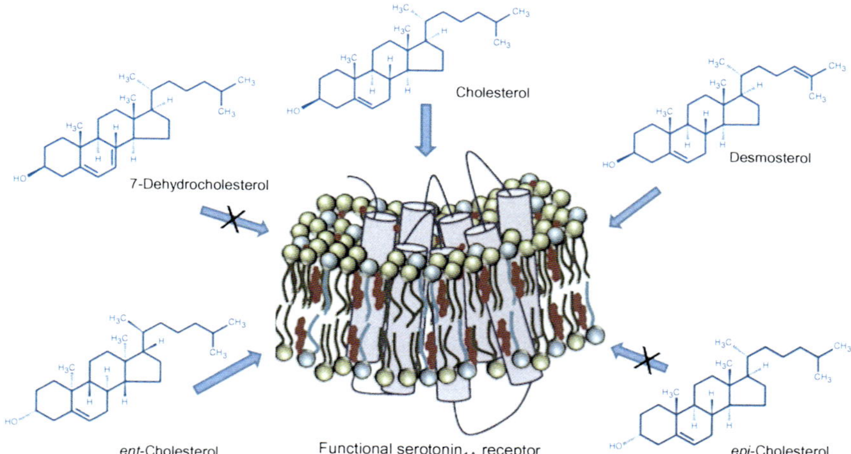

Amitabha Chattopadhyay *et al.*, Figure 5 A schematic representation showing the role of various sterols in supporting the function of the reconstituted serotonin$_{1A}$ receptor. The serotonin$_{1A}$ receptor is shown in purple, and the replenished sterol molecules are shown in maroon. The ligand-binding activity of the reconstituted serotonin$_{1A}$ receptor was supported upon replenishment with cholesterol, desmosterol, and *ent*-cholesterol, whereas replenishment with 7-DHC and *epi*-cholesterol could not support the function of the receptor. This brings out the important message that even subtle changes in sterol structure could be detrimental for receptor function, thereby implying stringent receptor–lipid interaction.

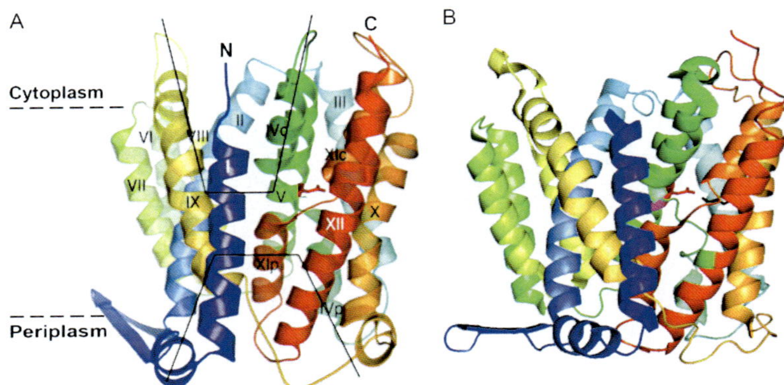

Etana Padan and Manish Dwivedi, Figure 1 The crystal structure of NhaA (A). NhaA was originally crystallized in 2005 (Hunte et al., 2005). (B) In 2007, NhaA was crystallized again, with very similar results, but with different assignment of TM X (Lee, Yashiro, et al. in PDB:4AU5, 2014). It is not yet known whether TM X was misassigned in the original structure or whether the second structure represents a different conformation. The figures were illustrated using Pymol (rainbow, ribbon presentation). In (A), black lines indicate funnels, and the TMs are marked with roman numerals.

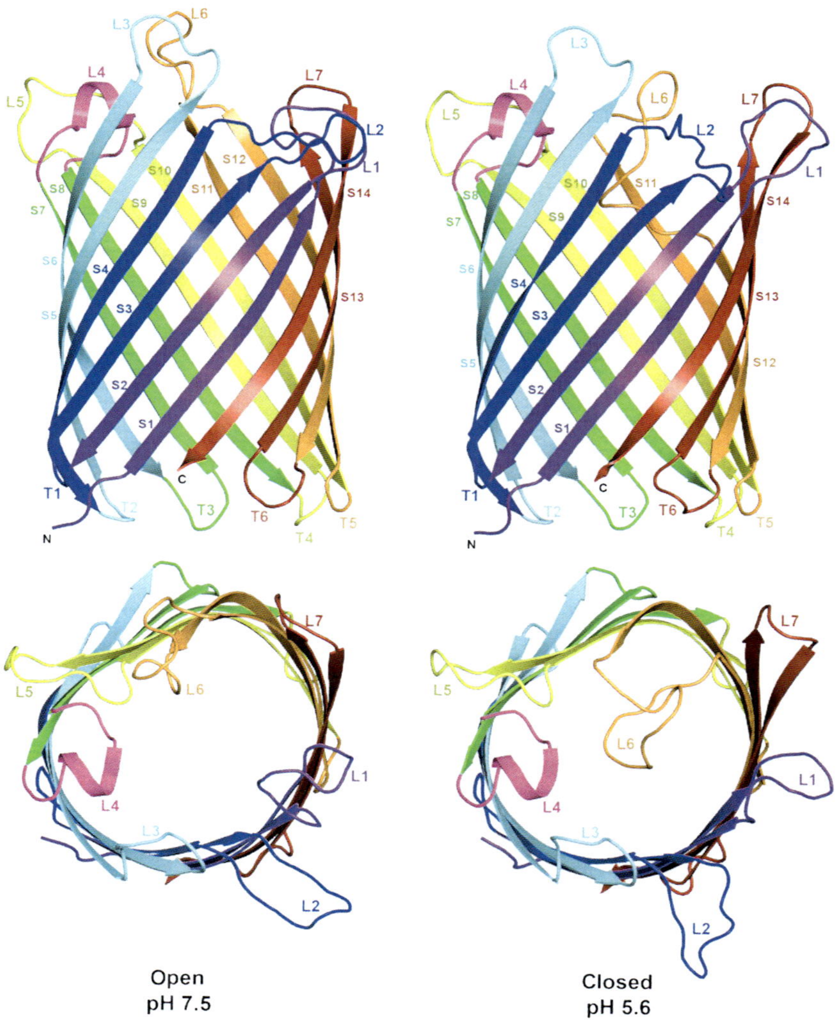

Stefan Köster *et al.*, Figure 1 Overall structure of OmpG. The X-ray structure shows the 14-stranded, monomeric β-barrel in the open and closed conformation (Yildiz, Vinothkumar, Goswami, & Kühlbrandt, 2006) viewed along the membrane plane (top) and from the extracellular side (bottom). *Adapted from Yildiz et al. (2006) with permission from John Wiley & Sons.*

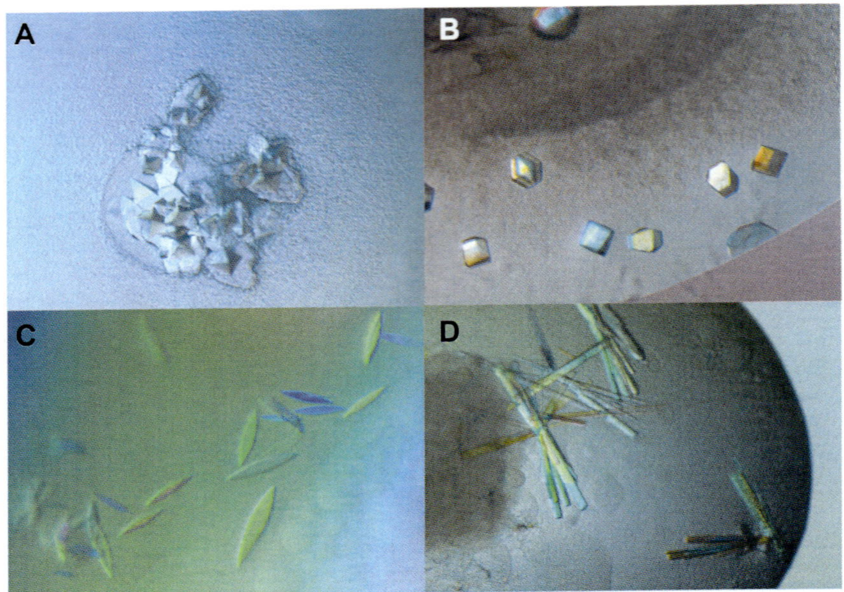

Stefan Köster *et al.*, Figure 6 Initial 3D crystals of OmpG. Initial 3D crystals grown in 96-well plates were typically small (A–C) or clustered (D). Often, the crystals were surrounded by viscose reservoir (A), precipitated protein (B), or phase separation (C, D).

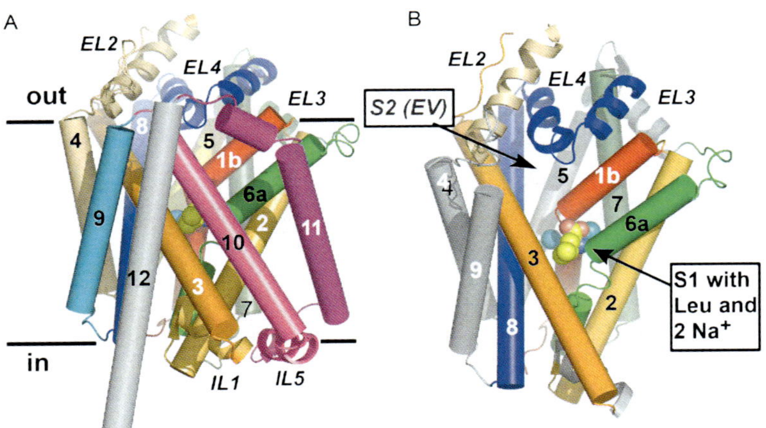

Satinder K. Singh and Aritra Pal, Figure 3 (A) LeuT in the membrane plane. Transmembrane helices (TM) and extracellular/intracellular loops are depicted as cylinders and coils, respectively. Heavy black lines demarcate the membrane boundaries. (B) Same view as (A) except tilted ~15° toward the reader and with TMs 10–12 removed to expose the substrates Leu (in yellow [carbon]/red [oxygen] spheres) and the two sodium ions (cyan spheres), all bound near the unwound sections of TMs 1 and 6. *Adapted from Singh (2008).*

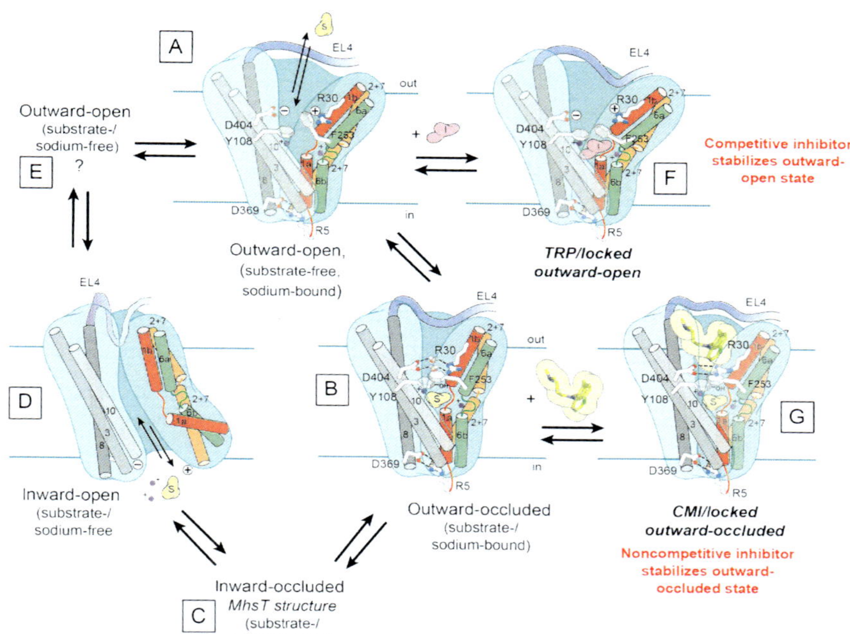

Satinder K. Singh and Aritra Pal, Figure 13 Schematic of transport and inhibition based exclusively on crystal structures and steady-state kinetic data of LeuT except for the substrate/sodium-bound inward-occluded state, which is based on MhsT, and the apo outward-open state, for which no representative structure is yet available. Conformational changes associated with isomerization from (A) sodium-bound, substrate-free outward-open to (B) substrate/sodium-bound outward-occluded state to (C) sodium/substrate-bound inward-occluded (MhsT) to (D) apo inward-open to (E) sodium/substrate-free outward-open. Panel (F) and (G), respectively, depict structures of the competitively inhibited, stabilized outward-open Trp/Na^+-bound and the noncompetitively inhibited, stabilized outward-occluded Leu/Na^+/CMI-bound states. TM5 is not shown for clarity, although, as mentioned, it does play an important role in the translocation cycle.

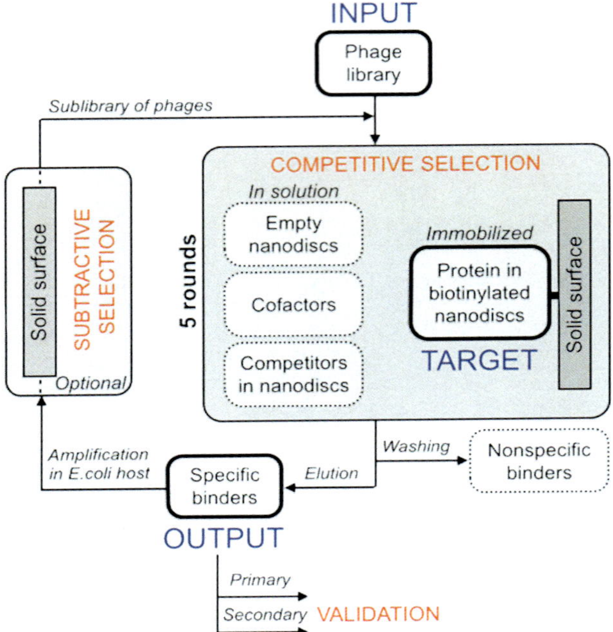

Pawel K. Dominik and Anthony A. Kossiakoff, Figure 4 Competitive and subtractive phage display selection. The input phage library is incubated with the target membrane protein embedded in biotinylated nanodiscs, which are immobilized on a solid surface. In solution, the excess of empty nanodiscs, competitors (undesired conformations or regions of target protein), as well as necessary cofactors are used to create a desired competitive selection environment. Nonspecific or competitor-specific phage particles are washed away, while specific "binders" are recovered by elution in detergent. Subsequently, the specific phage particles are propagated in *E. coli* for another selection round. Optionally, prior to the next round, a subtractive selection step is performed by short incubation of sublibrary of phage particles with biotinylated competitors or solid surface alone to remove additional nonspecific binders. The leftover phage pool is used as a new input. In total, five consecutive rounds of binding selection are performed using decreasing concentrations of specific antigen as a selection pressure. As a result, specific phage clones are amplified and used for primary and secondary validation.

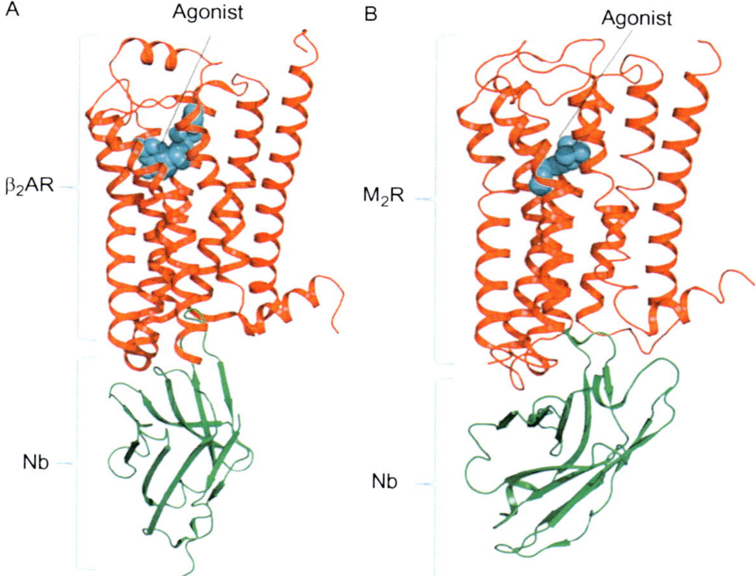

Arun K. Shukla *et al.*, Figure 2 Nanobody-mediated trapping of active GPCR conformations. (A) Crystal structure of a high-affinity agonist-bound β_2AR in a fully active conformation in complex with a nanobody. The diagram represents the cartoon depiction of the structure (PDB ID 3POG) generated using PyMol software. The β_2AR is shown in red color, nanobody (Nb) in green and the agonist (BI-167107) is shown in cyan. (B) Crystal structure of the agonist-bound human muscarinic M_2 receptor (M_2R) in complex with an active state selective nanobody. The diagram represents the cartoon depiction of the structure (PDB ID 4MQS) generated using PyMol software. The M_2R is shown in red color, nanobody (Nb) in green, and the agonist (iperoxo) is shown in cyan.

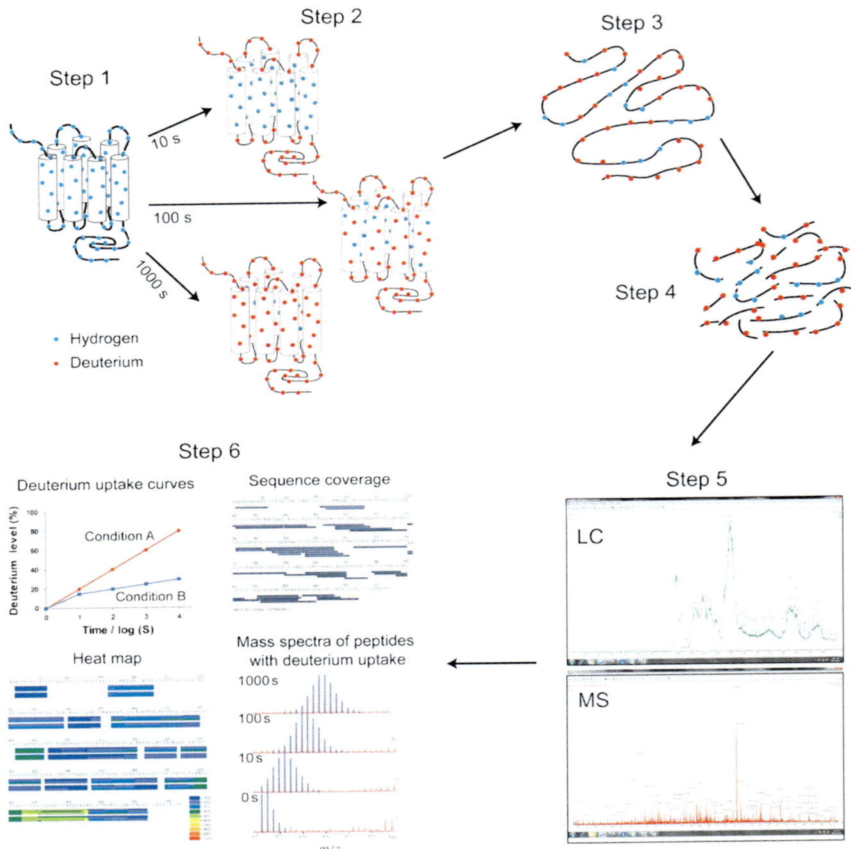

Sheng Li *et al.*, Figure 1 Theory and overall step-by-step procedure for HDX-MS regarding GPCR analysis.

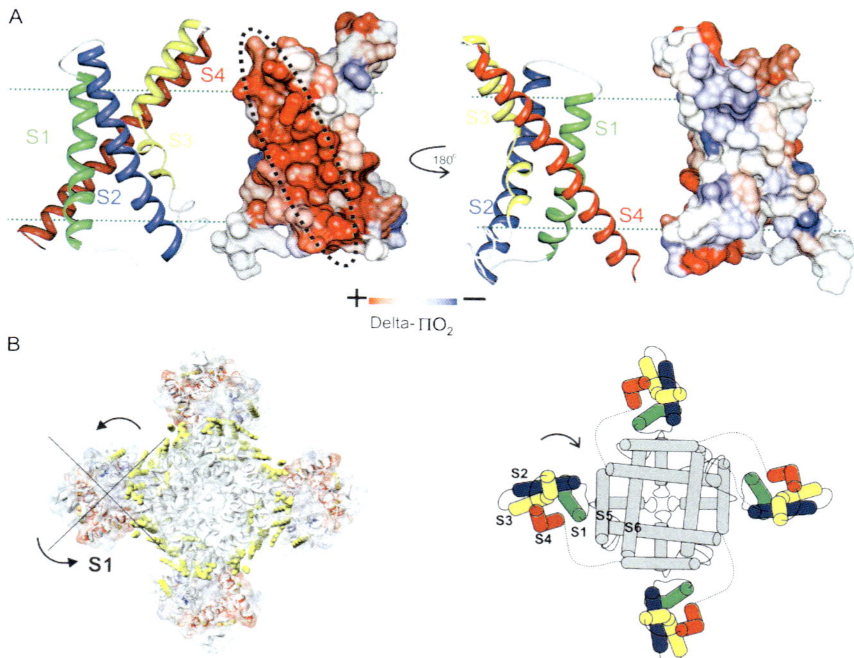

Sudha Chakrapani, Figure 2 Mapping the molecular interacting surface of the VSD with the PD. Fractional differences in ΠO_2 mapped on the sensor structure and color coded with a gradient from red–white–blue. Red denotes increase and blue represents decrease in the environmental parameter. The surface of highest impact (S1 and S2) is marked by dotted black line. (B) Evaluation of the open-state structures for the K_v channels. The delta values mapped on to the coordinates of Kv1.2–Kv2.1 chimera structure (2R9R.pdb). A view from the intracellular part of the channel is shown. The EPR data from KvAP in the open-inactivated state predict a rotation of the VSD (~70–100°) to place S1 and S2 in the proximity of the pore as shown in the cartoon. *This figure is adapted from the originally published data in Chakrapani et al. (2008).*

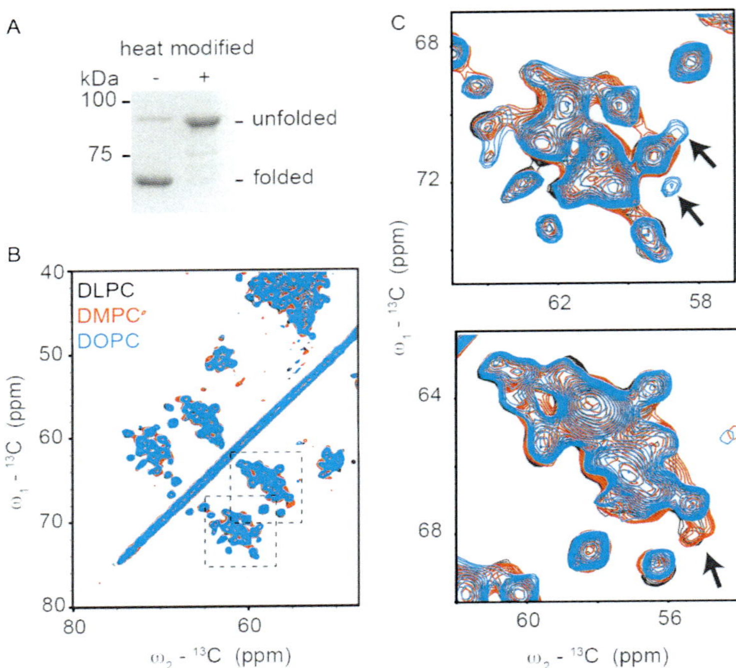

Lindsay A. Baker *et al.*, Figure 3 Proteins with β-barrel folds such as BamA (β-barrel assembly machinery A, UniProt ID: P0A940) from *E. coli* can be studied by ssNMR. (A) Folding of BamA analyzed on seminative SDS-PAGE. BamA reconstituted in DOPC vesicles at a molar lipid: protein ratio of 25:1 is shown. Native, folded BamA shows a characteristic shift of the electrophoretic mobility from 88 to 70 kDa, compared to heat-denatured protein. (B) and (C) Comparison of BamA reconstituted in different lipid bilayers. Shown in (B) is a section of a 2D ^{13}C,^{13}C correlation experiment with 30 ms PARIS mixing of BamA in DLPC (black), DMPC (red), and DOPC (blue) at a molar lipid: protein ratios of 25:1. (C) Close-ups of the boxed regions in (B) reveal small differences in certain regions of the spectra.

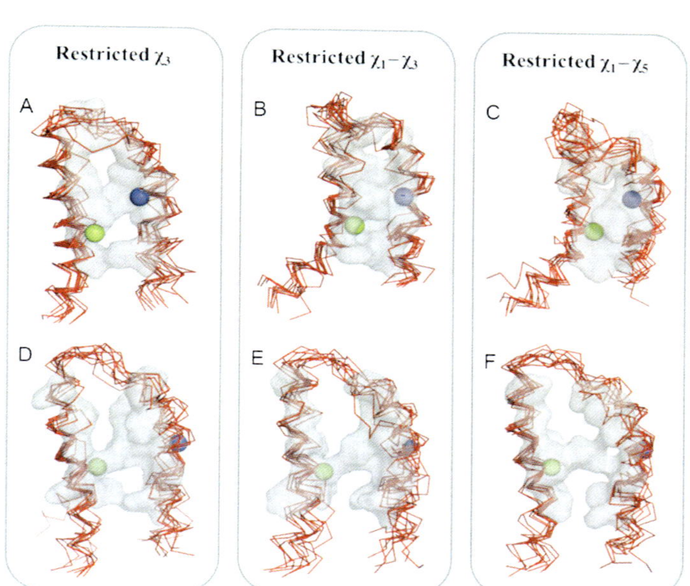

Linda Columbus and Brett Kroncke, Figure 3 Lowest energy structures (10 lowest) of TM0026. Structures were calculated with 35 hydrogen bond (identified from amide proton NOEs) and PRE restraints. PRE restraints were obtained from A13R1 (a tertiary contact site) and V15R1 (a detergent-exposed site) (A–C; 87 PRE restraints) or from only A13R1 (D–F; 55 PRE restraints). Each set of structures (A and D, B and E, and C and F) was calculated using a different restriction on the conformation of the spin label as labeled. The rmsds for each set of structures, calculated from the most average structure, are shown in Table 2. Light green and blue spheres show the beta carbon of A13 and V43, respectively, to aid in helical twist assessment. A transparent gray surface shows the interface between the two helices (and correlates with the solvent accessible surface area).

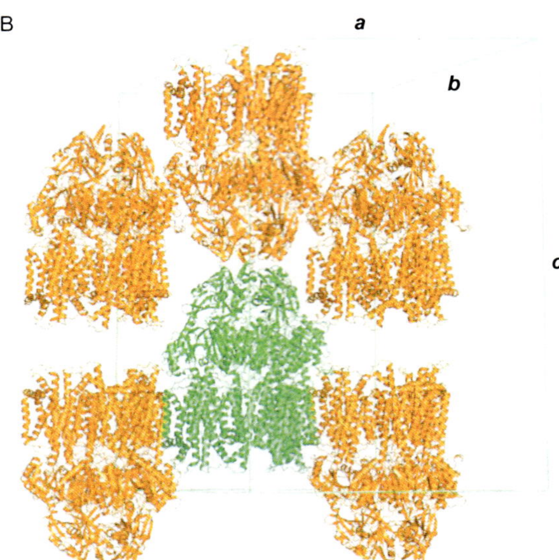

Jared A. Delmar *et al.*, Figure 4 Crystal structure of the *E. coli* inner membrane heavy-metal efflux pump CusA. (B) Packing diagram of the CusA crystal viewed orthogonal to the long axis of the unit cell.

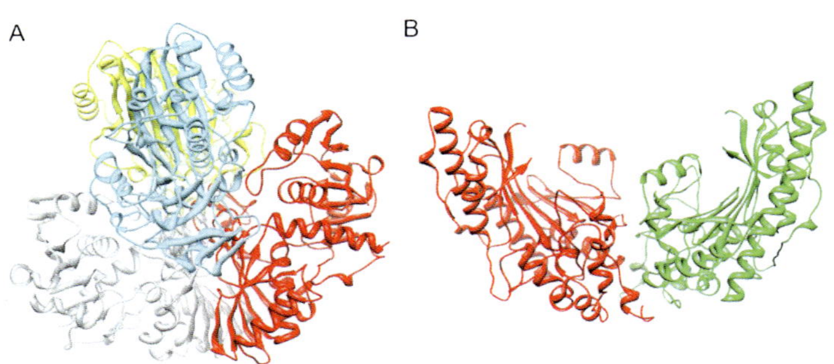

Bo Y. Baker *et al.*, Figure 4 Ribbon representations of bovine GAPDH (left panel) solved to 1.93 Å and bovine CK-B (right panel) solved to 1.65 Å. Individual subunits are presented in different colors.

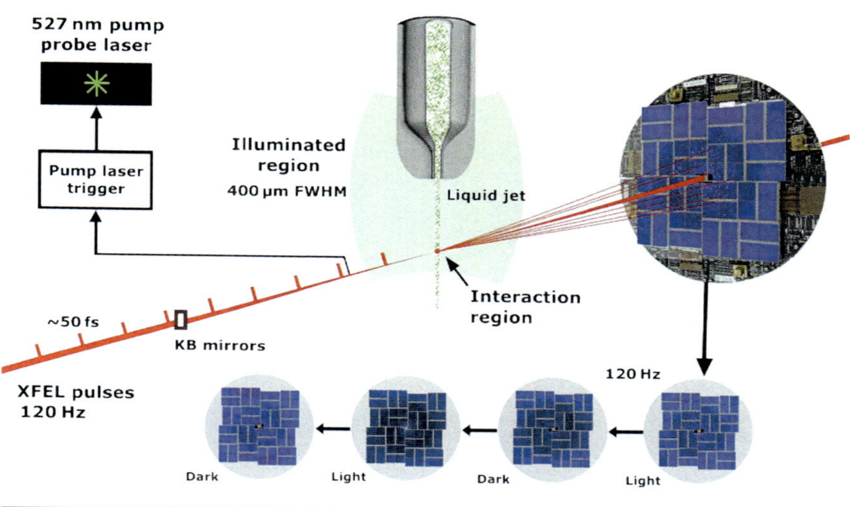

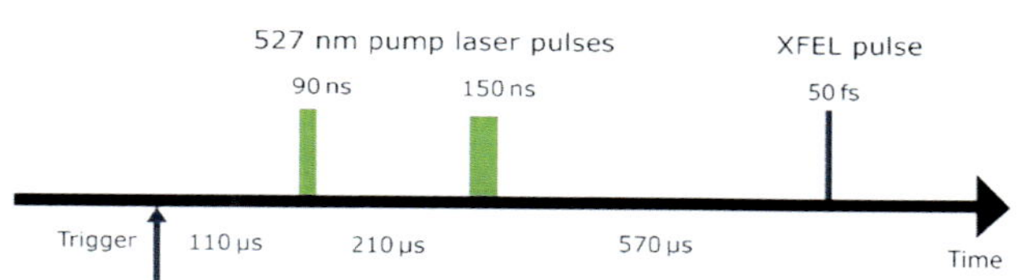

Jesse Coe *et al.*, Figure 3 Diagram of the experimental setup for the time-resolved SFX experiment in which the S_1 and S_3 states are probed alternately at 60 Hz each. The optical pump laser scheme is shown at the bottom, indicating delay times of 210 µs between the first and second pump, and 570 µs between the second pump and interaction with the FEL beam, allowing population of the S_2 and S_3 states, respectively. *Figure originally published in Kupitz, Basu, et al. (2014).*

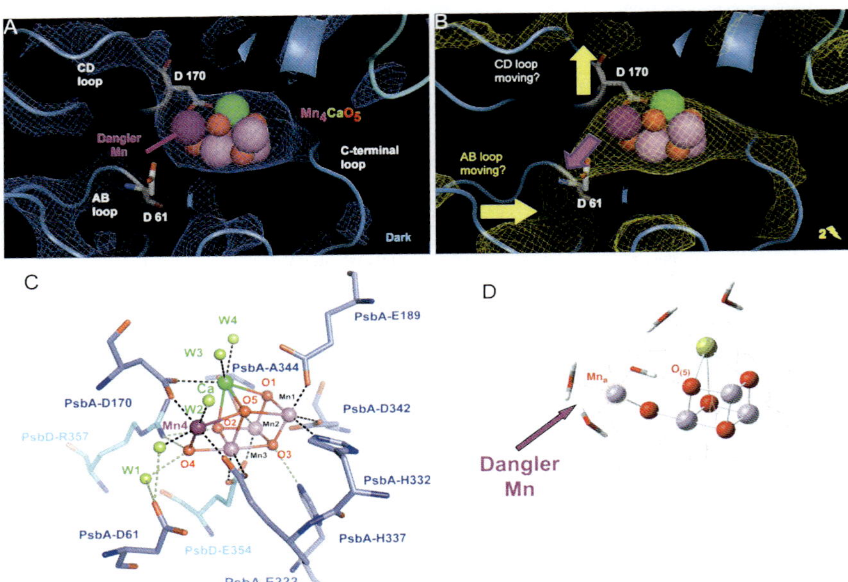

Jesse Coe *et al.*, Figure 6 Simulated annealing omit maps of the OEC. (A) and (B) Exhibit the S_1 and S_3 states, respectively, at 1.5σ. In (B), the blue represents the S_1 state, whereas the yellow represents the S_3 state for comparison. (C) Shows the crystal structure from the 1.9 Å structure of the OEC from Umena et al. (2011). (D) Shows the proposed S_3 state derived from DFT calculations performed in Isobe et al. (2012). *Figure originally published in Kupitz, Basu, et al. (2014).*

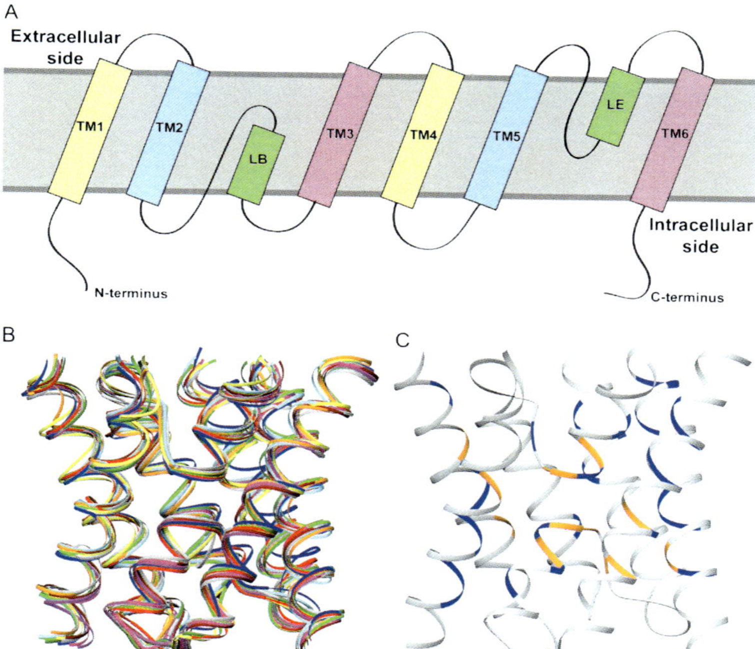

Ravi Kumar Verma *et al.*, Figure 1 (A) Membrane topology diagram of an MIP channel. Each MIP channel has six transmembrane helical segments (TM1–TM6) and two half-helices formed by the loops LB and LE. The transmembrane region is shown in dark gray color. (B) Superposition of 11 unique MIP structures determined to date. These are AQP0, AQP1, AQP2, AQP4, AQP5, AqpZ, GlpF, SoPIP2;1, Aqy1, AqpM, and PfAQP. Their respective PDB IDs are 2B6O, 1J4N, 4NEF, 3GD8, 3D9S, 1RC2, 1FX8, 1Z98, 2W1P, 2F2B, and 3C02. Only the transmembrane region (TM1–TM6 helices and LB and LE loop regions) was considered for the superposition. All MIP structures from diverse sources adopt the unique hourglass helical fold. (C) Positions of 13 highly conserved residues (orange) and those 27 positions in which small and weakly polar residues exhibit a high level of group conservation (blue) are shown in the ribbon diagram of the hourglass fold. The plant MIP structure SoPIP2;1 (PDB ID: 1Z98) was used as a reference structure to illustrate the highly conserved positions within the transmembrane region.

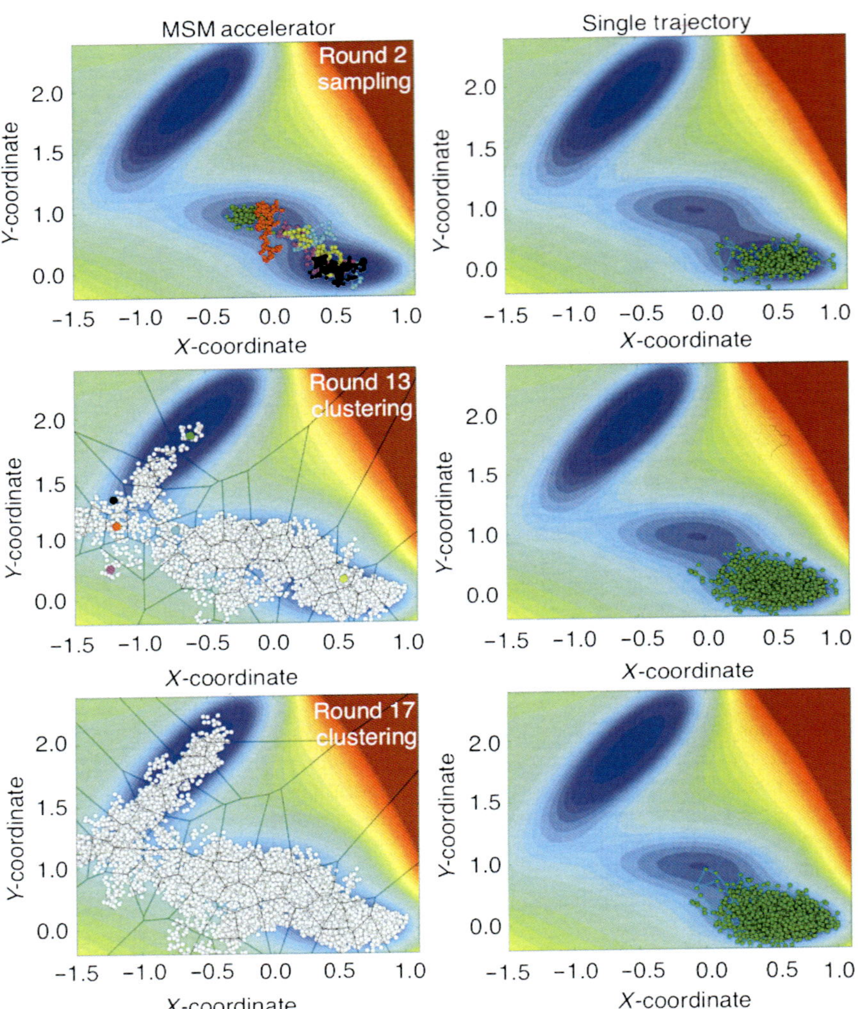

Diwakar Shukla *et al.*, Figure 3 Two-dimensional demo of MSM accelerator, running an adaptive sampling algorithm to efficiently sample space. Here, we show the final rounds, where the MSM has fully converged with much less computer power needed to simulate the event using traditional molecular dynamics (right pane).

CPI Antony Rowe
Eastbourne, UK
May 07, 2015